Angular Momentum
in Quantum Physics
Theory and Application

GIAN-CARLO ROTA, *Editor*
ENCYCLOPEDIA OF MATHEMATICS AND ITS APPLICATIONS

	Volume	Section
1	LUIS A. SANTALÓ **Integral Geometry and Geometric Probability,** 1976 (2nd printing, with revisions, 1979)	Probability
2	GEORGE E. ANDREWS **The Theory of Partitions,** 1976 (2nd printing, 1981)	Number Theory
3	ROBERT J. McELIECE **The Theory of Information and Coding** A Mathematical Framework for Communication, 1977 (2nd printing, with revisions, 1979)	Probability
4	WILLARD MILLER, Jr. **Symmetry and Separation of Variables,** 1977	Special Functions
5	DAVID RUELLE **Thermodynamic Formalism** The Mathematical Structures of Classical Equilibrium Statistical Mechanics, 1978	Statistical Mechanics
6	HENRYK MINC **Permanents,** 1978	Linear Algebra
7	FRED S. ROBERTS **Measurement Theory** with Applications to Decisionmaking, Utility, and the Social Sciences, 1979	Mathematics and the Social Sciences
8	L. C. BIEDENHARN and J. D. LOUCK **Angular Momentum in Quantum Physics:** Theory and Application, 1981	Mathematics of Physics
9	L. C. BIEDENHARN and J. D. LOUCK **The Racah-Wigner Algebra in Quantum Theory**, 1981	Mathematics of Physics

GIAN-CARLO ROTA, *Editor*
ENCYCLOPEDIA OF MATHEMATICS AND ITS APPLICATIONS

	Volume	Section
10	JOHN D. DOLLARD and CHARLES N. FRIEDMAN **Product Integration** with Application to Differential Equations, 1979	Analysis
11	WILLIAM B. JONES and W. J. THRON **Continued Fractions:** Analytic Theory and Applications, 1980	Analysis
12	NATHANIEL F. G. MARTIN and JAMES W. ENGLAND **Mathematical Theory of Entropy,** 1981	Real Variables
13	GEORGE A. BAKER, Jr., and PETER R. GRAVES-MORRIS **Padé Approximants** **Part I: Basic Theory,** 1981	Mathematics of Physics
14	GEORGE A. BAKER, Jr., and PETER R. GRAVES-MORRIS **Padé Approximants** **Part II: Extensions and Applications,** 1981	Mathematics of Physics
15	E. C. BELTRAMETTI and G. CASSINELLI **The Logic of Quantum Mechanics,** 1981	Mathematics of Physics
16	G. D. JAMES and A. KERBER **The Representation Theory of the Symmetric Group,** 1981	Algebra

Other volumes in preparation

ENCYCLOPEDIA OF MATHEMATICS and Its Applications

GIAN-CARLO ROTA, *Editor*
ENCYCLOPEDIA OF MATHEMATICS AND ITS APPLICATIONS
Volume 8

Section: Mathematics of Physics
Peter A. Carruthers, *Section Editor*

Angular Momentum in Quantum Physics
Theory and Application

L. C. Biedenharn
Physics Department
Duke University
Durham, North Carolina

J. D. Louck
Los Alamos National Laboratory
University of California
Los Alamos, New Mexico

With a Foreword by
Peter A. Carruthers
Los Alamos National Laboratory
University of California
Los Alamos, New Mexico

1981

Addison-Wesley Publishing Company
Advanced Book Program
Reading, Massachusetts

London • Amsterdam • Don Mills, Ontario • Sydney • Tokyo

Library of Congress Cataloging in Publication Data

Biedenharn, L C
Angular momentum in quantum physics.

(Encyclopedia of mathematics and its applications;
v. 8: Section, Mathematics and physics)
Bibliography: p.
Includes indexes.
1. Angular momentum. 2. Quantum theory.
I. Louck, James D., joint author. II. Title.
III. Series: Encyclopedia of mathematics and its applications; v. 8.
QC174.17.A53B53 539.7′2 80-36893
ISBN 0-201-13507-8

American Mathematical Society (MOS) Subject Classification Scheme (1980): 81A60, 81A66, 81A72, 81A78, 81A81, 22E70, 15A03, 15A18, 15A72, 47D10, 47D15, 33A50, 33A65, 33A75, 05A10, 05A15, 78A40, 78A45, 20C35

Manufactured in the United States of America

ABCDEFGHIJ-HA-8987654321

EUGENE P. WIGNER

(Courtesy AIP Niels Bohr Laboratory)

To
Eugene P. Wigner

Contents

PART II

Contents of Volume 9

Editor's Statement

A large body of mathematics consists of facts that can be presented and described much like any other natural phenomenon. These facts, at times explicitly brought out as theorems, at other times concealed within a proof, make up most of the applications of mathematics, and are the most likely to survive changes of style and of interest.

This ENCYCLOPEDIA will attempt to present the factual body of all mathematics. Clarity of exposition, accessibility to the non-specialist, and a thorough bibliography are required of each author. Volumes will appear in no particular order, but will be organized into sections, each one comprising a recognizable branch of present-day mathematics. Numbers of volumes and sections will be reconsidered as times and needs change.

It is hoped that this enterprise will make mathematics more widely used where it is needed, and more accessible in fields in which it can be applied but where it has not yet penetrated because of insufficient information.

GIAN-CARLO ROTA

Foreword

The study of the symmetries of physical systems remains one of the principal contemporary theoretical activities. These symmetries, which basically express the geometric structure of the physical system in question, must be clearly analyzed in order to understand the dynamical behavior of the system. The analysis of rotational symmetry, and the behavior of physical quantities under rotations, is the most common of such problems. Accordingly, every professional physicist must achieve a good working knowledge of the "theory of angular momentum."

In addition, the theory of angular momentum is the prototype of continuous symmetry groups of many types now found useful in the classification of the internal symmetries of elementary particle physics. Much of the intuition and mathematical apparatus developed in the theory of angular momentum can be transferred with little change to such research problems of current interest.

If there is a single essential book in the arsenal of the physicist, it is a good book on the theory of angular momentum. I have worn out several earlier texts on this subject and have spent much time checking signs and Clebsch-Gordan coefficients. Such books are the most borrowed and least often returned. I look forward to a long association with the present fine work.

A good book on the theory of angular momentum needs to be thoroughly reliable yet must develop the material with insight and good taste in order to lay bare the elegant texture of the subject. Originality should not be erected in opposition to current practices and conventions if the text is to be truly useful.

The present text, written by two well-known contributors to the field, satisfies all these criteria and more. Subtleties and scholarly comments are presented clearly yet unobtrusively. Moreover, the footnotes contain fascinating historical material of which I was previously unaware. The two chapters on the "theory of turns" and "boson calculus" are significant new additions to the pedagogical literature on angular momentum. Much of the theory of turns presented here was developed by the authors. By means of this approach the concept of "double group" is made very clear. The development of the boson calculus employs Gel'fand patterns in an essential way, in addition to the more traditional Young tableaux. This section provides an excellent prototype for the analysis of all compact groups.

The representation theory is developed in the complete detail required for physical applications. This exposition of the lore of rotation matrices is especially thorough, including the Euler angle parametrization as well as others of practical value.

The text ends with a long chapter on applications well chosen to illustrate the power of the general techniques. The book concludes with a masterly development of the group theoretical description of the spectra of spherical top molecules. To my mind the recent experimental confirmation of this theory in high resolution laser spectrometry experiments is one of the most spectacular confirmations of quantum theory.

The present text is really a book for physicists. Nevertheless, the theory generates substantial material of interest for mathematicians. Recent research (for example in non-Abelian gauge field theory) has produced topics of common interest to both mathematicians and physicists. Some of the more interesting mathematical outgrowths of the theory of angular momentum are developed in the companion volume currently in press.

PETER A. CARRUTHERS
General Editor, Section on Mathematics of Physics

Preface

"The art of doing mathematics," Hilbert[1] has said, "consists in finding that *special* case which contains all the germs of *generality*." In our view, angular momentum theory plays the role of that "special case," with symmetry—one of the most fruitful themes of modern mathematics and physics—as the "generality." We would only amend Hilbert's phrase to include physics as well as mathematics. In the Preface to the second edition of his famous book *Group Theory and Its Applications to the Quantum Mechanics of Atomic Spectra*, Wigner[2] records von Laue's view of how remarkable it is that "almost all the rules of [atomic] spectroscopy follow from the symmetry of the problem." The symmetry at issue is *rotational symmetry*, and the spectroscopic rules are those implied by *angular momentum conservation*. In this monograph, we have tried to expand on these themes.

The fact that this monograph is part of an encyclopedia imposes a responsibility that we have tried to take seriously. This responsibility is rather like that of a library. It has been said that a library must satisfy two disparate needs: One should find the book one is looking for, but one should also find books that one had no idea existed. We believe that much the same sort of thing is true of an encyclopedia, and we would be disappointed if the reader did not have both needs met in the present work. To accomplish this objective, we have found it necessary to split our monograph into two volumes, one dealing with the "standard" treatment of angular momentum theory and its applications, the other dealing in depth with the fundamental concepts of the subject and the interrelations of angular momentum theory with other areas of mathematics.

Fulfilling this responsibility further, we have made an effort to address readers who seek *very* detailed answers on *specific* points—hence, we have a large index, and many notes and appendices—as well as readers who seek an overview of the subject, especially a description of its unique and appealing aspects. This accounts for the uneven level of treatment which varies from chapter to chapter, or even within a chapter, quite unlike a

[1]Quoted in M. Kac, "Wiener and integration in function spaces," *Bull. Amer. Math. Soc.* **72** (1966), p. 65. (The italics are in the original; Kac notes that the statement may be apocryphal.)

[2]E. P. Wigner, *Group Theory and Its Applications to the Quantum Mechanics of Atomic Spectra*. p.v. Academic Press, New York, 1959. (We have added in brackets the word "atomic," since this was clearly von Laue's intended meaning.)

textbook with its uniformly increasing levels of difficulty. The variation in the treatment applies particularly to the Remarks. Quite often these Remarks contain material that has not been developed or explained earlier. Such material is intended for the advanced reader, and it can be disregarded by others. We urge the reader to browse and skip, rather than trying, at first, any more systematic approach.

These considerations apply also to the applications. Some applications may be almost too elementary, whereas others are at the level of current research. The field of applications is so broad that we have surely failed to do justice in many cases, but we do hope that the treatment of some applications is successful.

In discussing a particular subject, we have given more detail than is usual in mathematical books, where terseness is considered the cardinal virtue. Here we have followed the precepts of Littlewood[3] who points out that "*two* trivialities omitted can add up to an *impasse*."

Let us acknowledge one idiosyncrasy of our treatment: We have not explicitly used the methods of group theory, per se, but have proceeded algebraically so that the group theory, if it appears at all, appears naturally as a subject is developed. No doubt this method of technique is an overreaction to the censure—now disappearing—with which many physicists greeted the *Gruppenpest*[4]. In any event, we think that this treatment does make the material more accessible to some readers.

Let us make some brief suggestions as to how to use the first volume, *Angular Momentum in Quantum Physics* (AMQP). Part I: (*i*) Chapters 2 and 3 and parts of Chapter 6 constitute the standard treatment of angular momentum theory and will suffice for many readers who wish to learn the mechanics of the subject. The methods used are elementary (but by no means imprecise), and the whole treatment flows from the fundamental commutation relations of angular momentum. (*ii*) Chapters 4 and 5 are recommended to the reader who wishes a general overview of the subject with methods capable of great generalization. Paradoxically, although these two chapters contain much new material, this material also belongs to the very beginnings of the subject—in the multiplication of forms of Clebsch and Gordan, and in the ξ-η calculus of Weyl—all of which are now incorporated under the rubric of the "boson calculus." Part II: The applications given in Chapter 7 are totally independent of one another and can be understood from the results given in Chapter 3.

The second volume, *Racah-Wigner Algebra in Quantum Theory* (RWA), is also presented in two parts. Part I: In Chapters 2, 3, and 4 the algebra of the

[3] J. E. Littlewood, *A Mathematician's Miscellany*, p. 30. Methuen, London, 1953, (The italics are in the original.)

[4] B. G. Wybourne, "The Gruppenpest yesterday, today, and tomorrow," *International Journal of Quantum Chemistry*, Symposium No. 7 (1973), pp. 35–43.

operators associated with the two basic quantities in angular momentum theory—the Wigner and Racah coefficients—is developed within the framework of the algebra of bounded operators acting in Hilbert space. These chapters are intended to rephrase the concept of a "Wigner operator" (tensor operator) in algebraic terms, using methods from Gel'fand's development of Banach algebras. This approach to angular momentum theory is rather new, and is intended for the reader who wishes to pursue the subject from the viewpoint of mathematics. Part II: The twelve topics developed in Chapter 5 establish diverse interrelations between concepts in angular momentum theory and other areas of mathematics. These topics are independent of one another, but do draw for their development on the material of Chapter 3 of AMQP, and to a lesser extent on Chapters 2–4 of RWA. This material should be of interest to both mathematicians and physicists.

L. C. BIEDENHARN
J. D. LOUCK

Acknowledgments

This monograph could not have been completed without the extensive help of friends and colleagues. Professor L. P. Horwitz performed the vital chore of a thorough reading and criticism of the entire two volumes; his help is gratefully acknowledged.

Other colleagues who have helped us by reading and criticizing particular chapters, the applications, or topics in *Angular Momentum in Quantum Physics* and in *Racah-Wigner Algebra in Quantum Theory* are: (*i*) AMQP. D. Giebink, Chapters 2–4; Professors H. Bacry and B. Wolf, Chapter 4; Professor R. Rodenberg and Dr. M. Danos, Chapters 2–6; Professor B. R. Judd, Chapter 7, Section 5; Drs. H. W. Galbraith, C. W. Patterson, and B. J. Krohn, Chapter 7, Section 10; (*ii*) RWA. Professor L. Michel, Chapters 2–4; Drs. H. Ruch and R. Petry, Topic 2; Dr. M. M. Nieto, Professors M. Reed, N. Mukunda, and H. Bacry, Topic 7; Professor T. Regge, Topic 9; Dr. B. J. Krohn, Topic 10; Dr. C. W. Patterson and Professor J. Paldus, Topic 12. Dr. W. Holman read both volumes to assist us with the indexing and suggested many improvements. Further acknowledgment of help from those not mentioned here is indicated in the relevant chapters.

In a more general way, we are indebted also to Professors H. van Dam, E. Merzbacher, A. Bohm, and Dr. N. Metropolis for discussions and help extending over several years.

This monograph is dedicated to Professor Eugene P. Wigner, whose picture (courtesy of the Niels Bohr Library of the American Institute of Physics) appears as the frontispiece. We not only acknowledge the inspiration of Wigner's research, but also record his personal encouragement and help, and we are honored that he has accepted this dedication. We wish also to acknowledge our great indebtedness to the late Professor Giulio Racah, who encouraged our work at its most critical time, the very beginning. Professor Ugo Fano also helped us greatly in this same period.

Special thanks are due to Professor Gian-Carlo Rota, editor of this Encyclopedia, for encouraging us to write at length on the subject of this monograph.

Most monographs begin as course notes and lectures series. The present monograph is no exception and evolved from such notes and lectures given over the years. We are particularly indebted to Dr. H. William Koch for urging us to write up the lectures on angular momentum theory (based on the concept of the turn) given at the National Bureau of Standards. This was finally achieved many years later, when a much expanded version was presented at Canterbury University (Christchurch, New Zealand, in 1973) and supported by an Erskine Fellowship arranged by Professor Brian Wybourne; part of the actual writing of the monograph was carried out during the tenure of a von Humboldt Fellowship (1976) under the sponsorship of Professor Walter Greiner at the University of Frankfurt (am Main).

Without the patience and helpful attitude of the editorial and production staffs of the Advanced Book Program of Addison-Wesley, and the free-lance copy editor, Dorothea Thorburn, it would not have been possible to split our original manuscript into the present two volumes.

The entire typing of the original manuscript and several of its revisions were carried through, without flinching, by Nancy Simon. Her loyalty to this task, extending over several years, is especially appreciated. Thanks are also given to Lena Diehl and Julia Clark for typing many of the final revisions. We are also indebted to Graphic Arts and Illustration Services of the Los Alamos National Laboratory for the figures and reproductions.

The writing of a monograph can be traumatic to others besides the authors. We wish to thank our wives, Sarah Biedenharn and Marge Louck, for their unfailing support and encouragement in keeping us at our task.

L. C. BIEDENHARN
J. D. LOUCK

CHAPTER 1

Introduction

The quantum theory of angular momentum, as developed principally by Eugene Wigner and Giulio Racah, has become an indispensable discipline for the working physicist, irrespective of his field of specialization, be it solid-state physics, molecular-, atomic-, nuclear-, or even hadronic-structure physics. Angular momentum theory is no less valuable to theoretical physicists, mathematical physicists, and mathematicians, as one of the prime examples of the far-reaching power and great beauty of the symmetry approach that underlies, and generalizes, angular momentum theory.

The concept of angular momentum, defined initially as the moment of momentum ($\vec{L} = \vec{r} \times \vec{p}$), originated very early in classical mechanics. (Kepler's second law, in fact, contains precisely this concept.) Nevertheless, angular momentum had, for the development of classical mechanics, nothing like the central role this concept enjoys in quantum physics. Wigner [1] notes, for example, that most books on mechanics written around the turn of the century (and even later) do not mention the general theorem of the conservation of angular momentum. In fact, Cajori's well-known *History of Physics* [2] (1929 edition) gives exactly *half a line* to angular momentum conservation.

That the concept of angular momentum may be of greater importance in quantum mechanics is almost self-evident. The Planck quantum of action, h, has precisely the dimensions of an angular momentum, and, moreover, the Bohr quantization hypothesis[1] specified the unit of (orbital) angular momentum to be $\hbar \equiv h/2\pi$. Angular momentum theory and quantum physics are thus clearly linked.

---

[1] We discuss this in more detail in the Note to Section 3 of Chapter 7.

ENCYCLOPEDIA OF MATHEMATICS and Its Applications, Gian-Carlo Rota (ed.).
Vol. 8: L. C. Biedenharn and J. D. Louck, Angular Momentum in
Quantum Physics: Theory and Application

One may discern two principal reasons underlying this change in the importance of angular momentum for quantum mechanics. The first reason—which applies to some extent also to classical mechanics—is the change in the conceptual framework that gradually occurred as the Newtonian laws of motion were replaced by the Lagrangian (and later, Hamiltonian) equations. The second is peculiar to quantum mechanics and concerns the quantal concept of state. Let us discuss these two reasons briefly, even though a fully satisfactory treatment lies outside the purview of this monograph.

The shift in emphasis from the Newtonian laws of motion to the Lagrangian formulation was to some extent pragmatic (to eliminate holonomic constraints) but primarily motivated toward conceptual generalization (for example, to eliminate the particular role of Cartesian coordinates). This trend continues even today, for, as Lichnerowicz [3] has pointed out, the analytic mechanics of Lagrange, by providing the first examples of natural differentiable manifolds of arbitrary dimension, began what has now become differential geometry (Sternberg [4], Abraham [5], MacLane [6]).

To appreciate the significance of this change in viewpoint, let us recall that, originally, angular momentum was considered important because it was (for central forces) a *conserved quantity*, just as the analogous conservation of linear momentum gave content to that concept. But the true import and depth of the conceptual change became apparent only when Hamel [7] in 1904—using the Lagrangian–Hamiltonian formulation of mechanics—established that *there is a connection between the conservation laws of linear and angular momentum and the fundamental symmetries of time and space* (see Note 1).

For an isolated system one necessarily has the symmetry implied by the homogeneity (spatial translational invariance) and isotropy (rotational invariance) of space; thus, the Lagrangian is invariant to translations and to rotations, and hence one obtains, in this way, the conservation laws of linear and angular momentum.

This fundamental relationship between symmetry principles and conservation laws culminated in the classical general theorem of Noether [8] (see Note 1) relating the general invariance properties of the Lagrangian both to the conserved quantities and to their associated conservation laws (Roman [9]).

These results greatly increased the conceptual importance of angular momentum, and this lesson has, by now, been so thoroughly absorbed that "angular momentum conservation" and "invariance under rotation" are currently viewed as virtually synonymous.

It is important to note, however, that such a strong view is not completely justified, for, as Wigner [10] has emphasized, *the invariance of the equations of motion to rotations (say) does not imply the conservation of angular momentum.* The existence of a Lagrangian (or, correspondingly, a Hamiltonian) is, in fact, *essential* in establishing the connection between symmetry (invariance) and conservation laws. Conversely, we see from this the importance of the conceptual change from Newtonian mechanics (as exemplified by the Newtonian equations of motion).

Remarks. (*a*) There is a nice illustration of the point made in the last paragraph. Consider the motion of a particle in a viscous medium. The energy of such a particle is not conserved, although the equations of motion are indeed invariant under time displacement. (This example is taken from Ref. [10]; an example for angular momentum is given in Ref. [11].) (*b*) When one considers the conservation laws of momentum and angular momentum for particle systems interacting via the electromagnetic field with charges and magnets, there can be many apparent paradoxes, even at the classical, nonrelativistic level. A nice discussion of several such pseudo-paradoxes is given in the survey by Furry [12], where, in particular, the classical Trouton–Noble experiment and the more recent Shockley–James examples on "hidden" momentum are discussed.

The second reason for the greater importance of angular momentum in quantal physics than in classical physics is, as mentioned, the change in the concept of state. For quantum mechanics, the states of a system are vectors (more precisely, rays) in a separable Hilbert space; accordingly, one has (axiomatically) the possibility of linear superposition for state vectors. The validity of the superposition principle is one of the characteristic features of quantum mechanics and is qualitatively different, as Dirac [13] emphasized, from any kind of classical superposition (as might occur in a classical linear system).

To appreciate the profound difference that the superposition of quantal states makes possible, let us consider as an example the Kepler orbit of an electron moving around the proton in a hydrogen atom. For classical physics, the rotational symmetry of the problem (taking the proton to define the origin) merely implies that the rotated orbit for any arbitrary orientation is also an allowed motion having the same energy. This "new" information is rather obvious, and not very useful.[1]

By contrast, in quantum physics the fact that the rotated state vector is also an allowed state of the same energy is now implemented by the superposition principle: The original state may be superposed with the set of states generated by rotations to produce, in general, *new* states having simple rotational properties (for example, rotational invariance). It is this possibility that is fundamentally new in quantum theory.

[1] The fact that classically the orbit could have any orientation contains implicitly a paradox (for the classical hydrogen atom) that is resolved only by quantum mechanics (Wigner [14, p. 798]). Namely, there must be vanishing entropy connected with these infinitely many orientations, if they exist.

It is thus the hypothesis that in quantum physics the states form a linear manifold that changes the role of symmetry from being an interesting observation, as in classical physics, into an important constructive technique of great power and flexibility. Equally significant for this monograph is the fact that symmetry techniques are important not only for the state vectors of quantum physics, but also for quantum mechanical operators. (This is apparent from the observation that the set of linear transformations of a vector space is itself a vector space.) Thus, the operators of quantum mechanics, which are necessarily linear by the association of physical measurements to operators, can be classified by their symmetry, a concept that leads to the notion of tensor operators (treated in detail in Chapter 3 *et seq.*).

The fact that in quantum mechanics the state vectors form a linear manifold leads to a remarkable strengthening of the connection between conservation laws and invariance principles: *The conservation laws now follow from the basic kinematic concepts* (Wigner [1, p. 20]). Symmetry transformations are found to play a dual role in quantum mechanics in that the generators of infinitesimal symmetry transformations not only generate the transformations, but are themselves the conserved quantities. It is in this latter aspect of symmetry in quantum mechanics that physicists find their predilection for algebraic concepts (the algebra of the symmetry generators such as the angular momentum operator **L**) as opposed to the global (group) concept of the symmetries themselves. This local, algebraic, viewpoint is the viewpoint generally taken in this monograph.

We are now in a position to understand the central role played by angular momentum in quantum physics, as implied by the title of this monograph, for we see that the importance of angular momentum is based directly on kinematic concepts taken to characterize quantum physics itself. Thus, for nonrelativistic physics based on the kinematics of Newtonian relativity, we find the Galilei group as the group of invariance transformations; and for relativistic physics we have Einsteinian relativity with the Poincaré group as the group of invariance transformations. These kinematic symmetries, as discussed above, then imply the conserved quantities of the theory; in particular, in both relativities angular momentum is a conserved quantity. (See Schwinger [15] for a readable and authoritative account of this point of view.)

It follows from these results that *angular momentum plays a fundamental role in physics, a role that is preserved in going from Newtonian relativity to Einsteinian relativity. Angular momentum is, in both relativities, a conserved quantity related to the isotropy of space–time*. (See Notes 2 and 3.)

Notes

1. A brief survey of the literature tracing the development of the connection between conservation laws and the invariance of the Lagrangian from Jacobi

(1842) to Bessel-Hagen (1921) is given in the review by Houtappel *et al.* [11] (see their footnote 20). A thorough discussion of invariance and symmetry (restricted, however, to classical physics) has been given by Caratù *et al.* [16].

2. Having laid such stress on the importance of the superposition principle in implementing symmetry, it is surprising, and perhaps ironic, that angular momentum also furnishes an example of the contrary result: the splitting of the Hilbert space of quantum states into *noncombining* sets. (This is the concept of a *superselection rule.*) The example furnished by angular momentum is the splitting into states having half-integer angular momenta and states having integer angular momenta. This splitting, from the Pauli–Lüders spin-statistics theorem, is associated with the split into distinct types of particles having bosonic versus fermionic properties. (This superselection rule is discussed in Topic 1, RWA.)

3. Internal angular momentum is a classical, nonrelativistic concept (this follows from the *additivity* of the generators of the Galilean symmetry). The existence of discrete internal angular momentum is a standard result in classical, nonrelativistic field theory, as is clear from the existence of scalar waves (no internal angular momentum) and vector waves with longitudinal and/or transverse polarization (unit internal angular momentum).

Hepp and Jensen [17] have shown that spin-$\frac{1}{2}$ internal angular momentum is also a classical nonrelativistic field-theoretic concept. Arguing that the actual historical development of the particle picture for cathode rays (electrons) was in a large part a result of inadequately sensitive experimental devices, they reconstruct from the experimental facts a logically admissible classical field-theoretic interpretation, which is then quantized by the Jordan–Klein–Wigner procedure (so-called second quantization). In so doing, Hepp and Jensen clear up several misconceptions on the classical role of spinor fields.

The systematic use of the Galilean symmetry approach, such as that given by Lévy-Leblond [18], is an equally valid way to clarify these same concepts. In both approaches it is shown, for example, that the approximate g-factor of the electron ($g_S = 2$) is in no way a (Poincaré) relativistic result, but can readily result from a linear Galilean covariant wave equation. [Alternatively (as noted by Feynman) one may write $\vec{p}^2/2m$ (the nonrelativistic kinetic energy) in spinorial form, $(\vec{\sigma} \cdot \vec{p})^2/2m$, where $\vec{\sigma}$ is the Pauli spin vector. Then $\vec{p} \to \vec{p} - e\vec{A}/c$ yields $g_S = 2$.]

An informal survey (by Lévy-Leblond) found that some forty out of forty-six physics texts examined confused spin-$\frac{1}{2}$ and/or $g_S = 2$ as a special relativistic result.

The Galilean symmetry approach leads in a natural way to the Schrödinger equation. The invariance group of the Schrödinger equation itself, however, leads to a larger symmetry structure (see Niederer [19] and references cited there).

References

1. E. P. Wigner, *Symmetries and Reflections*, p. 14. Indiana Univ. Press, Bloomington and London, 1967. This book reprints the original reference: "Symmetry and conservation laws," *Proc. Natl. Acad. Sci. U.S.* **51** (1964), 956–965.

2. F. Cajori, *History of Physics*. Macmillan, New York, 1929.
3. A. Lichnerowicz, "Differential geometry and physical theories," in *Perspectives in Geometry and Relativity* (B. Hoffmann, ed.), pp. 1–6. Indiana Univ. Press, Bloomington, Ind., 1966.
4. S. Sternberg, *Celestial Mechanics*. Benjamin, New York, 1969.
5. R. Abraham, *Foundations of Mechanics*. Benjamin, New York, 1967.
6. S. MacLane, *Geometrical Mechanics*, Lectures in Department of Mathematics, Part I, pp. 1–124; Part II, pp. 1–111. University of Chicago, 1968.
7. G. P. Hamel, *Theoretische Mechanik*. Teubner, Stuttgart, 1912.
8. E. Noether, *Göttinger Nachrichten* (1918), 235.
9. P. Roman, *Theory of Elementary Particles*, 2nd ed., Chapter IV, Section 1a. North-Holland, Amsterdam, 1961.
10. E. P. Wigner, "Conservation laws in classical and quantum physics," *Prog. Theoret. Phys.* **11** (1954), 437–440.
11. R. M. F. Houtappel, H. van Dam, and E. P. Wigner, "The conceptual basis and use of the geometric invariance principles," *Rev. Mod. Phys.* **37** (1965), 595–632.
12. W. H. Furry, "Examples of momentum distributions in the electromagnetic field and in matter," *Amer. J. Phys.* **37** (1969), 621–636.
13. P. A. M. Dirac, *Principles of Quantum Mechanics*, 1st ed. Oxford Univ. Press, London, 1930.
14. E. P. Wigner, "Symmetry principles in old and new physics," Bull. Amer. Math. Soc. **74** (1968), 793–815.
15. J. Schwinger, *Particles, Sources, and Fields*. Addison-Wesley, Reading, Mass., 1970.
16. G. Caratù, G. Marmo, E. J. Saletan, A. Simoni, and B. Vitale, *Invariance and Symmetry in Classical Mechanics*. Istituto di Fisica Teorica, Naples, Italy, 1977. (To be published by Wiley, New York, 1982.)
17. H. Hepp and H. Jensen, "Klassische Feldtheorie der polarisierten Kathodenstrahlung und ihre Quantelung," *Sitzber. Heidelberg. Akad. Wiss. Math.-Naturw. Kl.* 4 Abh. (1971), 89–122.
18. J.-M. Lévy-Leblond, "Galilei group and Galilean invariance," in *Group Theory and Its Applications* (E. M. Loebl, ed.), Vol. II, pp. 222–299. Academic Press, New York, 1971.
19. U. Niederer, "The maximal kinematical invariance group of the free Schrödinger equation," *Helv. Phys. Acta* **45** (1972), 802–810; "The maximal kinematical invariance groups of Schrödinger equations with arbitrary potentials," *ibid.* **47** (1974), 167–172.

CHAPTER 2

The Kinematics of Rotations

1. Introduction

It is the purpose of this section to discuss the kinematics of rotations; by this term we mean *the description of physical objects under rotation*. Despite the apparent simplicity of the subject, the kinematics of rotations will prove to have some subtleties (see Sections 3 and 4).

The essential element in the description of physical systems lies in the association of a mathematical model to the underlying space (Mackey [1]). This association is necessarily postulated, and ultimately is an assumption as to the validity of a given system of relativity. Let us begin with the spatial concepts postulated in Newtonian relativity. The mathematical model to be associated with this physics is that physical (mass) points are to be identified with points belonging to a three-dimensional Euclidean space, $E(3)$. "Three-dimensional" means that a point is a triple of real numbers, point $\leftrightarrow \mathbf{x} \equiv (x_1, x_2, x_3)$, where $x_i \in \mathbb{R}$; "Euclidean space" means that under spatial symmetries belonging to the relativity group all distances between points are preserved (hence, all lengths and all angles are preserved).

The symmetries of Euclidean space can be composed from two special symmetries: (*a*) *translations*, which displace all points similarly: $\mathbf{x} \rightarrow \mathbf{x}' = \mathbf{x} + \mathbf{a}$, $\mathbf{a} = (a_1, a_2, a_3)$; and (*b*) *rotations and reversals*, which leave one point fixed – rotations preserve orientation, whereas reversals (Cartan's [2] term) reverse orientation.

2. Properties of Rotations

It is well known that *a Euclidean symmetry leaving one point fixed leaves all points along some line through this point fixed*. Accordingly, a *rotation* is a

---

ENCYCLOPEDIA OF MATHEMATICS and Its Applications, Gian-Carlo Rota (ed.).
Vol. 8: L. C. Biedenharn and J. D. Louck, Angular Momentum in
Quantum Physics: Theory and Application

Euclidean symmetry characterized by *an axis* (denoted by a unit vector $\hat{n}$), *an angle* of rotation, ϕ, and a *sense* of rotation about the axis (taken always as the "right-hand screw" rule). We denote a rotation by $\mathscr{R}(\phi, \hat{n})$, or by $\mathscr{R}$, if the meaning is clear.

From geometric considerations, one can obtain a general result for the transformation of a given *vector* $\vec{r}$, [defined as the directed line from the origin $(0, 0, 0)$ to the point (x, y, z)] under a rotation[1] (leaving the origin fixed). The component of $\vec{r}$ along $\hat{n}$ must remain unchanged: $(\hat{n} \cdot \vec{r})\hat{n} \to (\hat{n} \cdot \vec{r})\hat{n}$. The perpendicular component, $\vec{r} - (\hat{n} \cdot \vec{r})\hat{n}$, rotates through an angle ϕ and, like every vector perpendicular to $\hat{n}$, can be expressed, after rotation, as a linear combination of $\vec{r} \times \hat{n}$ and $\vec{r} - (\hat{n} \cdot \vec{r})\hat{n}$. One finds

$$\vec{r} - (\hat{n} \cdot \vec{r})\hat{n} \to [\vec{r} - (\hat{n} \cdot \vec{r})\hat{n}] \cos \phi + (\hat{n} \times \vec{r}) \sin \phi,$$

so that

$$\mathscr{R}(\phi, \hat{n}) \colon \vec{r} \to \vec{r}\,' = \mathscr{R}(\phi, \hat{n})\vec{r} = \vec{r} \cos \phi + \hat{n}(\hat{n} \cdot \vec{r})(1 - \cos \phi) + (\hat{n} \times \vec{r}) \sin \phi. \tag{2.1}$$

This is a useful relation for discussing rotations of vectors.

This result may be used to obtain the characteristic vectors of a rotation. Thus, one verifies directly that

$$\mathscr{R}(\phi, \hat{n})(\hat{p} + i\hat{q}) = e^{-i\phi}(\hat{p} + i\hat{q}),$$

$$\mathscr{R}(\phi, \hat{n})(\hat{p} - i\hat{q}) = e^{i\phi}(\hat{p} - i\hat{q}),$$

$$\mathscr{R}(\phi, \hat{n})\hat{n} = \hat{n}, \tag{2.2}$$

where $\hat{p}$ and $\hat{q}$ are any unit vectors perpendicular to $\hat{n}$ such that $(\hat{n}, \hat{p}, \hat{q})$ constitute a right-handed triad of perpendicular unit vectors. Thus, the eigenvalues of the rotation $\mathscr{R}(\phi, \hat{n})$ are $1, e^{i\phi}, e^{-i\phi}$.

The geometric point of view shows that a rotation $\mathscr{R}$ is a linear transformation of vectors with the two properties

$$\mathscr{R} \colon \ \vec{r} \cdot \vec{s} \to \mathscr{R}\vec{r} \cdot \mathscr{R}\vec{s} = \vec{r} \cdot \vec{s},$$

$$\mathscr{R} \colon \ \vec{r} \times \vec{s} \to \mathscr{R}\vec{r} \times \mathscr{R}\vec{s} = \mathscr{R}(\vec{r} \times \vec{s}). \tag{2.3}$$

[1] The terms "rotate" and "rotation" are sometimes used in a looser sense than we mean here. For example, a reuleaux is said to be able "to rotate inside a square" (Santaló [2a, p. 9]). This is *not* a rotation in the sense we use the term since the axis is *not* constant during the motion. (It is true that a reuleaux can be continuously put in a square in all positions.)

Conversely, any linear transformation of Euclidean space with these properties is a rotation.

Rotations form a group. We have introduced rotations as Euclidean symmetries. It is necessary only to observe that two rotations having a common center form a combined rotation (with the same center) to conclude that the rotations themselves form a group. The "product" of two rotations $\mathscr{R}_1$ and $\mathscr{R}_2$ (denoted $\mathscr{R}_1\mathscr{R}_2$) is the transformation (symmetry) resulting from acting *first* with $\mathscr{R}_2$, *then* with $\mathscr{R}_1$. This product is (as shown directly from geometric considerations or as inherited from the defining Euclidean group) *associative*: $(\mathscr{R}_1\mathscr{R}_2)\mathscr{R}_3 = \mathscr{R}_1(\mathscr{R}_2\mathscr{R}_3)$. The inverse to $\mathscr{R}(\phi, \hat{n})$ is $\mathscr{R}(-\phi, \hat{n})$, also denoted $\mathscr{R}^{-1}(\phi, \hat{n})$. The identity rotation is the zero rotation ($\phi = 0$—that is, no rotation at all).

Rotations having a *common axis* form an *abelian* (commuting) *subgroup*:

$$\mathscr{R}(\phi_1, \hat{n})\mathscr{R}(\phi_2, \hat{n}) = \mathscr{R}(\phi_2, \hat{n})\mathscr{R}(\phi_1, \hat{n}) = \mathscr{R}(\phi_1 + \phi_2, \hat{n}). \tag{2.4}$$

In general, rotations do not commute.

Rotations having a *common angle* obey the *class angle* relation (Wigner [3, p. 150]), $\mathscr{S}\mathscr{R}(\phi, \hat{n})\mathscr{S}^{-1} = \mathscr{R}(\phi, \mathscr{S}\hat{n})$, where $\mathscr{S}$ is a rotation carrying the rotation axis $\hat{n}$ into a new rotation axis $\hat{n}' = \mathscr{S}\hat{n}$.

A useful application of this result is the proof of the following identity between two sequences of rotations:

$$\mathscr{R}_n \cdots \mathscr{R}_2\mathscr{R}_1 = \mathscr{R}'_1\mathscr{R}'_2 \cdots \mathscr{R}'_n, \tag{2.5}$$

where $\mathscr{R}_k$ is a rotation of ϕ_k about $\hat{n}_k$, and $\mathscr{R}'_k$ is a rotation of ϕ_k about $\hat{n}'_k = (\mathscr{R}_{k-1} \cdots \mathscr{R}_2\mathscr{R}_1)^{-1}\hat{n}_k$ $(k = 2, \ldots, n)$ and $\hat{n}'_1 = \hat{n}_1$.

The geometric viewpoint described above focuses attention on the points as entities, and the transformations they undergo. A less intuitive but more powerful procedure is the analytic approach, which focuses attention on the three measure numbers correlated to any point. In this approach a triple is a relation between *two* structures: the *point* designated by the triple, and the *frame* by which the numbers are defined. Specifically, the *frame*, denoted $F = (\hat{e}_1, \hat{e}_2, \hat{e}_3)$, consists of vectors, $\hat{e}_i$ $(i = 1, 2, 3)$, that have a common origin; are of unit length $(\hat{e}_i \cdot \hat{e}_i = 1)$; are mutually orthogonal $(\hat{e}_i \cdot \hat{e}_j = 0, i \neq j)$; and are oriented $(\hat{e}_1 = \hat{e}_2 \times \hat{e}_3)$. Using the frame F, the triple $\mathbf{x} = (x_1, x_2, x_3)$ associated to the point $\mathbf{x}$ has $x_i \equiv \vec{x} \cdot \hat{e}_i$, $\vec{x}$ denoting the vector from the origin to the point $\mathbf{x}$.

In the analytic approach one distinguishes two alternative meanings of a transformation: (*a*) *Active meaning*. Each point is transformed into a new point. The physical significance is that the physical (mass) points are moved into new positions. (*b*) *Passive meaning*. The coordinate frame is changed, so that the (unchanged) points receive new labels. The physical significance is clearly that of a *re-description* without any physical changes occurring.[1]

Rotations as matrices. A rotation $\mathscr{R}$ has a well-defined action on each of the unit vectors defining a frame. A 3×3 matrix R can thus be associated with the rotation $\mathscr{R}$ by taking the number $R_{ij} \equiv \hat{e}_i \cdot \mathscr{R}\hat{e}_j$ to be the element belonging to row i and the column j of R. To each rotation $\mathscr{R}$ there corresponds a matrix R.

The action of $\mathscr{R}$ on an arbitrary vector, $\vec{x} = \sum_i x_i \hat{e}_i$, can now be written as $\mathscr{R}\vec{x} = \sum_i (\sum_j R_{ij} x_j)\hat{e}_i$. By writing the vectors $\vec{x}$ and $\mathscr{R}\vec{x} = \vec{x}'$ as matrices with one column, $\mathbf{x} = \mathrm{col}(x_1, x_2, x_3)$ and $\mathbf{x}' = \mathrm{col}(x_1', x_2', x_3')$, one obtains the matrix equation, $\mathbf{x}' = R\mathbf{x}$. In this way one obtains a correspondence $\mathscr{R} \to R$ between rotations and matrices such that the product of two rotations $\mathscr{R}_1\mathscr{R}_2$ corresponds to the matrix product $R_1 R_2$. The correspondence $\mathscr{R} \to R$ is said to be a *representation* of the group of rotations.

Reformulating rotations in these terms yields the matrix result:

$$\mathscr{R}(\phi, \hat{n}) \to R(\phi, \hat{n}) = \mathbb{1}_3 + N \sin\phi + N^2(1 - \cos\phi). \tag{2.6}$$

In this result, $\mathbb{1}_3$ is the 3×3 identity matrix, and N is a skew-symmetric matrix determined by the axis, $\hat{n}$, of the rotation: $N_{ij} = -e_{ijk} n_k$, where $e_{ijk} = +1$ and -1, respectively, for even and odd permutations (i, j, k) of $(1, 2, 3)$, and is otherwise 0.

This result for the matrix $R(\phi, \hat{n})$ shows that the set of 3×3 matrices representing rotations are *real*, *orthogonal* ($\tilde{R} = R^{-1}$, where the tilde denotes matrix transposition), and *proper* ($\det R = 1$). We denote this group of matrices by $SO(3)$. [That every real, proper, orthogonal matrix has the form (2.6) is proved in Section 6.]

Let us emphasize that the representation of a rotation $\mathscr{R}$ by a 3×3 real, proper, orthogonal matrix is defined relative to a fixed reference frame, $F = (\hat{e}_1, \hat{e}_2, \hat{e}_3)$. We shall consider in the next section the more difficult question as to the converse—that is, whether or not such a matrix determines a unique rotation.

3. Dirac's Construction

We wish to discuss in this section a seemingly simple question of kinematics: Is the mapping of rotations onto proper, orthogonal matrices $\mathscr{R} \to R$ one-to-

[1] It is common to express this distinction by the Latin terms "alias" versus "alibi" transformations. To correlate these with the previous terms, note that "alias" comes from *alias dictus* (L.), meaning "otherwise called," so that an "alias" transformation is passive, whereas "alibi" comes from *ali-ubi* (L.), meaning "other-where" or "elsewhere;" this clearly agrees with the active transformations.

one or not? Put in a different way, this question becomes: Is a rotation by $\phi = 2\pi$ about any axis *physically equivalent* to the identity rotation?

It is certainly true, from Eq. (2.1), that $\mathscr{R}(\phi + 2k\pi, \hat{n}) = \mathscr{R}(\phi, \hat{n})$ $(k = 0, \pm 1, \pm 2, \ldots)$, but let us recall that this is actually an identity for rotations acting on vectors: $\mathscr{R}(\phi + 2k\pi, \hat{n})\vec{r} = \mathscr{R}(\phi, \hat{n})\vec{r}$ for an arbitrary vector $\vec{r}$. Thus, a 2π rotation is equivalent to the identity rotation for a collection of physical mass points whose properties are determined by the corresponding collection of instantaneous position vectors. We can always map the rotations of such mass points onto proper, orthogonal matrices by introducing a frame F. But this leaves open the question as to whether a rotation by 2π is equivalent to the identity for all conceivable physical objects and, hence, as to whether the group of rotations is equivalent to the group of proper, orthogonal matrices.

Dirac[1] has given a very ingenious construction to demonstrate that, *for solid bodies*, a rotation by 2π is *not* equivalent to the identity, but that a rotation by 4π *is* equivalent to the identity. This fact will also be shown to be valid for *coordinate frames* as well as for solid bodies, thereby defining (as will be discussed) a quite unsuspected kinematic attribute of a coordinate frame, its *version* (Misner *et al.* [5]).

The concept of a *solid body* (nonpenetrable region of space), in contrast to that of a *rigid body* (fixed distances between mass points), is essential to this construction, as we shall show.

Consider an arbitrary solid body in free space; for simplicity, take it to be a cube (marked on its faces so as to be asymmetric). Dirac's procedure is to

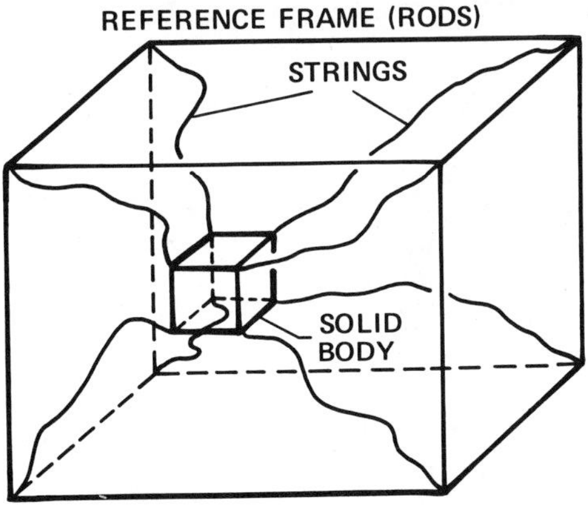

Figure 2.1.

[1] This construction is part of the more arcane lore in physics and has circulated widely, although it was never explicitly published by Dirac. (Newman [4] is the earliest published citation we have found.) We are indebted to Professor E. Merzbacher for first informing us of this construction.

identify points on the cube by attaching thin, perfectly extensible strings whose other end points are attached to a fixed reference frame. For definiteness, we use the arrangement shown in Fig. 2.1.

Now, perform a rotation of the solid body by 2π about *any* axis. The strings become hopelessly tangled up; no manipulation of the strings can untangle them. (The body and frame are to remain fixed, but *the strings must not be passed through the body or rods.*) If, however, we rotate an additional 2π (that is, start all over and rotate 4π about any axis), it is a remarkable fact that

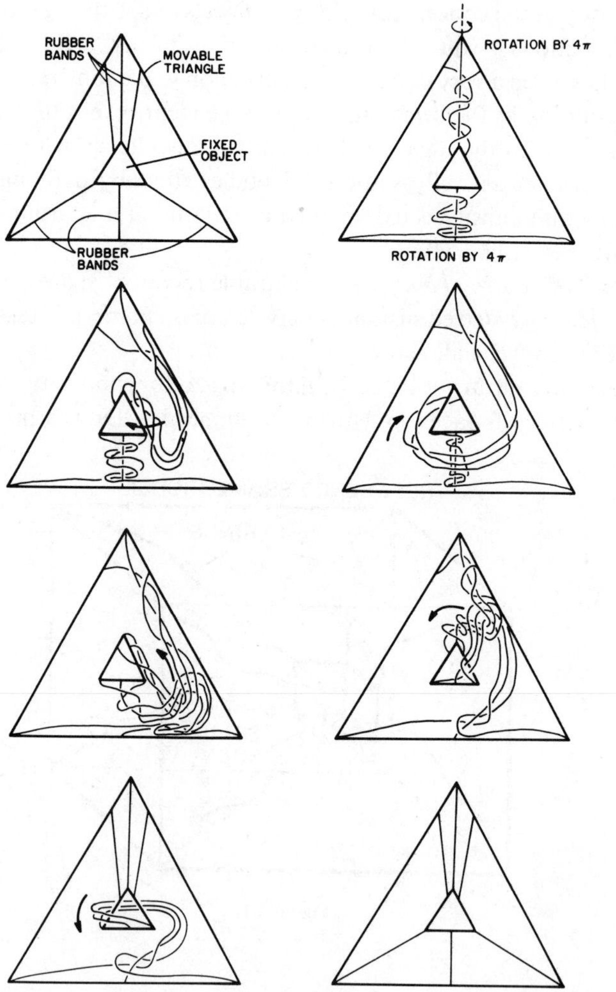

Figure 2.2.

suitable manipulation of the strings *does* lead to disentanglement recovering the initial configuration precisely.[1]

In Fig. 2.2 the operation of disentangling six strings after a rotation of 4π is illustrated in a sequence of diagrams. The solid body is here represented by the fixed triangle, and the reference frame by the movable triangle. One sees from these diagrams that the essential operation [as Footnote 1 below also shows] is that of carrying the strings *around* the solid body.

How many strings does one need? We discuss in Note 1 the analysis of this construction, which leads to the following conclusion: *The minimum number of strings for the construction is three.* It is easily verified that for fewer than three strings the entanglement due to a rotation by 2π can always be untangled (Fig. 2.3).

On the basis of this analysis, we can now assert the conclusion we draw from Dirac's construction: *It is a kinematic property of the real world that (a) a coordinate frame under rotation by 2π about any axis is in principle distinguishable from an unrotated coordinate frame; and (b) a coordinate frame under*

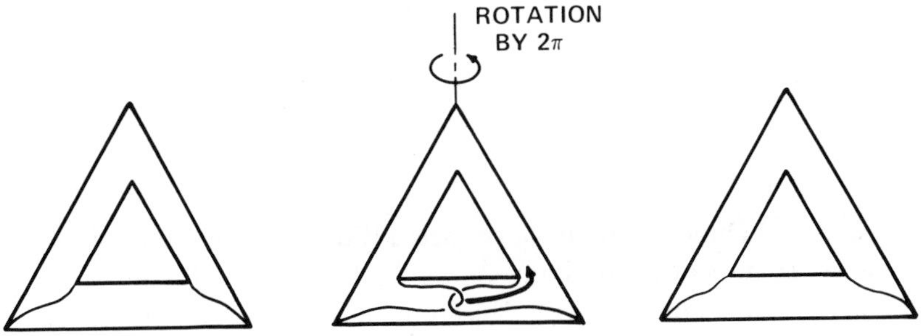

Figure 2.3.

Figure 2.4.

[1] The reader is invited to carry this experiment out for himself. An equivalent version of this construction uses a belt (= bundle of strings). Rotate the belt by 4π about an axis, say, along the length of the belt, and, holding the orientations of the belt fixed, carry the frame end around the buckle.

rotation by 4π *about any axis is indistinguishable from an unrotated frame*. This attribute of a frame we call its *version* (as denoted by Misner *et al.* [5, p. 1148]).

We have emphasized, in the discussion above, that the concept of a solid body (in the sense we use the term) is essential to the Dirac construction. As an illustration of this point, we show in Fig. 2.4 how one may untangle the strings, after a rotation by 2π, for a rigid body (fixed distances between mass points with *penetrable* regions between) containing three mass points. A second property is also brought out by consideration of Fig. 2.4: If we join the mass points by a line (impenetrable), it is still possible to untangle the strings, as illustrated in Fig. 2.5. Thus, *the solid body in the Dirac construction is an impenetrable region of space having a finite size* (3-simplex). (See also Weyl [6, pp. 165ff].)

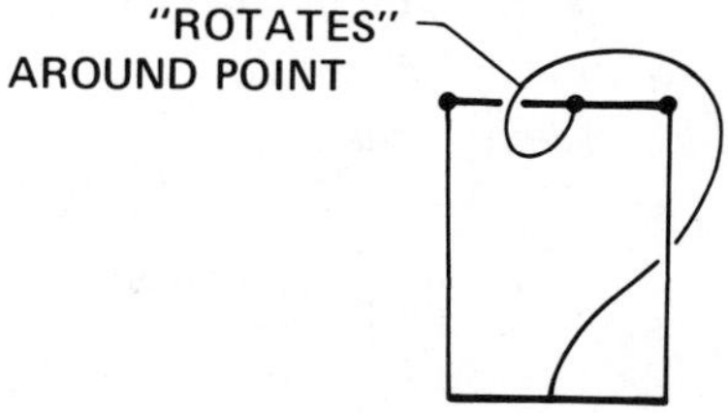

Figure 2.5.

Remarks. (*a*) Dirac's result must be carefully distinguished from the similar behavior of spinors under rotation. As we shall discuss (in Section 4), spinors are *point* objects, in contrast to the objects in Dirac's construction, which must have a *finite size*. (*b*) If we now turn to our original question, we see that the mapping, rotations $\{\mathscr{R}\} \to 3 \times 3$ matrices $\{R\}$, is 1:1 if we consider the rotations to be that of *points*, but clearly 2:1 if we consider the rotations to be that of a *physical* "solid body" (if such exists) or that of a *mathematical coordinate frame*. (*c*) The 2:1 mapping of frame rotations onto 3×3 matrices is conceptually the same (as we discuss in Section 6) as the 2:1 homomorphism of the $SU(2)$ group (multiplicative group of 2×2 unimodular unitary matrices) onto the $SO(3)$ group (multiplicative group of 3×3 proper, orthogonal, real matrices). It is an important contribution of Wigner [3] that $SU(2)$ is the group that enters quantum physics; it is the import of Dirac's construction that $SU(2)$ actually enters, conceptually, at the classical level (frame).

We discuss in Note 2 some of the implications of these results.

In the sequel we shall always mean by the phrase "rotation group" the group having $\mathscr{R}(4\pi, \hat{n}) = E =$ identity rotation, $\mathscr{R}(2\pi, \hat{n}) \neq E$—that is, the group $SU(2)$.

4. Cartan's Definition of a Spinor

Spinors were introduced into quantum mechanics, as the name indicates, to describe particles having half-integer intrinsic ("spin") angular momentum (Heisenberg and Jordan [7], Pauli [8]). In mathematics, however, spinors were discovered earlier[1], Cartan [9] having defined spinors in their most general form as early as 1913.

The great advantage of Cartan's approach is that it is purely geometric, and because of this geometric origin the Pauli matrices, familiar from quantum mechanics, appear of their own accord independently of any quantum mechanical considerations.

The key element in Cartan's approach is to consider[2] an *isotropic vector* (a vector of zero length) in three-dimensional (complex) Euclidean space. The components of such a vector satisfy $x_1^2 + x_2^2 + x_3^2 = 0$, and we can associate to this vector two complex numbers ξ_0, ξ_1 given by

$$x_1 = \xi_0^2 - \xi_1^2, \qquad x_2 = i(\xi_0^2 + \xi_1^2), \qquad x_3 = -2\xi_0\xi_1. \tag{2.7}$$

Alternatively, we can define ξ_0 and ξ_1 in terms of the isotropic vector $\mathbf{x}$ by

$$\xi_0 = \pm[\tfrac{1}{2}(x_1 - ix_2)]^{\frac{1}{2}}, \qquad \xi_1 = \pm[-\tfrac{1}{2}(x_1 + ix_2)]^{\frac{1}{2}}. \tag{2.8}$$

Cartan observed that it is not possible to assign a consistent sign to the equations above, *since under a* 2π *rotation* ξ_0, ξ_1 *reverse sign, but the isotropic vector returns to its original position.* (The relative sign of ξ_0, ξ_1 can be made definite, however.) Such behavior is the defining characteristic of a spinor [defined to be the pair (ξ_0, ξ_1)] in that a 4π rotation, and not 2π, is equivalent to the identity rotation. A spinor is a Euclidean tensor (Cartan [2, p. 42]).

This elegantly simple construction by Cartan points up two significant aspects: (*a*) It is necessary to generalize to complex numbers; (*b*) a (Cartan) spinor is a *point spinor* necessarily having *zero* length. This latter feature is in sharp contrast to the spinorial aspect of a solid body, which necessarily has *finite* size, as we saw in the previous section.

---

[1] Spinors are actually implicit in Hamilton's [10–12] quaternions (see also Cayley [13, 14]) and in the homographic transformations of Klein [15, 16] and Cayley [17], but to make the concept explicit is not trivial. To define spinors from quaternions is equivalent to defining a wave function from the density matrix (see Chapter 7, Section 7), or, in algebraic terms, to determining the maximal proper ideals of an algebra (see Chapter 2, RWA).

[2] For historical accuracy let us note that Klein and Sommerfeld [18] had considered defining a spinor by an isotropic three-vector in their amazingly foresighted treatment. This specific discussion is in Volume I, Section 3, p. 23 ff. They consider spinors as points lying on an imaginary cone of second order; they, of course, do not use the language of spinors or isotropic three-vectors.

Let us determine now how a spinor transforms under rotation by using the known transformation properties of *vectors* under rotation. The spinor (ξ_0, ξ_1) is associated with the isotropic vector (x_1, x_2, x_3); under a rotation R we have $\mathbf{x} \to \mathbf{x}' = R\mathbf{x}$, and (since lengths are preserved) the isotropic vector $\mathbf{x}'$ determines the transformed spinor (ξ_0', ξ_1'). Thus, for $(\xi_0')^2$ one finds from Eq. (2.7) that

$$\begin{aligned}(\xi_0')^2 &= \tfrac{1}{2}[(R_{11} - iR_{21})x_1 + (R_{12} - iR_{22})x_2 + (R_{13} - iR_{23})x_3] \\ &= \tfrac{1}{2}(R_{11} - iR_{21} + iR_{12} + R_{22})\xi_0^2 - (R_{13} - iR_{23})\xi_0\xi_1 \\ &\quad + \tfrac{1}{2}(-R_{11} + iR_{21} + iR_{12} + R_{22})\xi_1^2. \end{aligned} \tag{2.9}$$

It is remarkable that the right-hand side of this expression is a perfect square (the discriminant vanishes in consequence of the orthogonality of the matrix R). The expression for $(\xi_1')^2$ is also a perfect square. Thus, both ξ_0' and ξ_1' are determined *linearly* in terms of ξ_0 and ξ_1, to within an overall sign $\pm$. The *relative* sign of ξ_0' and ξ_1' is *not* ambiguous, but is determined [by Eq. (2.7)] from the sign of x_3'.

The explicit form of the linear transformation induced in a spinor by the rotation of the corresponding isotropic vector $\mathbf{x}$ will be developed in Section 5. We show there that the rotation $\mathbf{x} \to \mathbf{x}' = R\mathbf{x}$ of isotropic vectors induces the rotation $\boldsymbol{\xi}' = U\boldsymbol{\xi}$ of spinors, $\boldsymbol{\xi} = \begin{pmatrix}\xi_0 \\ \xi_1\end{pmatrix}$, where U is a unimodular unitary matrix obtained from the elements R_{ij} of R.

Remark. The association of isotropic vectors $\mathbf{x}$ to spinors $\boldsymbol{\xi}$ given by Eq. (2.7) will be called the *Cartan mapping*. (It has several interesting applications, as discussed in Chapter 6, Section 16.) A spinorial mass point is described by its real *position vector* $\vec{r}$ (in Euclidean space $\mathbb{R}^3$) and a *spinor* $\boldsymbol{\xi}$ (in two-dimensional complex space $\mathbb{C}^2$), where the spinor space defines, via the Cartan map, the space of *isotropic vectors* $\{\vec{x}\}$. Under a rotation both $\vec{r}$ and $\vec{x}$ are transformed as *vectors*. Observe that a spinorial mass point is a classical object and that it is quite distinct from a solid body in the Dirac construction.

Let us continue now with Cartan's construction. Equations (2.7) for the isotropic vector $\mathbf{x}$ associated with a spinor $\boldsymbol{\xi}$ yield the ratio ξ_0/ξ_1 in the form $\xi_0/\xi_1 = -(x_1 - ix_2)/x_3 = x_3/(x_1 + ix_2)$. The relationship expressed by this ratio may also be written in matrix notation as

$$(\mathbf{x} \cdot \boldsymbol{\sigma})\boldsymbol{\xi} = 0, \tag{2.10}$$

where $\mathbf{x}\cdot\boldsymbol{\sigma}$ denotes $x_1\sigma_1 + x_2\sigma_2 + x_3\sigma_3$, and the σ_i $(i = 1, 2, 3)$ denote the Pauli [8] matrices:

$$\sigma_1 = \begin{pmatrix} 0 & 1 \\ 1 & 0 \end{pmatrix}, \qquad \sigma_2 = \begin{pmatrix} 0 & -i \\ i & 0 \end{pmatrix}, \qquad \sigma_3 = \begin{pmatrix} 1 & 0 \\ 0 & -1 \end{pmatrix}. \tag{2.11}$$

Thus, $\mathbf{x}\cdot\boldsymbol{\sigma}$ is the 2×2 matrix given by

$$\mathbf{x}\cdot\boldsymbol{\sigma} = \begin{pmatrix} x_3 & x_1 - ix_2 \\ x_1 + ix_2 & -x_3 \end{pmatrix}. \tag{2.12}$$

The matrix (2.12) (hence, also the Pauli matrices) arises naturally in the Cartan map, giving the isotropic vector $\mathbf{x}$ associated with the spinor ξ.

Cartan also noted that the matrix (2.12) provides a general way of associating a 2×2 complex matrix with an *arbitrary* vector, and conversely:

$$\mathbf{x} = (x_1, x_2, x_3) \leftrightarrow \begin{pmatrix} x_3 & x_1 - ix_2 \\ x_1 + ix_2 & -x_3 \end{pmatrix}. \tag{2.13}$$

It is this association that we now use to develop the relationship between rotations $[SU(2)]$ and proper (real), orthogonal matrices.

5. Relation between $SU(2)$ and $SO(3)$ Rotations

Observe now that for $\mathbf{x}$ real the matrix $\mathbf{x}\cdot\boldsymbol{\sigma}$ is both Hermitian and traceless and that these properties are preserved by a unitary similarity transformation $[U^\dagger U = UU^\dagger = \sigma_0 = 2\times 2$ identity matrix, where $U^\dagger$ denotes the transposed $(\sim)$ and complex conjugated $(*)$ matrix$]$:

$$\mathbf{x}\cdot\boldsymbol{\sigma} \rightarrow \mathbf{x}'\cdot\boldsymbol{\sigma} = U(\mathbf{x}\cdot\boldsymbol{\sigma})U^\dagger. \tag{2.14}$$

[The corresponding equations $(\mathbf{x}\cdot\boldsymbol{\sigma})\xi = 0$ for the isotropic vector $\mathbf{x}'$ associated with the spinor $\xi' = U\xi$ become $(\mathbf{x}'\cdot\boldsymbol{\sigma})\xi' = 0$.] Thus, corresponding to the unitary transformation of spinors, $\xi' = U\xi$, we have the *linear* transformation of vectors $\mathbf{x}' = R\mathbf{x}$, as determined from Eq. (2.14):

$$x_i' = \tfrac{1}{2}\operatorname{tr}[\sigma_i U(\mathbf{x}\cdot\boldsymbol{\sigma})U^\dagger] = \sum_j R_{ij}x_j, \tag{2.15}$$

$$R_{ij} = \tfrac{1}{2}\operatorname{tr}(\sigma_i U \sigma_j U^\dagger), \tag{2.16}$$

where tr denotes the trace of a matrix.

An alternative and useful expression for the matrix $R = (R_{ij})$ may be obtained by rewriting the transformation (2.14) first as $\boldsymbol{\eta}' = (U \otimes U^*)\boldsymbol{\eta}$ (matrix direct product), where $\boldsymbol{\eta} = \text{col}(x_3, x_1 - ix_2, x_1 + ix_2, -x_3)$ and then as a relationship between the column matrices $\text{col}(x_1', x_2', x_3', \xi')$ and $\text{col}(x_1, x_2, x_3, \xi)$ (ξ an arbitrary variable). The result is

$$\begin{pmatrix} R & 0 \\ 0 & 1 \end{pmatrix} = S^\dagger(U \otimes U^*)S, \tag{2.17}$$

where $U \otimes U^*$ denotes the matrix direct product of U with U^*, and S is the unitary unimodular matrix given by

$$S = \frac{1}{\sqrt{2}}\begin{pmatrix} 0 & 0 & 1 & -i \\ 1 & -i & 0 & 0 \\ 1 & i & 0 & 0 \\ 0 & 0 & -1 & -i \end{pmatrix}. \tag{2.18}$$

Let $U(2)$ denote the group (matrix multiplication) of 2×2 unitary matrices:

$$U(2) = \{U \colon UU^\dagger = \sigma_0\}. \tag{2.19}$$

One may now prove, using either Eq. (2.16) or Eq. (2.17), that (*a*) R is *real*, *proper*, and *orthogonal* for each $U \in U(2)$; (*b*) if U is mapped to R by relation (2.16) (denoted $U \to R$ and $U' \to R'$), then $U'U \to R'R$; (*c*) the set of elements of $U(2)$, which is mapped to the identity of $SO(3)$ (kernel of the mapping) by relation (2.16), is $\{e^{i\phi}\sigma_0 \colon 0 \leqslant \phi < 2\pi\}$; and (*d*) there exists a $U \in U(2)$ such that $U \to R$ for each $R \in SO(3)$ (the mapping is onto).

It is useful for applications to have an explicit formula determining the set of solution matrices $U \in U(2)$ of Eq. (2.16) that correspond to a specified $R \in SO(3)$. For this purpose, we observe that each $U \in U(2)$ may be written in the form $U = e^{i\phi}U_0$, $0 \leqslant \phi < 2\pi$, where U_0 belongs to the group of 2×2 unitary unimodular matrices:

$$SU(2) = \{U_0 \colon U_0 U_0^\dagger = \sigma_0, \det U_0 = 1\}. \tag{2.20}$$

We next use the fact that each $U_0 \in SU(2)$ has the form

$$U_0 = U(\alpha_0, \boldsymbol{\alpha}) = \begin{pmatrix} \alpha_0 - i\alpha_3 & -i\alpha_1 - \alpha_2 \\ -i\alpha_1 + \alpha_2 & \alpha_0 + i\alpha_3 \end{pmatrix} = \alpha_0\sigma_0 - i\boldsymbol{\alpha}\cdot\boldsymbol{\sigma}, \tag{2.21}$$

where $(\alpha_0, \boldsymbol{\alpha}) = (\alpha_0, \alpha_1, \alpha_2, \alpha_3)$ are the Euler–Rodrigues (real) parameters, which define the surface of the unit sphere, S^3, in four-space—that is, $\alpha_0^2+\alpha_1^2+\alpha_2^2+\alpha_3^2=1$. The element $R=R(\alpha_0,\boldsymbol{\alpha})\in SO(3)$ corresponding to U_0 is found from Eq. (2.17) to be

$$R=\begin{pmatrix} \alpha_0^2+\alpha_1^2-\alpha_2^2-\alpha_3^2 & 2\alpha_1\alpha_2-2\alpha_0\alpha_3 & 2\alpha_1\alpha_3+2\alpha_0\alpha_2 \\ 2\alpha_1\alpha_2+2\alpha_0\alpha_3 & \alpha_0^2+\alpha_2^2-\alpha_3^2-\alpha_1^2 & 2\alpha_2\alpha_3-2\alpha_0\alpha_1 \\ 2\alpha_1\alpha_3-2\alpha_0\alpha_2 & 2\alpha_2\alpha_3+2\alpha_0\alpha_1 & \alpha_0^2+\alpha_3^2-\alpha_1^2-\alpha_2^2 \end{pmatrix}. \tag{2.22}$$

A solution of Eq. (2.22) for $(\alpha_0,\boldsymbol{\alpha})$ in terms of the matrix elements R_{ij} of R is now easily found:

$$\operatorname{tr} R\neq -1:\quad \alpha_0\equiv \tfrac{1}{2}(1+\operatorname{tr} R)^{\frac{1}{2}},\qquad \alpha_i\equiv -(R_{jk}-R_{kj})/4\alpha_0,\qquad i,j,k \text{ cyclic},$$

$$\operatorname{tr} R=-1:\quad \alpha_0=0,\qquad \alpha_i=(\operatorname{sign}\alpha_i)[(1+R_{ii})/2]^{\frac{1}{2}},\quad \operatorname{sign}\alpha_1=1,$$

$$\operatorname{sign}\alpha_2=\operatorname{sign}R_{12},\quad \operatorname{sign}\alpha_3=\operatorname{sign}R_{13}. \tag{2.23}$$

In summary, there are just two unitary unimodular matrices that solve Eq. (2.16) for each specified R—namely, the U_0 given by Eq. (2.21) with parameters (2.23), and the negative, $-U_0$, of this same matrix: $\{U_0,-U_0\}\to R$. The set of all solution matrices $U\in U(2)$ to Eq. (2.16) is thus $\{e^{i\phi}U_0: 0\leqslant\phi<2\pi\}$.

Writing 2×2 unitary unimodular matrices in the form (2.21) shows clearly that the elements of the group $SU(2)$ are in one-to-one correspondence with the set of points S^3, which define the surface of the unit sphere in four-space. As $(\alpha_0,\boldsymbol{\alpha})$ runs over all points of the unit sphere, S^3, the matrices (2.21) and (2.22) enumerate, respectively, all elements (exactly once) of $SU(2)$ and all elements (exactly twice) of $SO(3)$, it being evident from Eq. (2.22) that $(\alpha_0,\boldsymbol{\alpha})\in S^3$ and the diametrically opposite point $(-\alpha_0,-\boldsymbol{\alpha})\in S^3$ define the same proper, orthogonal matrix R.

Remark. Schwinger [19] apparently gave the first published result inverting Eq. (2.16); Eqs. (2.23) agree with his answer. Schwinger did not, however, discuss the singular case, $\operatorname{tr} R=-1$, occurring for the exceptional ($R=\tilde{R}$, $R\neq I$) proper, orthogonal matrices.

6. Parametrizations of the Group of Rotations

The representation of rotations by 3×3 real, proper, orthogonal matrices and 2×2 unitary unimodular matrices can be implemented by various parametrizations of importance in physical applications. In addition to the

Euler–Rodrigues parameters $(\alpha_0, \boldsymbol{\alpha})$ introduced in the last section, there are three common parametrizations[1] of rotations found in the literature: (*a*) the $(\phi, \hat{n})$ parameters, which Euler introduced for characterizing the direction and angle of a rotation; (*b*) the Cayley–Klein parameters; and (*c*) the Euler angles. We now discuss each of these parametrizations briefly and refer to the literature for more extensive discussions of the topological properties of the corresponding "group spaces."

The $(\phi, \hat{n})$ *parameters.* We have used these parameters in our initial discussion in Section 2, in defining the action of a rotation on a vector. The associated matrix $R(\phi, \hat{n})$ is given by Eq. (2.6). An exponential form can also be obtained for this matrix by noting that $N^3 = -N$ [see Eq. (2.6)]:

$$R(\phi, \hat{n}) = e^{\phi N}, \qquad 0 \leqslant \phi \leqslant \pi, \qquad \hat{n} \cdot \hat{n} = 1. \tag{2.24}$$

The domain of definition of the parameters, $0 \leqslant \phi \leqslant \pi$, $\hat{n} \cdot \hat{n} = 1$, covers the group elements of $SO(3)$ exactly once, except that, for $\phi = \pi$, $\hat{n}$ and $-\hat{n}$ determine the same rotation. [The elements of $SO(3)$ are thus in one-to-one correspondence (Wigner [3], Speiser [22]) with the points of a solid sphere of radius π, where diametrically opposite points on the surface are identified.]

The inverse relationship, expressing ϕ and $\hat{n}$ in terms of the elements R_{ij} of a real, proper, orthogonal matrix R, may be found directly by solving Eq. (2.6) for ϕ and N. The results are

$$\cos\phi = (-1 + \operatorname{tr} R)/2, \tag{2.25}$$

$$n_i \sin\phi = -(R_{jk} - R_{kj}), \qquad i, j, k \text{ cyclic.} \tag{2.26}$$

Since each proper, orthogonal matrix R has $-1 \leqslant \operatorname{tr} R \leqslant 3$, relation (2.25) determines a unique angle in the interval $0 \leqslant \phi \leqslant \pi$. With ϕ determined, we find n_i from relation (2.26), unless $\phi = \pi$. For $\phi = \pi$, we have the two solutions, $\pm\hat{n}$, for the direction, where $\hat{n} = \boldsymbol{\alpha}$ is obtained from Eqs. (2.23).

The set of unitary unimodular matrices in the $(\phi, \hat{n})$ parametrization is obtained from Eqs. (2.21), (2.23), (2.25), and (2.26):

$$U = U(\psi, \hat{n}) = \sigma_0 \cos(\psi/2) - i(\hat{n} \cdot \boldsymbol{\sigma}) \sin(\psi/2) = e^{-i\psi(\hat{n} \cdot \boldsymbol{\sigma})/2}, \tag{2.27}$$

where the last equality is proved by direct expansion, using $(\hat{n} \cdot \boldsymbol{\sigma})^2 = \sigma_0$. The domain of definition of the parameters $(\psi, \hat{n})$ is now

---

[1] A fourth parametrization, the Cayley [20] rational parameters, is primarily of technical importance (Weyl [21, p. 56]) and has not found much use in the practical applications of angular momentum. We mention it here for completeness, but shall not use it in this monograph.

$$\hat{n}\cdot\hat{n}=1, \qquad 0\leqslant\psi\leqslant 2\pi, \tag{2.28}$$

so that the set of points $\{(\alpha_0,\boldsymbol{\alpha})=(\cos\frac{1}{2}\psi,\hat{n}\sin\frac{1}{2}\psi)\}$ covers the sphere S^3 exactly once.

To make explicit the relation of $U(\psi,\hat{n})$ to $R(\phi,\hat{n})$, we split the points on the unit sphere S^3 into two open hemispheres:

$$\left\{\left(\cos\frac{\phi}{2},\ \hat{n}\sin\frac{\phi}{2}\right):\ 0\leqslant\phi<\pi,\quad \hat{n}\cdot\hat{n}=1\right\},$$

$$\left\{\left(\cos\frac{2\pi-\phi}{2},\ \hat{n}\sin\frac{2\pi-\phi}{2}\right):\ 0\leqslant\phi<\pi,\quad \hat{n}\cdot\hat{n}=1\right\}. \tag{2.29}$$

The set of points $\{(0,\hat{n}):\hat{n}\cdot\hat{n}=1\}$ corresponding to $\phi=\pi$ (the unit sphere S^2) constitutes the common boundary of the two hemispheres. The pair of unitary unimodular matrices, $U(\phi,\hat{n})$ and $U(2\pi-\phi,-\hat{n})=-U(\phi,\hat{n})$, corresponding to a point in the first hemisphere and the diametrically opposite point in the second hemisphere, map via Eq. (2.16) to $R(\phi,\hat{n})$.

The Cayley–Klein parameters. Stereographic projection of the (three-space) sphere S^2 to the complex plane leads to a relation between rotations and unitary matrices that is equivalent to the one we have already obtained using Cartan's spinors. Rotation of points on the sphere correspond to *homographic transformations* of the plane (also called *linear fractional transformations*, or *bilinear transformations*). This relationship was discovered by Klein [15] and developed fully by Klein [16] and Cayley [17].

A correspondence between points (x_1,x_2,x_3) on the sphere $x_1^2+x_2^2+x_3^2=1$ and points (ξ,η) of the equatorial plane is established by *stereographic projection* from the north pole N at $(0,0,1)$. Thus, the straight line through the north pole N and any point (x_1,x_2,x_3) on the sphere, except the pole N, determines a point (ξ,η) in the equatorial plane; conversely, the straight line through any finite point (ξ,η) of the equatorial plane and the north pole determines a point (x_1,x_2,x_3) on the sphere.

Geometric considerations between similar triangles yield the relations $\xi=x_1/(1-x_3)$, $\eta=x_2/(1-x_3)$; that is,

$$\zeta=\xi+i\eta=\frac{x_1+ix_2}{1-x_3}=\frac{1+x_3}{x_1-ix_2}. \tag{2.30}$$

We now state the principal theorem for homographic transformations. (See Weyl [6], Klein [16], Gel'fand *et al.* [23], and DuVal [24] for the proof):

Let $\mathbf{x} \to \mathbf{x}' = R\mathbf{x}$ *denote a proper orthogonal transformation of the sphere. The* $\xi\eta$*-plane then undergoes the transformation determined by the homographic transformation of* $\zeta = \xi + i\eta$,

$$\zeta \to \zeta' = \frac{u_{11}^* \zeta + u_{12}^*}{u_{21}^* \zeta + u_{22}^*}, \tag{2.31}$$

where the u_{ij} *are the elements of a unitary matrix* $U = (u_{ij})$, *which is related to* $R = (R_{ij})$ *by*

$$R_{ij} = \tfrac{1}{2} \operatorname{tr}(\sigma_i U \sigma_j U^\dagger). \tag{2.32}$$

Remarks. (*a*) Observe that the unitary matrices U and λU ($|\lambda| = 1$) determine the same homographic transformation of the plane. We may therefore choose U to be unimodular, in which case the two unimodular matrices U and $-U$ determine the same transformation. (*b*) The proper orthogonal transformations of the sphere and the transformations of the $\xi\eta$-plane are one-to-one; it is the identification of the 2×2 unitary unimodular matrices with the homographic transformations that then yields the two-to-one relationship with these transformations, hence with the proper orthogonal matrices. (*c*) The homographic transformation (2.31) $\zeta \to \zeta'$ of the complex plane corresponding to the rotation $\mathbf{x} \to \mathbf{x}' = R\mathbf{x}$ of the sphere was obtained for the first time by Cayley [17] in the form (see Note 3)

$$\zeta' = \frac{(\alpha_0 + i\alpha_3)\zeta + (i\alpha_1 - \alpha_2)}{(i\alpha_1 + \alpha_2)\zeta + (\alpha_0 - i\alpha_3)}, \tag{2.33}$$

where $\alpha_i \equiv n_i \sin \frac{1}{2}\psi$, $\alpha_0 = \cos \frac{1}{2}\psi$, $\hat{n} \cdot \hat{n} = 1$, $0 \leqslant \psi \leqslant 2\pi$. The four *complex numbers* $(a, b, c, d) = (\alpha_0 + i\alpha_3, i\alpha_1 - \alpha_2, i\alpha_1 + \alpha_2, \alpha_0 - i\alpha_3)$, or slight variants of these, are now called the *Cayley–Klein parameters*, whereas the four *real numbers* $(\alpha_0, \boldsymbol{\alpha})$ defining a point on the surface of the unit sphere in four-space, S^3, are known as the *Euler–Rodrigues parameters* (alternatively, the three ratios α_i/α_0 form the "homogeneous" or "symmetric" Euler parameters (see Klein and Sommerfeld [18, p. 60] and Cayley [25]).

The Euler angles. The three angles introduced by Euler for characterizing rotations are perhaps the most widely used in physical applications. This may explain the existence in the literature of several variations in their definition.

Geometrically, the Euler angles $(\alpha\beta\gamma)$ are described by a sequence of three rotations of vectors, which we now describe in terms of a fixed frame $(\hat{e}_1, \hat{e}_2, \hat{e}_3)$ (see Fig. 2.6):

$\mathscr{R}_1$: rotation of angle α about $\hat{n}_1 = (0, 0, 1)$,

$\mathscr{R}_2$: rotation of angle β about $\hat{n}_2 = (-\sin\alpha, \cos\alpha, 0)$,

$\mathscr{R}_3$: rotation of angle γ about $\hat{n}_3 = (\cos\alpha \sin\beta, \sin\alpha \sin\beta, \cos\beta)$. (2.34)

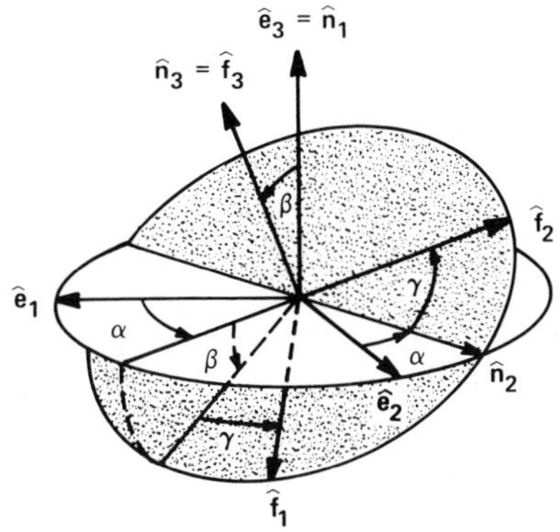

Figure 2.6. Euler angles. The three Euler angles $(\alpha\beta\gamma)$ are defined by a sequence of three rotations: α about $\hat{n}_1$; β about $\hat{n}_2$; and γ about $\hat{n}_3$.

Alternatively, using the identity Eq. (2.5), the Euler angles may be described in terms of three other rotations:

$$\mathscr{R}(\alpha\beta\gamma) = \mathscr{R}_3\mathscr{R}_2\mathscr{R}_1 = \mathscr{R}'_1\mathscr{R}'_2\mathscr{R}'_3, \tag{2.35}$$

$\mathscr{R}'_3$: rotation of γ about $\hat{e}_3 = (0, 0, 1)$,

$\mathscr{R}'_2$: rotation of β about $\hat{e}_2 = (0, 1, 0)$,

$\mathscr{R}'_1$: rotation of α about $\hat{e}_3 = (0, 0, 1)$. (2.36)

If we consider that the rotations are active (as opposed to passive) transformations, then $\mathscr{R}(\alpha\beta\gamma)$ maps an arbitrary vector $\vec{x}$ to the new vector $\vec{x}' = \mathscr{R}(\alpha\beta\gamma)\vec{x}$. The components of $\vec{x}'$ are related to those of $\vec{x}$ [in the fixed $(\hat{e}_1, \hat{e}_2, \hat{e}_3)$ frame] by

$$\mathbf{x}' = R(\alpha\beta\gamma)\mathbf{x},$$

$$R(\alpha\beta\gamma) = \begin{pmatrix} \cos\alpha & -\sin\alpha & 0 \\ \sin\alpha & \cos & 0 \\ 0 & 0 & 1 \end{pmatrix} \begin{pmatrix} \cos\beta & 0 & \sin\beta \\ 0 & 1 & 0 \\ -\sin\beta & 0 & \cos\beta \end{pmatrix} \begin{pmatrix} \cos\gamma & -\sin\gamma & 0 \\ \sin\gamma & \cos\gamma & 0 \\ 0 & 0 & 1 \end{pmatrix}$$

$$= \begin{pmatrix} \cos\alpha\cos\beta\cos\gamma - \sin\alpha\sin\gamma & -\cos\alpha\cos\beta\sin\gamma - \sin\alpha\cos\gamma & \cos\alpha\sin\beta \\ \sin\alpha\cos\beta\cos\gamma + \cos\alpha\sin\gamma & -\sin\alpha\cos\beta\sin\gamma + \cos\alpha\cos\gamma & \sin\alpha\sin\beta \\ -\sin\beta\cos\gamma & \sin\beta\sin\gamma & \cos\beta \end{pmatrix}. \tag{2.37}$$

The domain of definition of the Euler angles in Eq. (2.37) is

$$0 \leqslant \alpha < 2\pi, \qquad 0 \leqslant \beta \leqslant \pi, \qquad 0 \leqslant \gamma < 2\pi. \tag{2.38}$$

Distinct sets of numbers $(\alpha\beta\gamma)$ lying in these intervals correspond to different rotations except for $\beta = 0$ or $\beta = \pi$. If $\beta = 0$, the rotation is through the angle $\alpha + \gamma$ about $(0, 0, 1)$; if $\beta = \pi$, the rotation is through the angle $\alpha - \gamma$ about $(0, 0, 1)$. In these cases distinct values of α and γ may determine the same rotation.

The inverse to $R(\alpha\beta\gamma)$ is

$$R^{-1}(\alpha\beta\gamma) = R(2\pi - \gamma, \pi - \beta, 2\pi - \alpha). \tag{2.39}$$

In Fig. 2.6 we show the frame $(\hat{f}_1, \hat{f}_2, \hat{f}_3)$ obtained from $(\hat{e}_1, \hat{e}_2, \hat{e}_3)$ by application of the sequence of rotations $\mathscr{R}(\alpha\beta\gamma) = \mathscr{R}_3\mathscr{R}_2\mathscr{R}_1$ of Eq. (2.34). In this interpretation (passive) of rotations the elements of the matrix $R(\alpha\beta\gamma)$ appear as the direction cosines between axes: $R_{ij}(\alpha\beta\gamma) = \hat{e}_i \cdot \hat{f}_j$.

The Euler angle parametrization of $U \in SU(2)$ is given by

$$U(\alpha\beta\gamma) = e^{-i\alpha\sigma_3/2} e^{-i\beta\sigma_2/2} e^{-i\gamma\sigma_3/2}$$

$$= \begin{pmatrix} e^{-i\alpha/2}\cos(\beta/2)\, e^{-i\gamma/2} & -e^{-i\alpha/2}\sin(\beta/2)\, e^{i\gamma/2} \\ e^{i\alpha/2}\sin(\beta/2)\, e^{-i\gamma/2} & e^{i\alpha/2}\cos(\beta/2)\, e^{i\gamma/2} \end{pmatrix}, \tag{2.40}$$

where the domain of definition of $(\alpha\beta\gamma)$ is now extended to

$$0 \leqslant \alpha < 2\pi, \quad 0 \leqslant \beta \leqslant \pi \quad \text{or} \quad 2\pi \leqslant \beta \leqslant 3\pi, \quad 0 \leqslant \gamma < 2\pi. \tag{2.41}$$

Noting that $U(\alpha, \beta + 2\pi, \gamma) = -U(\alpha\beta\gamma)$, we see that both $U(\alpha\beta\gamma)$ and $U(\alpha, \beta + 2\pi, \gamma)$, $0 \leqslant \beta \leqslant \pi$, map to the same $R(\alpha\beta\gamma)$.

It is very common in the physics literature to find incomplete, or even incorrect, usage of the Euler angles when half-integral angular momenta [$SU(2)$] are discussed because of an improper covering of the unit sphere in four-space. (This problem is discussed in De Vries and Jonker [26].)

7. Notes

1. The Dirac construction. There is a fundamental group structure associated with such systems of strings (see p. 10), discovered by Artin [27–30] (in considering the theory of knots) and named the *braid(s) group* (*Zöpfegruppe*). The braid group of order n considers n strings attached as shown (for $n=5$) in Fig. 2.7. The strings run continuously downward (that is, they are not allowed to loop back). There are two elementary operations (a) σ_i: string i crosses *over* string $i+1$ (labeled from the left in the order of the strings as existing before the operation); and (b) σ_i^{-1}: string i crosses *under* string $i+1$. With these operations, the system forms a group; the group is clearly generated by the $n-1$ operators $\sigma_1, \sigma_2, \ldots, \sigma_{n-1}$. The system is greatly simplified by the results:

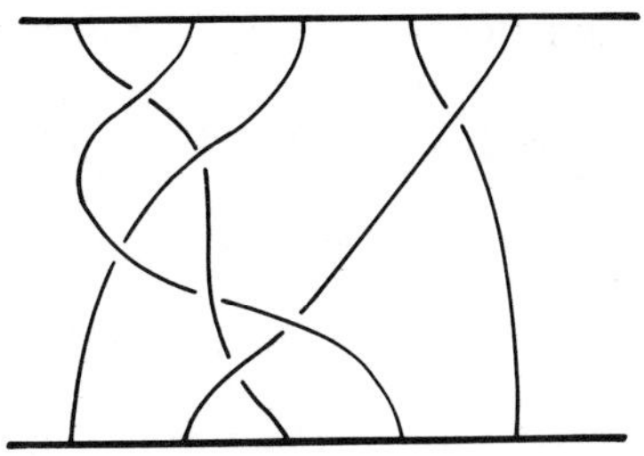

Figure 2.7.

(a) σ_1 and σ_0 generate the group:

$$\sigma_0 \equiv \sigma_1\sigma_2\cdots\sigma_{n-1}, \qquad \sigma_i = (\sigma_0)^{i-1}\sigma_1(\sigma_0)^{-i+1},$$

(b) $\sigma_i\sigma_j = \sigma_j\sigma_i$ if j is at least $i+2$,
(c) $\sigma_i\sigma_{i+1}\sigma_i = \sigma_{i+1}\sigma_i\sigma_{i+1}$.

The braid group is intrinsically interesting on two grounds: mathematically, because it is more fundamental than the permutation group; physically, because it is the natural tool to implement and analyze many path-dependent physical problems. Important examples are the Feynman "sum over paths" technique and Mandelstam's gauge-independent, but path-dependent wave functions.

In the case of Dirac's construction, there is an additional allowed operation in the braid group – strings may be carried *around the frame*. It was precisely this problem (Dirac's construction) that has been analyzed algebraically by Newman [4], using the theory of braids. Newman has proved that the *minimum number* of strings such that a rotation by 4π and not 2π is equivalent to the identity rotation is *three*.

2. The finite-size spinor. We quote the following opening paragraph from the introduction to Casimir's [31] dissertation:

Classical mechanics knows two classes of systems: "material points" and "rigid bodies." In building up a general mathematical theory applying to any system, one usually starts from the equations of motion of a material point. From a physical point of view, however, the idea of "rigid body" is as fundamental a conception as the idea of "material point."

We interpret "rigid body" as used here by Casimir to mean *solid body*, in the sense introduced in Section 3, for which we have demonstrated the 4π (and not 2π) rotation to be the identity rotation. As we have shown in Section 3, this object is necessarily of finite size and is associated with an impenetrable region of space. We propose to call this fundamental concept of a solid body a *finite-size spinor*. (A possible relation of this object to hadrons and quarks has been explored in Ref. [32].)

Even if we associate the minimum of three distinguished points with this object, owing to the requirement of the impenetrability of the "space" between the distinguished points, this object is conceptually distinct from a rigid body composed of three mass points separated by ordinary Euclidean space. The argument of Bopp and Haag [33] (see also Chapter 7, Section 10) shows conclusively that only integral angular momenta (hence, a rotation of 2π is the identity) can occur in the latter case. This result of Bopp and Haag is not in contradiction to the concept of a finite-size spinor, since, as we have shown, the impenetrability of the space between the three or more distinguished points composing a finite-size spinor is essential to its spinorial character.

3. Transformations of the spinor plane. It is a curious fact, which Klein [16] notes, that the homographic transformation (2.33) splits, under the substitution $\zeta = \zeta_1/\zeta_2$, into the two separate operations,

$$\zeta_1 \to \zeta_1' = (\alpha_0 + i\alpha_3)\zeta_1 + (i\alpha_1 + \alpha_2)\zeta_2,$$

$$\zeta_2 \to \zeta_2' = (i\alpha_1 - \alpha_2)\zeta_1 + (\alpha_0 - i\alpha_3)\zeta_2,$$

but fails to note, as also does Cayley [17], that the *quaternionic composition rule* (see Chapter 4, Section 3)

$$\alpha_0'' = \alpha_0'\alpha_0 - \boldsymbol{\alpha}' \cdot \boldsymbol{\alpha},$$

$$\boldsymbol{\alpha}'' = \alpha_0\boldsymbol{\alpha}' + \alpha_0'\boldsymbol{\alpha} + \boldsymbol{\alpha}' \times \boldsymbol{\alpha}, \tag{2.42}$$

obtained by them from the composition of the two homographic transformations, $\zeta \to \zeta'$ followed by $\zeta' \to \zeta''$, is, in fact, the rule of *matrix multiplication* (Hawkins [34]) applied to unitary matrices written in the form (2.21)

$$U^*(\alpha_0'', \boldsymbol{\alpha}'') = U^*(\alpha_0', \boldsymbol{\alpha}')U^*(\alpha_0, \boldsymbol{\alpha}).$$

Although one usually associates Hamilton [11] with quaternions, Klein [16] credits Rodrigues with the discovery of the composition rule (2.42).

References

1. G. W. Mackey, *Induced Representations of Groups and Quantum Mechanics*. Benjamin, New York, 1968.
2. E. Cartan, *The Theory of Spinors*. MIT Press, Cambridge, Mass., 1966. Translation of the first French edition printed in 1937 from the lecture notes of Elie Cartan.

2a. L. A. Santaló, *Integral Geometry and Geometric Probability. Encyclopedia of Mathematics and Its Applications* (G.-C. Rota, ed.), Vol. 1, Addison-Wesley, Reading, Mass., 1976.

3. E. P. Wigner, *Gruppentheorie und ihre Anwendung auf die Quantenmechanik der Atomspektren.* Vieweg, Braunschweig, 1931; reprinted by Edwards, Ann Arbor, Mich., 1944. Translated by J. J. Griffin as *Group Theory and Its Application to the Quantum Mechanics of Atomic Spectra.* Academic Press, New York, 1959. The translated version has been expanded by three additional chapters: "Racah coefficients" (Chapter 24); "Time inversion" (Chapter 26); and "Physical interpretation and classical limits of representation coefficients, three and six-j symbols" (Chapter 27).
4. M. H. A. Newman, "On a string problem of Dirac," *J. London Math. Soc.* **17** (1942), 173–177.
5. C. W. Misner, K. S. Thorne, and J. A. Wheeler, *Gravitation.* Freeman San Francisco, 1973.
6. H. Weyl, *Gruppentheorie und Quantenmechanik.* Hirzel, Leipzig, 1st ed., 1928; 2nd ed., 1931. Translated by H. P. Robertson, as *The Theory of Groups and Quantum Mechanics.* Methuen, London, 1931; reissued by Dover, New York, 1949.
7. W. Heisenberg and P. Jordan, "Anwendung der Quantenmechanik auf das Problem der anomalen Zeemaneffekte," *Z. Physik* **37** (1926), 263–277.
8. W. Pauli, "Zur Quantenmechanik des magnetischen Elektrons," *Z. Physik* **37** (1927), 601–623.
9. E. Cartan, "Les groupes projectifs qui ne laissent invariante aucune multiplicité plane," *Bull. Soc. Math. France* **41** (1913), 53–96.
10. W. R. Hamilton, *Quaternions,* Mathematical Papers III, pp. 103–105. Cambridge Univ. Press, London, 1967. (Notebook entry for 16 October 1843.)
11. W. R. Hamilton, "Letter to Graves on quaternions; or a new system of imaginaries in algebra," *Phil. Mag.* **25** (1844), 489–495. [Math. Papers III, 106–110.]
12. W. R. Hamilton, "On a new species of imaginary quantities connected with the theory of quaternions," *Proc. Roy. Irish. Acad.* **II** (1844), 424–434. [Math. Papers III, 111–116.]
13. A. Cayley, "On certain results relating to quaternions," *Phil. Mag.* **26** (1845), 141–145. [Collected Math. Papers I, 123–126.]
14. A. Cayley, "On the application of quaternions to the theory of rotation," *Phil. Mag.* **3** (1848), 196–200. [Collected Math. Papers I, 405–409.]
15. F. Klein, "Ueber binäre Formen mit linearen Transformationen in sich selbst," *Math. Ann.* **9** (1875), 183.
16. F. Klein, *The Icosahedron.* Dover, New York, 1956. Republication of the English translation (in 1913 by G. G. Morrice) of Klein's book published in 1884.
17. A. Cayley, "On the correspondence of homographies and rotations," *Math. Ann.* **15** (1879), 238–240. [Collected Math. Papers X, 153–154.]
18. F. Klein and A. Sommerfeld, *Über die Theorie des Kreisels,* Vol. 1. Teubner, Leipzig, 1897. This is the first of four volumes. See Vol. 4 for corrections.
19. J. Schwinger, "Unitary operator bases," *Proc. Natl. Acad. Sci. U.S.* **46** (1960), 570–579.
20. A. Cayley, "Sur quelques propriétés des déterminants gauches," *J. reine angew. Math.* **32** (1846), 119–123. [Collected Math. Papers I, 332–336.]
21. H. Weyl, *The Classical Groups.* Princeton Univ. Press, Princeton, N. J., 1946.
22. D. Speiser, "Theory of compact Lie groups and some applications to elementary particle physics," in *Group Theoretical Concepts and Methods in Elementary Particle Physics* (F. Gürsey, ed.), pp. 201–216. Gordon and Breach, New York, 1962.
23. I. M. Gel'fand, R. A. Minlos, and Z. Ya. Shapiro, *Representations of the Rotation and Lorentz Groups and Their Applications.* Macmillan, New York, 1963. Translated from the Russian by G. Cummins and T. Boddington.
24. P. DuVal, *Homographies, Quaternions, and Rotations.* Oxford Univ. Press, London, 1964.
25. A. Cayley, "On the motion of rotation of a solid body," *Cambridge Math. J.* **3** (1843), 224–232. [Collected Math. Papers I, 28–35.]
26. E. De Vries and J. E. Jonker, "A note on orthogonality and completeness of the rotation matrices," *Nucl. Phys.* **A105** (1967), 621–626.

27. E. Artin, "Theorie der Zöpfe," *Hamburg Univ. Math. Sem., Abh.* **4** (1926), 47–72.
28. E. Artin, "Theory of braids," *Ann. of Math.* **48** (1947), 101–126.
29. E. Artin, "Braids and permutations," *Ann. of Math.* **48** (1947), 643–649.
30. E. Artin, "The theory of braids," *Amer. Scientist* **38** (1950), 112–119.
31. H. B. G. Casimir, *Rotation of a Rigid Body in Quantum Mechanics.* Thesis, University of Leyden, Wolters, Groningen, 1931. [*Koninkl. Ned. Akad. Wetenschap, Proc.* **34** (1931), 844.]
32. L. C. Biedenharn, R. Y. Cusson, M. Y. Han, and J. D. Louck, "A kinematical model for quarks and hadrons," *Found. of Phys.* **2** (1972), 149–159.
33. F. Bopp and R. Haag, "Über die Möglichkeit von Spinmodellen," *Z. Naturforsch.* **5a** (1950), 644–653.
34. T. Hawkins, "The theory of matrices in the 19th century," *Proc. Int. Congr. Math., Vancouver* (1974), 561–570.

CHAPTER 3

Standard Treatment of Angular Momentum in Quantum Mechanics

1. Overview

It is a remarkable fact that in the very first papers establishing the matrix form of quantum mechanics[1] (Heisenberg [1], Born and Jordan [2], Born *et al.* [3], Dirac [4]), the operators for angular momentum were introduced and the decisive consequences of quantized angular momentum deduced. The techniques developed for this treatment are astonishingly close to, but wholly independent of, those developed by Cartan [6] in his definitive treatment of (semisimple) Lie groups of which angular momentum [$SU(2)$] is a prototype.

This material has become standard in all textbooks of quantum mechanics; we shall discuss it, albeit synoptically, in this chapter. Our purpose in doing so is to establish a common ground based on the way all physicists are trained to look at angular momentum. In doing so we shall see that the standard treatment is, in fact, quite rigorous (in essence) and establishes fundamental results capable of far-reaching generalizations (which we shall indicate in due course).

2. Definition of the Angular Momentum Operators

In classical mechanics, the angular momentum, $\vec{L}$, of a (point) particle is defined to be the moment of momentum:

$$\vec{L} = \vec{r} \times \vec{p}, \tag{3.1}$$

[1] References [1–3] appear in English translation in a book edited by van der Waerden [5].

ENCYCLOPEDIA OF MATHEMATICS and Its Applications, Gian-Carlo Rota (ed.).
Vol. 8: L. C. Biedenharn and J. D. Louck, Angular Momentum in
Quantum Physics: Theory and Application

where $\vec{r}$ is the position of the particle (relative to a fixed origin) and $\vec{p}$ is the (linear) momentum of the particle. Angular momentum is an *additive quantity*, like momentum, and the total angular momentum is defined to be the sum, $\vec{L}_{\text{total}} \equiv \sum_{\alpha=1}^{n} \vec{L}^{\alpha}$, of the angular momenta of the constituents.

It is natural to extend this definition to quantum mechanics by interpreting the symbols as operators; that is, we redefine the angular momentum of a particle to be an *operator*, $\vec{L} \to \mathbf{L}^{\text{op}} \equiv \mathbf{r}^{\text{op}} \times \mathbf{p}^{\text{op}}$, where the components of the position operator $\mathbf{r}^{\text{op}}$ and the momentum operator $\mathbf{p}^{\text{op}}$ obey the (Heisenberg) commutation rule, $[p_i^{\text{op}}, x_j^{\text{op}}] = -i\hbar\delta_{ij}$.

It follows directly that the angular momentum operators obey the fundamental commutation rule, which is expressed by

$$[L_i, L_j] = i\hbar e_{ijk} L_k, \tag{3.2}$$

where

$$e_{ijk} = \begin{cases} +1 \text{ for } (i,j,k) \text{ an even permutation of } (1,2,3), \\ -1 \text{ for } (i,j,k) \text{ an odd permutation of } (1,2,3), \\ 0 \text{ otherwise.} \end{cases} \tag{3.3}$$

Vectorially, this result is expressed by

$$\mathbf{L} \times \mathbf{L} = i\hbar\mathbf{L}. \tag{3.4}$$

Remark. If we define the skew-Hermitian operators $\mathfrak{G}_i = -iL_i/\hbar$, then the commutation relations of Eq. (3.2) take the form: $[\mathfrak{G}_i, \mathfrak{G}_j] = e_{ijk}\mathfrak{G}_k$. This identifies the algebra generated by the $\{\mathfrak{G}_i\}$ over $\mathbb{R}$ as the Lie algebra $A_1 = su(2)$ of the simple Lie group $SU(2)$.

One appeals next to kinematics (the independence of the description of the motion of each particle) to deduce that the operators for distinct particles α and β commute (dropping now the superscript "op"):

$$[x_i^{\alpha}, x_j^{\beta}] = [x_i^{\alpha}, p_j^{\beta}] = [p_i^{\alpha}, p_j^{\beta}] = 0, \qquad \alpha \neq \beta. \tag{3.5}$$

The total angular momentum operator $\mathbf{L}$ defined by

$$\mathbf{L} = \sum_{\alpha} \mathbf{L}^{\alpha}, \tag{3.6}$$

as well as the angular momentum of each particle $\mathbf{L}^{\alpha}$, obeys the same commutation rule, Eq. (3.4).

Remarks. (*a*) A striking feature of the commutation rule (3.4) is that it is *not* invariant to scale changes ($\mathbf{L} \to \lambda\mathbf{L}$)—unlike the classical Poisson brackets—and, accordingly, *there is an absolute scale for angular momenta.* It is customary to choose $\hbar$ as the unit for angular momentum and to replace $\mathbf{L}$ by $\mathbf{L}/\hbar$. *Henceforth all angular momenta will be scaled in this way.*

(*b*) Since the components of angular momentum do not commute, we must accordingly have [from Eq. (3.2)] an uncertainty principle:

$$(\Delta L_i)^2(\Delta L_j)^2 \geqslant \tfrac{1}{4}\langle L_k\rangle^2, \qquad i \neq j \neq k. \tag{3.7}$$

States for which any one component of the angular momentum is "sharp" [meaning exactly known: $(\Delta L_i)^2 = 0$] necessarily have the remaining two components "unsharp" (randomly distributed)—except, of course, for the trivial case where all components vanish. This poses an interesting conceptual problem, for the angular momentum—considered as a vector—can never "have" a sharp direction; this behavior contrasts markedly with that of the momentum and position vectors, each of which can define a precise direction. [We discuss this further in connection with the "vector model" (p. 36) and develop the subject in RWA.]

(*c*) It is customary in physics at this point to ignore the derivation of the commutation rules and simply postulate that an angular momentum operator $\mathbf{J}$ is an observable (that is, Hermitian operator) satisfying the rule:

$$\mathbf{J} \times \mathbf{J} = i\mathbf{J}. \tag{3.8}$$

The purpose of this step is to accommodate "spin" angular momentum (see below), at least heuristically. A more satisfactory procedure (Mackey [7]) is to examine critically the concept of a particle, thereby arriving at a group-theoretic definition in terms of imprimitivity and an associated motion group (Poincaré group or Galilei group; see Note 3, Chapter 1).

3. The Angular Momentum Multiplets

Let us assume now that we are dealing with the standard formulation of quantum mechanics—that is, a Hilbert space over the complex numbers with observables represented by Hermitian linear operators. We denote the vectors of the Hilbert space by kets, $|\phi\rangle$, and the vectors of the dual space by bras, $\langle\psi|$. The inner product is denoted by the bracket: $\langle\psi|\phi\rangle \in \mathbb{C}$. We seek to deduce consequences from the commutation relations for $\mathbf{J}$.

The strategy chosen by Cartan [6] and by Born *et al.* [3] is to convert the problem into a finite-dimensional matrix eigenvalue problem. To do this one first observes that the operator $\mathbf{J}^2$,

$$\mathbf{J}^2 \equiv J_1^2 + J_2^2 + J_3^2, \tag{3.9}$$

commutes with $\mathbf{J}$:

$$[\mathbf{J}^2, J_i] = 0.$$

The eigenvalue problem is then to find the set of eigenkets[1] such that $\mathbf{J}^2$ and J_3 are diagonal:

$$\mathbf{J}^2|J^{2\prime}, J_3'\rangle = J^{2\prime}|J^{2\prime}, J_3'\rangle,$$

$$J_3|J^{2\prime}, J_3'\rangle = J_3'|J^{2\prime}, J_3'\rangle, \tag{3.10}$$

where $J^{2\prime}$ is a real, nonnegative number and J_3' is a real number, these properties following from the assumption that the J_i are Hermitian operators.

The next step is not so straightforward. One defines the *non-Hermitian* operators (complex extension of the Lie algebra):

$$J_\pm \equiv J_1 \pm iJ_2, \qquad J_+^\dagger = J_-, \tag{3.11}$$

where the dagger (†) denotes Hermitian conjugation. These new operators satisfy the rules

$$[\mathbf{J}^2, J_\pm] = 0, \qquad [J_3, J_\pm] = \pm J_\pm, \qquad [J_+, J_-] = 2J_3, \tag{3.12}$$

which are an immediate consequence of $\mathbf{J} \times \mathbf{J} = i\mathbf{J}$.

We assume that there exists an eigenket that satisfies Eqs. (3.10), and we form the new kets $J_\pm|J^{2\prime}, J_3'\rangle \equiv |\pm\rangle$. Using the commutation relations, Eqs. (3.12), one finds that the ket $|+\rangle$, if it is nonzero, has the eigenvalues

$$\mathbf{J}^2|+\rangle = J^{2\prime}|+\rangle,$$

$$J_3|+\rangle = (J_3' + 1)|+\rangle.$$

---

[1] The notation $|J^{2\prime}, J_3'\rangle$ for an eigenket suppresses the set of quantum numbers $(\alpha) = (\alpha_1, \alpha_2, \ldots)$ that describe the eigenstates of a set A of observables, which together with $\mathbf{J}^2$ and J_3 constitute a *complete set of commuting observables* for a physical system. (The prime denotes "eigenvalue" – that is, $\mathbf{J}^2 \to J^{2\prime} \in \mathbb{R}$.)

We quote Dirac [8, p. 57] for a definition of the latter concept: "Let us define a complete set of commuting observables to be a set of observables which all commute with one another and for which there is only one simultaneous eigenstate belonging to any set of eigenvalues."

In discussing abstract angular momentum, we shall generally suppress the labels (α), since they add nothing to our understanding of the angular momentum itself. One must always keep in mind that in any physical problem the notation $\{|jm\rangle\}$ [see Eq. (3.16)] must be supplied with these extra quantum numbers (α).

Similarly $|-\rangle$, if it is nonzero, has the eigenvalues $J^{2\prime}$ and $J_3' - 1$. (This behavior accounts for the names "*raising*" operator for J_+ and "*lowering*" operator for J_-.)

The process can be repeated. One apparently generates eigenkets all belonging to the eigenvalue $J^{2\prime}$, but having J_3 eigenvalues $J_3' + k$, where k is an integer (positive, zero, or negative).

The raising process must terminate. Proof. The norm of $|+\rangle$ is given by

$$\langle +|+\rangle = \langle J^{2\prime}, J_3'|J_-J_+|J^{2\prime}, J_3'\rangle$$

$$= [J^{2\prime} - J_3'(J_3' + 1)]\,\langle J^{2\prime}, J_3'|J^{2\prime}, J_3'\rangle \geqslant 0.$$

Clearly, if J_3' can be raised indefinitely, we contradict the fact that the norm of a vector in Hilbert space is nonnegative. Hence, for some $J'_{3,\max}$ we must have $J_+|J^{2\prime}, J'_{3,\max}\rangle = 0$, $|J^{2\prime}, J'_{3,\max}\rangle \neq 0$, and therefore $J^{2\prime} = J'_{3,\max}(J'_{3,\max} + 1)$. ■

The lowering process must terminate. Proof. The norm of $|-\rangle$ is given by

$$\langle -|-\rangle = \langle J^{2\prime}, J_3'|J_+J_-|J^{2\prime}, J_3'\rangle$$

$$= [J^{2\prime} - J_3'(J_3' - 1)]\langle J^{2\prime}, J_3'|J^{2\prime}, J_3'\rangle \geqslant 0.$$

By the same reasoning used in the raising process, it follows that there exists a $J'_{3,\min}$ such that $J_-|J^{2\prime}, J'_{3,\min}\rangle = 0$, $|J^{2\prime}, J'_{3,\min}\rangle \neq 0$, and therefore $J^{2\prime} = J'_{3,\min}(J'_{3,\min} - 1)$. ■

The fact that the process must terminate *above* and *below* quantizes the eigenvalue of $\mathbf{J}^2$ and J_3. To see this, recall that we assumed that one simultaneous eigenket, $|J^{2\prime}, J_3'\rangle$, existed with eigenvalues $J^{2\prime}$ and J_3'. Let the raising terminate in k steps, and the lowering in l steps—that is, k is the smallest nonnegative integer such that

$$J_+^k|J^{2\prime}, J_3'\rangle \neq 0, \qquad \text{but} \qquad J_+^{k+1}|J^{2\prime}, J_3'\rangle = 0,$$

while l is the smallest nonnegative integer such that

$$J_-^l|J^{2\prime}, J_3'\rangle \neq 0, \qquad \text{but} \qquad J_-^{l+1}|J^{2\prime}, J_3'\rangle = 0.$$

Then $J'_{3,\max} = J_3' + k$, $J'_{3,\min} = J_3' - l$, and $J^{2\prime} = (J_3' + k)(J_3' + k + 1) = (J_3' - l)\,(J_3' - l - 1)$, which requires $(k + l + 1)(2J_3') = (k + l + 1)(l - k)$

—that is,

$$J_3' = \frac{l-k}{2}, \qquad J^{2\prime} = \left(\frac{k+l}{2}\right)\left(\frac{k+l}{2}+1\right),$$

$$J'_{3,\max} = \left(\frac{k+l}{2}\right), \qquad J'_{3,\min} = -\left(\frac{k+l}{2}\right). \tag{3.13}$$

Thus, the existence of the single nonzero simultaneous eigenket, $|J^{2\prime}, J_3'\rangle$, of $\mathbf{J}^2$ and J_3 implies the existence of a *set* of nonzero simultaneous eigenkets:

$$J_+^k|J^{2\prime}, J_3'\rangle, \ldots, J_+|J^{2\prime}, J_3'\rangle, |J^{2\prime}, J_3'\rangle,$$

$$J_-|J^{2\prime}, J_3'\rangle, \ldots, J_-^l|J^{2\prime}, J_3'\rangle. \tag{3.14}$$

Each eigenket in this set has the same $\mathbf{J}^2$ eigenvalue,

$$J^{2\prime} = \left(\frac{k+l}{2}\right)\left(\frac{k+l}{2}+1\right),$$

and the J_3 eigenvalues are

$$J'_{3,\max} = \frac{k+l}{2}, \ldots, \frac{l-k}{2} + 1,$$

$$\frac{l-k}{2} = J_3', \frac{l-k}{2} - 1, \ldots, -\frac{k+l}{2} = J'_{3,\min}. \tag{3.15}$$

The standard method for enumerating these results is to introduce $j = (k+l)/2$ and let m denote any number in the set $\{j, j-1, \ldots, -j\}$. The notation $|jm\rangle$ is also introduced to designate a simultaneous normalized eigenket of $\mathbf{J}^2$ and J_3. Then the results enumerated in Eqs. (3.13)–(3.15) may be summarized by

$$\mathbf{J}^2|jm\rangle = j(j+1)|jm\rangle,$$

$$J_3|jm\rangle = m|jm\rangle, \tag{3.16}$$

where j belongs[1] to the set $\{0, \frac{1}{2}, 1, \frac{3}{2}, 2, \ldots\}$ and, for a given j, the value of m ranges from j to $-j$ in steps of unity.

---

[1] Brownstein [8a] has given an unusual derivation of the eigenvalues of $\mathbf{J}^2$ in which J_3 plays no role; the derivation requires, however, the concept of adding two angular momenta (see Section 11).

Observe that the single eigenket $|J^{2\prime}, J_3'\rangle$ from which the set of eigenkets $\{|jm\rangle : m = j, j-1, \ldots, -j\}$ was generated is denoted in the new notation by $|j, j-k\rangle$. However, having proved the existence of the entire set $\{|jm\rangle$: $m = j, j-1, \ldots, -j\}$ (for $|j, j-k\rangle \neq 0$), we can now generate the set from the eigenket $|jj\rangle$. The defining properties of $|jj\rangle \neq 0$ are

$$\mathbf{J}^2|jj\rangle = j(j+1)|jj\rangle, \qquad J_3|jj\rangle = j|jj\rangle,$$

$$J_+|jj\rangle = 0, \qquad J_-^{2j+1}|jj\rangle = 0,$$

$$J_-^{j-m}|jj\rangle \neq 0, \qquad m = j, j-1, \ldots, -j. \tag{3.17}$$

The normalized eigenkets $|jm\rangle$ (for $|jj\rangle$ normalized) are now given by

$$|jm\rangle = \left[\frac{(j+m)!}{(2j)!(j-m)!}\right]^{\frac{1}{2}} J_-^{j-m}|jj\rangle, \tag{3.18}$$

where the normalization factor is easily obtained by using the identity

$$J_+^k J_-^k = \prod_{s=1}^{k} [\mathbf{J}^2 - (J_3 - s)(J_3 - s + 1)]. \tag{3.19}$$

An arbitrary phase factor in the normalization factor is chosen by convention in the physics literature to be $+1$ (see Note 1 for further discussion of the phase).

The action of J_+ and J_- on the general eigenket, Eq. (3.18), is now easily calculated:

$$J_+|jm\rangle = [(j-m)(j+m+1)]^{\frac{1}{2}}|jm+1\rangle,$$

$$J_-|jm\rangle = [(j+m)(j-m+1)]^{\frac{1}{2}}|jm-1\rangle. \tag{3.20}$$

Finally, we observe that the eigenkets corresponding to distinct values of j and distinct values of m are, in consequence of the Hermitian property of the J_i, orthogonal:

$$\langle j'm'|jm\rangle = \delta_{j'j}\delta_{m'm}. \tag{3.21}$$

Summary. On a Hilbert space for which the J_i are Hermitian operators, the only eigenvalues of $\mathbf{J}^2$ that can occur are of the form $j(j+1)$, where j belongs to the set $\{0, \frac{1}{2}, 1, \frac{3}{2}, 2, \ldots\}$; for each j there exist $2j+1$ orthonormal eigenkets of J_3 having eigenvalues $m = j, j-1, \ldots, -j$.

It is important to remark that when one is presented with a particular Hilbert space on which there is defined a set of three Hermitian operators $\{J_i\}$ satisfying $[J_i, J_j] = ie_{ijk}J_k$, the theory developed above makes no prediction as to *which* values $j \in \{0, \frac{1}{2}, 1, \frac{3}{2}, \ldots\}$ will occur, nor does it predict what their multiplicities (repetitions) will be.

Despite the simplicity of deriving these results, they are of fundamental importance, and all further developments are direct consequences of them.

Remarks. (*a*) The eigenvalue j is called the angular momentum; the eigenvalue m is called the z-component of the angular momentum, or sometimes the projection quantum number. In mathematical usage, the eigenvalue m is called a weight.

Operationally, one determines j to be the *maximum* eigenvalue of J_3 and/or the negative of the *minimum* eigenvalue.

(*b*) Determining j from the quadratic equation $J^{2'} = j(j+1)$ leads to a second solution: $-j-1$. This "negative angular momentum" defines an important reflection symmetry of the coupling coefficients of angular momentum theory (Wigner and Racah coefficients; see Sections 12 and 18), as is discussed in detail in RWA.

(*c*) The eigenvalues $\{(j, m)\}$ divide into two sets: integers and half-integers. *Orbital angular momentum* refers to the spatial motion of particles and has the form (3.1); it always has integral angular momentum (see Chapter 6, Section 5). *Spin angular momentum* refers to an intrinsic property of a particle and may be either integral or half-integral (half an odd integer).

(*d*) The fact that the magnitude, $[j(j+1)]^{\frac{1}{2}}$, of the angular momentum vector $\mathbf{J}$ is not equal to the magnitude of its maximal component, j, is a consequence of the uncertainty principle, which forbids any complete alignment of $\mathbf{J}$ along a given direction.

One may obtain a more accurate uncertainty relation of the form

$$(\Delta J_1)^2 + (\Delta J_2)^2 = j(j+1) - m^2, \tag{3.22}$$

which shows that the minimum fluctuations (perpendicular to the z-axis) occur for $m = \pm j$.

These peculiar results led to the concept of "spatial quantization," which is pictorially represented by the "vector model" of an angular momentum (Sommerfeld [9]). In this model an angular momentum has length $[j(j+1)]^{\frac{1}{2}}$ and assumes *discrete* orientations $\cos\theta = m/\sqrt{j(j+1)}$ as shown in Fig. 3.1.

To account for the uncertainty in direction, the vector is considered randomly distributed on a cone around the z-axis.

The vector model, though necessarily imprecise, is a qualitatively correct picture of a quantized vector, which becomes precise in the (classical) limit $j \gg 1$.

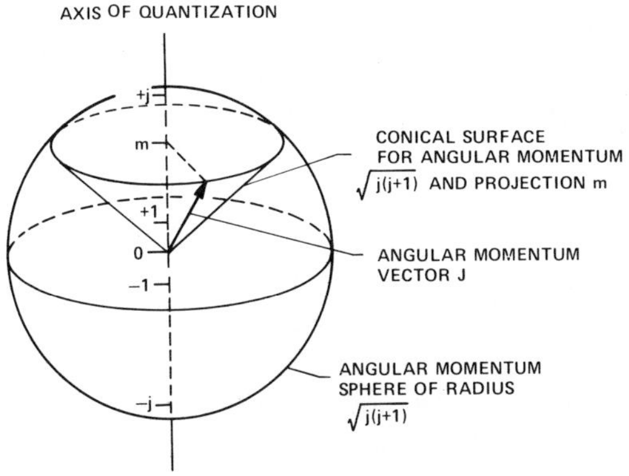

Figure 3.1.

(*e*) The letter m was originally chosen from *magnetic* quantum number, based on the Zeeman effect (Chapter 7, Section 3) used to unravel spectra.

(*f*) Feynman [10] has given a neat way to "explain" the value $J^{2\prime} = j(j+1)$. Let us assume that we know that the eigenvalue m of J_3 lies between $m = j$ and $m = -j$, as in the vector model. An average over all values of m is equivalent to an average over all spatial directions; hence, the averages $\langle J_1^2 \rangle$, $\langle J_2^2 \rangle$, $\langle J_3^2 \rangle$ are all equal. It follows that

$$\langle \mathbf{J}^2 \rangle = 3\langle J_3^2 \rangle/(2j+1) = 3 \sum_{m=-j}^{j} m^2/(2j+1).$$

Using $\sum m^2 = j(j+1)(2j+1)/3$, we find $\langle \mathbf{J}^2 \rangle = j(j+1)$. Since $\mathbf{J}^2$ is a scalar (rotational invariant), it follows that $J^{2\prime} = j(j+1)$.

This method of multiplet averaging can be exploited in more general problems; it was a frequent tool in spectroscopy before the development of more formal angular momentum techniques (Breit and Wills [11], Slater [12]).

4. Matrices of the Angular Momentum

In the previous section, we found that the angular momentum $\mathbf{J}$ is quantized and that the magnitude $\langle \mathbf{J}^2 \rangle^{\frac{1}{2}} \equiv \langle jm|\mathbf{J}^2|jm \rangle^{\frac{1}{2}} = [j(j+1)]^{\frac{1}{2}}$ has the eigenvalues given by $j = 0, \frac{1}{2}, 1, \ldots$; for each value of j, there are $2j+1$ orientations of the angular momentum operator $\mathbf{J}$, specified by the J_3 eigenvalues, $m = j, j-1, \ldots, -j+1, -j$.

Correspondingly, for each j there are $2j+1$ eigenkets labeled $|jm\rangle$, $m = j, \ldots, -j$.

The general matrix element of the angular momentum operator **J** is given by Eqs. (3.16) and (3.20):

$$\langle j'm'|J_\pm|jm\rangle = [(j \mp m)(j \pm m + 1)]^{\frac{1}{2}}\delta_{j'j}\delta_{m',m\pm 1},$$

$$\langle j'm'|J_3|jm\rangle = m\delta_{j'j}\delta_{m'm}. \tag{3.23}$$

Using these matrix elements, one can write out the corresponding matrices, $J_i^{(j)}$ (representing J_i), for the first few cases (the convention for labeling rows and columns is as follows: The column labels are $m = j, j-1, \ldots, -j$ when read from left to right, and the row labels are $m = j, j-1, \ldots, -j$ when read from top to bottom):

$j = 0$: $\quad J^{(0)} = 0;$

$j = \frac{1}{2}$: $\quad J_1^{(\frac{1}{2})} = \frac{1}{2}\begin{pmatrix} 0 & 1 \\ 1 & 0 \end{pmatrix}, \qquad J_2^{(\frac{1}{2})} = \frac{1}{2}\begin{pmatrix} 0 & -i \\ i & 0 \end{pmatrix},$

$$J_3^{(\frac{1}{2})} = \frac{1}{2}\begin{pmatrix} 1 & 0 \\ 0 & -1 \end{pmatrix};$$

$j = 1$: $\quad J_1^{(1)} = \frac{1}{\sqrt{2}}\begin{pmatrix} 0 & 1 & 0 \\ 1 & 0 & 1 \\ 0 & 1 & 0 \end{pmatrix}, \qquad J_2^{(1)} = \frac{1}{\sqrt{2}}\begin{pmatrix} 0 & -i & 0 \\ i & 0 & -i \\ 0 & i & 0 \end{pmatrix},$

$$J_3^{(1)} = \begin{pmatrix} 1 & 0 & 0 \\ 0 & 0 & 0 \\ 0 & 0 & -1 \end{pmatrix}. \tag{3.24}$$

(The 2×2 matrices are the familiar Pauli matrices, $\boldsymbol{\sigma}$, that is, $\mathbf{J}^{(\frac{1}{2})} = \frac{1}{2}\boldsymbol{\sigma}$.)

The angular momentum matrices (3.24) satisfy the characteristic equations $(J_i^{(\frac{1}{2})})^2 = \frac{1}{4}\mathbb{1}^{(\frac{1}{2})}$, $(J_i^{(1)})^3 = J_i^{(1)}$, and, generally,

$$\prod_{m=-j}^{j} (J_i^{(j)} - m\mathbb{1}^{(j)}) = 0, \tag{3.25}$$

where $\mathbb{1}^{(j)}$ denotes the unit matrix of dimension $2j+1$.

Remark. Observe that the finite-dimensional vector space (dimension $2j+1$) of all complex column matrices z of length $2j+1$ equipped with the scalar product $z^\dagger z$ is a Hilbert space on which the operators defined by $J_i\colon z \to J_i^{(j)}z$

are Hermitian and satisfy $[J_i, J_j] = ie_{ijk}J_k$. In this rather trivial sense, there always exist Hilbert spaces and Hermitian operators J_i that realize each value $0, \frac{1}{2}, 1, \ldots$ of j given by the abstract angular momentum theory developed above.

It is clear from these examples that the (abstract, Hermitian) angular momentum operator $\mathbf{J}$, satisfying the (defining) relation $\mathbf{J} \times \mathbf{J} = i\mathbf{J}$, can be represented by its action on the eigenkets of any physical system *in block form*:

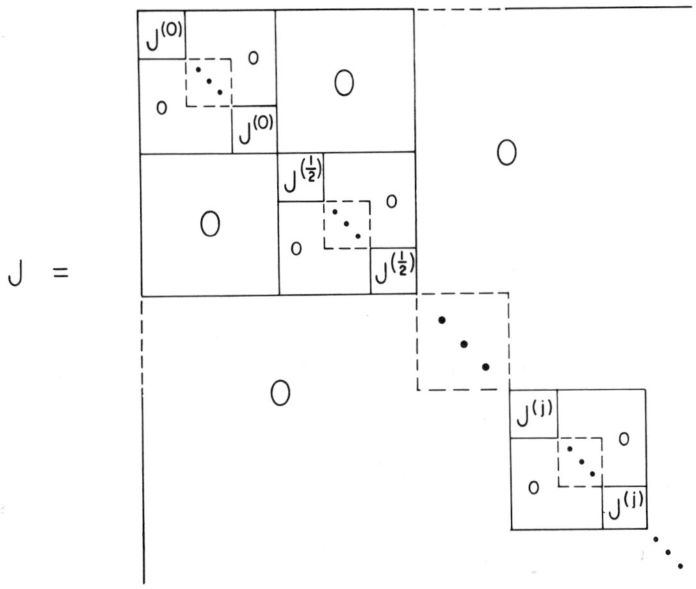

(3.26)

In general, the block matrix $\mathbf{J}^{(j)}$ will occur with repetition n_{2j+1}, different for each angular momentum j. (We have arbitrarily ordered the blocks by size.)

The distinction between the operator $\mathbf{J}$ and its realization by one or more blocks is not always made clear in the literature. The orbital angular momentum operator $\hbar\mathbf{L} = \mathbf{x} \times \mathbf{p}$ is limited to blocks having j an integer—that is, $n_2 = n_4 = n_6 = \cdots = 0$. *Spin* angular momentum can denote a block with any j, integer or half-integer.

5. The Rotation Matrices (General Properties)

In Chapter 2 we found two unitary matrix representations of the group of rotations $[SU(2)]$—namely, the set of 2×2 matrices:

$$\{e^{-i\psi(\hat{n}\cdot\boldsymbol{\sigma})/2} : \hat{n}\cdot\hat{n} = 1,\ 0 \leqslant \psi < 2\pi\} \tag{3.27}$$

and the set of 3×3 matrices:

$$\{e^{\psi N}: 0 \leqslant \psi < 2\pi\}, \tag{3.28}$$

where N is the 3×3 skew-symmetric matrix defined in Eq. (2.6).

The matrix N is unitarily equivalent to

$$-i\hat{n} \cdot \mathbf{J}^{(1)} \equiv -i(n_1 J_1^{(1)} + n_2 J_2^{(2)} + n_3 J_3^{(3)}),$$

the explicit transformation being given by

$$A^{\dagger} N A = -i\hat{n} \cdot \mathbf{J}^{(1)}, \tag{3.29}$$

where

$$A = \begin{pmatrix} -1/\sqrt{2} & 0 & 1/\sqrt{2} \\ -i/\sqrt{2} & 0 & -i/\sqrt{2} \\ 0 & 1 & 0 \end{pmatrix}. \tag{3.30}$$

In this result, A is the matrix transforming (x_1, x_2, x_3) to the *spherical basis*

$$\left[-\frac{1}{\sqrt{2}}(x_1 + ix_2),\ x_3,\ \frac{1}{\sqrt{2}}(x_1 - ix_2)\right].$$

Transforming $e^{\psi N}$ by A, we can thus bring Eq. (3.28) to a form analogous to Eq. (3.27):

$$\{e^{-i\psi\hat{n} \cdot \mathbf{J}^{(1)}}: \hat{n} \cdot \hat{n} = 1,\ 0 \leqslant \psi < 2\pi\}. \tag{3.31}$$

The 2×2 and 3×3 unitary matrices given by Eqs. (3.27) and (3.31) are both instances of the $(2j + 1) \times (2j + 1)$ unitary matrix defined by

$$D^j(\psi, \hat{n}) \equiv e^{-i\psi\hat{n} \cdot \mathbf{J}^{(j)}}, \qquad 0 \leqslant \psi < 2\pi. \tag{3.32}$$

We are accordingly led to study the significance of these matrices for general $j \in \{0, \frac{1}{2}, 1, \dots\}$. Observe that there is no problem as to the existence of these matrices (using the power series for the exponential), since the angular momentum matrices $J_i^{(j)}$ are finite.

In principle, the matrices $D^j(\psi, \hat{n})$ are fully determined by the definition (3.32), from knowledge of the angular momentum matrices; one uses the characteristic equation for $\hat{n} \cdot J^{(j)}$ [Eq. (3.25)] to reduce the exponential expansion to a finite sum. This is practical for $j = \frac{1}{2}$ and 1, but it becomes increasingly difficult for larger j. (We discuss methods that give the explicit evaluation in Section 6.)

Let us next observe that the unitary matrix (3.32) is the matrix representation on the space $\mathscr{H}_j$ with basis $\{|jm\rangle : m = j, j-1, \ldots, -j\}$ of the unitary operator defined by

$$\mathscr{U}(\psi, \hat{n}) \equiv e^{-i\psi\hat{n}\cdot\mathbf{J}}, \qquad \hat{n}\cdot\mathbf{J} \equiv n_1J_1 + n_2J_2 + n_3J_3. \tag{3.33}$$

There is no difficulty in exponentiating the angular momentum operator $\mathbf{J}$ itself, since its action always splits into a direct sum of actions on subspaces of the type $\mathscr{H}_j$.

The action of $\mathscr{U}(\psi, \hat{n})$ on a given eigenket $|jm\rangle$ is given by

$$\begin{aligned}\mathscr{U}(\psi, \hat{n})\colon |jm\rangle \to |jm\rangle' &\equiv \mathscr{U}(\psi, \hat{n})|jm\rangle\\ &= \sum_{m'} \langle jm'|\mathscr{U}(\psi, \hat{n})|jm\rangle |jm'\rangle\\ &= \sum_{m'} D^j_{m'm}(\psi, \hat{n})|jm'\rangle,\end{aligned} \tag{3.34}$$

where we have now obtained the elements of the matrix $D^j(\psi, \hat{n})$ in the form

$$D^j_{m'm}(\psi, \hat{n}) = \langle jm'|e^{-i\psi\hat{n}\cdot\mathbf{J}}|jm\rangle. \tag{3.35}$$

To interpret the physical significance of the unitary operators $\mathscr{U}(\psi, \hat{n})$, we consider the transformation of the angular momentum $\mathbf{J}$ corresponding to the unitary change of basis given by Eq. (3.34). Under the action of $\mathscr{U}(\psi, \hat{n})$, we have

$$\begin{aligned}\mathscr{U}(\psi, \hat{n})\colon J_i|jm\rangle \to (J_i|jm\rangle)' &= \mathscr{U}(\psi, \hat{n})(J_i|jm\rangle)\\ &= \mathscr{U}(\psi, \hat{n})J_i\mathscr{U}^{-1}(\psi, \hat{n})|jm\rangle'.\end{aligned}$$

Thus, the transformation of J_i is

$$\mathscr{U}(\psi, \hat{n})\colon J_i \to J_i' = \mathscr{U}(\psi, \hat{n})J_i\mathscr{U}^{-1}(\psi, \hat{n}). \tag{3.36}$$

To evaluate the transform of $\mathbf{J}$, we once again use the series expansion and rearrange the terms (valid, since absolutely convergent) into the Baker–Campbell–Hausdorff form:

$$\mathscr{U}(\psi, \hat{n})\mathbf{J}\mathscr{U}^{-1}(\psi, \hat{n}) = \sum_{k=0}^{\infty} \frac{1}{k!}[-i\psi\hat{n}\cdot\mathbf{J}, \mathbf{J}]_{(k)}, \tag{3.37}$$

where we denote the multiple commutator by

$$\begin{aligned} &[A, B]_{(k)} \equiv [A, [A, B]_{(k-1)}], \\ &[A, B]_{(1)} \equiv [A, B] = AB - BA, \\ &[A, B]_{(0)} \equiv B. \end{aligned} \tag{3.38}$$

Note now an important technical feature: *The evaluation of the multiple commutators depends only on the commutation properties of the operators* J_i, *and this is independent of which subspace* $\mathscr{H}_j$ *has been singled out for evaluating the action of each* J_i *in Eq.* (3.37).

It follows from the commutation relations for $\mathbf{J}$ that

$$[\mathbf{a}\cdot\mathbf{J}, \mathbf{J}] = -i\mathbf{a}\times\mathbf{J}, \tag{3.39}$$

where $\mathbf{a}$ is a vector that commutes with the J_i. Thus, we find for the transform

$$\begin{aligned} \mathbf{J}' &= e^{-i\psi\hat{n}\cdot\mathbf{J}}\mathbf{J}e^{i\psi\hat{n}\cdot\mathbf{J}} \\ &= \mathbf{J}\cos\psi + \hat{n}(\hat{n}\cdot\mathbf{J})(1 - \cos\psi) - (\hat{n}\times\mathbf{J})\sin\psi. \end{aligned} \tag{3.40}$$

This result gives an explicit evaluation of the transformation, Eq. (3.36).

Remarks. (*a*) If we now compare Eq. (3.40)—the transformation of the vector operator $\mathbf{J}$—with the rotation of a vector as given by Eq. (2.1), we see that the two results differ by the sign of the $\sin\psi$ term [set $\phi = \psi$ in Eq. (2.1)]. This difference in sign is not an error, but is a consequence of an important distinction in the meaning of the two transformations.

In Eq. (2.1) we made an active rotation of the vector $\vec{x}$; that is, we kept the coordinate frame fixed and moved the vector to a new position. Writing the vector $\vec{x}$ in terms of coordinates (x_1, x_2, x_3), we have $\vec{x} = \sum_i x_i\hat{e}_i$, where $(\hat{e}_1, \hat{e}_2, \hat{e}_3)$ are the unit vectors of a fixed right-handed coordinate frame. The transformation $\vec{x} \to \vec{x}'$ is effected by

$$x_i = \sum_j R_{ij}(\psi, \hat{n})x_j \equiv x_i', \tag{3.41}$$

so that the coordinates (x_1, x_2, x_3) were changed to (x_1', x_2', x_3') by the rotation.

By contrast, the transformation of the vector operator $\mathbf{J} = (J_1, J_2, J_3)$ given by Eq. (3.40) is

$$J_i \to \sum_j R_{ji}(\psi, \hat{n})J_j \equiv J'_i = \sum_j R_{ij}(\psi, -\hat{n})J_j, \tag{3.42}$$

and this transformation was induced from that of the Hilbert space eigenket, $|jm\rangle \to |jm\rangle'$, given by Eq. (3.34). This latter transformation, which will be taken to be standard throughout this monograph [see Remark (*d*) at the end of Section 14 for further discussion], changes the ket vectors $\{|jm\rangle\}$ and is dual to a transformation changing the bra vectors $\{\langle(x_1, x_2, x_3)|\}$ labeled by coordinates.

Thus we see that in Eq. (2.1) the coordinates $\{(x_1, x_2, x_3)\}$ were changed under the rotation, whereas in Eq. (3.40) the kets $\{|jm\rangle\}$ were changed. These two possible interpretations of a rotation are dual to each other, and this duality accounts for the fact that $(\psi, \hat{n})$ appears in Eq. (3.41) and $(\psi, -\hat{n})$ in Eq. (3.42).

(*b*) The result given in Eq. (3.40) has another interesting aspect: this equation is the prototype for the general definition of a tensor operator in terms of its transformation properties. This subject is developed more fully in Section 14.

(*c*) In view of the interpretation of $\mathscr{U}(\psi, \hat{n})$ as corresponding to a rotation of a physical system, it is customary in physics to call the operator $\mathscr{U}(\psi, \hat{n})$ a *rotation operator*. The matrix representation $D^j(\psi, \hat{n})$ is correspondingly called a *rotation matrix*.

Group property of the rotation matrices. We have already remarked that the sets of rotation matrices corresponding to $j = \frac{1}{2}$ and $j = 1$ are unitary matrix representations of the group $SU(2)$. We next prove this property for arbitrary $j \in \{0, \frac{1}{2}, 1, \ldots\}$.

We must show that the $(2j + 1) \times (2j + 1)$ unitary matrices (3.32) multiply in the same way as the 2×2 unitary matrices $U(\psi, \hat{n}) = \exp(-i\psi\hat{n} \cdot \boldsymbol{\sigma}/2)$—that is,

$$U(\psi', \hat{n}')U(\psi, \hat{n}) = U(\psi'', \hat{n}'') \tag{3.43}$$

implies

$$D^j(\psi', \hat{n}')D^j(\psi, \hat{n}) = D^j(\psi'', \hat{n}'') \tag{3.44}$$

for general j.

Proof. A direct proof of this result, using the definition (3.32), is fairly cumbersome and difficult. Let us proceed indirectly, using the similarity transformation (3.40).

Let $\mathbf{a} = (a_1, a_2, a_3)$ denote a vector that commutes with $\mathbf{J}$. Then it follows from Eq. (3.40) and Eq. (2.1) that

$$\mathscr{U}(\psi, \hat{n})(\mathbf{a} \cdot \mathbf{J})\mathscr{U}^{-1}(\psi, \hat{n}) = \mathbf{b} \cdot \mathbf{J}, \tag{3.45}$$

where in column matrix notation

$$\mathbf{b} = R(\psi, \hat{n})\mathbf{a}. \tag{3.46}$$

It follows at once that

$$\mathscr{U}(\psi, \hat{n})\mathscr{U}(\phi, \hat{a})\mathscr{U}^{-1}(\psi, \hat{n}) = \mathscr{U}(\psi, \hat{b}) \tag{3.47}$$

is a valid operator identity on each subspace $\mathscr{H}_j$. [Observe that this result generalizes the class angle relation (Chapter 2, p. 9) to general j.]

Transforming this result by $\mathscr{U}(\psi', \hat{n}')$ yields

$$\mathscr{U}(\psi', \hat{n}')\mathscr{U}(\psi, \hat{n})\mathscr{U}(\phi, \hat{a})[\mathscr{U}(\psi', \hat{n}')\mathscr{U}(\psi, \hat{n})]^{-1} = \mathscr{U}(\phi, \hat{c}), \tag{3.48}$$

where

$$\mathbf{c} = R(\psi', \hat{n}')\mathbf{b} = R(\psi', \hat{n}')R(\psi, \hat{n})\mathbf{a} = R(\psi'', \hat{n}'')\mathbf{a}, \tag{3.49}$$

the last step following from the fact (proved in Chapter 2) that the 3×3 orthogonal matrices multiply in the same way as the corresponding 2×2 unitary matrices.

In the final step, we now transform Eq. (3.48) by $\mathscr{U}^{-1}(\psi'', \hat{n}'')$:

$$[\mathscr{U}^{-1}(\psi'', \hat{n}'')\mathscr{U}(\psi', \hat{n}')\mathscr{U}(\psi, \hat{n})]\mathscr{U}(\phi, \hat{a})[\mathscr{U}^{-1}(\psi'', \hat{n}'')\mathscr{U}(\psi', \hat{n}')\mathscr{U}(\psi, \hat{n})]^{-1}$$
$$= \mathscr{U}(\phi, \hat{a}).$$

Thus, the transform by

$$\mathscr{U}^{-1}(\psi'', \hat{n}'')\mathscr{U}(\psi', \hat{n}')\mathscr{U}(\psi, \hat{n}) \tag{3.50}$$

leaves an arbitrary unitary operator $\mathscr{U}(\phi, \hat{a})$ invariant. Since the operators $\{J_i\}$ leave no vector of the space $\mathscr{H}_j$ invariant, it follows that the operator (3.50) is a multiple α of the unit operator. Since the determinant of $D^j(\psi, \hat{n})$ is 1, it follows also that $\alpha = 1$. Thus, we have proved that the result

$$\mathscr{U}(\psi', \hat{n}')\mathscr{U}(\psi, \hat{n}) = \mathscr{U}(\psi'', \hat{n}'') \tag{3.51}$$

follows from the corresponding relation (3.43) for the 2×2 unitary matrices.

The multiplication property (3.44) of the matrices $D^j(\psi, \hat{n})$ now follows from the definition (3.35). ■

The preceding method of proof of the group property may also be applied to the Euler angle parametrization of rotations. Since this parametrization is a popular one for applications, we introduce now the relevant notations.

The unitary *rotation operator* is defined by

$$\mathscr{U}(\alpha\beta\gamma) = e^{-i\alpha J_3} e^{-i\beta J_2} e^{-i\gamma J_3} \tag{3.52}$$

[parameter domains given by Eq. (2.41)], and the corresponding *rotation matrices* are given by

$$D^j(\alpha\beta\gamma) = e^{-i\alpha J_3^{(j)}} e^{-i\beta J_2^{(j)}} e^{-i\gamma J_3^{(j)}} \tag{3.53}$$

with *matrix elements*

$$D^j_{m'm}(\alpha\beta\gamma) = \langle jm'|\mathscr{U}(\alpha\beta\gamma)|jm\rangle. \tag{3.54}$$

The proof that the multiplication of rotations [see Eq. (2.40)]

$$U(\alpha'\beta'\gamma')U(\alpha\beta\gamma) = U(\alpha''\beta''\gamma'') \tag{3.55}$$

implies the general multiplication property for operators

$$\mathscr{U}(\alpha'\beta'\gamma')\mathscr{U}(\alpha\beta\gamma) = \mathscr{U}(\alpha''\beta''\gamma'') \tag{3.56}$$

is a repetition of the preceding proof given for the $(\psi, \hat{n})$ parametrization. The multiplication property for the matrices (3.53) then follows from Eq. (3.54):

$$D^j(\alpha'\beta'\gamma')D^j(\alpha\beta\gamma) = D^j(\alpha''\beta''\gamma''). \tag{3.57}$$

The essential element in the proof of the multiplication laws (3.44) and (3.57) for matrices is the demonstration of the multiplication laws (3.51) and (3.56) for operators. For the parameters $(\psi, \hat{n})$ and $(\alpha\beta\gamma)$, we are able to prove the desired operator relations by using directly the properties of exponential operators of the type $\exp(-i\psi\hat{n} \cdot \mathbf{J})$ and the known action of the J_i on the space $\mathscr{H}_j$. Thus, the results of this section are independent of the particular realization of the space $\mathscr{H}_j$.

In the subsequent sections we shall be discussing various explicit Hilbert spaces $\mathscr{H}_j$ on which the standard actions of the angular momentum operators are realized. In these explicit realizations, verification that the correspondence $U \to \mathscr{U}$ implies $\mathscr{U}'\mathscr{U} = \mathscr{U}''$ for $U'U = U''$ is often an elementary result, following directly from the definition of $\mathscr{U}$, as a mapping on the space $\mathscr{H}_j$, making no direct use of the exponential operators. These methods (see Chapter 5, Section 4) are independent of the parametrization of U, and yield directly the rotation matrices, which are denoted by

$$D^j(U) \equiv D^j\begin{pmatrix} u_{11} & u_{12} \\ u_{21} & u_{22} \end{pmatrix}, \tag{3.58}$$

where it is convenient to introduce the notation $f(U)$ to denote the values of a function f defined on the four elements u_{ij} of U.

We turn next to the description of the explicit forms of the rotation matrices in terms of the more common parametrizations of a rotation.

6. The Rotation Matrices (Explicit Forms)

Rotation matrices in terms of Euler angles. Equation (3.53) [equivalently, Eq. (3.54)] gives the complete expression for the rotation matrices. Using the fact that $J_3^{(j)}$ is diagonal, we find

$$D^j_{m'm}(\alpha\beta\gamma) = e^{-im'\alpha} d^j_{m'm}(\beta) e^{-im\gamma}, \tag{3.59}$$

where $d^j_{m'm}(\beta)$ is defined by

$$d^j_{m'm}(\beta) \equiv \langle jm'|e^{-i\beta J_2}|jm\rangle; \tag{3.60}$$

that is,

$$d^j(\beta) = e^{-i\beta J_2^{(j)}}. \tag{3.61}$$

It is straightforward to obtain $d^j(\beta)$ for $j = \frac{1}{2}$ and 1 by using the characteristic equation (3.25) and the series expansion of the exponential:[1]

$$d^{\frac{1}{2}}(\beta) = \begin{pmatrix} \cos\dfrac{\beta}{2} & -\sin\dfrac{\beta}{2} \\ \sin\dfrac{\beta}{2} & \cos\dfrac{\beta}{2} \end{pmatrix},$$

[1] The labeling of rows and columns follows the standard convention discussed above Eq. (3.24).

$$d^1(\beta) = \begin{pmatrix} \dfrac{1+\cos\beta}{2} & \dfrac{-\sin\beta}{\sqrt{2}} & \dfrac{1-\cos\beta}{2} \\ \dfrac{\sin\beta}{\sqrt{2}} & \cos\beta & \dfrac{-\sin\beta}{\sqrt{2}} \\ \dfrac{1-\cos\beta}{2} & \dfrac{\sin\beta}{\sqrt{2}} & \dfrac{1+\cos\beta}{2} \end{pmatrix}. \tag{3.62}$$

The rotation matrices for $j = \frac{1}{2}$ and 1 are, of course, just the original 2×2 unimodular unitary matrix $U(\alpha\beta\gamma)$ [Eq. (2.40)] and the original 3×3 orthogonal matrix $R(\alpha\beta\gamma)$ [Eq. (2.37)], respectively, the latter being transformed by the matrix A of Eq. (3.30):

$$D^{\frac{1}{2}}(\alpha\beta\gamma) = U(\alpha\beta\gamma),$$

$$D^1(\alpha\beta\gamma) = A^\dagger R(\alpha\beta\gamma)A. \tag{3.63}$$

There are several methods available for obtaining the elements $d^j_{m'm}(\beta)$ of the matrix $d^j(\beta)$ for general j. One finds, for example, the following techniques used: (*a*) differential equations; (*b*) a rotation of J_2 into J_3; (*c*) recursion relations; (*d*) characteristic equation, using the principal idempotents of $J_2^{(j)}$, and (*e*) boson operator techniques.

Since the differential equation approach is the most common, it is developed in some detail in Section 8. Methods (*b*) and (*c*) are discussed below, and method (*d*) is sketched in Note 2.

The most elegant procedure uses boson operator techniques; these techniques are of sufficient general interest that we develop this theme in detail in Chapter 5. It is useful to avail ourselves at this point of the essential results for $D^j(U)$ – for the purpose of discussing the structure of the results – and defer the actual derivation to that chapter.

Let ζ denote a parametrization $U(\zeta)$ of U; then the D-matrix in the parametrization ζ is obtained (abusing the notation) as

$$D^j(\zeta) \equiv D^j(U(\zeta)). \tag{3.64}$$

Using the Euler angle parametrization given by the matrix (2.40) for $U(\alpha\beta\gamma)$, we obtain (Wigner [13])

$$d^j_{m'm}(\beta) = [(j+m')!(j-m')!(j+m)!(j-m)!]^{\frac{1}{2}}$$

$$\times \sum_s \frac{(-1)^{m'-m+s}\left(\cos\dfrac{\beta}{2}\right)^{2j+m-m'-2s}\left(\sin\dfrac{\beta}{2}\right)^{m'-m+2s}}{(j+m-s)!s!(m'-m+s)!(j-m'-s)!}, \tag{3.65}$$

where the summation index s extends over all integral values such that the factorials in the denominator are nonnegative.

Equation (3.65) expresses $d^j_{m'm}(\beta)$ as a polynomial in $\cos(\beta/2)$ and $\sin(\beta/2)$. It is found, however, by factoring out the highest possible (positive) powers of $\cos(\beta/2)$ and $\sin(\beta/2)$, that what remains in the summation is actually a polynomial in the squares, $\cos^2(\beta/2)$ and $\sin^2(\beta/2)$. Thus, by putting this explicitly in evidence,

$$\frac{1+x}{2} = \cos^2\frac{\beta}{2}, \qquad \frac{1-x}{2} = \sin^2\frac{\beta}{2} \tag{3.66}$$

(yielding $x = \cos\beta$), the form of Eq. (3.65) becomes that of a polynomial in x multiplied by powers of $\cos(\beta/2)$ and $\sin(\beta/2)$. These polynomials are the *Jacobi polynomials*, which we now discuss, paying particular attention to the subtleties involved in making this identification.

The Jacobi polynomials. The Jacobi polynomials $P_n^{(\alpha,\beta)}(x)$ are polynomials of degree n defined by (Szegö [14], Erdélyi *et al.* [15])

$$P_n^{(\alpha,\beta)}(x) = \sum_s \binom{n+\alpha}{s}\binom{n+\beta}{n-s}\left(\frac{x-1}{2}\right)^{n-s}\left(\frac{x+1}{2}\right)^s, \tag{3.67}$$

where (α, β) are parameters in this standard notation,[1] n is a nonnegative integer, and $\binom{z}{a}$ is defined, for arbitrary complex z, by

$$\binom{z}{a} = \begin{cases} \dfrac{z(z-1)\cdots(z-a+1)}{a!} & \text{for } a = 1, 2, \ldots \\ 1 & \text{for } a = 0 \\ 0 & \text{for } a = -1, -2, \ldots \end{cases} \tag{3.68}$$

In our use of these polynomials, the numbers

$$n,\ n+\alpha,\ n+\beta,\ n+\alpha+\beta \tag{3.69}$$

are always nonnegative integers, and we may rewrite definition (3.67) in the form

[1] The reader is cautioned against confusing the two standard notations using (α, β)—Euler angles in (2.40) and parameters in (3.67).

$$P_n^{(\alpha,\beta)}(x) = (n+\alpha)!(n+\beta)!\sum_s \frac{\left(\frac{x-1}{2}\right)^{n-s}\left(\frac{x+1}{2}\right)^{s}}{s!(n+\alpha-s)!(\beta+s)!(n-s)!}, \qquad (3.70)$$

where the summation on s extends over all integral values for which the arguments of the factorials are nonnegative.

If either α or β is a negative integer, or if both are, it turns out that the polynomials (3.70), subject to the parameter restrictions (3.69), are simply related to Jacobi polynomials having only nonnegative (α', β') parameters.[1] We note these formulas explicitly because *the same phenomenon makes its appearance again in the decomposition of a Wigner coefficient (or a Racah coefficient) into analogous forms in which square root factors multiply a polynomial.* This view concerning the Wigner and Racah coefficients is developed in RWA.

The following three formulas, useful for relating Jacobi polynomials having negative parameters (α, β) to those having positive parameters, may be verified directly from the definition, Eq. (3.70):

$$P_n^{(\alpha,\beta)}(x) = \frac{(n+\alpha)!(n+\beta)!}{n!(n+\alpha+\beta)!}\left(\frac{x+1}{2}\right)^{-\beta} P_{n+\beta}^{(\alpha,-\beta)}(x),$$

$$P_n^{(\alpha,\beta)}(x) = \frac{(n+\alpha)!(n+\beta)!}{n!(n+\alpha+\beta)!}\left(\frac{x-1}{2}\right)^{-\alpha} P_{n+\alpha}^{(-\alpha,\beta)}(x),$$

$$P_n^{(\alpha,\beta)}(x) = \left(\frac{x-1}{2}\right)^{-\alpha}\left(\frac{x+1}{2}\right)^{-\beta} P_{n+\alpha+\beta}^{(-\alpha,-\beta)}(x). \qquad (3.71)$$

Rotation matrices and Jacobi polynomials. The relation of the functions $d^j_{m'm}(\beta)$ to the Jacobi polynomials follows directly from the definition (3.70), upon multiplying in the necessary factors and comparing with the definition (3.65):

$$d^j_{m'm}(\beta) = \left[\frac{(j+m)!(j-m)!}{(j+m')!(j-m')!}\right]^{\frac{1}{2}} \left(\sin\frac{\beta}{2}\right)^{m-m'} \left(\cos\frac{\beta}{2}\right)^{m'+m}$$

$$\times\, P_{j-m}^{(m-m',m+m')}(\cos\beta). \qquad (3.72)$$

[1] In mathematical discussions (Erdélyi *et al.* [15]) of the orthogonality properties of the Jacobi polynomials, it is assumed that $\alpha > -1, \beta > -1$ in order that the weight function $(1-x)^\alpha(1+x)^\beta$ have no negative powers and be integrable. Such restrictions are not directly of interest, for applications of the rotation group in physics, for two reasons: (*a*) any orthogonality properties are to be implied by the rotation matrices themselves, and (*b*) more general negative indices, such as are allowed by (3.69), occur naturally in physical applications.

This result is valid for the full range of the projection quantum numbers: $-j \leqslant m \leqslant j$; $-j \leqslant m' \leqslant j$. *Negative powers of the sine and cosine factors multiplying the Jacobi polynomial are always canceled by factors coming out of the polynomial because of properties* (3.71). It is useful to make these polynomial forms fully explicit, since the same problem, as mentioned above, arises in the context of Wigner coefficients. These results are given below.

Let k be the smallest of the four (nonnegative) integers:

$$k \equiv \min(j+m, j-m, j+m', j-m'). \tag{3.73}$$

Define the (nonnegative) parameters μ, ν, and λ—corresponding to the four possible cases for k—by means of the following table:

k	μ	ν	λ
$j+m$	$m'-m$	$-m'-m$	$m'-m$
$j-m$	$m-m'$	$m'+m$	0
$j+m'$	$m-m'$	$-m'-m$	0
$j-m'$	$m'-m$	$m'+m$	$m'-m$

In terms of these nonnegative parameters the following expression for the rotation function is always in polynomial form:

$$d^j_{m'm}(\beta) = (-1)^\lambda \left[\frac{k!(2j-k)!}{(k+\mu)!(k+\nu)!}\right]^{\frac{1}{2}} \left(\sin\frac{\beta}{2}\right)^\mu \left(\cos\frac{\beta}{2}\right)^\nu P_k^{(\mu,\nu)}(\cos\beta). \tag{3.74}$$

Determination of $d^j(\beta)$ by rotating J_3 into J_2. The rotation of eigenkets

$$|jm\rangle \rightarrow |jm\rangle' = \mathscr{U}\left(\frac{\pi}{2}, -\hat{e}_1\right)|jm\rangle \tag{3.75}$$

induces the rotation of $\pi/2$ about $\hat{e}_1$ of the angular momentum $\vec{J} = J_1\hat{e}_1 + J_2\hat{e}_2 + J_3\hat{e}_3$—that is, $\mathbf{J} \rightarrow \mathbf{J}' = J_1\hat{e}_1 - J_3\hat{e}_2 + J_2\hat{e}_3$. Correspondingly, the transformation for the J_3 angular momentum component [using Eq. (3.36)] is given by

$$J_2 = \mathscr{U}\left(\frac{\pi}{2}, -\hat{e}_1\right) J_3 \mathscr{U}\left(\frac{\pi}{2}, \hat{e}_1\right). \tag{3.76}$$

Hence, we also have

$$e^{-i\beta J_2} = \mathscr{U}\left(\frac{\pi}{2}, -\hat{e}_1\right) e^{-i\beta J_3} \mathscr{U}\left(\frac{\pi}{2}, \hat{e}_1\right). \tag{3.77}$$

Taking matrix elements of this result, we obtain an expression for the elements of $d^j(\beta)$ in terms of exponentials:

$$d^j_{m'm}(\beta) = \sum_\mu D^{j*}_{\mu m'}\left(\frac{\pi}{2}, \hat{e}_1\right) D^j_{\mu m}\left(\frac{\pi}{2}, \hat{e}_1\right) e^{-i\mu\beta}, \tag{3.78}$$

where we have used the unitary property of the D-matrices in obtaining this result. The numerical coefficient $D^j_{m'm}(\pi/2, \hat{e}_1)$ is given by[1]

$$D^j_{m'm}\left(\frac{\pi}{2}, \hat{e}_1\right) = \frac{(-i)^{m'-m}}{2^j}[(j+m)!(j-m)!(j+m')!(j-m')!]^{\frac{1}{2}}$$

$$\times \sum_s \frac{(-1)^s}{(j+m-s)!(j-m'-s)!(m'-m+s)!s!}. \tag{3.79}$$

Simple relations among the $d^j_{m'm}(\beta)$. A variety of useful relations among the matrix elements of $d^j(\beta)$ may be proved by direct inspection of the definition (3.65) (Edmonds [16], Rose [17]). For completeness, we state several of these relations. (A comprehensive development of the relations among the elements of the rotation matrix itself is given in Chapter 5, Appendix C):

$$d^j_{m'm}(\beta) = d^j_{mm'}(-\beta),$$

$$d^j_{m'm}(-\beta) = (-1)^{m'-m} d^j_{m'm}(\beta),$$

$$d^j_{m'm}(\beta) = d^j_{-m,-m'}(\beta). \tag{3.80}$$

Combining the first of these relations with the second one yields

$$d^j_{m'm}(\beta) = (-1)^{m'-m} d^j_{mm'}(\beta). \tag{3.81}$$

This relation, when combined with the third one of Eqs. (3.80), gives

$$d^j_{m'm}(\beta) = (-1)^{m'-m} d^j_{-m',-m}(\beta). \tag{3.82}$$

Recursion relations and their source. In addition to satisfying the group multiplication rule discussed in the last section, the rotation matrices also obey a number of rules, called *coupling rules*, which follow from a study of the *direct product* of two rotation matrices; these relations involve *Wigner coefficients* of the rotation group. These relations are developed in detail in Section 13.

[1] This result comes from Eq. (3.89).

It is not always realized that these coupling laws are the source of most of the so-called recursion relations for the $d^j_{m'm}(\beta)$. As might be expected, since there exists a vast mathematical literature developing the properties of Jacobi polynomials, these relations may also be proved by using classical mathematical formulas, without making direct reference to groups and Wigner coefficients.

In order that we may complete the discussion of the properties of the $d^j_{m'm}(\beta)$ in this section, and at the same time verify the statement above, we shall avail ourselves of the general result, Eq. (3.188), here. We need use only the simplest instance of that result, namely $j_2 = \frac{1}{2}$. There are four relations coming from Eq. (3.188) for $j_2 = \frac{1}{2}$, and upon supplying the explicit coefficients and $d^{\frac{1}{2}}(\beta)$ matrix elements (the exponential factors cancel out), the four *basic relations* are found to be

$$\begin{aligned}
&(j-m+1)^{\frac{1}{2}}\left(\cos\frac{\beta}{2}\right)d^{j+\frac{1}{2}}_{m'-\frac{1}{2},m-\frac{1}{2}}(\beta) + (j+m+1)^{\frac{1}{2}}\left(\sin\frac{\beta}{2}\right)d^{j+\frac{1}{2}}_{m'-\frac{1}{2},m+\frac{1}{2}}(\beta)\\
&\quad = (j-m'+1)^{\frac{1}{2}}d^j_{m'm}(\beta),\\
&-(j-m+1)^{\frac{1}{2}}\left(\sin\frac{\beta}{2}\right)d^{j+\frac{1}{2}}_{m'+\frac{1}{2},m-\frac{1}{2}}(\beta) + (j+m+1)^{\frac{1}{2}}\left(\cos\frac{\beta}{2}\right)d^{j+\frac{1}{2}}_{m'+\frac{1}{2},m+\frac{1}{2}}(\beta)\\
&\quad = (j+m'+1)^{\frac{1}{2}}d^j_{m'm}(\beta),\\
&(j+m)^{\frac{1}{2}}\left(\cos\frac{\beta}{2}\right)d^{j-\frac{1}{2}}_{m'-\frac{1}{2},m-\frac{1}{2}}(\beta) - (j-m)^{\frac{1}{2}}\left(\sin\frac{\beta}{2}\right)d^{j-\frac{1}{2}}_{m'-\frac{1}{2},m+\frac{1}{2}}(\beta)\\
&\quad = (j+m')^{\frac{1}{2}}d^j_{m'm}(\beta),\\
&(j+m)^{\frac{1}{2}}\left(\sin\frac{\beta}{2}\right)d^{j-\frac{1}{2}}_{m'+\frac{1}{2},m-\frac{1}{2}}(\beta) + (j-m)^{\frac{1}{2}}\left(\cos\frac{\beta}{2}\right)d^{j-\frac{1}{2}}_{m'+\frac{1}{2},m+\frac{1}{2}}(\beta)\\
&\quad = (j-m')^{\frac{1}{2}}d^j_{m'm}(\beta).
\end{aligned}\tag{3.83}$$

As remarked above, these four equations are just the explicit expression of the reduction of the direct product $D^j \otimes D^{\frac{1}{2}}$ into $D^{j+\frac{1}{2}}$ and $D^{j-\frac{1}{2}}$.

To illustrate the use of Eqs. (3.83), we indicate the derivation of two recursion relations that have been found to be particularly useful in numerical calculations of the $d^j_{m'm}(\beta)$.

Multiplying the third equation of (3.83) by $(j+m')^{\frac{1}{2}}$, and using the second equation (with $j \to j-\frac{1}{2}$ and $m' \to m'-\frac{1}{2}$) to elevate all angular momenta to a common j, yields the relation (Gel'fand and Shapiro [18], Fox and Ozier [19])

$$[(j-m)(j+m+1)]^{\frac{1}{2}} \sin\beta \, d^j_{m',m+1}(\beta) + [(j+m)(j-m+1)]^{\frac{1}{2}} \sin\beta \, d^j_{m',m-1}(\beta) = 2(m\cos\beta - m')d^j_{m'm}(\beta). \tag{3.84}$$

Similar manipulations may be used to obtain (Altmann and Bradley [20])

$$[(j+m)(j-m+1)]^{\frac{1}{2}} d^j_{m',m-1}(\beta) + [(j+m')(j-m'+1)]^{\frac{1}{2}} d^j_{m'-1,m}(\beta) = (m-m')\cot\frac{\beta}{2}\, d^j_{m'm}(\beta). \tag{3.85}$$

[Alternatively, this second result may be obtained directly from the general result (3.188), upon choosing $j_1 = j, j_2 = 1$.]

The general rotation matrices. We shall subsequently derive [Chapter 5, Eq. (5.44)] a form of the rotation matrices that is independent of parametrizations. We present this form here, not only for completeness, but also because it displays, more than any other form, the structural simplicity of the rotation matrices. The form is easily comprehended (hence, remembered) and from a practical point of view may be used to obtain the rotation matrices in any parametrization. This latter property is particularly useful for applications to physical problems involving nonstandard parametrizations (Gilmore [21]).

A preliminary remark on notation may be in order. We have seen in connection with the Jacobi polynomials the importance of nonnegative parameters [see Eq. (3.73) and the table given there]. In combinatorics the problem often arises of counting the number of rectangular arrays that can be constructed subject to the conditions that (a) the entries in the array are nonnegative integers, and (b) the sum of the integers in each row and in each column is specified (fixed). (This is a generalization of the magic square.) A notation based on such considerations proves very valuable here.

In terms of the elements u_{ij} of a 2×2 unitary unimodular matrix U, the elements of the rotation matrix may be written as

$$D^j_{m'm}(U) = [(j+m)!(j-m)!(j+m')!(j-m')!]^{\frac{1}{2}} \times \sum_{\boxed{\alpha}} \frac{(u_{11})^{\alpha_{11}}(u_{12})^{\alpha_{12}}(u_{21})^{\alpha_{21}}(u_{22})^{\alpha_{22}}}{\alpha_{11}!\alpha_{12}!\alpha_{21}!\alpha_{22}!}, \tag{3.86}$$

where the notation $\boxed{\alpha}$ symbolizes a 2×2 array of nonnegative integers,

$$\begin{array}{|cc|l} \hline \alpha_{11} & \alpha_{12} & j+m' \\ \alpha_{21} & \alpha_{22} & j-m' \, . \\ \hline j+m & j-m & \end{array} \tag{3.87}$$

In this array the α_{ij} are nonnegative integers subject to the row and column constraints (sums) indicated by the (nonnegative) integers $j \pm m$, $j \pm m'$. (Explicitly, $\alpha_{11} + \alpha_{12} = j + m'$, $\alpha_{21} + \alpha_{22} = j - m'$, $\alpha_{11} + \alpha_{21} = j + m$, $\alpha_{12} + \alpha_{22} = j - m$.) The summation is over all such arrays. (Observe that any one of the α_{ij} may serve as a single summation index if one wishes to eliminate the redundancy inherent in the square-array notation.)

The group multiplication property of the rotation matrices is now expressed in the form

$$D^j(U)D^j(U') = D^j(UU'). \tag{3.88}$$

Rotation matrices in terms of the Euler–Rodrigues parameters. In many physical problems (notably in molecular physics, crystal field theory, and electron paramagnetic resonance) one encounters the problem of reducing the representation $D^j(U)$ of the rotation group into the irreducible representations of a finite subgroup (point group) of the rotation group. The rotational elements of the point group are often most simply described as rotations about fixed axes. It is therefore useful to know explicitly the rotation matrices as expressed in terms of the Euler–Rodrigues parameters.

The desired form of the rotation matrix is obtained by substituting U as given by Eq. (2.21) into the general form (3.86) (we choose $\alpha_{21} = s$ as the single summation index):

$$D^j_{m'm}(\alpha_0, \boldsymbol{\alpha}) = [(j+m')!(j-m')!(j+m)!(j-m)!]^{\frac{1}{2}}$$
$$\times \sum_s \frac{(\alpha_0 - i\alpha_3)^{j+m-s}(-i\alpha_1 - \alpha_2)^{m'-m+s}(-i\alpha_1 + \alpha_2)^s(\alpha_0 + i\alpha_3)^{j-m'-s}}{(j+m-s)!(m'-m+s)!s!(j-m'-s)!}. \tag{3.89}$$

The additional substitution $\alpha_0 = \cos(\psi/2)$, $\alpha_i = n_i \sin(\psi/2)$ then yields also the explicit form of $D^j_{m'm}(\psi, \hat{n})$. (In many applications it is useful to have the trace of the matrix $D^j(\psi, \hat{n})$. This formula is developed in Note 3.)

Observe that the multiplication property of the rotation matrices (3.89) is now given by

$$D^j(\alpha_0', \boldsymbol{\alpha}')D^j(\alpha_0, \boldsymbol{\alpha}) = D^j(\alpha_0'', \boldsymbol{\alpha}''), \tag{3.90}$$

where the parameters $(\alpha_0'', \boldsymbol{\alpha}'')$ are obtained from $(\alpha_0', \boldsymbol{\alpha}')$ and $(\alpha_0, \boldsymbol{\alpha})$ by the quaternionic multiplication rules (2.42). [In Note 4, the properties of the functions $D^j_{m'm}$ are developed directly in terms of the parameters $(\alpha_0, \boldsymbol{\alpha})$. See also Note 6 for further discussion of properties of the rotation matrices.]

Remark. We have abused the mapping notation for a function f—that is, $f: X \to \mathbb{C}$, or equivalently, $f: x \to \alpha = f(x)$ for $\alpha \in \mathbb{C}$ and $x \in X$, where X denotes the domain of definition of f and $\mathbb{C}$ denotes the set of complex numbers—in using the same symbol $f = D^j_{m'm}$ to denote the various functions with values $D^j_{m'm}(\alpha\beta\gamma)$, $D^j_{m'm}(\psi, \hat{n})$, $D^j_{m'm}(U)$, $D^j_{m'm}(\alpha_0, \boldsymbol{\alpha})$, etc. This defect in the notation is particularly apparent when one notes that $f = D^j_{m'm}$ in Eq. (3.86) is a real polynomial (coefficients in $\mathbb{R}$) in the complex variables $(u_{11}, u_{12}, u_{21}, u_{22})$, while $f' = D^j_{m'm}$ in Eq. (3.89) is a complex polynomial (coefficients in $\mathbb{C}$) in the real variables $(\alpha_0, \boldsymbol{\alpha})$. In a strict usage, we should introduce a different symbol for the function in each parametrization of U, for example, $\mathscr{D}^j_{m'm}(\alpha\beta\gamma) = D^j_{m'm}(U(\alpha\beta\gamma))$, etc. (or, alternatively, use the standard notation for the composition of two functions). Following this latter course would, however, complicate unnecessarily the discussion of the rotation matrices, since the use of the same symbol, $D^j_{m'm}$, usually presents no difficulty. There is a precaution, however, and this is to be exercised in discussing the complex conjugate of a function versus the complex conjugate of the set of values of the function: If the domain of definition X of f is a set of real variables, there is no difficulty with the notation, since the complex conjugate of f, denoted f^*, is defined by $f^*(x) = [f(x)]^*$; in the only other case above for the D-functions, namely, for the notation $D^j_{m'm}(U)$, where $f = D^j_{m'm}$ is a real polynomial in the elements u_{ij} of U, we will avoid any ambiguity by using the convention that the symbol $D^{j*}_{m'm}(U)$ *always* denotes the complex conjugated set of values of $D^j_{m'm}(U)$—that is, $D^{j*}_{m'm}(U) \equiv [D^j_{m'm}(U)]^* = D^j_{m'm}(U^*)$, the last equality following from the fact that $D^j_{m'm}$ in Eq. (3.86) is a real polynomial.

7. Wave Functions for Angular Momentum Systems

In the discussion above, we have consistently based our work on the viewpoint that the kets, $\{|jm\rangle\}$, are to be the basis vectors for the elementary ("sharp") angular momentum systems. It is customary in physics to use an alternative viewpoint, that of *wave functions* (or *probability amplitudes*), to accord with the viewpoint taken for the Schrödinger equation. In this alternative view, the state labels denote functions, and the arguments of the function denote the position labels of the particle (or collection of particles) in configuration space. This shift in viewpoint is not trivial and will be discussed in detail here.

The wave functions for angular momentum systems can all be obtained in a uniform manner in the following way. Consider the product law for the rotation matrices, $D^j(UU') = D^j(U)D^j(U')$; that is,

$$D^j_{mm'}(UU') = \sum_{m''} D^j_{mm''}(U)D^j_{m''m'}(U'). \tag{3.91}$$

We seek to interpret this result as a transformation induced by the rotation $\mathscr{U}$ acting on the "system" described by the wave function $D^j_{m''m'}(U')$.

As the prototype for the interpretation, consider this same action $\mathscr{U}$ on a ket vector $|jm\rangle$:

$$\mathscr{U}: |jm\rangle \rightarrow |jm\rangle' \equiv \sum_{m'} D^j_{m'm}(U)|jm'\rangle. \tag{3.92}$$

It is the placement of the (m', m) indices in this result that is crucial in interpreting Eq. (3.91) as a wave function transformation.

Let us regard the function $D^j_{mm'}(UU')$, appearing on the left-hand side of Eq. (3.91), as the "transformed system"—that is,

$$[D^j_{mm'}(U')]' \equiv D^j_{mm'}(UU'). \tag{3.93}$$

We find that the product law, Eq. (3.91), then implies

$$\mathscr{U}: D^j_{mm'}(U') \rightarrow [D^j_{mm'}(U')]' \equiv \sum_{m''} D^j_{mm''}(U)D^j_{m''m'}(U'). \tag{3.94}$$

The indices in this equation do not accord with the standard form (3.92) *for transforming ket vectors.* To remedy this defect we must preserve the product law (since it is correct), but (somehow) change its form. To do this we use the fact that the D^j are unitary, replace U by U^{-1}, and take the complex conjugate of the equation. We then obtain an equation having the desired interpretation:

$$\mathscr{U}: D^{j*}_{mm'}(U') \rightarrow [D^{j*}_{mm'}(U')]' = \sum_{m''} D^j_{m''m}(U)D^{j*}_{m''m'}(U'). \tag{3.95}$$

This equation shows that it is the functions $D^{j*}_{mm'}(U')$—as opposed to the $D^j_{mm'}$ themselves—that transform properly as *state vectors* carrying angular momentum labels (j, m), this result being true for each $m' = j, j-1, \ldots, -j$. *The use of the complex conjugated rotation matrices is not a perverse convention, but a necessity for overall consistency.*

Before proceeding, it is useful to clarify the notation further. In the interpretation above, the angles of U' play no role. To be notationally consistent, the transformation law should be written without U' in a clear ket-like manner. Thus, one should write

$$\mathscr{U}: |_{m}{}^{j}{}_{m'}\rangle \to |_{m}{}^{j}{}_{m'}\rangle' \equiv \sum_{m''} D^{j}_{m''m}(U)|_{m''}{}^{j}{}_{m'}\rangle. \tag{3.96}$$

Here $|_{m}{}^{j}{}_{m'}\rangle$ denotes explicitly the ket vectors [which are implicit in Eq. (3.95)]. Note that the label m' is, so far, *a free index* and, as yet, uninterpreted. Linear combinations over this free index would, in fact, leave the results discussed here unchanged.

Having identified the ket vectors, we can now reintroduce without confusion the rotation U' that appears in Eq. (3.95). Letting U' be denoted by U, we find

$$\langle U|_{m}{}^{j}{}_{m'}\rangle = D^{j*}_{mm'}(U) = \langle jm'|\mathscr{U}^{-1}|jm\rangle. \tag{3.97}$$

Before we can discuss the significance of these results—and in particular how they yield all spherical functions—, we must first obtain the differential equations for the rotation matrices, since it is on this that the standard treatment of spherical harmonics is based.

8. Differential Equations for the Rotation Matrices

Since we have chosen to determine first the finite bases for angular momentum systems, and then to define the (unitary) rotation matrices by exponentiation, it is clear that the functions so defined possess derivatives of all orders; we can safely proceed in an elementary way, with no loss of generality.

Consider the rotation matrix using the Euler angle parametrization:

$$D^{j}(\alpha\beta\gamma) \equiv e^{-i\alpha J_3^{(j)}} e^{-i\beta J_2^{(j)}} e^{-i\gamma J_3^{(j)}}. \tag{3.98}$$

Differentiating with respect to α, β, and γ yields the following results:

$$\frac{\partial}{\partial\alpha} D^{j}(\alpha\beta\gamma) = -iJ_3^{(j)} D^{j}(\alpha\beta\gamma),$$

$$\begin{aligned}\frac{\partial}{\partial\beta} D^{j}(\alpha\beta\gamma) &= -i(e^{-i\alpha J_3^{(j)}} J_2^{(j)} e^{i\alpha J_3^{(j)}}) D^{j}(\alpha\beta\gamma) \\ &= -i(-J_1^{(j)} \sin\alpha + J_2^{(j)} \cos\alpha) D^{j}(\alpha\beta\gamma),\end{aligned}$$

$$\begin{aligned}\frac{\partial}{\partial\gamma} D^{j}(\alpha\beta\gamma) &= -i[D^{j}(\alpha\beta\gamma) J_3^{(j)} (D^{j}(\alpha\beta\gamma))^{-1}] D^{j}(\alpha\beta\gamma) \\ &= -i(J_1^{(j)} \cos\alpha \sin\beta + J_2^{(j)} \sin\alpha \sin\beta + J_3^{(j)} \cos\beta) D^{j}(\alpha\beta\gamma),\end{aligned} \tag{3.99}$$

where we have used the transformation of angular momentum, Eq. (3.40), in obtaining these forms.

One can now invert these results to express the action of the matrix operators $J_i^{(j)}$ on the matrices $D^j(\alpha\beta\gamma)$ as *differential operators* on the matrix functions $D^j(\alpha\beta\gamma)$. One finds the result

$$J_i^{(j)} D^j(\alpha\beta\gamma) = -\mathscr{J}_i D^j(\alpha\beta\gamma), \tag{3.100}$$

where the $\mathscr{J}_i$ are the differential operators defined by

$$\mathscr{J}_1 = i\cos\alpha\cot\beta\,\frac{\partial}{\partial\alpha} + i\sin\alpha\,\frac{\partial}{\partial\beta} - i\,\frac{\cos\alpha}{\sin\beta}\,\frac{\partial}{\partial\gamma},$$

$$\mathscr{J}_2 = i\sin\alpha\cot\beta\,\frac{\partial}{\partial\alpha} - i\cos\alpha\,\frac{\partial}{\partial\beta} - i\,\frac{\sin\alpha}{\sin\beta}\,\frac{\partial}{\partial\gamma},$$

$$\mathscr{J}_3 = -i\,\frac{\partial}{\partial\alpha}. \tag{3.101}$$

There is only one point in this derivation of the $\mathscr{J}_i$ that requires explanation: Why was a minus sign inserted in Eq. (3.100)? The necessity for this sign can be seen in two ways. (*a*) The sign is essential in order that the differential operators $\mathscr{J}_i$ obey the commutation relation, $\boldsymbol{\mathscr{J}} \times \boldsymbol{\mathscr{J}} = i\boldsymbol{\mathscr{J}}$; (*b*) The sign has its origin in the use of $D^{j*}_{mm'}$ as wave functions, as discussed in the previous section.

We must also be careful and recall that the $J_i^{(j)}$ in Eq. (3.100) are finite matrices, and that this equation is strictly valid only for each fixed j-basis. Only when the completeness of the direct sum of spaces $(\mathscr{H}_0 \oplus \mathscr{H}_{\frac{1}{2}} \oplus \mathscr{H}_1 \oplus \cdots)$ has been shown can we conclude—as we shall—that the differential operators $\mathscr{J}_i$ are valid operator realizations of a set of angular momentum operators satisfying $\boldsymbol{\mathscr{J}} \times \boldsymbol{\mathscr{J}} = i\boldsymbol{\mathscr{J}}$.

Since we have found for each angular momentum subspace $j = 0, \frac{1}{2}, 1, \ldots$ that the differential operators $\mathscr{J}_i$ may be represented in the subspace in terms of the matrices $J_i^{(j)}$ and that the form of this realization is *independent of* j, we may assert that the differential operators are, in fact, a realization of a set of angular momentum operators, in general, and not merely on the basis D^j (we use this result repeatedly in this section).

It is useful to rewrite Eqs. (3.101) in the form

$$\mathscr{J}_+ = \mathscr{J}_1 + i\mathscr{J}_2 = e^{i\alpha}\left(i\cot\beta\frac{\partial}{\partial\alpha} + \frac{\partial}{\partial\beta} - \frac{i}{\sin\beta}\frac{\partial}{\partial\gamma}\right),$$

$$\mathscr{J}_- = \mathscr{J}_1 - i\mathscr{J}_2 = e^{-i\alpha}\left(i \cot\beta \frac{\partial}{\partial\alpha} - \frac{\partial}{\partial\beta} - \frac{i}{\sin\beta}\frac{\partial}{\partial\gamma}\right),$$

$$\mathscr{J}_3 = -i\frac{\partial}{\partial\alpha}. \tag{3.102}$$

We next complex conjugate Eq. (3.100), taking into account the relations

$$\mathscr{J}_i^* = -\mathscr{J}_i, \qquad J_1^{(j)*} = J_1^{(j)}, \qquad J_2^{(j)*} = -J_2^{(j)}, \qquad J_3^{(j)*} = J_3^{(j)}, \tag{3.103}$$

and take matrix elements of the resulting equation. We obtain the standard action of the differential operators $\mathscr{J}_i$:

$$\mathscr{J}_\pm D^{j*}_{m'm}(\alpha\beta\gamma) = [(j \mp m')(j \pm m' + 1)]^{\frac{1}{2}} D^{j*}_{m'\pm 1,m}(\alpha\beta\gamma),$$

$$\mathscr{J}_3 D^{j*}_{m'm}(\alpha\beta\gamma) = m' D^{j*}_{m'm}(\alpha\beta\gamma). \tag{3.104}$$

The differential equations for the rotation matrices expressed by Eq. (3.104) are almost, but not quite, the desired objective of this section. The missing element, to which we now turn, is the interpretation of the index m.

Let us proceed physically. Suppose we had a single mass point moving in an orbit about a (fixed) origin. In this case, the angular momentum is $\vec{L} = (\vec{r} \times \vec{p})$, where the components of the position vector $\vec{r}$ of the particle may be taken to be

$$x_1 = r\cos\alpha\sin\beta,$$

$$x_2 = r\sin\alpha\sin\beta,$$

$$x_3 = r\cos\beta. \tag{3.105}$$

The angles (α, β) are thus identified as the azimuthal and polar angles of the position vector of the particle. We now obtain the differential operator for the angular momentum (using $\mathscr{L} = \hbar^{-1}\vec{r} \times \vec{p} = -i\vec{r} \times \nabla$):

$$\mathscr{L}_1 = i\cos\alpha\cot\beta\frac{\partial}{\partial\alpha} + i\sin\alpha\frac{\partial}{\partial\beta},$$

$$\mathscr{L}_2 = i\sin\alpha\cot\beta\frac{\partial}{\partial\alpha} - i\cos\alpha\frac{\partial}{\partial\beta},$$

$$\mathscr{L}_3 = -i\frac{\partial}{\partial\alpha}. \tag{3.106}$$

These equations are precisely the expressions for the $\mathscr{J}_i$, given by Eqs. (3.101), in which *we have deleted the term containing the derivative* $\partial/\partial\gamma$. Since the choice $m = 0$ removes the angle γ from the rotation matrix, we see that in the case of a single particle it is uniquely $\mathscr{J}$ of Eqs. (3.101) that is the physical angular momentum. Therefore $\mathscr{J}$ must be the physical angular momentum, in general. (We shall again see that this identification is correct in the more detailed discussion of the rigid rotator given in Chapter 7, Section 10.)

Observe that the choice $m = 0$ implies that j is integral, so that the generalization represented by $\mathscr{J}$ of Eq. (3.101) is that *half-integer* values of angular momenta are allowed, in contrast to $\mathscr{L}$, for which only integer values are allowed. (Actually, our argument shows that $m = 0$ is sufficient to remove γ from the theory for a single particle. But is it necessary – that is, is the existence of half-integer orbital angular momentum a possibility? The answer is that half-integer orbital angular momenta cannot exist, a result we shall prove in Chapter 6, Section 5.)

We can draw one further conclusion from these results. The fact that it is consistent to delete $\partial/\partial\gamma$ from the $\mathscr{J}$ in Eq. (3.102) implies that $\partial/\partial\gamma$ *must commute with the* $\mathscr{J}$. It follows that the Hermitian operator $i(\partial/\partial\gamma)$ is the missing element that completes the set of differential equations defining the rotation functions $D^{j*}_{m'm}(\alpha\beta\gamma)$. Accordingly, we add to Eqs. (3.102) the equations[1]

$$\mathscr{P}_3 \equiv -i\frac{\partial}{\partial\gamma},$$

$$\mathscr{P}_3 D^{j*}_{m'm}(\alpha\beta\gamma) = m D^{j*}_{m'm}(\alpha\beta\gamma). \tag{3.107}$$

To interpret these results, note now that we can solve Eqs. (3.101) for the operator $\partial/\partial\gamma$, and hence for $\mathscr{P}_3$. The result is

$$\mathscr{P}_3 = \sin\beta\cos\alpha\,\mathscr{J}_1 + \sin\beta\sin\alpha\,\mathscr{J}_2 + \cos\beta\,\mathscr{J}_3,$$

or equivalently, but more suggestively,

$$\mathscr{P}_3 = \sum_i R_{i3}(\alpha\beta\gamma)\mathscr{J}_i, \tag{3.108}$$

where $R(\alpha\beta\gamma)$ is the proper orthogonal matrix defined by Eq. (2.37). We are

---

[1] The physical interpretation of $(\mathscr{P}_1, \mathscr{P}_2, \mathscr{P}_3)$, demonstrated below, is that $\mathscr{P}_i$ is the component of the total angular momentum $\mathscr{J}$ of a rigid or solid body projected on the ith axis of a right-handed frame *fixed* in the body. There is no uniformity in the literature as to the notation for these components. We have adopted one used frequently by molecular spectroscopists (Nielsen [22]).

thus led to introduce the three operators $\mathscr{P}_j$ $(j = 1, 2, 3)$ defined by

$$\mathscr{P}_j \equiv \sum_i R_{ij}(\alpha\beta\gamma)\mathscr{J}_i, \tag{3.109}$$

and ask for their physical significance.[1]

The physical interpretation of the angular momentum operators $\mathscr{P}_i$ *and* $\mathscr{J}_i$. We proceed geometrically to obtain the physical interpretation of $\mathscr{P}_j$. Let $(\hat{e}_1, \hat{e}_2, \hat{e}_3)$ denote an inertial frame, and let $(\hat{f}_1, \hat{f}_2, \hat{f}_3)$ denote the frame (noninertial) whose instantaneous orientation, relative to the inertial frame, is described by the Euler angles:

$$\hat{f}_j = \sum_i R_{ij}(\alpha\beta\gamma)\hat{e}_i. \tag{3.110}$$

Since the operator $\vec{\mathscr{J}} \equiv \hat{e}_1\mathscr{J}_1 + \hat{e}_2\mathscr{J}_2 + \hat{e}_3\mathscr{J}_3$ [$\mathscr{J}_i$ defined by Eq. (3.101)] has previously been identified as the total angular momentum of a physical system, it is correct that $\mathscr{P}_j$, defined by

$$\mathscr{P}_j = \hat{f}_j \cdot \mathscr{J} = \sum_i R_{ij}(\alpha\beta\gamma)\mathscr{J}_i, \tag{3.111}$$

is the component of the total angular momentum referred to the moving axis $\hat{f}_j$.

The physical significance of the operators $\mathscr{P}_j$ has thus been established partially; we still must identify more closely the significance of the total angular momentum $\mathscr{J}$ itself.

The key relation is found in the commutator of the $\mathscr{J}_i$ with the direction cosines $R_{jl}(\alpha\beta\gamma)$ defining the frame $(\hat{f}_1, \hat{f}_2, \hat{f}_3)$:

$$[\mathscr{J}_i, R_{jl}] = ie_{ijk}R_{kl}, \qquad i, j, k \text{ cyclic} \tag{3.112}$$

---

[1] One must resist the temptation to interpret Eq. (3.109) by using the transformation (3.40), which, in terms of Euler angles, assumes the form

$$J'_j = \mathscr{U}(\alpha\beta\gamma)J_j\mathscr{U}^{-1}(\alpha\beta\gamma) = \sum_i R_{ij}(\alpha\beta\gamma)J_i.$$

The operators (J'_1, J'_2, J'_3) and (J_1, J_2, J_3) appearing in this result are the components of the angular momentum referred to the axes of two different frames, $(\hat{e}'_1, \hat{e}'_2, \hat{e}'_3)$ and $(\hat{e}_1, \hat{e}_2, \hat{e}_3)$, which stand in the relation $\hat{e}'_j = \sum_i R_{ij}(\alpha\beta\gamma)\hat{e}_i$ to each another, but the operators J_i themselves are *not* related to the position or motion of the $(\hat{e}'_1, \hat{e}'_2, \hat{e}'_3)$ frame—that is, *they are independent of the Euler angles*, whereas the operators $\mathscr{J}_i$ defined by Eqs. (3.101), and appearing in Eq. (3.109), do depend on the Euler angles and, accordingly, operate on the direction cosines $R_{ij}(\alpha\beta\gamma)$. *The* $\mathscr{P}_j$ *cannot be obtained by a unitary similarity transformation of the operators* $\mathscr{J}_i$.

for each $l = 1, 2, 3$. Using this result and the definition of the rotating frame, Eq. (3.110), we find[1]

$$[\hat{n}\cdot\mathscr{J}, \hat{f}_j] = -i(\hat{n}\times\hat{f}_j). \tag{3.113}$$

We now exponentiate this last result, using the same procedure that gave Eq. (3.40), to obtain the transformation [see Eq. (2.1) and Remark (a), p. 42]

$$\hat{f}'_j \equiv e^{-i\psi\hat{n}\cdot\mathscr{J}}\hat{f}_j e^{i\psi\hat{n}\cdot\mathscr{J}} = \mathscr{R}(\psi, -\hat{n})\hat{f}_j, \tag{3.114}$$

in which $\hat{n} = n_1\hat{e}_1 + n_2\hat{e}_2 + n_3\hat{e}_3$ is an arbitrary direction (specified with respect to the inertial frame), and ψ is an arbitrary angle. It is this result, the transformation (3.114), that yields the interpretation of $\mathscr{J}$: The angular momentum $\mathscr{J}$ is the generator of the rotations of the body-fixed frame itself; hence, *$\mathscr{J}$ is to be identified as the total angular momentum of a solid or rigid body whose instantaneous orientation is specified by the frame $(\hat{f}_1, \hat{f}_2, \hat{f}_3)$, which is itself fixed (no relative motion) in the body.*

Relations satisfied by the angular momentum operators of a solid body. We summarize here some of the relations satisfied by the angular momentum operators $\{\mathscr{J}_i\}$ and $\{\mathscr{P}_i\}$:

(*a*) Standard commutation of the $\mathscr{J}_i$:

$$[\mathscr{J}_i, \mathscr{J}_j] = i\mathscr{J}_k, \qquad i, j, k \text{ cyclic}; \tag{3.115}$$

(*b*) $\mathscr{J}_i$ can stand to either side:

$$\mathscr{P}_j = \sum_i R_{ij}(\alpha\beta\gamma)\mathscr{J}_i = \sum_i \mathscr{J}_i R_{ij}(\alpha\beta\gamma); \tag{3.116}$$

(*c*) The famous factor of $-i$ in the commutation of the $\mathscr{P}_i$:

$$[\mathscr{P}_i, \mathscr{P}_j] = -i\mathscr{P}_k, \qquad i, j, k \text{ cyclic}; \tag{3.117}$$

(*d*) Mutual commutivity of the $\{\mathscr{J}_i\}$ and $\{\mathscr{P}_i\}$:

$$[\mathscr{P}_i, \mathscr{J}_j] = 0, \qquad i, j, k = 1, 2, 3; \tag{3.118}$$

[1] Each of the vectors $\hat{f}_i$ defining the frame $(\hat{f}_1, \hat{f}_2, \hat{f}_3)$ is a vector operator with respect to the angular momentum $\mathscr{J}$ – see Section 15.

(*e*) Same invariant (squared) total angular momentum:

$$\mathscr{P}_1^2 + \mathscr{P}_2^2 + \mathscr{P}_3^2 = \mathscr{J}_1^2 + \mathscr{J}_2^2 + \mathscr{J}_3^2 = \mathscr{J}^2$$

$$= -\csc^2 \beta \left(\frac{\partial^2}{\partial \alpha^2} + \frac{\partial^2}{\partial \gamma^2} - 2\cos\beta \frac{\partial^2}{\partial \alpha\, \partial \gamma} \right) - \frac{\partial^2}{\partial \beta^2} - \cot \beta \frac{\partial}{\partial \beta}. \tag{3.119}$$

Let us observe again the surprising result for the commutation relations $[\mathscr{P}_i, \mathscr{P}_j] = -i\mathscr{P}_k$, that is, it is the operators $-\mathscr{P}_i$ that satisfy standard angular momentum commutation relations. (This result is a direct consequence of the fact that the $\mathscr{J}_i$ do not commute with the rotated axes $\hat{f}_i$ that appear in the definition $\mathscr{P}_i = \hat{f}_i \cdot \mathscr{J}$.) Observe also from $\mathscr{P}_i = \hat{f}_i \cdot \mathscr{J}$ that the $\mathscr{P}_i$ are *invariants* with respect to the rotations generated by $\mathscr{J}$, a result that follows geometrically from the fact that the rotations generated by $\mathscr{J}$ rotate both $\mathscr{J}$ and $\hat{f}_i$ *simultaneously*, thereby leaving the scalar product invariant.

Let us summarize the results so far obtained: The rotation matrices $D^{j*}_{m'm}(\alpha\beta\gamma)$ are simultaneous eigenfunctions of the three mutually commuting operators $\mathscr{J}^2$, $\mathscr{J}_3$, and $\mathscr{P}_3$:

$$\mathscr{J}^2 D^{j*}_{m'm}(\alpha\beta\gamma) = j(j+1) D^{j*}_{m'm}(\alpha\beta\gamma),$$

$$\mathscr{J}_3 D^{j*}_{m'm}(\alpha\beta\gamma) = m' D^{j*}_{m'm}(\alpha\beta\gamma),$$

$$\mathscr{P}_3 D^{j*}_{m'm}(\alpha\beta\gamma) = m D^{j*}_{m'm}(\alpha\beta\gamma). \tag{3.120}$$

The $\mathscr{P}_i$ are the components of the angular momentum $\mathscr{J}$ referred to the rotating frame—that is, $\mathscr{P}_i = \hat{f}_i \cdot \mathscr{J}$. The action of $\mathscr{J}_\pm$ on the $D^{j*}_{m'm}(\alpha\beta\gamma)$ is the standard action given by Eqs. (3.104); we have yet to determine the action of the operators $\mathscr{P}_i$ $(i = 1, 2)$ on the $D^{j*}_{m'm}(\alpha\beta\gamma)$.

Action of the body-referred angular momentum operators on the rotation matrices. The action of the operators $\mathscr{P}_i$ on the rotation matrices $D^j(\alpha\beta\gamma)$ can be obtained by transposing the matrix relation, Eq. (3.100), using the properties $\tilde{J}_1^{(j)} = J_1^{(j)}$, $\tilde{J}_2^{(j)} = -J_2^{(j)}$, $\tilde{J}_3^{(j)} = J_3^{(j)}$, and $\tilde{D}^j(\gamma, -\beta, \alpha) = D^j(\alpha\beta\gamma)$. [That the matrices $J_1^{(j)}$ and $J_3^{(j)}$ are symmetric, whereas $J_2^{(j)}$ is skew-symmetric, follows directly from Eq. (3.20). The property $\tilde{D}^j(\gamma, -\beta, \alpha) = D^j(\alpha\beta\gamma)$ then follows from Eq. (3.53).] The result of this calculation is

$$D^j(\alpha\beta\gamma) J_i^{(j)} = -\mathscr{P}_i D^j(\alpha\beta\gamma), \tag{3.121}$$

where $\mathscr{P}_1$, $\mathscr{P}_2$, and $\mathscr{P}_3$ are obtained from $\mathscr{J}_1$, $-\mathscr{J}_2$, and $\mathscr{J}_3$, respectively, by interchanging α and γ and replacing β by $-\beta$. These remarks show that the operators $\mathscr{P}_i$ take the explicit forms

$$\mathscr{P}_1 + i\mathscr{P}_2 = e^{-i\gamma}\left(-i\cot\beta\frac{\partial}{\partial\gamma} + \frac{\partial}{\partial\beta} + \frac{i}{\sin\beta}\frac{\partial}{\partial\alpha}\right),$$

$$\mathscr{P}_1 - i\mathscr{P}_2 = e^{i\gamma}\left(-i\cot\beta\frac{\partial}{\partial\gamma} - \frac{\partial}{\partial\beta} + \frac{i}{\sin\beta}\frac{\partial}{\partial\alpha}\right),$$

$$\mathscr{P}_3 = -i\frac{\partial}{\partial\gamma}. \tag{3.122}$$

[These results agree with those obtained directly, but more laboriously, from Eq. (3.109).]

To obtain the action of the operators $\mathscr{P}_i$ on the rotation matrix $D^{j*}(\alpha\beta\gamma)$, we complex conjugate Eq. (3.121) and take matrix elements, arriving at the following results:

$$(\mathscr{P}_1 - i\mathscr{P}_2)D^{j*}_{m'm}(\alpha\beta\gamma) = [(j-m)(j+m+1)]^{\frac{1}{2}}D^{j*}_{m',m+1}(\alpha\beta\gamma),$$

$$(\mathscr{P}_1 + i\mathscr{P}_2)D^{j*}_{m'm}(\alpha\beta\gamma) = [(j+m)(j-m+1)]^{\frac{1}{2}}D^{j*}_{m',m-1}(\alpha\beta\gamma),$$

$$\mathscr{P}_3 D^{j*}_{m'm}(\alpha\beta\gamma) = mD^{j*}_{m'm}(\alpha\beta\gamma). \tag{3.123}$$

Observe that—because of the minus sign in the commutation relations (3.117)—it is $\mathscr{P}_1 - i\mathscr{P}_2$ and $\mathscr{P}_1 + i\mathscr{P}_2$, that act now as step-up and step-down ("ladder") operators, respectively.

The physical interpretation of these results is that the wave functions $D^{j*}_{m'm}(\alpha\beta\gamma)$ are the wave functions of a rotating symmetric top (a solid body with center of mass fixed in space); and that $\mathscr{J}_3$ is the z-component of the angular momentum referred to space-fixed axes, while $\mathscr{P}_3$ is the angular momentum referred to the body-fixed z-axis (axis of symmetry) (Wigner [13], Casimir [23]).

This physical interpretation is discussed in detail in Chapter 7, Section 10.

Remark. This discussion of the operators $\mathscr{P}_i$ defined by Eq. (3.111) has been confined to *proper* orthogonal transformations $R(\alpha\beta\gamma)$ [right-handed frames $(\hat{e}_1, \hat{e}_2, \hat{e}_3)$ and $(\hat{f}_1, \hat{f}_2, \hat{f}_3)$]. We may extend the discussion to *improper* orthogonal transformations by noting that under spatial inversion the unit vectors $\hat{f}_i$ reverse ($\hat{f}_i \to -\hat{f}_i$, $i = 1, 2, 3$), this property being a consequence of the definition of a body-fixed frame (see Chapter 7, Section 10). [Here we do not invert the inertial frame $(\hat{e}_1, \hat{e}_2, \hat{e}_3)$.] Defining $R_{ij} = \hat{e}_i \cdot \hat{f}_j$, we find that the commutation relations (3.117) are now changed, and become $[\mathscr{P}_i, \mathscr{P}_j] = -ie_{ijk}(\det R)\mathscr{P}_k$; equivalently, one has $[(\det R)\mathscr{P}_i, (\det R)\mathscr{P}_j] = -ie_{ijk}(\det R)\mathscr{P}_k$. Thus, if we generalize to include frames $(\hat{e}_1, \hat{e}_2, \hat{e}_3)$ and $(\hat{f}_1, \hat{f}_2, \hat{f}_3)$ related by both proper and improper rotations, then it is the operators $(\det R)\mathscr{P}_i$ $(i = 1, 2, 3)$ that obey the commutation relations (3.117). [This result is of importance for molecular spectroscopy—see Chapter 7, Section 10.]

Generation of the rotation matrices by differential operators. Since each operator in the set $\{\mathscr{J}_i\}$ commutes with each operator in the set $\{\mathscr{P}_i\}$, we may generate the functions $D^{j*}_{m'm}$ by a double application of the lowering operator technique of Eq. (3.18). It follows that:

$$D^{j*}_{m'm}(\alpha\beta\gamma) = \left[\frac{(j+m')!(j+m)!}{(2j)!(j-m')!(2j)!(j-m)!}\right]^{\frac{1}{2}}$$

$$\times (\mathscr{J}_-)^{j-m'}(\mathscr{P}_1 + i\mathscr{P}_2)^{j-m} D^{j*}_{jj}(\alpha\beta\gamma). \tag{3.124}$$

In this result, D^{j*}_{jj} is itself uniquely determined by the *first-order* differential equations:

$$\mathscr{J}_3 D^{j*}_{jj}(\alpha\beta\gamma) = jD^{j*}_{jj}(\alpha\beta\gamma),$$

$$\mathscr{J}_+ D^{j*}_{jj}(\alpha\beta\gamma) = 0,$$

$$\mathscr{P}_3 D^{j*}_{jj}(\alpha\beta\gamma) = jD^{j*}_{jj}(\alpha\beta\gamma),$$

$$(\mathscr{P}_1 - i\mathscr{P}_2) D^{j*}_{jj}(\alpha\beta\gamma) = 0. \tag{3.125}$$

Using the explicit forms (3.102) and (3.122) for the differential operators, we find that Eqs. (3.125) require

$$D^{j*}_{jj}(\alpha\beta\gamma) = N_j e^{ij\alpha} d^j_{jj}(\beta) d^{ij\gamma},$$

where $d^j_{jj}(\beta)$ satisfies

$$\left(\frac{\partial}{\partial\beta} + j\tan\frac{\beta}{2}\right) d^j_{jj}(\beta) = 0,$$

yielding

$$d^j_{jj}(\beta) = \left(\cos\frac{\beta}{2}\right)^{2j}.$$

The arbitrary constant N_j is evaluated by using $D^j(0,0,0)$ = unit matrix, yielding $N_j = 1$. Thus,

$$D^{j*}_{jj}(\alpha\beta\gamma) = e^{ij\alpha}\left(\cos\frac{\beta}{2}\right)^{2j} e^{ij\gamma}. \tag{3.126}$$

This result is now substituted into Eq. (3.124) to obtain a fully defined form of the wave functions $D^{j*}_{m'm}(\alpha\beta\gamma)$. It is fairly tedious to carry out the details of applying the differential operators $\mathscr{J}_-$ and $\mathscr{P}_1 + i\mathscr{P}_2$ repeatedly to obtain the answer in the form involving the Jacobi polynomials, as discussed previously in Section 6. We refer to the literature (Shaffer [24]) for these details.

9. Orthogonality of the Rotation Matrices

The orthogonality of the rotation matrices follows from the fact that the angular momentum operators $\{\mathscr{J}_i\}$ and $\{\mathscr{P}_i\}$ are Hermitian with respect to the scalar product

$$(\Psi, \Phi) \equiv \int d\Omega\, \Psi^*(\mathbf{x})\Phi(\mathbf{x}), \tag{3.127}$$

where the integration is carried out over all points $\{\mathbf{x} = (x_1, x_2, x_3, x_4)\}$ of the unit sphere in four-space:

$$\int d\Omega = 2\pi^2. \tag{3.128}$$

Let us indicate briefly the explicit forms that the surface element (invariant measure) $d\Omega$ and the orthogonality relations take in terms of various parametrizations.

(*a*) *Spherical polar coordinates.* We may express the Euler–Rodrigues parameters $(\alpha_0, \boldsymbol{\alpha})$ (point on the unit sphere in four-space) in terms of the usual spherical polar coordinates:

$$\begin{aligned}
\alpha_1 &= \sin\theta\cos\phi\sin\chi, && 0 \leqslant \phi < 2\pi,\\
\alpha_2 &= \sin\theta\sin\phi\sin\chi, && 0 \leqslant \theta \leqslant \pi,\\
\alpha_3 &= \cos\theta\sin\chi, && 0 \leqslant \chi \leqslant \pi,\\
\alpha_0 &= \cos\chi.
\end{aligned} \tag{3.129}$$

Then $d\Omega$ is given by

$$d\Omega = d\omega \sin^2 \chi \, d\chi, \tag{3.130}$$

where $d\omega$ is itself the differential surface area of the unit sphere in three-space:

$$d\omega = d\phi \, \sin\theta \, d\theta. \tag{3.131}$$

(*b*) *The* $(\psi, \hat{n})$ *parameters*. In this case the point $(\alpha_0, \boldsymbol{\alpha})$ is parametrized by

$$\alpha_0 = \cos\frac{\psi}{2}, \qquad 0 \leqslant \psi \leqslant 2\pi,$$

$$\boldsymbol{\alpha} = \hat{n} \sin\frac{\psi}{2}, \qquad \hat{n}\cdot\hat{n} = 1,$$

$$d\Omega = dS(\hat{n}) \sin^2\frac{\psi}{2} \frac{d\psi}{2}, \tag{3.132}$$

where $dS(\hat{n})$ denotes the differential surface area at the point $n_1^2 + n_2^2 + n_3^2 = 1$ of the unit sphere [for $n = (\sin\theta\cos\phi, \sin\theta\sin\phi, \cos\theta)$, we have $dS(\hat{n}) = d\omega$].

(*c*) *Euler angles*. In this case the point $(\alpha_0, \boldsymbol{\alpha})$ is parametrized by

$$\alpha_0 = \cos\frac{\beta}{2} \cos\frac{1}{2}(\gamma + \alpha),$$

$$\alpha_1 = \sin\frac{\beta}{2} \sin\frac{1}{2}(\gamma - \alpha),$$

$$\alpha_2 = \sin\frac{\beta}{2} \cos\frac{1}{2}(\gamma - \alpha),$$

$$\alpha_3 = \cos\frac{\beta}{2} \sin\frac{1}{2}(\gamma + \alpha), \tag{3.133}$$

where we note that the upper hemisphere ($\alpha_0 \geqslant 0$) is covered by $0 \leqslant \alpha < 2\pi$, $0 \leqslant \beta \leqslant \pi$, $0 \leqslant \gamma < 2\pi$, while the lower hemisphere ($\alpha_0 \leqslant 0$) is covered by $0 \leqslant \alpha < 2\pi$, $2\pi \leqslant \beta \leqslant 3\pi$, $0 \leqslant \gamma < 2\pi$:

$$d\Omega = \tfrac{1}{8} d\alpha \, d\gamma \sin\beta \, d\beta. \tag{3.134}$$

The two most often used forms of the orthogonality relations of the rotation matrices may now be stated:

$$\int dS(\hat{n}) \int_0^{2\pi} \frac{d\psi}{2} \left(\sin \frac{\psi}{2}\right)^2 D^{j*}_{m'm}(\psi, \hat{n}) D^{j'}_{\mu'\mu}(\psi, \hat{n}) = \frac{2\pi^2}{2j+1} \delta_{jj'}\delta_{m'\mu'}\delta_{m\mu}, \tag{3.135}$$

$$\frac{1}{8}\int_0^{2\pi} d\alpha \int_0^{2\pi} d\gamma \int_0^{\pi} d\beta \sin \beta \, D^{j*}_{m'm}(\alpha\beta\gamma) D^{j'}_{\mu'\mu}(\alpha\beta\gamma)$$
$$+ \frac{1}{8}\int_0^{2\pi} d\alpha \int_0^{2\pi} d\gamma \int_{2\pi}^{3\pi} d\beta \sin \beta \, D^{j*}_{m'm}(\alpha\beta\gamma) D^{j'}_{\mu'\mu}(\alpha\beta\gamma) = \frac{2\pi^2}{2j+1} \delta_{jj'}\delta_{m'\mu'}\delta_{m\mu}. \tag{3.136}$$

For j and j' *both integral*, the integration over the lower hemisphere is the same as that over the upper, and, in particular, the orthogonality relation (3.136) reduces to

$$\int_0^{2\pi} d\alpha \int_0^{2\pi} d\gamma \int_0^{\pi} d\beta \sin \beta \, D^{j*}_{m'm}(\alpha\beta\gamma) D^{j'}_{\mu'\mu}(\alpha\beta\gamma) = \frac{8\pi^2}{2j+1} \delta_{jj'}\delta_{m'\mu'}\delta_{m\mu}. \tag{3.137}$$

10. Spherical Harmonics

We showed earlier [discussion circa Eq. (3.106)] that setting $m=0$ in $D^j_{m'm}(\alpha\beta\gamma)$ was sufficient to remove γ from these functions and that in this case we obtain the theory of orbital angular momentum of a single particle. For completeness we summarize the results obtained by specializing the functions $D^j_{mm'}(\alpha\beta\gamma)$ to $D^j_{m0}(\alpha\beta\gamma)$, where we now denote j by l to indicate the restriction to integral values:

(*a*) Relation of spherical harmonics to rotation matrices:

$$Y_{lm}(\beta\alpha) = \left(\frac{2l+1}{4\pi}\right)^{\frac{1}{2}} D^{l*}_{m0}(\alpha\beta\gamma). \tag{3.138}$$

(*b*) Orthogonality of spherical harmonics:

$$\int_0^{2\pi} d\alpha \int_0^{\pi} d\beta \sin \beta \, Y^*_{l'm'}(\beta\alpha) Y_{lm}(\beta\alpha) = \delta_{l'l}\delta_{m'm}. \tag{3.139}$$

(c) Action of orbital angular momentum operators on the spherical harmonics:

$$\mathscr{L}_{\pm} Y_{lm}(\beta\alpha) = [(l \mp m)(l \pm m + 1)]^{\frac{1}{2}} Y_{l,m\pm 1}(\beta\alpha),$$

$$\mathscr{L}_3 Y_{lm}(\beta\alpha) = m Y_{lm}(\beta\alpha),$$

$$\mathscr{L}^2 Y_{lm}(\beta\alpha) = l(l+1) Y_{lm}(\beta\alpha). \tag{3.140}$$

Explicit forms of the spherical harmonics. Using the form (3.138) for the spherical harmonics yields

$$Y_{lm}(\beta\alpha) = \left(\frac{2l+1}{4\pi}\right)^{\frac{1}{2}} e^{im\alpha} d^l_{m0}(\beta). \tag{3.141}$$

By using the Wigner form (3.65), the form (3.72) containing the Jacobi polynomials, or the exponential form (3.78), we may express Eq. (3.141) in a variety of ways.

It is common practice to express the spherical harmonics in terms of *associated Legendre functions*. These functions may be defined in terms of the Jacobi polynomials by

$$P^m_l(\cos\beta) \equiv \frac{(l+m)!}{l!} \left(\frac{\sin\beta}{2}\right)^m P^{(m,m)}_{l-m}(\cos\beta). \tag{3.142}$$

Using this result, and relations (3.71) and (3.72), we find

$$d^l_{m0}(\beta) = (-1)^m \left[\frac{(l-m)!}{(l+m)!}\right]^{\frac{1}{2}} P^m_l(\cos\beta), \tag{3.143}$$

$$Y_{lm}(\beta\alpha) = (-1)^m \left[\frac{(2l+1)(l-m)!}{4\pi(l+m)!}\right]^{\frac{1}{2}} P^m_l(\cos\beta) e^{im\alpha}. \tag{3.144}$$

Observe that relation (3.143) is valid for both positive and negative values of m; the relation

$$P^{-m}_l(\cos\beta) = (-1)^m \frac{(l-m)!}{(l+m)!} P^m_l(\cos\beta) \tag{3.145}$$

follows from Eqs. (3.71). Hence, we also obtain

$$Y^*_{lm}(\beta\alpha) = (-1)^m Y_{l,-m}(\beta\alpha). \tag{3.146}$$

Let us also note that setting $m = 0$ in $Y_{lm}(\beta\alpha)$ removes α from these functions:

$$Y_{l0}(\beta\alpha) = \left(\frac{2l+1}{4\pi}\right)^{\frac{1}{2}} d^l_{00}(\beta) = \left(\frac{2l+1}{4\pi}\right)^{\frac{1}{2}} P_l(\cos\beta), \tag{3.147}$$

where

$$P_l(\cos\beta) \equiv P_l^0(\cos\beta) \tag{3.148}$$

defines the *Legendre polynomials* (in $\cos\beta$).

There is still another popular method of obtaining the spherical harmonics without the necessity of appealing to the rotation matrices or of referring directly to the functions of classical mathematics. This method utilizes the general formula (3.18) to which we now turn.

Generation of spherical harmonics by lowering the highest-weight eigenstate. The highest-weight ($m = l$) eigenstate is obtained by solving the first-order differential equations,

$$\mathscr{L}_+ Y_{ll}(\beta\alpha) = 0,$$

$$\mathscr{L}_3 Y_{ll}(\beta\alpha) = l Y_{ll}(\beta\alpha), \tag{3.149}$$

where $\mathscr{L}_+$ and $\mathscr{L}_3$ are defined by Eqs. (3.106).

The normalized [in the sense of Eq. (3.139)] solution to these equations is

$$Y_{ll}(\beta\alpha) = \frac{(-1)^l}{2^l l!}\left[\frac{(2l+1)!}{4\pi}\right]^{\frac{1}{2}} (e^{i\alpha}\sin\beta)^l, \tag{3.150}$$

where an arbitrary phase (the so-called Condon and Shortley convention) has been chosen to obtain agreement with our previous results.

The general spherical harmonic is then obtained by the standard lowering formula, Eq. (3.18):

$$Y_{lm}(\beta\alpha) = \left[\frac{(l+m)!}{(2l)!(l-m)!}\right]^{\frac{1}{2}} \mathscr{L}_-^{l-m} Y_{ll}(\beta\alpha). \tag{3.151}$$

Repeated application of the lowering operator $\mathscr{L}_-$ to $Y_{ll}(\beta\alpha)$ in this result leads to the following form of the spherical harmonics:

$$Y_{lm}(\beta\alpha) = (-1)^m \left[\frac{2l+1}{4\pi}(l+m)!(l-m)!\right]^{\frac{1}{2}} e^{im\alpha}$$

$$\times \sum_k \frac{(-1)^k(\sin\beta)^{2k+m}(\cos\beta)^{l-2k-m}}{2^{2k+m}(k+m)!k!(l-m-2k)!}. \tag{3.152}$$

This last form is of interest because it displays (most obviously) the relation of the spherical harmonics to the *solid harmonics* (homogeneous polynomials of degree l in x_1, x_2, x_3, which solve Laplace's equation). Recognizing that

$$e^{im\alpha}(\sin\beta)^{2k+m} = (e^{i\alpha}\sin\beta)^{k+m}(e^{-i\alpha}\sin\beta)^k$$

$$= (x_1 + ix_2)^{k+m}(x_1 - ix_2)^k,$$

we see from Eq. (3.152) that the solid harmonics $\mathscr{Y}_{lm}(\mathbf{x})$ are given by

$$\mathscr{Y}_{lm}(\mathbf{x}) = \left[\frac{2l+1}{4\pi}(l+m)!(l-m)!\right]^{\frac{1}{2}}$$

$$\times \sum_k \frac{(-x_1 - ix_2)^{k+m}(x_1 - ix_2)^k x_3^{l-2k-m}}{2^{2k+m}(k+m)!k!(l-m-2k)!}, \tag{3.153}$$

in which the point $\mathbf{x} = (x_1, x_2, x_3)$ is no longer required to be on the unit sphere.

The solid harmonics satisfy the standard equations (3.140), where the angular momentum operators are now to be left in their original Cartesian form – that is,

$$\mathscr{L}_i = -i\left(x_j \frac{\partial}{\partial x_k} - x_k \frac{\partial}{\partial x_j}\right), \qquad i, j, k \text{ cyclic}. \tag{3.154}$$

(The solid harmonics up to degree four and the corresponding spherical harmonics are tabulated in Table 4 in the Appendix of Tables. The properties of the spherical harmonics and related spherical functions are developed further in Chapter 6.)

11. The Addition of Angular Momentum

We have determined in Section 3 all possible eigenspaces $\mathscr{H}_j$ (standard basis of ket vectors $|jm\rangle$) corresponding to the maximal (commuting) observables of angular momentum $[\mathbf{J}^2 \rightarrow j(j+1), J_3 \rightarrow m]$. As noted at the beginning, we may regard these states as a description of the *total* angular momentum of a composite system or, equally well, as the states available to describe the angular momentum of a single particle. (This freedom of interpretation exists because (*a*) kinematic independence implies that $\mathbf{J} = \sum_{i=1}^{n} \mathbf{J}(i)$ obeys the same commutation relations as the $\mathbf{J}(i)$ separately, and (*b*) conceptually, the total angular momentum of the system exists because of the invariance of the description of the composite system to rotations.)

It is natural now to examine these two interpretations simultaneously and consider the problem of adding the angular momentum of two (independent) particles (or systems). The underlying physics specifies the necessary assumptions, which are as follows:

(*a*) The angular momenta of the two particles are to be simultaneously observable; hence, we have the commutation relation:

$$[J_i(1), J_k(2)] = 0, \qquad i, k = 1, 2, 3. \tag{3.155}$$

Correspondingly, the ket vectors of the composite system are products of the ket vectors of the individual particles.

(*b*) The two particles belong to sharp angular momentum eigenspaces $\mathscr{H}_{j_i}(i)$ $(i = 1, 2)$ with bases given by

$$\{|j_i m_i\rangle : j_i \text{ is fixed}, j_i \geqslant m_i \geqslant -j_i\}. \tag{3.156}$$

(*c*) The total angular momentum is the sum $\mathbf{J} = \mathbf{J}(1) + \mathbf{J}(2)$. Since $\mathbf{J}(k) \times \mathbf{J}(k) = i\mathbf{J}(k)$ and the components of $\mathbf{J}(1)$ commute with those of $\mathbf{J}(2)$, we see that

$$\mathbf{J} \times \mathbf{J} = i\mathbf{J}. \tag{3.157}$$

These assumptions imply that the ket vectors on which the total angular momentum operator $\mathbf{J}$ acts is the *tensor product space* $\mathscr{H}_{j_1}(1) \otimes \mathscr{H}_{j_2}(2)$ spanned by the set of orthonormal product basis vectors:[1]

$$\{|j_1 m_1\rangle |j_2 m_2\rangle \equiv |j_1 m_1\rangle \otimes |j_2 m_2\rangle : j_1 \geqslant m_1 \geqslant -j_1, j_2 \geqslant m_2 \geqslant -j_2\}. \tag{3.158}$$

(Tensor product spaces are discussed in Note 5.)

The action of the total angular momentum operator $J_i = J_i(1) + J_i(2)$ on the space $\mathscr{H}_{j_1}(1) \otimes \mathscr{H}_{j_2}(2)$ is defined by

[1] This notation suppresses the additional labels (α_1) and (α_2) required to complete the bases $\{|(\alpha_1) j_1 m_1\rangle\}$ and $\{|(\alpha_2) j_2 m_2\rangle\}$ (see footnote, p. 32).

$$J_i|j_1m_1\rangle|j_2m_2\rangle = (J_i(1)|j_1m_1\rangle)|j_2m_2\rangle + |j_1m_1\rangle(J_i(2)|j_2m_2\rangle). \quad (3.159)$$

[More explicitly, this action leads to the $(2j_1 + 1)(2j_2 + 1)$ dimensional matrix, M_i, of the total angular momentum J_i given by

$$J_i \to M_i = J^{(j_1)} \otimes \mathbb{1}^{(j_2)} + \mathbb{1}^{(j_1)} \otimes J^{(j_2)}, \quad (3.160)$$

where $\mathbb{1}^{(j)}$ denotes the unit matrix of dimension $2j+1$ and $\otimes$ designates (for matrices) the matrix direct product.]

One sees that, for the choice of basis (3.158), the eigenvalues of J_3 are $m = m_1 + m_2$. The possible values of m for $j_1 \geqslant m_1 \geqslant -j_1$ and $j_2 \geqslant m_2 \geqslant -j_2$ are determined by the lattice points displayed in Fig. 3.2: (m_1, m_2), $m = m_1 + m_2$.

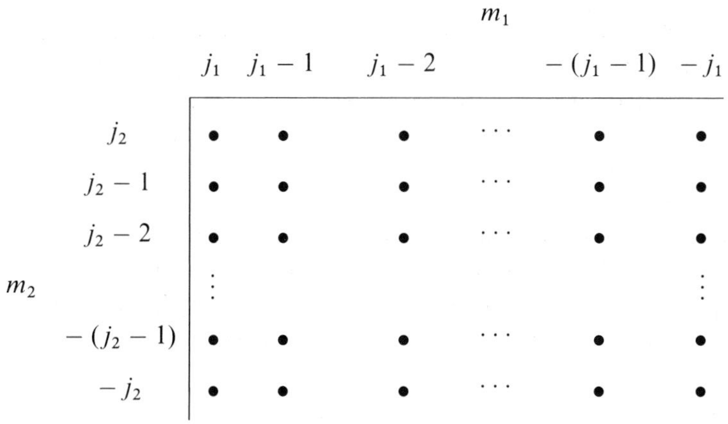

Figure 3.2.

The problem is to determine the angular momentum multiplets of **J** that occur in the product space $\mathscr{H}_{j_1}(1) \otimes \mathscr{H}_{j_2}(2)$.

The method of solution was known to the early spectroscopists, and is based on the spectrum of m (implicitly contained in Fig. 3.2) and the general properties of multiplets given in Section 3. The basic property taken from multiplet theory is that, whenever a *particular* eigenvalue of J_3, say m_0, occurs, *all* members of a multiplet $j_0, j_0 - 1, \ldots, m_0, \ldots, -j_0$ for some $j_0 \in \{0, \frac{1}{2}, 1, \ldots\}$ must occur with m_0. (This is the essence of the raising and lowering operator technique given in Section 3, this process detailing the explicit construction of the full j_0-multiplet.)

Examining Fig. 3.2, we see that the largest value of m that occurs is $j_1 + j_2$, this value occurring *exactly once*. Since no larger value of m occurs, the m-value

$j_1 + j_2$ must be a member of the multiplet $j_1 + j_2, j_1 + j_2 + 1, \ldots, -(j_1 + j_2)$. The ket vector having this maximum value of J_3 is clearly $|j_1 j_1\rangle |j_2 j_2\rangle$, and the members of the $(j_1 + j_2)$-multiplet are generated explicitly by applying the lowering operator J_- sequentially [Eq. (3.18)].

Examining Fig. 3.2 again, we see that $m = j_1 + j_2 - 1$ occurs exactly twice. One of these m-values occurs in the $(j_1 + j_2)$-multiplet whose construction is described in the preceding paragraph. The other m-value, $j_1 + j_2 - 1$, must therefore belong to a multiplet $j_1 + j_2 - 1, j_1 + j_2 - 2, \ldots, -(j_1 + j_2 - 1)$. The first member $(j = m = j_1 + j_2 - 1)$ of this multiplet is determined (uniquely, up to normalization) as that linear combination of the two $m = j_1 + j_2 - 1$ ket vectors, $|j_1 j_1 - 1\rangle |j_2 j_2\rangle$ and $|j_1 j_1\rangle |j_2 j_2 - 1\rangle$ that is annihilated by J_+ (equivalently, it is the linear combination that is perpendicular to $J_- |j_1 j_1\rangle |j_2 j_2\rangle$). The remaining members of the $(j_1 + j_2 - 1)$-multiplet are now generated by applying the lowering operator J_- sequentially [Eq. (3.18)].

The process iterates, since from Fig. 3.2 one sees that there are exactly k ket vectors having $m = j_1 + j_2 - k + 1$ $[k \leqslant 1 + 2\min(j_1, j_2)]$, and $k - 1$ linear combinations of these vectors are orthogonal to one another and are, in fact, the $m = j_1 + j_2 - k + 1$ members of the j-multiplets corresponding to $j = j_1 + j_2,\ j_1 + j_2 - 1, \ldots, j_1 + j_2 - k + 2$. Thus, there exists exactly one more linear combination of the $m = j_1 + j_2 - k + 1$ vectors, orthogonal to the previous ones, which is uniquely determined by the requirement that it be annihilated by J_+. This ket vector is then the $j = m = j_1 + j_2 - k + 1$ member of the $(j_1 + j_2 - k + 1)$-multiplet, the remaining members of this multiplet being generated by applying the lowering operator J_- sequentially [Eq. (3.18)].

The process must terminate[1] at a value $k = k_0 \geqslant 0$ such that the number of independent states in the various multiplets equals the dimension $(2j_1 + 1)(2j_2 + 1)$ of the space $\mathscr{H}_{j_1}(1) \otimes \mathscr{H}_{j_2}(2)$. This condition is

$$\sum_{j=k_0}^{j_1+j_2} (2j + 1) = (2j_1 + 1)(2j_2 + 1). \tag{3.161}$$

Thus, we must have $(j_1 + j_2 + 1)^2 - k_0^2 = (2j_1 + 1)(2j_2 + 2)$; that is, $k_0^2 = (j_1 - j_2)^2$, and therefore $k_0 = |j_1 - j_2|$.

This procedure has demonstrated a special case of the well-known result of Clebsch and Gordan, namely: *The addition of two (independent) angular*

[1] Geometrically one sees that the identification and removal of one multiplet is the dissection of Fig. 3.2 by gnomons ¬, which remove sequentially the top-most row and right-most column of dots. The use of gnomons is familiar in the proofs of finite enumeration relations in classical Greek mathematics ["geometric numbers" – for example, Eq. (3.161)].

momenta, having values j_1 *and* j_2*, yields for the total angular momenta the values* $j = j_1 + j_2, j_1 + j_2 - 1, \ldots, |j_1 - j_2|$*, each value occurring once and only once.*

This result is of fundamental importance in physics and was obtained at the very beginning of quantum mechanics by Born *et al.* [3]. The content was embodied in the "vector model," which schematically portrayed the addition as *quantized* (see Fig. 3.3; another aspect of the vector model has been mentioned in connection with Fig. 3.1, p. 37.)

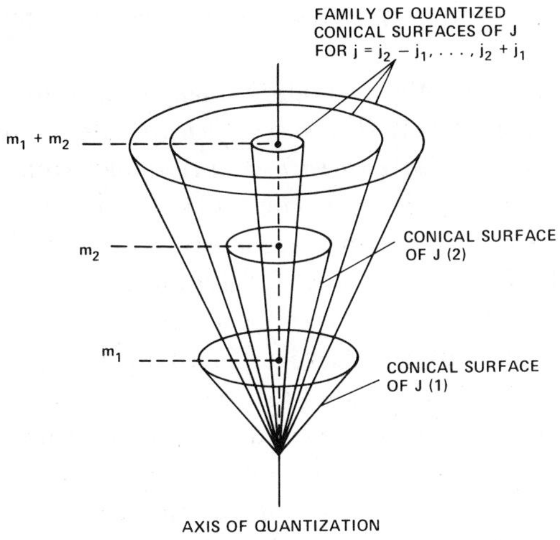

Figure 3.3.

12. The Wigner Coefficients[1]

We have seen above how the addition of two angular momenta defines a composite space that may be uniquely decomposed into eigenspaces of the

---

[1] This designation is not standard in the physics literature, and the terms "vector-addition coefficients", "vector-coupling coefficients", or "Clebsch-Gordan coefficients" are frequently used as synonyms. The work of Clebsch [24a, Section 8] and Gordan [24b, Section 2] on the invariant theory of algebraic forms can be seen, in retrospect, to be an equivalent formulation of the coupling problem for angular momentum. In particular, the reduction, by Clebsch and Gordan, of products of binary forms is exactly the ξ-η calculus used by Weyl [24c] to develop the Clebsch-Gordan *series*. (It should be noted that Weyl does not mention the Clebsch-Gordan coefficients in his book.)

We have chosen the designation "Wigner coefficients" since the significance of these coefficients in relation to the quantum theory of angular momentum and the rotation matrices appears first in Wigner [25, 26].

We would like to thank Professor D.-N. Verma for his remarks on the work of Clebsch and Gordan, and Professor P. J. Brussaard for bringing to our attention his historical paper (Brussaard [24d]).

total angular momentum. The contribution of Wigner [13, 25, 26] to this problem was the determination of the explicit coefficients for the set of eigenvectors belonging to this decomposition. [Probably the more important contribution was the demonstration by Wigner [13] and by Eckart [27] of the importance of the physical applications of this result (see Section 15).]

Wigner's original derivation of these coefficients used their relation to the rotation matrices, to be discussed in Section 13. This method is usually regarded as being more group-theoretic than some of the later methods that are based on using the algebraic properties of the angular momentum operators themselves.

Let us mention some of the methods that may be used for calculating these coefficients (hereafter called *Wigner coefficients*; also known as *Clebsch–Gordan and vector-coupling coefficients*): (*a*) Implementation of the construction of the previous section; (*b*) iteration of recursion relations that the Wigner coefficients can be shown to satisfy (Edmonds [16], Rose [17], Condon and Shortley [28], Racah [29], Louck [30]); (*c*) utilization of properties of particular realizations of the angular momentum operators and of the product space (van der Waerden [31], Kramers [32], Brinkman [33, 34], Hamermesh [35], Nachamkin [36]; see also Chapter 5, Section 8); (*d*) extension of method (*c*) to obtain generating functions (Schwinger [37], Regge [38], Bargmann [39], Bincer [40]); utilization of projection operator techniques (Löwdin [41]); (*e*) exploitation of the relation of Wigner coefficients to hypergeometric functions (Majumdar [42]); and (*f*) factorization of the group elements (Gel'fand and Graev [43]).

These many approaches to the calculation of Wigner coefficients are indicative of the broad scope of interpretations and viewpoints that can be ascribed to the mathematical apparatus of angular momentum theory.

In this section, we elect to carry to completion the construction given in Section 11, principally because it is purely algebraic, is elementary, and requires no further development of concepts.[1]

The detailed expression for the Wigner coefficients depends on the method of derivation. (We discuss this further in Note 6.) All such forms are equivalent and (for all presently known forms) can be transformed one into the other by symmetry transformations and/or a transformation method introduced by Racah [29] and discussed in Appendix A.

We shall give explicitly three forms for the Wigner coefficients – that of Wigner [13], that of van der Waerden [31], and that of Racah [29].

The method of highest weights for determining the Wigner coefficients. This is the method described in Section 11, but now with all the details implemented.

[1] An interesting symbolic variation of this method is given by Gel'fand *et al.* [44]; see also Hagedorn [45].

The ket vector in a multiplet $|(\alpha)jm\rangle$, $m = j, j-1, \ldots, -j$, for which the weight m is maximal (or highest—that is, $J_+|(\alpha)jm\rangle = 0$ for $m = j$), is called a vector of *highest weight*. Once the highest-weight vector is determined, the full multiplet is generated by the sequential application of the lowering operator J_-.

Let us denote the ket vectors in one of the j-multiplets discussed in the last section by $|(j_1 j_2)jm\rangle$ (see footnote 1, p. 32). The highest-weight vector is then denoted by $|(j_1 j_2)jj\rangle$, and the j-multiplet is given by

$$|(j_1 j_2)jm\rangle = \left[\frac{(j+m)!}{(2j)!(j-m)!}\right]^{\frac{1}{2}} J_-^{j-m}|(j_1 j_2)jj\rangle. \tag{3.162}$$

It is clear from the discussion in Section 11 that the highest-weight vector, $|(j_1 j_2)jj\rangle$, for each $j = j_1 + j_2, j_1 + j_2 - 1, \ldots, |j_1 - j_2|$, is uniquely determined (up to normalization) by (a) belonging to the J_3 eigenvalue j, and (b) being annihilated by the raising operator J_+:

$$|(j_1 j_2)jj\rangle = \sum_{\substack{m_1 m_2 \\ m_1 + m_2 = j}} A_{m_1 m_2} |j_1 m_1\rangle |j_2 m_2\rangle,$$

$$J_+|(j_1 j_2)jj\rangle = 0.$$

Using $J_+ = J_+(1) + J_+(2)$, and the standard action of the operator $J_+(i)$ on the ket vector $|j_i m_i\rangle$, one now derives a two-term recursion relation for the coefficients $\{A_{m_1 m_2}\}$, the solution of which is found by iteration to be

$$A_{m_1 m_2} = \delta(m_1 + m_2, j)(-1)^{j_1 - m_1} \left[\frac{(j_1 + m_1)!(j_2 + m_2)!}{(j_1 - m_1)!(j_2 - m_2)!}\right]^{\frac{1}{2}} A,$$

where A is a constant that is independent of m_1 and m_2.

The normalization condition

$$\sum_{\substack{m_1 m_2 \\ m_1 + m_2 = j}} |A_{m_1 m_2}|^2 = 1$$

and the summation formula

$$\sum_{\substack{m_1 m_2 \\ m_1 + m_2 = j}} \frac{(j_1 + m_1)!(j_2 + m_2)!}{(j_1 - m_1)!(j_2 - m_2)!}$$

$$= \frac{(j + j_1 + j_2 + 1)!(j + j_1 - j_2)!(j + j_2 - j_1)!}{(2j + 1)!(j_1 + j_2 - j)!} \tag{3.163}$$

now give us the desired result for the highest-weight j:

$$|(j_1 j_2)jj\rangle = \sum_{m_1 m_2} C^{j_1 j_2 j}_{m_1 m_2 j} |j_1 m_1\rangle |j_2 m_2\rangle, \tag{3.164}$$

where the coefficient in this result is defined by

$$C^{j_1 j_2 j}_{m_1 m_2 j} \equiv \delta(m_1 + m_2, j) \left[\frac{(2j+1)!(j_1 + j_2 - j)!}{(j + j_1 + j_2 + 1)!(j + j_1 - j_2)!(j + j_2 - j_1)!}\right]^{\frac{1}{2}}$$

$$\times (-1)^{j_1 - m_1} \left[\frac{(j_1 + m_1)!(j_2 + m_2)!}{(j_1 - m_1)!(j_2 - m_2)!}\right]^{\frac{1}{2}}. \tag{3.165}$$

(An arbitrary phase factor has been chosen to be unity to obtain agreement with the standard Wigner form of these coefficients. There are a great many different notations for the Wigner coefficients. We discuss these in Note 7.)

The j-multiplet (3.162) may now be written in the form

$$|(j_1 j_2)jm\rangle = \sum_{m_1 m_2} C^{j_1 j_2 j}_{m_1 m_2 m} |j_1 m_1\rangle |j_2 m_2\rangle, \tag{3.166}$$

where

$$C^{j_1 j_2 j}_{m_1 m_2 m} \equiv \left[\frac{(j+m)!}{(2j)!(j-m)!}\right]^{\frac{1}{2}} \langle j_1 m_1 ; j_2 m_2 | J_-^{j-m} | (j_1 j_2) jj\rangle$$

$$= \left[\frac{(j+m)!}{(2j)!(j-m)!}\right]^{\frac{1}{2}} \langle (j_1 j_2) jj | J_+^{j-m} | j_1 m_1 ; j_2 m_2\rangle, \tag{3.167}$$

in which

$$|j_1 m_1 ; j_2 m_2\rangle \equiv |j_1 m_1\rangle |j_2 m_2\rangle. \tag{3.168}$$

One next expands $J_+^{j-m} = [J_+(1) + J_+(2)]^{j-m}$, using the binomial theorem, and calculates the action of $[J_+(i)]^s$ on $|j_i m_i\rangle$ from Eq. (3.20). These steps are straightforward, and Eq. (3.168) yields the following expression for the Wigner coefficients ("Racah's first form"):

$$C^{j_1 j_2 j}_{m_1 m_2 m} = \delta(m_1 + m_2, m)$$

$$\times \left[\frac{(2j+1)(j_1 + j_2 - j)!(j_1 - m_1)!(j_2 - m_2)!(j - m)!(j + m)!}{(j_1 + j_2 + j + 1)!(j + j_1 - j_2)!(j + j_2 - j_1)!(j_1 + m_1)!(j_2 + m_2)!}\right]^{\frac{1}{2}} \times$$

$$\times \sum_t (-1)^{j_1-m_1+t}\left[\frac{(j_1+m_1+t)!(j+j_2-m_1-t)!}{t!(j-m-t)!(j_1-m_1-t)!(j_2-j+m_1+t)!}\right]. \tag{3.169}$$

Using the transformation described in Appendix A, Racah was able to bring the right-hand side of Eq. (3.169) to the symmetric form (van der Waerden [31]):

$$C^{j_1j_2j}_{m_1m_2m} = \delta(m_1+m_2,m)\left[\frac{(2j+1)(j_1+j_2-j)!(j+j_1-j_2)!(j+j_2-j_1)!}{(j_1+j_2+j+1)!}\right]^{\frac{1}{2}}$$

$$\times \sum_z \frac{(-1)^z[(j_1+m_1)!(j_1-m_1)!(j_2+m_2)!(j_2-m_2)!(j+m)!(j-m)!]^{\frac{1}{2}}}{z!(j_1+j_2-j-z)!(j_1-m_1-z)!(j_2+m_2-z)!(j-j_2+m_1+z)!(j-j_1-m_2+z)!}. \tag{3.170}$$

Wigner's method of calculation (described below) gave the coefficients in the form

$$C^{j_1j_2j}_{m_1m_2m} = \delta(m_1+m_2,m)$$

$$\times \left[\frac{(2j+1)(j+j_1-j_2)!(j-j_1+j_2)!(j_1+j_2-j)!}{(j+j_1+j_2+1)!}\right]^{\frac{1}{2}}$$

$$\times \left[\frac{(j+m)!(j-m)!}{(j_1+m_1)!(j_1-m_1)!(j_2+m_2)!(j_2-m_2)!}\right]^{\frac{1}{2}}$$

$$\times \sum_s \frac{(-1)^{j_2+m_2+s}(j_2+j+m_1-s)!(j_1-m_1+s)!}{s!(j-j_1+j_2-s)!(j+m-s)!(j_1-j_2-m+s)!}. \tag{3.171}$$

Remark. It is curious that it has never been pointed out in the literature that Racah's first form, Eq. (3.169), may be obtained from Wigner's form, Eq. (3.171), by the following two transformations:

(*a*) $m_1 \to -m_1, \quad m_2 \to -m_2, \quad m \to -m;$

(*b*) $j_1+m_1 \to j_1+m_1, \quad j_2+m_2 \to j_2+m_2, \quad j-m \to j-m$

$$j_1-m_1 \leftrightarrow j+j_2-j_1, \quad j_2-m_2 \leftrightarrow j+j_1-j_2, \quad j+m \leftrightarrow j_1+j_2-j. \tag{3.172}$$

In other words, the fact that the Wigner coefficients may be written in the two forms, Eq. (3.169) and (3.171), *proves* that the transformation (*b*) is a symmetry transformation of the Wigner coefficients [since (*a*) is readily verified to be such a symmetry]. The symmetry transformation (*b*) was not observed until 1958 by Regge [38] (by a completely different method), although a direct comparison of Wigner's form and Racah's first form would have revealed it much earlier.

Systematic discussion of the symmetries of the Wigner coefficients is deferred to Chapter 5; a summary is given below.

Orthogonality of the Wigner coefficients. Since the ket vectors in the set

$$\{|(j_1 j_2) jm\rangle : j = j_1 + j_2, j_1 + j_2 - 1, \ldots, |j_1 - j_2|;\ m = j, j-1, \ldots, -j\} \tag{3.173}$$

are orthonormal, the Wigner coefficients (by convention, taken to be real numbers[1]) are the elements of a real, orthogonal matrix C of dimension $(2j_1 + 1)(2j_2 + 1)$, expressing a change of basis of the tensor product space $\mathscr{H}_{j_1}(1) \otimes \mathscr{H}_{j_2}(2)$. This change of basis splits the angular momentum matrices M_i given by Eq. (3.160) into the basic angular momentum matrices $J^{(j)}$:

$$C(J_i^{(j_1)} \otimes \mathbb{1}^{(j_2)} + \mathbb{1}^{(j_1)} \otimes J_i^{(j_2)})\tilde{C} = \sum_j \oplus J_i^{(j)}$$

$$= \begin{pmatrix} J_i^{(j_1+j_2)} & 0 & \cdots & 0 \\ 0 & J_i^{(j_1+j_2-1)} & & 0 \\ \vdots & & \ddots & \vdots \\ 0 & 0 & \cdots & J_i^{(|j_1-j_2|)} \end{pmatrix}. \tag{3.174}$$

The elements of C are identified explicitly to be

$$(C)_{jm;m_1m_2} = C^{j_1j_2j}_{m_1m_2m},$$

where the index pairs (jm) and (m_1m_2) enumerate rows and columns,

[1] It is important to make this observation explicit. The Wigner coefficients can be taken, without loss of generality, to be unitary. The freedom of phase on state vectors in quantum mechanics (see Note 1 and Chapter 5, Topic 1 in RWA) allows one to choose arbitrarily a phase in each (j, m) subspace. (The Wigner coefficients remain unitary under this rephasing.) The standard (Wigner [13, p. 188]) phase choice is to define $C^{j_1j_2j}_{j_1,-j_2,j_1-j_2}$ to be *real and positive*. Since the phases of the angular momentum multiplets are defined to make the matrix elements of the raising-lowering operators $J_\pm$ real (and positive), the explicit calculations given above show that *all* Wigner coefficients in this convention are *real*, as asserted.

respectively. Thus, the $(2j_1 + 1)(2j_2 + 1)$ rows and columns are enumerated [in an order appropriate to the right-hand side of Eq. (3.174)], respectively, by

$$(jm): \qquad j = j_1 + j_2, \ldots, |j_1 - j_2|;\ m = j, \ldots, -j,$$

$$(m_1 m_2): \quad m_1 = j_1, \ldots, -j_1;\ m_2 = j_2, \ldots, -j_2.$$

The complete statements of the orthogonality relations of C are the following:

(*a*) Orthogonality of rows:

$$\sum_{m_1 m_2} C^{j_1 j_2 j}_{m_1 m_2 m} C^{j_1 j_2 j'}_{m_1 m_2 m'} = \delta_{jj'}\delta_{mm'}. \tag{3.175}$$

(*b*) Orthogonality of columns:

$$\sum_{jm} C^{j_1 j_2 j}_{m_1 m_2 m} C^{j_1 j_2 j}_{m_1' m_2' m} = \delta_{m_1 m_1'}\delta_{m_2 m_2'}. \tag{3.176}$$

Remark. It should be noted that these orthogonality relations refer to the collection of elements of a very specific matrix C. It will later prove valuable to extend the concept of a Wigner coefficient to values of the quantum numbers $(j_1 m_1; j_2 m_2; jm)$ that lie outside those defining the orthogonal matrix C. Caution must be exercised in interpreting the orthogonality relations for such extended coefficients. For example, a Wigner coefficient is often defined to be zero unless the "triangle conditions" are satisfied [see Eq. (3.191) below]. In this case Eqs. (3.175) and (3.176) are valid properties of the extended coefficients when we modify Eq. (3.175) by inserting the factor $\varepsilon_{j_1 j_2 j}$ in the right-hand side. Here $\varepsilon_{j_1 j_2 j}$ is defined to be unity when the triangle conditions are satisfied; otherwise, it is zero [see Eq. (3.272)].

Owing to the fact that many of the elements of C are zero, the statement of the orthogonality relations may be substantially simplified by removing trivial $0 = 0$ relations. All such relations are a consequence of

$$C^{j_1 j_2 j}_{m_1 m_2 m} = 0 \quad \text{for } m_1 + m_2 \neq m, \tag{3.177}$$

which is itself a consequence of the fact that $J_3 = J_3(1) + J_3(2)$ has eigenvalue m when acting on $|j_1 m_1\rangle|j_2 m_2\rangle$.

Accounting for these zeros in the matrix C, the orthogonality of rows reduces to

$$\sum_{\substack{m_1 m_2 \\ m_1 + m_2 = m}} C^{j_1 j_2 j}_{m_1 m_2 m} C^{j_1 j_2 j'}_{m_1 m_2 m} = \delta_{jj'}, \tag{3.178}$$

while the orthogonality of columns is best written in the two forms

$$\sum_j C^{j_1 j_2 j}_{m_1, m-m_1, m} C^{j_1 j_2 j}_{m_1', m-m_1', m} = \delta_{m_1 m_1'},$$

$$\sum_j C^{j_1 j_2 j}_{m-m_2, m_2, m} C^{j_1 j_2 j}_{m-m_2', m_2', m} = \delta_{m_2 m_2'}. \qquad (3.179)$$

In the preparation of tables of the Wigner coefficients, it is customary to assume $j_1 \geqslant j_2$ and to display an array of coefficients with the row and column headings j and m_2, respectively. Equations (3.178) and (3.179) then express the orthogonality of the rows and columns of this matrix of dimension $2j_2 + 1$. (Table 1 in the Appendix of Tables gives the Wigner coefficients for $j_2 = \frac{1}{2}, 1, \frac{3}{2}, 2, \frac{5}{2}, 3$.)

Symmetry properties of the Wigner coefficients. Explicit knowledge of the symmetry properties of the Wigner coefficients is important not only for understanding structural interrelationships, but also for practical applications where symmetry transformations on the Wigner coefficients prove to be among the most frequently used properties. We therefore summarize these very useful results here.

In Chapter 5, Appendix C, we derive these symmetries without the benefit of explicit forms for the coefficients. It is true, however, that all these symmetries may be verified directly, by inspection, from the van der Waerden form (3.170).[1,2]

Discussions of the symmetry properties of the Wigner coefficients were given in the original work of Wigner [13] and Racah [29], and in the thesis of Eisenbud [46]; the result was a symmetry group of twelve elements (identified below). It came as a surprise to learn from Regge's [38] work that the symmetry group is, in fact, much larger—a 72-element group. (An excellent exposition of Regge's result is given by Bargmann [39].)

All the symmetry relations are generated from the first four of the following six relations (the last two relations have been included here for convenience of reference):

$$C^{j_1 j_2 j}_{m_1 m_2 m} = (-1)^{j_1 + j_2 - j} C^{j_1 j_2 j}_{-m_1, -m_2, -m},$$

$$C^{j_1 j_2 j}_{m_1 m_2 m} = (-1)^{j_1 + j_2 - j} C^{j_2 j_1 j}_{m_2 m_1 m},$$

---

[1] This method, of course, does not reveal the underlying origin of these symmetries, which is discussed in Chapter 5, Appendix C.

[2] Indeed, Kleima and van Wageningen [45a] did find the Regge symmetry directly from the van der Waerden form (independently, but several years after Regge's paper).

$$C^{j_1 j_2 j}_{m_1, m_2, m_1+m_2} = C^{\frac{1}{2}(j_1+j_2+m_1+m_2)\,\frac{1}{2}(j_1+j_2-m_1-m_2)\,j}_{\frac{1}{2}(j_1-j_2+m_1-m_2),\frac{1}{2}(j_1-j_2-m_1+m_2),j_1-j_2},$$

$$C^{j_1 j_2 j}_{m_1 m_2 m} = (-1)^{j_2+m_2}[(2j+1)/(2j_1+1)]^{\frac{1}{2}} C^{j j_2 j_1}_{-m, m_2, -m_1},$$

$$C^{j_1 j_2 j}_{m_1 m_2 m} = (-1)^{j_1-m_1}[(2j+1)/(2j_2+1)]^{\frac{1}{2}} C^{j_1 j j_2}_{m_1, -m, -m_2},$$

$$C^{j_1 j_2 j}_{m_1 m_2 m} = (-1)^{j_2+m_2}[(2j+1)/(2j_1+1)]^{\frac{1}{2}} C^{j_2 j j_1}_{-m_2, m, m_1}. \qquad (3.180)$$

Temporarily ignoring phase changes and dimensionality factors, we see that the four basic symmetries (the top four relations) generate a group of order 72. The simplest way to see this is to arrange the six labels of a Wigner coefficient into a *Regge array*:

$$\begin{bmatrix} j_1+m_1 & j_2+m_2 & j-m \\ j_1-m_1 & j_2-m_2 & j+m \\ j-j_1+j_2 & j+j_1-j_2 & -j+j_1+j_2 \end{bmatrix}. \qquad (3.181)$$

The nine elements of this array are redundant, since the elements in each row and each column sum to a fixed value, j_1+j_2+j. (The array is thus a "magic square.")

The transformations of the quantum numbers $(j_1 m_1 j_2 m_2 j m)$ induced by the top four relations, Eqs. (3.180), correspond, respectively, to the following operations on the Regge array: (*a*) interchange rows 1 and 2; (*b*) interchange columns 1 and 2; (*c*) transpose; and (*d*) interchange columns 1 and 3. These four operations generate all six permutations of the columns of the array and all six permutations of the rows, which, together with transposition, yield 72 relations among the Wigner coefficients.

To take into account now the phase changes and dimensionality factors, let us define a new symbol that corresponds, as closely as possible, to the symmetries exhibited by the Regge array. This is the 3-j symbol, introduced by Wigner [47], and defined by

$$\begin{pmatrix} j_1 & j_2 & j \\ m_1 & m_2 & -m \end{pmatrix} \equiv (-1)^{j_1-j_2+m}(2j+1)^{-\frac{1}{2}} C^{j_1 j_2 j}_{m_1 m_2 m}$$

$$\equiv R\begin{bmatrix} j_1+m_1 & j_2+m_2 & j-m \\ j_1-m_1 & j_2-m_2 & j+m \\ j-j_1+j_2 & j+j_1-j_2 & -j+j_1+j_2 \end{bmatrix}. \qquad (3.182)$$

The second line of Eq. (3.182) defines the Regge notation for the 3-j symbol in terms of the Regge array.

The symmetry properties now have an elegant expression in terms of the symmetries of the determinant of the Regge array: *Thus, the Regge symbol is invariant under all even permutations of its rows or columns, and under transposition of the array, and is multiplied by the factor* $(-1)^{j_1+j_2+j}$ *under odd permutations of its rows or columns.*

The 3-j symbol clearly optimizes the symmetry properties inherent in the Wigner coefficients. The actual origin of these symbols, however, may be traced in Wigner's work to the construction of rotational invariants (discussed in Section 20).

The most often used symmetries of the 3-j symbol (and hence Wigner coefficients) are the twelve "classical" ones: (a) *even permutations* of the columns leave the coefficient invariant; (b) *odd permutations* of the columns multiply the coefficient by $(-1)^{j_1+j_2+j}$; and (c) *sign reversal* of all the projection quantum numbers multiplies the coefficient by $(-1)^{j_1+j_2+j}$.

13. Relations between Rotation Matrices and Wigner Coefficients

We have emphasized that, under rotation of a physical system, it is the total angular momentum operator that generates the transformations of the quantal states of the system. (In the passive viewpoint one rotates the frame; in the active viewpoint, *all* constituents of the system, considered as a composite object, are rotated *simultaneously*. See Note 8 for the case when the "parts" of a physical system may be rotated *independently*.) Thus, under a rotation $\mathscr{U}(\psi,\hat{n})$, the eigenket $|(j_1j_2)jm\rangle$ undergoes the transformation

$$\mathscr{U}\colon |(j_1j_2)jm\rangle \to |(j_1j_2)jm\rangle' \equiv \mathscr{U}|(j_1j_2)jm\rangle$$

$$= e^{-i\psi\hat{n}\cdot\mathbf{J}}|(j_1j_2)jm\rangle = \sum_{m'} D^j_{m'm}(\psi,\hat{n})|(j_1j_2)jm'\rangle. \tag{3.183}$$

On the other hand, since we may write $\mathscr{U}$ as

$$\mathscr{U} = e^{-i\psi\hat{n}\cdot\mathbf{J}} = e^{-i\psi\hat{n}\cdot\mathbf{J}(1)} \otimes e^{-i\psi\hat{n}\cdot\mathbf{J}(2)},$$

the product ket vector $|j_1m_1\rangle \otimes |j_2m_2\rangle$ undergoes the transformation

$$\mathscr{U}\colon |j_1m_1\rangle \otimes |j_2m_2\rangle \to (|j_1m_1\rangle \otimes |j_2m_2\rangle)'$$

$$\equiv (e^{-i\psi\hat{n}\cdot\mathbf{J}(1)}|j_1m_1\rangle) \otimes (e^{-i\psi\hat{n}\cdot\mathbf{J}(2)}|j_2m_2\rangle)$$

$$= \sum_{m_1'm_2'} [D^{j_1}(\psi,\hat{n}) \otimes D^{j_2}(\psi,\hat{n})]_{m_1'm_2';m_1m_2}|j_1m_1'\rangle \otimes |j_2m_2'\rangle, \tag{3.184}$$

where

$$[D^{j_1}(\psi,\hat{n}) \otimes D^{j_2}(\psi,\hat{n})]_{m_1'm_2';m_1m_2}$$

denotes an element of the direct product matrix

$$D^{j_1}(\psi,\hat{n}) \otimes D^{j_2}(\psi,\hat{n}).$$

Using the relationship between the two bases $\{|(j_1j_2)jm\rangle\}$ and $\{|j_1m_1\rangle \otimes |j_2m_2\rangle\}$ of the product space $\mathscr{H}_{j_1}(1) \otimes \mathscr{H}_{j_2}(2)$,

$$|(j_1j_2)jm\rangle = \sum_{m_1m_2} C^{j_1j_2j}_{m_1m_2m}|j_1m_1\rangle \otimes |j_2m_2\rangle, \tag{3.185}$$

we now derive from Eqs. (3.183) and (3.184) the basic result:

$$C[D^{j_1}(U) \otimes D^{j_2}(U)]\tilde{C} = \sum_j \oplus D^j(U)$$

$$\equiv \begin{pmatrix} D^{j_1+j_2}(U) & 0 & \cdots & 0 \\ 0 & D^{j_1+j_2-1}(U) & \cdots & 0 \\ \vdots & & & \vdots \\ 0 & \cdots & & D^{|j_1-j_2|}(U) \end{pmatrix}, \tag{3.186}$$

in which we have now replaced the rotation parametrized by $(\psi, \hat{n})$ by an arbitrary unitary unimodular matrix U. [Relation (3.186) is independent of parametrization. Observe that the exponentiation of Eq. (3.174) is just Eq. (3.186).]

In terms of matrix elements, Eq. (3.186) takes the form

$$\sum_{\substack{m_1'm_2'\\ m_1m_2}} C^{j_1j_2j'}_{m_1'm_2'm'} C^{j_1j_2j}_{m_1m_2m} D^{j_1}_{m_1'm_1}(U) D^{j_2}_{m_2'm_2}(U) = \delta_{j'j} D^j_{m'm}(U). \tag{3.187}$$

Equation (3.186) [or (3.187)] expresses the fact that the Wigner coefficients are elements of a real, orthogonal matrix, which splits (or reduces) the direct product representation $D^{j_1}(U) \otimes D^{j_2}(U)$ of a rotation U into the basic rotation matrices $D^j(U)$, $j = j_1 + j_2, \ldots, |j_1 - j_2|$.

Using the orthogonality properties of the Wigner coefficients, Eq. (3.187) may be written in several other useful forms:

$$\sum_{m_1m_2} C^{j_1j_2j}_{m_1m_2m} D^{j_1}_{m_1'm_1}(U) D^{j_2}_{m_2'm_2}(U) = C^{j_1j_2j}_{m_1',m_2',m_1'+m_2'} D^j_{m_1'+m_2',m}(U), \tag{3.188}$$

$$D^{j_1}_{m_1'm_1}(U)D^{j_2}_{m_2'm_2}(U)$$

$$= \sum_j C^{j_1 j_2 j}_{m_1',m_2',m_1'+m_2'} C^{j_1 j_2 j}_{m_1,m_2,m_1+m_2} D^j_{m_1'+m_2',m_1+m_2}(U). \tag{3.189}$$

Remark. Wigner [25] based the calculation of the coefficients $C^{j_1 1 j}_{m_1 m_2 m}$ on the product property, Eq. (3.189). (In that paper the term $j = j_1$ was accidentally omitted. In correcting this oversight, Wigner [26] gave the explicit $j_2 = 1$ coefficients; this, to our knowledge, constitutes the first time these coefficients appear in print.) The standard phase convention was fixed in Wigner's general derivation (Wigner [13]), yielding Eq. (3.171) based on the integral property expressed by Eq. (3.190) below.

Integrals over three rotation matrices. The orthogonality relations given in Section 9 may now be used, together with the product property (3.189), to derive

$$\int d\Omega\, D^{j*}_{m'm}(U)D^{j_1}_{m_1'm_1}(U)D^{j_2}_{m_2'm_2}(U) = \frac{2\pi^2}{2j+1} C^{j_1 j_2 j}_{m_1' m_2' m'} C^{j_1 j_2 j}_{m_1 m_2 m}, \tag{3.190}$$

in which the angular momenta j_1, j_2, j may assume the values $0, \frac{1}{2}, 1, \ldots$, and *we now extend the domain of definition* of a Wigner coefficient by defining

$$C^{j_1 j_2 j}_{m_1 m_2 m} = 0 \qquad \text{unless } j \in \{j_1 + j_2, j_1 + j_2 - 1, \ldots, |j_1 - j_2|\}. \tag{3.191}$$

These conditions are known as the *triangle conditions* on $(j_1 j_2 j)$. This result (Eq. 3.190) is valid in any parametrization of $U \in SU(2)$ that covers exactly once the unit sphere in four-space. [We indicate in greater detail in Chapter 5, Appendix C, how Eq. (3.190) may be used to calculate the Wigner coefficients.]

Equation (3.190) may be particularized, both in the choice of parametrization and in the choice of the projection quantum numbers, to obtain a number of useful results. The best known of these relations are obtained by using the Euler angle parametrization, to which we now turn.

Gaunt's [48] integral over three spherical harmonics is obtained by using Eq. (3.138), relating rotation matrices and spherical harmonics:

$$\int_0^{2\pi} d\alpha \int_0^{\pi} \sin\beta\, d\beta\, Y^*_{lm}(\beta\alpha) Y_{l_1 m_1}(\beta\alpha) Y_{l_2 m_2}(\beta\alpha)$$

$$= \left[\frac{(2l_1+1)(2l_2+1)}{4\pi(2l+1)}\right]^{\frac{1}{2}} C^{l_1 l_2 l}_{0\,0\,0} C^{l_1 l_2 l}_{m_1 m_2 m}. \tag{3.192}$$

Specializing this result still further, using Eq. (3.147), and carrying out the integration over α, yields the integral over three Legendre functions:

$$\int_0^{\pi} \sin\beta \, d\beta \, P_l(\cos\beta)\, P_{l_1}(\cos\beta)\, P_{l_2}(\cos\beta) = \frac{2}{2l+1}\,(C^{l_1 l_2 l}_{0\,0\,0})^2. \qquad (3.193)$$

The special Wigner coefficient $C^{l_1 l_2 l}_{0\,0\,0}$ appearing in this result can be expressed as a single term (no summation). (In general, it is not possible to reduce a Wigner coefficient to a single term, except for the *boundary* or *extremal* cases, where one of the projection quantum numbers equals its maximum or minimum value.) Racah [29] was able to carry out the summation for this special case $m_1 = m_2 = m = 0$:

$$C^{l_1 l_2 l}_{0\,0\,0} = \left[\frac{(2l+1)(l_1+l_2-l)!(l_1-l_2+l)!(-l_1+l_2+l)!}{(l_1+l_2+l+1)!}\right]^{\frac{1}{2}}$$

$$\times \frac{(-1)^{L-l}L!}{(L-l_1)!(L-l_2)!(L-l)!}, \qquad (3.194)$$

where

$$L \equiv (l_1+l_2+l)/2 \qquad \text{for } l_1+l_2+l \text{ even.}$$

It is a consequence of the first symmetry relation [Eq. (3.180)] that one must have

$$C^{l_1 l_2 l}_{0\,0\,0} = 0 \qquad \text{for } l_1+l_2+l \text{ odd.} \qquad (3.195)$$

There is a physical interpretation of the result expressed by Eq. (3.195). One adjoins to the rotations, the *parity operation* $\mathscr{I}$ (inversion through the coordinate origin), so that the group becomes the group of rotations and reversals. Under the parity operation, $\mathscr{I}$: $\mathbf{x} \to -\mathbf{x}$, the solid harmonics [Eq. (3.153)], and hence the spherical harmonics [Eq. (3.152)], are multiplied by $(-1)^l$. It follows that under the parity operation the integrand of Eq. (3.192) undergoes the phase change $(-1)^{l_1+l_2+l}$; since the integral is invariant, one concludes that $l_1 + l_2 + l$ must be even for the integral to be nonvanishing. Equation (3.195) accordingly expresses the *conservation of parity*.

14. Concept of a Tensor Operator

Survey. The concept of angular momentum and the related concept of rotational symmetry have led us, from the viewpoint of physics, to the study of the characteristic states $\{|(\alpha)jm\rangle\}$ and their transformation properties. The next step is to examine the action of physically important operators on these characteristic states; that is to say, we must characterize the matrix elements

$$\text{M.E.} = \langle \text{final state}|\text{operator}|\text{initial state}\rangle \in \mathbb{C}$$

that determine the (observable) *transition probabilities* $|\text{M.E.}|^2$ for transitions from one characteristic state to another induced by the operator (perturbation). Attention thus shifts in physics from the states themselves, $\{|(\alpha)jm\rangle\}$, to transformations acting *on* these states. It is this need for additional structure that is responsible for physical developments in angular momentum theory that go beyond results available in the mathematical literature. It is our view that these physically oriented concepts can, and should, have mathematical significance. We shall show in this chapter that these developments center on two structures: the basis-dependent *Wigner operators*, and the basis-independent (invariant) *Racah operators.*

Definition of a tensor operator. Let us consider a physical system having rotational symmetry and, hence, characterized by states of sharp angular momentum $|(\alpha)jm\rangle$. Since there can be, and generally is, additional structure besides the angular momentum variables, we have used a labeling set (α) to denote a set of quantum labels. (These labels necessarily correspond to operators that commute with the angular momenta – see footnote on p. 32.)

Consider now an arbitrary operator $\mathcal{O}$ that acts on the system. We can describe the action of the operator, in complete detail, by the set of matrix elements

$$\{\langle(\alpha')j'm'|\mathcal{O}|(\alpha)jm\rangle\}. \tag{3.196}$$

Since these matrix elements are *probability amplitudes*, and since the physical probabilities must necessarily be invariant to the choice of coordinate frame, we must have

$$|\langle(\alpha')j'm'|\mathcal{O}|(\alpha)jm\rangle|^2 = \text{invariant to rotations of coordinate frames.} \tag{3.197}$$

The implications of this constraint are contained in the fundamental theorem on symmetry (Wigner [13, 49]):

The invariance of the physical probability (3.197) *under a symmetry implies that either* (*a*) *the probability amplitude*, (3.196) *is invariant, or* (*b*) *the probability amplitude transforms under the symmetry into its complex conjugate.*

As shown by Wigner, the latter case corresponds to time reversal, and the result for rotational symmetry is that we have *invariance of the probability amplitude itself* (the proof of the Wigner theorem is given in RWA):

$$\langle(\alpha')j'm'|\mathcal{O}|(\alpha)jm\rangle = \text{invariant to rotations of coordinate frames.} \tag{3.198}$$

Since this relation is valid for all ket vectors $\{|(\alpha)jm\rangle\}$ and for all bra vectors $\{\langle(\alpha)jm|\}$ (vectors in the dual Hilbert space), we conclude that

$$\langle\Psi|\mathcal{O}|\Phi\rangle = \text{invariant to rotations of coordinate frames,} \tag{3.199}$$

where $|\Phi\rangle$ and $\langle\Psi|$ denote arbitrary vectors of the Hilbert space and its dual, respectively.

The action of the rotational symmetry operator $\mathcal{U} = \exp(-i\psi\hat{n}\cdot\mathbf{J})$ on the ket vector $|\Phi\rangle$ is the unitary transformation

$$|\Phi\rangle \to |\Phi'\rangle \equiv \mathcal{U}|\Phi\rangle \tag{3.200}$$

with a corresponding unitary transformation on the bra vector $\langle\Psi|$,

$$\langle\Psi| \to \langle\Psi'| \equiv \langle\Psi|\mathcal{U}^{-1}. \tag{3.201}$$

The explicit statement of the invariance (3.199) now takes the form

$$\langle\Psi|\mathcal{O}|\Phi\rangle = \langle\Psi'|\mathcal{O}'|\Phi'\rangle, \tag{3.202}$$

where $\mathcal{O}'$ denotes the operator $\mathcal{O}$ relative to the rotated frame. Thus,

$$\langle\Psi|\mathcal{O}|\Phi\rangle = \langle\Psi|\mathcal{U}^{-1}\mathcal{O}'\mathcal{U}|\Phi\rangle \tag{3.203}$$

for arbitrary ket vectors $|\Phi\rangle$ and arbitrary bra vectors $\langle\Psi|$. We conclude: $\mathcal{O} = \mathcal{U}^{-1}\mathcal{O}'\mathcal{U}$, or, equivalently,

$$\mathcal{O}' = \mathcal{U}\mathcal{O}\mathcal{U}^{-1}. \tag{3.204}$$

Under infinitesimal transformations, $\mathcal{U} = \exp(-i\delta\boldsymbol{\omega}\cdot\mathbf{J})$, where $\boldsymbol{\omega} = \psi\hat{n}$, we may write

$$\mathcal{O}' = \mathcal{O} - i\delta\boldsymbol{\omega}\cdot[\mathbf{J}, \mathcal{O}]$$

and

$$\delta\mathcal{O} \equiv \mathcal{O}' - \mathcal{O} = -i\delta\boldsymbol{\omega} \cdot [\mathbf{J}, \mathcal{O}]. \tag{3.205}$$

The appearance of the commutator of the general operator $\mathcal{O}$ with the angular momentum $\mathbf{J}$ motivates the definition of an *irreducible tensor operator* of rank $J (J = 0, \frac{1}{2}, 1, \ldots)$. An irreducible tensor operator of rank J (denoted by $\mathbf{T}^J$) is defined to be *a set of linear operators* $\{T^J_M: M = J, J-1, \ldots, -J\}$, where each operator is a mapping of the Hilbert space of ket vectors into itself and the commutator action of the angular momentum $\mathbf{J}$ on the set of operators is to map the set into itself according to the following specific (defining) rules:

$$[J_+, T^J_M] = [(J - M)(J + M + 1)]^{\frac{1}{2}} T^J_{M+1},$$

$$[J_-, T^J_M] = [(J + M)(J - M + 1)]^{\frac{1}{2}} T^J_{M-1},$$

$$[J_3, T^J_M] = MT^J_M. \tag{3.206}$$

The linear operators T^J_M are themselves often called tensor operators, but they are more correctly referred to as the *components* of the tensor operator $\mathbf{T}^J$.

The conditions expressed by the commutation relations (3.206) are that the irreducible tensor operator components T^J_M transform under the *commutation action* of the components J_i of $\mathbf{J}$ as if they possessed sharp angular momentum J, M:

$$\sum_i [J_i, [J_i, T^J_M]] = J(J + 1)T^J_M,$$

$$[J_3, T^J_M] = MT^J_M. \tag{3.207}$$

A second useful result, which follows from the defining relations (3.206), shows how one may generate all the components $(T^J_J, \ldots, T^J_{-J})$ of a tensor operator of rank J from the *highest component* T^J_J by multiple commutation with the lowering operator J_-:

$$T^J_M = \left[\frac{(J + M)!}{(2J)!(J - M)!}\right]^{\frac{1}{2}} [J_-, T^J_J]_{(J-M)}. \tag{3.208}$$

It is often convenient to write the defining relations, Eqs. (3.206), in a notational form in which the Wigner coefficients $C^{J\,1\,J}_{M,m,M+m}$ are introduced explicitly (see Table 1 in Appendix of Tables). One then introduces the so-called *spherical components* of the angular momentum $\mathbf{J}$ by

$$J_{+1} \equiv -(J_1 + iJ_2)/\sqrt{2},$$

$$J_0 \equiv J_3,$$

$$J_{-1} \equiv (J_1 - iJ_2)/\sqrt{2}, \tag{3.209}$$

and obtains a uniform expression[1] for Eqs. (3.206):

$$[J_m, T^J_M] = [J(J+1)]^{\frac{1}{2}} C^{J\,1\,J}_{M,m,M+m} T^J_{M+m}. \tag{3.210}$$

The definition of a tensor operator has been so constructed that we can find the effect of finite transformations on T^J_M by applying the multiple commutator [see Eq. (3.37)] result:

$$\mathscr{U}(\boldsymbol{\omega}) T^J_M \mathscr{U}^{-1}(\boldsymbol{\omega}) = \sum_{n=0}^{\infty} [-i\boldsymbol{\omega}\cdot\mathbf{J}, T^J_M]_{(n)}/n!. \tag{3.211}$$

To evaluate the multiple commutator, observe that the tensor operator property may be written

$$[J_i, T^J_M] = \sum_{M'} \langle JM'|J_i|JM\rangle T^J_{M'}. \tag{3.212}$$

This result implies that the multiple commutator becomes a matrix product:

$$[-i\boldsymbol{\omega}\cdot\mathbf{J}, T^J_M]_{(n)} = \sum_{M'} \langle JM'|(-i\boldsymbol{\omega}\cdot\mathbf{J})^n|JM\rangle T^J_{M'}. \tag{3.213}$$

Equation (3.213) shows that we can sum the multiple commutator to find

$$\mathscr{U}(\boldsymbol{\omega}) T^J_M \mathscr{U}^{-1}(\boldsymbol{\omega}) = \sum_{M'} \langle JM'|e^{-i\boldsymbol{\omega}\cdot\mathbf{J}}|JM\rangle T^J_{M'}$$

$$= \sum_{M'} D^J_{M'M}(\boldsymbol{\omega}) T^J_{M'}. \tag{3.214}$$

It follows that the components of a tensor operator of rank J transform under the *similarity action* of $\mathscr{U}$ in the same way as a ket vector does under the standard action of $\mathscr{U}$:

$$\mathscr{U}\colon |(\alpha)JM\rangle \to |(\alpha)JM\rangle' \equiv \mathscr{U}|(\alpha)JM\rangle$$

$$= \sum_{M'} D^J_{M'M}(U)|(\alpha)JM'\rangle, \tag{3.215}$$

---

[1] The phase conventions in Eq. (3.209) should be noted carefully. See Appendix D.

$$\mathscr{U}: T^J_M \to (T^J_M)' = \mathscr{U} T^J_M \mathscr{U}^{-1}$$

$$= \sum_{M'} D^J_{M'M}(U) T^J_{M'}. \tag{3.216}$$

Remarks. (*a*) Irreducible tensor operators are the operator analogs of the basic ket vectors in the multiplet $\{|(\alpha)jm\rangle: m = j, \ldots, -j\}$, and, like these multiplets, *physical tensor operators* usually carry additional labels that specify the physical properties of the operator.

(*b*) The basic ket vectors $\{|(\alpha)jm\rangle: m = j, \ldots, -j\}$ may themselves be considered as operators acting on the Hilbert space of states of a physical system. Let us formulate this concept in the language of wave functions and differential operators, since it usually arises this way in physical problems. Let $\Phi_{(\alpha)jm}$ denote a wave function of sharp angular momentum, and Φ and Ψ denote arbitrary functions in the Hilbert space. Then, if the angular momentum operators $\mathscr{J}_i$ possess the *derivation property*,

$$\mathscr{J}_i(\Phi\Psi) = (\mathscr{J}_i\Phi)\Psi + \Phi(\mathscr{J}_i\Psi), \tag{3.217}$$

we may write

$$[\mathscr{J}_i, \Phi_{(\alpha)jm}]\Psi = (\mathscr{J}_i\Phi_{(\alpha)jm})\Psi,$$

that is,

$$[\mathscr{J}_i, \Phi_{(\alpha)jm}] = (\mathscr{J}_i\Phi_{(\alpha)jm}) \tag{3.218}$$

is a valid operator identity. Hence, Eqs. (3.206) hold for wave functions having sharp angular momentum. The necessity of the derivation property for the interpretation of wave functions as tensor operators has often been overlooked in the literature (Aebersold and Biedenharn [50]).

(*c*) The definition of a tensor operator, Eqs. (3.206), may appear overly restrictive. An apparently more general definition of a tensor operator **T** is that its components $(T_1, T_2, \ldots, T_n)$ should be linear operators that are transformed linearly among themselves under the commutation action of the angular momentum operators J_i:

$$[J_i, T_\lambda] = \sum_{\mu=1}^{n} a^{(i)}_{\mu\lambda} T_\mu. \tag{3.219}$$

In Note 9 we discuss conditions under which it is possible, by taking linear combinations of the T_λ $(\lambda = 1, 2, \ldots, n)$, to split **T** into subsets of operators,

$$\{T^J_M\colon M = J, \ldots, -J\}, \qquad J = J_1, J_2, \ldots, \tag{3.220}$$

that satisfy the rules (3.206). [The term *irreducible* means that no further split is possible for operators satisfying Eqs. (3.206).]

Sets of operators satisfying the defining relations, Eqs. (3.206), occur frequently in physical problems, thus elevating the study of irreducible tensor operators to a position of fundamental importance in angular momentum theory. Examples of tensor operators are given in Note 10.

(*d*) We have noted earlier [Remark (*a*), p. 42] that the active transformation of a vector [as given by Eq. (2.1) or (3.41)] and the standard transformation of a vector operator [as given by Eq. (3.40) or (3.42)] differ by R versus R^{-1}. This difference, to repeat, is not an error, but – in the final analysis – the result of conventions made in the definition of a tensor operator. The present remark is intended to clarify this situation.

In order to be explicit let us compare the general tensor operator transformation (3.216) and the transformations (3.41) and (3.42) by rewriting the latter in terms of spherical components, using the relation $D^1(U) = A^\dagger RA$ [see Eqs. (3.28)–(3.32) and Eq. (3.209)]. The results are

$$x_\mu \to x'_\mu \equiv \sum_\nu D^{1*}_{\mu\nu}(U)x_\nu, \tag{3.41$'$}$$

$$J_\mu \to J'_\mu \equiv \sum_\nu D^1_{\nu\mu}(U)J_\nu. \tag{3.42$'$}$$

We thus see from Eq. (3.42′) that the angular momentum operator transforms according to the definition (3.216) of an irreducible tensor operator of rank $J = 1$, whereas the (active) transformation (3.41′) of a vector $\mathbf{x} \in \mathbb{R}^3$ is inverse to that of Eq. (3.42′), since $[D^1(U)]^{-1} = [D^1(U)]^\dagger$.

Let us next note that the tensor operator concept involves two distinct spaces on which transformations occur: The first is the Hilbert space which undergoes the transformation $|\Psi\rangle \to \mathscr{U}|\Psi\rangle$; and the second is the index space $\{(J, M)\}$, which labels the operators. The tensor operator concept links these two spaces by requiring that the induced transformation $T^J_M \to \mathscr{U}\, T^J_M \mathscr{U}^{-1}$ be the same as the index transformation $|JM\rangle \to \sum_{M'} D^J_{M'M}(U)|JM'\rangle$.

The essence of the tensor operator concept is this linking together of distinct transformations in separate spaces. (Michel [50a, p. 48], in his definitive discussion, makes this same point, provocatively, by asserting that "tensor operators on $\mathscr{H}$ are not operators on $\mathscr{H}$!".)

It is important to note that there is a *free choice* in this association of the two transformations: instead of the standard convention (wherein tensor operators transform as ket vectors), one could equally well have chosen bra vector transformations. This would be a perverse and very unnatural convention, which, in practice, would lead to untold confusion because of the unfortunate ambiguity in the matrix element notation $\langle\Psi|\mathcal{O}|\Phi\rangle$ for non-Hermitian operators – does $\mathcal{O}$ act to the left or to the right?

We conclude that the duality between Eq. (3.41′) and (3.42′) is a *convention*, which is based on different meanings associated with vector as opposed to vector operator.

15. The Wigner–Eckart Theorem

The general concept of a *vector operator* (defined by its commutation relations with the total angular momentum) was introduced into physics by Güttinger and Pauli [51] (see also Dirac [52], Born and Jordan [53]) for the purpose of deriving from the new quantum mechanics (matrix mechanics) the relative intensities of the components of Zeeman patterns, arising in complex spectra (Hönl–Kronig intensity formulas). (The standard textbook treatments of vector operators are those of Wigner [13], Condon and Shortley [28], and Born and Jordan [53].)

By definition, the Cartesian components V_j $(j = 1, 2, 3)$ of a vector operator $\mathbf{V}$ obey the same commutation relations with the components J_i $(i = 1, 2, 3)$ of $\mathbf{J}$ as do the J_j themselves – that is,

$$[J_i, V_j] = ie_{ijk}V_k, \tag{3.221}$$

which implies[1]

$$\mathbf{J}\times\mathbf{V} + \mathbf{V}\times\mathbf{J} = 2i\mathbf{V}. \tag{3.222}$$

To show that Eq. (3.221) defines an irreducible tensor operator of rank 1, we rewrite the commutation relations (3.221) in terms of J_+, J_-, and J_3, thus finding

$$\begin{aligned}[][J_\pm, V_\mu] &= [(1 \mp \mu)(2 \pm \mu)]^{\frac{1}{2}} V_{\mu\pm 1},\\ [J_3, V_\mu] &= \mu V_\mu,\end{aligned} \tag{3.223}$$

where

$$\begin{aligned} V_{+1} &= -(V_1 + iV_2)/\sqrt{2},\\ V_0 &= V_3,\\ V_{-1} &= (V_1 - iV_2)/\sqrt{2}.\end{aligned} \tag{3.224}$$

[1] Equation (3.221) implies (3.222), but it is interesting that the converse is not true.

Since the result, Eq. (3.223), agrees with the definition, Eqs. (3.206), we conclude that **V** is an irreducible tensor of rank 1. The V_μ ($\mu = +1, 0, -1$) are called *spherical components* of **V** (see Appendix D).

Remark. Aside from **J** itself, we have already encountered several examples of vector operators: (*a*) Each of the basis vectors $\hat{f}_1, \hat{f}_2, \hat{f}_3$ [see Eq. (3.113)], defining the orientation of a solid body, is a vector operator with respect to the total angular momentum $\mathcal{J}$; (*b*) each of the angular momenta **J**(1) and **J**(2) of two kinematically independent systems is a vector operator with respect to the total angular momentum $\mathbf{J} = \mathbf{J}(1) + \mathbf{J}(2)$; and (*c*) both the position vector **r** and the linear momentum **p** are vector operators with respect to the orbital angular momentum $\mathbf{L} = \mathbf{r} \times \mathbf{p}$ of a single particle.

Using only algebraic consequences of the defining relations, Eqs. (3.223), the explicit results for the nonvanishing matrix elements, $\langle(\alpha')j'm'|V_\mu|(\alpha)jm\rangle$, were obtained in the very first discussions of the problem (Condon and Shortley [28], Güttinger and Pauli [51], Born and Jordan [53]). We shall not give this derivation here, since these early results for vector operators pointed the way to the general result for an irreducible tensor operator of arbitrary rank (Wigner [13], Eckart [27]) and to a much simpler derivation of the form of the nonvanishing matrix elements.

Eckart [27] and Wigner [13] recognized that the results developed for a vector operator were a special case of the general result for irreducible tensor operators.

Remark. Eckart's contribution[1] here was in observing that the form of the result for the coupling of two angular momenta,

$$\langle(\alpha j_1 j_2)jm|\alpha_1 j_1 m_1; \alpha_2 j_2 m_2\rangle \propto C^{j_1 j_2 j}_{m_1 m_2 m},$$

where

$$|\alpha_1 j_1 m_1; \alpha_2 j_2 m_2\rangle \equiv |(\alpha_1)j_1 m_1\rangle\, |(\alpha_2)j_2 m_2\rangle$$

would remain essentially correct when the eigenkets $|(\alpha_i)j_i m_i\rangle$ were replaced by operators $T_{(\alpha_i)j_i m_i}$ having the same transformation properties under rotations as the eigenkets themselves.

Wigner [13] gave the explicit definition, Eq. (3.216), of an irreducible tensor operator, and then determined the nonvanishing matrix elements by the procedure that we now give.

---

[1] Eckart was explicit only in the case of vector operators, but the general result is clearly implied.

Wigner-Eckart Theorem: Each matrix element in an angular momentum basis of an irreducible tensor operator is a product of two factors: a purely angular momentum dependent factor ("Wigner coefficient") and a factor that is independent of projection quantum numbers ("reduced matrix element").

Proof. Consider the probability amplitude, $\langle(\alpha')j'm'|T^J_M|(\alpha)jm\rangle$. Under the transformation $\mathscr{U}$, we have $|(\alpha)jm\rangle \to \mathscr{U}|(\alpha)jm\rangle$, $\langle(\alpha')j'm'| \to \langle(\alpha')j'm'|\mathscr{U}^{-1}$, and $T^J_M \to \mathscr{U}T^J_M\mathscr{U}^{-1}$. Since the probability amplitude is invariant to these transformations, we obtain

$$\langle(\alpha')j'm'|T^J_M|(\alpha)jm\rangle$$

$$= \sum_{\mu'M'\mu} D^{j'*}_{\mu'm'}(U)D^{j}_{\mu m}(U)D^{J}_{M'M}(U)\langle(\alpha')j'\mu'|T^J_{M'}|(\alpha)j\mu\rangle. \quad (3.225)$$

Since this result is valid for arbitrary rotations, we can integrate over all rotations, using the integral over the rotation matrices given by Eq. (3.190). We thus obtain

$$\langle(\alpha')j'm'|T^J_M|(\alpha)jm\rangle = \langle(\alpha')j'||\mathbf{T}^J||(\alpha)j\rangle C^{jJj'}_{mMm'}, \quad (3.226)$$

where we have defined the reduced matrix element by

$$\langle(\alpha')j'||\mathbf{T}^J||(\alpha)j\rangle \equiv (2j'+1)^{-1} \sum_{\mu'M'\mu} C^{jJj'}_{\mu M'\mu'}\langle(\alpha')j'\mu'|T^J_{M'}|(\alpha)j\mu\rangle. \quad (3.227)$$

An alternative form for the reduced matrix element is given by

$$\langle(\alpha')j'||\mathbf{T}^J||(\alpha)j\rangle = \sum_{\mu M'} C^{jJj'}_{\mu M'\mu'}\langle(\alpha')j'\mu'|T^J_{M'}|(\alpha)j\mu\rangle. \quad (3.227')$$

This result follows directly from Eq. (3.226) and the orthogonality of the Wigner coefficients (the summation over μ and M' yields a coefficient that is independent of μ'). [Observe that the extended definition (3.191) of the Wigner coefficients occurs in these results.] Notice also that

$$\sum_{\mu M'} C^{jJj''}_{\mu M'\mu'}\langle(\alpha')j'\mu'|T^J_{M'}|(\alpha)j\mu\rangle = 0, \qquad j'' \neq j', \quad (3.228)$$

this result following by direct substitution of Eq. (3.226) into the second identity, Eq. (3.227′). ■

The quantity

$$\langle(\alpha')j'||\mathbf{T}^J||(\alpha)j\rangle \quad (3.229)$$

is Condon and Shortley's notation for a reduced matrix element and is, as its definition shows, independent of all projection quantum numbers and hence a rotational invariant.

Remarks. (*a*) The most striking feature of the Wigner–Eckart theorem is the clear-cut separation of the generic (group-theoretic) aspects of an operator from its particularities—the reduced matrix element—that relate to the physical measurement in question. Conventionally, one expresses this by saying that one has separated the "geometric aspects" from the "physics of the problem." However one expresses it, it is this separation that accounts for the wide applicability of general angular momentum structures in physical problems.

(*b*) The justification of the unusual notation $\langle(\alpha')j'\|\mathbf{T}^J\|(\alpha)j\rangle$ [Eq. (3.229)] for a reduced matrix element that has $\langle(\alpha')j'\|$ and $\|(\alpha)j\rangle$ appearing as "reduced bra–ket vectors" has been given by Condon and Shortley: They observe that this notational scheme allows one to treat reduced matrix elements like the elements of a matrix (thus justifying also the term "reduced matrix element") when transforming from one complete set of basis states $\{|(\alpha)jm\rangle\}$ to another $\{|(\beta)jm\rangle\}$. Thus, under the change of basis given by

$$|(\beta)jm\rangle = \sum_{(\alpha)} \langle(\alpha)j|(\beta)j\rangle|(\alpha)jm\rangle, \tag{3.230}$$

one obtains the reduced matrix elements in the new basis as

$$\langle(\beta')j'\|\mathbf{T}^J\|(\beta)j\rangle = \sum_{(\alpha')(\alpha)} \langle(\beta')j'|(\alpha')j'\rangle\langle(\alpha')j'\|\mathbf{T}^J\|(\alpha)j\rangle\langle(\alpha)j|(\beta)j\rangle.$$

(The transformation coefficient $\langle(\alpha)j|(\beta)j\rangle$ is independent of m because the action of the angular momentum operator $\mathbf{J}$ is the same on both basis sets.)

16. The Coupling of Tensor Operators

There is a second aspect of the Wigner coefficients that is equally as important as the property expressed by the Wigner–Eckart theorem—this is the coupling aspect: *The Wigner coefficients* $\{C^{j_1j_2j}_{m_1m_2m}\}$ *effect a coupling in the space of tensor operators.*

Let us be explicit. Consider two operators $\mathbf{S}^{k_1}$ and $\mathbf{T}^{k_2}$, each of which is an irreducible tensor operator with respect to the *total angular momentum* $\mathbf{J}$ of a physical system having the basis kets $\{|(\alpha)jm\rangle\}$. Thus, by definition, each of the sets of operators $\{S^{k_1}_{\mu_1}\}$ and $\{T^{k_2}_{\mu_2}\}$ obey the commutation relations (3.206) with the components of $\mathbf{J}$. Alternatively, under the rotation generated by $\mathbf{J}$, each of these sets of operators undergoes the unitary transformation given by Eq. (3.216).

We now introduce the symbol

$$[\mathbf{S}^{k_1} \mathbf{x} \mathbf{T}^{k_2}]^k \tag{3.231}$$

to denote the set of operators with components

$$[\mathbf{S}^{k_1} \mathbf{x} \mathbf{T}^{k_2}]^k_\mu, \qquad \mu = k, k-1, \ldots, -k, \tag{3.232}$$

which are defined by the coupling of the components of $\mathbf{S}^{k_1}$ and $\mathbf{T}^{k_2}$:

$$[\mathbf{S}^{k_1} \mathbf{x} \mathbf{T}^{k_2}]^k_\mu \equiv \sum_{\mu_1 \mu_2} C^{k_1 k_2 k}_{\mu_1 \mu_2 \mu} S^{k_1}_{\mu_1} T^{k_2}_{\mu_2}. \tag{3.233}$$

We must prove that this operator, $[\mathbf{S}^{k_1} \mathbf{x} \mathbf{T}^{k_2}]^k$, is indeed an irreducible tensor with the angular momentum properties designated by the index k – that is, it is of rank k.

Proof. By assumption, the action of the rotation operator $\mathscr{U}$ on the components of $\mathbf{S}^{k_1}$ and $\mathbf{T}^{k_2}$ is

$$\mathscr{U}: \mathscr{U} S^{k_1}_{\mu_1} \mathscr{U}^{-1} = \sum_{\mu_1'} D^{k_1}_{\mu_1' \mu_1}(U) S^{k_1}_{\mu_1'},$$

$$\mathscr{U} T^{k_2}_{\mu_2} \mathscr{U}^{-1} = \sum_{\mu_2'} D^{k_2}_{\mu_2' \mu_2}(U) T^{k_2}_{\mu_2'}.$$

Hence, the action of $\mathscr{U}$ on the product $S^{k_1}_{\mu_1} T^{k_2}_{\mu_2}$ is

$$\begin{aligned} \mathscr{U}: \mathscr{U} S^{k_1}_{\mu_1} T^{k_2}_{\mu_2} \mathscr{U}^{-1} &= (\mathscr{U} S^{k_1}_{\mu_1} \mathscr{U}^{-1})(\mathscr{U} T^{k_2}_{\mu_2} \mathscr{U}^{-1}) \\ &= \sum_{\mu_1' \mu_2'} D^{k_1}_{\mu_1' \mu_1}(U) D^{k_2}_{\mu_2' \mu_2}(U) S^{k_1}_{\mu_1'} T^{k_2}_{\mu_2'}. \end{aligned}$$

Using this result and Eq. (3.188), we find

$$\mathscr{U}[\mathbf{S}^{k_1} \mathbf{x} \mathbf{T}^{k_2}]^k_\mu \mathscr{U}^{-1} = \sum_{\mu'} D^k_{\mu' \mu}(U) [\mathbf{S}^{k_1} \mathbf{x} \mathbf{T}^{k_2}]^k_{\mu'}.$$

This result shows that $[\mathbf{S}^{k_1} \mathbf{x} \mathbf{T}^{k_2}]^k$ is also an irreducible tensor of rank k with respect to the angular momentum $\mathbf{J}$. ■

Remarks. (*a*) Note that the coupling is entirely kinematic in its properties; that is, the space on which the component operators act is not directly relevant to the coupling; the commutation properties of the tensor operators $S^{k_1}_{\mu_1}$ and $T^{k_2}_{\mu_2}$ with each other does not affect the coupling; and the order in which the component operators act is preserved in the coupling.

(*b*) The possibility of *combining tensor operators to form new tensor operators* is a far-reaching conceptual extension, which shows that the underlying structure is a generalization of the ring structure familiar in algebra. [It is a generalization in that many distinct products are defined – see Remark (*c*).] Tensor operators, as linear operators, may be added as well as multiplied by numerical constants and invariant operators; the resulting structure obeys the necessary distributive laws (see RWA).

(*c*) Using the coupling structure, one can effect several distinct types of multiplication of tensor operators: A scalar product, coupling two tensor operators into a rotational invariant, is effected by using the Wigner coefficients

$$C^{jk0}_{m\mu 0} = \delta_{jk}\delta_{m,-\mu}\frac{(-1)^{j-m}}{\sqrt{2j+1}}. \tag{3.234}$$

(This scalar coupling has all the properties of a *metric*, as is discussed in Section 20.) A coupling into a vector is effected by using the Wigner coefficients, $C^{jk1}_{m,\mu,m+\mu}$; clearly, further tensorial couplings of higher rank may be effected.

(*d*) A basic property, inherited from the Hilbert space structure, is that the tensor coupling operation is *associative*. This property is fundamental: It implies the important B–E identity (see Section 18), a result that is proved in RWA.

17. Applications of the Wigner–Eckart Theorem

General coupling law for reduced matrix elements. Let us apply the Wigner–Eckart theorem to the coupled tensor operator $[\mathbf{S}^{k_1}\mathbf{x}\,\mathbf{T}^{k_2}]^k$:

$$\begin{aligned}&\langle(\alpha')j'm'|[\mathbf{S}^{k_1}\mathbf{x}\,\mathbf{T}^{k_2}]^k_\mu|(\alpha)jm\rangle\\ &\qquad = \langle(\alpha')j'\|[\mathbf{S}^{k_1}\mathbf{x}\,\mathbf{T}^{k_2}]^k\|(\alpha)j\rangle C^{jkj'}_{m\mu m'},\end{aligned} \tag{3.235}$$

where the reduced matrix element is determined from Eqs. (3.227) and (3.233) to be[1]

[1] The reduced matrix element is obtained by setting $j''' = j'$ in Eq. (3.236). It is useful, however, to incorporate the Kronecker delta factor, coming from Eq. (3.228), into this result in order to exhibit correctly the coupling to zero of the Wigner coefficients themselves [see Eq. (3.240)].

$$\delta_{j'j'''}\langle(\alpha')j'\|[\mathbf{S}^{k_1}\times\mathbf{T}^{k_2}]^{k}\|(\alpha)j\rangle$$

$$= \sum_{\mu_1\mu_2\mu m} C^{k_1k_2k}_{\mu_1\mu_2\mu} C^{jkj'''}_{m\mu m'} \langle(\alpha')j'm'|S^{k_1}_{\mu_1} T^{k_2}_{\mu_2}|(\alpha)jm\rangle, \qquad (3.236)$$

where m' $(-j' \leqslant m' \leqslant j')$ is arbitrary.

The right-hand side of Eq. (3.236) may be simplified still further, since $\mathbf{S}^{k_1}$ and $\mathbf{T}^{k_2}$ are irreducible tensor operators with respect to $\mathbf{J}$. Hence, the Wigner–Eckart theorem applies, yielding the following result:

$$\langle(\alpha')j'm'|S^{k_1}_{\mu_1}|(\alpha)jm\rangle = \langle(\alpha')j'\|\mathbf{S}^{k_1}\|(\alpha)j\rangle C^{jk_1j'}_{m\mu_1m'}, \qquad (3.237)$$

$$\langle(\alpha')j'm'|T^{k_2}_{\mu_2}|(\alpha)jm\rangle = \langle(\alpha')j'\|\mathbf{T}^{k_2}\|(\alpha)j\rangle C^{jk_2j'}_{m\mu_2m'}. \qquad (3.238)$$

Supplying the intermediate states $|(\alpha'')j''m''\rangle$ in the expansion of the matrix element

$$\langle(\alpha')j'm'|S^{k_1}_{\mu_1} T^{k_2}_{\mu_2}|(\alpha)jm\rangle,$$

and using Eqs. (3.237) and (3.238), we bring the reduced matrix element (3.236) to the form

$$\langle(\alpha')j'\|[\mathbf{S}^{k_1}\times\mathbf{T}^{k_2}]^{k}\|(\alpha)j\rangle$$

$$= (-1)^{k_1+k_2-k} \sum_{(\alpha'')j''} [(2j''+1)(2k+1)]^{\frac{1}{2}} W(j'k_1jk_2;j''k)$$

$$\times \langle(\alpha')j'\|\mathbf{S}^{k_1}\|(\alpha'')j''\rangle\langle(\alpha'')j''\|\mathbf{T}^{k_2}\|(\alpha)j\rangle, \qquad (3.239)$$

where the coefficients $W(j'k_1jk_2;j''k)$ are defined by a sum over four Wigner coefficients:

$$(-1)^{k_1+k_2-k}\delta_{j'j'''}[(2j''+1)(2k+1)]^{\frac{1}{2}} W(j'k_1jk_2;j''k)$$

$$\equiv \sum_{\mu_1\mu_2\mu m m''} C^{k_1k_2k}_{\mu_1\mu_2\mu} C^{jkj'''}_{m\mu m'} C^{j''k_1j'}_{m''\mu_1m'} C^{jk_2j''}_{m\mu_2m''}. \qquad (3.240)$$

[m' may be chosen arbitrarily $(-j \leqslant m' \leqslant j)$ in this result.]

The coefficients $W(j'k_1jk_2;j''k)$ were introduced into angular momentum theory by Racah [29] and are known as *Racah coefficients*. Similar coefficients, called 6-j coefficients, were also introduced by Wigner [47]. The relation between the Wigner 6-j coefficients (the curly bracket symbol) and

Racah coefficients is

$$\begin{Bmatrix} j' & k_1 & j'' \\ k_2 & j & k \end{Bmatrix} = (-1)^{j'+k_1+j+k_2} W(j'k_1jk_2;j''k). \tag{3.241}$$

Equation (3.239) expresses the general coupling law for reduced matrix elements. Its form should be compared with the coupling law, Eq. (3.233), for the tensor operators themselves. In the latter, it is the Wigner coefficients that are used to couple two tensor operators to form a third; in the former, the coupling law for reduced matrix elements, it is the Racah coefficients that play the role of coupling coefficients.

Remarks. (*a*) That the Racah coefficients themselves, rather than the Wigner coefficients, may be considered as the basic entities of angular momentum theory is not apparent from their definition (3.240) in terms of Wigner coefficients. (This viewpoint will be clarified in Section 18; it is developed in detail in RWA.) It is this result – that the Racah coefficients exist "on their own" – that elevates the coupling law (3.239) for reduced matrix elements to a principal theorem in angular momentum theory. The special consequences of Eq. (3.239) as developed in the remainder of this section are the main tools of spectroscopy. (*b*) Historically, Racah [29; II] introduced the coefficients $[(2j'' + 1)(2k + 1)]^{\frac{1}{2}} W(j'k_1jk_2;j''k)$, through Eq. (3.240), for the expression of the reduced matrix elements of an invariant interaction in a coupled angular momentum basis [Eq. (3.260) below]. He subsequently (Racah [29; III]) recognized that these coefficients were also the transformation coefficients between different coupling schemes for the addition of three angular momenta, a subject that was developed much further in Ref. [54]. (This important subject is developed in detail in RWA.) (*c*) Wigner [47] arrived at a definition of the 6-*j* symbol [equivalent to Eq. (3.240)] by considering the consequences of the associative law of multiplication of three representation matrices [Eq. (3.189)]. He then showed how the 6-*j* symbols were related to group integrals (see Note 11), thus providing yet another method of "freeing the definition of a Racah coefficient from that of a Wigner coefficient."

The Wigner–Eckart theorem, the coupling law [Eq. (3.233)] for tensor operators, and the coupling law [Eq. (3.239)] for reduced matrix elements represent the three major properties of tensor operators (and find many applications in physical problems). Accordingly, the study and interpretation of these results becomes an important task. We summarize in the next section, Section 18, the more useful properties of the Racah coefficients, but defer to RWA the deeper and more difficult problem of interpreting these three principal tensor operator properties.

We emphasize again the generality of the coupling law, Eq. (3.239): *It is the most general form for the coupling of the reduced matrix elements of two tensor operators with respect to an angular momentum* **J** *having eigenkets* $\{|(\alpha)jm\rangle\}$.

Particular cases of Eq. (3.239) are sufficiently important to be noted explicitly. We therefore now specialize the general coupling law, Eq. (3.239), to a useful particular case.

Coupling laws for reduced matrix elements of tensor operators acting in kinematically independent spaces. A particular case of the coupling law (3.239) is obtained by choosing the tensor operators to have the tensor product forms

$$S^{k_1}_{\mu_1} = T^{k_1}_{\mu_1}(1) \otimes \mathbb{1}(2),$$

$$T^{k_2}_{\mu_2} = \mathbb{1}(1) \otimes T^{k_2}_{\mu_2}(2), \tag{3.242}$$

where $\mathbb{1}(i)$ denotes the unit operator and $\mathbf{T}^{k_i}(i)$ $(i = 1, 2)$ is an irreducible tensor operator with respect to the angular momentum $\mathbf{J}(i)$ of part i of a physical system composed of two parts $(i = 1, 2)$, and thus having total angular momentum $\mathbf{J} = \mathbf{J}(1) + \mathbf{J}(2)$.

One then verifies that $\mathbf{S}^{k_1}$ and $\mathbf{T}^{k_2}$ are tensor operators with respect to **J**.

The eigenkets $|(\alpha)jm\rangle$ appearing in the Wigner–Eckart theorem, Eq. (3.235), become the coupled angular momentum eigenkets,

$$|(\alpha)jm\rangle = |(\alpha_1\alpha_2 j_1 j_2)jm\rangle$$

$$\equiv \sum_{m_1 m_2} C^{j_1 j_2 j}_{m_1 m_2 m} |(\alpha_1)j_1 m_1\rangle \otimes |(\alpha_2)j_2 m_2\rangle, \tag{3.243}$$

where

$$(\alpha) = (\alpha_1\alpha_2 j_1 j_2). \tag{3.244}$$

Our problem now is to express the reduced matrix elements

$$\langle(\alpha_1'\alpha_2' j_1' j_2')j'\|\mathbf{T}^{k_1}(1) \otimes \mathbb{1}(2)\|(\alpha_1\alpha_2 j_1 j_2)j\rangle$$

occurring in Eq. (3.239) in terms of the reduced matrix elements

$$\langle(\alpha_1')j_1'\|\mathbf{T}^{k_1}(1)\|(\alpha_1)j_1\rangle$$

occurring in the Wigner–Eckart theorem as applied to the tensor operator $\mathbf{T}^{k_1}(1)$ [with respect to $\mathbf{J}(1)$]:

$$\langle(\alpha_1')j_1'm_1'|T^{k_1}_{\mu_1}(1)|(\alpha_1)j_1m_1\rangle$$
$$= \langle(\alpha_1')j_1'\|\mathbf{T}^{k_1}(1)\|(\alpha_1)j_1\rangle C^{j_1k_1j_1'}_{m_1\mu_1m_1'}. \tag{3.245}$$

[We must also solve a similar problem for $\mathbb{1}(1) \otimes \mathbf{T}^{k_2}(2)$.]

We assert (and prove below) that

$$\langle(\alpha_1'\alpha_2'j_1'j_2')j'\|\mathbf{T}^{k_1}(1) \otimes \mathbb{1}(2)\|(\alpha_1\alpha_2j_1j_2)j\rangle$$
$$= (-1)^{j_1'+k_1+j_2+j}[(2j_1'+1)(2j+1)]^{\frac{1}{2}}\begin{Bmatrix} j' & j_2 & j_1' \\ j_1 & k_1 & j \end{Bmatrix}$$
$$\times\ \delta_{\alpha_2'\alpha_2}\delta_{j_2'j_2}\langle(\alpha_1')j_1'\|\mathbf{T}^{k_1}(1)\|(\alpha_1)j_1\rangle;$$

$$\langle(\alpha_1'\alpha_2'j_1'j_2')j'\|\mathbb{1}(1) \otimes \mathbf{T}^{k_2}(2)\|(\alpha_1\alpha_2j_1j_2)j\rangle$$
$$= (-1)^{j'+k_2+j_1+j_2}[(2j_2'+1)(2j+1)]^{\frac{1}{2}}\begin{Bmatrix} j' & j_1 & j_2' \\ j_2 & k_2 & j \end{Bmatrix}$$
$$\times\ \delta_{\alpha_1'\alpha_1}\delta_{j_1'j_1}\langle(\alpha_2')j_2'\|\mathbf{T}^{k_2}(2)\|(\alpha_2)j_2\rangle. \tag{3.246}$$

Proof. Consider the proof of the first relation. The reduced matrix element in question is obtained from Eq. (3.227′) to be

$$\langle(\alpha_1'\alpha_2'j_1'j_2')j'\|\mathbf{T}^{k_1}(1) \otimes \mathbb{1}(2)\|(\alpha_1\alpha_2j_1j_2)j\rangle$$
$$= \sum_{m\mu_1} C^{jk_1j'}_{m\mu_1m'}\langle(\alpha_1'\alpha_2'j_1'j_2')j'm'|T^{k_1}_{\mu_1}(1) \otimes \mathbb{1}(2)|(\alpha_1\alpha_2j_1j_2)jm\rangle$$
$$= \delta_{\alpha_2'\alpha_2}\delta_{j_2'j_2}\sum_{m_1m_2m\mu_1} C^{j_1j_2j}_{m_1m_2m}C^{jk_1j'}_{m\mu_1m'}C^{j_1'\,j_2\,j'}_{m'-m_2,m_2,m'}$$
$$\times\ \langle(\alpha_1')j_1'm_1'|T^{k_1}_{\mu_1}(1)|(\alpha_1)j_1m_1\rangle$$
$$= \delta_{\alpha_2'\alpha_2}\delta_{j_2'j_2}\sum_{m_1m_2m\mu_1} C^{j_1j_2j}_{m_1m_2m}C^{jk_1j'}_{m\mu_1m'}C^{j_1'\,j_2\,j'}_{m'-m_2,m_2,m'}C^{j_1k_1j_1'}_{m_1\mu_1m_1'}$$
$$\times\ \langle(\alpha_1')j_1'\|\mathbf{T}^{k_1}(1)\|(\alpha_1)j_1\rangle. \tag{3.247}$$

In obtaining this result, we have used the coupled eigenkets, Eq. (3.243), and the matrix elements given by Eq. (3.245). Using the second symmetry relation in Eqs. (3.180), the summation over Wigner coefficients in Eq. (3.247) is found from Eq. (3.240) to be

$$(-1)^{j_1'-j'+j-j_1}[(2j_1'+1)(2j+1)]^{\frac{1}{2}}W(j'j_2k_1j_1;j_1'j).$$

Expressing the Racah coefficient in terms of the 6-j coefficient [Eq. (3.241)], we obtain the desired result. A similar procedure establishes the second of Eqs. (3.246). ■

The next general relation is obtained by substituting the reduced matrix elements given by Eqs. (3.246) into the general formula (3.239). Writing the final result in terms of 6-j coefficients, we obtain the coupling law for reduced matrix elements of tensor operators acting in independent spaces:

$$\langle(\alpha_1'\alpha_2'j_1'j_2')j'\|[\mathbf{T}^{k_1}(1)\times\mathbf{T}^{k_2}(2)]^k\|(\alpha_1\alpha_2j_1j_2)j\rangle$$

$$=\begin{bmatrix} j_1 & j_2 & j \\ k_1 & k_2 & k \\ j_1' & j_2' & j' \end{bmatrix}\langle(\alpha_1')j_1'\|\mathbf{T}^{k_1}(1)\|(\alpha_1)j_1\rangle\langle(\alpha_2')j_2'\|\mathbf{T}^{k_2}(2)\|(\alpha_2)j_2\rangle, \tag{3.248}$$

where

$$[\mathbf{T}^{k_1}(1)\times\mathbf{T}^{k_2}(2)]_\mu^k \equiv \sum_{\mu_1\mu_2} C_{\mu_1\mu_2\mu}^{k_1k_2k}T_{\mu_1}^{k_1}(1)\otimes T_{\mu_2}^{k_2}(2), \tag{3.249}$$

$$\begin{bmatrix} j_1 & j_2 & j \\ k_1 & k_2 & k \\ j_1' & j_2' & j' \end{bmatrix} \equiv [(2j_1'+1)(2j_2'+1)(2j+1)(2k+1)]^{\frac{1}{2}}\begin{Bmatrix} j_1 & j_2 & j \\ k_1 & k_2 & k \\ j_1' & j_2' & j' \end{Bmatrix}, \tag{3.250}$$

$$\begin{Bmatrix} j_1 & j_2 & j \\ k_1 & k_2 & k \\ j_1' & j_2' & j' \end{Bmatrix} \equiv (-1)^{j_1+j_2+j+k_1+k_2+k+j_1'+j_2'+j'}\sum_{j''}(-1)^{2j''}(2j''+1)$$

$$\times\begin{Bmatrix} j' & k_1 & j'' \\ k_2 & j & k \end{Bmatrix}\begin{Bmatrix} j' & j_2' & j_1' \\ j_1 & k_1 & j'' \end{Bmatrix}\begin{Bmatrix} j'' & j_1 & j_2' \\ j_2 & k_2 & j \end{Bmatrix}. \tag{3.251}$$

Equation (3.251) defines the Wigner 9-j coefficients, the properties of which are summarized in Section 19. We note here a relation of the 9-j coefficients to the Wigner coefficients themselves, since this result also is derived directly from Eqs. (3.235) and (3.249) and the definitions of coupled states and coupled tensors:

$$\sum_{m_1'm_2'm_1m_2\mu_1\mu_2} C_{m_1'm_2'm'}^{j_1'j_2'j'}C_{m_1m_2m}^{j_1j_2j}C_{\mu_1\mu_2\mu}^{k_1k_2k}C_{m_1\mu_1m_1'}^{j_1k_1j_1'}C_{m_2\mu_2m_2'}^{j_2k_2j_2'}$$

$$=\begin{bmatrix} j_1 & j_2 & j \\ k_1 & k_2 & k \\ j_1' & j_2' & j' \end{bmatrix}C_{m\mu m'}^{jkj'}. \tag{3.252}$$

The coupling law Eq. (3.248) is often derived in the literature independently of the coupling law Eq. (3.239). The derivation given here has emphasized the greater generality of the latter coupling law.

Remark. The reduced matrix elements given by Eqs. (3.246) are regained from the result, Eq. (3.248), upon choosing $T^{k_2}_{\mu_2}(2) = T^0_0(2) = \mathbb{1}(2)$ and $T^{k_1}_{\mu_1}(1) = T^0_0(1) = \mathbb{1}(1)$.

In addition to Eqs. (3.246), there is another special case of Eq. (3.248), which we note explicitly. This is the case $k = 0$. This is the coupling of two tensor operators to a rotational invariant. Explicitly, this invariant is

$$[\mathbf{S}^{k_1} \times \mathbf{T}^{k_2}]^0_0 = \delta_{k_1k_2}[\mathbf{S}^{k_1} \times \mathbf{T}^{k_1}]^0_0, \tag{3.253}$$

where, using a change of notation and the special Wigner coefficient (3.234), we find

$$[\mathbf{S}^k \times \mathbf{T}^k]^0_0 = \sum_\mu \frac{(-1)^{k-\mu}}{(2k+1)^{\frac{1}{2}}} S^k_\mu T^k_{-\mu}. \tag{3.254}$$

In the case represented by Eqs. (3.242), the results above are denoted by

$$[\mathbf{T}^k(1) \times \mathbf{T}^k(2)]^0_0 = \sum_\mu \frac{(-1)^{k-\mu}}{(2k+1)^{\frac{1}{2}}} T^k_\mu(1) \otimes T^k_{-\mu}(2)$$

$$= \sum_\mu \frac{(-1)^{k-\mu}}{(2k+1)^{\frac{1}{2}}} T^k_\mu(1) T^k_{-\mu}(2). \tag{3.255}$$

Specializing Eq. (3.235) to the invariant coupling yields

$$\langle(\alpha')j'm'|[\mathbf{S}^k \times \mathbf{T}^k]^0_0|(\alpha)jm\rangle = \delta_{j'j}\delta_{m'm}\langle(\alpha')j\|[\mathbf{S}^k \times \mathbf{T}^k]^0\|(\alpha)j\rangle, \tag{3.256}$$

where

$$\langle(\alpha')j\|[\mathbf{S}^k \times \mathbf{T}^k]^0\|(\alpha)j\rangle = \sum_{(\alpha'')j''} (-1)^{k+j''-j}\left[\frac{(2j''+1)}{(2k+1)(2j+1)}\right]^{\frac{1}{2}}$$

$$\times\, \varepsilon_{kjj''}\langle(\alpha')j\|\mathbf{S}^k\|(\alpha'')j''\rangle\langle(\alpha'')j''\|\mathbf{T}^k\|(\alpha)j\rangle. \tag{3.257}$$

[We have used the following special Racah coefficient in obtaining Eq. (3.257):

$$W(j'k_1jk_2;j''0) = (-1)^{j'+k_1+j+k_2}\begin{Bmatrix} j' & k_1 & j'' \\ k_2 & j & 0 \end{Bmatrix}$$

$$= \delta_{j'j}\delta_{k_1k_2}\varepsilon_{k_1jj''}\frac{(-1)^{k_1+j-j''}}{[(2k_1+1)(2j+1)]^{\frac{1}{2}}}. \tag{3.258}$$

This result itself is obtained directly from the definition, Eq. (3.240), using the coefficient (3.234), and the symmetry and orthogonality relations for the Wigner coefficients.]

The further specialization of Eq. (3.257) to tensor operators of the form given by Eq. (3.242) leads to the result

$$\langle(\alpha_1'\alpha_2'j_1'j_2')j'm'|[\mathbf{T}^k(1)\times\mathbf{T}^k(2)]_0^0|(\alpha_1\alpha_2j_1j_2)jm\rangle$$

$$= \delta_{j'j}\delta_{m'm}\langle(\alpha_1'\alpha_2'j_1'j_2')j\|[\mathbf{T}^k(1)\times\mathbf{T}^k(2)]^0\|(\alpha_1\alpha_2j_1j_2)j\rangle, \tag{3.259}$$

where the reduced matrix is obtained from Eq. (3.248):

$$\langle(\alpha_1'\alpha_2'j_1'j_2')j\|[\mathbf{T}^k(1)\times\mathbf{T}^k(2)]^0\|(\alpha_1\alpha_2j_1j_2)j\rangle$$

$$= (-1)^{j_2'+k+j_1+j}\left[\frac{(2j_1'+1)(2j_2'+1)}{2k+1}\right]^{\frac{1}{2}}\begin{Bmatrix} j_1 & j_2 & j \\ j_2' & j_1' & k \end{Bmatrix}$$

$$\times\ \langle(\alpha_1')j_1'\|\mathbf{T}^k(1)\|(\alpha_1)j_1\rangle\langle(\alpha_2')j_2'\|\mathbf{T}^k(2)\|(\alpha_2)j_2\rangle. \tag{3.260}$$

In order to present this result in final form, we have used a relation between 9-j coefficients and 6-j coefficients, which is proved in Section 19 [Eq. (3.325)]:

$$\begin{Bmatrix} j_1 & j_2 & j \\ k_1 & k_2 & 0 \\ j_1' & j_2' & j' \end{Bmatrix} = \delta_{k_1k_2}\delta_{jj'}\frac{(-1)^{j_1+j_2'+j+k_1}}{[(2j+1)(2k_1+1)]^{\frac{1}{2}}}\begin{Bmatrix} j_1 & j_2 & j \\ j_2' & j_1' & k_1 \end{Bmatrix}. \tag{3.261}$$

Summary. We have obtained in this section two of the principal coupling laws for the reduced matrix elements of tensor operators, Eqs. (3.239) and (3.248) (for further constructions of this type, see Jucys *et al.* [55]; see also Note 10 for an application of these coupling laws to the spherical harmonics, and Table 8 in the Appendix of Tables for a summary of results).

18. Racah Coefficients

We have introduced the Racah coefficients through their role in tensor operator theory. The coefficients have appeared in three related contexts:

(*a*) as factors in the matrix elements of the tensor product operator

$$[\mathbf{S}^{k_1} \times \mathbf{T}^{k_2}]^k \tag{3.262}$$

taken in the space $\mathscr{H}_j$; (*b*) as coupling coefficients in the space of reduced matrix elements; and (*c*) as matrix elements of the invariant operator

$$[\mathbf{T}^k(1) \times \mathbf{T}^k(2)]^0 \tag{3.263}$$

taken in the tensor product space $\mathscr{H}_{j_1}(1) \otimes \mathscr{H}_{j_2}(2)$.

From the point of view of physics, Racah coefficients enter in still another significant way: They are the recoupling coefficients for the addition of three angular momenta (Biedenharn *et al.* [54]), and, more important, they constitute a "basis" for all recoupling coefficients in the addition of an arbitrary number of angular momenta (this is developed in RWA).

From the point of view of mathematics, the study of Racah coefficients is interesting for several reasons: (*a*) They are basis-independent objects (independent of the projection quantum numbers); (*b*) the Wigner coefficients themselves may be obtained as the limit of the Racah coefficients [a result proved below – see Eq. (3.300)]; (*c*) they occur as generalized hypergeometric series; (*d*) they are related to projective geometry; and (*e*) they define an algebra. [Topics (*c*), (*d*), and (*e*) are developed in RWA.]

It is thus apparent that Racah coefficients merit study on their own, quite aside from the definition in terms of Wigner coefficients as given by Eq. (3.240). Indeed, one may adopt the viewpoint that it is the Racah coefficients that are the basic objects to be studied in angular momentum theory, since all other "angular momentum entities" may be obtained from them.

It is essential toward implementing this viewpoint to present the properties of the Racah coefficients in a form that "frees" them from their definition in terms of Wigner coefficients. We initiate this program in this section, but in order to make the machinery of using Racah coefficients accessible to a larger audience, we shall state, without always giving the proof, many of their properties, leaving the sometimes technical proofs to RWA.

Notation and domain of definition. The Racah coefficient[1] $W(abcd;ef)$ is defined for all values of a, b, c, d, e, f that belong to the set $\{0, \frac{1}{2}, 1, \frac{3}{2}, \ldots\}$. However, in consequence of Eq. (3.240), expressing the Racah coefficients in terms of Wigner coefficients, and of the extended definition (3.191) of the latter, we have

[1] We follow Racah [29] in introducing the Roman letters $a, b, c, \ldots$ to denote angular momenta and Greek letters $\alpha, \beta, \gamma, \ldots$ to denote their projections.

$W(abcd;ef) = 0$ unless the triples of nonnegative integers and half-integers (abe), (cde), (acf), (bdf) satisfy the triangle conditions. (3.264)

[Observe that, despite the apparent asymmetry in the triangle conditions on (abc) given by

$$c \in \{a+b, a+b-1, \ldots, |a-b|\}, \qquad (3.265)$$

these conditions are, in fact, symmetric in a, b, c in the sense that, if conditions (3.265) hold, then also the triangle conditions are valid for all permutations of a, b, c.]

Relations between Racah coefficients and Wigner coefficients. Various other forms of the defining relation, Eq. (3.240), are useful both for determining algebraic forms for the Racah coefficients and in applications. We give below four basic relations (which are interpreted algebraically in RWA), after first rewriting the defining relation in a form that agrees with Racah's expression. [This redefinition uses the symmetry relation $W(abcd;ef) = W(cdab;ef)$, which is easily proved directly from Eq. (3.240) when this latter relation is summed over γ.]:

$$\begin{aligned}&\delta_{cc'}[(2e+1)(2f+1)]^{\frac{1}{2}} W(abcd;ef)\\ &\quad = \sum_{\beta\delta} C^{bdf}_{\beta,\delta,\beta+\delta}\, C^{edc}_{\gamma-\delta,\delta,\gamma}\, C^{abe}_{\gamma-\beta-\delta,\beta,\gamma-\delta}\, C^{afc'}_{\gamma-\beta-\delta,\beta+\delta,\gamma},\end{aligned} \qquad (3.266)$$

$$\begin{aligned}&\sum_{\beta\delta} C^{bdf}_{\beta\delta\gamma}\, C^{edc}_{\alpha+\beta,\delta,\alpha+\gamma}\, C^{abe}_{\alpha,\beta,\alpha+\beta}\\ &\quad = [(2e+1)(2f+1)]^{\frac{1}{2}} W(abcd;ef)\, C^{afc}_{\alpha,\gamma,\alpha+\gamma},\end{aligned} \qquad (3.267)$$

$$\begin{aligned}&\sum_{\beta\delta e} [(2e+1)(2f+1)]^{\frac{1}{2}} W(abcd;ef)\\ &\quad \times C^{bdf'}_{\beta\delta\gamma}\, C^{edc}_{\alpha+\beta,\delta,\alpha+\gamma}\, C^{abe}_{\alpha,\beta,\alpha+\beta} = \delta_{ff'}\, C^{afc}_{\alpha,\gamma,\alpha+\gamma},\end{aligned} \qquad (3.268)$$

$$\begin{aligned}&\sum_{f} [(2e+1)(2f+1)]^{\frac{1}{2}} W(abcd;ef)\, C^{bdf}_{\beta,\delta,\beta+\delta}\, C^{afc}_{\alpha,\beta+\delta,\alpha+\beta+\delta}\\ &\quad = C^{edc}_{\alpha+\beta,\delta,\alpha+\beta+\delta}\, C^{abe}_{\alpha,\beta,\alpha+\beta},\end{aligned} \qquad (3.269)$$

$$\begin{aligned}&\sum_{e} [(2e+1)(2f+1)]^{\frac{1}{2}} W(abcd;ef)\, C^{edc}_{\alpha+\beta,\delta,\alpha+\beta+\delta}\, C^{abe}_{\alpha,\beta,\alpha+\beta}\\ &\quad = C^{bdf}_{\beta,\delta,\beta+\delta}\, C^{afc}_{\alpha,\beta+\delta,\alpha+\beta+\delta}.\end{aligned} \qquad (3.270)$$

[Let us note that the index γ in Eq. (3.266) can be chosen arbitrarily ($-c \leqslant \gamma \leqslant c$), as noted earlier in Eq. (3.240).]

Let us sketch a proof of these results. To obtain Eq. (3.267), we replace γ by $\gamma + \alpha$ in the defining relation, Eq. (3.266), multiply by $C^{afc'}_{\alpha,\gamma,\alpha+\gamma}$ and sum over c', and use orthogonality relation (3.179). The result is Eq. (3.267).

Equation (3.267) now becomes the key relation for proving the orthogonality relation for Racah coefficients:

$$\sum_f (2e+1)(2f+1)W(abcd;ef)W(abcd;e'f) = \delta_{ee'}\varepsilon_{abe}\varepsilon_{cde}, \tag{3.271}$$

where[1]

$$\varepsilon_{abc} = \begin{cases} 1 & \text{for all triples } a, b, c \in \{0, \tfrac{1}{2}, \ldots\} \text{ that} \\ & \text{satisfy the triangle conditions,} \\ 0 & \text{otherwise.} \end{cases} \tag{3.272}$$

To obtain this important orthogonality relation, we first replace γ by $\gamma - \alpha$ in Eq. (3.267) and write out a corresponding equation with e replaced by e'. We then multiply together the two equations and sum first on α, then on f, using the orthogonality relations for Wigner coefficients. The result of these operations is Eq. (3.271).

The orthogonality relation (3.271) may also be written in the form

$$\sum_e (2e+1)(2f+1)W(abcd;ef)W(abcd;ef') = \delta_{ff'}\varepsilon_{acf}\varepsilon_{bdf} \tag{3.273}$$

upon noting the symmetry relation

$$W(abcd;ef) = W(acbd;fe).$$

This relation itself is a consequence of the definition, Eq. (3.266), and the symmetry of the Wigner coefficients [Eqs. (3.180)]. (A systematic discussion of the symmetries of the Racah coefficients is given below.)

The orthogonality relations (3.271) and (3.273) of the Racah coefficients, together with those for the Wigner coefficients, Eq. (3.175) and (3.176), may now be used to prove the remaining relations, Eq. (3.268)–(3.270).

There are three more basic relations satisfied by the Racah coefficients, not involving Wigner coefficients. These relations will next be stated with a brief description of their significance.

---

[1] The introduction of this symbol into the right-hand side of Eq. (3.271) is necessary so that one obtains $0 = 0$ when the left-hand side vanishes [see Eq. (3.264)].

The Racah sum rule. Racah [29] gave the following relation between the coefficients $W(abcd;ef)$:

$$\sum_f (-1)^{b+d-f}(2f+1)W(abcd;ef)W(adcb;gf) = (-1)^{e+g-a-c}W(bacd;eg). \tag{3.274}$$

Racah's original derivation of this result uses (*a*) the definition of the W-coefficient in terms of Wigner coefficients; (*b*) the symmetry relation $C^{abc}_{\alpha\beta\gamma} = (-1)^{a+b-c} C^{bac}_{\beta\alpha\gamma}$ of the Wigner coefficients; and (*c*) the orthogonality relations of the Wigner coefficients.

Racah's sum rule[1] expresses a fundamental property of a mapping diagram associated with the transformations between sets of basis vectors that occur in the coupling of three angular momenta–, namely, that the diagram is commutative. This somewhat technical result is established in detail in Topic 12, RWA.

Biedenharn–Elliott (B–E) identity. Using recoupling theory for four angular momenta, Biedenharn [56] and Elliott [57] derived the following identity between Racah coefficients:

$$W(a'ab'b;c'e)W(a'ed'd;b'c)$$
$$= \sum_f (2f+1)W(abcd;ef)W(c'bd'd;b'f)W(a'ad'f;c'c). \tag{3.275}$$

We give in RWA two derivations of this result, (*a*) one using recoupling theory, and (*b*) the other showing that *this result is a consequence of the associative law of multiplication for unit tensor operators*. (This latter result shows that it is the B–E identity that has the significance for the "algebra of tensor operators" that the Jacobi identity has for the structure constants of a Lie algebra.)

We shall spend a great deal of effort in interpreting Eq. (3.275). *It is the key relationship for elevating the study of Racah coefficients to a position that is independent of the concept of a Wigner coefficient.*

The triangle sum rule. For each triple (*abc*) obeying the triangle conditions, we define a *triangle coefficient*:

$$\Delta(abc) = \left[\frac{(a+b-c)!(a-b+c)!(-a+b+c)!}{(a+b+c+1)!}\right]^{\frac{1}{2}}. \tag{3.276}$$

[1] Rose and Yang [55a] interpreted the Racah sum rule as an eigenvalue equation and made an interesting application of this interpretation to the Pomeranchuk relation for asymptotic cross-sections.

The triangle coefficients and Racah coefficients satisfy the following relation:

$$[\Delta(acf)\Delta(bdf)]^{-1} = (2f+1)\sum_e [\Delta(abe)\Delta(cde)]^{-1} W(abcd;ef). \tag{3.277}$$

This result (Biedenharn and Louck [58]) is a consequence of the associative law of multiplication of *symplecton polynomials*, as discussed in RWA.

Alternative notation for Racah coefficients. We introduce here an alternative notation for the Racah coefficients.[1] [Our purpose in introducing this (nonconventional) notation here is to make available to the reader who does not proceed to RWA the close relationships that exist between Racah and Wigner coefficients.] We introduce the symbol,

$$W^{abc}_{\rho\sigma\tau}(j),$$

which is defined for all $j = 0, \frac{1}{2}, 1, \ldots$; for all integer and half-integer triples of nonnegative numbers (abc) that satisfy the triangle conditions $|a-b| \leqslant c \leqslant a+b$; and for all $\rho = a, a-1, \ldots, -a$; $\sigma = b, b-1, \ldots, -b$; $\tau = c, c-1, \ldots, -c$. The definition is

(*a*) $W^{abc}_{\rho\sigma\tau}(j) \equiv 0$ unless $\rho + \sigma = \tau$;

(*b*) $W^{abc}_{\rho\sigma\tau}(j) \equiv 0$ unless each of the triples $(j-\tau, a, j-\sigma)$, $(j-\sigma, b, j)$, $(j-\tau, c, j)$ consists of nonnegative integers and half-integers that satisfy the triangle conditions;

(*c*) $W^{abc}_{\rho\sigma\tau}(j) \equiv [(2c+1)(2j-2\sigma+1)]^{\frac{1}{2}} W(j-\tau, a, j, b; j-\sigma, c)$, otherwise. (3.278)

Since we also have the relation

$$[(2e+1)(2f+1)]^{\frac{1}{2}} W(abcd;ef) = W^{bdf}_{e-a,c-e,c-a}(c), \tag{3.279}$$

where $b \geqslant |e-a|$, $d \geqslant |c-e|$, and $f \geqslant |c-a|$, *each Racah coefficient is expressible in the new notation.*

We show later [Eq. (3.300)] that

$$\lim_{j\to\infty} W^{abc}_{\rho\sigma\tau}(j) = C^{abc}_{\rho\sigma\tau}. \tag{3.280}$$

[1] This notation is suggested naturally in the algebraic theory of Wigner operators developed in RWA.

It is this relation that provides the most persuasive reason for our choice of notation.

In terms of this notation, the orthogonality relations for the Racah coefficients take a form very similar to those for the Wigner coefficients:

$$\sum_{\rho\sigma} W^{abc}_{\rho\sigma\tau}(j) W^{abd}_{\rho\sigma\tau}(j) = \delta_{cd}\varepsilon_{j-\tau,j,c},$$

$$\sum_{c} W^{abc}_{\rho,\tau-\rho,\tau}(j) W^{abc}_{\rho',\tau-\rho',\tau}(j) = \delta_{\rho\rho'}\varepsilon_{j-\tau,a,j-\tau+\rho}\varepsilon_{j,b,j-\tau+\rho}. \tag{3.281}$$

[It is implicit in these relations that the triples (abc) and (abd) satisfy the triangle conditions.]

Explicit algebraic forms. The calculation of explicit forms of the Racah coefficients is quite difficult. Racah [29] obtained the general form by setting $m' = j'$ in Eq. (3.240) and reducing the resulting multiple summation to one over a single index by a series of "tricks." Schwinger [37] (and also Bargmann [39], using the same ideas) obtained the general form using a generating function (see Chapter 5, Appendix D). Biedenharn [56] gave recursion formulas for generating, by iteration, the Racah coefficients. Both Racah's and Schwinger's methods yield the same expression, but the general iteration of the recursion relations yields a different (and less symmetric) general expression. Thus, at present we can give no simple, elegant derivation (in contrast to the case for Wigner coefficients) of Racah's form of these coefficients within the framework of concepts developed so far in our presentation.

Having emphasized the general significance of the Racah coefficients for angular momentum theory, it is essential to our purpose to give a derivation of the Racah form. We have elected to follow Racah's original derivation, giving the details, however, in Appendix A. We state here the result of that derivation after noting some special cases, which are more easily found.

Up to phase, the result,

$$W(abc0;ef) = \frac{\delta_{bf}\delta_{ce}\varepsilon_{abe}}{[(2e+1)(2f+1)]^{\frac{1}{2}}}, \tag{3.282}$$

is determined by the triangle conditions, Eq. (3.264), and the orthogonality relations, Eqs. (3.271). The phase is fixed by the convention

$$W^{a0a}_{\rho 0 \rho}(j) = \varepsilon_{j-\rho,a,j}, \tag{3.283}$$

which is equivalent to the phase convention used for Wigner coefficients.

Further values of the Racah coefficients for particular arguments may be obtained in a variety of ways. For example, to evaluate the "stretched" coefficients, as these are called,[1] we may first choose $\gamma = f$ and $\alpha = c - f$ in Eq. (3.267) and obtain

$$[(2e+1)(2f+1)]^{\frac{1}{2}} W(abcd;ef) C^{a\,f\,c}_{c-f,f,c}$$
$$= \sum_{\delta} C^{b\,d\,f}_{f-\delta,\delta,f}\, C^{e\,d\,c}_{c-\delta,\delta,c}\, C^{a\,b\,e}_{c-f,f-\delta,c-\delta}. \tag{3.284}$$

If we now set $f = b + d$, then the summation over δ reduces to the single term $\delta = d$. Thus,

$$[(2e+1)(2b+2d+1)]^{\frac{1}{2}} W(abcd;e,b+d)\, C^{a\,b+d\,c}_{c-b-d,b+d,c}$$
$$= C^{b\,d\,b+d}_{b,d,b+d}\, C^{e\,d\,c}_{c-d,d,c}\, C^{a\,b\,e}_{c-b-d,b,c-d}. \tag{3.285}$$

Using the special Wigner coefficient

$$C^{a\,b\,c}_{\gamma-b,b,\gamma} = (-1)^{a+b-c} v(abc)[(2c+1)(2b)!]^{\frac{1}{2}}$$
$$\times \left[\frac{(a+b-\gamma)!(c+\gamma)!}{(a-b+\gamma)!(c-\gamma)!}\right]^{\frac{1}{2}}, \tag{3.286}$$

where

$$v(abc) = [(a-b+c)!/(a+b+c+1)!(a+b-c)!(-a+b+c)!]^{\frac{1}{2}}, \tag{3.287}$$

we obtain

$$W(abcd;e,b+d) = \left[\frac{(2b)!(2d)!}{(2b+2d+1)!}\right]^{\frac{1}{2}} \times \frac{v(abe)v(edc)}{v(a,b+d,c)}. \tag{3.288}$$

We now turn to the calculation of the general coefficient. We follow Racah's procedure and set $\gamma = c$ in Eq. (3.266). It is convenient to introduce the triangle coefficient $\Delta(abc)$ defined by Eq. (3.276), noting that

[1] Stretching means that one of the four angular momentum triangles degenerates to a line (see Fig. 3.4); this situation is also termed a "boundary" Racah coefficient from its position in a tabular array (see Table 2 in Appendix of Tables).

$$C^{abc}_{c-\beta,\beta,c} = (-1)^{a-c+\beta}\frac{\Delta(abc)}{(a-b+c)!(-a+b+c)!}$$

$$\times\left[\frac{(2c+1)!(a+c-\beta)!(b+\beta)!}{(a-c+\beta)!(b-\beta)!}\right]^{\frac{1}{2}}. \qquad (3.289)$$

Using next the general form, Eq. (3.170), for a Wigner coefficient, we obtain the following expression for a Racah coefficient:

$$W(abcd;ef) = \Delta(bdf)\Delta(edc)\Delta(abe)\Delta(afc)w(abcd;ef), \qquad (3.290)$$

where $w(abcd;ef)$ is defined by

$$w(abcd;ef) = \frac{(2c+1)!}{(a-f+c)!(-a+f+c)!(e-d+c)!(-e+d+c)!}$$

$$\times\sum_{\beta\delta rs}\frac{(-1)^{a-e+\beta+r+s}(a+c-\beta-\delta)!(f+\beta+\delta)!(e+c-\delta)!(d+\delta)!}{(e-b+c-\beta-\delta+r)!(a-c+\beta+\delta-r)!(d+\delta-s)!(f-b-\delta+s)!}$$

$$\times\left[\frac{(b-\beta)!}{(a+b-e-r)!(e-a-\beta+r)!s!(b-\beta-s)!}\right]$$

$$\times\left[\frac{(b+\beta)!}{(b+\beta-r)!r!(f-d+\beta+s)!(b+d-f-s)!}\right]. \qquad (3.291)$$

Racah was able to reduce this four-index summation expression to one involving but a single summation index, thus obtaining a form that is essentially no more complicated than a Wigner coefficient. Racah's procedure is given in Appendix A, where the following result is proved:

$$W(abcd;ef) = \Delta(abe)\Delta(cde)\Delta(acf)\Delta(bdf)$$

$$\times\sum_{z}\frac{(-1)^{a+b+c+d+z}(z+1)!}{(z-a-b-e)!(z-c-d-e)!(z-a-c-f)!(z-b-d-f)!}$$

$$\times\frac{1}{(a+b+c+d-z)!(a+d+e+f-z)!(b+c+e+f-z)!}.$$

$$(3.292)$$

Fundamental Racah coefficients. The four spin-$\frac{1}{2}$ Racah coefficients $W^{a\,\frac{1}{2}\,a\pm\frac{1}{2}}_{\rho,\pm\frac{1}{2},\rho\pm\frac{1}{2}}(j)$ are called *fundamental coefficients.* We give the values of these coefficients in the $W^{abc}_{\rho\sigma\tau}(j)$ notation for purposes of comparison with the fundamental (spin-$\frac{1}{2}$) Wigner coefficients:

$$W^{a\,\frac{1}{2}\,a+\frac{1}{2}}_{\rho,\frac{1}{2},\rho+\frac{1}{2}}(j) = \left[\frac{(a+\rho+1)(2j+a-\rho+1)}{(2a+1)(2j+1)}\right]^{\frac{1}{2}},$$

$$W^{a\,\frac{1}{2}\,a+\frac{1}{2}}_{\rho,-\frac{1}{2},\rho-\frac{1}{2}}(j) = \left[\frac{(a-\rho+1)(2j-a-\rho+1)}{(2a+1)(2j+1)}\right]^{\frac{1}{2}},$$

$$W^{a\,\frac{1}{2}\,a-\frac{1}{2}}_{\rho,\frac{1}{2},\rho+\frac{1}{2}}(j) = -\left[\frac{(a-\rho)(2j-a-\rho)}{(2a+1)(2j+1)}\right]^{\frac{1}{2}},$$

$$W^{a\,\frac{1}{2}\,a-\frac{1}{2}}_{\rho,-\frac{1}{2},\rho-\frac{1}{2}}(j) = \left[\frac{(a+\rho)(2j+a-\rho+2)}{(2a+1)(2j+1)}\right]^{\frac{1}{2}}. \qquad (3.293)$$

The interesting and important property, Eq. (3.280), which we shall later show to be general, is here exhibited by the fundamental Racah coefficients—in the limit of large j, they become Wigner coefficients:

$$\lim_{j\to\infty} W^{a\,\frac{1}{2}\,a+\tau}_{\rho,\sigma,\rho+\sigma}(j) = C^{a\,\frac{1}{2}\,a+\tau}_{\rho,\sigma,\rho+\sigma} \qquad (3.294)$$

for $\sigma, \tau = -\frac{1}{2}, \frac{1}{2}$.

Structural consequences of the B–E identity. Of all the relations between coefficients arising in angular momentum theory, the B–E identity is one of the most fundamental. This view is based on a remarkable structural result, which we shall demonstrate in this section. *The B–E identity, Eq.* (3.275), *together with the orthogonality properties, Eqs.* (3.271) *and* (3.273), *and the fundamental spin-*$\frac{1}{2}$ *Racah coefficients imply the complete structure of angular momentum theory.*[1]

To prove this assertion, we cast the B–E identity, Eq. (3.275), in several other useful forms.

Multiplying Eq. (3.275) by $(2e+1)W(abcd;eg)$, summing over e, and using orthogonality relation (3.273) yields

$$W(c'bd'd;b'g)W(a'ad'g;c'c)$$

$$= \sum_e (2e+1)W(abcd;eg)W(a'ab'b;c'e)W(a'ed'd;b'c). \qquad (3.295)$$

[1] This result was first sketched by Racah [private communication, unpublished].

In this result, we rename b' to be e', multiply by $(2e' + 1)(2f + 1) \times W(c'bd'd;e'f)$, sum over e', and use orthogonality relation (3.273), thus obtaining

$$\delta_{fg} W(a'ad'f;c'c) = \sum_{e'e} (2f + 1)(2e' + 1)(2e + 1) W(abcd;eg)$$

$$\times W(c'bd'd;e'f) W(a'ae'b;c'e) W(a'ed'd;e'c). \qquad (3.296)$$

For the purpose of proving below the limit relation (3.280), it is useful to display this result in the notation $W^{abc}_{\rho\sigma\tau}(j)$. We put $a' = j - \lambda - \rho$, $c' = j - \lambda$, $d' = j$, and $e' = j - \lambda + \sigma$ to obtain

$$\delta_{fg} W^{afc}_{\rho,\lambda,\rho+\lambda}(j)$$

$$= \sum_{\sigma e} [(2e + 1)(2g + 1)]^{\frac{1}{2}} W(abcd;eg) W^{bdf}_{\sigma,\lambda-\sigma,\lambda}(j) W^{edc}_{\rho+\sigma,\lambda-\sigma,\rho+\lambda}(j)$$

$$\times W^{abe}_{\rho,\sigma,\rho+\sigma}(j - \lambda + \sigma), \qquad (3.297)$$

where we have left the coefficient that does not depend on j in standard Racah form [compare with Eq. (3.268)].

Consider now this result particularized to $g = f$, $d = \frac{1}{2}$, $b = f - \frac{1}{2}$. Using the explicit fundamental Racah coefficients in the result leads to the following five-term recursion relation (we have renamed some of the indices):

$$W^{bdf}_{\beta,\delta,\beta+\delta}(j)$$

$$= \left[\frac{(b+d-f)(b+f-d+1)(d-\delta)(2j-d-\delta+1)(f+\beta+\delta+1)(2j+f-\beta-\delta+2)}{(2d)(2f+1)(2d)(2j+1)(2f+2)(2j+1)}\right]^{\frac{1}{2}}$$

$$\times W^{b\,d-\frac{1}{2}\,f+\frac{1}{2}}_{\beta,\delta+\frac{1}{2},\beta+\delta+\frac{1}{2}}(j + \tfrac{1}{2})$$

$$- \left[\frac{(b+d-f)(b+f-d+1)(d+\delta)(2j+d-\delta+1)(f-\beta-\delta+1)(2j-f-\beta-\delta)}{(2d)(2f+1)(2d)(2j+1)(2f+2)(2j+1)}\right]^{\frac{1}{2}}$$

$$\times W^{b\,d-\frac{1}{2}\,f+\frac{1}{2}}_{\beta,\delta-\frac{1}{2},\beta+\delta-\frac{1}{2}}(j - \tfrac{1}{2})$$

$$+ \left[\frac{(d+f-b)(b+d+f+1)(d+\delta)(2j+d-\delta+1)(f+\beta+\delta)(2j+f-\beta-\delta+1)}{(2d)(2f+1)(2d)(2j+1)(2f)(2j+1)}\right]^{\frac{1}{2}}$$

$$\times W^{b\,d-\frac{1}{2}\,f-\frac{1}{2}}_{\beta,\delta-\frac{1}{2},\beta+\delta-\frac{1}{2}}(j - \tfrac{1}{2}) +$$

$$+\left[\frac{(d+f-b)(b+d+f+1)(d-\delta)(2j-d-\delta+1)(f-\beta-\delta)(2j-f-\beta-\delta+1)}{(2d)(2f+1)(2d)(2j+1)(2f)(2j+1)}\right]^{\frac{1}{2}}$$

$$\times\, W^{b\,d-\frac{1}{2}\,f-\frac{1}{2}}_{\beta,\delta+\frac{1}{2},\beta+\delta+\frac{1}{2}}(j+\tfrac{1}{2}). \tag{3.298}$$

In standard Racah notation, this recursion relation becomes

$$\begin{aligned}
(2c+1)(2d)(2f+1)W(abcd;ef)&\\
&= [(b+d-f)(b+f-d+1)(d+e-c)(c+e-d+1)\\
&\quad\times(c+f-a+1)(a+c+f+2)]^{\frac{1}{2}}W(a,b,c+\tfrac{1}{2},d-\tfrac{1}{2};e,f+\tfrac{1}{2})\\
&\quad-[(b+d-f)(b+f-d+1)(c+d-e)(c+d+e+1)\\
&\quad\times(a+c-f)(a+f-c+1)]^{\frac{1}{2}}W(a,b,c-\tfrac{1}{2},d-\tfrac{1}{2};e,f+\tfrac{1}{2})\\
&\quad+[(d+f-b)(b+d+f+1)(c+d-e)(c+d+e+1)\\
&\quad\times(c+f-a)(a+c+f+1)]^{\frac{1}{2}}W(a,b,c-\tfrac{1}{2},d-\tfrac{1}{2};e,f-\tfrac{1}{2})\\
&\quad+[(d+f-b)(b+d+f+1)(d+e-c)(c+e-d+1)\\
&\quad\times(a+f-c)(a+c-f+1)]^{\frac{1}{2}}W(a,b,c+\tfrac{1}{2},d-\tfrac{1}{2};e,f-\tfrac{1}{2}).
\end{aligned} \tag{3.299}$$

The five-term recursion relation, Eq. (3.298), is valid for all allowed arguments of the Racah coefficients. Furthermore, starting with the initial value $d=1$ and the spin-$\frac{1}{2}$ coefficients (3.293), it generates recursively all Racah coefficients. This result demonstrates that *all Racah coefficients are completely determined by the fundamental spin-$\frac{1}{2}$ coefficients and the Biedenharn–Elliott relation* [and the orthogonality relations that were used to bring Eq. (3.275) to the form (3.297)].

Equation (3.298) also may be used to prove the general validity of the relation

$$\lim_{j\to\infty} W^{abc}_{\rho\sigma\tau}(j) = C^{abc}_{\rho\sigma\tau}. \tag{3.300}$$

The proof is given in two steps: We first note from Eqs. (3.298) that the coefficient

$$\bar{C}^{abc}_{\rho\sigma\tau} \equiv \lim_{j\to\infty} W^{abc}_{\rho\sigma\tau}(j) \tag{3.301}$$

satisfies the recursion relation:

$$\begin{aligned}
&\bar{C}^{b\,d\,f}_{\beta,\delta,\beta+\delta} \\
&= \left[\frac{(b+d-f)(b+f-d+1)(d-\delta)(f+\beta+\delta+1)}{(2d)(2f+1)(2d)(2f+2)}\right]^{\frac{1}{2}} \bar{C}^{b\,d-\frac{1}{2}\,f+\frac{1}{2}}_{\beta,\delta+\frac{1}{2},\beta+\delta+\frac{1}{2}} \\
&\quad - \left[\frac{(b+d-f)(b+f-d+1)(d+\delta)(f-\beta-\delta+1)}{(2d)(2f+1)(2d)(2f+2)}\right]^{\frac{1}{2}} \bar{C}^{b\,d-\frac{1}{2}\,f+\frac{1}{2}}_{\beta,\delta-\frac{1}{2},\beta+\delta-\frac{1}{2}} \\
&\quad + \left[\frac{(d+f-b)(b+d+f+1)(d+\delta)(f+\beta+\delta)}{(2d)(2f+1)(2d)(2f)}\right]^{\frac{1}{2}} \bar{C}^{b\,d-\frac{1}{2}\,f-\frac{1}{2}}_{\beta,\delta-\frac{1}{2},\beta+\delta-\frac{1}{2}} \\
&\quad + \left[\frac{(d+f-b)(b+d+f+1)(d-\delta)(f-\beta-\delta)}{(2d)(2f+1)(2d)(2f)}\right]^{\frac{1}{2}} \bar{C}^{b\,d-\frac{1}{2}\,f-\frac{1}{2}}_{\beta,\delta+\frac{1}{2},\beta+\delta+\frac{1}{2}}.
\end{aligned} \tag{3.302}$$

On the other hand, if in Eq. (3.268) we choose $d = \frac{1}{2}$, $b = f - \frac{1}{2}$, and substitute in the spin-$\frac{1}{2}$ Racah and Wigner coefficients, we obtain exactly (after renaming labels) the recursion relation (3.302), now satisfied by $C^{b\,d\,f}_{\beta,\delta,\beta+\delta}$.

Since (starting from $d = 1$) the recursion relation (3.302) evidently determines all coefficients $\bar{C}^{b\,d\,f}_{\beta,\delta,\beta+\delta}$ from the fundamental ones, and since

$$\bar{C}^{b\,\frac{1}{2}\,f}_{\beta,\delta,\beta+\delta} = C^{b\,\frac{1}{2}\,f}_{\beta,\delta,\beta+\delta},$$

we find that

$$\bar{C}^{b\,d\,f}_{\beta,\delta,\beta+\delta} = C^{b\,d\,f}_{\beta,\delta,\beta+\delta} \tag{3.303}$$

in true generality, thus proving Eq. (3.300).

We may now assert: *All Wigner coefficients may be obtained as the asymptotic limit of Racah coefficients.*

Thus, we have shown that the B-E identity, the orthogonality properties of the Racah coefficients, and the spin-$\frac{1}{2}$ Racah coefficients determine all Racah coefficients and all Wigner coefficients. Since the Wigner coefficients also give the angular momentum **J** as a special case (by specifying the matrix elements), hence, also the commutation relations $\mathbf{J} \times \mathbf{J} = i\mathbf{J}$, we see that the group structure is implied. Moreover, as shown in Section 8, Chapter 5 there is an asymptotic limit of the Racah coefficients that yields the rotation matrices themselves, or, alternatively, the rotation matrices may be obtained from the representation of the Wigner coefficients as discretized rotation matrices [Eq.

(5.89)]. It is in this sense that we conclude that the complete structure of angular momentum theory [$SU(2)$] is determined as asserted on p. 115.

The method above of finding limits can be taken one step further. Thus, in Eq. (3.302), we set $b = j$, $d = J$, $f = j + \Delta$, $\beta = j - \alpha$, $\delta = M$, and take the limit $j \to \infty$ to obtain

$$2J \lim_{j\to\infty} C^{jJ\,j+\Delta}_{j-\alpha,M,j-\alpha+M}$$
$$= [(J-\Delta)(J-M)]^{\frac{1}{2}} \lim_{j\to\infty} C^{jJ-\frac{1}{2}\,j+\Delta+\frac{1}{2}}_{j-\alpha,M+\frac{1}{2},j-\alpha+M+\frac{1}{2}}$$
$$+ [(J+\Delta)(J+M)]^{\frac{1}{2}} \lim_{j\to\infty} C^{jJ-\frac{1}{2}\,j+\Delta-\frac{1}{2}}_{j-\alpha,M-\frac{1}{2},j-\alpha+M-\frac{1}{2}}. \quad (3.304)$$

Since

$$\lim_{j\to\infty} C^{j\frac{1}{2}\,j+\Delta}_{j-\alpha,M,j-\alpha+M} = \delta_{\Delta M},$$

it follows by induction on J in Eq. (3.304) that

$$\lim_{j\to\infty} C^{jJ\,j+\Delta}_{j-\alpha,M,j-\alpha+M} = \delta_{\Delta M}. \quad (3.305)$$

We exhibit in RWA [see Eqs. (4.98) and (3.32)] the Racah and Wigner coefficients in forms that show explicitly the limit properties (3.280) and (3.305).

A variety of other recursion relations may be produced by starting from one or another of the forms of the B–E identity. A typical three-term relation may be obtained by choosing $a' = j - \rho - \sigma - \frac{1}{2}$, $b' = j - \frac{1}{2}$, $c' = j - \sigma - \frac{1}{2}$, $d' = j$, $d = \frac{1}{2}$, $c = e + \frac{1}{2}$ in Eq. (3.275). Using the values of the fundamental Racah coefficients in the resulting equation then leads to the following relation:

$$W^{a\,b+\frac{1}{2}\,c+\frac{1}{2}}_{\rho,\sigma+\frac{1}{2},\rho+\sigma+\frac{1}{2}}(j)$$
$$= (2b+1)\left[\frac{(2c+2)(c+\rho+\sigma+1)(2j+c-\rho-\sigma+1)}{(2c+1)(b+\sigma+1)(b+c-a+1)(a+b+c+2)(2j+b-\sigma+1)}\right]^{\frac{1}{2}}$$
$$\times W^{abc}_{\rho,\sigma,\rho+\sigma}(j-\tfrac{1}{2})$$
$$-\left[\frac{(b-\sigma)(a+b-c)(a+c-b+1)(2j-b-\sigma)}{(b+\sigma+1)(a+b-c+1)(a+b+c+2)(2j+b-\sigma+1)}\right]^{\frac{1}{2}}$$
$$\times W^{a\,b-\frac{1}{2}\,c+\frac{1}{2}}_{\rho,\sigma+\frac{1}{2},\rho+\sigma+\frac{1}{2}}(j). \quad (3.306)$$

This recursion relation is invalid for either $\sigma = -b - 1$ or $c = a - b - 1$ (or both). However, one easily develops the expressions for $W^{abc}_{\rho,-b,\rho-b}(j)$ and $W^{a\,b\,a-b}_{\rho,\sigma,\rho+\sigma}(j)$ so that Eq. (3.306), when augmented with these special cases, may be used to generate the Racah coefficients. These features are typical of three-term recursion relations. (Taking the limit $j \to \infty$ also gives a correct relation between Wigner coefficients.)

It might be thought that relation (3.299) or (3.306) would provide a useful and simple method for deriving Racah's expression, Eq. (3.292). This is not the case! An uncritical iteration (one in which no attention is paid to algebraic cancellations) of either of these expressions produces a different (and unsymmetric) form of the coefficients. In a critical iteration, one finds that considerable algebraic manipulation must be done to bring the result to Racah's form. That this is so may be seen as follows: An iteration process is an inductive process whereby one obtains the answer by iterating sufficiently many times as required to guess the general answer, which is then proved rigorously (inductively) by demonstrating that it *solves* the recursion relation (and the initial data). It is extremely difficult to show directly that $W(abcd;ef)$, given by Eq. (3.292), solves either of Eqs. (3.299) or (3.306). This means that this answer is not the "natural" one that emerges from the iteration.

These recursion relations (and others) are useful for general proofs, for calculating the first few coefficients, and for generating numerical values. They do not seem to provide a simple derivation of Racah's result. Nonetheless, the B–E identity does provide an *in principle* method for generating all Racah coefficients (hence, all Wigner coefficients) from the fundamental ones, as shown earlier.

In Appendix B, we show how the fundamental Racah coefficients are themselves determined by (*a*) a sign convention, (*b*) the orthogonality relations, and (*c*) a symmetry relation.

Remark. By using the limit property (3.300) in a relationship between Racah coefficients [after rewriting some, or all, of the coefficients in terms of the notation $W^{abc}_{\rho\sigma\tau}(j)$], one can reproduce various relations between Racah coefficients and Wigner coefficients. For example, the orthogonality relations for Racah coefficients go into those for Wigner coefficients; Eq. (3.297) goes into Eq. (3.268), etc. In this way, one can obtain results for Wigner coefficients as limiting cases of relations between Racah coefficients alone.

Symmetries. The Racah coefficients possess a large group (144 elements) of symmetries, the physical origin of which is obscure.

The easiest way to obtain the symmetries of the Racah coefficients is by direct inspection of the explicit form, and this is one reason we required a derivation of the expression for Racah coefficients in the form given by Eq.

(3.292). [See Ref. [77, p. 140] for an equivalent form of this result due to Lehrer-Ilamed (replace $2q_i$ by q_i to correct a typographical error).] An alternative way of deriving the symmetries, which uses a generating function (see Chapter 5, Appendix D), has been given by Schwinger [37] and Bargmann [39]. Both ways leave the origin of the symmetry obscure.

Some of the symmetries of expression (3.292) are more obvious than others. We begin by looking for symmetries under permutations of the six labels *abcdef*. Furthermore, we restrict our attention to those permutations that leave each term z under the summation invariant. The term

$$(k-e-f-z)!(k-b-c-z)!(k-a-d-z)!, \tag{3.307}$$

where

$$k = a+b+c+d+e+f,$$

is the most restrictive (a symmetry of this factor is a symmetry of the other factors), and we consider it first. The term (3.307) is invariant to the $3! = 6$ permutations of the three pairs

$$(ad), (bc), (ef) \tag{3.308}$$

as well as to the interchange of the two entries in a given pair. Thus, there are $6 \times 2 \times 2 \times 2 = 48$ permutations of *abcdef* that leave the term (3.307) invariant. However, one observes that the interchange of entries in just one pair (or all three pairs), for example,

$$\{(ad), (bc), (ef)\} \to \{(da), (bc), (ef)\},$$

does not leave the triangle factors in Eq. (3.292) invariant. Thus, the group of permutations that leaves the Racah coefficient (3.292) termwise invariant can contain no more than twenty-four elements, corresponding to the permutations of the three pairs $(ad), (bc), (ef)$ among themselves and the interchange of the order of the entries in either none or two of the three pairs. Let us now prove that all twenty-four of these permutations are symmetries of the Racah coefficient.

Consider the two permutation operators (cycle notation) given by

$$P = (ac)(bd), \qquad Q = (afde)(bc). \tag{3.309}$$

The actions of these permutation operators on the pairs (3.308) are

$$P\colon \{(ad), (bc), (ef)\} \to \{(cb), (da), (ef)\},$$

$$Q\colon \{(ad), (bc), (ef)\} \to \{(fe), (cb), (ad)\}. \tag{3.310}$$

Since $\Delta(rst)$ is invariant under the six permutations of its arguments, one now readily verifies that P and Q each maps the four factors

$$\frac{\Delta(rst)}{(z-r-s-t)!}, \qquad (rst) = (abe), (cde), (acf), (bdf)$$

among themselves (that is, leaves the product invariant). Thus the group of permutations generated by P and Q leaves a Racah coefficient invariant, except possibly for a change of phase.

By direct multiplication, one verifies that

$$P^2 = Q^4 = (PQ)^3 = \text{identity}. \tag{3.311}$$

This result uniquely identifies P and Q as the generators of a group that is isomorphic to the symmetric group S_4.

It is proved in RWA that the 6-j and 9-j symbols (indeed, all 3n-j symbols) are associated with *planar cubic graphs* (see Jucys *et al.* [55]). In particular, the planar cubic graph corresponding to the 6-j symbol may also be represented by the regular tetrahedron. *It is the group T_d of rotation–inversions that carry the regular tetrahedron into itself that provides the geometric realization of the group S_4 of symmetries of the Racah coefficients.*

That the *tetrahedron* is the natural geometric object for depicting the symmetries of the Racah coefficients is also suggested by the four triangles of a tetrahedron and the four sets of triangle conditions on the Racah coefficients:

$$(abe), (cde), (acf), (bdf).$$

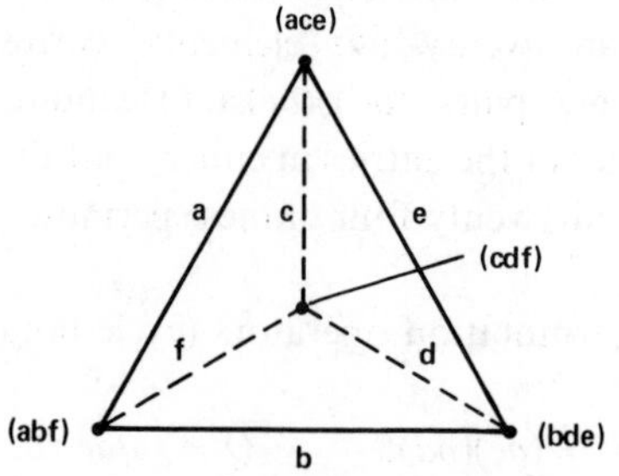

Figure 3.4.

Thus, we label the edges of one of the faces (see Fig. 3.4) with *abe*, and then label the edges of the remaining three faces with the triples (*cde*), (*acf*), and (*bdf*) such that common letters are assigned to shared edges. Observe then that the meaning of the pairs

$$(ad), (bc), (ef)$$

is that they label the three sets of opposite (and perpendicular) edges.

The group of twenty-four rotation and inversion operations that map the tetrahedron onto itself may now be shown to induce the twenty-four permutations of *abcdef* that leave the Racah coefficients invariant (except for phase). Alternatively, one can obtain this same group of permutations by labeling each of the four vertices of the tetrahedron by the triple of labels on the edges emanating from the vertex (see Fig. 3.4). We then make all possible twenty-four assignments of the four symbols (*abf*), (*cdf*), (*ace*), (*bde*) to the four vertices, labeling an edge by the letter common to the two vertex symbols at its ends. These operations again produce the permutations of *abcdef* corresponding to symmetries of the Racah coefficients.

The *complete quadrilateral* provides still another method[1] (noted by Fano and Racah [59]) of depicting these permutational symmetries. The complete quadrilateral is the planar figure obtained from the tetrahedron by the mapping "face" → "line" and "edge" → "point." Thus, a complete quadrilateral has four lines each of which intersects the other three (no common intersection points) as shown in Fig. 3.5. The permutational symmetries of the Racah coefficients are obtained from the complete quadrilateral by making all possible assignments of *abcdef* to the six points such that the three letters in each of the triples (*abe*), (*cde*), (*acf*), and (*bdf*) fall on the same straight line.

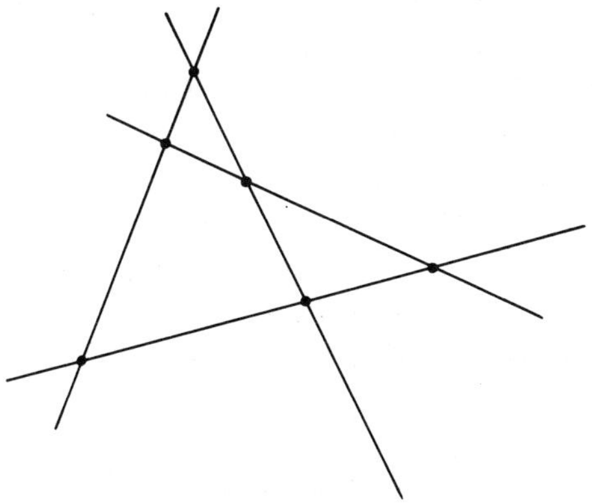

Figure 3.5.

[1] A third realization is discussed in RWA, Topic 9.

The preceding two methods (tetrahedron ↔ complete quadrilateral) are interesting geometric methods of realizing the permutational symmetries of the Racah coefficients. However, the quickest way to write out these symmetries is based on the notation for the 6-j symbol, which takes advantage of the three pairs (3.308) (Wigner [47], Erdélyi [60]) corresponding to opposite edges of the tetrahedron. Thus, we identify the top row of the 6-j symbol with a triangle of the tetrahedron, say (abe), and the vertical columns with the pairs of letters labeling opposite edges:

$$\begin{Bmatrix} a & b & e \\ d & c & f \end{Bmatrix}.$$

In order to have coefficients that are truly invariant under tetrahedral permutations, the definition of the 6-j symbol removes the phase factor from the Racah coefficient (3.292):

$$\begin{Bmatrix} a & b & e \\ d & c & f \end{Bmatrix} \equiv (-1)^{a+b+c+d} W(abcd;ef). \tag{3.312}$$

The *permutational symmetries* are now very simply expressed: The 6-j symbol is invariant under all permutations of its columns and under the exchange of any *pair* of elements in the top row with the corresponding pair in the bottom row.

The permutational symmetries of these coefficients were already noted by Racah [29]. After Regge discovered the additional symmetries of the Wigner coefficients, he also reexamined the Racah coefficients and found further symmetries.

The new symmetries discovered by Regge [61] are still termwise symmetries of the form (3.292), but involve now *linear transformations* on $abcdef$, which define new triangles. The proof of these symmetries is again by direct verification. If we reexamine Eq. (3.292), we see that it will be termwise invariant (ignoring the phase) under any transformation that simultaneously (a) transforms $\Delta(abe)\Delta(cde)\Delta(acf)\Delta(bdf)$ into itself; (b) transforms the sums $a+b+e$, $c+d+e$, $a+c+f$, $b+d+f$ among themselves; (c) transforms the sums $a+d$, $b+c$, $e+f$ among themselves. It is straightforward now to verify the equality of the following 6-j symbols, giving rise to a total of 144 symmetry relations:

$$\begin{Bmatrix} a & b & e \\ d & c & f \end{Bmatrix} = \begin{Bmatrix} a & \dfrac{b+c+e-f}{2} & \dfrac{b+e+f-c}{2} \\ d & \dfrac{b+c+f-e}{2} & \dfrac{c+e+f-b}{2} \end{Bmatrix}$$

$$= \begin{Bmatrix} \frac{a+d+e-f}{2} & b & \frac{a+e+f-d}{2} \\ \frac{a+d+f-e}{2} & c & \frac{d+e+f-a}{2} \end{Bmatrix}$$

$$= \begin{Bmatrix} \frac{a+b+d-c}{2} & \frac{a+b+c-d}{2} & e \\ \frac{a+c+d-b}{2} & \frac{b+c+d-a}{2} & f \end{Bmatrix}$$

$$= \begin{Bmatrix} \frac{a+b+d-c}{2} & \frac{b+c+e-f}{2} & \frac{a+e+f-d}{2} \\ \frac{a+c+d-b}{2} & \frac{b+c+f-e}{2} & \frac{d+e+f-a}{2} \end{Bmatrix}$$

$$= \begin{Bmatrix} \frac{a+d+e-f}{2} & \frac{a+b+c-d}{2} & \frac{b+e+f-c}{2} \\ \frac{a+d+f-e}{2} & \frac{b+c+d-a}{2} & \frac{c+e+f-b}{2} \end{Bmatrix}. \tag{3.313}$$

Bargmann [39] gives the 6-j symbol in the form of a 4×3 array (see also Shelepin [62]), which we arrange in the following form:

$$\begin{Bmatrix} a & b & e \\ d & c & f \end{Bmatrix} = \begin{bmatrix} d+f-b & c+f-a & c+d-e \\ a+f-c & b+f-d & a+b-e \\ d+e-c & b+e-a & b+d-f \\ a+e-b & c+e-d & a+c-f \end{bmatrix}. \tag{3.314}$$

Our reason for writing Bargmann's array in this form is that under the identifications $a = j - \gamma$, $b = b$, $c = j$, $d = d$, $e = j - \beta$, $f = f$, $\gamma = \alpha + \beta$, the first three rows are just a rearrangement of the entries of the Regge array (3.182), with $(j_1, m_1) = (b, \alpha)$, $(j_2 m_2) = (d, \beta)$, and $(j, m) = (f, \gamma)$, so that one has

$$W^{bdf}_{\alpha\beta\gamma}(j) = (-1)^{b+d+2j-\gamma}[(2j - 2\beta + 1)(2f + 1)]^{\frac{1}{2}}$$

$$\times \begin{bmatrix} d+f-b & f+\gamma & d+\beta \\ f-\gamma & b+f-d & b-\alpha \\ d-\beta & b+\alpha & b+d-f \\ 2j-b-\beta-\gamma & 2j-d-\beta & 2j-f-\gamma \end{bmatrix}. \tag{3.315}$$

This is a useful result for the discussion of the relationship between the symmetries of Racah coefficients and Wigner coefficients (see Appendix C).

In Bargmann's discussion of the symmetries of the Racah coefficients (based on the known Regge result and Schwinger's generating function), it emerges quite naturally that *the* 144 *symmetries of the Racah coefficients correspond to the transformations induced on abcdef by the row and column operations on the array* (3.314).

The twenty-four permutations[1] of the rows of the array (3.314) induce the transformations on *abcdef* that correspond to the twenty-four permutations of the four symbols

$$(a + b + e,\ c + d + e,\ a + c + f,\ b + d + f)$$

and that leave each of the three symbols in the triple

$$(a + d, b + c, e + f)$$

invariant. Similarly, the six permutations of the columns of the array (3.314) induce the transformations on *abcdef* that correspond to the six permutations of the three symbols

$$(a + d, b + c, e + f)$$

and that leave each of the four symbols in

$$(a + b + e, c + d + e, a + c + f, b + d + f)$$

invariant. Thus, an alternative way of expressing the invariance group of the 6-j symbol may be given: It is the group of transformations on *abcdef* induced by the group of transformations of the form

$$\begin{aligned}(a + b + e, c + d + e, a + c + f, b + d + f)&\\ \rightarrow (a' + b' + e', c' + d' + e', a' + c' + f', b' + d' + f'),&\end{aligned}$$

$$(a + d, b + c, e + f) \rightarrow (a' + d', b' + c', e' + f'),$$

where the symbols on the right are rearrangements of the symbols on the left. Observe that this group is the direct product group, $S_4 \times S_3$.

[1] It should be noted that the twenty-four tetrahedral permutations do not correspond to the twenty-four row permutations of the array.

An interesting question concerning the symmetries of the Racah and Wigner coefficients is posed by the existence of the limit relation (3.280). In this limit, which of the symmetries of the Racah coefficients remain and become symmetries of the Wigner coefficients? This question is answered in Appendix C.

19. 9-j Coefficients

Since the 9-j coefficients reduce in special cases to 6-j coefficients [see Eqs. (3.325)], it might appear that the former are even more "basic" than the latter. However, the fundamental theorem of recoupling theory states that every recoupling coefficient ($3n$-j coefficients, $n = 3, 4, \ldots$) is expressible as a summation over products of Racah coefficients (6-j coefficients). This result is proved in RWA. *It is this "basis property" that establishes the Racah coefficient as the fundamental object in the theory of rotational invariants* (hence, in recoupling theory).

We summarize here a few of the properties of the 9-j symbols, referring the reader to the literature and RWA for the proofs and for additional relations not included here (Edmonds [16], Jucys *et al.* [55], Jahn and Hope [63], Arima *et al.* [64], Danos [65], Wu [66], El Baz and Castel [67]).

Expression in terms of 3-j *and* 6-j *symbols.* The expression relating the 9-j symbols to Wigner coefficients is given by Eqs. (3.250) and (3.252). We identify $(j_{11}, j_{21}, j_{31}) = (k_1, k_2, k)$, $(j_{12}, j_{22}, j_{32}) = (j_1, j_2, j)$, and $(j_{13}, j_{23}, j_{33}) = (j_1', j_2', j')$ in that result, express the Wigner coefficients in terms of 3-j symbols [Eq. (3.182)], and finally use the symmetries of the 3-j symbols, thus obtaining the following relation in which $\phi \equiv \sum_{kl} j_{kl}$:

$$(-1)^{\phi} \begin{Bmatrix} j_{12} & j_{22} & j_{32} \\ j_{11} & j_{21} & j_{31} \\ j_{13} & j_{23} & j_{33} \end{Bmatrix} \begin{pmatrix} j_{31} & j_{32} & j_{33} \\ m_{31} & m_{32} & m_{33} \end{pmatrix}$$

$$= \sum_{\text{all } (m_{1i})(m_{2i})} \begin{pmatrix} j_{11} & j_{21} & j_{31} \\ m_{11} & m_{21} & m_{31} \end{pmatrix} \begin{pmatrix} j_{12} & j_{22} & j_{32} \\ m_{12} & m_{22} & m_{32} \end{pmatrix}$$

$$\times \begin{pmatrix} j_{13} & j_{23} & j_{33} \\ m_{13} & m_{23} & m_{33} \end{pmatrix} \begin{pmatrix} j_{11} & j_{12} & j_{13} \\ m_{11} & m_{12} & m_{13} \end{pmatrix} \begin{pmatrix} j_{21} & j_{22} & j_{23} \\ m_{21} & m_{22} & m_{23} \end{pmatrix}. \tag{3.316}$$

Using the symmetries of the 3-j symbols, one may now prove the symmetry relation (the full group of symmetries is discussed below):

$$\begin{Bmatrix} j_{11} & j_{12} & j_{13} \\ j_{21} & j_{22} & j_{23} \\ j_{31} & j_{32} & j_{33} \end{Bmatrix} = (-1)^{\phi} \begin{Bmatrix} j_{12} & j_{22} & j_{32} \\ j_{11} & j_{21} & j_{31} \\ j_{13} & j_{23} & j_{33} \end{Bmatrix}. \tag{3.317}$$

Using this symmetry relation and the orthogonality of the Wigner 3-j coefficients in Eq. (3.316), we obtain the following symmetric form for the 9-j coefficients

$$\begin{Bmatrix} j_{11} & j_{12} & j_{13} \\ j_{21} & j_{22} & j_{23} \\ j_{31} & j_{32} & j_{33} \end{Bmatrix}$$

$$= \sum_{\text{all } m_{ij}} \begin{pmatrix} j_{11} & j_{12} & j_{13} \\ m_{11} & m_{12} & m_{13} \end{pmatrix} \begin{pmatrix} j_{21} & j_{22} & j_{23} \\ m_{21} & m_{22} & m_{23} \end{pmatrix} \begin{pmatrix} j_{31} & j_{32} & j_{33} \\ m_{31} & m_{32} & m_{33} \end{pmatrix}$$

$$\times \begin{pmatrix} j_{11} & j_{21} & j_{31} \\ m_{11} & m_{21} & m_{31} \end{pmatrix} \begin{pmatrix} j_{12} & j_{22} & j_{32} \\ m_{12} & m_{22} & m_{32} \end{pmatrix} \begin{pmatrix} j_{13} & j_{23} & j_{33} \\ m_{13} & m_{23} & m_{33} \end{pmatrix}. \tag{3.318}$$

On the other hand, Eq. (3.251) expresses the 9-j symbols directly in terms of the 6-j symbols. Using the symmetries of the 6-j symbols and Eq. (3.317), we may reexpress this result in the symmetric form:

$$\begin{Bmatrix} j_{11} & j_{12} & j_{13} \\ j_{21} & j_{22} & j_{23} \\ j_{31} & j_{32} & j_{33} \end{Bmatrix}$$

$$= \sum_k (-1)^{2k}(2k+1) \begin{Bmatrix} j_{11} & j_{21} & j_{31} \\ j_{32} & j_{33} & k \end{Bmatrix} \begin{Bmatrix} j_{12} & j_{22} & j_{32} \\ j_{21} & k & j_{23} \end{Bmatrix} \begin{Bmatrix} j_{13} & j_{23} & j_{33} \\ k & j_{11} & j_{12} \end{Bmatrix}. \tag{3.319}$$

This relation shows [as does Eq. (3.316)] that a 9-j coefficient is zero unless the triangle conditions are fulfilled by the entries in each row and each column of the symbol.

Orthogonality relations. The orthogonality of the 6-j coefficients is expressed by

$$\sum_f (2e+1)(2f+1) \begin{Bmatrix} a & b & e \\ d & c & f \end{Bmatrix} \begin{Bmatrix} a & b & e' \\ d & c & f \end{Bmatrix} = \delta_{ee'}\, \varepsilon_{abe}\, \varepsilon_{dce}. \tag{3.320}$$

This result follows directly from the orthogonality relation (3.271) for the Racah coefficients and the definition (3.312) of the 6-j symbol. The orthogonality relation for the 9-j coefficient follows directly from Eq. (3.319), the symmetries of the 6-j symbol, and the orthogonality relation (3.320):

$$\sum_{hi}(2c+1)(2f+1)(2h+1)(2i+1)\begin{Bmatrix} a & b & c \\ d & e & f \\ h & i & j \end{Bmatrix}\begin{Bmatrix} a & b & c' \\ d & e & f' \\ h & i & j \end{Bmatrix} = \delta_{cc'}\,\delta_{ff'}. \tag{3.321}$$

Let us note explicitly that this result is to be applied only to triples (abc), (def), (cfj), (abc'), (def'), and $(c'f'j)$ that satisfy the triangle conditions, since we have *not* included the $\varepsilon_{abc}, \ldots$ factors necessary for Eq. (3.321) to have a more general validity.

Symmetries of the 9-j *symbol.* There are seventy-two known symmetry relations for the 9-j symbols. Each of these symmetries is a *termwise* symmetry of expression (3.318) and is a consequence of the symmetries of the 3-j symbols. One easily verifies that the 9-j symbol

$$\begin{Bmatrix} j_{11} & j_{12} & j_{13} \\ j_{21} & j_{22} & j_{23} \\ j_{31} & j_{32} & j_{33} \end{Bmatrix}$$

is invariant under even permutation of its rows, even permutation of its columns, and under the interchange of rows and columns (matrix transposition). It is multiplied by the factor

$$(-1)^{\sum_{kl} j_{kl}} \tag{3.322}$$

under odd permutations of its rows or columns.

Special values of the 9-j *symbols.* Equation (3.319) is particularly useful for evaluating certain 9-j symbols. If, for example, we have $j_{33} = 0$, then

$$\begin{Bmatrix} j_{11} & j_{12} & j_{13} \\ j_{21} & j_{22} & j_{23} \\ j_{31} & j_{32} & 0 \end{Bmatrix} = \frac{(-1)^{j_{12}+j_{13}+j_{21}+j_{31}}\delta_{j_{13}j_{23}}\delta_{j_{31}j_{32}}\begin{Bmatrix} j_{12} & j_{22} & j_{32} \\ j_{21} & j_{11} & j_{23} \end{Bmatrix}}{[(2j_{31}+1)(2j_{13}+1)]^{\frac{1}{2}}} \tag{3.323}$$

in consequence of

$$\begin{Bmatrix} a & b & 0 \\ d & c & f \end{Bmatrix} = \begin{Bmatrix} a & f & c \\ d & 0 & b \end{Bmatrix} = \frac{(-1)^{a+c+f}\delta_{ab}\delta_{cd}}{[(2a+1)(2c+1)]^{\frac{1}{2}}}. \tag{3.324}$$

The symmetry relations may now be used to move the zero in Eq. (3.323) to any desired position in the 3×3 array, thus yielding the following formulas:

$$\begin{Bmatrix} a & b & e \\ c & d & e \\ f & f & 0 \end{Bmatrix} = \begin{Bmatrix} 0 & e & e \\ f & d & b \\ f & c & a \end{Bmatrix} = \begin{Bmatrix} e & 0 & e \\ c & f & a \\ d & f & b \end{Bmatrix}$$

$$= \begin{Bmatrix} f & f & 0 \\ d & c & e \\ b & a & e \end{Bmatrix} = \begin{Bmatrix} f & b & d \\ 0 & e & e \\ f & a & c \end{Bmatrix} = \begin{Bmatrix} a & f & c \\ e & 0 & e \\ b & f & d \end{Bmatrix}$$

$$= \begin{Bmatrix} b & a & e \\ f & f & 0 \\ d & c & e \end{Bmatrix} = \begin{Bmatrix} e & d & c \\ e & b & a \\ 0 & f & f \end{Bmatrix} = \begin{Bmatrix} c & e & d \\ a & e & b \\ f & 0 & f \end{Bmatrix}$$

$$= \frac{(-1)^{b+c+e+f}}{[(2e+1)(2f+1)]^{\frac{1}{2}}} \begin{Bmatrix} a & b & e \\ d & c & f \end{Bmatrix}. \tag{3.325}$$

We conclude this introduction to the 9-j symbol by giving its algebraic expression in the simplest known form (Jucys and Bandzaitis [67a]):

$$\begin{Bmatrix} a & b & c \\ d & e & f \\ h & i & j \end{Bmatrix} = (-1)^{c+f-j} \frac{(d\,a\,h)(b\,e\,i)(j\,h\,i)}{(d\,e\,f)(b\,a\,c)(j\,c\,f)}$$

$$\times \sum_{xyz} \frac{(-1)^{x+y+z}}{x!y!z!} \times \frac{(2f-x)!(2a-z)!}{(2i+1+y)!(a+d+h+1-z)!}$$

$$\times \frac{(d+e-f+x)!(c+j-f+x)!}{(e+f-d-x)!(c+f-j-x)!} \times \frac{(e+i-b+y)!(h+i-j+y)!}{(b+e-i-y)!(h+j-i-y)!}$$

$$\times \frac{(b+c-a+z)!}{(a+d-h-z)!(a+c-b-z)!(a+d+h+1-z)!}$$

$$\times \frac{(a+d+j-i-y-z)!}{(d+i-b-f+x+y)!(b+f-a-j+x+z)!}, \tag{3.326}$$

where

$$(a\,b\,c) \equiv \left[\frac{(a-b+c)!(a+b-c)!(a+b+c+1)!}{(b+c-a)!}\right]^{\frac{1}{2}}.$$

Numerous special cases of the 9-j coefficients may be found in Varshalovich *et al.* [67b].

20. Rotationally Invariant Products

Physical applications of angular momentum theory often involve the construction of rotational invariants from two or more tensor operators. The general procedure for this construction is based on the notion of a *conjugate irreducible tensor operator* and the unitary property of the rotation matrices.

The tensor operator coupling, $[\mathbf{S}^{k_1} \mathbf{x} \mathbf{S}^{k_2}]^k$, has already led us to the construction of a rotational invariant:

$$[\mathbf{S}^J \mathbf{x} \mathbf{T}^J]_0^0 = \sum_M \frac{(-1)^{J-M}}{(2J+1)^{\frac{1}{2}}} S_M^J T_{-M}^J. \tag{3.327}$$

The essential structural element in this construction is the transformation property, under rotations, of the set of operators $\{\bar{T}_M^J : M = J, \ldots, -J\}$, where

$$\bar{T}_M^J \equiv (-1)^{J-M} T_{-M}^J. \tag{3.328}$$

One finds that these operators are transformed linearly among themselves with coefficients that are the elements of the *complex conjugate rotation matrix*:

$$\mathscr{U}: \bar{T}_M^J \to \mathscr{U} \bar{T}_M^J \mathscr{U}^{-1} = \sum_{M'} D_{M'M}^{J*}(U) \bar{T}_{M'}^J. \tag{3.329}$$

Equivalently, the definition may be given by specifying the commutation action of the total angular momentum operator:

$$\begin{aligned}
[J_+, \bar{T}_M^J] &= -[(J+M)(J-M+1)]^{\frac{1}{2}} \bar{T}_{M-1}^J, \\
[J_-, \bar{T}_M^J] &= -[(J-M)(J+M+1)]^{\frac{1}{2}} \bar{T}_{M+1}^J, \\
[J_3, \bar{T}_M^J] &= -M \bar{T}_M^J.
\end{aligned} \tag{3.330}$$

Proof. The proof of Eq. (3.329) follows directly from a property of the rotation matrix,

$$D_{-M',-M}^J(U) = (-1)^{M'-M} D_{M'M}^{J*}(U), \tag{3.331}$$

which is proved in Chapter 5, Appendix C. ■

The transformation property (3.329) of the operators (3.328) is *generic*; that is, any set of operators $\{\bar{T}_M^J\}$ possessing this transformation property may be used to define a rotational invariant. We call such a set of operators a *conjugate irreducible tensor operator*.

We are thus led to the following result: *Let* $\mathbf{S}^J$ *and* $\bar{\mathbf{T}}^J$ *denote tensor and conjugate tensor operators, respectively. Then the quantity*

$$\sum_M S_M^J \bar{T}_M^J \tag{3.332}$$

is a rotational invariant (as also is $\sum_M \bar{S}_M^J T_M^J$).

The proof of this result may be given by using the transformations, Eqs. (3.216) and (3.329), to show that $\mathscr{U}$ commutes with the form (3.332). Alternatively, the commutation relations, Eqs. (3.206) and (3.330), may be used directly to show that the total angular momentum $\mathbf{J}$ commutes with the form (3.332).

Examples of conjugate irreducible tensor operators are (*a*) the *complex conjugate* $\Phi^*_{(\alpha)JM}$ of a sharp angular momentum wave function (see, however, the discusssion on p. 92); (*b*) the *Hermitian conjugate* of an irreducible tensor operator – that is, the operator $\mathbf{T}^{J\dagger}$ defined by

$$\langle(\alpha)jm|T_M^{J\dagger}|(\alpha')j'm'\rangle = \langle(\alpha')j'm'|T_M^J|(\alpha)jm\rangle^*; \tag{3.333}$$

and (*c*) the operator $\bar{\mathbf{T}}^J$ defined by Eq. (3.328) multiplied by an arbitrary factor that is independent of M. For example, one often finds

$$(\bar{T}_M^J)' \equiv (-1)^{J+M} T_{-M}^J \tag{3.334}$$

used in the literature.

The coupling of three tensor operators (or wave functions, or a combination) to a rotational invariant occurs frequently in applications of the theory of angular momentum and, indeed, may be viewed as a basic structure for determining the Wigner coefficients themselves [van der Waerden [31], Schwinger [37], Regge [38], Kramers [68]). This coupling is given by

$$\begin{aligned}
[[\mathbf{A}^a \times \mathbf{B}^b]^c \times \mathbf{C}^c]_0^0 &= \sum_\gamma [\mathbf{A}^a \times \mathbf{B}^b]_\gamma^c \bar{C}_\gamma^c \\
&= \sum_{\alpha\beta\gamma} \frac{(-1)^{c-\gamma}}{\sqrt{2c+1}} C_{\alpha\beta\gamma}^{abc} A_\alpha^a B_\beta^b C_{-\gamma}^c \\
&= (-1)^{a-b+c} \sum_{\alpha\beta\gamma} \begin{pmatrix} a & b & c \\ \alpha & \beta & \gamma \end{pmatrix} A_\alpha^a B_\beta^b C_\gamma^c,
\end{aligned} \tag{3.335}$$

since

$$\begin{pmatrix} a & b & c \\ \alpha & \beta & \gamma \end{pmatrix} = \frac{(-1)^{a-b-\gamma}}{\sqrt{2c+1}} C_{\alpha,\beta,-\gamma}^{abc}.$$

We see from Eq. (3.335) that the 3-j coefficients occur naturally in effecting the coupling of three tensor operators to an invariant. The invariant expression

$$\sum_{\alpha\beta\gamma} \begin{pmatrix} a & b & c \\ \alpha & \beta & \gamma \end{pmatrix} A^a_\alpha B^b_\beta C^c_\gamma \tag{3.336}$$

has a variety of applications, both in the tensor operator interpretation and in the wave function interpretation.

As noted above, if the Hermitian conjugate to $\mathbf{C}^c$ is defined, then it is also true that the form

$$\sum_{\alpha\beta\gamma} C^{bac}_{\beta\alpha\gamma} A^a_\alpha B^b_\beta C^{c\dagger}_\gamma \tag{3.337}$$

is a rotational invariant, and, in general, the two invariants, given Eqs. (3.336) and (3.337), are not the same.

The metric interpretation of the scalar coupling Wigner coefficients $C^{ab0}_{\alpha\beta 0}$ is associated with the conjugation operation defined by Eq. (3.328). These coefficients may be regarded as effecting the transformation from a set of irreducible tensor operator components $\{A^a_\alpha\}$ to the conjugate set $\{\bar{A}^a_\alpha\}$:

$$A^a_\alpha \to \bar{A}^a_\alpha \equiv \sum_{a'\alpha'} C^{aa'0}_{\alpha\alpha'0} A^{a'}_{\alpha'} = \frac{(-1)^{a-\alpha}}{\sqrt{2a+1}} A^a_{-\alpha}. \tag{3.338}$$

It is this result and the property

$$\sum_\alpha A^a_\alpha \bar{A}^a_\alpha = \text{rotational invariant} \tag{3.339}$$

that admit a metric interpretation for the scalar coupling Wigner coefficients (see Wigner [13, p. 292]).

21. Operators Associated with Wigner, Racah, and 9-j Coefficients

In this section we define operators associated with Wigner, Racah, and 9-j coefficients. The operators defined here are all bounded linear operators acting on a separable Hilbert space $\mathscr{H}$. We shall assume that $\mathscr{H}$ has been fully split into irreducible subspaces with respect to the angular momentum $\mathbf{J}$, and, furthermore, that each such subspace $\mathscr{H}_j$ $(j = 0, \frac{1}{2}, 1, \ldots)$ occurs exactly once and is spanned by a set of orthonormal basis vectors $\{|jm\rangle,\ m = j, j-1, \ldots, -j\}$ on which $\mathbf{J}$ has the standard action given by Eqs. (3.16) and

(3.20).[1] Examples of such spaces are given in RWA, where the algebraic relationships given here are developed in detail in three chapters. Since these results are not part of the standard textbook literature, it is useful to summarize some of the basic definitions and algebraic relations in order that these concepts be accessible for use in the applications in Chapter 7.

There are four types of operators defined below: *Wigner operators* (also called unit tensor operators), *Racah invariants, Racah operators, and* 9-*j invariants.* The Wigner operators and Racah invariants are defined by giving their actions on the angular momentum basis of $\mathscr{H}$. The Racah operators and 9-j invariants are defined by giving their actions on the coupled angular momentum basis of the tensor product space $\mathscr{H} \otimes \mathscr{H}$. We give a number of properties of Wigner operators, since these operators have been used most often in the applications. Only a few properties of the Racah operators are given, but let us note that these operators possess properties analogous to those of Wigner operators (see RWA). Racah invariants and 9-j invariants arise naturally in the coupling rules for Wigner and Racah operators.

Wigner Operators

Notation (Gel'fand patterns; see Chapter 5, Appendix A):

$$\left\langle \begin{matrix} & J+\Delta & \\ 2J & & 0 \\ & J+M & \end{matrix} \right\rangle, \qquad \begin{matrix} M, \Delta = J, J-1, \ldots, -J \\ 2J = 0, 1, 2, \ldots \end{matrix} \tag{3.340}$$

Definition (shift action):

$$\left\langle \begin{matrix} & J+\Delta & \\ 2J & & 0 \\ & J+M & \end{matrix} \right\rangle |jm\rangle = C^{j\,J\,j+\Delta}_{m,M,m+M} |j+\Delta, m+M\rangle$$

$$\text{for all } j = 0, \tfrac{1}{2}, \ldots;\ m = j, j-1, \ldots, -j. \tag{3.341}$$

Conjugation:

$$\left\langle \begin{matrix} & J+\Delta & \\ 2J & & 0 \\ & J+M & \end{matrix} \right\rangle^{\dagger} |jm\rangle = C^{j-\Delta\,J\,j}_{m-M,M,m} |j-\Delta, m-M\rangle. \tag{3.342}$$

[1] We discuss in Note 12 how such abstract spaces are used in applications.

Orthogonality:

$$\sum_{M}\left\langle\begin{matrix} & J+\Delta' & \\ 2J & & 0 \\ & J+M & \end{matrix}\right\rangle\left\langle\begin{matrix} & J+\Delta & \\ 2J & & 0 \\ & J+M & \end{matrix}\right\rangle^{\dagger}|jm\rangle=\delta_{\Delta'\Delta}\varepsilon_{j-\Delta,J,j}|jm\rangle,$$

$$\sum_{\Delta}\left\langle\begin{matrix} & J+\Delta & \\ 2J & & 0 \\ & J+M' & \end{matrix}\right\rangle^{\dagger}\left\langle\begin{matrix} & J+\Delta & \\ 2J & & 0 \\ & J+M & \end{matrix}\right\rangle|jm\rangle=\delta_{M'M}|jm\rangle,$$

$$\sum_{m}\langle jm|\left\langle\begin{matrix} & J'+\Delta' & \\ 2J' & & 0 \\ & J'+M' & \end{matrix}\right\rangle\left\langle\begin{matrix} & J+\Delta & \\ 2J & & 0 \\ & J+M & \end{matrix}\right\rangle^{\dagger}|jm\rangle=\frac{2j+1}{2J+1}\,\delta_{J'J}\delta_{M'M}\delta_{\Delta'\Delta}. \tag{3.343}$$

Tensor operator property [see Eq. (3.216)]:

$$\mathscr{U}\left\langle\begin{matrix} & J+\Delta & \\ 2J & & 0 \\ & J+M & \end{matrix}\right\rangle\mathscr{U}^{-1}=\sum_{M'}D^{J}_{M'M}(U)\left\langle\begin{matrix} & J+\Delta & \\ 2J & & 0 \\ & J+M & \end{matrix}\right\rangle. \tag{3.344}$$

Basis property (Wigner–Eckart theorem):

$$T^{J}_{M}|jm\rangle=\left(\sum_{\Delta}\langle j+\Delta\|\mathbf{T}^{J}\|j\rangle\left\langle\begin{matrix} & J+\Delta & \\ 2J & & 0 \\ & J+M & \end{matrix}\right\rangle\right)|jm\rangle. \tag{3.345}$$

Characteristic null space: The characteristic null space[1] of the Wigner operator defined by Eq. (3.341) is the set of irreducible subspaces $\mathscr{H}_j \subset \mathscr{H}$ given by

$$\{\mathscr{H}_j\colon 2j=0,1,\ldots,J-\Delta-1\}. \tag{3.346}$$

Coupling law:

$$\sum_{\alpha\beta}C^{abc}_{\alpha\beta\gamma}\left\langle\begin{matrix} & b+\sigma & \\ 2b & & 0 \\ & b+\beta & \end{matrix}\right\rangle\left\langle\begin{matrix} & a+\rho & \\ 2a & & 0 \\ & a+\alpha & \end{matrix}\right\rangle=\mathbf{W}^{a\;b\;c}_{\rho,\sigma,\rho+\sigma}\left\langle\begin{matrix} & c+\rho+\sigma & \\ 2c & & 0 \\ & c+\gamma & \end{matrix}\right\rangle. \tag{3.347}$$

---

[1] The distinction between null space (or kernel) and the characteristic null space is that the latter space is constituted of whole irrep spaces and not just individual vectors. This concept and its implications for the structure of Wigner operators and Racah operators is developed in detail in Chapters 3 and 4 of RWA.

In this result, $\mathbf{W}^{abc}_{\rho\sigma\tau}$ is an invariant operator (commutes with $\mathbf{J}$) and is called a *Racah invariant*, since it has eigenvalues on a generic state $|jm\rangle$ given by a Racah coefficient:

$$\mathbf{W}^{abc}_{\rho\sigma\tau}|jm\rangle = W^{abc}_{\rho\sigma\tau}(j)|jm\rangle. \tag{3.348}$$

In terms of the Condon and Shortley notation for reduced matrix elements (see p. 97), one has

$$W^{abc}_{\rho\sigma\tau}(j+\Delta) = \langle j+\Delta, m+M|\mathbf{W}^{abc}_{\rho\sigma\tau}|j+\Delta, m+M\rangle = \langle j+\Delta||\mathbf{W}^{abc}_{\rho\sigma\tau}||j\rangle. \tag{3.349}$$

Note that it is this latter coefficient with the shifted label $j+\Delta$ that appears in Eq. (3.347) when one takes matrix elements $\langle j+\Delta, m+M|\cdots|jm\rangle$.

Product law:

$$\left\langle \begin{matrix} & b+\sigma & \\ 2b & & 0 \\ & b+\beta & \end{matrix} \right\rangle \left\langle \begin{matrix} & a+\rho & \\ 2a & & 0 \\ & a+\alpha & \end{matrix} \right\rangle = \sum_c \mathbf{W}^{a\,b\,c}_{\rho,\sigma,\rho+\sigma} C^{a\,b\,c}_{\alpha,\beta,\alpha+\beta} \left\langle \begin{matrix} & c+\rho+\sigma & \\ 2c & & 0 \\ & c+\alpha+\beta & \end{matrix} \right\rangle. \tag{3.350}$$

Racah invariant:

$$\delta_{\tau'\tau}\mathbf{W}^{abc}_{\rho\sigma\tau} = \sum_{\alpha\beta\gamma} C^{abc}_{\alpha\beta\gamma} \left\langle \begin{matrix} & b+\sigma & \\ 2b & & 0 \\ & b+\beta & \end{matrix} \right\rangle \left\langle \begin{matrix} & a+\rho & \\ 2a & & 0 \\ & a+\alpha & \end{matrix} \right\rangle \left\langle \begin{matrix} & c+\tau & \\ 2c & & 0 \\ & c+\gamma & \end{matrix} \right\rangle^{\dagger}. \tag{3.351}$$

These Racah invariants satisfy orthogonality relations that correspond to Eqs. (3.281) for the Racah coefficients.

Taking matrix elements of the operator relations above with respect to the basis states of the underlying Hilbert space $\mathscr{H}$, one recovers the algebraic relations between Wigner and Racah coefficients as have already been developed in Sections 12 and 18 – for example, the orthogonality relations, Eqs. (3.178) and (3.179) and the set of relations given in Eqs. (3.266)–(3.270). These relations are valid without reference to the underlying physics. Indeed, the essence of the Wigner–Eckart theorem is the factorization of a physical tensor operator into two parts: a part "containing" the physics and associated with an operator (the reduced matrix elements), which is usually *unbounded*, and a part, the Wigner operator, which contains the implications of rotational symmetry, that is *bounded*. It is the properties of these bounded operators that constitute the *Racah–Wigner algebra* outlined above (and developed in detail in RWA).

The B–E identity, Eq. (3.275), is a consequence of the associative law for the product of three Wigner operators and may be proved by using the product law (3.350) (see RWA for this proof). The existence of the B–E identity is the essential result that allows one to define yet another algebra of bounded operators in angular momentum theory. The relevant underlying Hilbert space is the tensor product space $\mathscr{H} \otimes \mathscr{H}$. We assume that this space has been fully split into irreducible subspaces with respect to the total angular momentum $\mathbf{J} = \mathbf{J}(1) + \mathbf{J}(2)$ and, furthermore, that each such subspace is spanned by a set of orthonormal basis vectors $\{|(j_1 j_2)jm\rangle : m = j, \ldots, -j\}$, where

$$|(j_1 j_2)jm\rangle = \sum_{m_1 m_2} C^{j_1 j_2 j}_{m_1 m_2 m} |j_1 m_1\rangle \otimes |j_2 m_2\rangle. \qquad (3.352)$$

Racah Operators

Notation:

$$\left\{\begin{matrix} & a+\rho & \\ 2a & & 0 \\ & a+\sigma & \end{matrix}\right\}, \qquad \begin{matrix} \rho, \sigma = a, a-1, \ldots, -a \\ 2a = 0, 1, 2, \ldots \end{matrix}. \qquad (3.353)$$

Definition:

$$\left\{\begin{matrix} & a+\rho & \\ 2a & & 0 \\ & a+\sigma & \end{matrix}\right\} = (-1)^{a+\sigma} \sum_{\alpha} \left\langle \begin{matrix} & a+\rho & \\ 2a & & 0 \\ & a+\alpha & \end{matrix} \right\rangle \otimes \left\langle \begin{matrix} & a-\sigma & \\ 2a & & 0 \\ & a+\alpha & \end{matrix} \right\rangle^{\dagger}. \qquad (3.354)$$

Alternative definition:

$$\left\{\begin{matrix} & a+\rho & \\ 2a & & 0 \\ & a+\sigma & \end{matrix}\right\} |(j_1 j_2)jm\rangle = [(2j_1 + 2\rho + 1)(2j_2 + 1)]^{\frac{1}{2}} \times W(j, j_1, j_2 + \sigma, a; j_2, j_1 + \rho) \times |(j_1 + \rho, j_2 + \sigma)jm\rangle. \qquad (3.355)$$

Conjugation:

$$\left\{\begin{matrix} & a+\rho & \\ 2a & & 0 \\ & a+\sigma & \end{matrix}\right\}^{\dagger} |(j_1 j_2) jm\rangle = [(2j_1+1)(2j_2-2\sigma+1)]^{\frac{1}{2}}$$
$$\times\, W(j, j_1-\rho, j_2, a; j_2-\sigma, j_1)$$
$$\times\, |(j_1-\rho, j_2-\sigma) jm\rangle. \qquad (3.356)$$

Orthogonality:

$$\sum_{\sigma} \left\{\begin{matrix} & a+\rho' & \\ 2a & & 0 \\ & a+\sigma & \end{matrix}\right\} \left\{\begin{matrix} & a+\rho & \\ 2a & & 0 \\ & a+\sigma & \end{matrix}\right\}^{\dagger} |(j_1 j_2) jm\rangle = \delta_{\rho'\rho}\varepsilon_{j_1-\rho, a, j_1} |(j_1 j_2) jm\rangle,$$

$$\sum_{\rho} \left\{\begin{matrix} & a+\rho & \\ 2a & & 0 \\ & a+\sigma' & \end{matrix}\right\}^{\dagger} \left\{\begin{matrix} & a+\rho & \\ 2a & & 0 \\ & a+\sigma & \end{matrix}\right\} |(j_1 j_2) jm\rangle = \delta_{\sigma'\sigma}\varepsilon_{j_2+\sigma, a, j_2} |(j_1 j_2) jm\rangle,$$

$$\sum_{j=|j_1-j_2|}^{j_1+j_2} \langle (j_1 j_2) jm| (2j+1) \left\{\begin{matrix} & a'+\rho' & \\ 2a' & & 0 \\ & a'+\sigma' & \end{matrix}\right\} \left\{\begin{matrix} & a+\rho & \\ 2a & & 0 \\ & a+\sigma & \end{matrix}\right\}^{\dagger} |(j_1 j_2) jm\rangle$$

$$= [(2j_1+1)(2j_2-2\sigma+1)/(2a+1)]\delta_{a'a}\delta_{\rho'\rho}\delta_{\sigma'\sigma}\varepsilon_{j_1-\rho, j_1, a}\varepsilon_{j_2-\sigma, j_2, a}. \qquad (3.357)$$

Product law:

$$\left\{\begin{matrix} & b+\sigma & \\ 2b & & 0 \\ & b+\beta & \end{matrix}\right\} \left\{\begin{matrix} & a+\rho & \\ 2a & & 0 \\ & a+\alpha & \end{matrix}\right\} = \sum_{c} \bar{\mathbf{W}}^{a\,b\,c}_{\rho,\sigma,\rho+\sigma} \underline{\mathbf{W}}^{a\,b\,c}_{\alpha,\beta,\alpha+\beta} \left\{\begin{matrix} & c+\rho+\sigma & \\ 2c & & 0 \\ & c+\alpha+\beta & \end{matrix}\right\}.$$
$$(3.358)$$

In this result, $\bar{\mathbf{W}}^{abc}_{\rho\sigma\tau}$ and $\underline{\mathbf{W}}^{abc}_{\alpha\beta\gamma}$ denote Racah invariants with respect to the angular momenta $\mathbf{J}(1)$ and $\mathbf{J}(2)$, respectively, so that

$$\bar{\mathbf{W}}^{abc}_{\rho\sigma\tau} |(j_1 j_2) jm\rangle = W^{abc}_{\rho\sigma\tau}(j_1) |(j_1 j_2) jm\rangle,$$

$$\underline{\mathbf{W}}^{abc}_{\alpha\beta\gamma} |(j_1 j_2) jm\rangle = W^{abc}_{\alpha\beta\gamma}(j_2) |(j_1 j_2) jm\rangle.$$

In evaluating the action of the operator relation (3.358) on a basis vector $|(j_1 j_2) jm\rangle$, these two Racah invariants are to be evaluated on the shifted labels $j_1+\rho+\sigma$ and $j_2+\alpha+\beta$, respectively.

The product law, Eq. (3.358), is an operator version of the B–E identity, the latter being recovered from Eq. (3.358) by taking matrix elements $\langle(j_1+\rho+\sigma, j_2+\alpha+\beta)jm|\cdots|(j_1j_2)jm\rangle$. Thus, the B–E identity underlies the existence of the algebra of Racah operators (called W-algebra in RWA).

We have given above two equivalent definitions of a Racah operator – the first in terms of Wigner operators, the second in terms of Racah coefficients. The latter definition "frees" the definition of a Racah operator from that of a Wigner operator in the sense already described in Section 18.

9-j Invariant Operators

The definition of a Racah operator given by Eq. (3.354) is a special case of the general operator defined by the coupling

$$T^{abc}_{\rho\sigma\gamma} \equiv \sum_{\alpha\beta} C^{abc}_{\alpha\beta\gamma} \left\langle \begin{matrix} & a+\rho & \\ 2a & & 0 \\ & a+\alpha & \end{matrix} \right\rangle \otimes \left\langle \begin{matrix} & b+\sigma & \\ 2b & & 0 \\ & a+\beta & \end{matrix} \right\rangle. \tag{3.359}$$

Thus, one finds

$$\left\{ \begin{matrix} & a+\rho & \\ 2a & & 0 \\ & a+\sigma & \end{matrix} \right\} |(j_1j_2)jm\rangle = \left[\frac{(2a+1)(2j_2+1)}{(2j_2+2\sigma+1)}\right]^{\frac{1}{2}} T^{aa0}_{\rho\sigma 0}|(j_1j_2)jm\rangle. \tag{3.360}$$

The operator $T^{abc}_{\rho\sigma\gamma}$ is a tensor operator of type T^c_γ with respect to the total angular momentum $\mathbf{J}$. Accordingly, it has an expansion in terms of the Wigner operators with respect to $\mathbf{J}$ (basis property):

$$\left\langle \begin{matrix} & c+\tau & \\ 2c & & 0 \\ & c+\gamma & \end{matrix} \right\rangle.$$

These operators have the action on the coupled basis given by

$$\left\langle \begin{matrix} & c+\tau & \\ 2c & & 0 \\ & c+\gamma & \end{matrix} \right\rangle |(j_1j_2)jm\rangle = C^{j\,c\,j+\tau}_{m,\gamma,m+\gamma}|(j_1j_2)j+\tau, m+\gamma\rangle. \tag{3.361}$$

Thus, one must have a relation

$$T^{abc}_{\rho\sigma\gamma} = \sum_\tau \begin{bmatrix} abc \\ \rho\sigma\tau \end{bmatrix} \left\langle \begin{matrix} & c+\tau & \\ 2c & & 0 \\ & c+\gamma & \end{matrix} \right\rangle, \tag{3.362}$$

where

$$\begin{bmatrix} abc \\ \rho\sigma\tau \end{bmatrix} \tag{3.363}$$

is an invariant operator with respect to **J**. Using the Condon and Shortley notation for the reduced matrix elements of the operator (3.363), we find

$$\langle (j_1 + \rho, j_2 + \sigma)j + \tau \| \begin{bmatrix} abc \\ \rho\sigma\tau \end{bmatrix} \| (j_1 j_2)j \rangle$$

$$= [(2j + 1)(2c + 1)(2j_1 + 2\rho + 1)(2j_2 + 2\sigma + 1)]^{\frac{1}{2}}$$

$$\times \begin{Bmatrix} j_1 & j_2 & j \\ a & b & c \\ j_1 + \rho & j_2 + \sigma & j + \tau \end{Bmatrix}. \tag{3.364}$$

The operator $\begin{bmatrix} abc \\ \rho\sigma\tau \end{bmatrix}$ is called a 9-j *invariant operator*, since its matrix elements are (up to dimension factors) the 9-j coefficients.

Remark. The notation for Wigner and Racah operators introduced above is based on the notion of a Gel'fand pattern. This notation is a natural one for basis vectors and tensor operators for the group $U(2)$, as well as for $SU(2)$, and has an immediate generalization to $U(n)$ (see Chapter 5, Appendix A). A $U(2)$ Wigner operator, for example, is denoted by

$$\left\langle \begin{matrix} & \Gamma_{11} & \\ M_{12} & & M_{22} \\ & M_{11} & \end{matrix} \right\rangle, \tag{3.365}$$

where the entries in this pattern may be any integers (negative, zero, or positive) that satisfy

$$M_{12} \geqslant M_{11} \geqslant M_{22}; M_{12} \geqslant \Gamma_{11} \geqslant M_{22}. \tag{3.366}$$

When a $U(2)$ Wigner operator is restricted in its action to an $SU(2)$ basis vector $|jm\rangle$, it gives the same result as an $SU(2)$ Wigner operator:

$$\left\langle \begin{matrix} & \Gamma_{11} & \\ M_{12} & & M_{22} \\ & M_{11} & \end{matrix} \right\rangle \sim \left\langle \begin{matrix} & J + \Delta & \\ 2J & & 0 \\ & J + M & \end{matrix} \right\rangle, \tag{3.367}$$

where

$$J = (M_{12} - M_{22})/2,$$

$$M = M_{11} - (M_{12} + M_{22})/2,$$

$$\Delta = \Gamma_{11} - (M_{12} + M_{22})/2. \tag{3.368}$$

It is sometimes convenient to use the general $U(2)$ notation even in discussions of $SU(2)$ Wigner operators. Equations (3.367) and (3.368) then given the rule for mapping a $U(2)$ Wigner operator to its equivalent $SU(2)$ version.

A rule identical to Eqs. (3.367) and (3.368) also holds for Racah operators.

Tables 5–7 in the Appendix of Tables give further properties of some special Wigner and Racah operators, respectively.

22. Notes

1. *Phases of the angular momentum matrices.* A state of a physical system in quantum mechanics is determined only up to a phase (a state is a *ray* in Hilbert space). This implies that a phase transformation of the "standard" angular momentum state $|jm\rangle$ cannot alter the physical predictions of angular momentum theory:

$$|jm\rangle \to |jm\rangle' \equiv e^{i\alpha(j,m)}|jm\rangle, \tag{3.369}$$

where the real numbers

$$\alpha(j, j), \alpha(j, j-1), \ldots, \alpha(j, -j) \tag{3.370}$$

may be chosen arbitrarily.

The effect of this transformation on the "standard" relations, Eqs. (3.16) and (3.20), is to leave the eigenvalue equations unaltered but to modify the latter to

$$J_+|jm\rangle' = e^{i\delta(j,m)}[(j-m)(j+m+1)]^{\frac{1}{2}}|jm+1\rangle',$$

$$J_-|jm\rangle' = e^{-i\delta(j,m-1)}[(j+m)(j-m+1)]^{\frac{1}{2}}|jm-1\rangle', \tag{3.371}$$

where

$$\delta(j, j) \equiv \alpha(j, j),$$

$$\delta(j, m) \equiv \alpha(j, m) - \alpha(j, m+1),$$

$$m = j-1, \ldots, -j. \tag{3.372}$$

Conversely, the real numbers

$$\delta(j, j), \delta(j, j-1), \ldots, \delta(j, -j) \tag{3.373}$$

may be chosen arbitrarily in Eqs. (3.371), in which case these equations may be put in standard form by defining the state vectors $|jm\rangle$ by Eq. (3.369), where

$$\alpha(j, m) \equiv \sum_{\mu=m}^{j} \delta(j, \mu). \tag{3.374}$$

Equations (3.371) yield the most general form of the angular momentum matrices where the factors (3.370) are fully arbitrary, but, as we have shown, these equations may always be put in standard form.

An unusual choice of phase occurs in Schiff [69] in the theory of orbital angular momentum because of the way in which the spherical harmonics are defined. Using the Y_{lm} of Section 10 as standard, the Schiff phase choice in Eqs. (3.369) corresponds to

$$\delta(l, l) = l\pi, \qquad \delta(l, m) = -\pi \; (m = 0, 1, \ldots, l-1), \qquad \delta(l, m) = 0 \; (m < 0), \tag{3.375}$$

yielding

$$\alpha(l, m) = m\pi \; (m \geqslant 0), \qquad \alpha(l, m) = 0 \; (m < 0), \tag{3.376}$$

$$Y_{lm}^{\text{Schiff}} = \begin{cases} (-1)^m Y_{lm}, & m \geqslant 0 \\ Y_{lm}, & m < 0. \end{cases} \tag{3.377}$$

2. *The method of principal idempotents and projection operators.* Let us first show how the method of principal idempotents may be used to determine the matrices $d^j(\beta)$. The $2j+1$ Hermitian matrices (Perlis [70]) defined by

$$E_m^{(j)} \equiv \prod_{\substack{\mu=-j \\ \mu \neq m}}^{j} \frac{J_2^{(j)} - \mu \mathbb{1}^{(j)}}{m - \mu}, \qquad m = j, j-1, \ldots, -j, \tag{3.378}$$

are called the *principal idempotents* of $J_2^{(j)}$. They possess the multiplication properties

$$E_{m'}^{(j)} E_m^{(j)} = \delta_{m'm} E_m^{(j)}; \tag{3.379}$$

are a resolution of the unit matrix $\mathbb{1}^{(j)}$ of dimension $2j+1$,

$$\mathbb{1}^{(j)} = \sum_m E_m^{(j)}; \tag{3.380}$$

and $J_2^{(j)}$ itself may be written (a reexpression of the characteristic equation) as

$$J_2^{(j)} = \sum_m m E_m^{(j)}. \tag{3.381}$$

This result and the multiplication property (3.379) may now be used to prove

$$F(J_2^{(j)}) = \sum_m F(m)E_m^{(j)} \tag{3.382}$$

for F any well-defined function of one variable, say, ζ.

In particular, for $F(\zeta) = e^{-i\beta\zeta}$, we obtain

$$e^{-i\beta J_2^{(j)}} = \sum_\mu e^{-i\mu\beta} E_\mu^{(j)}, \tag{3.383}$$

which expresses $d^j(\beta)$ as a polynomial of degree $2j$ in the matrix $J_2^{(j)}$.

We thus obtain

$$d^j_{m'm}(\beta) = \sum_\mu (E_\mu^{(j)})_{m'm} e^{-i\mu\beta}. \tag{3.384}$$

Comparison of this result with Eq. (3.78) shows that the element in row m' and column m of the matrix $E_\mu^{(j)}$ defined by Eq. (3.378) is

$$(E_\mu^{(j)})_{m'm} = D^{j*}_{\mu m'}\left(\frac{\pi}{2}, \hat{e}_1\right) D^j_{\mu m}\left(\frac{\pi}{2}, \hat{e}_1\right).$$

The method of principal idempotents is closely related to the *projection operator* technique used by Löwdin [41] and Calais [71] (see also Slater [72, p. 89], Bethe and Jackiw [73, p. 108], and Lehrer-Ilamed [73a]). The approach outlined above has been considered in more detail by Hull [74].

Projection operator techniques employing the rotation matrices are also used, in a very different way, to construct wave functions of sharp angular momentum (Peierls and Yoccoz [74a]). This method uses the operator P^j_{mk} defined by

$$P^j_{mk} \equiv \frac{2j+1}{8\pi^2} \int d\Omega D^{j*}_{mk}(\alpha\beta\gamma)\, \mathscr{U}(\alpha\beta\gamma),$$

where $\mathscr{U}(\alpha\beta\gamma)$ is the rotation operator given by Eq. (3.52). These operators satisfy, however, the rules for projection operators (each operator is idempotent and distinct operators yield perpendicular projections) *only for the case* $m = k$. Disregard of this restriction has led to numerous errors in the literature as discussed by Lamme and Boeker [74b] (see also, Sorensen [74c]). Bohr and Mottelson [74d, pp. 90–91], for example, incorrectly use the D-matrices to project states of sharp (J, M, K) from the intrinsic wave functions.

3. *Trace of the rotation matrix*. Let $G = \{g_1, g_2, \ldots, g_n\}$ denote a finite subgroup (of order n) of $SU(2)$ [or $SO(3)$], and let the element g_i be a rotation with angle ψ_i about direction $\hat{n}_i$. Then the correspondence $g_i \to D^j(\psi_i, \hat{n}_i)$ is a representation of G by unitary matrices of dimension $2j + 1$,—that is, $g_i g_k = g_l$ implies the corresponding multiplication property for the matrices: $D^j(\psi_i, \hat{n}_i)D^j(\psi_k, \hat{n}_k) = D^j(\psi_l, \hat{n}_l)$. This representation is (generally) reducible.

In the theory of finite groups, one finds general formulas for determining which irreducible representations of a group will occur, and with what multiplicity, in a given reducible representation. These formulas require knowledge of the trace (character) of the reducible representation, which in our case is tr $D^j(\psi_i, \hat{n}_i)$. We give here the proof of the general result:

$$\mathrm{tr} D^j(\psi, \hat{n}) = \frac{\sin(j + \frac{1}{2})\psi}{\sin(\psi/2)}. \tag{3.385}$$

The proof is given by observing [see Eq. (3.45)] that the Hermitian matrix $\hat{n} \cdot \mathbf{J}^{(j)}$ is unitarily equivalent (by a unitary similarity transformation) to $J_3^{(j)}$; hence $D^j(\psi, \hat{n})$ is unitarily equivalent to $e^{-i\psi J_3^{(j)}}$. Since the trace of a matrix is invariant under unitary similarity transformations, we obtain

$$\mathrm{tr}\, D^j(\psi, n) = \sum_{m=-j}^{j} e^{-im\psi},$$

which is reduced to Eq. (3.385) by using standard trigonometric formulas.

4. *Rotation matrices are spherical harmonics in four-space.* Casimir [23] points out that the rotation matrices are solutions to Laplace's equation in four-space (spherical harmonics). This result is not surprising in view of the fact that the Euler–Rodrigues parameters define a point on the unit sphere in four-space.

Casimir's observation becomes particularly evident when examined from the viewpoint of the quaternionic multiplication rule (2.42) discussed in Note 3 of Chapter 2.

The occurrence of the vector product in the quaternionic multiplication rule, together with the definition (3.1) of orbital angular momentum, suggests that we consider the multiplication of the "position quaternion," $(x_0, \mathbf{x})$, with the negative of the conjugate of the "momentum quaternion," $(-p_0, \mathbf{p})$. This multiplication gives the results

$$(x_0, \mathbf{x})(-p_0, \mathbf{p}) = (2i\mathscr{J}_0, 2\mathscr{J}), \tag{3.386}$$

where $\mathscr{J}_0$ and $\mathscr{J}$ are defined by

$$\mathscr{J}_0 \equiv \frac{i}{2}(x_0 p_0 + \mathbf{x} \cdot \mathbf{p}),$$

$$\mathscr{J} \equiv \tfrac{1}{2}[\mathbf{x} \times \mathbf{p} - (\mathbf{x} p_0 - x_0 \mathbf{p})]. \tag{3.387}$$

[The factors $2i$ and 2 have been included in Eq. (3.386) to fix conveniently certain properties of $\mathscr{J}_0$ and $\mathscr{J}$ – for example, the commutation relation, Eq. (3.391), for $\mathscr{J}$ and the eigenvalue relation, Eq. (3.397), for $\mathscr{J}_0$.]

A similar multiplication of the negative of the conjugate position quaternion with the momentum quaternion gives the results

$$(-x_0, \mathbf{x})(p_0, \mathbf{p}) = (2i\mathscr{J}_0, 2\mathscr{K}), \tag{3.388}$$

where $\mathscr{K}$ is defined by

$$\mathscr{K} \equiv \tfrac{1}{2}[\mathbf{x} \times \mathbf{p} + (\mathbf{x} p_0 - x_0 \mathbf{p})]. \tag{3.389}$$

We next identify the momentum quaternion $(p_0, \mathbf{p})$ with the operators

$$p_0 = -i\frac{\partial}{\partial x_0}, \qquad \mathbf{p} = -i\boldsymbol{\nabla}, \tag{3.390}$$

and derive, by direct calculation, the following properties of $\mathscr{J}$ and $\mathscr{K}$:

(*a*) Standard angular momentum commutation relations:

$$\begin{aligned} \mathscr{J} \times \mathscr{J} &= i\mathscr{J}, \\ \mathscr{K} \times \mathscr{K} &= i\mathscr{K}; \end{aligned} \tag{3.391}$$

(*b*) Commutivity of $\mathscr{J}$ and $\mathscr{K}$:

$$[\mathscr{J}_i, \mathscr{K}_j] = 0, \qquad i, j = 1, 2, 3; \tag{3.392}$$

(*c*) Common invariant:

$$\mathscr{J}^2 = \mathscr{J} \cdot \mathscr{J} = \mathscr{K} \cdot \mathscr{K} = -\tfrac{1}{4}R_4^2\boldsymbol{\nabla}_4^2 + \mathscr{J}_0^2 + \mathscr{J}_0, \tag{3.393}$$

where $R_4^2 = x_0^2 + \mathbf{x} \cdot \mathbf{x}$, and $\boldsymbol{\nabla}_4^2$ is the Laplacian in four-space,

$$\boldsymbol{\nabla}_4^2 = \frac{\partial^2}{\partial x_0^2} + \boldsymbol{\nabla}^2. \tag{3.394}$$

(*d*) $\mathscr{J}$ and $\mathscr{K}$ related:

$$-\mathscr{K}_j = \sum_i R_{ij}\mathscr{J}_i = \sum_i \mathscr{J}_i R_{ij}, \tag{3.395}$$

where R_{ij} is an element of a real, proper, orthogonal matrix $R = (R_{ij})$, which is defined for each quaternion $(x_0, \mathbf{x}) \neq (0, 0, 0, 0)$ by

$$R_{ij} \equiv [(x_0^2 - \mathbf{x} \cdot \mathbf{x})\delta_{ij} - 2e_{ijk}x_0x_k + 2x_ix_j]/(x_0^2 + \mathbf{x} \cdot \mathbf{x}). \tag{3.396}$$

This is a mapping of all points of four-space (except the origin) into the group of proper, orthogonal matrices; for $x_0^2 + \mathbf{x} \cdot \mathbf{x} = 1$, it is just the Euler–Rodrigues parametrization, Eq. (2.22).

Consider now the functions obtained from $D^{j*}_{m'm}(\alpha_0, \boldsymbol{\alpha})$ [see Eq. (3.89)] by identifying $x_0 = \alpha_0$, $\mathbf{x} = \boldsymbol{\alpha}$, and extending the domain of definition to all points $(x_0, \mathbf{x})$ of four-space. Then the $D^{j*}_{m'm}(x_0, \mathbf{x})$ are *homogeneous harmonic polynomials of degree* $2j$—that is,

$$\begin{aligned} \mathscr{J}_0 D^{j*}_{m'm}(x_0, \mathbf{x}) &= jD^{j*}_{m'm}(x_0, \mathbf{x}), \\ \boldsymbol{\nabla}_4^2 D^{j*}_{m'm}(x_0, \mathbf{x}) &= 0. \end{aligned} \tag{3.397}$$

Furthermore, the action of $\mathscr{J}$ and $\mathscr{P} \equiv -\mathscr{K}$ on $D^{j*}_{m'm}(x_0, \mathbf{x})$ is precisely that obtained from Eqs. (3.104) and (3.123) by replacing the Euler angle variables $(\alpha\beta\gamma)$ by $(x_0, \mathbf{x})$:

$$\mathscr{J}_{\pm} D^{j*}_{m'm}(x_0, \mathbf{x}) = [(j \mp m')(j \pm m' + 1)]^{\frac{1}{2}} D^{j*}_{m' \pm 1, m}(x_0, \mathbf{x}),$$

$$\mathscr{J}_3 D^{j*}_{m'm}(x_0, \mathbf{x}) = m' D^{j*}_{m'm}(x_0, \mathbf{x});$$

$$(-\mathscr{K}_1 \pm i\mathscr{K}_2) D^{j*}_{m'm}(x_0, \mathbf{x}) = [(j \mp m)(j \pm m + 1)]^{\frac{1}{2}} D^{j*}_{m', m \pm 1}(x_0, \mathbf{x}),$$

$$-\mathscr{K}_3 D^{j*}_{m'm}(x_0, \mathbf{x}) = m D^{j*}_{m'm}(x_0, \mathbf{x});$$

$$\mathscr{J}^2 D^{j*}_{m'm}(x_0, \mathbf{x}) = j(j+1) D^{j*}_{m'm}(x_0, \mathbf{x}). \tag{3.398}$$

Let us sketch the proof of these results, Eqs. (3.397)–(3.398): One finds the highest-weight $D^{j*}_{jj}(x_0, \mathbf{x})$ by requiring it to be a *harmonic polynomial* of degree $2j$, which is annihilated by $\mathscr{J}_+$ and $-\mathscr{K}_1 + i\mathscr{K}_2$ and having j as the eigenvalue of $\mathscr{J}_3$ and $-\mathscr{K}_3$. The solution is uniquely

$$D^{j*}_{jj}(x_0, \mathbf{x}) = (x_0 + ix_3)^{2j}, \tag{3.399}$$

where $D^{j*}_{jj}(1, 0, 0, 0) = 1$ determines a multiplying constant to be unity. The general function is now generated by the lowering procedure:

$$D^{j*}_{m'm}(x_0, \mathbf{x}) = \left[\frac{(j+m')!(j+m)!}{(2j)!(j-m')!(2j)!(j-m)!}\right]^{\frac{1}{2}}$$

$$\times (\mathscr{J}_-)^{j-m'}(-\mathscr{K}_1 - i\mathscr{K}_2)^{j-m}(x_0 + ix_3)^{2j}. \tag{3.400}$$

Carried out explicitly, this result yields the complex conjugate of Eq. (3.89) in which we replace α_0 by x_0 and $\boldsymbol{\alpha}$ by $\mathbf{x}$.

The harmonic property, Eq. (3.397), follows from the fact that the Laplacian, ∇_4^2, commutes with $\mathscr{J}$ and $\mathscr{K}$; hence, in its action on $D^{j*}_{m'm}(x_0, \mathbf{x})$, as defined by Eq. (3.400), one may move ∇_4^2 through the lowering operators so that it acts on the highest weight, giving zero. Alternatively, the same result follows from Eq. (3.393) and the eigenvalue equations for $\mathscr{J}_0$ and $\mathscr{J}^2$.

The structure outlined here thus generalizes the theory of the solid body to arbitrary quaternionic variables $(x_0, \mathbf{x})$.

One may recover the theory of the solid body, *in any parametrization*, simply by transforming the operators $\mathscr{J}$, $\mathscr{K}$, $\mathscr{J}_0$, ∇_4^2, and the functions $D^{j*}_{m'm}(x_0, \mathbf{x})$ from $(x_0, \mathbf{x})$ to the desired parameters [for example, the transformation (3.133) of the results of this Note will reproduce the equations of Section 8].

5. *The product space of kinematically independent systems is the tensor product space.* Let $\mathscr{H}$ and $\mathscr{K}$ denote vector spaces (over the same field, which we always consider to be $\mathbb{C}$) with bases $\phi_1, \phi_2, \ldots, \phi_n$ and $\psi_1, \psi_2, \ldots, \psi_m$, respectively. Suppose we are given a mapping f, which assigns to every pair of vectors (ϕ, ψ), $\phi \in \mathscr{H}$, $\psi \in \mathscr{K}$, a vector, denoted by $\phi \otimes \psi$, which belongs to a third vector space $\mathscr{L}$:

$$f\colon (\phi, \psi) \to \phi \otimes \psi \in \mathscr{L}. \tag{3.401}$$

We suppose further that the mapping f has the properties that (a) it is linear in ϕ and ψ separately (bilinear)—that is,

$$(\phi + \phi') \otimes \psi = \phi \otimes \psi + \phi' \otimes \psi,$$

$$\phi \otimes (\psi + \psi') = \phi \otimes \psi + \phi \otimes \psi',$$

$$(\alpha\phi) \otimes \psi = \phi \otimes (\alpha\psi) = \alpha(\phi \otimes \psi); \qquad (3.402)$$

and (*b*) it maps the set of nm pairs (ϕ_i, ψ_j) $(i = 1, 2, \ldots, n; j = 1, 2, \ldots, m)$ into a set

$$\phi_i \otimes \psi_j \qquad (i = 1, 2, \ldots, n;\ j = 1, 2, \ldots, m) \qquad (3.403)$$

of linearly independent vectors (vectors in $\mathscr{L}$).

The vector space (subspace of $\mathscr{L}$, which may be $\mathscr{L}$ itself) obtained by forming all linear combinations,

$$\sum_{i=1}^{n} \sum_{j=1}^{m} a_{ij}\phi_i \otimes \psi_j, \qquad (3.404)$$

is denoted by $\mathscr{H} \otimes \mathscr{K}$ and is called the *tensor product space* of $\mathscr{H}$ and $\mathscr{K}$.

Observe from the bilinear properties (3.402) that

$$\phi \otimes \psi = \sum_{ij} a_i b_j \phi_i \otimes \psi_j \qquad (3.405)$$

for $\phi = \sum_i a_i\phi_i$ and $\psi = \sum_j b_j\psi_j$. Thus, the vector space $\mathscr{H} \otimes \mathscr{K}$ contains vectors that are not of the form $\phi \otimes \psi$ – that is, vectors that are not mappings (by f) of any pair (ϕ, ψ). (See Weyl [24c, p. 92] for an interpretation of this result in terms of "pure" and "mixed" states in quantum mechanics.)

In the prototypical construction of a tensor product space, one first maps the spaces $\mathscr{H}$ and $\mathscr{K}$, to the vector spaces $C^{(n)}$ and $C^{(m)}$ of column matrices composed from the components $(a_1, a_2, \ldots, a_n)$ and $(b_1, b_2, \ldots, b_m)$ of $\phi = \sum_i a_i\phi_i$ and $\psi = \sum_j b_j\psi_j$ relative to bases of the respective spaces. Thus, putting $\mathbf{a} = \text{col}(a_1, a_2, \ldots, a_n)$ and $\mathbf{b} = \text{col}(b_1, b_2, \ldots, b_m)$, the vector $\mathbf{a} \otimes \mathbf{b}$ is defined to be the vector in $C^{(nm)}$ having a_ib_j as its $[(i-1)n + j]$th component, $i = 1, 2, \ldots, n; j = 1, 2, \ldots, m$.

Using this definition of $\mathbf{a} \otimes \mathbf{b}$, one may now verify the bilinear properties (3.402). Furthermore, if $\mathbf{e}_1, \mathbf{e}_2, \ldots, \mathbf{e}_n$ and $\mathbf{f}_1, \mathbf{f}_2, \ldots, \mathbf{f}_m$ are bases of $C^{(n)}$ and $C^{(m)}$, respectively, then $\{\mathbf{e}_i \otimes \mathbf{f}_j : i = 1, 2, \ldots, n;\ j = 1, 2, \ldots, m\}$ is a basis of $C^{(nm)}$. Thus, $C^{(n)} \otimes C^{(m)} = C^{(nm)}$.

Let S and T denote operators defined on the spaces $\mathscr{H}$ and $\mathscr{K}$, respectively:

$$S\colon \phi \to S\phi, \qquad \text{each } \phi \in \mathscr{H},$$

$$T\colon \psi \to T\psi, \qquad \text{each } \psi \in \mathscr{K}. \qquad (3.406)$$

We denote by $S \otimes T$ the operator that is defined on $\mathscr{H} \otimes \mathscr{K}$ by the mapping

$$S \otimes T\colon \phi \otimes \psi \to S\phi \otimes T\psi, \qquad \text{each } \phi \in \mathscr{H},\ \psi \in \mathscr{K}. \qquad (3.407)$$

In the physics literature, it is customary, when no confusion can result, to denote the tensor product $\phi \otimes \psi$ by juxtaposition, $\phi\psi$, and to denote $S\phi \otimes T\psi$ by

$$(ST)\phi\psi \equiv (S\phi)(T\psi). \tag{3.408}$$

Finally, let us note that if $\mathscr{H}$ and $\mathscr{K}$ are equipped with scalar products $(\ ,\ )_1$ and $(\ ,\ )_2$, respectively, then one can also define a scalar product on $\mathscr{H} \otimes \mathscr{K}$. Thus, for each pair of elements of $\mathscr{H} \otimes \mathscr{K}$,

$$\chi = \sum_{ij} a_{ij}\phi_i \otimes \psi_j, \qquad \chi' = \sum_{ij} a'_{ij}\phi_i \otimes \psi_j,$$

we define a scalar product by

$$(\chi, \chi') \equiv \sum_{iji'j'} a^*_{ij}a'_{i'j'}(\phi_i, \phi_{i'})_1(\psi_j, \psi_{j'})_2.$$

6. *Variety of forms of the rotation matrices and the Wigner coefficients.* ▶ Rotation matrices. There is no standard definition (or notation) for the elements of the rotation matrix (see Sections 5 and 6). It is impractical to summarize the many definitions of these functions that one finds in the literature; however, it is useful to discuss some of the reasons for this disparity. These include: (1) Different definitions of Euler angles. [Here one must often finds that the second rotation in the description (2.36) is replaced by a rotation β about the axis $\hat{e}_1 = (1, 0, 0)$, the other two remaining the same. Denoting the corresponding Euler angles by $(\alpha'\beta'\gamma')$, one finds $\alpha' = \alpha + (\pi/2)$, $\beta' = \beta$, $\gamma' = \gamma$.] (2) Different definitions of the unitary rotation operator (acting in Hilbert space) that corresponds to a rotation in $\mathbb{R}^3$. (This relationship is determined only up to unitary equivalence.) (3) Different definitions of the homomorphism of the group of unitary transformations [$SU(2)$] onto the group of proper orthogonal transformations [$SO(3)$]. (4) Different definitions of the rotation operator (acting in Hilbert space) that corresponds to $SU(2)$-rotations in $\mathbb{C}^2$ and $SO(3)$-rotations in $\mathbb{R}^3$.

Let us recall the conventions used in this monograph. For this discussion it is convenient to use the $(\psi, \hat{n})$ parametrization:

$$U(\psi, \hat{n}) \to R(\psi, \hat{n}) \to \mathscr{U}(\psi, \hat{n}) \equiv e^{-i\psi\hat{n}\cdot\mathbf{J}},$$

$$\mathscr{U}(\psi, \hat{n})|jm\rangle = \sum_{m'} D^j_{m'm}(\psi, \hat{n})|jm'\rangle.$$

Here $U(\psi, \hat{n}) \in SU(2)$ and $R(\psi, \hat{n}) \in SO(3)$ are related by the homomorphism (2.16), which may also be expressed as $e^{-i\psi\hat{n}\cdot\boldsymbol{\sigma}/2} \to e^{-i\psi\hat{n}\cdot\mathbf{M}}$, where $\mathbf{M} \equiv A\mathbf{J}^{(1)}A^\dagger$ [see Eq. (3.30)]; alternatively, $U(\psi, \hat{n})$ and $R(\psi, \hat{n})$ are given explicitly by Eqs. (2.27) and (2.6). Moreover, the angular momentum $\mathbf{J}$ has the standard action [see Eqs. (3.16) and (3.20)] on the basis $\{|jm\rangle\}$, and the rotation matrix elements, $D^j_{m'm}(\psi, \hat{n})$, are given explicitly by Eq. (3.89). In this monograph all specific realizations of the unitary operator $\mathscr{U}(\psi, \hat{n})$ and the Hilbert space $\mathscr{H}$ (spanned by the basis vectors $\{|jm\rangle\}$) are chosen to conform to the above conventions.

For comparison with other conventions, it is useful to give the explicit differential operator realization of the operator $\mathscr{U}(\phi, \hat{n})$ $(0 \leqslant \phi \leqslant \pi; \hat{n} \cdot \hat{n} = 1)$ and its action on (a suitable class of) functions defined on $\mathbb{R}^3$:

$$e^{-i\phi\hat{n}\cdot\mathbf{L}}F(\mathbf{x}) = F(R^{-1}(\phi, \hat{n})\mathbf{x}), \tag{$*$}$$

where $\mathbf{L}$ denotes the differential operator, $\mathbf{L} = -i\mathbf{x} \times \nabla$, and the point $\mathbf{x}' \equiv R^{-1}(\phi, \hat{n})\mathbf{x}$ has components $x'_i = \sum_j R_{ji}(\phi, \hat{n})x_j$. The solid harmonics $\mathscr{Y}_{lm}(\mathbf{x})$ are the basis functions that undergo the standard transformation:

$$e^{-i\phi\hat{n}\cdot\mathbf{L}}\mathscr{Y}_{lm}(\mathbf{x}) = \sum_{m'} D^l_{m'm}(\phi, \hat{n})\mathscr{Y}_{lm'}(\mathbf{x}).$$

It is useful to note that Eq. $(*)$ is a mathematical identity (for a suitably defined class of differentiable functions) that expresses the action of a differential operator on a function defined on $\mathbb{R}^3$ in terms of the same function evaluated at a new point of $\mathbb{R}^3$.

The property $(*)$ is the analog, for rotations, of the Taylor series expansion for displacements:

$$e^{-i\mathbf{a}\cdot\mathbf{p}/\hbar}F(\mathbf{x}) = F(\mathbf{x} - \mathbf{a}),$$

where $\mathbf{p}/\hbar = -i\nabla$ and $\mathbf{a} = (a_1, a_2, a_3)$ is a numerical vector.

There are many incorrect versions of Eq. $(*)$ in the physics literature. For example, in Ref. [16] one finds the relation [combine Eqs. (4.1.2), (4.1.8), and set $L_y = L_2$, $L_z = L_3$]:

$$e^{i\alpha L_3}e^{i\beta L_2}e^{i\gamma L_3}f(\mathbf{x}) = f(\mathbf{x}'). \tag{$**$}$$

Here (following Ref. [16]) $\mathbf{x}$ and $\mathbf{x}'$ are the coordinates of the same point P referred to reference frames F and F', which are related by Euler angles $(\alpha\beta\gamma)$. Accordingly, these coordinates are related by $\mathbf{x}' = R^{-1}(\alpha\beta\gamma)\mathbf{x}$ (column vector notation, where $R(\alpha\beta\gamma)$ is given by Eq. (2.37). This is an incorrect result for $\mathbf{x}'$ in Eq. $(**)$, as we now show.

The correct result for $\mathbf{x}'$ in Eq. $(**)$ is obtained by three applications of Eq. $(*)$, which yield: $\mathbf{x}' = S^{-1}(\alpha\beta\gamma)\mathbf{x}$, where

$$\begin{aligned} S(\alpha\beta\gamma) &\equiv R(\alpha, -\hat{e}_3)R(\beta, -\hat{e}_2)R(\gamma, -\hat{e}_3) = [R(\gamma, \hat{e}_3)R(\beta, \hat{e}_2)R(\alpha, \hat{e}_3)]^{-1} \\ &= R^{-1}(\gamma\beta\alpha) = TR(\alpha\beta\gamma)T, \end{aligned}$$

in which T is the diagonal matrix with elements $T_{11} = 1$, $T_{22} = T_{33} = -1$. Thus, the unitary operator $D(\alpha\beta\gamma) \equiv e^{i\alpha L_3}e^{i\beta L_2}e^{i\gamma L_3}$ effects, on functions defined on $\mathbb{R}^3$, the rather unusual transformation of coordinates involving a numerical similarity transformation (since $T^{-1} = T$) of the Euler angle matrix $R(\alpha\beta\gamma)$ by the matrix T.

We may use the homomorphism (2.16) between the groups $U(2)$ and $SO(3)$ to obtain $\sigma_1 \to T$ and $\sigma_1 U(\alpha\beta\gamma)\sigma_1 \to TR(\alpha\beta\gamma)T$ [see Eq. (2.40)]. From this result we find that the representation matrix $\mathscr{D}^{(j)}(\alpha\beta\gamma)$ used in Ref. [16] [which is defined by the matrix element relation, Eq. (4.1.10) (Ref. [16]), and *not* by the incorrect relation discussed above] is unitarily equivalent to the representation matrix $D^j(\alpha\beta\gamma)$ used in this monograph. Explicitly, one finds

$$\mathscr{D}^{(j)}(\alpha\beta\gamma) = D^j(\sigma_1)D^j(\alpha\beta\gamma)D^j(\sigma_1),$$

where the representation matrix $D^j(\sigma_1)$ is given by Eq. (3.86) as the matrix with elements $D^j_{m'm}(\sigma_1) = \delta_{m',-m}$. [We have conveniently used here the fact that the functions (3.86) are also representation functions of the group $U(2)$ (see Appendix E, Chapter 5); this is a detail that could have been avoided by careful use of $i\sigma_1 \in SU(2)$, and the inverse $(i\sigma_1)^{-1} = -i\sigma_1$, in the homomorphism (2.16).]

Using the relations and techniques given in this note, one can usually determine the explicit equivalence relation between (properly defined) rotation matrices used by various authors.

▶ Wigner coefficients. The van der Waerden form of the Wigner coefficients, Eq. (3.170), is important because it exhibits *termwise* the symmetry properties discussed in Section 12. (Our derivation of this symmetrical form is given in Chapter 5, Section 8.) Any expression for the Wigner coefficients that does not exhibit this termwise symmetry will be transformed into a new form by application of one or more of the seventy-two symmetries. It is thus apparent that there are, superficially, many "different" expressions for these coefficients. As is evident from the literature, the particular expression obtained depends on the method of derivation. This is discussed further by Smorodinskiĭ and Shelepin [75] and by Barut and Wilson [75a].

7. *Notations for Wigner and related coefficients.* Many notations have been introduced in the literature to denote the Wigner and related coefficients. We list here the more common notations together with some of the authors who have used these symbols.

Wigner Coefficients

$S^{(j_1 j_2)}_{j m_1 m_2}$	Wigner [13]
$(j_1 j_2 m_1 m_2 \mid j_1 j_2 j m)$	Condon and Shortley [28] Racah [29] Biedenharn *et al.* [54] Biedenharn [56] Arima *et al.* [64] Edmonds [16] Jucys *et al.* [55]
$(j_1 m_1 j_2 m_2 \mid j m)$	Hamermesh [35]
$(j m \mid m_1 m_2)$ $(j_1 j_2 j m \mid j_1 m_1, j_2 m_2)$	Fano and Racah [59]
$\langle j_1 m_1, j_2 m_2 \mid (j_1 j_2) j m\rangle$	Fano [76]
$(j_1 m_1 j_2 m_2 \mid j_1 j_2 j m)$	de-Shalit and Talmi [77]
$(j_1 j_2 m_1 m_2 \mid j m)$	Heine [78], Judd [79] Davydov [80] Shelepin [62]

$\langle j_1 j_2 m_1 m_2 \| jm \rangle$	Brink and Satchler [81]
$\langle j_1 m_1 j_2 m_2 \| jm \rangle$	Bohr and Mottelson [82]
$\langle j_1 j_2 m_1 m_2 \| j_1 j_2 jm \rangle$	Merzbacher [83] Messiah [84] Hagedorn [45] Abragam and Bleany [85]
$C(j_1 m_1 j_2 m_2, jm)$	Bethe and Jackiw [73]
$C_{j_1 j_2}(jm; m_1 m_2)$	Blatt and Weisskopf [86]
$C^{jm}_{j_1 m_1 j_2 m_2}$	Jahn [87] Alder [88]
$C(j_1 j_2 j; m_1 m_2 m)$	Rose [17]
$C^{j}_{m_1 m_2}$	van der Waerden [31] Landau and Lifshitz [89]
$c^{j_1 j_2}_{jm}(m_1, m_2)$	Fock [90]
$A^{jj_1 j_2}_{mm_1 m_2}$	Eckart [27]
$A^{j_1 j_2 j}_{m_1 m_2 m}$	Tinkham [91]
$C^{j_1 j_2 j}_{m_1 m_2 m}$	Biedenharn [92] Lomont [93] Redmond [94]
$B^{jm}_{j_1 m_1 j_2 m_2}$	Gel'fand *et al.* [44]
$X(j, m, j_1, j_2, m_1)$	Boys and Sahni [95]

Wigner 3-j Symbol

$\begin{pmatrix} j_1 & j_2 & j_3 \\ m_1 & m_2 & m_3 \end{pmatrix}$	Wigner [47]
$(-1)^{j_2+j_3-j_1} V(j_1 j_2 j_3; m_1 m_2 m_3)$	Racah [29]
$(-1)^{j_1-j_2+j_3} S_{j_1 m_1; j_2 m_2; j_3 m_3}$	Landau and Lifshitz [89]
$X(j_1 j_2 j_3; m_1 m_2 m_3)$	Schwinger [37]
$\begin{bmatrix} j_2+j_3-j_1 & j_3+j_1-j_2 & j_1+j_2-j_3 \\ j_1-m_1 & j_2-m_2 & j_3-m_3 \\ j_1+m_1 & j_2+m_2 & j_3+m_3 \end{bmatrix}$	Regge [38]
$(-1)^{j_1+j_2+j}\bar{V}\begin{pmatrix} j_1 & j_2 & j_3 \\ m_1 & m_2 & m_3 \end{pmatrix}$	Fano and Racah [59]

Wigner 6-j Symbol

$\begin{Bmatrix} j_1 & j_2 & j_3 \\ l_1 & l_2 & l_3 \end{Bmatrix}$	Wigner [47]

$(-1)^{j_1+j_2+l_1+l_2}W(j_1j_2l_2l_1;j_3l_3)$	Racah [29]
$(-1)^{j_1+j_2+l_1+l_2}[(2j_3+1)(2l_3+1)]^{\frac{1}{2}} \times U(j_1j_2l_2l_1;j_3l_3)$	Jahn [87]
$[(2j_3+1)(2l_3+1)]^{\frac{1}{2}}U\begin{pmatrix} j_1j_2l_1l_2 \\ j_3l_3 \end{pmatrix}$	Boys and Sahni [95]
$(-1)^{j_1+j_2+l_1+l_2}\begin{bmatrix} j_1 & j_2 & j_3 \\ l_2 & l_1 & \\ & l_3 & \end{bmatrix}$	Banerjee and Saha [96]
$\overline{W}\begin{pmatrix} j_1 & j_2 & j_3 \\ l_1 & l_2 & l_3 \end{pmatrix}$	Fano and Racah [59]
$\begin{Vmatrix} j_1 & j_1 & l_1 & l_1 \\ j_2 & l_2 & l_2 & j_1 \\ j_3 & l_3 & j_3 & l_3 \end{Vmatrix}$	Shelepin [62]

Wigner 9-j Symbol

$\begin{Bmatrix} j_1 & j_2 & j_3 \\ k_1 & k_2 & k_3 \\ l_1 & l_2 & l_3 \end{Bmatrix}$	Wigner [47]
$X\begin{pmatrix} j_1 & j_2 & j_3 \\ k_1 & k_2 & k_3 \\ l_1 & l_2 & l_3 \end{pmatrix}$	Fano [76] Fano and Racah [59]
$(-1)^{j_1+k_2-j_3-l_2}S(j_1j_2k_1k_2;j_3k_3l_1l_2;l_3)$	Schwinger [37]
$[(2j_3+1)(2k_3+1)(2l_1+1)(2l_2+1)]^{\frac{1}{2}} \times \chi(j_1j_2k_1k_2;j_3k_3;l_1l_2;l_3)$	Jahn and Hope [63]
$[(2j_3+1)(2l_3+1)(2l_1+1)(2l_2+1)]^{\frac{1}{2}} \times A\begin{pmatrix} j_1 & j_2 & j_3 \\ k_1 & k_2 & k_3 \\ l_1 & l_2 & l_3 \end{pmatrix}$	Kennedy and Cliff [97]
$U\begin{pmatrix} j_1 & j_2 & j_3 \\ k_1 & k_2 & k_3 \\ l_1 & l_2 & l_3 \end{pmatrix}$	Arima *et al.* [64]

8. *Rotations of kinematically independent physical systems.* Properties of complex composite physical systems may often be approximated (using perturbation theory) by the properties of a number of simpler systems, which are considered as weakly interacting. Each of the simpler systems may then be rotated (active viewpoint) independently of the others without substantially altering the properties of the whole system. (In the passive viewpoint, independent inertial frames, one for each system, are rotated.) It is therefore

useful to consider a collection of n physical systems, each of which may be rotated without influence on the other systems.

Let the quantum eigenkets of sharp angular momentum for the system a ($a = 1, 2, \ldots, n$) be denoted by

$$|(\alpha_a)j_a m_a\rangle. \tag{3.409}$$

Under a rotation of this physical system, these eigenkets undergo the transformation

$$\mathscr{U}_a\colon |(\alpha_a)j_a m_a\rangle \to \mathscr{U}_a|(\alpha_a)j_a m_a\rangle = \sum_{m_a'} D^j_{m_a' m_a}(U_a)|(\alpha_a)j_a m_a'\rangle, \tag{3.410}$$

where for a rotation of ψ_a about direction $\hat{n}_a$, we have

$$\mathscr{U}_a = \exp -i[\psi_a \hat{n}_a \cdot \mathbf{J}(a)], \tag{3.411}$$

in which $\mathbf{J}(a)$ is the generator of rotations of system a.

The state vector space for the composite system is taken (in perturbation theory) to be the tensor product space with basis

$$|(\alpha_1)j_1 m_1\rangle \otimes |(\alpha_2)j_2 m_2\rangle \otimes \cdots \otimes |(\alpha_n)j_n m_n\rangle. \tag{3.412}$$

The operator (see Note 5) denoted by

$$\mathscr{U}_1 \otimes \mathscr{U}_2 \otimes \cdots \otimes \mathscr{U}_n \tag{3.413}$$

then acts on the basis ket (3.412) according to the rule

$$\begin{aligned}
(\mathscr{U}_1 \otimes \cdots \otimes \mathscr{U}_n)&|(\alpha_1)j_1 m_1\rangle \otimes \cdots \otimes |(\alpha_n)j_n m_n\rangle \\
&= (\mathscr{U}_1|(\alpha_1)j_1 m_1\rangle) \otimes \cdots \otimes (\mathscr{U}_n|(\alpha_n)j_n m_n\rangle) \\
&= \sum_{m_1' \cdots m_n'} [D^{j_1}(U_1) \otimes \cdots \otimes D^{j_n}(U_n)]_{m_1' \cdots m_n';\, m_1 \cdots m_n} \\
&\quad \times |(\alpha_1)j_1 m_1'\rangle \otimes \cdots \otimes |(\alpha_n)j_n m_n'\rangle,
\end{aligned} \tag{3.414}$$

where

$$[D^{j_1}(U_1) \otimes \cdots \otimes D^{j_n}(U_n)]_{m_1' \cdots m_n';\, m_1 \cdots m_n} \equiv D^{j_1}_{m_1' m_1}(U_1) \cdots D^{j_n}_{m_n' m_n}(U_n) \tag{3.415}$$

denotes the element in row $(m_1' \cdots m_n')$ and column $(m_1 \cdots m_n)$ of the *matrix direct product*

$$D^{j_1}(U_1) \otimes \cdots \otimes D^{j_n}(U_n). \tag{3.416}$$

[The result given by Eq. (3.414) is obtained directly from the property (extended to n products) of the tensor product space expressed by Eq. (3.405).]

In group-theoretic terms, one says that the $\prod_{a=1}^{n}(2j_a + 1)$-dimensional tensor product space $\mathscr{H}_{j_1}(1) \otimes \cdots \otimes \mathscr{H}_{j_n}(n)$ with basis

$$|(\alpha_1)j_1m_1\rangle \otimes \cdots \otimes |(\alpha_n)j_nm_n\rangle, \qquad m_a = j_a, \ldots, -j_a, \qquad a = 1, \ldots, n \tag{3.417}$$

is the carrier space of a matrix representation of the *direct product group* denoted by

$$SU(2) \times \cdots \times SU(2). \tag{3.418}$$

The elements of this group are the (ordered) n-tuples

$$(U_1, \ldots, U_n), \qquad \text{each } U_a \in SU(2), \tag{3.419}$$

where the multiplication rule for these n-tuples is

$$(U'_1, \ldots, U'_n)(U_1, \ldots, U_n) = (U'_1U_1, \ldots, U'_nU_n). \tag{3.420}$$

The space $\mathscr{H}_{j_1}(1) \otimes \cdots \otimes \mathscr{H}_{j_n}(n)$ then carries the matrix representation $D^{j_1}(U_1) \otimes \cdots \otimes D^{j_n}(U_n)$ of $SU(2) \times \cdots \times SU(2)$.

A rotation of the composite system as a whole means that we choose all the rotations U_a to be the same: $U_1 = U_2 = \cdots = U_n = U$. In this case the group elements $(U, \ldots, U)$ of $SU(2) \times \cdots \times SU(2)$ are seen to form a subgroup, the so-called *diagonal subgroup*, which is abstractly the same as (*isomorphic to*) $SU(2)$ itself.

Thus, the action of a rotation on a composite physical system, each of whose parts has independent rotational symmetry, is to transform the basis states given by Eq. (3.417) among themselves with coefficients that are the elements of the *reducible rotation matrix*

$$D^{j_1}(U) \otimes \cdots \otimes D^{j_n}(U). \tag{3.421}$$

From the point of view of group theory, the construction of states of sharp total angular momentum (j, m) for composite systems in which the individual angular momenta $j_1, \ldots, j_n$ of the parts are conserved is described as the problem of reducing the representation (3.421) of the rotation U into the (irreducible) representations $D^j(U)$, $j = 0, \frac{1}{2}, \ldots$.

Our discussion here serves to show how the two structures—independent rotations of the parts of a physical system, and rotations of the composite whole—are subsumed under a common structure.

9. *The irreducible tensor operator multiplets.* We suppose, as given, a set of linearly independent (over $\mathbb{C}$) linear operators $T_1, T_2, \ldots, T_n$, such that each T_i has a well-defined action on the basic angular momentum eigenkets $\{|(\alpha)jm\rangle\}$ of a physical system. We further assume:

(*a*) Under the commutation action of the angular momentum operators, J_i, the operators T_λ $(\lambda = 1, \ldots, n)$ are transformed linearly (with coefficients from $\mathbb{C}$) among themselves—that is,

$$[J_i, T_\lambda] = \sum_\mu a^{(i)}_{\mu\lambda} T_\mu. \tag{3.422}$$

(*b*) For each pair of operators $S \equiv \sum_\lambda s_\lambda T_\lambda$ and $T \equiv \sum_\lambda t_\lambda T_\lambda$ $(s_\lambda, t_\lambda \in \mathbb{C})$, there is defined a *rotationally invariant scalar product*, denoted by (S,T), such that

$$([J_i, S], T) = (S, [J_i, T]). \tag{3.423}$$

We seek now to classify the operators $T_1, T_2, \ldots, T_n$ with respect to their transformation properties under rotations. The strategy is to cast the problem into the form of the angular momentum multiplet problem already solved in Section 3. The properties expressed by Eqs. (3.207) provide the key to this reformulation.

Let us introduce the *commutation operators* C_i $(i = 1, 2, 3)$ defined by[1]

$$C_i T \equiv [J_i, T]. \tag{3.424}$$

These operators are:

(*a*) Linear

$$C_i(\alpha S + \beta T) = \alpha(C_i S) + \beta(C_i T). \tag{3.425}$$

(*b*) Derivations

$$C_i(ST) = (C_i S)T + S(C_i T). \tag{3.426}$$

One also defines the product $C_i C_j$ by

$$(C_i C_j)T \equiv C_i(C_j T) = [J_i, [J_j, T]]. \tag{3.427}$$

Using this definition, and the Jacobi identity for commutators, we find

$$[C_i, C_j] = i e_{ijk} C_k. \tag{3.428}$$

With this result, we now see that the C_i $(i = 1, 2, 3)$ have the action of angular momentum operators when acting on the set of operators

$$\left\{ \sum_\lambda a_\lambda T_\lambda \right\}. \tag{3.429}$$

Furthermore, Eq. (3.423) expresses the property

$$(C_i S, T) = (S, C_i T); \tag{3.430}$$

that is, the C_i are Hermitian operators on the n-dimensional inner product space with "vectors" $S, T, \ldots$ and scalar (inner) product (S, T).

All the assumptions used in the construction of the basic angular momentum multiplets given in Section 3 are thus seen to be satisfied by the C_i. The details of that construction may now be applied to split the space spanned by the linear combinations $\{\sum_\lambda a_\lambda T_\lambda\}$ into irreducible tensor operator multiplets:

---

[1] The operators C_i are called ad_i in the mathematical literature.

$$T^J_M = \sum_\lambda A^J_{M\lambda} T_\lambda, \qquad M = J, \ldots, -J$$

and

$$J \in \{0, \tfrac{1}{2}, 1, \ldots\}. \tag{3.431}$$

(These multiplets obey, of course, all the standard relations of angular momentum theory under the action of the C_i.)

We conclude this note with one further observation. The operator Ω defined by

$$\Omega T \equiv [\mathbf{J}^2, T] \tag{3.432}$$

commutes with all the C_i:

$$[\Omega, C_i] = 0, \qquad i = 1, 2, 3. \tag{3.433}$$

Ω is also Hermitian. Thus, Ω, C^2, and C_3 may be simultaneously diagonalized on the space spanned by the linear combinations $\{\sum_\lambda a_\lambda T_\lambda\}$.

10. *Examples of tensor operators.* We note here the properties of several types of irreducible tensor operators that occur frequently in applications.

(*a*) The solid harmonics as tensor operators. The orbital angular momentum operators $\vec{L} = \vec{r} \times \vec{p}$ possess the derivation property; hence, the solid harmonics

$$\{\mathscr{Y}_{k\mu}(\mathbf{x}) : \mu = k, \ldots, -k\} \tag{3.434}$$

are the components of a tensor operator $\boldsymbol{\mathscr{Y}}_k$ (with respect to $\vec{L}$).

Since the eigenkets of $\vec{L}$ are themselves the solid harmonics

$$\{\mathscr{Y}_{lm}(\mathbf{x}) : m = l, \ldots, -l\}, \tag{3.435}$$

application of the Wigner–Eckart theorem yields the result

$$\mathscr{Y}_{k\mu}(\mathbf{x}) \mathscr{Y}_{lm}(\mathbf{x}) = \sum_{l'} \langle l' \| \boldsymbol{\mathscr{Y}}_k \| l \rangle \, C^{lkl'}_{m,\mu,m+\mu} \mathscr{Y}_{l',m+\mu}(\mathbf{x}), \tag{3.436}$$

where

$$\langle l' \| \boldsymbol{\mathscr{Y}}_k \| l \rangle \equiv r^{l+k-l'} \left[\frac{(2l+1)(2k+1)}{4\pi(2l'+1)} \right]^{\frac{1}{2}} C^{lkl'}_{000}. \tag{3.437}$$

This expression for the reduced matrix element is obtained from the homogeneity of the solid harmonics and the Gaunt integral (3.192). [Expressions of the type (3.436) for Bessel functions may be derived using properties of the Euclidean group in two-space; this is discussed in Note 13.]

In the form and notation expressed by Eq. (3.436) one is thinking of the tensor operator $\mathscr{Y}_{k\mu}(\mathbf{x})$ as acting on the orbital angular momentum eigenkets

$|lm\rangle$. Thus, from this viewpoint, one has

$$\mathscr{Y}_{k\mu}|lm\rangle = \sum_{l'} \langle l'\|\mathscr{Y}_k\|l\rangle\, C^{lkl'}_{m,\mu,m+\mu}|l', m+\mu\rangle. \tag{3.438}$$

Yet another interpretation of Eq. (3.436) may be given in terms of the tensor product:

$$[\mathscr{Y}_{k_1}(\mathbf{x}) \times \mathscr{Y}_{k_2}(\mathbf{x})]_{k\mu} \equiv \sum_{\mu_1\mu_2} C^{k_1k_2k}_{\mu_1\mu_2\mu}\, \mathscr{Y}_{k_1\mu_1}(\mathbf{x})\mathscr{Y}_{k_2\mu_2}(\mathbf{x}), \tag{3.439}$$

so that the result given by Eq. (3.436) is

$$[\mathscr{Y}_{k_1}(\mathbf{x}) \times \mathscr{Y}_{k_2}(\mathbf{x})]_{k\mu} = \langle k\|\mathscr{Y}_{k_1}\|k_2\rangle \mathscr{Y}_{k\mu}(\mathbf{x}). \tag{3.440}$$

In other words, *the coupled tensor product of two solid harmonics* (of the same argument) *is again a solid harmonic up to an invariant multiplier.*

The simplification given by Eq. (3.440) for solid harmonics of the same argument (acting in the same space $\{|(\alpha)lm\rangle\}$ of eigenkets of $\mathbf{L}$) is not realized for the tensor product $[\mathscr{Y}_{k_1}(\mathbf{x}^1) \times \mathscr{Y}_{k_2}(\mathbf{x}^2)]_{k\mu}$ of solid harmonics $\mathscr{Y}_{k_1\mu_1}(\mathbf{x}^1)$ and $\mathscr{Y}_{k_2\mu_2}(\mathbf{x}^2)$ acting in *different* spaces $\{|(\alpha_1)l_1m_1\rangle\}$ and $\{|(\alpha_2)l_2m_2\rangle\}$. One must, in this case, appeal to the general result given by Eq. (3.249).

(*b*) Irreducible tensors obtained as forms on the angular momentum components. We observed in Section 15 that the angular momentum operator $\mathbf{J}$ is itself a tensor operator of rank 1. This result may be generalized.

Using the notation of Note 9, one has

$$C_+ J_+^k = 0; \qquad k = 0, 1, 2, \ldots . \tag{3.441}$$

Consider the operators $\mathscr{T}^k_\mu$ defined by

$$\mathscr{T}^k_\mu \equiv a_k \left[\frac{(k+\mu)!}{(2k)!(k-\mu)!}\right]^{\frac{1}{2}} C_-^{k-\mu} J_+^k \tag{3.442}$$

for $\mu = k, k-1, \ldots, -k$, where a_k is an arbitrary constant (independent of μ). One finds by a straightforward application of the commutation properties of C_+, C_-, and C_3 that the $\mathscr{T}^k_\mu$ $(\mu = k, \ldots, -k)$ obey the defining relations, Eqs. (3.206), of a tensor of rank k.

Carrying out the repeated commutation with J_- indicated in Eq. (3.442) clearly produces a *polynomial form* (with real coefficients) in J_+, J_-, and J_3. These polynomials are given for $k = 1, 2, 3, 4$ in Table 5 in the Appendix of Tables.

The Wigner–Eckart theorem when applied to the tensor operator $\mathscr{T}^k$ yields

$$\langle(\alpha')j'm'|\mathscr{T}^k_\mu|(\alpha)jm\rangle = \delta_{(\alpha')(\alpha)}\delta_{j'j}\langle j\|\mathscr{T}^k\|j\rangle C^{jkj}_{m\mu m'}, \tag{3.443}$$

where the reduced matrix element is given by

$$\langle j\|\mathscr{T}^k\|j\rangle = a_k(-1)^k \left[\frac{(2j+k+1)!k!k!}{(2j+1)(2j-k)!(2k)!}\right]^{\frac{1}{2}}. \tag{3.444}$$

This result may be obtained by evaluating the matrix elements of J_+^k for $m = j - k$, $\mu = k$, and $m' = j$, and using

$$C^{jkj}_{-k+j,k,j} = (-1)^k \left[\frac{(2j+1)!(2k)!}{(2j+k+1)!k!} \right]^{\frac{1}{2}}.$$

The tensor operators $\mathscr{Y}_k$ and $\mathscr{T}^k$, discussed in (*a*) and (*b*) above, are examples of a more general tensor operator $\mathscr{V}^k$ whose components $\mathscr{V}^k_\mu$ are the polynomial forms defined on the components of an arbitrary vector operator $\mathbf{V}$ by

$$\mathscr{V}^k_\mu \equiv \left[\frac{(k+\mu)!}{(2k)!(k-\mu)!} \right]^{\frac{1}{2}} C_-^{k-\mu} V_+^k. \tag{3.445}$$

(*c*) Other constructions. One may apply the construction (3.445) to tensor operators of arbitrary rank. For example, a spin-$\frac{1}{2}$ tensor operator is, by definition, a pair of operators, u and v, that satisfy

$$\begin{aligned} C_+u &= 0, \qquad & C_-u &= v, \qquad & C_3u &= u/2, \\ C_+v &= u, \qquad & C_-v &= 0, \qquad & C_3v &= -v/2. \end{aligned} \tag{3.446}$$

One may now verify, using the method of (*b*) above, that S^k_μ defined by

$$S^k_\mu \equiv \left[\frac{(k+\mu)!}{(2k)!(k-\mu)!} \right]^{\frac{1}{2}} C_-^{k-\mu} u^k \tag{3.447}$$

is the μth component of a tensor operator $\mathbf{S}^k$ of rank $k/2$ ($k = 1, 2, \ldots$).

S^k_μ is clearly a polynomial form in the components u and v (which may themselves be noncommuting).

11. *Expressions for the 6-j symbol in terms of group integrals.* A Racah invariant $\mathbf{W}^{abc}_{\rho\sigma\tau}$ commutes with the total angular momentum $\mathbf{J}$ and is accordingly invariant under unitary transformations of the space $\mathscr{H}$ (see Section 21) generated by $\mathbf{J}$. Since the angular momentum label j is also preserved under such transformations, the Racah coefficients $W^{abc}_{\rho\sigma\tau}(j)$ are invariants. Equivalently, the 6-j symbols are invariants. It is reasonable, therefore, to expect that there exist relationships between 6-j symbols and group integrals over characters, since such integrals are also invariants. The key result for establishing such relationships is Eq. (3.190). Notice, however, that the triangle of angular momentum labels $(j_1 j_2 j)$ occurs twice in Eq. (3.190). It will be necessary then to consider products of 6-j symbols that contain an even number of identical triangles. Wigner [47] has given four such relations between 6-j coefficients and group integrals over characters (see Note 3). Sharp [98] has given the detailed proofs of each of these four relations together with a geometric interpretation in terms of paths in group space. Sharp has also discussed the freedom of phase remaining in the 6-j symbol when they are considered as defined by Eqs. (3.448)–(3.451) below. We state Wigner's results, proving only the first relation ($V = 2\pi^2$):

$$\begin{Bmatrix} a & b & e \\ d & c & f \end{Bmatrix}^2 = V^{-3} \int d\Omega_U \, d\Omega_V \, d\Omega_W \chi^a(U)\chi^b(V)\chi^e(W)\chi^d(VW^{-1}) \times \chi^c(WU^{-1})\chi^f(UV^{-1}), \tag{3.448}$$

$$\begin{Bmatrix} a & b & e \\ a & b & f \end{Bmatrix} = \frac{(-1)^{2f}}{V^2} \int d\Omega_U \, d\Omega_V \chi^a(UV)\chi^b(U^{-1}V)\chi^e(V)\chi^f(U), \tag{3.449}$$

$$\begin{Bmatrix} a & b & e \\ c & b & f \end{Bmatrix}\begin{Bmatrix} a & b & d \\ c & b & e \end{Bmatrix}\begin{Bmatrix} a & b & f \\ c & b & d \end{Bmatrix} = \frac{(-1)^{2a}}{V^4} \int d\Omega_U \, d\Omega_V \, d\Omega_W \, d\Omega_X \chi^f(U)\chi^e(V)\chi^d(W) \times \chi^a(XV)\chi^b(XWVUX)\chi^c(UXW), \tag{3.450}$$

$$\begin{Bmatrix} a & b & e \\ d & c & f \end{Bmatrix}\begin{Bmatrix} a & c & f \\ a & b & e \end{Bmatrix}\begin{Bmatrix} d & b & f \\ a & b & e \end{Bmatrix}\begin{Bmatrix} d & c & e \\ a & b & e \end{Bmatrix} = \frac{(-1)^{2d}}{V^5} \int d\Omega_U \, d\Omega_V \, d\Omega_W \, d\Omega_X \, d\Omega_Y \chi^d(U)\chi^c(V)\chi^f(W) \times \chi^a(X)\chi^a(Y^{-1}WV)\chi^b(X^{-1}YUW)\chi^e(YXVU). \tag{3.451}$$

Proof of Eq. (3.448). We proceed directly from the definition (3.266) of Racah coefficients, first summing that result over γ. Squaring the resulting equation and identifying the various products of Wigner coefficients with the integral (3.190), we obtain

$$\begin{aligned} W^2(abcd;ef) = {} & V^{-4} \sum_{\substack{\text{all} \\ \text{subscripts}}} \int d\Omega_U D^{f*}_{\rho'\rho}(U) D^b_{\beta'\beta}(U) D^d_{\delta'\delta}(U) \\ & \times \left(\int d\Omega_V D^{c*}_{\gamma'\gamma}(V) D^e_{\varepsilon'\varepsilon}(V) D^d_{\delta'\delta}(V) \right)^* \\ & \times \left(\int d\Omega_W D^{e*}_{\varepsilon'\varepsilon}(W) D^a_{\alpha'\alpha}(W) D^b_{\beta'\beta}(W) \right)^* \\ & \times \int d\Omega_X D^{c*}_{\gamma'\gamma}(X) D^a_{\alpha'\alpha}(X) D^f_{\rho'\rho}(X) \\ = {} & V^{-4} \int d\Omega_U \, d\Omega_V \, d\Omega_W \, d\Omega_X \\ & \times \chi^a(W^{-1}X)\chi^d(V^{-1}U)\chi^e(V^{-1}W) \\ & \times \chi^f(U^{-1}X)\chi^c(X^{-1}V)\chi^b(W^{-1}U). \end{aligned} \tag{3.452}$$

The relation

$$\sum_{m'm} D^{j*}_{m'm}(U_1)D^{j}_{m'm}(U_2) = \chi^j(U_1^{-1}U_2) \tag{3.453}$$

has been used repeatedly in obtaining the right-hand side of Eq. (3.452). We now put $U_1 = W^{-1}X$, $U_2 = W^{-1}U$, and $U_3 = V^{-1}W$ in Eq. (3.452) and use the invariance of the integral to translations, thus obtaining

$$\begin{aligned} W^2(abcd;ef) = V^{-4} \int d\Omega_{U_1}\, d\Omega_{U_2}\, d\Omega_{U_3}\, d\Omega_X \\ \times \chi^a(U_1)\chi^b(U_2)\chi^e(U_3) \\ \times \chi^d(U_3U_2)\chi^c(U_1^{-1}U_3^{-1})\chi^f(U_2^{-1}U_1). \end{aligned} \tag{3.454}$$

Setting $U = U_1^{-1}$, $V = U_2^{-1}$, $W = U_3$, using the property $\chi^j(U) = \chi^j(U^{-1})$ of characters, and integrating over $d\Omega_X$, we obtain the desired result, Eq. (3.448). ■

12. *Expansion of physical tensor operators in terms of Wigner operators.* The transformations effected in angular momentum spaces (spaces of sharp j and m) by physical interaction operators may often be expressed concisely in terms of Wigner operators (or Racah operators). (For applications of this concept, see Chapter 7, Sections 5, 7, 9, 10.) For later reference it is useful to discuss here the underlying operator relations.

Let us assume that one is given an operator T with an action defined on a physical state vector space that is spanned by an angular momentum basis $\{|(\alpha)jm\rangle\}$ (angular momentum $\mathbf{J}$). It is often possible to represent T by an expansion in terms of a set of (physical) irreducible tensor operators (with respect to $\mathbf{J}$), say, $\{T^k_\mu : k = 0, \frac{1}{2}, 1, \ldots, \mu = k, k-1, \ldots, -k\}$. Let us therefore assume an expansion of the form

$$T = \sum_{k\mu} t^k_\mu T^k_\mu, \qquad t^k_\mu \in \mathbb{C}. \tag{3.455}$$

The operator T then has the matrix elements given by

$$\langle(\alpha')j'm'|T|(\alpha)jm\rangle = \sum_{k\mu} t^k_\mu \langle(\alpha')j'\|\mathbf{T}^k\|(\alpha)j\rangle C^{jkj'}_{m\mu m'}. \tag{3.456}$$

In many physical applications (for example, perturbation theory) one is interested only in the set of matrix elements of the operator T that connect particular angular momentum subspaces, say, the $(2j'+1)$-dimensional space $\mathscr{H}_{j'}(\alpha')$ spanned by the vectors $\{|(\alpha')j'm'\rangle : m' = j', \ldots, -j'\}$ and the $(2j+1)$-dimensional space $\mathscr{H}_j(\alpha)$ spanned by the basis vectors $\{|(\alpha)jm\rangle : m = j, \ldots, -j\}$. Using the matrix elements of the operator T, we may associate with T the mapping $T(\alpha', \alpha): \mathscr{H}_{j'}(\alpha') \to \mathscr{H}_j(\alpha)$ defined by

$$T(\alpha',\alpha)|(\alpha)jm\rangle \equiv \sum_{j'm'} \langle(\alpha')j'm'|T|(\alpha)jm\rangle\, |(\alpha')j'm'\rangle. \tag{3.457}$$

Using the explicit matrix elements (3.456) and the definition (3.341) of a Wigner operator, we may write Eq. (3.457) in the form

$$T(\alpha',\alpha) = \sum_{k\mu\Delta} \mathbf{T}^{k,\Delta}_{\mu}(\alpha',\alpha) \left\langle \begin{matrix} & k+\Delta & \\ 2k & & 0 \\ & k+\mu & \end{matrix} \right\rangle . \tag{3.458}$$

In this relation $\mathbf{T}^{k,\Delta}_{\mu}(\alpha',\alpha)$ is an invariant operator which together with the Wigner operator effects the transformation

$$\mathbf{T}^{k,\Delta}_{\mu}(\alpha',\alpha) \left\langle \begin{matrix} & k+\Delta & \\ 2k & & 0 \\ & k+\mu & \end{matrix} \right\rangle |(\alpha)jm\rangle$$

$$= t^{k,\Delta}_{\mu}(\alpha',\alpha,j+\Delta) C^{jk\,j+\Delta}_{m\mu\,m+\mu} |(\alpha')j+\Delta, m+\mu\rangle, \tag{3.459}$$

where $t^{k,\Delta}_{\mu}(\alpha',\alpha,j+\Delta)$ is a numerical coefficient defined by

$$t^{k,\Delta}_{\mu}(\alpha',\alpha,j+\Delta) = t^{k}_{\mu}\langle(\alpha')j+\Delta\|\mathbf{T}^{k}\|(\alpha)j\rangle. \tag{3.460}$$

The coefficients $t^{k,\Delta}_{\mu}(\alpha',\alpha,j+\Delta)$ may also be expressed in terms of $T(\alpha',\alpha)$ by using the orthogonality of the Wigner operators under the (partial) trace operation [see Eqs. (3.343)]. Thus, defining

$$\operatorname{tr}\mathcal{O} = \sum_{m} \langle(\alpha)jm|\mathcal{O}|(\alpha)jm\rangle \tag{3.461}$$

for an arbitrary operator $\mathcal{O}: \mathcal{H} \to \mathcal{H}$, where $\mathcal{H} = \sum_{(\alpha)j} \oplus \mathcal{H}_j(\alpha)$, one finds

$$\operatorname{tr}\left(\left\langle \begin{matrix} & k'+\Delta' & \\ 2k' & & 0 \\ & k'+\mu' & \end{matrix} \right\rangle \left\langle \begin{matrix} & k+\Delta & \\ 2k & & 0 \\ & k+\mu & \end{matrix} \right\rangle^{\dagger} \right) = \delta_{k'k}\delta_{\mu'\mu}\delta_{\Delta'\Delta} \frac{2j+1}{2k+1}. \tag{3.462}$$

Using this result in Eq. (3.458) yields

$$t^{k,\Delta}_{\mu}(\alpha',\alpha,j) = \frac{2k+1}{2j+1} \operatorname{tr}\left(T(\alpha',\alpha) \left\langle \begin{matrix} & k+\Delta & \\ 2k & & 0 \\ & k+\mu & \end{matrix} \right\rangle^{\dagger} \right). \tag{3.463}$$

It is Eq.(3.458)—which expresses the operator mapping (3.457) in terms of Wigner operators—and Eq.(3.463) for the expansion coefficients that find numerous applications in physical problems. (Similar relations hold for Racah operators.)

13. *Sharp's formula for the integral of three Bessel functions.* By considering the irreducible unitary representations and 3-j coefficients of the *extended Euclidean group* $E'(2)$ [the simply reducible group obtained by adjoining the reflection in the x-axis to the Euclidean group $E(2)$], Sharp [98, p. 305] derived

an elegant formula for the product of three Bessel functions of integral order:

$$\int_0^\infty J_{k_1}(p_1 r) J_{k_2}(p_2 r) J_{k_3}(p_3 r) r\, dr = \frac{\cos(k_1\tau_1 - k_2\tau_2)}{2\pi\Delta(p_1 p_2 p_3)}.$$

The symbols in this expression have the following definitions: (1) The k_i denote arbitrary integers that sum to zero – that is, $k_1 + k_2 + k_3 = 0$; (2) the p_i denote the lengths of the sides of a triangle, which we denote as $(p_1 p_2 p_3)$; (3) the τ_i denote the exterior angles of the triangle $(p_1 p_2 p_3)$ – that is, $\tau_i = \pi - \theta_i$, where θ_i $(0 \leqslant \theta_i \leqslant \pi, i = 1, 2, 3)$ denotes the interior angle opposite to side p_i; and (4) $\Delta(p_1 p_2 p_3)$ denotes the area of the triangle $(p_1 p_2 p_3)$, and, by definition, is infinite (so that the right-hand of the above expression is zero) if the lengths p_i do not form a triangle.

Sharp's formula is of interest here because it is the analog, for Bessel functions, of Gaunt's formula (3.192) for the integral of the product of three spherical harmonics in terms of the 3-j coefficients of the rotation group. Moreover, this formula is indicative of relations that may be obtained for special functions by considering the Racah-Wigner algebra of simply reducible groups (Wigner [47, 99]) of which the rotation group is the prototype (see also Vilenkin [100] and Talman [101]).

23. Appendices

A. Algebraic Form of the Racah Coefficients

We reproduce in this Appendix Racah's method (Racah [29]) of reducing the fourfold index summation appearing in the definition of $w(abcd;ef)$ given by Eq. (3.291) to a single index summation. Racah's own derivation is quite terse, and by supplying more of the details we hope to make the proof more accessible.

We first bring Eq. (3.291) to the form given by Racah. To accomplish this we introduce new summation indices α, γ, t, and u by $\alpha = c - \delta - \beta$, $\gamma = c - \delta$, $t = e - b + c - \beta - \delta + r$, and $u = d + \delta - s$. Noting also that r is integral in Eq. (3.291), hence, $(-1)^{a-e+\beta+r+s} = (-1)^{a-e+\beta-r+s} = (-1)^{a-b+c+d}(-1)^{t+u}$, we obtain

$$w(abcd;ef) = \frac{(-1)^{a-b+c+d}(2c+1)!}{(a-f+c)!(-a+f+c)!(e-d+c)!(-e+d+c)!}$$

$$\times \sum_{tu} \frac{(-1)^{t+u} w_{tu}(abcd;ef)}{t!(a+e-b-t)!u!(d+f-b-u)!}, \qquad \text{(A-1)}$$

where

$$w_{tu}(abcd;ef) = \sum_{\alpha\gamma} (a+\alpha)!(c+f-\alpha)!(e+\gamma)!(c+d-\gamma)! \times$$

$$\times\left[\frac{(b+\alpha-\gamma)!}{(a+\alpha-t)!(b-a-\gamma+t)!(c+d-\gamma-u)!(b-c-d+\alpha+u)!}\right]$$

$$\times\left[\frac{(b+\gamma-\alpha)!}{(e+\gamma-t)!(b-e-\alpha+t)!(c+f-\alpha-u)!(b-c-f+\gamma+u)!}\right]. \tag{A-2}$$

Racah was able to simplify Eq. (A-1) by taking the unlikely step of replacing the double summation in the defining equation for w_{tu} by a quadruple summation; that is, he proved that, for $0 \leqslant t \leqslant a+e-b$, $0 \leqslant u \leqslant d+f-b$, one can write

$$w_{tu}(abcd;ef)/t!(a+e-b-t)!u!(d+f-b-u)!$$

$$=\sum_{vw}\frac{(a+c+f+1)!(e+d+c+1)!(v-w)!(e+d+c-v)!}{w!(a+f-d-e+w)![(e+d+c+1-w)!]^2(b-c-e-f+u+v-w)!(b-a+e+t-v)!}$$

$$\times\left[\sum_r\frac{(b-a-e+u+v-w)!}{r!(a+e-b-t-r)!(u-r)!(b-a-e+v-w+r)!}\right]$$

$$\times\left[\sum_s\frac{(b+c+e+t-f-v)!}{s!(d+f-b-u-s)!(t-s)!(b+c+e-f-v+s)!}\right]. \tag{A-3}$$

The proof of the equality of Eqs. (A-2) and (A-3) is quite easy (after the fact), requiring only the use of the binomial coefficient sum rule:

$$\binom{x+y}{k}=\sum_{m+n=k}\binom{x}{m}\binom{y}{n}, \tag{A-4}$$

where x and y are indeterminates, k is a nonnegative integer, and the sum is over all nonnegative integer pairs (m,n) that add to k. The proof will be given by bringing each of Eqs. (A-2) and (A-3) to the common form:

$$w_{tu}(abcd;\,ef)$$

$$=\sum_{vw}\frac{(a+c+f+1)!(e+d+c+1)!(v-w)!(e+d+c-v)!}{w!(a+f-d-e+w)![(e+d+c+1-w)!]^2(b-c-e-f+u+v-w)!(b-a+e+t-v)!}$$

$$\times\left[\frac{(u+v-t-w)!(e+d+c+t-u-v)!}{(v-t-w)!(e+d+c-u-v)!}\right]. \tag{A-5}$$

Consider Eq. (A-3) first. The summation over r and the summation over s appearing in Eq. (A-3) are carried out by using the binomial coefficient sum rule (A-4) (for the summation over r, identify $k=u$, $m=r$, $n=u-r$, $x=a+e-b-t$, $y=b-a-e+u+v-w$; for the summation over s, identify $k=t$, $m=s$, $n=t-s$, $x=d+f-b-u$, $y=b+c+e+t-f-v$). Thus, the result is that Eqs. (A-3) and (A-5) are equivalent.

We next show the equivalence of Eqs. (A-2) and (A-5). The binomial coefficient sum rule is first used to replace each of the two terms in square brackets in Eq. (A-2) by the following summations, respectively:

$$\sum_v [1/(v-e-\gamma)!(c+d+e-u-v)!(b-a+e+t-v)!(a-c-d-e+\alpha-t+u+v)!],$$

$$\sum_w [1/(v-t-w)!(c+f-\alpha-u-v+t+w)!(e+\gamma-v+w)!(b-c-e-f+u+v-w)!].$$

(To obtain the first expression, choose $m = v - e - \gamma$, $n = c + d + e - u - v$, $k = c + d - \gamma - u$, $x = b - a - \gamma + t$, $y = a + \alpha - t$; to obtain the second, choose $m = c + f - \alpha - u - v + t + w$, $n = v - t - w$, $k = c + f - \alpha - u$, $x = b - e - \alpha + t$, $y = e + \gamma - t$.) When these replacements are made, Eq. (A-2) becomes

$$w_{tu}(abcd;ef)$$

$$= \sum_{vw} [1/(c+d+e-u-v)!(b-a+e+t-v)!(v-t-w)!(b-c-e-f+u+v-w)!]$$

$$\times \sum_\alpha \frac{(a+\alpha)!(c+f-\alpha)!}{(a-c-d-e+\alpha-t+u+v)!(c+f-\alpha-u-v+t+w)!}$$

$$\times \sum_\gamma \frac{(e+\gamma)!(c+d-\gamma)!}{(e+\gamma-v+w)!(v-e-\gamma)!}.$$

The summations over α and γ may now be carried out by an application of the binomial coefficient sum rule applied to $x = -\xi$ and $y = -\eta$:

$$\binom{\xi+\eta+k-1}{k} = \sum_{m+n=k} \binom{\xi+m-1}{m}\binom{\eta+n-1}{n}. \tag{A-6}$$

(We choose $m = a - c - d - e + \alpha - t + u + v$, $n = c + f - \alpha - u - v + t + w$, $k = a + f - d - e + w$, $\xi = c + d + e + t - u - v + 1$, $\eta = u + v - t - w + 1$ to obtain the α-summation; and $m = e + \gamma - v + w$, $n = v - e - \gamma$, $k = w$, $\xi = v - w + 1$, $\eta = c + d + e - v + 1$ to obtain the γ-summation.) Thus, Eq. (A-2) is equivalent to Eq. (A-5).

The next step is to substitute the quadruple summation expression (A-3) into Eq. (A-1), thus obtaining an expression for $w(abcd;ef)$ involving six summations. Remarkably, one can now carry out five of the internal summations (using only the binomial coefficient sum rule), leaving a single summation for the final expression. The order in which the summations are carried out is: sum t, sum u, sum v, sum w; set $r + s = z$, eliminate s, and sum r. The explicit summations involved are:

(1) *sum over t*:

$$\sum_t \frac{(-1)^t(b+c+e-f-v+t)!}{(t-s)!(b-a+e+t-v)!(a+e-b-t-r)!} =$$

$$= \frac{(-1)^{a+e-b-r}(b+c+e-f-v+s)!(a+c-f)!}{(a+e-b-s-r)!(2e-v-r)!(b+c-e-f+s+r)!},$$

(2) *sum over u*:

$$\sum_u \frac{(-1)^u(b-a-e+u+v-w)!}{(u-r)!(b-c-e-f+u+v-w)!(d+f-b-u-s)!}$$

$$= \frac{(-1)^{f+d-b-s}(b-a-e+v-w+r)!(c+f-a)!}{(f+d-b-r-s)!(d-c-e+v-w-s)!(b+c-a-d+r+s)!}.$$

These two summations use the following variant of the binomial coefficient sum rule:

$$(-1)^k \binom{\xi-y+k-1}{k} = \sum_{m+n=k} (-1)^m \binom{\xi+m-1}{m}\binom{y}{n}.$$

(3) *sum over v*:

$$\sum_v \frac{(v-w)!(e+d+c-v)!}{(d-c-e+v-w-s)!(2e-v-r)!}$$

$$= \frac{(c+d+e+1-w)!(c+e-d+s)!(c+d-e+r)!}{(d+e-c-w-r-s)!(2c+r+s+1)!},$$

(4) *sum over w*:

$$\sum_w \frac{1}{w!(a+f-d-e+w)!(e+d+c+1-w)!(d+e-c-w-r-s)!}$$

$$= \frac{(a+d+e+f+1-r-s)!}{(e+d-c-r-s)!(a+f+c+1)!(e+d+c+1)!(a+f-c-r-s)!},$$

(5) *sum over r*:

$$\sum_r \frac{(c+d-e+r)!(c+e-d+z-r)!}{r!(z-r)!} = \frac{(2c+z+1)!(c+d-e)!(c+e-d)!}{z!(2c+1)!}.$$

The result of carrying out the above five operations is

$$w(abcd;ef) = (-1)^{b+c-e-f}$$

$$\times \sum_z \frac{(-1)^z(a+d+e+f+1-z)!}{z!(a+f-c-z)!(a+e-b-z)!(d+e-c-z)!(d+f-b-z)!}$$

$$\times \frac{1}{(b+c-a-d+z)!(b+c-e-f+z)!}. \qquad \text{(A-7)}$$

We finally shift the summation index z to $z' = a + d + e + f - z$ (and then rename z' to be z) to obtain the following standard expression for the Racah coefficients:

$$W(abcd;ef) = \Delta(abe)\Delta(cde)\Delta(acf)\Delta(bdf)$$

$$\times \sum_z \frac{(-1)^{a+b+c+d+z}(z+1)!}{(z-a-b-e)!(z-c-d-e)!(z-a-c-f)!(z-b-d-f)!}$$

$$\times \frac{1}{(a+b+c+d-z)!(a+d+e+f-z)!(b+c+e+f-z)!}, \tag{A-8}$$

where, for convenience, we repeat the definition of a triangle coefficient:

$$\Delta(abc) = \left[\frac{(a+b-c)!(a-b+c)!(-a+b+c)!}{(a+b+c+1)!}\right]^{\frac{1}{2}}. \tag{A-9}$$

B. Determination of the Fundamental Racah Coefficients

We have demonstrated in Section 18 that the Racah coefficients are uniquely determined by the orthogonality conditions, the B–E relation, and the four fundamental (spin-$\frac{1}{2}$) coefficients. It is of interest therefore to give a set of assumptions that allow us to calculate directly and uniquely the fundamental coefficients. We assert: *The fundamental Racah coefficients are uniquely determined by the following conditions*:

(*a*) The sign convention:

$$\text{sign } W^{a\frac{1}{2}c}_{\rho\sigma\tau}(j) = \text{sign}(c - a - \sigma), \tag{B-1}$$

where, by definition, sign $0 \equiv +$.

(*b*) The orthogonality relations applied to the fundamental coefficients.

(*c*) The symmetry relation:

$$W^{a\frac{1}{2}c}_{\rho\sigma\tau}(j) = (-1)^{a-c+\sigma}\left[\frac{(2c+1)(2j-2\sigma+1)}{(2a+1)(2j+1)}\right]^{\frac{1}{2}} W^{c\frac{1}{2}a}_{\tau,-\sigma,\rho}(j-\sigma), \tag{B-2}$$

which in standard Racah notation is

$$W(afe\tfrac{1}{2};cb) = (-1)^{b+c-e-f}W(abc\tfrac{1}{2};ef). \tag{B-3}$$

Proof. Let us introduce a more concise notation:

$$A^a_\tau(j) = W^{a\frac{1}{2}a+\frac{1}{2}}_{\tau,\frac{1}{2},\tau+\frac{1}{2}}(j) \geqslant 0,$$

$$B^a_\tau(j) = W^{a\frac{1}{2}a+\frac{1}{2}}_{\tau+1,-\frac{1}{2},\tau+\frac{1}{2}}(j) \geqslant 0,$$

$$C^a_\tau(j) = W^{a\frac{1}{2}a-\frac{1}{2}}_{\tau,\frac{1}{2},\tau+\frac{1}{2}}(j) \leqslant 0,$$

$$D^a_\tau(j) = W^{a\frac{1}{2}a-\frac{1}{2}}_{\tau+1,-\frac{1}{2},\tau+\frac{1}{2}}(j) \geqslant 0,$$

where the values follow from sign assumption (*a*).

The orthogonality relations then require that each of the 2 × 2 matrices

$$\begin{pmatrix} A^a_\tau(j) & B^a_\tau(j) \\ C^a_\tau(j) & D^a_\tau(j) \end{pmatrix}, \qquad \tau = a-1, \ldots, -a, \tag{B-4}$$

be orthogonal. [For $\tau = a$, the orthogonality relations and sign conventions imply $A^a_a(j) = \varepsilon_{j-a-\frac{1}{2},j,a+\frac{1}{2}}$, the remaining three coefficients being zero because the triangle conditions are violated; for $\tau = -a-1$, the orthogonality relations and sign convention imply $B^a_{-a-1}(j) = 1$ (all j), the remaining three coefficients being zero because the triangle conditions are violated.] It follows then that each of the matrices (B-4) must have the form

$$\begin{pmatrix} A^a_\tau(j) & B^a_\tau(j) \\ -B^a_\tau(j) & A^a_\tau(j) \end{pmatrix}, \qquad \tau = a-1, \ldots, -a. \tag{B-5}$$

Thus,

$$\begin{aligned} D^a_\tau(j) &= A^a_\tau(j), \\ C^a_\tau(j) &= -B^a_\tau(j) \end{aligned} \tag{B-6}$$

for each $\tau = a-1, \ldots, -a$. The effect of relation (B-2) is to turn Eqs. (B-6) into recursion relations. Thus, conditions (c) yield

$$\begin{aligned} D^a_\tau(j) &= \left[\frac{(2a)(2j+2)}{(2a+1)(2j+1)}\right]^{\frac{1}{2}} A^{a-\frac{1}{2}}_{\tau+\frac{1}{2}}(j+\tfrac{1}{2}), \\ C^a_\tau(j) &= -\left[\frac{(2a)(2j)}{(2a+1)(2j+1)}\right]^{\frac{1}{2}} B^{a-\frac{1}{2}}_{\tau-\frac{1}{2}}(j-\tfrac{1}{2}), \end{aligned} \tag{B-7}$$

so that the desired recursion relations are

$$\begin{aligned} A^a_\tau(j) &= \left[\frac{(2a)(2j+2)}{(2a+1)(2j+1)}\right]^{\frac{1}{2}} A^{a-\frac{1}{2}}_{\tau+\frac{1}{2}}(j+\tfrac{1}{2}), \\ B^a_\tau(j) &= \left[\frac{(2a)(2j)}{(2a+1)(2j+1)}\right]^{\frac{1}{2}} B^{a-\frac{1}{2}}_{\tau-\frac{1}{2}}(j-\tfrac{1}{2}), \end{aligned} \tag{B-8}$$

where $\tau = a-1, \ldots, -a$ in each of these recursion formulas, and $A^a_a(j) = \varepsilon_{j-a-\frac{1}{2},j,a+\frac{1}{2}}$, $B^a_{-a-1}(j) = 1$. Iterating the first of Eqs. (B-8) (in a) a number of times equal to the difference $a - \tau$ yields

$$A^a_\tau(j) = \left[\frac{(a+\tau+1)(2j+a-\tau+1)}{(2a+1)(2j+1)}\right]^{\frac{1}{2}} \tag{B-9}$$

for each $j \geqslant \tau + \frac{1}{2}$, and zero otherwise. A similar iteration of the second of Eqs. (B-8) yields

$$B^a_\tau(j) = \left[\frac{(a-\tau)(2j-a-\tau)}{(2a+1)(2j+1)}\right]^{\frac{1}{2}} \tag{B-10}$$

for each $j \geqslant \tau + \frac{1}{2}$, and zero otherwise. Equations (B-9) and (B-10) give the unique set of fundamental Racah coefficients determined by the rules (*a*)–(*c*). ■

C. Relationship between Symmetries of Racah and Wigner Coefficients

We have seen that a Wigner coefficient may be obtained as a limit of a Racah coefficient – that is,

$$C^{bdf}_{\alpha\beta\gamma} = \lim_{j\to\infty} W^{bdf}_{\alpha\beta\gamma}(j). \tag{C-1}$$

Because of this relation we may pose the question: Which symmetries of the Racah coefficients remain in this limit and become symmetries of the Wigner coefficients? The answer is obtained by determining those *row and column* interchanges on the array (3.315) that induce a transformation on $(b, \alpha, d, \beta, f, \gamma)$ that is also induced by *row, column, and transposition* operations on the Regge array:

$$\begin{pmatrix} b+\alpha & d+\beta & f-\gamma \\ b-\alpha & d-\beta & f+\gamma \\ d+f-b & b+f-d & b+d-f \end{pmatrix}. \tag{C-2}$$

For example, the interchange of columns 1 and 2 followed by the interchange of rows 1 and 2 in array (C-2) induces the same transformation ($b \to d$, $d \to b$, $f \to f$, $\alpha \to -\beta$, $\beta \to -\alpha$, $\gamma \to -\gamma$) as does the interchange of columns 1 and 2 followed by the interchange of rows 1 and 2 in array (3.315). Indeed, we see that any permutation of the columns of array (3.315) followed by the same permutation of the rows is allowed and corresponds to those symmetries of $C^{bdf}_{\alpha\beta\gamma}$ (with appropriate phase and dimension factors) that have the following relabelings of (b, α), (d, β), and (f, γ):

$$\begin{array}{lll} (b,\alpha) & (d,\beta) & (f,\gamma) \\ (d,\beta) & (f,-\gamma) & (b,-\alpha) \\ (f,-\gamma) & (b,\alpha) & (d,-\beta) \\ (d,-\beta) & (b,-\alpha) & (f,-\gamma) \\ (b,-\alpha) & (f,\gamma) & (d,\beta) \\ (f,\gamma) & (d,-\beta) & (b,\alpha) \end{array} \tag{C-3}$$

The interchange of rows 1 and 2 in array (3.315) induces the same transformation as does transposition of array (C-2). Thus, the transformation from (b, α), (d, β), (f, γ) to

$$\left(\frac{b+d+\gamma}{2}, \frac{b-d+\alpha-\beta}{2}\right), \left(\frac{b+d-\gamma}{2}, \frac{b-d-\alpha+\beta}{2}\right), (f, b-d) \tag{C-4}$$

must be included. Combining this transformation with those of (C-3) now yields 12 common transformations of $(b, \alpha, d, \beta, f, \gamma)$ induced by row exchanges and column exchanges of array (3.315) and row exchanges, column exchanges, and transposition of array (C-2). There can be no more common transformations. This is proved as follows: The transformations on the 9 entries of the first three rows of array (3.315) induced by row exchanges and column exchanges is a subgroup G_1 of order 36 of the symmetric group S_9. Similarly, the transformations of these same 9 entries induced by row exchanges, column exchanges, and transposition of array (C-2) is a subgroup G_2 of order 72 of S_9. Furthermore, $G_1 \cap G_2$ is a subgroup of G_1 and G_2, and we have already demonstrated that $G_1 \cap G_2$ contains a subgroup of 12 elements. Hence, $G_1 \cap G_2$ contains $12n$ elements for some positive integer n, and $12n$ divides both 36 and 72. Thus, $n = 1$ or 3. If $n = 3$, then every element of G_1 belongs to $G_1 \cap G_2$, which is false, since individual row and column exchanges of the array (3.315) do not induce transformations of the 9 entries corresponding to any element of G_2. Thus, $n = 1$.

In going to the limit (C-1), we lose all but 12 of the 144 symmetries of the Racah coefficients, these 12 remaining as symmetries of the Wigner coefficients. The remaining 60 symmetries of the Wigner coefficients then arise independently of the Racah coefficient symmetries. The interesting conclusion is as follows: We cannot deduce the Wigner coefficient symmetries from the Racah coefficient symmetries and the limit relation (C-1) (short of examining explicitly the Wigner coefficients).

D. Properties of Spherical Basis Vectors

One usually understands that the angular momentum operator $\vec{J}$ denotes the quantity

$$\vec{J} = J_1\hat{e}_1 + J_2\hat{e}_2 + J_3\hat{e}_3, \tag{D-1}$$

where the J_i are the operators satisfying the standard commutation relations $[J_i, J_j] = ie_{ijk}J_k$ and the $(\hat{e}_1, \hat{e}_2, \hat{e}_3)$ are (real) unit vectors constituting a right-handed triad (commuting with the angular momentum):

$$\begin{aligned} \hat{e}_i \cdot \hat{e}_j &= \delta_{ij}, \\ \hat{e}_i \times \hat{e}_j &= ie_{ijk}\hat{e}_k. \end{aligned} \tag{D-2}$$

If we define complex vectors by

$$\begin{aligned} \hat{\xi}_{+1} &\equiv -(\hat{e}_1 + i\hat{e}_2)/\sqrt{2}, \\ \hat{\xi}_0 &\equiv \hat{e}_3, \\ \hat{\xi}_{-1} &\equiv (\hat{e}_1 - i\hat{e}_2)/\sqrt{2}, \end{aligned} \tag{D-3}$$

then Eq. (D-1) may also be written in the form

$$\vec{J} = J_{+1}\hat{\xi}^*_{+1} + J_0\hat{\xi}^*_0 + J_{-1}\hat{\xi}^*_{-1}. \tag{D-4}$$

The J_μ ($\mu = +1, 0, -1$) in this result are the operators

$$J_{+1} \equiv -\frac{1}{\sqrt{2}} J_+, \qquad J_0 \equiv J_3, \qquad J_{-1} \equiv \frac{1}{\sqrt{2}} J_-, \tag{D-5}$$

where $J_\pm \equiv J_1 \pm iJ_2$. (This confusing notation is, unfortunately, conventional.)

The complex vectors defined by Eq. (D-3) are called *spherical basis vectors*, and the operators J_μ ($\mu = +1, 0, -1$) are called *spherical components* of $\vec{J}$. This nomenclature and the sign choices in the definitions (D-3) and (D-4) are patterned after the expressions for the spherical (solid) harmonics of degree 1 (see Table 4 in Appendix of Tables).

It is instructive to recall here the transformation properties of the preceding quantities. The standard transformation of basis kets used throughout this monograph [see the *Remarks* pp. 42, 43, and 93] is given by

$$\mathscr{U}: |(\alpha)j\mu\rangle \to |(\alpha)j\mu\rangle' = \sum_\nu D^j_{\nu\mu}(U)|(\alpha)j\nu\rangle, \tag{D-6}$$

and this transformation of basis states induces the transformation of spherical angular momentum components given by

$$\mathscr{U}: J_\mu \to J'_\mu = \mathscr{U} J_\mu \mathscr{U}^{-1} = \sum_\nu D^1_{\nu\mu}(U) J_\nu. \tag{D-7}$$

This latter result is equivalent to

$$\mathscr{U}: J_j \to J'_j = \mathscr{U} J_j \mathscr{U}^{-1} = \sum_i R_{ij} J_i, \tag{D-8}$$

where U and R are related by Eq. (2.16). Corresponding to this transformation of the components J_i of the vector operator $\mathbf{J} = (J_1, J_2, J_3)$, we have the transformation of the basis vectors $(\hat{e}_1, \hat{e}_2, \hat{e}_3) \to (\hat{e}'_1, \hat{e}'_2, \hat{e}'_3)$ and $(\hat{\xi}_{+1}, \hat{\xi}_0, \hat{\xi}_{-1}) \to (\hat{\xi}'_{+1}, \hat{\xi}'_0, \hat{\xi}'_{-1})$ as determined from $\vec{J} = \sum_i J_i \hat{e}_i = \sum_i J'_i \hat{e}'_i = \sum_\mu J_\mu \hat{\xi}^*_\mu = \sum_\mu J'_\mu \hat{\xi}'^*_\mu$:

$$\begin{aligned} \hat{e}_j &\to \hat{e}'_j = \sum_i R_{ij} \hat{e}_i, \\ \hat{\xi}_\mu &\to \hat{\xi}'_\mu = \sum_\nu D^1_{\nu\mu}(U) \hat{\xi}_\nu, \\ \hat{\xi}^*_\mu &\to \hat{\xi}'^*_\mu = \sum_\nu D^{1*}_{\nu\mu}(U) \hat{\xi}^*_\nu. \end{aligned} \tag{D-9}$$

Let us next observe that the complex vectors (D-3) satisfy the orthogonality relations (Hermitian scalar product)

$$\hat{\xi}^*_\mu \cdot \hat{\xi}_\nu = \delta_{\mu\nu}, \tag{D-10}$$

and therefore the spherical component J_μ of $\vec{J}$ is obtained as

$$J_\mu = \vec{J} \cdot \hat{\xi}_\mu. \tag{D-11}$$

It follows that the complex vectors $\hat{\xi}_\mu^*$ ($\mu = +1, 0, -1$) are a basis of complex three-space (as are the ξ_μ).

A second characteristic feature of the set of spherical vectors (D-3) is expressed by the dot products

$$\hat{\xi}_{+1} \cdot \hat{\xi}_{+1} = \hat{\xi}_{-1} \cdot \hat{\xi}_{-1} = 0, \qquad \hat{\xi}_0 \cdot \hat{\xi}_0 = 1. \tag{D-12}$$

Thus, two of the vectors are *isotropic*, while the third is not.

The properties noted in Eqs. (D-10) and (D-12) suggest a general definition of a spherical basis: *Three vectors denoted by $\hat{\eta}_{+1}, \hat{\eta}_0, \hat{\eta}_{-1}$ are called a spherical basis if and only if they obey the rules*

$$\begin{gathered} \hat{\eta}_\mu^* \cdot \hat{\eta}_\nu = \delta_{\mu\nu}, \qquad \mu, \nu = +1, 0, -1, \\ \hat{\eta}_{+1} \cdot \hat{\eta}_{+1} = \hat{\eta}_{-1} \cdot \hat{\eta}_{-1} = 0, \\ \hat{\eta}_0 = \alpha_0 \hat{n}, \qquad |\alpha_0| = 1, \qquad \hat{n} \cdot \hat{n} = 1. \end{gathered} \tag{D-13}$$

The orthogonality property expressed by the first of these equations allows one to express any (real or complex) vector $\vec{A}$ in the form

$$\vec{A} = A_{+1}\hat{\eta}_{+1}^* + A_0\hat{\eta}_0^* + A_{-1}\hat{\eta}_{-1}^*, \tag{D-14}$$

where

$$A_\mu = \vec{A} \cdot \hat{\eta}_\mu. \tag{D-15}$$

The principal result, which can now be proved, is as follows: *Every spherical basis has the form*

$$\hat{\eta}_\mu = \alpha_\mu(\mathscr{R}\hat{\xi}_\mu), \qquad \mu = +1, 0, -1, \tag{D-16}$$

where α_μ is a complex number of unit modulus, $|\alpha_\mu| = 1$, and $\mathscr{R}$ is a rotation of the basis $(\hat{e}_1, \hat{e}_2, \hat{e}_3)$.

Proof. The proof of this result may be given by using the three defining properties: (*a*) $\hat{\eta}_{+1}$ and $\hat{\eta}_{-1}$ are perpendicular to the real unit vector $\hat{n}$; (*b*) $\hat{\eta}_{+1}$ and $\hat{\eta}_{-1}$ are isotropic vectors; and (*c*) $\hat{\eta}_{+1}$ and $\hat{\eta}_{-1}$ are perpendicular—that is, $\hat{\eta}_{+1}^* \cdot \hat{\eta}_{-1} = 0$. These properties imply that $\hat{\eta}_{+1}$ and $\hat{\eta}_{-1}$ have the forms

$$\begin{aligned} \hat{\eta}_{+1} &= -\frac{\alpha_{+1}}{\sqrt{2}}(\hat{p} + i\hat{q}), \\ \hat{\eta}_{-1} &= \frac{\alpha_{-1}}{\sqrt{2}}(\hat{p} - i\hat{q}), \end{aligned} \tag{D-17}$$

where $(\hat{p}, \hat{q}, \hat{n})$ constitute a right-handed triad of (real) perpendicular unit vectors (Cartesian basis). Hence, there exists a rotation $\mathscr{R}$ of basis vectors such that $\hat{p} = \mathscr{R}\hat{e}_1$, $\hat{q} = \mathscr{R}\hat{e}_2$, $\hat{n} = \mathscr{R}\hat{e}_3$, thus proving Eq. (D-16). ■

Stated somewhat differently, the result we have found above is that every spherical basis is a rotation of the particular spherical basis given by

$$\begin{aligned}\hat{\zeta}_{+1} &\equiv -\frac{\alpha_{+1}}{\sqrt{2}}(\hat{e}_1 + i\hat{e}_2),\\ \hat{\zeta}_0 &\equiv \alpha_0 \hat{e}_3,\\ \hat{\zeta}_{-1} &\equiv \frac{\alpha_{-1}}{\sqrt{2}}(\hat{e}_1 - i\hat{e}_2)\end{aligned} \qquad \text{(D-18)}$$

for some choice of the complex numbers α_μ of unit modulus. *As we let* $(\hat{e}_1, \hat{e}_2, \hat{e}_3)$ *run over all right-handed Cartesian frames and* $(\alpha_{+1}, \alpha_0, \alpha_{-1})$ *over all complex numbers of unit modulus, the vectors* $(\hat{\zeta}_{+1}, \hat{\zeta}_0, \hat{\zeta}_{-1})$ *enumerate all spherical bases sets.*

There is no censensus in the literature as to the choice of factors α_μ in defining a *standard basis set*, although $\hat{\xi}_{+1}, \hat{\xi}_0, \hat{\xi}_{-1}$ is a popular choice. Danos [65], however, has suggested a choice of spherical basis vectors that offers distinct advantages for transcribing vector relations expressed in terms of dot and vectors products to standard coupling relations.

The idea is a simple one: Just as we use the vector product of Cartesian basis sets $(\hat{e}_1, \hat{e}_2, \hat{e}_3)$ to distinguish between left- and right-handed frames, we use the properties of the products $\hat{\zeta}_\mu \times \hat{\zeta}_\nu$ of spherical basis sets in order to partition further the set of all spherical bases.

By direct calculation, we obtain

$$\begin{aligned}\hat{\zeta}_0 \times \hat{\zeta}_\mu &= i\alpha_0 \mu \hat{\zeta}_\mu, \qquad \mu = +1, 0, -1\\ \hat{\zeta}_{+1} \times \hat{\zeta}_{-1} &= i\alpha_{+1}\alpha_{-1}\alpha_0^* \hat{\zeta}_0.\end{aligned} \qquad \text{(D-19)}$$

These results may be expressed concisely as a single equation by using the $C^{111}_{\mu\nu\lambda}$ Wigner coefficients:

$$\hat{\zeta}_\mu \times \hat{\zeta}_\nu = i\sqrt{2}\,\alpha_\mu \alpha_\nu \alpha^*_{\mu+\nu} C^{111}_{\mu,\nu,\mu+\nu} \hat{\zeta}_{\mu+\nu}. \qquad \text{(D-20)}$$

The dot product has a similar form:

$$\hat{\zeta}_\mu \cdot \hat{\zeta}_\nu = -\sqrt{3}\,\alpha_\mu \alpha_\nu C^{110}_{\mu\nu 0} = \alpha_\mu \alpha_{-\mu} (-1)^\mu \delta_{\mu,-\nu}. \qquad \text{(D-21)}$$

Danos suggested that one define a standard basis $\hat{\eta}_{+1}, \hat{\eta}_0, \hat{\eta}_{-1}$ by requiring the signs in front of the Wigner coefficients in Eqs. (D-20) and (D-21) to be real and positive:

$$\begin{aligned}\hat{\eta}_\mu \times \hat{\eta}_\nu &= \sqrt{2}\,C^{111}_{\mu,\nu,\mu+\nu} \hat{\eta}_{\mu+\nu},\\ \hat{\eta}_\mu \cdot \hat{\eta}_\nu &= \sqrt{3}\,C^{110}_{\mu\nu 0}.\end{aligned} \qquad \text{(D-22)}$$

This requires $\alpha_0 = -i$ and $\alpha_{-1} = -\alpha^*_{+1}$. Thus, the general form of this spherical basis is

$$\hat{\eta}_{+1} = -\frac{\alpha}{\sqrt{2}}(\hat{e}_1 + i\hat{e}_2),$$

$$\hat{\eta}_0 = -i\hat{e}_3,$$

$$\hat{\eta}_{-1} = -\frac{\alpha^*}{\sqrt{2}}(\hat{e}_1 - i\hat{e}_2), \tag{D-23}$$

still containing an arbitrary factor α of unit modulus (Danos chooses $\alpha = -i$, but we see that this choice is arbitrary). [If we order the spherical basis vectors and the Cartesian basis vectors by $\eta_{+1}, \eta_0, \eta_{-1}$ and $\hat{e}_1, \hat{e}_2, \hat{e}_3$, respectively, then the determinant of the transformation (D-18) is $-i\alpha_{+1}\alpha_0\alpha_{-1}$, which assumes the value $+1$ for Eqs. (D-23).]

The merit of using a spherical basis of the form of Eqs. (D-23) is that for

$$\vec{A} = \sum_\mu A^1_\mu \hat{\eta}^*_\mu,$$

$$\vec{B} = \sum_\mu B^1_\mu \hat{\eta}^*_\mu, \tag{D-24}$$

one has

$$\vec{A} \cdot \vec{B} = \sqrt{3}[\mathbf{A}^1 \times \mathbf{B}^1]^0,$$

$$\vec{A} \times \vec{B} = \sqrt{2}[\mathbf{A}^1 \times \mathbf{B}^1]^1, \tag{D-25}$$

where the symbols on the right refer to the coupling of tensor operators discussed in Section 16.

Spherical basis vectors also make their appearance in still another role – as the eigenvectors of a rotation [see Eq. (2.2)]. To understand this result from another point of view, we interpret the formation of the vector (cross) product of vectors as an operator acting on vectors.

With each vector $\vec{A}$ (real or complex) one may associate an operator $\Gamma_{\vec{A}} \equiv \vec{A}\,\mathbf{x}$, which is defined by specifying its action on an arbitrary vector (real or complex) $\vec{B}$:

$$\Gamma_{\vec{A}}\vec{B} \equiv \vec{A} \times \vec{B}. \tag{D-26}$$

The product of two such operators, $\Gamma_{\vec{A}}\Gamma_{\vec{B}}$ is defined by $(\Gamma_{\vec{A}}\Gamma_{\vec{B}})\vec{C} = \Gamma_{\vec{A}}(\Gamma_{\vec{B}}\vec{C})$. We also define the action of $\Gamma_{\vec{A}}$ on a complex number a by

$$\Gamma_{\vec{A}}a \equiv 0. \tag{D-27}$$

The set of operators $\{\Gamma_{\vec{A}}\}$ possesses a number of useful properties, which may be proved from the definition (D-26) and well-known properties of the vector product:

$$(a)\qquad \Gamma_{\vec{A}}(b\vec{B} + c\vec{C}) = b(\Gamma_{\vec{A}}\vec{B}) + c(\Gamma_{\vec{A}}\vec{C}),$$

$$(b)\qquad \Gamma_{a\vec{A}+b\vec{B}} = a\Gamma_{\vec{A}} + b\Gamma_{\vec{B}},$$

$$(c)\qquad \Gamma_{\vec{A}}(\vec{B}\cdot\vec{C}) = 0 = (\Gamma_{\vec{A}}\vec{B})\cdot\vec{C} + \vec{B}\cdot(\Gamma_{\vec{A}}\vec{C}),$$

$$(d)\qquad \Gamma_{\vec{A}}(\vec{B}\times\vec{C}) = (\Gamma_{\vec{A}}\vec{B})\times\vec{C} + \vec{B}\times(\Gamma_{\vec{A}}\vec{C}), \tag{D-28}$$

where a, b, c and $\hat{A}$, $\hat{B}$, $\hat{C}$ are arbitrary complex numbers and vectors, respectively.

That there exists a relation between the operators $\{\Gamma_{\vec{A}}\}$ and rotations becomes apparent when we determine properties of the special operators defined by $\Gamma_k \equiv i\hat{e}_k\,\mathbf{x}$ for $k = 1, 2, 3$:

$$\Gamma_i\hat{e}_j = ie_{ijk}\hat{e}_k,$$

$$[\Gamma_i, \Gamma_j] = ie_{ijk}\Gamma_k, \tag{D-29}$$

where we note that $\Gamma^2 \equiv \Gamma_1^2 + \Gamma_2^2 + \Gamma_3^2 = 2$. These results express the fact that the Γ_i $(i = 1, 2, 3)$ are a realization of angular momentum operators having $j = 1$.

A second useful result follows directly from the properties of rotations given by Eq. (2.3):

$$\mathscr{R}\Gamma_{\vec{A}}\mathscr{R}^{-1} = \Gamma_{\mathscr{R}\vec{A}}. \tag{D-30}$$

This relation expresses the fact that under the similarity action of a rotation the set of operators $\{\Gamma_{\vec{A}}\}$ is transformed into itself (tensor operator property).

As may be anticipated from the fact that the Γ_i $(i = 1, 2, 3)$ are angular momentum operators with $j = 1$, spherical basis sets arise when one considers the eigenvalue problem for $\Gamma_{i\hat{n}}$, where $\hat{n}$ is an arbitrary unit (real) vector $(\hat{n}\cdot\hat{n} = 1)$. A simple derivation shows that $\Gamma_{i\hat{n}}(\Gamma_{i\hat{n}} + 1)(\Gamma_{i\hat{n}} - 1) = 0$, thus demonstrating that the eigenvalues of $\Gamma_{i\hat{n}}$ are $+1, 0, -1$. Futhermore, with respect to the *Hermitian scalar product* defined by

$$(\vec{A}, \vec{B}) \equiv \vec{A}^* \cdot \vec{B}, \tag{D-31}$$

one sees that $\Gamma_{i\hat{n}}$ is Hermitian. It follows that the eigenvectors $\hat{u}_{+1}, \hat{u}_0, \hat{u}_{-1}$ of $\Gamma_{i\hat{n}}$ are perpendicular and normalizable. Lastly, one may use the derivation property (c), Eq. (D-28), to prove that $\hat{u}_{+1}$ and $\hat{u}_{-1}$ are isotropic. Thus, the eigenvectors of $\Gamma_{i\hat{n}}$ define a spherical basis set:

$$\Gamma_{i\hat{n}}\hat{u}_\nu = \nu\hat{u}_\nu, \qquad \nu = +1, 0, -1. \tag{D-32}$$

As will be recognized, the theory above is just a particularization of the abstract theory of angular momentum to the case $j = 1$ with z-axis chosen to be $\hat{n}$, and the arbitrary phase factors in Eq. (D-18) are a consequence of the general freedom one has in the problem as discussed in Note 1. Furthermore, the rotation operator for rotations of ϕ about direction $\hat{n}$ now assumes the very simple form

$$\mathscr{R}(\phi, \hat{n}) = e^{-i\phi\hat{n}\cdot\mathbf{\Gamma}} = e^{-i\phi\Gamma_{\hat{n}}}, \qquad \text{(D-33)}$$

since

$$i\Gamma_{\hat{n}} = \Gamma_{i\hat{n}} = i(n_1\Gamma_1 + n_2\Gamma_3 + n_3\Gamma_3) = i\hat{n}\cdot\mathbf{\Gamma}.$$

Remark. The presentation above, developing the properties of $\Gamma_{\vec{A}}$, is the physicist's version [for $SU(2)$] of the general ad A operator in the mathematician's treatment of Lie algebras (the vector product is replaced by the bracket operation for pairs of elements of an abstract Lie algebra). Just as $\Gamma_{\vec{A}}$ may be used for developing the properties of the $j = 1$ (3×3 matrices) representation of the Lie algebra of the rotation group, the ad A operator is used for developing the properties of the *adjoint representation* of a Lie algebra.

References

1. W. Heisenberg, "Über quantentheoretische Umdeutung kinematischer und mechanischer Beziehungen," *Z. Physik* **33** (1925), 879–893.
2. M. Born and P. Jordan, "Zur Quantenmechanik," *Z. Physik* **34** (1925), 858–888.
3. M. Born, W. Heisenberg, and P. Jordan, "Zur Quantenmechanik II," *Z. Physik* **35** (1926), 557–615.
4. P. A. M. Dirac, "The fundamental equations of quantum mechanics," *Proc. Roy. Soc.* **A109** (1925), 642–653.
5. B. L. van der Waerden, *Sources of Quantum Mechanics*. North-Holland, Amsterdam, 1967.
6. E. Cartan, *Sur la Structure des Groupes Finis and Continus*. Thesis, Paris, Nony, 1894. [*Ouevres Complète*, Part 1, pp. 137–287. Gauthier-Villars, Paris, 1952.]
7. G. W. Mackey, *The Mathematical Foundations of Quantum Mechanics*. Benjamin, New York, 1963.
8. P. A. M. Dirac, *The Principles of Quantum Mechanics*. Oxford Univ. Press, London, 1st ed., 1930; 4th ed., 1958.

8a. K. R. Brownstein, "Eigenvalues of $\mathbf{J}^2$," *Amer. J. Phys.* **47** (1979), 809–810.

9. A. Sommerfeld, "Zur Quantentheorie der Spektrallinien," *Ann. Physik* **51** (1916), 1–94, 125–167.
10. R. P. Feynman, *Feynman Lectures on Physics*, Chapter 34. Addison-Wesley, Reading, Mass., 1963.
11. G. Breit and L. A. Wills, "Hyperfine structure in intermediate coupling," *Phys. Rev.* **44** (1933), 470–490.
12. J. C. Slater, "The theory of complex spectra," *Phys. Rev.* **34** (1929), 1293–1322.
13. E. P. Wigner, *Group Theory and Its Application to the Quantum Mechanics of Atomic Spectra*. Academic Press, New York, 1959. Translation by J. J. Griffin of the 1931 German edition.
14. G. Szegö, *Orthogonal Polynomials*. Edwards, Ann Arbor, 1948.
15. A. Erdélyi, W. Magnus, F. Oberhettinger, and G. F. Tricomi, *Higher Transcendental Functions*, Vol. I. McGraw-Hill, New York, 1953.
16. A. R. Edmonds, *Angular Momentum in Quantum Mechanics*. Princeton Univ. Press, Princeton, N. J., 1957.
17. M. E. Rose, *Elementary Theory of Angular Momentum*. Wiley, New York, 1957.

18. I. M. Gel'fand and Z. Ya. Shapiro, "Representations of the group of rotations of 3-dimensional space and their applications," *Amer. Math. Soc. Transl.* **2** (1956), 207–316.
19. K. Fox and I. Ozier, "Construction of tetrahedral harmonics," *J. Chem. Phys.* **52** (1970), 5044–5056.
20. S. L. Altmann and C. J. Bradley, "A note on the calculation of the matrix elements of the rotation group," *Phil. Trans. Roy. Soc.* **A255** (1962), 193–198.
21. R. Gilmore, *Lie Groups, Lie Algebras, and Some of Their Applications.* Wiley, New York, 1974.
22. H. H. Nielsen, "The vibration–rotation energies of molecules," *Rev. Mod. Phys.* **13** (1951), 90–136.
23. H. B. G. Casimir, *Rotation of a Rigid Body in Quantum Mechanics.* Thesis, University of Leyden, Wolters, Groningen, 1931. [*Koninkl. Ned. Akad. Wetenschap, Proc.* **34** (1931), 844.]
24. W. H. Shaffer, "Operational derivation of wave functions for a symmetrical rigid rotator," *J. Mol. Spectrosc.* **1** (1957), 69–80.
24a. A. Clebsch, "Theorie der binären algebraischen Formen," Teubner, Leipzig (1872).
24b. P. Gordan, "Über das Formensystem binärer Formen," Teubner, Leipzig (1875).
24c. H. Weyl, *Gruppentheorie und Quantenmechanik.* Hirzel, Leipzig, 1st ed., 1928; 2nd ed., 1931. Translated by H. P. Robertson as *The Theory of Groups and Quantum Mechanics.* Methuen, London, 1931; reissued by Dover, New York, 1949.
24d. P. J. Brussaard, "Clebsch-Gordan of Wigner-Coefficienten," *Ned. T. Natuurk.* **33** (1967), 202–222.
25. E. P. Wigner, "Einige Folgerungen aus der Schrödingerschen Theorie für die Termstrukturen," *Z. Physik* **43** (1927), 624–652.
26. E. P. Wigner, "Berichtigung zu der Arbeit: Einige Folgerungen aus der Schrödingerschen Theorie für die Termstrukturen," *Z. Physik* **45** (1927), 601–602.
27. C. Eckart, "The application of group theory to the quantum dynamics of monatomic systems," *Rev. Mod. Phys.* **2** (1930), 305–380.
28. E. U. Condon and G. H. Shortley, *The Theory of Atomic Spectra.* Cambridge Univ. Press, London, 1935.
29. G. Racah, "Theory of complex spectra. II," *Phys. Rev.* **62** (1942), 438–462; III, *ibid.* **63** (1943), 367–382.
30. J. D. Louck, "New recursion relation for the Clebsch–Gordan coefficients," *Phys. Rev.* **110** (1958), 815–816.
31. B. L. van der Waerden, *Die gruppentheoretische Methode in der Quantenmechanik.* Springer, Berlin, 1932.
32. H. A. Kramers, "Zur Ableitung der quantenmechanischen Intensitätformeln," *Proc. Amst. Acad.* **33** (1930), 953–958; "Die Multiplettaufspaltung bei Koppelung zweier Vektoren," *ibid.* **34** (1931), 965–976.
33. H. C. Brinkman, *Zur Quantenmechanik der Multipolstrahlung.* Dissertation, Utrecht, Groningen, 1932; "Die Multiplettaufspaltung der Spectren von Atomen mit zwei Leuchtelektronen," *Z. Physik* **79** (1932), 753–775.
34. H. C. Brinkman, *Applications of Spinor Invariants in Atomic Physics.* North-Holland, Amsterdam, 1956.
35. M. Hamermesh, *Group Theory and Its Applications to Physical Problems.* Addison-Wesley, Reading, Mass., 1962.
36. J. Nachamkin, "Direct use of Young tableau algebra to generate the Clebsch–Gordan coefficients of $SU(2)$," *J. Math. Phys.* **16** (1975), 2391–2394.
37. J. Schwinger, *On Angular Momentum.* U.S. Atomic Energy Commission Report NYO-3071, 1952, unpublished. Published in *Quantum Theory of Angular Momentum* (L. C. Biedenharn and H. van Dam, eds.), pp. 229–279. Academic Press, New York, 1965.

38. T. Regge, "Symmetry properties of Clebsch–Gordan coefficients," *Nuovo Cimento* [10] **10** (1958), 544–545.
39. V. Bargmann, "On the representations of the rotation group," *Rev. Mod. Phys.* **34** (1962), 829–845.
40. A. Bincer, "Interpretation of the symmetry of the Clebsch–Gordan coefficients discovered by Regge," *J. Math. Phys.* **11** (1970), 1835–1844.
41. P.-O. Löwdin, "Expansion theorems for the total wave function and extended Hartree–Fock schemes," *Rev. Mod. Phys.* **32** (1960), 328–334.
42. S. D. Majumdar, "The Clebsch–Gordan coefficients,"*Prog. Theoret. Phys.* **20** (1958), 798–803.
43. I. M. Gel'fand and M. I. Graev, "Finite-dimensional irreducible representations of the unitary and full linear groups, and related special functions," *Izv. Akad. Nauk SSSR Ser. Mat.* **29** (1965), 1329–1356. [*Am. Math. Soc. Transl.* **64** (1967), Ser. 2, 116–146.]
44. I. M. Gel'fand, R. A. Minlos, and Z. Ya. Shapiro, *Representations of the Rotation and Lorentz Groups and Their Applications.* Macmillan, New York, 1963. Translated from the Russian by G. Cummins and T. Boddington.
45. R. Hagedorn, *Selected Topics on Scattering Theory*: *Part IV. Angular Momentum.* Lectures given at the Max-Planck-Institut für Physik, Munich, 1963.
45a. D. Kleima and R. van Wageningen, "On the Regge symmetries of the $3j$-symbol," *Nucl. Phys.* **73** (1965), 625–630.
46. L. Eisenbud, Ph.D. thesis, Princeton University, 1948.
47. E. P. Wigner, "On the matrices which reduce the Kronecker products of representations of S. R. groups," 1940, unpublished. Published in *Quantum Theory of Angular Momentum* (L. C. Biedenharn and H. van Dam, eds.), pp. 87–133. Academic Press, New York, 1965.
48. J. A. Gaunt, "IV. The triplets of helium," *Trans. Roy. Soc.* **A228** (1929), 151–196.
49. E. P. Wigner, "Über die Operation der Zeitumkehr in der Quantenmechanik," *Göttinger Nachrichten, Math.-Phys.* (1932), 546–559.
50. D. Aebersold and L. C. Biedenharn, "Cautionary remark on applying symmetry techniques to the Coulomb problem," *Phys. Rev.* **A15** (1977), 441–443.
50a. L. Michel, "Application of group theory to quantum physics: Algebraic aspects," in *Lecture Notes in Physics*: *Group Representations in Mathematics and Physics, Battelle Recontres*, 1969 (V. Bargmann, ed.), pp. 36–143. Springer, Berlin, 1970.
51. P. Güttinger and W. Pauli, "Zur Hyperfeinstruktur von Li^+. Teil II, Mathematischer Anhang," *Z. Physik* **67** (1931), 754–765.
52. P. A. M. Dirac, "The elimination of the nodes in quantum mechanics," *Proc. Roy. Soc.* **A111** (1926), 281–305.
53. M. Born and P. Jordan, *Elementare Quantenmechanik*. Springer, Berlin, 1930.
54. L. C. Biedenharn, J. M. Blatt, and M. E. Rose, "Some properties of the Racah and associated coefficients," *Rev. Mod. Phys.* **24** (1952), 249–257.
55. A. P. Jucys, I. B. Levinson, and V. V. Vanagas, *The Theory of Angular Momentum.* (*Mathematicheskii apparat teorii momenta kolichestva dvizheniya.*) Vilnius, USSR. (1960). Translated from the Russian by A. Sen and A. R. Sen, Jerusalem, Israel (1962). Available from Office of Technical Services, U.S. Department of Commerce, Washington, D.C. Of particular interest are the graphical methods and the explicit consideration of ($3n$-j) symbols up to $n = 6$.
55a. M. E. Rose and C. N. Yang, "Eigenvalues and eigenvectors of a symmetric matrix of $6j$ symbols," *J. Math. Phys.* **3** (1962), 106.
56. L. C. Biedenharn, "An identity satisfied by Racah coefficients," *J. Math. Phys.* **31** (1953), 287–293.

57. J. P. Elliott, "Theoretical studies in nuclear structure V: The matrix elements of non-central forces with an application to the 2p-shell," *Proc. Roy. Soc.* **A218** (1953), 345–370.
58. L. C. Biedenharn and J. D. Louck, "An intrinsically self-conjugate boson structure: The symplecton," *Ann. Phys.* **63** (1971), 459–475.
59. U. Fano and G. Racah, *Irreducible Tensorial Sets.* Academic Press, New York, 1959.
60. A. Erdélyi, *Math. Rev.* **14** (1957), 642; review of Ref. [56].
61. T. Regge, "Symmetry properties of Racah's coefficients," *Nuovo Cimento* [10] **11** (1959), 116–117.
62. L. A. Shelepin, "Calculus of Clebsch–Gordan coefficients and its physical applications," in *Group-Theoretical Methods in Physics* (D. V. Skobel'tsyn, ed.), pp. 1–109, 1975. A special research report translated from the Russian by Consultants Bureau, New York and London.
63. H. A. Jahn and J. Hope, "Symmetry properties of the Wigner 9*j*-symbol," *Phys. Rev.* **93** (1954), 318–321.
64. A. Arima, H. Horie, and Y. Tanabe, "Generalized Racah coefficient and its application," *Prog. Theoret. Phys.* **11** (1954), 143–154.
65. M. Danos, "Fully consistent phase conventions in angular momentum theory," *Ann. Phys.* **63** (1971), 319–334.
66. A. C. T. Wu, "Structure of the Wigner 9*j* coefficients in the Bargmann approach," *J. Math. Phys.* **13** (1972), 84–90.
67. E. A. El Baz and B. Castel, *Graphical Methods of Spin Algebras in Atomic, Nuclear, and Particle Physics.* Dekker, New York, 1972.
67a. A. P. Jucys and A. A. Bandzaitis, *Angular Momentum Theory in Quantum Physics.* Mokslas, Vilnius, 1977.
67b. D. A. Varshalovich, A. N. Moskalev, and V. K. Khersonskii, *Quantum Theory of Angular Momentum.* Nauka, Leningrad, 1975 (in Russian).
68. H. A. Kramers, *Quantum Mechanics.* North-Holland, Amsterdam, 1957. Translation by D. ter Haar of Kramer's monograph published in the *Hand- und Jahrbuch der chemischen Physik*, 1937.
69. L. Schiff, *Quantum Mechanics.* McGraw-Hill, New York, 1949.
70. S. Perlis, *Theory of Matrices.* Addison-Wesley, Cambridge, Mass., 1952.
71. J.-L. Calais, "Derivation of the Clebsch–Gordan coefficients by means of projection operators," Technical Note 25 [AF-61(514)-1200], 1959, unpublished.
72. J. C. Slater, *Quantum Theory of Atomic Structure*, Vol. 2. McGraw-Hill, New York, 1960.
73. H. A. Bethe and R. W. Jackiw, *Intermediate Quantum Mechanics.* Benjamin, New York, 1968.
73a. Y. Lehrer-Ilamed, "On the direct calculations of the representations of the three-dimensional rotation group," *Proc. Camb. Phil. Soc.* **60** (1964), 61–66.
74. P. G. Hull, *Some Properties of Angular Momentum Eigenfunctions.* Thesis, Auburn University, Auburn, Ala., 1962.
74a. R. E. Peierls and J. Yoccoz, "The collective model of nuclear motion," *Proc. Phys. Soc.* **A70** (1957), 381–387.
74b. M. A. Lamme and E. Boeker, "Exact and approximate angular momentum projection for light nuclei," *Nucl. Phys.* **A111** (1968), 492–512.
74c. R. A. Sorensen, "Problems of angular momentum projection in nuclear physics," *Nucl. Phys.* **A281** (1977), 475–485.
74d. A. Bohr and B. R. Mottelson, *Nuclear Structure.* Vol. II: *Nuclear Deformations.* Benjamin, New York, 1975.
75. Ya. A. Smorodinskiĭ and L. A. Shelepin, "Clebsch-Gordan coefficients viewed from different sides," *Sov. Phys. Usp.* **15** (1972), 1–24 [*Usp. Fiz. Nauk* **106** (1972), 3–45].

75a. A. O. Barut and R. Wilson, "Some new identities of Clebsch-Gordan coefficients and representation functions of $SO(2,1)$ and $SO(4)$," *J. Math. Phys.* **17** (1976), 900–915.
76. U. Fano, *Statistical Matrix Techniques and Their Application to the Directional Correlations of Radiations*. U.S. National Bureau of Standards Report 1214, 1951, unpublished.
77. A. de-Shalit, and I. Talmi, *Nuclear Shell Theory* (Pure and Applied Physics Series, Vol. 14). Academic Press, New York, 1963.
78. V. Heine, *Group Theory and Quantum Mechanics; An Introduction to Its Present Usage*. Pergamon, New York, 1960.
79. R. Judd, *Operator Techniques in Atomic Spectroscopy*. McGraw-Hill, New York, 1963.
80. A. S. Davydov, *Quantum Mechanics*. Pergamon, London, Addison-Wesley, Reading, Mass., 1965. Translation from the Russian of *Kvantovaya Mekhanika*; published in Moscow, 1963, with revisions and additions by D. ter Haar.
81. D. M. Brink and G. R. Satchler, *Angular Momentum*. Oxford Univ. Press, London, 1962.
82. A. Bohr and B. R. Mottelson, *Nuclear Structure*. Vol. I: *Single-Particle Motion*. Benjamin, New York, 1969.
83. E. Merzbacher, *Quantum Mechanics*. Wiley, New York, 1961.
84. A. Messiah, *Quantum Mechanics*, Vol. II. North-Holland, Amsterdam, 1965. Translated by J. Potter from the French of *Mécanique Quantique*, Vol. II.
85. A. Abragam and B. Bleany, *Electron Paramagnetic Resonance of Transition Ions*. Clarendon, Oxford, 1970.
86. J. M. Blatt and V. F. Weisskopf, *Theoretical Nuclear Physics*. Wiley, New York, 1952.
87. H. A. Jahn, "Theoretical studies in nuclear structure II. Nuclear d^2, d^3, and d^4 configurations. Fractional parentage coefficients and central force matrix elements," *Proc. Roy. Soc.* **A205** (1951), 192–237.
88. K. Alder, "Beiträge zur Theorie der Richtungskorrelation," *Helv. Phys. Acta* **25** (1952), 235–258.
89. L. D. Landau and E. M. Lifshitz, *Quantum Mechanics. Nonrelativistic Theory*. Addison-Wesley, Reading, Mass. 1958. Translated from the Russian by J. B. Sykes and J. S. Bell.
90. V. A. Fock, *Zhur. Eksper. i Teoret. Fiz.* **10** (1940), 383 (in Russian).
91. M. Tinkham, *Group Theory and Quantum Physics*. McGraw-Hill, New York, 1964.
92. L. C. Biedenharn, *Tables of Racah Coefficients*. Oak Ridge National Laboratory Report ORNL-1098, 1952.
93. J. S. Lomont, *Applications of Finite Groups*. Academic Press, New York, 1959.
94. P. J. Redmond, "An explicit formula for the calculation of fractional parentage coefficients," *Proc. Roy. Soc.* **A222** (1954), 84–93.
95. S. F. Boys and R. C. Sahni, "Electronic wave functions XII. The evaluation of the general vector-coupling coefficients by automatic computation," *Phil. Trans. Roy. Soc.* **A246** (1954), 463–479.
96. M. K. Banerjee and A. K. Saha, "Shape factors for β-decay," *Proc. Roy. Soc.* **A224** (1954), 473–487.
97. J. M. Kennedy and M. J. Cliff, *Transformation Coefficients between LS and jj Coupling*. Report AECL-224, Atomic Energy of Canada, Chalk River, Ontario, 1955.
98. W. T. Sharp, *Racah Algebra and the Contraction of Groups*. Thesis, Princeton University, 1960; issued as Report AECL-1098, Atomic Energy of Canada, Chalk River, Ontario, 1960.
99. E. P. Wigner, *Application of Group Theory to the Special Functions of Mathematical Physics*. Lecture notes, unpublished, Princeton University, Princeton, N. J., 1955.
100. N. Vilenkin, *Special Functions and the Theory of Group Representations* (transl. from the Russian; Amer. Math. Soc. Transl., Vol. 22). Amer. Math. Soc., Providence, R. I., 1968.
101. J. D. Talman, *Special Functions: A Group Theoretic Approach*. Benjamin, New York, 1968. Based on E. P. Wigner's lectures; see Ref. [99].

CHAPTER 4

The Theory of Turns Adapted from Hamilton

1. An Alternative Approach to Rotations

In the preceding chapters, we have discussed the standard view of the quantum theory of angular momentum. The present chapter is concerned with developing a new view of the subject starting from the beginning with a fresh look at rotations based on a viewpoint adapted primarily from Hamilton [1], but also partly from Klein and Sommerfeld [2] and from Wigner [3] (see Note 1). This new survey will be rewarded by the concept of a novel geometric entity – the *turn* – which in its quantal version (Chapter 5) will be realized by an elementary operator structure. In discussing this alternative viewpoint, we are primarily interested in motivation and in suggesting concepts. Accordingly, we shall proceed, at first, intuitively.

When reduced to fundamentals, the concept of an elementary (point) particle in physics (as discussed in Chapter 1) becomes synonymous with the fundamental symmetries assumed to characterize physical space. We have seen that, from the homogeneity and isotropy assumed for space, one induces the symmetries of spatial displacements (translations) and rotations; that is, one assumes that space is Euclidean and three-dimensional, having the isometry group, $E(3)$, of rotations and translations.

It is quite easy to characterize translations, since three-dimensional translations necessarily (Artin [4]) form an abelian (commutative) group. For physics, the translation generator is the (Hermitian) momentum operator $\mathbf{p}$, and the associated operator generating finite displacements is realized by $U(\mathbf{a}) = \exp(-i\mathbf{a} \cdot \mathbf{p}/\hbar)$. This operator displaces the system by the (numerical) vector $\mathbf{a}$.

ENCYCLOPEDIA OF MATHEMATICS and Its Applications, Gian-Carlo Rota (ed.).
Vol. 8: L. C. Biedenharn and J. D. Louck, Angular Momentum in
Quantum Physics: Theory and Application

[To verify this statement, let the operator $U(\mathbf{a})$ act on the function $f(\mathbf{x})$. We find

$$(U(\mathbf{a})f)(\mathbf{x}) = \sum_{n=0}^{\infty} \frac{1}{n!}(-i\mathbf{a} \cdot \mathbf{p}/\hbar)^n f(\mathbf{x}), \tag{4.1}$$

where $(\mathbf{p}f)(\mathbf{x}) = -i\hbar\nabla f(\mathbf{x})$. This latter step uses the Schrödinger realization of the Heisenberg commutation rule: $[p_i, x_j] = -i\hbar\delta_{ij}$. Thus the action of $U(\mathbf{a})$ on functions $f(\mathbf{x})$, from the Taylor series, takes the form of a displacement:

$$(U(\mathbf{a})f)(\mathbf{x}) = f(\mathbf{x} - \mathbf{a}), \tag{4.2}$$

as stated. To be fully explicit we interpret Eq. (4.2) to mean that under the action of the displacement operator $U(\mathbf{a})$ the function f becomes the new function $U(\mathbf{a})f$, and the point $\mathbf{x}$ becomes the displaced point $\mathbf{x} - \mathbf{a}$, such that the new function evaluated at the old point is the same as the old function evaluated at the displaced point.]

This is elementary, but fundamental. The reason it all works so smoothly is as follows:

(*a*) *Technically*: One has the uniqueness of the (unitary) displacement operator $U(\mathbf{a})$ as the content of the Stone [5]–von Neumann [6] theorem (Jauch [7]).

(*b*) *Conceptually*: The primitive concept of a translation has been thoroughly understood, using the derived concept of a "vector" defined to be an ordered pair of points in $E(3)$ (initial point = "tail of vector," final point = "head of vector"). The fact that one and the same translation is defined by the displacement of any point expresses the equivalence of all displacement vectors under parallel transport.

The composition of two displacements is now found to be the addition of vectors, using the freedom of parallel transport to effect a joining of the two vectors head (of first)-to-tail (of second). The commutativity of translations is vectorially realized as the parallelogram law.

This is all thoroughly familiar, *for translations*. But when it comes to *rotations*, it is quite clear that we lack a primitive model – the rotational analog to vectors – capable of making the rotational structure intuitively obvious.

To find the proper model, let us begin by asking: Is there a geometric way to realize both translations and rotations in a uniform way? There is indeed: *Both translations and rotations can be obtained by successive reflections in two planes* (see Note 2).

Let us validate this observation:

If the two planes are identical, then we get the identity transformation.

If the two planes are *distinct* and *parallel*, we get a translation. Consider the diagram shown in Fig. 4.1. Geometrically, one sees that the translation so generated displaces all points by *twice* the distance between the planes in a direction perpendicular to the planes *from* plane 1 *to* plane 2. We may describe the translation by the two reflection planes and the vector $\vec{a}/2$ (half the actual displacement), and verify that this description has all the desired properties of a translation (including independence of the translation to parallel displacement of the planes and vector $\vec{a}/2$). Thus, *translation is the operation of two successive reflections through parallel planes.*

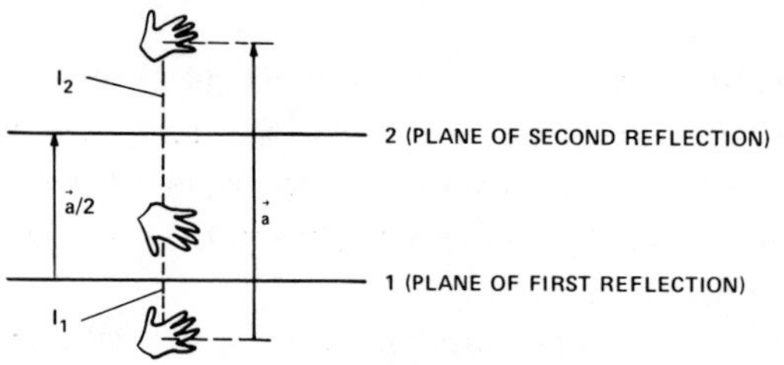

Figure 4.1. A translation $\vec{a}$ is generated by two successive reflections $\mathbf{1}_1$ and $\mathbf{1}_2$ through two parallel planes separated by $\vec{a}/2$.

If the two planes are *distinct*, but *not* parallel, then the two reflections generate a rotation. Consider the diagram shown in Fig. 4.2. Geometrically one sees that the rotation $\mathscr{R}(\phi, \hat{n}) = \mathbf{1}_2\mathbf{1}_1$ has (*a*) the axis $\hat{n}$ defined by the intersection of the two planes, (*b*) the rotation angle ϕ, *twice* the dihedral angle of the two planes, and (*c*) the sense defined by the order of reflecting *first* in plane 1 and *then* in plane 2. Thus, *a rotation is the operation of two successive reflections through two intersecting planes.*

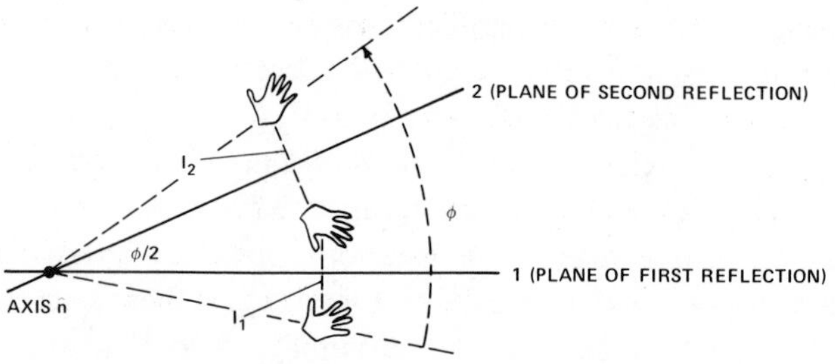

Figure 4.2. A rotation of ϕ about $\hat{n}$ is generated by two successive reflections $\mathbf{1}_1$ and $\mathbf{1}_2$ through two intersecting planes with dihedral angle $\phi/2$.

The importance of this uniform presentation of translations and rotations is that it *suggests* that the proper analogy to pursue is between $\vec{a}/2$ (*half the displacement vector*) and $\vec{\phi}/2$ (*half the directed arc of the rotation*). For translations, a convention using $\vec{a}/2$ in place of $\vec{a}$ changes nothing, but for rotations (see Remark (*h*), p. 192) the use of half-angles is all important.

Let us focus attention on those rotations that all leave one special point (the center) fixed. The pairs of planes by which we describe these rotations must then all pass through this center. To describe the rotation $\mathscr{R}(\phi, \hat{n})$, we use a pair of planes whose intersection is the axis $\hat{n}$, and use the dihedral angle $\phi/2$ defined by the planes (with $\phi/2 \leqslant \pi$), with the sense of rotation taken to be *from* plane 1 *to* plane 2 (equivalently the right-hand rule for orienting $\hat{n}$). (See Fig. 4.3.)

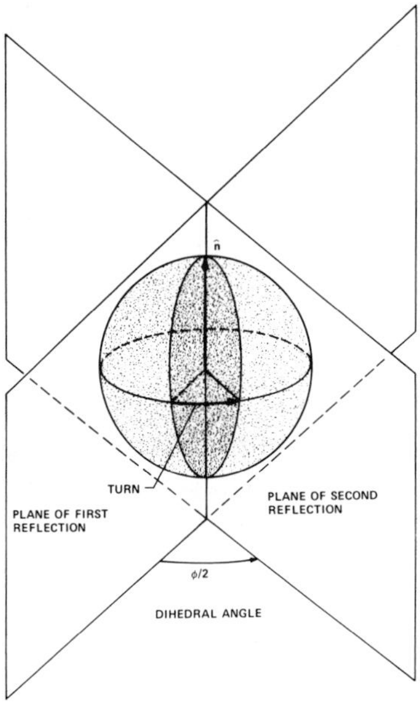

Figure 4.3. A turn is an ordered pair of points shown here, geometrically, as a directed arc length of a great circle. Each turn defines a rotation by specifying two intersecting planes.

We now have our model: *The primitive object for describing rotations is the turn*[1], *half the directed arc of the rotation parametrized as an ordered pair of points on the surface of a unit sphere, modulo great circle transport.* The order is given by denoting one point as the "tail of the turn," the other point as the "head of the turn." A rotation is generated by reflection first in the plane

[1] The name *turn* is adapted from Klein and Sommerfeld [2], who used a related (but not identical) concept (*Wendestreckung*).

perpendicular to the tail of the turn, followed by reflection in the plane perpendicular to the head of the turn. This generates a rotation about the axis $\hat{n}$ (intersection of the planes) by an angle ϕ, which is *twice* the arc length of the turn, the sense of the rotation being determined by the sense of the turn (equivalent to the right-hand rule).

2. Properties of Turns (Geometric View)

In accord with our aim of developing an intuitive understanding, let us now accept the viewpoint that the concept of a turn (for rotations) is the proper analog to that of a vector (for translations) and use this concept to find out the properties of rotations. The geometry of turns will lead us quite directly to the essence of rotations.

Intuitively, it may help to think of the turn as the directed arc of the great circle joining the two points, with the beginning of the arc being the tail of the turn, and the end, the head of the turn, but the more precise definition[1] (as an ordered pair of points) has technical advantages. To see this, consider the points that lie on the surface of our unit sphere (Fig. 4.3). Any two distinct points, Q and P, that are not diametrically opposite define a unique great circle containing the two points. The ordered pair (QP) defines a unique turn, whereas there are *two* directed arcs joining Q to P. Hence the definition in terms of ordered pairs is technically much neater.

We seek to abstract the geometric rules for turns from the analogy between turns and vectors in the plane. Following this guide, one notes that in order to implement the rule for adding vectors it is first necessary to specify the equivalence class of vectors (parallel transport), so that two arbitrary vectors may be placed in the head-to-tail relationship required for adding. What is to be the equivalence relationship for turns? To answer this we appeal to the relation between our model and rotations: Any two turns of the same length and direction on the same great circle produce the same rotation. Hence, we abstract the result: *Two turns are equivalent if they can be superposed by displacing either one along a great circle containing it.* The phrase "modulo great circle transport" is included in the definition of a turn to specify just this equivalence relation.

Consider now any two distinct, but diametrically opposite, points, Q and P. The geometric fact that a great circle containing these two points is not unique shows that one can in fact superpose the turn (QP) with any pair of diametrically opposite points, using great circle transport. Hence, we con-

[1] We are indebted to Prof. H. Bacry (Université d'Aix-Marseille, Marseille, France) and Prof. K. Baclawski (Haverford College, Haverford, Pennsylvania) for pointing out (independently) these advantages (private communications).

clude: *Turns defined by any pair of diametrically opposite points are all equivalent*. We denote this equivalence class by $\mathbf{T}_\pi$. Note that this implies that the addition of $\mathbf{T}_\pi$ *commutes with all turns* (since we are free to choose $\mathbf{T}_\pi$ to have the same great circle as the given turn, and the addition of turns on the same great circle is commutative). This property justifies our calling $\mathbf{T}_\pi$ a scalar turn.

Let us now give the law of combination and discuss its properties.

(*a*) The *sum of two turns* is defined geometrically in analogy to the sum of two vectors. Let $\mathbf{T}_1$ and $\mathbf{T}_2$ denote the two turns. The sum of $\mathbf{T}_1$ and $\mathbf{T}_2$ (first $\mathbf{T}_2$, then $\mathbf{T}_1$) is denoted by $\mathbf{T}_1 + \mathbf{T}_2$ and is defined by the following rule: Choose *either of the two points* where the great circle of $\mathbf{T}_1$ and the great circle of $\mathbf{T}_2$ intersect. Place the head of $\mathbf{T}_2$ and the tail of $\mathbf{T}_1$ at the point of intersection (using the equivalence relation to move each turn along its great circle). *The turn going from the tail of* $\mathbf{T}_2$ *to the head of* $\mathbf{T}_1$ *is defined to be* $\mathbf{T}_1 + \mathbf{T}_2$. (The choice of the opposite intersection point produces the same sum.)

(*b*) *The addition of turns is noncommutative*: $\mathbf{T}_1 + \mathbf{T}_2 \neq \mathbf{T}_2 + \mathbf{T}_1$, in general. This basic result can be demonstrated geometrically, as illustrated in Fig. 4.4. Note that the arc lengths of the two sums are the same, but that their great circles differ, in general.

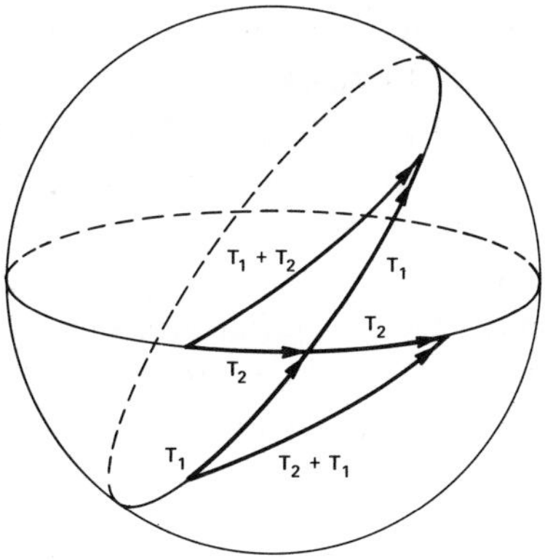

Figure 4.4. Addition of turns is noncommutative.

(*c*) *The addition of turns is an associative operation*. To verify this geometrically, let us consider the addition of three turns, adding first $\mathbf{T}_3$ to $\mathbf{T}_2$, then $\mathbf{T}_1$ to the sum, $\mathbf{T}_{23} = \mathbf{T}_2 + \mathbf{T}_3$. The law of combination requires the three turns to be in the head-to-tail arrangement, shown in Fig. 4.5. This geometric arrangement shows directly that the resultant, $\mathbf{T}_{\text{sum}}$, is independent of any bracketings:

$$\mathbf{T}_{\text{sum}} = \mathbf{T}_1 + \mathbf{T}_{23} = \mathbf{T}_{12} + \mathbf{T}_3 = \mathbf{T}_1 + \mathbf{T}_2 + \mathbf{T}_3;$$

that is, the resultant position (tail and head, defining $\mathbf{T}_{\text{sum}}$) is quite independent of the specific path (specific bracketing). (An algebraic verification of the associative law will be given in the next section.)

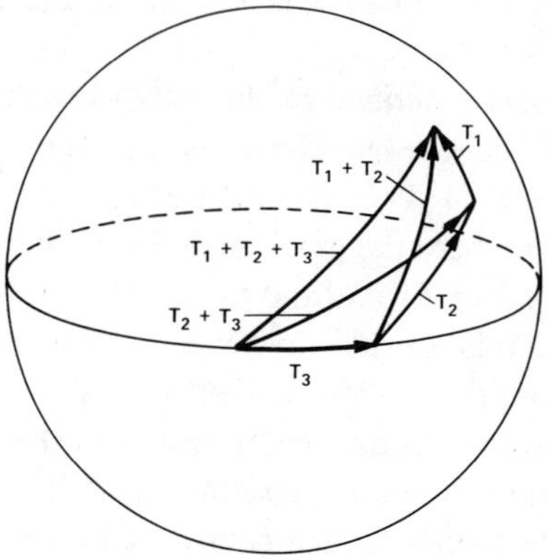

Figure 4.5. Addition of turns is associative.

(*d*) It is also clear geometrically that the *identity operation of turns is the turn of zero length*: $\mathbf{T}_0$. [Note that, from the definition of a turn, the turns $\mathbf{T}_{2k\pi}$ ($k = 0, 1, \ldots$) are equivalent to $\mathbf{T}_0$.]

(*e*) Using the law of combination, one sees that the *inverse* to a given turn $\mathbf{T}$ is the turn $-\mathbf{T}$ having the same great circle and length, but opposite sense. This is clear from the definition of a turn as an ordered pair.

These geometric results establish the (unsurprising) fact that the turns under addition constitute a group. We call this group *Hamilton's group of turns*, or simply the group of turns; this group will be shown to be isomorphic to the quantal rotation group, $SU(2)$. Note that the equivalence classes of turns are (for the generic case) three parameter objects.

There are two involutions in the group of turns. The first is the operation defining the inverse: $\mathbf{T} \to -\mathbf{T}$. The second involution is defined by using $\mathbf{T}_\pi$:

$$\mathbf{T} \to \mathbf{T}^c \equiv \mathbf{T} + \mathbf{T}_\pi = \mathbf{T}_\pi + \mathbf{T}.$$

Clearly, $(\mathbf{T}^c)^c = \mathbf{T}$. Figure 4.6 illustrates these two involutions geometrically.

We have seen that to any pair of distinct, nondiametrical points there is associated a unique equivalence class of turns whose (arc) length, l, lies in the interval $0 < l < \pi$. To each such equivalence class there is associated a

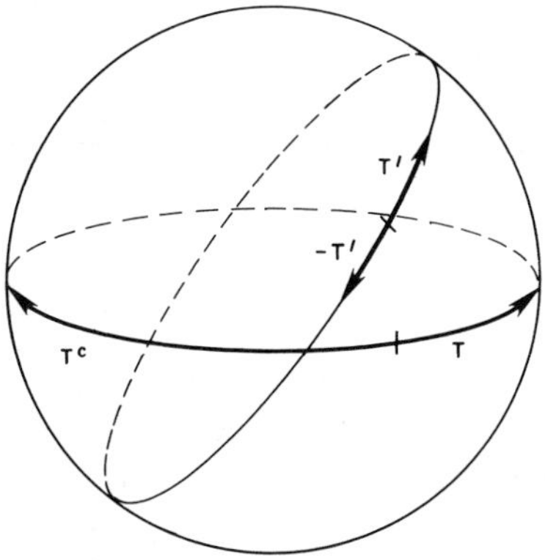

Figure 4.6. The two involution operators: $\mathbf{T} \to \mathbf{T}^c$ and $\mathbf{T}' \to -\mathbf{T}'$.

direction $\hat{n}$ (the axis of the great circle through the points) and a sense. The special (scalar) turns, $\mathbf{T}_0$ and $\mathbf{T}_\pi$, are associated with pairs of identical points or diametrical points, respectively, and correspond to equivalence classes of turns of length 0 and length π, respectively, and *no* direction $\hat{n}$ or sense. This exhausts the set of (equivalence classes of) turns belonging to Hamilton's group of turns.

The special role of turns of length $\pi/2$. Turns of length $\pi/2$ are of particular interest. We denote a generic turn of length $\pi/2$ by $\mathbf{E}$. The set of turns of length $\pi/2$ is the subset of the group of turns characterized by the property $\mathbf{E}^c = -\mathbf{E}$, or, equivalently, by $\mathbf{E} + \mathbf{E} = \mathbf{T}_\pi$.

An arbitrary turn can always be decomposed (nonuniquely) into a sum of two turns of length $\pi/2$; that is, for each $\mathbf{T}$ there exist turns $\mathbf{E}$ and $\mathbf{E}'$ of length $\pi/2$ such that

$$\mathbf{T} = \mathbf{E}' + \mathbf{E}. \tag{4.3}$$

Proof. Let $\hat{n}$ denote the normal to the great circle of $\mathbf{T}$, and P the point of intersection of $\hat{n}$ with the unit sphere. Then we may choose $\mathbf{E}$ to be the turn going from the tail of $\mathbf{T}$ to P, and $\mathbf{E}'$ is then the turn going from P to the head of $\mathbf{T}$. ∎

Turns of length $\pi/2$ possess a second important property. Let $\hat{e}$ denote the unit vector perpendicular (right-hand rule) to the great circle of $\mathbf{E}$, and let $\hat{n}$ denote the unit vector perpendicular to the great circle of an arbitrary nonzero turn $\mathbf{T}$. Then from geometric considerations, we find[1]

[1] We thank Professor Pieter Brussaard (Fysisch Laboratorium der Rijksuniversiteit, Utrecht) for pointing out an error in our initial discussion.

$$\hat{e} \cdot \hat{n} = \cos\|\mathbf{T} - \mathbf{E}\|/\sin\|\mathbf{T}\|, \tag{4.4}$$

where $\|\mathbf{T}\|$ denotes the length l of the turn $\mathbf{T}$.

Let us consider now three turns, $(\mathbf{E}_1, \mathbf{E}_2, \mathbf{E}_3)$, each of length $\pi/2$, which are chosen to satisfy the condition

$$\mathbf{E}_3 + \mathbf{E}_2 + \mathbf{E}_1 = \mathbf{T}_0. \tag{4.5}$$

Such a set of three turns is provided by the spherical triangle covering an octant of the unit sphere. (Equivalently, these three turns define a frame, an ordered triple of orthogonal unit vectors as discussed earlier in Chapter 2, Section 2. This is discussed further in Remark (*c*) below.)

Using the defining relation, Eq. (4.5), and the fact that the $\mathbf{E}_i$ have length $\pi/2$, it follows that the four turns $\{\mathbf{T}_0, \mathbf{E}_1, \mathbf{E}_2, \mathbf{E}_3\}$ and their conjugates $\{\mathbf{T}_\pi, \mathbf{E}_i^c = -\mathbf{E}_i\}$ form a finite group of eight elements under addition of turns. *This group is none other than Hamilton's quaternion group.*

To make this result evident, it is convenient firstly to replace the additive notation for turns by a multiplicative notation (juxtaposition), and secondly to denote the turns belonging to the quaternion group by more suggestive labels:

$$\begin{aligned} \mathbf{T}_0 &\to 1 & &\text{and} & \mathbf{T}_\pi &\to -1 \\ \mathbf{E}_1 &\to i & & & \mathbf{E}_1^c &\to i^{-1} \\ \mathbf{E}_2 &\to j & & & \mathbf{E}_2^c &\to j^{-1} \\ \mathbf{E}_3 &\to k & & & \mathbf{E}_3^c &\to k^{-1}. \end{aligned} \tag{4.6}$$

In this new notation, one finds that

(*a*) $i^2 = j^2 = k^2 = -1$ (since the turns $\mathbf{E}_i$ are of length $\pi/2$),

and

$$\begin{aligned} &(b)\ ij = -ji = k \\ &\quad jk = -kj = i \qquad \text{from relation (4.5)} \\ &\quad ki = -ik = j. \end{aligned} \tag{4.7}$$

These are the well-known identities satisfied by the quaternion group. As noted by Coxeter [8], the least redundant definition is $i^2 = j^2 = (ij)^2$.

The fact that we have been led—using our alternative approach to rotations—directly to a realization of the quaternion group is not surprising, in view of the intimate relationship that exists abstractly between these two group structures. What does deserve emphasis, however, is that we have achieved, by means of the concept of a turn, a geometric realization of the quantal rotation group, and this implies that a geometric realization has been achieved not only for the quaternion group, but for all unit quaternions as well. Put differently, by using the proper primitive object, the turn, we have achieved the proper physical picture for rotations. This picture then includes a geometric realization for all finite subgroups of the quantal rotation group, among which the quaternion group is but a special case. Let us note explicitly that the turns thus afford a physical realization of the double groups as well – as emphasized by Harter and dos Santos [9]. This physical picture afforded by the turns is, moreover, the analog to the physical picture afforded by vectors for the group of translations.

Remarks. (*a*) The scalar turn $\mathbf{T}_\pi$ (called scalar because it commutes with all other turns) plays a special role in the group of turns. If we recall how turns were introduced – relating rotations to a product of two reflections – it is clear that the turn $\mathbf{T}_\pi$ requires two reflections *in the same plane*. This operation must, by definition, be the identity. *The geometry of turns that distinguishes* $\mathbf{T}_\pi$ *from the identity* $\mathbf{T}_0$ *is distinct from the geometry of the model from which we started*! This happy accident is not unusual in physics. Feynman [10] has emphasized that it is precisely for this reason that it is so important to develop all possible views of a subject, since in the process of generalizing, or abstracting, one model may be favored in this way over another.

(*b*) If one imposes the requirement that $\mathbf{T}_\pi$ and $\mathbf{T}_0$ be identified, the resulting structure is still a group – namely, the proper orthogonal group [$SO(3)$], which is a factor group of the quantal rotation group [$SU(2)$]. That is, $SU(2)/Z_2 \cong SO(3)$. $\mathbf{T}_0$ and $\mathbf{T}_\pi$ commute with all turns and generate the two-element normal subgroup Z_2.

This identification of $\mathbf{T}_\pi$ and $\mathbf{T}_0$ reduces the group space of the group of turns [$SU(2)$] by half to the group space of $SO(3)$. (This has been discussed in Chapter 2.)

(*c*) It is of interest to make explicit the (homomorphic) mapping of the group of turns [$SU(2)$] onto the group of 3×3 proper orthogonal matrices [$SO(3)$], noted in Remark (*b*). We introduce a right-handed reference frame $(\hat{e}_1, \hat{e}_2, \hat{e}_3)$ by associating each turn of length $\pi/2$ in the triple $(\mathbf{E}_1, \mathbf{E}_2, \mathbf{E}_3)$, $\mathbf{E}_3 + \mathbf{E}_2 + \mathbf{E}_1 = \mathbf{T}_0$, with its normal vector $\hat{e}_i$. [There is thus a one-to-one map between the set of all right-handed reference frames $\{(\hat{e}_1, \hat{e}_2, \hat{e}_3)\}$ and the set of all triples of turns of length $\pi/2$, $\{(\mathbf{E}_1, \mathbf{E}_2, \mathbf{E}_3) : \mathbf{E}_3 + \mathbf{E}_2 + \mathbf{E}_1 = \mathbf{T}_0\}$.]

Consider the turns $\mathbf{E}'_j$ defined by

$$\mathbf{E}'_j = -\mathbf{T} + \mathbf{E}_j + \mathbf{T}, \qquad j = 1, 2, 3, \text{ with } \mathbf{T} \text{ being a given turn.} \qquad (4.8)$$

The normals $\hat{e}'_1, \hat{e}'_2, \hat{e}'_3$ to the great circles of $\mathbf{E}'_1, \mathbf{E}'_2, \mathbf{E}'_3$, respectively, are given by $\hat{e}'_i = \mathscr{R}\hat{e}_i$, where $\mathscr{R}$ is the rotation associated with the turn $\mathbf{T}$. Using $\mathbf{E}'_j$ in place of $\mathbf{T}$ and $\mathbf{E}_i$ in place of $\mathbf{E}$ in Eq. (4.4), we obtain

$$R_{ij} = \hat{e}_i \cdot (\mathscr{R}\hat{e}_j) = \cos\|-\mathbf{T} + \mathbf{E}_j + \mathbf{T} - \mathbf{E}_i\|. \tag{4.9}$$

Equation (4.9) is the desired mapping of the group of turns onto 3×3 proper orthogonal matrices, $\mathbf{T} \to R \equiv (R_{ij})$. Note that Eq. (4.9) yields the same result when $\mathbf{T}$ is replaced by $\mathbf{T}^c$, so that the mapping is 2 : 1, both $\mathbf{T}$ and $\mathbf{T}^c$ mapping onto the same orthogonal matrix.

Equation (4.9) is an interesting formula geometrically. It asserts that *the elements of each proper orthogonal matrix can be obtained as the cosines of a specific set of arc lengths on the unit sphere.*

(*d*) Let us reconsider the Dirac string construction discussed in Chapter 2, Section 3, using now the concept of a turn. For a solid body with one fixed point the state of the system is specified by giving the orientation of a body-fixed frame $\{\hat{b}_i\}$ with respect to a space-fixed frame $\{\hat{e}_i\}$, both frames having the same origin (the fixed point in the solid body). [The phrase "state of the system" is to be understood kinematically and in the sense of classical (and not quantal) physical concepts.]

Using the results of Remark (*c*), we find from Eq. (4.9) the relation between the two frames to be

$$\hat{b}_j = \sum_{i=1}^{3} B_{ij}\hat{e}_i, \qquad j = 1, 2, 3,$$

where

$$B_{ij} = \cos\|-\mathbf{B} + \mathbf{E}_j + \mathbf{B} - \mathbf{E}_i\|.$$

In this expression for the $\{B_{ij}\}$ – the matrix elements of an orthogonal matrix – we have denoted by $\{\mathbf{E}_i\}$ any triple of turns of length $\pi/2$ that satisfy $\mathbf{E}_3 + \mathbf{E}_2 + \mathbf{E}_1 = \mathbf{T}_0$. The turn $\mathbf{B}$ is to be associated with the state of the solid body.

To interpret these results let us specialize $\mathbf{B}$ to be the zero turn $\mathbf{T}_0$. Then $B_{ij} = \delta_{ij}$, so that the two frames $\{\hat{e}_i\}$ and $\{\hat{b}_i\}$ coincide.

Note, however, that the turn $\mathbf{B}$ is not uniquely specified, since $\mathbf{B} \to \mathbf{B} + \mathbf{T}_\pi$ leaves each B_{ij} unchanged. It is clear that the B_{ij} determine exactly two turns, $\mathbf{B}$ and $\mathbf{B} + \mathbf{T}_\pi \equiv \mathbf{B}^c$. Thus, *the associations*

State of system $\leftrightarrow$ frame $\{\hat{b}_i\}$ relative to frame $\{\hat{e}_i\}$ $\leftrightarrow$ turns $(\mathbf{B}, \mathbf{B}^c)$

are intrinsically ambiguous even for coinciding frames. Expressed in different words, *the version of a frame is inherently undefined*.

This situation is the precise analog of the familiar situation in quantum mechanics wherein one recognizes that the phase of a wave function is undefined; for the case of the solid body the (two-valued) version plays the role of the phase [$U(1)$ or circle group].

We can now understand the significance of the Dirac string construction. Working entirely within classical physical structures, the Dirac string construction shows that, although the version of any frame is undefined, *the relative version between a frame and the same frame after rotation is indeed well-defined*. Expressed differently, the turn $\mathbf{B}$ to be associated with the state of a solid body is undefined (by $\mathbf{T}_\pi$) but *under rotation by* $\mathbf{R}$ *the turn* $\mathbf{B}$ *becomes* $\mathbf{R} + \mathbf{B}$, *and this addition is well-defined* (*not* ambiguous by $\mathbf{T}_\pi$, since the system is distinguishable from the same system rotated by 2π). (See Note 3.)

We remark that for the quantum mechanical analog the fact that differences of phase for wave functions are well-defined, but not the phase itself, is the starting point for developing interactions based on gauge symmetry. (Dirac [10a], Boya [10b]).

It is quite interesting conceptually that the description of the state of a solid body by a turn $\mathbf{B}$ is so "quantum mechanical" in flavor, with a "measurement" being the determination of the matrix elements B_{ij}. These considerations show quite clearly the distinction between a finite-sized spinor (for which the "measurement" yields the three angles of a general 3×3 orthogonal matrix) and a point spinor (for which a measurement yields only two angles; see Chapter 7, Section 7). Despite this quantum mechanical flavor, our considerations on the solid body have been classical, since a correct quantal treatment must consider the uncertainty principle (see RWA). We remark that in the limit of large (massive, nonrelativistic) rotators these classical concepts can be validated in principle.

(*e*) The concept of a turn allows one to give a geometric realization of the individual elements of the quantal rotation group. Although this realization is of most importance in discussing the characteristic functions of this space (the rotation matrices), one can, however, adapt this realization to accord with the various subspaces (coset spaces).

The coset space (G/K) [where G is the group of turns and K is the group $SU(1)$ generated by J_3; see Chapter 3] can be described as the space of turns, *in which all* (*nonscalar*) *turns having the same great circle are identified*. Thus, the elements in the coset space $SU(2)/SU(1)$ are in one-to-one correspondence with the set of all unit vectors (the normal vectors to the great circles), which, in turn, defines the surface of the unit sphere, S^2. The appropriate spherical functions are the spherical harmonics Y_{lm}. (The fact that l is *integral* results from the identification: $\mathbf{T} \sim \mathbf{T}^c$.)

The double coset space $(K/G/K)$ is the subspace of (G/K) where points having the same polar angle, but different azimuths, are identified. The appropriate spherical functions are the Legendre functions P_l, as discussed in Chapter 3, Section 11. (This discussion of coset spaces is rather brief. A nice discussion is given in the text by Dym and McKean [11].)

(*f*) The concept of a turn is most useful for the intuitive understanding of rotations. Let us illustrate this by considering the familiar remark, "Infinitesimal rotations commute." One has become so accustomed to this statement that it probably no longer seems puzzling (if it ever was), but there *is* a concealed problem: Why should infinitesimals differ *qualitatively* from finite quantities in such a basic structural relation?

Turns show quite clearly the origin of this qualitative change. *There is an intrinsic scale for turns, set by the (unit) radius of the sphere.* The addition of infinitesimal turns means that one is considering, in effect, a very large sphere or, equivalently, the tangent plane to the unit sphere. In this limit, infinitesimal turns correspond to genuine vectors in the tangent space, and commutativity is obvious.

The existence of an intrinsic scale appears also in the commutation relations for angular momentum, which are not invariant to scale changes.

If one considers an infinitesimal turn, and (using the mode of expression of Klein–Sommerfeld) forms the "differential quotient" by dividing by an infinitesimal of time dt, one obtains a rotational velocity $\boldsymbol{\omega}$. This quantity is accordingly a genuine vector, as follows from the corresponding vectorial nature of infinitesimal turns. But it is not surprising that this velocity is not an exact differential (one of the peculiarities of rotational motion in three dimensions), since, intuitively speaking, the use of infinitesimal turns (tangent plane) has lost information.

(*g*) One further insight is afforded by turns. We may consider the group of translations (in the tangent plane) to be a *limiting structure* for the group of rotations (defined by turns on a sphere of "large radius"). This affords an intuitive understanding of the so-called *contraction limit* (Inönü and Wigner [12, 13], Segal [14], Saletan [15], which relates the compact quantal rotation group $SU(2)$ to the *noncompact* Euclidean group $E(2)$. In terms of eigenfunctions, this is the root cause for the asymptotic relationship that exists between spherical harmonics and Bessel functions (Sharp [16], Wigner [17], Talman [18]).

(*h*) Our discussion of the geometry of turns would be incomplete if we failed to mention the basic geometric theorem that Hamilton [1] used in his research on quaternions, (see also Klein and Sommerfeld [2]). This is the theorem determining (in effect) the product of two given rotations. The theorem states: *Let the (dihedral) angles at the vertices of a spherical triangle PQR be denoted P,*

Q, and R (in a given, positive sense). Then the product of rotations through angles $2P, 2Q, 2R$ about the radial axes $\hat{P}, \hat{Q}, \hat{R}$ is the identity.

It is not difficult to prove this theorem; one simply expresses the given rotations as products (in pairs) of reflections in the planes that form the sides of the spherical triangle: $RP, PQ; PQ, QR; QR, RP$. The resulting product is the identity. (This theorem, and the proof stated above, appear in Coxeter [8, pp. 53, 54].)

The importance of this theorem lies in the fact that it makes evident the fundamental role played by *half-angles* in the kinematics of rotations.

We have learned, however, in Remarks (*a*) and (*d*) above, to be aware that proofs of this sort may actually be in error by "rotations of 2π." We shall find that this is indeed the case here. *The resulting product is not the identity, but the scalar turn* $\mathbf{T}_\pi$.

This fact can be verified by transcribing the theorem directly into the language of turns. A much nicer way to proceed is to recognize that a conveniently chosen special case can settle the issue. Let us take the limiting case of an infinitesimal spherical triangle. Then all the rotations are about a single axis, and the sum of the rotation angles is clearly 2π. The stated result follows (that is, the theorem actually leads to $\mathbf{T}_\pi$ and not to the identity $\mathbf{T}_0$).

It is significant that this subtle distinction (the concept of version of a frame or solid body) was not recognized prior to the concept of a spinor, and is neglected in the references cited in this remark. Note that the use of a spherical triangle whose sides are interpreted as *turns* (in constrast to the interpretation of the theorem stated above) avoids the subtle error of changing the "version." (Misner *et al.* [19] give the standard interpretation of the theorem on p. 1139 and introduce version on p. 1148 as the concept of "orientation–entanglement relation." See our Chapter 2, Section 3.)

3. Properties of Turns (Algebraic View)

In the discussion above, we have presented the theory of turns geometrically in order to exploit the analogy between vectors (for translations) and turns (for rotations). We shall depart now from the purely geometric view; our objective is to find an algebraic description.

We have noted earlier that a given turn is defined in terms of an ordered pair of points (head and tail of the turn) on a unit sphere. To give an algebraic description of the turn, we replace the two points by unit vectors $\hat{\alpha}, \hat{\beta}$ from the center of the unit sphere to the ends of the turn. Two such unit vectors define a scalar, $s \equiv \hat{\alpha} \cdot \hat{\beta} = \cos\|\mathbf{T}\|$, and a vector, $\mathbf{v} = \hat{\alpha} \times \hat{\beta} = \hat{n} \sin\|\mathbf{T}\|$, where $\hat{n}$ is the unit vector along the axis of the rotation associated with the turn $\mathbf{T}$. The sense of the turn is implied by the right-hand rule of the cross product.

We shall call these two quantities, s and $\mathbf{v}$, the parameters of the turn $\mathbf{T}$, and write

$$\mathbf{T} = T(s, \mathbf{v}), \tag{4.10}$$

where $s = \cos\|\mathbf{T}\|$ is the scalar parameter and $\mathbf{v} = \hat{n} \sin\|\mathbf{T}\|$ is the vector parameter. (It is easily seen that equivalent turns have the same parameters, s and $\mathbf{v}$.)

The parameters of a turn necessarily satisfy the relation

$$s^2 + \mathbf{v} \cdot \mathbf{v} = 1. \tag{4.11}$$

We may interpret this relation as identifying $(s, \mathbf{v})$ *with a point on the surface of the unit sphere, S^3, in Euclidean four-space.* As we let the length of the turn $\|\mathbf{T}\|$ assume any value in the interval $0 \leqslant \|\mathbf{T}\| \leqslant \pi$, and $\hat{n}$ be any direction in three-space, the parameters associated with the set of turns ranges over all the points of S^3, and conversely. We conclude: *The set of turns may be put in one-to-one correspondence with the set of points of S^3.*

Thus, each turn has the algebraic representation

$$\mathbf{T} = T(s, \mathbf{v}) \to (s, v_1, v_2, v_3), \tag{4.12}$$

where (s, v_1, v_2, v_3) is any point of S^3. We obtain the geometric turn corresponding to a point of S^3 by solving for the length $\|\mathbf{T}\|$ $(0 \leqslant \|\mathbf{T}\| \leqslant \pi)$ of the turn from $\cos\|\mathbf{T}\| = s$ and solving for $\hat{n}$ from $(n_1, n_2, n_3)\sin\|\mathbf{T}\| = (v_1, v_2, v_3)$. The identity turn has the algebraic representation

$$\mathbf{T}_0 = T(1, \mathbf{0}) \to (1, 0, 0, 0), \tag{4.13}$$

and the turn $\mathbf{T}_\pi$ has the algebraic representation

$$\mathbf{T} = T(-1, \mathbf{0}) \to (-1, 0, 0, 0). \tag{4.14}$$

The algebraic representation of the inverse turn is

$$-\mathbf{T} = T(s, -\mathbf{v}), \tag{4.15}$$

whereas for the conjugate turn one has

$$\mathbf{T}^c = T(-s, -\mathbf{v}). \tag{4.16}$$

It is important to note, from Eqs. (4.15) and (4.16), that $T(-s, -\mathbf{v}) \neq -T(s, \mathbf{v})$. Thus, the parameters of a turn do not obey the rules of four-space vectors. Turns are, in fact, not vectors in four-space. The difficulty is only apparent, and is an artifact of using + to denote the composition law for mapping the set of ordered pairs of turns onto the set of turns itself. The *addition of turns* [with unit $T(1, \mathbf{0})$] is to be distinguished from the very different operation *addition of vectors* [with unit (0, 0, 0, 0)].

The task now is to deduce, using only the geometric rule of addition of turns, the combination law for the parameters of a turn. Consider the addition of two turns, $\mathbf{T}' + \mathbf{T}$. Let $\hat{\alpha}$, $\hat{\beta}$, and $\hat{\gamma}$ denote, respectively, unit vectors from the center of the unit sphere to the tail of $\mathbf{T}$, the head of $\mathbf{T}$ (tail of $\mathbf{T}'$), and the head of $\mathbf{T}'$. Geometric considerations suffice to show that

$$\hat{\beta} = s\hat{\alpha} + \mathbf{v} \times \hat{\alpha}, \tag{4.17}$$

$$\hat{\gamma} = s'\hat{\beta} + \mathbf{v}' \times \hat{\beta}, \tag{4.18}$$

where $\mathbf{T} = T(s, \mathbf{v}), \mathbf{T}' = T(s', \mathbf{v}')$, $s = \hat{\alpha} \cdot \hat{\beta}$, $\mathbf{v} = \hat{\alpha} \times \hat{\beta}$, $s' = \hat{\beta} \cdot \hat{\gamma}$, and $\mathbf{v}' = \hat{\beta} \times \hat{\gamma}$. Substituting $\hat{\beta}$ into Eq. (4.18) and using the vector triple product rule,

$$\mathbf{v}' \times (\mathbf{v} \times \hat{\alpha}) = (\mathbf{v}' \times \mathbf{v}) \times \hat{\alpha} - (\mathbf{v} \cdot \mathbf{v}')\hat{\alpha},$$

we obtain

$$\hat{\gamma} = s''\hat{\alpha} + \mathbf{v}'' \times \hat{\alpha}, \tag{4.19}$$

where

$$s'' = ss' - \mathbf{v} \cdot \mathbf{v}', \tag{4.20}$$

$$\mathbf{v}'' = s\mathbf{v}' + s'\mathbf{v} + \mathbf{v}' \times \mathbf{v}. \tag{4.21}$$

Observing that $\hat{\alpha} \cdot \mathbf{v}'' = 0$, we obtain from Eq. (4.19) the relations

$$s'' = \hat{\alpha} \cdot \hat{\gamma}, \qquad \mathbf{v}'' = \hat{\alpha} \times \hat{\gamma}, \tag{4.22}$$

which show that $(s'', \mathbf{v}'')$ are the parameters of the composite turn:

$$T(s'', \mathbf{v}'') = T(s', \mathbf{v}') + T(s, \mathbf{v}). \tag{4.23}$$

We conclude: *Corresponding to the geometric rule for the addition of turns,*

$$\mathbf{T}'' = \mathbf{T}' + \mathbf{T}, \tag{4.24}$$

one has the law of combination of parameters given by

$$T(ss' - \mathbf{v}' \cdot \mathbf{v}, s\mathbf{v}' + s'\mathbf{v} + \mathbf{v}' \times \mathbf{v}) = T(s', \mathbf{v}') + T(s, \mathbf{v}) \tag{4.25}$$

This law of combination, Eq. (4.25), can be recognized as identical to the product law for quaternions. If we write the quaternion q having the components $(s, \mathbf{v})$ in the form

$$q = s1 + v_1 i + v_2 j + v_3 k, \tag{4.26}$$

and similarly for q' with components $(s', \mathbf{v}')$, we find that the quaternion product, $q'q$, has – using the quaternion multiplication laws given in Eq. (4.7) – precisely the components given in Eq. (4.25). Thus, the correspondences $q \to T(s, \mathbf{v})$ and $q' \to T(s', \mathbf{v}')$ imply $q'q \to T(s', \mathbf{v}') + T(s, \mathbf{v})$.

The conjugation (involution) for turns defined by $\mathbf{T} \to -\mathbf{T}$ or, equivalently, $T(s, \mathbf{v}) \to T(s, -\mathbf{v})$ allows us to define for quaternions the *conjugate quaternion*:

$$q \to \bar{q} = s1 - v_1 i - v_2 j - v_3 k, \tag{4.27}$$

and hence a norm for quaternions:

$$N(q) \equiv q\bar{q} = s^2 + \mathbf{v} \cdot \mathbf{v} \geqslant 0. \tag{4.28}$$

We can now recognize that the concept of a *turn* is no more and no less than a geometric realization for the multiplication of quaternions of unit norm (unimodular). The unimodularity results from Eq. (4.11). *The group of turns* [quantal rotation group, $SU(2)$] *is a geometric realization of the multiplicative group of unimodular quaternions.*

Remarks. (*a*) Hamilton's achievement went much beyond the recognition of the structural properties of rotations as realized by unimodular quaternions. He recognized that quaternions form a division algebra with the remarkable composition property for the norm, $N(q)N(q') = N(qq')$, and he subsequently devoted much effort toward a general quaternion calculus. This aspect of his work, in contrast to the rotational aspects of quaternions, has not gained general acceptance in the physics literature. We shall accordingly not discuss quaternions (as such, apart from turns) further in this work, referring for these purely quaternionic aspects to the physically oriented, elegant survey by Ehlers *et al.* [20].

(*b*) It is customary in physics to use the Pauli matrices instead of the quaternion units. As *basis units*, one has the correspondence

$$1 \to \sigma_0,$$

$$i \to -\sqrt{-1}\,\sigma_1,$$

$$j \to -\sqrt{-1}\,\sigma_2,$$

$$k \to -\sqrt{-1}\,\sigma_3,$$

where σ_i are the 2×2 matrices given earlier [Eqs. (3.24)]. This is not simply a notational convenience, and we must explain why the change occurs.

Note first that the two structures are very different. The quaternion units, $\pm 1, \pm i, \pm j, \pm k$, form the eight-element *quaternion group*, which has a unique faithful irreducible representation (irrep) of matrix dimension two. By contrast, the *Pauli group* has 16 elements and two inequivalent (conjugate) irreps of dimension two. The two groups are quite distinct.

There are two reasons why physicists use the Pauli matrices rather than the quaternions. The first reason is purely technical. The generators of infinitesimal transformations (displacements, rotations) have a direct physical significance as observables, and hence Hermitian operators. If, however, the infinitesimal transformations are integrated to finite transformations, the resulting transformation must be unitary to conserve probability. Physicists meet these two requirements by explicitly adjoining the imaginary unit $\sqrt{-1}$ (as in the Pauli matrices), whereas mathematicians, concerned as they are to distinguish the real field from the complex field, avoid the explicit imaginary unit by taking the generators as anti-Hermitian.

The second reason why physicists use the Pauli matrices, and not the quaternion units, can also be understood from quantum mechanics. Quantum mechanics uses ket vectors (Hilbert space vectors), $\{|\psi\rangle\}$, and bra vectors (dual Hilbert space vectors) $\{\langle x|\}$, and ascribes to the inner product, $\langle x|\psi\rangle$, the meaning of probability amplitude. From an algebraic point of view, the space of ket vectors is a left ideal, and a pure state is associated with a primitive idempotent $[\rho_\psi^2 = \rho_\psi, \operatorname{tr}\rho_\psi = 1]$.

If we apply these algebraic concepts to the quaternions, we see that for quantum mechanical applications one needs to adjoin the imaginary unit $\sqrt{-1}$, exactly as in the Pauli group.

Thus, for example, a pure state is associated with the idempotent $\begin{pmatrix} 1 & 0 \\ 0 & 0 \end{pmatrix}$, which is given by the algebraic element

$$\tfrac{1}{2}(1 + \sigma_3).$$

These idempotents associated with pure states – called density matrices, ρ, in quantum mechanics – have the general form

$$\rho = \tfrac{1}{2}(1 + \boldsymbol{\sigma} \cdot \hat{P}),$$

where $\hat{P}$ is a unit three-vector.

We shall discuss density matrices systematically in Chapter 7, Section 7. Our purpose here was to explain the underlying algebraic reasons why the Pauli matrices – and not quaternionic units – appear naturally in quantum mechanics.

4. The Space of Turns as a Carrier Space

Let us denote an individual turn [an element of $SU(2)$, the quantal rotation group] by τ. There is a standard way, due to Cayley, whereby one can obtain from a given group a realization of this group by transformations. This realization associates with each group element g the transformation induced on the set of all group elements $\{\tau\}$ by *left translation*:

$$g : \tau \to \tau' \equiv L_g(\tau) = g + \tau. \tag{4.29}$$

There is clearly a second standard realization by *right translation*:

$$g : \tau \to \tau' \equiv R_g(\tau) = \tau + g^{-1} = \tau + (-g). \tag{4.30}$$

From the original group laws for the turns, one finds the composition rules for left and right translations of the group of turns:

$$L_{g'}(L_g(\tau)) = L_{g'+g}(\tau), \tag{4.31}$$

where the operations are, *first*, transform by g, and *then* by g'. Similarly, one has

$$R_{g'}(R_g(\tau)) = R_{g'+g}(\tau). \tag{4.32}$$

[Note that the desire to achieve the same formal structure in Eqs. (4.31) and (4.32) accounts for the use of g^{-1} in Eq. (4.30).]

We can thus express the realization by left and right translations as a mapping

$$\tau \to L_\tau \qquad \text{or} \qquad \tau \to R_\tau \tag{4.33}$$

such that the group law is preserved:

$$\tau + \tau' \rightarrow L_{\tau+\tau'} \qquad \text{or} \qquad \tau + \tau' \rightarrow R_{\tau+\tau'}. \tag{4.34}$$

This latter property defines the mapping to be a representation.

The essential point now is to observe that – in consequence of associativity in combining turns – *left and right translations commute*:

$$R_g L_{g'} = L_{g'} R_g : \tau \rightarrow \tau' = g' + \tau + g^{-1} \tag{4.35}$$

Since the left translations taken by themselves define a representation of the group of turns [$SU(2)$], and similarly for the right translations by themselves, we conclude that *the space of turns taken as a carrier space for translations realizes a representation of the direct product group* $SU(2) \times SU(2)$.

This result becomes clearer if we make use of the mapping of the group of turns onto the group of unimodular quaternions or, equivalently, onto the group of 2×2 unimodular unitary matrices. Specifically, this mapping is [see Eqs. (4.25) and (4.26)]

$$T(x_0, \mathbf{x}) \rightarrow q(x_0, \mathbf{x}) \rightarrow U(x_0, \mathbf{x}), \tag{4.36}$$

where

$$q(x_0, \mathbf{x}) = x_0 1 + x_1 i + x_2 j + x_3 k, \tag{4.37}$$

$$U(x_0, \mathbf{x}) = x_0 \sigma_0 - i(\mathbf{x} \cdot \boldsymbol{\sigma}), \tag{4.38}$$

in which $(x_0, \mathbf{x})$ denotes a generic point of the three-sphere, S^3. Using now the same notation $g, g', \tau, \tau', \ldots$ for corresponding turns, quaternions, and matrices [given by the mapping (4.36)], one obtains the transformations of quaternions or matrices corresponding to those of turns [Eqs. (4.29)–(4.35)] simply by changing the additive notation to a multiplicative one. In particular, the transformation of turns, Eq. (4.35), becomes

$$R_g L_{g'} : \tau \rightarrow \tau' = g' \tau g^{-1} \tag{4.39}$$

in which g, g', τ, τ' denote either unimodular quaternions or the corresponding unimodular matrices.

The realization of the transformations L_g and R_g as differential operators acting on the set of differentiable functions $\{f\}$ of $(x_0, \mathbf{x})$ that are square integrable over the sphere S^3 has already been obtained in Chapter 3, Note 4, under a different guise. The realizations of L_g and R_g in this function space are given, respectively, by the unitary operators $\mathscr{L}_g$ and $\mathscr{R}_g$ defined by

$$(\mathscr{L}_g f)(\tau) = f(L_{g^{-1}}(\tau)),$$

$$(\mathscr{R}_g f)(\tau) = f(R_{g^{-1}}(\tau)), \tag{4.40}$$

where $\tau = \tau(x_0, \mathbf{x})$ and $f(\tau) = f(x_0, \mathbf{x})$.

The differential operators generating these transformations may be found by considering infinitesimal left and right translations. Thus, the infinitesimal left translation of τ by $g = (1, \varepsilon\hat{n}/2)$ gives

$$L_{(1, -\varepsilon\hat{n}/2)}(\tau) = \tau + \delta\tau,$$

where the parameters of $\delta\tau$ are $\varepsilon(\hat{n}\cdot\mathbf{x}, -\hat{n}x_0 - \hat{n}\times\mathbf{x})/2$. The corresponding infinitesimal transformation in function space is thus found to be

$$(\mathscr{L}_g f)(\tau) = (1 - i\varepsilon\hat{n}\cdot\mathscr{J} + \cdots)f(\tau),$$

where $\mathscr{J}$ is the differential operator

$$\mathscr{J} = -\frac{i}{2}\left[\mathbf{x}\times\nabla - \left(\mathbf{x}\frac{\partial}{\partial x_0} - x_0\nabla\right)\right]. \tag{4.41}$$

Similarly, an infinitesimal right translation leads to

$$(\mathscr{R}_g f)(\tau) = (1 - i\varepsilon n\cdot\mathscr{K} + \cdots)f(\tau),$$

where $\mathscr{K}$ is the differential operator

$$\mathscr{K} = -\frac{i}{2}\left[\mathbf{x}\times\nabla + \left(\mathbf{x}\frac{\partial}{\partial x_0} - x_0\nabla\right)\right]. \tag{4.42}$$

These results are precisely the same as the results obtained in Chapter 3 for the rotation matrices (viewed as wave functions), and hence for the quantum mechanics of the symmetric top. Stated differently, we have shown that *translations realized on the space of turns are a fundamental representation of the quantum theory of the symmetric top*. The turn is the primitive element in this realization.

We shall not develop the consequences of these results further at this point, preferring instead to introduce a new technique – the Jordan mapping – in the next chapter, which, when combined with the turns, will lead us to an elegant way to develop both representation theory and the Wigner coefficients.

5. Notes

1. It has been our experience that the use of turns is the most intuitive and accessible way to approach rotations and to achieve the realization of angular momentum theory by 2×2 matrices, or, equivalently, to achieve Hamilton's approach to rotations via quaternions. [For a discussion of the usefulness of this concept see Harter [9].) This is the reason for the viewpoint chosen in the present chapter, but it must be admitted that Hamilton himself never explicitly introduced the concept of a turn, lacking (to our knowledge) the essential idea of relating a turn directly to a rotation via two reflections. This Note attempts to set these ideas in reasonable historical perspective (without, however, any claim to definitiveness).

Hamilton's way of introducing quaternions, extensively developed in his "Lectures on Quaternions" [1], is curiously roundabout (and appeared so to his contemporaries[1]). Hamilton defined quaternions as "*quotients of vectors*" (vectors in three-space, see [1]; cf. Vol. I, Book II, Chapter 1, pp. 107ff.). We shall show that *this definition is, in fact, completely equivalent to that of a turn.* Our argument is an extension of some results given by Klein and Sommerfeld.

First let us recall that an arbitrary three-vector can be considered as a quaternion having zero scalar part; the quaternion v corresponding to the vector $\mathbf{V} = (v_1, v_2, v_3)$ may be written as

$$\mathbf{V} \to v = v_1 i + v_2 j + v_3 k.$$

It is a fundamental result[2] that a rotation of the vector $\mathbf{V}$ into a vector $\mathbf{V}' = (v_1', v_2', v_3') \to v' = v_1' i + v_2' j + v_3' k$ is given by the quaternionic multiplication

$$\mathbf{V}' \to v' = qvq^{-1}, \tag{4.43}$$

where, for

$$q = \left(\cos\frac{\phi}{2}\right)1 + \left(\sin\frac{\phi}{2}\right)(n_1 i + n_2 j + n_3 k),$$

the vector $\mathbf{V}'$ is given in terms of $\mathbf{V}$ by $\mathbf{V}' = R(\phi, \hat{n})\mathbf{V}$ [column matrix notation, see Eq. (2.6).]

Consider now the particular case of Eq. (4.43) for $\hat{n} \cdot \mathbf{V} = 0$. It then follows that $vq^{-1} = qv$, so that Eq. (4.43) may be written as $v' = q^2 v$; that is,

$$v'/v = q^2.$$

This relation has several consequences:

---

[1] The discussion by Klein and Sommerfeld [2], (see §7 beginning on p. 55) is a remarkably clear and informative one. The background of Hamilton's ideas is also discussed by MacDuffee [21].

[2] This result was discovered independently by Hamilton [22, 23] and Cayley [24]. See the footnotes on p. 361 of Ref. [22] and p. 391 of Ref. [23] in which Hamilton acknowledges Cayley's independent discovery (and prior publication) and points out that he had exhibited his formula to Graves in early October, 1844. In his Collected Papers, Vol. I, p. 586, Cayley notes that Hamilton's discovery preceded by some months the date of his (Cayley's) paper.

(*a*) Since q^2 is itself a quaternion, we see that Hamilton's definition of a quaternion "as the ratio of two vectors" is quite meaningful and general. (Our discussion required that q^2 be unimodular, but if we allow **V** and **V**′ to have different lengths, the unimodular restriction can be removed.)

(*b*) It is meaningful to take the square root of this relation and solve for the unimodular quaternion q:

$$q = (v'/v)^{\frac{1}{2}},$$

where the positive sign is chosen by identifying $q = 1$ with the zero rotation. *The square root introduces the all-important half-angle* $\phi/2$.

We have thus shown that the concept of a turn **T** *of length* $\phi/2$ *and axis* $\hat{n}$ *agrees in all details with Hamilton's definition of the unimodular quaternion* q *associated with the same rotation.*

What is lacking in Hamilton's approach is the direct association of the turn (the q above, for example) with the rotation itself. The use of the basic relation, Eq. (4.43), does not really suffice, for this relation defines a faithful representation of the *group* $SO(3)$ and not of the group of turns [$SU(2)$]. It is the direct association of a turn with the rotation induced by two reflections that supplies the desired $SU(2)$ group; this result (to our knowledge, see Klein and Sommerfeld [2, p. 938]) was first stated by Schoenfliess [25] in 1910. (It was independently used by Wigner [3] in his famous paper on the Poincaré group.)

We have gone into this historical background rather more than is customary because these basic ideas have become part of the folklore of physics, and inaccuracies abound. We first learned of the use of two reflections in generating rotations from Prof. T. A. Welton, who attributed the idea (without claiming accuracy) to Hamilton.

The use of turns was developed intuitively, and (mistakenly) attributed by one of us to Hamilton, in [26]; the present Note is an attempt to make amends. We wish to acknowledge a correspondence with Dr. R. Hartung[1] concerning the origin of turns. (An earlier version of the present chapter was presented in lectures [27] given at the University of Canterbury in 1974.)

2. The use of reflections as the basic element in developing geometry is well known in the mathematical literature (see, for example, Bachmann [28]).

3. The problem of "rotations by 2π" is discussed in a rather different way in the mathematical literature.

One first considers the group space of $SO(3)$. This is the solid sphere of radius π in three-space. The generic rotation is described by a direction $\hat{n}$, and the angle of rotation θ is the radial distance ($0 \leqslant \theta \leqslant \pi$). For $SO(3)$ a rotation by 2π is *declared* to be the identity so that diametrically opposite points on the surface of the sphere are identified.

Next one considers closed paths in group space beginning and ending at the identity (homotopies). All such paths are continuously deformable into one of two distinct types: (*a*) a path equivalent to the null rotation (identity) or (*b*) a path equivalent to the rotation by 2π (Wigner [29]).

[1] Hartung [26a] has considered implications of 2π rotations (and turns) for the Pauli principle and geometry.

What does this imply for the physical realization of rotations? The concept of homotopy means, physically, that one considers a given physical object in three-space, and carries out any closed continuous sequence of rotations (that is, motions leaving one point of the object fixed and leading back to the initial state). The result, from the homotopy discussion above, is that any such motion is deformable into one of two motions: (*a*) no rotation at all, or (*b*) a rotation by 2π. *This latter is declared, according to the group $SO(3)$, to be equivalent, in its effect on the physical object, to the identity*! Physically, we have learned very little from this, for everything hinges on the unanswered question as to whether the physical object does or does not change under a 2π-rotation. (Weyl [30, p. 184] is unclear on just this point.)

The great advantage of the Dirac construction of Chapter 2, and of the turns (this chapter), is that it shows physically that 2π-rotations are *not* equivalent to the identity for a reference frame. Hence, the second type of homotopic path does not in fact return *this* physical object to its initial state.

We have devoted considerable space to this discussion, for it is a source of much confusion in the physics literature. We see, moreover, from the discussion above, that the physical and mathematical viewpoints have little in common.

The discussion of the problem in physics is made more difficult by the fact that a spinorial wave function under 2π-rotation suffers only a change in sign. A phase change in a wave function is, however, of no physical consequence. (This is discussed more fully in RWA.) With this in mind, Bacry [31] has stated two physical rules to clarify the physical problem of 2π-rotations: (1) It is impossible to distinguish the state of an *isolated* system from the state obtained after a rotation of $2n\pi$ ($n = 0, 1, \ldots$) regardless of whether or not the system is classical or quantal. (2) If the system is not isolated, one can determine at most whether or not the number of 2π-rotations is even or odd. We remark that the validity of rule (1) requires the validity of the Wigner superselection rule (see RWA).

References

1. W. R. Hamilton, *Lectures on Quaternions*. Dublin, 1853.
2. F. Klein and A. Sommerfeld, *Über die Theorie des Kreisels*, Vol. 1. Teubner, Leipzig, 1897.
3. E. P. Wigner, "On unitary representations of the inhomogeneous Lorentz group," *Ann. of Math.* **40** (1939), 149–204.
4. E. Artin, *Geometric Algebra*. Interscience, New York, 1964.
5. M. H. Stone, "Linear transformations in Hilbert space: I. Geometrical aspects," *Proc. Natl. Acad. Sci. U.S.* **15** (1929), 198–200; "Linear transformations in Hilbert space: II. Analytical aspects," *ibid.* 423–425.
6. J. von Neumann, "Allgemeine Eigenwerttheorie Hermitescher Funktionaloperatoren," *Math. Ann.* **102** (1930), 49–131; "Zur Algebra der Funktionaloperationen und Theorie der normalen Operatoren," *ibid.* 370–427.
7. J. M. Jauch, *Foundations of Quantum Mechanics*. Addison-Wesley, Reading, Mass., 1968.
8. H. S. M. Coxeter, *Regular Polytopes*. Methuen, London, 1948.
9. W. G. Harter and N. dos Santos, "Double-group theory on the half-shell and the two-level system. I. Rotation and half-integral spin states," *Am. J. Phys.* **46** (1978), 251–263.
10. R. P. Feynman, *The Character of Physical Law*. MIT Press, Cambridge, Mass., 1965. Messenger Lectures at Cornell University by R. P. Feynman, 1964.

10a. P. A. M. Dirac, "Quantized singularities in the electromagnetic field", *Proc. Roy. Soc.* **A130** (1931), 60–72.
10b. L. J. Boya, J. F. Carinena, and J. Mateos, "Homotopy and solitons", *Fortschr. Physik* **26** (1978), 175–214.
11. H. Dym and H. P. McKean, *Fourier Series and Integrals*. Academic Press, New York, 1972.
12. E. Inönü and E. P. Wigner, "On the contraction of groups and their representations," *Proc. Natl. Acad. Sci. U.S.* **39** (1953), 510–524.
13. E. Inönü, "Contraction of Lie groups and their representations," in *Group Theoretical Concepts and Methods in Elementary Particle Physics* (F. Gürsey, ed.), pp. 391–402. Gordon and Breach, New York, 1964.
14. I. E. Segal, "A class of operator algebras which are determined by groups," *Duke Math. J.* **18** (1951), 221–265.
15. E. J. Saletan, "Contraction of Lie groups," *J. Math. Phys.* **2** (1961), 1–21.
16. W. T. Sharp, *Racah Algebra and the Contraction of Groups*. Thesis, Princeton University, 1960; issued as Report AECL-1098, Atomic Energy of Canada, Chalk River, Ontario, 1960.
17. E. P. Wigner, *Application of Group Theory to the Special Functions of Mathematical Physics*. Lecture notes, unpublished, Princeton University, Princeton, N. J., 1955.
18. J. D. Talman, *Special Functions; A Group Theoretic Approach*. Benjamin, New York, 1968.
19. C. W. Misner, K. S. Thorne, and J. A. Wheeler, *Gravitation*. Freeman, San Francisco, 1973.
20. J. Ehlers, W. Rindler, and I. Robinson, "Quaternions, bivectors, and the Lorentz group," in *Perspectives in Geometry and Relativity* (B. Hoffmann, ed.), pp. 134–149. Indiana Univ. Press, Bloomington, Ind., 1966.
21. C. C. MacDuffee, "Algebra's debt to Hamilton," in *A Collection of Papers in Memory of Sir William Rowan Hamilton, Scripta Math.* **10** (1944), 25–36.
22. W. R. Hamilton, "On quaternions, or a new system of imaginaries in algebra; with some applications," *Proc. Roy. Irish Acad.* **III** (1847), 1–16. [Math. Papers III, 355–362.]
23. W. R. Hamilton, "On quaternions and the rotation of a solid body," *Proc. Roy. Irish Acad.* **IV** (1850), 38–56. [Math. Papers III, 381–391.]
24. A. Cayley, "On the application of quaternions to the theory of rotation," *Phil. Mag.* **3** (1848), 196–200. [Collected Math. Papers I, 405–409.]
25. A. Schoenfliess, *Rend. del Circolo Matematico di Palermo*, t. **29** (1910).
26. L. C. Biedenharn and H. van Dam, *Quantum Theory of Angular Momentum*. Academic Press, New York, 1965. A collection of reprints and original papers with a historical introduction.
26a. R. W. Hartung, "Pauli principle and Euclidean geometry," *Amer. J. Phys.* **47** (1979), 900–910.
27. L. C. Biedenharn and J. D. Louck, *Erskine Lectures*, Canterbury University, Christchurch, N. Z. 1974, unpublished.
28. E. Bachmann, *Aufbau der Geometrie aus dem Spiegelungsbegriff*. Springer, Berlin, 1959.
29. E. P. Wigner, *Group Theory and Its Application to the Quantum Mechanics of Atomic Spectra*. Academic Press, New York, 1959. Translation by J. J. Griffin of the 1931 German edition.
30. H. Weyl, *Gruppentheorie und Quantenmechanik*. Hirzel, Leipzig, 1st ed., 1928; 2nd ed., 1931. Translated by H. P. Robertson as *The Theory of Groups and Quantum Mechanics*. Methuen, London, 1931; reissued by Dover, New York, 1949.
31. H. Bacry, "La rotation des fermions," *La Recherche* **83** (1977), 1010.

CHAPTER 5

The Boson Calculus Applied to the Theory of Turns

1. Introduction

The space of turns was shown in the preceding chapter to be the carrier space of the defining (fundamental) representation of the group $SU(2) * SU(2)$ – the group of the symmetric top.[1] We shall show in the present chapter how the complete representation theory of this structure can be obtained from a mapping (the generalized Jordan map) that maps the generic turn into an operator (the matrix boson operator) acting in a Hilbert space. By means of this construction we shall obtain a unified presentation of the angular momentum states, the rotation matrices, and the Wigner coefficients as well. An interesting aspect of this technique is that it leads automatically to a view of the Wigner coefficients as a form of discretized rotation matrix (Section 8 below), which in turn implies a relationship between the Wigner coefficients and Jacobi polynomials (Gel'fand *et al.* [1, Supplement III]).

The key idea behind this unified presentation – the Jordan mapping of the turn into a matrix boson – is of more general validity than the application made here. To put this structure in a larger context we first develop, in some detail, the techniques of the *boson calculus*. It is hoped that this digression does not break the thread of the development too badly.

[1] The star denotes that this group is a subgroup of the direct product group, the subgroup being defined by the extra relation $\mathscr{J}^2 = \mathscr{K}^2$; see Chapter 3, Note 4.

ENCYCLOPEDIA OF MATHEMATICS and Its Applications, Gian-Carlo Rota (ed.).
Vol. 8: L. C. Biedenharn and J. D. Louck, Angular Momentum in
Quantum Physics: Theory and Application

2. Excursus on the Boson Calculus

The harmonic oscillator. The boson calculus originated in the treatment of a basic physical problem, the quantum mechanics of the linear harmonic oscillator. Because in quantum physics a (free) field—for example, the electromagnetic field—can be viewed as a collection of infinitely many harmonic oscillators, this elementary problem is of much greater applicability, and importance, than might first appear.

The Hamiltonian for such an oscillator, in one dimension, is given by

$$H = \frac{p^2}{2m} + \frac{m\omega^2}{2} q^2, \tag{5.1}$$

where q is the displacement of the oscillator (particle) from some fixed origin, m is the mass of the particle, and ω is the (classical) circular frequency of the oscillation.

Quantization is defined by taking p and q to be self-adjoint operators[1] acting on a Hilbert space that satisfy (on a dense domain) the Heisenberg commutation relations:

$$[q,p] = i\hbar 1. \tag{5.2}$$

The Hamiltonian thus becomes a self-adjoint operator, which can be recognized to be positive definite on all vectors $|\psi\rangle$ in its domain:

$$\langle\psi|H\psi\rangle = \frac{1}{2m}\langle p\psi|p\psi\rangle + \frac{m\omega^2}{2}\langle q\psi|q\psi\rangle > 0. \tag{5.3}$$

(The operator H is strictly positive, since $\langle\psi|H\psi\rangle = 0$ would imply $p|\psi\rangle = q|\psi\rangle = 0$, which contradicts the Heisenberg relation.)

Let us now introduce the *boson operators*, defined by

$$\text{“Creation operator”} \equiv a = (m\omega/2\hbar)^{\frac{1}{2}}q - i(2\hbar m\omega)^{-\frac{1}{2}}p, \tag{5.4}$$

$$\text{“Destruction operator”} \equiv \bar{a} = (m\omega/2\hbar)^{\frac{1}{2}}q + i(2\hbar m\omega)^{-\frac{1}{2}}p. \tag{5.5}$$

(Note that $\bar{a}$ is adjoint to a.) In consequence of the definition, the boson operators obey the commutation relation

$$[\bar{a}, a] = 1. \tag{5.6}$$

[1] The distinction between Hermitian (or symmetric) and self-adjoint operators is important when there can be domain problems.

The Hamiltonian H written in terms of boson operators takes the form

$$H = \frac{\hbar\omega}{2}(a\bar{a} + \bar{a}a) = \hbar\omega(a\bar{a} + \tfrac{1}{2}). \tag{5.7}$$

Let $|\psi_0\rangle$ be a normalized vector that satisfies the relation

$$\bar{a}|\psi_0\rangle = 0. \tag{5.8}$$

Using the Schrödinger realization of the operators p and q, which takes $p = -i\hbar(d/dq)$, Eq. (5.8) becomes a first-order linear differential equation whose solution is

$$\langle x|\psi_0\rangle \equiv \psi_0(x) = \pi^{-\frac{1}{4}} \exp(-x^2/2), \tag{5.9}$$

where x denotes the dimensionless coordinate

$$x \equiv (m\omega/\hbar)^{\frac{1}{2}} q.$$

The normalization condition on $|\psi_0\rangle$ is obtained from the inner product $\langle\phi|\psi\rangle = \int_{-\infty}^{\infty} \phi^*(x)\psi(x)\,dx$:

$$\langle\psi_0|\psi_0\rangle = \int_{-\infty}^{\infty} |\psi_0(x)|^2\,dx = 1, \tag{5.10}$$

which identifies the Hilbert space to be $L^2(-\infty, \infty)$.

The Hamiltonian H operating on the vector $|\psi_0\rangle$ thus has the eigenvalue $\hbar\omega/2$. It is customary to subtract this energy (the "zero point" or minimum energy) and define a new dimensionless operator, N, the "number" operator, by

$$(\hbar\omega)^{-1}H - \tfrac{1}{2} \equiv N = a\bar{a}. \tag{5.11}$$

The complete set of normalized eigenvectors, $\{|\psi_n\rangle\}$, can be constructed from $|\psi_0\rangle$ by using the definition

$$|\psi_n\rangle = (n!)^{-\frac{1}{2}}(a)^n|\psi_0\rangle. \tag{5.12}$$

It is easily seen that these vectors are normalized eigenvectors of the number operator N belonging to the eigenvalue n:

$$N|\psi_n\rangle = n|\psi_n\rangle. \tag{5.13}$$

The names "creation operator" for a and "destruction operator" for $\bar{a}$ come from interpreting Eq. (5.12). The operator a raises the eigenvalue n by one unit ("creates a quantum of excitation $\hbar\omega$"), and correspondingly $\bar{a}$ acting on $|\psi_n\rangle$ lowers the eigenvalue n by one unit ("destroys one quantum").

The analysis of the operator N, and correspondingly the Hamiltonian H, leads to the following significant result: *The number operator N has for its spectrum the nonnegative integers $\mathbb{Z}^+$; each eigenvalue is nondegenerate, and the eigenvectors $|\psi_n\rangle$ form a complete orthonormal basis for $\mathscr{H}$, a (separable) Hilbert space $L^2(-\infty, \infty)$ over the real line.*

A Hilbert space of analytic functions.[1] There is another way to view the quantum mechanics of the harmonic oscillator that emphasizes the classical aspects of the quantal structure most strikingly; this is the construction (originated by Schrödinger [10]) of wave functions that *minimize the Heisenberg uncertainty relation*: $\Delta p\,\Delta q \geqslant \hbar/2$. The existence of such "coherent" or "minimum uncertainty" states (which actually achieve the lower limit $\hbar/2$) is a peculiarity of the harmonic oscillator. (This subject is developed further in RWA in connection with the angular momentum uncertainty relations.)

An alternative way to view the minimal uncertainty states is to consider these states as *eigenstates of the destruction operator* $\bar{a}$. This approach will lead us very quickly to the construction of the coherent states.

To begin, let us note that the ground state $|\psi_0\rangle$ is itself an eigenstate of the destruction operator [from Eq. (5.8)] and belongs to the eigenvalue zero. The average value for the momentum and position operators for this state—$\langle\psi_0|p\psi_0\rangle \equiv p_{\text{ave}}$ and $\langle\psi_0|q\psi_0\rangle \equiv q_{\text{ave}}$—are, from Eq. (5.8), also zero. The dispersion in momentum, $\Delta p \equiv [\langle\psi_0|(p - p_{\text{ave}})^2\psi_0\rangle]^{\frac{1}{2}}$, and the dispersion in position, $\Delta q \equiv [\langle\psi_0|(q - q_{\text{ave}})^2\psi_0\rangle]^{\frac{1}{2}}$, are easily calculated to be

$$\Delta p = (\hbar m\omega/2)^{\frac{1}{2}}, \qquad \Delta q = (\hbar/2m\omega)^{\frac{1}{2}}, \tag{5.14}$$

so that $\Delta p\,\Delta q = \hbar/2$, verifying that the ground state $|\psi_0\rangle$ is indeed a state of minimum uncertainty. (The numerical factors can be checked by noting that these values lead to the energy $\hbar\omega/2$, as required for the ground state.)

To obtain all other minimum uncertainty states we simply displace the ground state using the (unitary) displacement operator:

$$\mathscr{U}(\xi, \eta) = \exp\{-i[(2/m\hbar\omega)^{\frac{1}{2}}\xi p - (2m\omega/\hbar)^{\frac{1}{2}}\eta q]\}, \tag{5.15}$$

[1] Important papers introducing and developing this subject are those of Dirac [2], Bergman [3], Bargmann [4], Segal [5], Klauder [6], Klauder and Sudarshan [7], and Mukunda and Sudarshan [7a]. For a generalization see Grossman [7b] and the review paper by Wolf [7c]. Applications to angular momentum theory are due to Schwinger [8] and Bargmann [9].

where ξ and η are dimensionless real numbers. From the Baker–Campbell–Hausdorff operator identity,

$$e^{A}Be^{-A} = \sum_{k=0}^{\infty} [A, B]_{(k)}/k!,$$

with $[A, B]_{(0)} \equiv B$, $[A, B]_{(1)} \equiv [A, B]$, and $[A, B]_{(k+1)} \equiv [A, [A, B]_{(k)}]$, one finds the result

$$\mathscr{U}(\xi, \eta)\, \bar{a} \mathscr{U}^{-1}(\xi, \eta) = \bar{a} - (\xi + i\eta). \tag{5.16}$$

This justifies the name displacement operator, since $\mathscr{U}(\xi, \eta)$ shifts the operator $\bar{a}$ by the complex number $\zeta \equiv \xi + i\eta$. In terms of a and $\bar{a}$ the shift operator $\mathscr{U}(\zeta)$ reads:

$$\mathscr{U}(\zeta) = \exp(\zeta a - \zeta^* \bar{a}) \tag{5.15'}$$

The desired set of minimal uncertainty states are now seen to be the set $\{\mathscr{U}(\zeta)|\psi_0\rangle\}$. These states are normalized, by construction, and form a family of unit vectors, one for each complex number ζ. Let us denote these vectors by

$$|\hat{e}_\zeta\rangle \equiv \mathscr{U}(\zeta)|\psi_0\rangle. \tag{5.17}$$

The defining characteristic (which follows from the construction) is that

$$\bar{a}|\hat{e}_\zeta\rangle = \zeta|\hat{e}_\zeta\rangle. \tag{5.18}$$

Expressed in words, the operator $\bar{a}$ acts on the unit vectors $\{|\hat{e}_\zeta\rangle\}$ as a multiplication operator: $\bar{a} \to \zeta$.

This family of unit vectors $\{|\hat{e}_\zeta\rangle\}$ is, however, *not orthonormal*. One finds the inner product to be

$$\langle \hat{e}_\alpha|\hat{e}_\beta\rangle = \langle \psi_0|\mathscr{U}^{-1}(\alpha)\mathscr{U}(\beta)|\psi_0\rangle = \exp(-\tfrac{1}{2}|\alpha|^2 - \tfrac{1}{2}|\beta|^2 + \alpha^*\beta), \tag{5.19}$$

which yields the magnitude

$$|\langle \hat{e}_\alpha|\hat{e}_\beta\rangle| = \exp(-\tfrac{1}{2}|\alpha - \beta|^2). \tag{5.20}$$

This expression clearly shows that *no two of these unit vectors are orthogonal*. In the physics literature (Klauder [6]), this situation is described as "an over-complete family of states."

The commutation relation $[\bar{a}, a] = 1$ suggests that, corresponding to the realization $\bar{a} \to \zeta$ on the $\{|\hat{e}_\zeta\rangle\}$, one seek to implement the action of a as a differential operator: $-a \to d/d\zeta$. To achieve this requires that one replace the normalized vectors $|\hat{e}_\zeta\rangle$ by the *unnormalized vectors* $|\mathbf{e}_\zeta\rangle \equiv e^{\frac{1}{2}|\zeta|^2}|\hat{e}_\zeta\rangle$. The most expedient procedure is to express these vectors in terms of the basis $|\psi_n\rangle = (n!)^{-\frac{1}{2}} a^n |\psi_0\rangle$. Using $\mathscr{U}(\zeta) = \exp(\zeta a - \zeta^* \bar{a})$ and property (5.8), one finds that

$$|\mathbf{e}_\zeta\rangle = \sum_{k=0}^{\infty} (k!)^{-\frac{1}{2}} \zeta^k |\psi_k\rangle = e^{\zeta a}|\psi_0\rangle. \tag{5.21}$$

In this form one easily validates the action $\bar{a} \to \zeta$, $-a \to d/d\zeta$ on the set $\{|\mathbf{e}_\zeta\rangle\}$. (Note, incidentally, that any eigenvector for a is necessarily the zero vector.)

The set of $\{|\mathbf{e}_\zeta\rangle\}$ are called (in the mathematical literature: Bargmann [4]) *a family of principal vectors*. In terms of these vectors we can now obtain a new Hilbert space $\mathscr{F}$ isomorphic to the Hilbert space $\mathscr{H}$ of states of the harmonic oscillator.

For each ket vector $|f\rangle$ in the space $\mathscr{H}$ let us define a function f of the complex variable ζ by

$$f(\zeta) = \langle f|\mathbf{e}_\zeta\rangle, \tag{5.22}$$

where the inner product on the right is in $\mathscr{H}$. Using the expression (5.21) for $|\mathbf{e}_\zeta\rangle$, we find

$$f(\zeta) = \sum_{k=0}^{\infty} \frac{A_k}{(k!)^{\frac{1}{2}}} \zeta^k, \tag{5.23}$$

with $A_k \equiv \langle f|\psi_k\rangle$. Thus, f is an entire analytic function of the complex variable ζ. Conversely, since each entire analytic function f may be expanded in the form (5.23), we may define a ket vector $|f\rangle \in \mathscr{H}$ by $|f\rangle = \sum_k A_k^* |\psi_k\rangle$, and this ket vector then corresponds to f. These results show that there is a one-to-one correspondence between the set of entire analytic functions of a complex variable ζ and the set of vectors of the Hilbert space $\mathscr{H}$.

Particular examples of the correspondence of vectors of $\mathscr{H}$ to entire analytic functions are given by

$$\begin{aligned}
&|\psi_k\rangle \to u_k, \qquad k = 0, 1, 2, \ldots, \\
&u_k(\zeta) = \langle \psi_k|\mathbf{e}_\zeta\rangle = \zeta^k/(k!)^{\frac{1}{2}}; \\
&|\mathbf{e}_\alpha\rangle \to \mathbf{e}_\alpha, \qquad \alpha \in \mathbb{C}, \\
&\mathbf{e}_\alpha(\zeta) = \langle \mathbf{e}_\alpha|\mathbf{e}_\zeta\rangle = e^{\alpha^* \zeta}.
\end{aligned} \tag{5.24}$$

The Hilbert space $\mathscr{F}$ is now defined to be the space of entire analytic functions with inner product of two functions $f, g \in \mathscr{F}$ given by

$$(f, g) = \frac{1}{\pi} \iint_{-\infty}^{\infty} d\xi \, d\eta \, e^{-|\zeta|^2} f^*(\zeta) g(\zeta). \tag{5.25}$$

By definition, $f \in \mathscr{F}$ if and only if f has finite norm, $(f, f)^{\frac{1}{2}} < \infty$.

Using the inner product (5.25), one finds that $(u_k, u_{k'}) = \delta_{kk'}$ and, from this result, that $(f, g) = \langle f | g \rangle^*$. The Hilbert spaces $\mathscr{H}$ and $\mathscr{F}$ are thus isomorphic.[1]

Remarks. The space $\mathscr{F}$ has some quite remarkable properties, which make it particularly convenient for applications in mathematical physics. We note here the following:

(*a*) Every element in $\mathscr{F}$ corresponds to precisely one function. (This is quite in contrast to the general situation in L^2-space, where an element of the space is an equivalence class of functions differing at most on sets of measure zero. Analytic functions differing on sets of measure zero are identical.)

(*b*) A sequence of functions f_n converging strongly to f implies pointwise convergence: $f_n(z) \to f(z)$. (This is another aspect of the strong correlation between local and global properties characteristic of analytic functions.)

(*c*) The analog of the Dirac delta function $\delta(q - q')$ for the space $\mathscr{F}$ is the reproducing kernel $\mathscr{K}(\omega, \zeta)$ defined by

$$f(\omega) = \frac{1}{\pi} \iint_{-\infty}^{\infty} d\xi \, d\eta \, e^{-|\zeta|^2} \mathscr{K}(\omega, \zeta) f(\zeta),$$

with $\mathscr{K}(\omega, \zeta) = (\mathbf{e}_\omega, \mathbf{e}_\zeta) = e^{\omega\zeta^*}$. Although the kernel $\mathscr{K}$ behaves as Dirac's delta function is supposed to do (because of the self-reproducing property), it does so without exhibiting any singularity whatsoever.

(*d*) The most remarkable of all the properties of $\mathscr{F}$ concerns the ease with which operators and their domains can be rigorously defined. In particular, *every linear operator in* $\mathscr{F}$ *having the* $\{\mathbf{e}_\alpha\}$ *as domain is an integral operator*.

The importance for mathematical physics of this remarkable result is that it provides a key example for which the use of Dirac's bra–ket notation is

---

[1] The most efficient way for evaluating the inner product is not by integration but by differentiation. For $|\psi\rangle = \psi(a)|\psi_0\rangle$, $\psi(a) = \sum_k \alpha_k a^k$, and $\psi^*(a) = \sum_k \alpha_k^* a^k$, one finds, using the commutation relation $[\bar{a}, a] = 1$ and $\bar{a}|\psi_0\rangle = 0$, that $\langle f | g \rangle = [f^*(\partial/\partial a) g(a)]_{a=0}$.

rigorously valid. In Dirac's ingenious notation, a *ket vector* $|\alpha\rangle$ (with α a label), is an element of the Hilbert space; a *bra vector*, written $\langle\beta|$, is an element of the dual Hilbert space, and the pairing $\{\langle\beta|, |\alpha\rangle\} \rightarrow \mathbb{C}$ is denoted $\langle\beta|\alpha\rangle$, and constitutes an inner product. So far this is essentially standard, but Dirac goes further to define the unit operator to be $\sum_\alpha |\alpha\rangle\langle\alpha|$, and moreover the general operator to be a linear combination of rank-one operators – that is,

$$\text{Operator} \equiv X = \sum_{\alpha,\beta} X_{\alpha\beta}|\alpha\rangle\langle\beta|. \tag{5.26}$$

This is equivalent to defining a general operator to be a matrix. For the space $\mathscr{F}$ these results are, remarkably enough, completely valid (Jauch [11]).

(*e*) Let us note that, just as for the product Hilbert space of n harmonic oscillators, we can define a Hilbert space $\mathscr{F}_n$ to be the direct product:

$$\mathscr{F}_n \equiv \mathscr{F} \otimes \cdots \otimes \mathscr{F}, \qquad n \text{ factors, } n < \infty.$$

(*f*) The importance of the Hilbert spaces $\mathscr{H}_n$ and $\mathscr{F}_n$ for the present volume lies in the fact that the special properties of these spaces, particularly that of remark (*d*), would allow one to justify fully our formal manipulations with boson operator techniques. We shall not carry this justification out, but it is essential for confidence in our results that the possibility of doing so be clearly in evidence.

3. The Jordan Mapping

Origin of the Jordan map. Physical phenomena in the relativistic energy domain are characterized by the creation and destruction of particles; the corresponding generalization of quantum mechanics to include Poincaré symmetry (Einsteinian relativity) is the *quantum theory of fields*, necessarily incorporating indefinite particle numbers. Jordan [12] observed (in 1935) that the basic technique of field quantization could be characterized algebraically quite simply, by what we shall call the *Jordan map.*[1] This is a mapping from a one-particle realization of the kinematic symmetry group (Jost [13]) into field operators of either boson or fermion type. We shall use the Jordan map into boson operators to implement the theory of turns, as a calculational device for angular momentum theory of the greatest utility.

---

[1] This mapping was used by Schwinger [8] in his famous (1952) report "On Angular Momentum." Accordingly, it is often called the "Jordan-Schwinger map." It is interesting to note that (in 1944) Dirac [12a] used a boson construction, which he called "expansors," to give unitary representations of the Lorentz group.

Jordan's observation may be expressed in this way: Consider the Hilbert space $\mathscr{H}_n$, corresponding to n kinematically independent bosons, and let X denote an $n \times n$ matrix.

The mapping $\mathscr{L}$ of $n \times n$ matrices into boson operators acting in $\mathscr{H}_n$ given by

$$\mathscr{L}: X = (X_{ij}) \rightarrow \sum_{i,j=1}^{n} X_{ij} a_i \bar{a}_j \equiv \mathscr{L}_X,$$

or

$$\mathscr{L}: X \rightarrow \mathscr{L}_X, \tag{5.27}$$

preserves the operation of commutation of matrices.

Proof. The proof follows directly from the boson commutation relations: $[\bar{a}_i, a_j] = \delta_{ij}$, $[\bar{a}_i, \bar{a}_j] = [a_i, a_j] = 0$. Let us denote by a the column vector $\operatorname{col}(a_1, \ldots, a_n)$, by $\bar{a}$ the column vector of adjoint operators $\operatorname{col}(\bar{a}_1, \ldots, \bar{a}_n)$, and by $\tilde{a}$ the row vector $(a_1, \ldots, a_n)$. Then, for any two $n \times n$ matrices X, Y we have

$$\mathscr{L}: X \rightarrow \tilde{a} X \bar{a} = \mathscr{L}_X, \qquad Y \rightarrow \tilde{a} Y \bar{a} = \mathscr{L}_Y.$$

Thus, for the commutator $[\mathscr{L}_X, \mathscr{L}_Y]$ we find

$$[\mathscr{L}_X, \mathscr{L}_Y] = [\tilde{a} X \bar{a}, \tilde{a} Y \bar{a}] = \tilde{a}[X, Y]\bar{a} = \mathscr{L}_{[X,Y]}. \qquad \blacksquare$$

The Jordan map thus has the property that

$$[\mathscr{L}_X, \mathscr{L}_Y] = \mathscr{L}_{[X,Y]}, \tag{5.28}$$

and is, moreover, linear over $\mathbb{C}$:

$$\lambda \mathscr{L}_X + \mu \mathscr{L}_Y = \mathscr{L}_{\lambda X + \mu Y}, \qquad \lambda, \mu \in \mathbb{C}. \tag{5.29}$$

The unit matrix in the ring of $n \times n$ matrices has the map

$$\mathbb{1} \rightarrow \mathscr{L}_{\mathbb{1}} = \tilde{a}\bar{a} = \sum_{i=1}^{n} a_i \bar{a}_i; \tag{5.30}$$

the operator $\mathscr{L}_{\mathbb{1}}$ clearly commutes with all $\{\mathscr{L}_X\}$. We shall call the operator $\mathscr{L}_{\mathbb{1}}$ *the Euler operator* (denoted by $\mathscr{E}$), since it is diagonal on the subspace of homogeneous polynomials in $\mathscr{H}_n$.

4. An Application of the Jordan Map

In order to appreciate the importance of the Jordan mapping for angular momentum, let us demonstrate now the ease with which many of the results of Chapter 3 can be obtained. In Chapter 3 we used Lie algebraic techniques applied to the generic angular momentum operators **J**. To apply the Jordan map we consider the simplest faithful realization, and hence we use the realization given by the turns, $T(\cos\frac{1}{2}\psi, \hat{n}\sin\frac{1}{2}\psi)$, of Chapter 4. We take ψ to be infinitesimal and, accordingly [from Eq. (4.38)], consider the Pauli matrices $\mathbf{J} \to \frac{1}{2}\boldsymbol{\sigma}$.

Consider then the Hilbert space $\mathscr{H}_2$ over two bosons, and the Jordan map:

$$\mathscr{L}: \tfrac{1}{2}\sigma_i \to J_i \equiv \mathscr{L}_{\frac{1}{2}\sigma_i} = \tfrac{1}{2}(\bar{a}\sigma_i\bar{a}). \tag{5.31}$$

The Jordan map assures us that the boson operators[1] J_i satisfy the commutation relation:

$$[J_i, J_j] = ie_{ijk}J_k. \tag{5.32}$$

Let $\mathscr{H}^{(2j)}$ denote the linear space of state vectors of the form $P(a_1, a_2)|0\rangle$ in which $P(a_1, a_2)$ is a homogeneous polynomial in (a_1, a_2) of degree $2j$ (an integer) and $|0\rangle$ denotes the ket vector such that $\bar{a}_1|0\rangle = \bar{a}_2|0\rangle = 0$. The space $\mathscr{H}^{(2j)}$ is invariant under the action of the boson operators J_i, so that the eigenvectors for angular momentum may be put in the form

$$|jm\rangle = P_{jm}(a_1, a_2)|0\rangle,$$

$$P_{jm}(a_1, a_2) = \frac{a_1^{j+m}a_2^{j-m}}{[(j+m)!(j-m)!]^{\frac{1}{2}}}. \tag{5.33}$$

This set of basis vectors for $\mathscr{H}^{(2j)}$ is enumerated by $m = j, j-1, \ldots, -j$; the space $\mathscr{H}^{(2j)}$ thus has dimension $2j+1$.

The operators $J_\pm \equiv J_1 \pm iJ_2, J_3$ have [from Eq. (5.31)] the explicit boson operator realization:

$$J_+ = a_1\bar{a}_2, \qquad J_- = a_2\bar{a}_1,$$

$$J_3 = \tfrac{1}{2}(a_1\bar{a}_1 - a_2\bar{a}_2). \tag{5.34}$$

[1] For notational simplicity we use the same symbol J_i to denote a boson operator that we have used previously to denote a generic angular momentum operator. These usages are logically distinct, but we believe no confusion should arise.

The action of these operators on the basis $\{|jm\rangle\}$ is easily worked out to be the standard action:

$$J_\pm|jm\rangle = [(j \mp m)(j \pm m + 1)]^{\frac{1}{2}}|j, m \pm 1\rangle,$$

$$J_3|jm\rangle = m|jm\rangle. \tag{5.35}$$

The Euler operator, $\mathscr{E} = \sum_i a_i\bar{a}_i = \tilde{a}\bar{a}$, is also diagonal on $\mathscr{H}^{(2j)}$:

$$\mathscr{E}|jm\rangle = 2j|jm\rangle. \tag{5.36}$$

These results show that by means of the Jordan map one obtains the set of *all* eigenspaces for angular momentum (each exactly once) most economically.

Remark. The verification of Eqs. (5.35) and (5.36) utilizes the properties $\bar{a}_1|0\rangle = \bar{a}_2|0\rangle = 0$, which imply $J_+|0\rangle = J_-|0\rangle = J_3|0\rangle = 0$, as well as $\mathscr{E}|0\rangle = 0$. If $\mathscr{O}$ is any polynomial in the a_i and $\bar{a}_i$ such that $\mathscr{O}|0\rangle = 0$, then the action of $\mathscr{O}$ on the state $|jm\rangle = P_{jm}(a_1, a_2)|0\rangle$ is given by $\mathscr{O}|jm\rangle = [\mathscr{O}, P_{jm}(a_1, a_2)]|0\rangle$. The basic commutators $[a_i, a_j] = [\bar{a}_i, \bar{a}_j] = 0$ and $[\bar{a}_i, a_j] = \delta_{ij}$ then allow one to replace the commutator $[\mathscr{O}, P_{jm}(a_1, a_2)]$ by another polynomial, acting on $|0\rangle$.

The finite transformations of the space $\mathscr{H}^{(2j)}$ (hence, of $\mathscr{H}_2$) generated by $\mathbf{J}$ are obtained by exponentiation of the Jordan map of $-i\psi\hat{n}\cdot\boldsymbol{\sigma}/2$. To a given turn $T(\cos\frac{1}{2}\psi, \hat{n}\sin\frac{1}{2}\psi)$ there corresponds the 2×2 unitary unimodular matrix $U(\psi, \hat{n}) = \exp(-i\psi\hat{n}\cdot\boldsymbol{\sigma}/2)$, which itself corresponds to a unitary boson operator via the Jordan map of $-i\psi\hat{n}\cdot\boldsymbol{\sigma}/2$:

$$T(\cos\tfrac{1}{2}\psi, \hat{n}\sin\tfrac{1}{2}\psi) \to \exp(-i\psi\hat{n}\cdot\boldsymbol{\sigma}/2) \to \exp(\mathscr{L}_{-i\psi\hat{n}\cdot\boldsymbol{\sigma}/2})$$

$$= \exp(-i\psi\hat{n}\cdot\mathscr{L}_{\boldsymbol{\sigma}/2}) = \exp(-i\psi\hat{n}\cdot\mathbf{J}). \tag{5.37}$$

Taking matrix elements in $\mathscr{H}^{(2j)}$ then leads to

$$\langle jm'|\exp(-i\psi\hat{n}\cdot\mathscr{L}_{\boldsymbol{\sigma}/2})|jm\rangle = D^j_{m'm}(\psi, \hat{n}). \tag{5.38}$$

We conclude that *the exponentiated Jordan map of* $(-i\psi\hat{n}\cdot\boldsymbol{\sigma}/2)$ *yields every irreducible representation of the group of turns*:

$$T(\cos\tfrac{1}{2}\psi, \hat{n}\sin\tfrac{1}{2}\psi) \to D^j(\psi, \hat{n}), \qquad j = 0, \tfrac{1}{2}, 1, \ldots. \tag{5.39}$$

Remark. One may also consider the Jordan map of the matrix $U = U(\psi, \hat{n})$: $U \to \mathscr{L}_U = \sum_{ij} u_{ij}a_i\bar{a}_j$. Notice, however, that

$$\mathscr{L}_{\exp(-i\psi\hat{n}\cdot\boldsymbol{\sigma}/2)} \neq \exp(\mathscr{L}_{-i\psi\hat{n}\cdot\boldsymbol{\sigma}/2}).$$

The significance of $\mathscr{L}_U$ is seen to be that of an element in the Lie algebra with basis $J_+, J_-, J_3, \mathscr{E}$, and not that of a finite transformation generated by $\mathbf{J}$.

The basic result, Eq. (5.38), allows us to give an alternative derivation of the rotation matrices, which avoids the direct evaluation of the exponentials as required in our original definition, Eq. (3.35). Thus, for $U = \exp(-i\psi\hat{n}\cdot\boldsymbol{\sigma}/2)$, we have, by definition of the rotation matrices:

$$\mathscr{T}_U|jm\rangle \equiv \sum_{m'} D^j_{m'm}(U)P_{jm'}(a_1, a_2)|0\rangle, \tag{5.40}$$

where

$$\mathscr{T}_U \equiv \exp(\mathscr{L}_{-i\psi\hat{n}\cdot\boldsymbol{\sigma}/2}). \tag{5.41}$$

On the other hand, noting Eq. (5.33) and $\mathscr{T}_U|0\rangle = |0\rangle$, we may write

$$\begin{aligned} \mathscr{T}_U|jm\rangle &= (\mathscr{T}_U P_{jm}(a_1, a_2)\mathscr{T}_{U^{-1}})\mathscr{T}_U|0\rangle \\ &= P_{jm}(a_1', a_2')|0\rangle, \end{aligned} \tag{5.42}$$

where $a_i' = \mathscr{T}_U a_i \mathscr{T}_{U^{-1}}$, yielding[1]

$$\begin{aligned} a_2' &\equiv u_{11}a_1 + u_{21}a_2, \\ a_2' &\equiv u_{12}a_1 + u_{22}a_2, \end{aligned} \tag{5.43}$$

with $U = (u_{ij})$. The procedure now is to substitute these transformed bosons into the monomials given by Eq. (5.33) and expand on the original monomial basis:

$$\begin{aligned} &\frac{(u_{11}a_1 + u_{21}a_2)^{j+m}(u_{12}a_1 + u_{22}a_2)^{j-m}}{[(j+m)!(j-m)!]^{\frac{1}{2}}}|0\rangle \\ &= [(j+m)!(j-m)!]^{\frac{1}{2}} \sum_{st} \frac{(u_{11})^{j+m-s}(u_{21})^s(u_{12})^{j-m-t}(u_{22})^t}{(j+m-s)!s!(j-m-t)!t!} \cdot a_1^{2j-s-t}a_2^{s+t}|0\rangle \\ &= \sum_{m'} D^j_{m'm}(U)P_{jm'}(a_1, a_2)|0\rangle, \end{aligned}$$

[1] It is a general result that an arbitrary unitary transformation of n bosons $a_1, \ldots, a_n$ given by $U: a_i \to a_i' = \sum_j u_{ji}a_j$ preserves the basic commutation relations between the bosons. This result may be used directly to prove that $\mathscr{T}_U$ is a unitary operator.

where

$$D^j_{m'm}(U) = [(j+m)!(j-m)!(j+m')!(j-m')!]^{\frac{1}{2}}$$

$$\times \sum_s \frac{(u_{11})^{j+m-s}(u_{21})^s(u_{12})^{m'-m+s}(u_{22})^{j-m'-s}}{(j+m-s)!s!(m'-m+s)!(j-m'-s)!}. \quad (5.44)$$

Remark. This derivation of the rotation matrices (representation functions) shows that we could consider an arbitrary linear transformation $a_1 \to z_{11}a_1 + z_{21}a_2, a_2 \to z_{12}a_1 + z_{22}a_2, z_{ij} \in \mathbb{C}$, of the bosons, thereby obtaining the functions $D^j_{m'm}(Z)$ in which the unitary elements u_{ij} are replaced by arbitrary complex numbers z_{ij}. The essential property of these functions, which the derivation validates, is that the matrices $D^j(Z)$ possess the *multiplication property* $D^j(Z)D^j(Z') = D^j(ZZ')$ for arbitrary complex matrices Z and Z' (or for any matrix Z of indeterminates that commute).

5. Generalization of the Jordan Map

The necessity of generalizing the Jordan map can be seen in two very different ways. From the point of view of turns, one can see that the Jordan map realizes only the angular momentum representations [$SU(2)$] and not the full structure carried by the turns [the group of the symmetric top, $SU(2) * SU(2)$]. Alternatively, one can note that the Jordan map, as a characterization of the field quantization procedure, was designed to treat only the two symmetry structures: totally symmetric (bosons) or totally antisymmetric (fermions). To afford the possibility of more general structures, including mixed symmetry (general Young frames; see Appendix A), it is useful to generalize the Jordan map from n bosons to n^2 bosons, or equivalently *to the $n \times n$ matrix boson A.*

Let A denote an $n \times n$ matrix whose elements $(A)_{ij} = a^j_i$ are boson operators satisfying the commutation relations

$$[\bar{a}^j_i, a^{j'}_{i'}] = \delta_{ii'}\,\delta^{jj'},$$

$$[\bar{a}^j_i, \bar{a}^{j'}_{i'}] = [a^j_i, a^{j'}_{i'}] = 0. \quad (5.45)$$

We shall denote by $\bar{A}$ the $n \times n$ matrix whose elements are the adjoint boson operators $(\bar{A})_{ij} = \bar{a}^j_i$.

Let X denote an arbitrary $n \times n$ matrix with numerical elements $(X)_{ij} = x_{ij}$. Then *the mapping of the $n \times n$ matrices $\{X\}$ into linear operators $\mathscr{L}_X$ given by*

$$X \to \mathscr{L}_X = \text{trace}(\tilde{A}X\bar{A}) = \sum_{i,j=1}^{n} x_{ij} \sum_{\alpha=1}^{n} a_i^\alpha \bar{a}_j^\alpha \tag{5.46}$$

preserves the operation of commutation of the $n \times n$ *matrices,* $\{X\}$.

Proof. The proof follows as before upon noting that the n^2 operators E_{ij} defined by

$$E_{ij} = \sum_{\alpha=1}^{n} a_i^\alpha \bar{a}_j^\alpha, \qquad i, j = 1, 2, \ldots, n \tag{5.47}$$

satisfy the same commutation relations as do the n^2 boson operators $a_i \bar{a}_j$; that is,

$$[E_{ij}, E_{kl}] = \delta_{jk} E_{il} - \delta_{il} E_{kj}. \tag{5.48}$$

■

The generalized Jordan map allows of a second mapping: *The mapping of* $n \times n$ *matrices* $\{Y\}$ *into linear operators* $\mathscr{R}_Y$ *given by*

$$Y \to \mathscr{R}_Y = \text{trace}(\tilde{Y}\tilde{A}\bar{A}) = \sum_{\alpha,\beta=1}^{n} y_{\alpha\beta} \sum_{i=1}^{n} a_i^\alpha \bar{a}_i^\beta \tag{5.49}$$

preserves the operation of commutation of the matrices $\{Y\}$.

Proof. The proof of this result follows immediately upon noting that the n^2 operators $E^{\alpha\beta}$ defined by

$$E^{\alpha\beta} = \sum_{i=1}^{n} a_i^\alpha \bar{a}_i^\beta, \qquad \alpha, \beta = 1, 2, \ldots, n \tag{5.50}$$

satisfy the commutation relations of the same type as the E_{ij} occurring in the mapping $\mathscr{L}_X$:

$$[E^{\alpha\beta}, E^{\gamma\delta}] = \delta^{\beta\gamma} E^{\alpha\delta} - \delta^{\alpha\delta} E^{\gamma\beta}. \tag{5.51}$$

■

It is an important property of these two mappings that *all the operators* $\mathscr{L}_X$ *commute with all the operators* $\mathscr{R}_Y$:

$$[\mathscr{L}_X, \mathscr{R}_Y] = 0, \tag{5.52}$$

for each (numerical) $n \times n$ matrix X and each (numerical) matrix Y. This result follows most quickly by showing that each of the n^2 operators E_{ij} commutes with each of the n^2 operators $E^{\alpha\beta}$.

The generalized Jordan map, and the associated operator structures, are the natural starting point for constructing and studying the family of unitary groups $\{U(n)\}$ (see Ref. [14] and the references cited therein).

It is perhaps useful to remark that the $m \times n$ matrix boson ($m \neq n$) leads to commuting $\mathscr{L}$ and $\mathscr{R}$ maps, by specializing the $n \times n$ results above.

6. Application of the Generalized Jordan Map

Let us indicate the motivation, and geometric significance, underlying the generalized Jordan mapping, before carrying out the actual applications. For the Jordan map applied in Section 4, we considered a geometric object, the turn, which mapped into an operator, realizing the $SU(2)$ representation structures on a pair of bosons (a_1, a_2). If we were to seek to generalize these concepts, it would be natural to attempt to map the turns—in their more general role as a carrier space—into the matrix boson A—on which the generalized Jordan operator mapping is to realize the representation structures of the larger group $[SU(2) * SU(2)]$. The turns, however, are three-parameter objects (unimodular quaternions), and the natural map into the matrix boson (four bosons) indicates that it is the general quaternion that should be considered. Geometrically this enlarges the group of turns to include, besides the rotations, also the multiplicative group of scale transformations (scaling the radius of the unit sphere by the modulus of the quaternion). The corresponding geometric view identifies the general quaternion as a *Wendestreckung* (Klein–Sommerfeld)—that is, a turn together with a stretching. From the point of view of the boson calculus, this generalization appears as the inclusion of a new object: a pair of (independent) two-component bosons coupled (antisymmetrically) to zero angular momentum: $a_{12}^{12} \equiv a_1^1 a_2^2 - a_2^1 a_1^2 = \det A$. These paired bosons, a_{12}^{12}, are the carrier space for the scale transformations.

Accordingly, the first step is to map the general quaternion q into the matrix boson:

$$q \to A = \begin{pmatrix} a_1^1 & a_1^2 \\ a_2^1 & a_2^2 \end{pmatrix}, \tag{5.53}$$

which maps the quaternion norm to

$$N(q) \equiv q\bar{q} \to \det A = a_{12}^{12}. \tag{5.54}$$

The next step is to implement the group of left and right translations acting on the quaternions (and hence the turns) as a carrier space:

$$L_u: q \to uq = L_u(q),$$

$$R_v: q \to q\tilde{v} = R_v(q), \tag{5.55}$$

where u and v are quaternions. For $v = (v_0, v_1, v_2, v_3)$ the quaternion $\tilde{v}$ is defined by $\tilde{v} = (v_0, v_1, -v_2, v_3)$. One then finds $\widetilde{uv} = \tilde{v}\,\tilde{u}$ so that $R_u(R_v(q)) = R_{uv}(q)$ as desired for the multiplication of right translations.[1]

Consider now the corresponding structure for the matrix boson A. The matrix boson A is to function as the carrier space [analogous to q in Eq. (5.55)], whereas the analog to the quaternion u (resp. v) generating the left (resp. right) translation is not a matrix boson but a 2×2 unitary matrix. It follows that the desired action on the matrix boson A is given by

$$L_U: A \to UA = L_U(A),$$

$$R_V: A \to A\tilde{V} = R_V(A), \tag{5.56}$$

where U and V are unitary matrices.

Let $P(A)$ denote a polynomial in the four bosons (a_i^j) of A, and let $|0\rangle$ denote the ket vector (in $\mathscr{H}_4 = \mathscr{H}_1 \otimes \mathscr{H}_1 \otimes \mathscr{H}_1 \otimes \mathscr{H}_1$) defined by $\bar{a}_i^j|0\rangle = 0$. (The ket $|0\rangle$ is the "vacuum" ket having zero quanta; see Section 2.)

We can now state the significant result: Let the unitary matrices U and V be written in the forms $U = e^{-iX}$ and $V = e^{-iY}$, respectively, where X and Y are Hermitian matrices. *Then the exponentiated Jordan maps,* $\mathscr{S}_U = \exp(-i\mathscr{L}_X)$ *and* $\mathscr{T}_V = \exp(-i\mathscr{R}_Y)$, *implement the left and right translations, respectively, as boson operators acting on* $P(A)|0\rangle$:

$$\mathscr{S}_U P(A)|0\rangle = P(L_{\tilde{U}}(A))|0\rangle = P(\tilde{U}A)|0\rangle,$$

$$\mathscr{T}_V P(A)|0\rangle = P(R_{\tilde{V}}(A))|0\rangle = P(AV)|0\rangle,$$

$$\mathscr{S}_U\mathscr{T}_V P(A)|0\rangle = \mathscr{T}_V\mathscr{S}_U P(A)|0\rangle = P(\tilde{U}AV)|0\rangle, \tag{5.57}$$

for all $U,\ V \in U(2)$.

---

[1] It is convenient to use $\tilde{v}$ in place of v^{-1} in the definition of a right translation, since it leads, for the matrix boson, to a symmetric relationship between the generators of left and right translations [see Eqs. (5.58) and (5.59)] whereby one set of generators is obtained from the other by interchanging superscripts and subscripts in the notation for the elements of the matrix boson.

Proof. We have $\mathscr{S}_U P(A)|0\rangle = (\mathscr{S}_U P(A)\mathscr{S}_{U^{-1}})|0\rangle = P(\mathscr{S}_U A \mathscr{S}_{U^{-1}})|0\rangle$, since $\mathscr{S}_U|0\rangle = |0\rangle$. A straightforward application of the Baker–Hausdorff–Campbell formula yields $\exp(-i\mathscr{L}_X)A\exp(i\mathscr{L}_X) = \tilde{U}A$. A similar procedure establishes the stated result for right translations. ■

Choosing $X = \frac{1}{2}(\text{tr } X)\sigma_0 + \frac{1}{2}\psi(\hat{n}\cdot\boldsymbol{\sigma})$ (and similarly for Y), we now obtain two sets of commuting angular momentum operators and the number operator:

Generators $\mathbf{J}$ of unimodular left translations ($\sigma_\pm \equiv \sigma_1 \pm i\sigma_2$):

$$J_+ = \tfrac{1}{2}\text{tr}(\tilde{A}\sigma_+\bar{A}) = E_{12} = \sum_k a_1^k \bar{a}_2^k,$$

$$J_- = \tfrac{1}{2}\text{tr}(\tilde{A}\sigma_-\bar{A}) = E_{21} = \sum_k a_2^k \bar{a}_1^k,$$

$$J_3 = \tfrac{1}{2}\text{tr}(\tilde{A}\sigma_3\bar{A}) = \tfrac{1}{2}(E_{11} - E_{22}) = \tfrac{1}{2}\sum_k (a_1^k\bar{a}_1^k - a_2^k\bar{a}_2^k). \tag{5.58}$$

Generators $\mathbf{K}$ of unimodular right translations:

$$K_+ = \tfrac{1}{2}\text{tr}(\tilde{\sigma}_+\tilde{A}\bar{A}) = E^{12} = \sum_k a_k^1 \bar{a}_k^2,$$

$$K_- = \tfrac{1}{2}\text{tr}(\tilde{\sigma}_-\tilde{A}\bar{A}) = E^{21} = \sum_k a_k^2 \bar{a}_k^1,$$

$$K_3 = \tfrac{1}{2}\text{tr}(\tilde{\sigma}_3\tilde{A}\bar{A}) = \tfrac{1}{2}(E^{11} - E^{22}) = \tfrac{1}{2}\sum_k (a_k^1\bar{a}_k^1 - a_k^2\bar{a}_k^2). \tag{5.59}$$

Generator of phase transformations:

$$\mathscr{E} = \text{tr}(\tilde{A}\bar{A}) = \sum_k E^{kk} = \sum_{ij} a_i^j \bar{a}_i^j, \tag{5.60}$$

which can be recognized as the operator corresponding to the total number of quanta.

This realization of the operators $\mathbf{J}$ and $\mathbf{K}$ has the important property that the two invariant operators $\mathbf{J}\cdot\mathbf{J}$ and $\mathbf{K}\cdot\mathbf{K}$ are identical: $\mathbf{J}\cdot\mathbf{J} = \mathbf{K}\cdot\mathbf{K}$.

These results show that the generalized Jordan map of the turns or unimodular quaternions realizes the group $SU(2) * SU(2)$ of the symmetric top, and not the general direct product group $SU(2) \times SU(2)$.

The most expeditious way to develop the characteristic vectors for this operator realization is to recognize that these vectors are essentially the elements of the rotation matrices D^j introduced in Section 4. We recall that the definition of the $D^j(U)$ is meaningful for any four indeterminates arranged in 2×2 matrix form: $Z = \begin{pmatrix} z_{11} & z_{12} \\ z_{21} & z_{22} \end{pmatrix}$. Moreover, the representation property,

$$D^j(Z_1)D^j(Z_2) = D^j(Z_1 Z_2), \tag{5.61}$$

is valid for these indeterminates as well. When the matrix Z is taken to be numerical, unitary, and unimodular, the D^j have their usual interpretations as irreps of $SU(2)$. (If we drop the unimodular restriction, we can obtain irreps of the group of real quaternions under multiplication.) More generally, by using the correspondence $Z \to (\det Z)^k D^j(Z)$, we have the representation property:

$$Z_1 Z_2 \to (\det Z_1)^k\, D^j(Z_1)(\det Z_2)^k\, D^j(Z_2) = (\det Z_1 Z_2)^k\, D^j(Z_1 Z_2). \tag{5.62}$$

Replacing the indeterminates Z by the matrix boson A, we obtain (for k a nonnegative integer) *the boson polynomials*, $(\det A)^k\, D^j_{m'm}(A)$, which are homogeneous polynomials in the boson operators (a^j_i) of degree $2(j + k)$.

The desired characteristic vectors for the group generated by the operators $(\mathbf{J}, \mathbf{K}, \mathscr{E})$ are thus given by the eigenkets:

$$|k,j;m,m'\rangle \equiv [(k - j)!(k + j + 1)!/(2j + 1)]^{-\frac{1}{2}} (\det A)^{k-j}\, D^j_{mm'}(A)|0\rangle, \tag{5.63}$$

where the determination of the normalization factor is given in Appendix A.

To be explicit let us write out the action of the generators [given by the Jordan map, Eqs. (5.58)–(5.60)] acting on this eigenket basis, Eq. (5.63):

$$J_\pm|k,j;m,m'\rangle = [(j \mp m)(j \pm m + 1)]^{\frac{1}{2}}|k,j;m \pm 1,m'\rangle,$$

$$J_3|k,j;m,m'\rangle = m|k,j;m,m'\rangle; \tag{5.64}$$

$$K_\pm|k,j;m,m'\rangle = [(j \mp m')(j \pm m' + 1)]^{\frac{1}{2}}|k,j;m,m' \pm 1\rangle,$$

$$K_3|k,j;m,m'\rangle = m'|k,j;m,m'\rangle; \tag{5.65}$$

$$\mathscr{E}|k,j;m,m'\rangle = 2k|k,j;m,m'\rangle; \tag{5.66}$$

in which $k = 0, \frac{1}{2}, \ldots;\ j = 0, \frac{1}{2}, \ldots, k;$ and $m, m' = j, j-1, \ldots, -j$.

7. Application of the Generalized Jordan Map To Determine the Wigner Coefficients

There is another way in which we may consider the space of homogeneous polynomials over the matrix boson A. Instead of considering the two pairs of bosons as forming a single entity (the matrix boson A), we may consider each boson pair as a separate entity to which there is associated a (commuting) Jordan map. The first realization corresponds to "coupled boson pairs," the second realization to "uncoupled boson pairs"; the interrelation between the two realizations defines, as we shall show, the Wigner coefficients in an elegant way.

Let us denote the two pairs of bosons by $i = 1, 2$; that is, $\mathbf{a}^1 \equiv (a_1^1, a_2^1)$ for $i = 1$ and $\mathbf{a}^2 \equiv (a_1^2, a_2^2)$ for $i = 2$. The Jordan map (Section 3) applied to each pair yields the angular momentum generators:

$$
\begin{aligned}
J_+(i) &= \tilde{a}^i \sigma_+ \bar{a}^i/2 = a_1^i \bar{a}_2^i, \\
J_-(i) &= \tilde{a}^i \sigma_- \bar{a}^i/2 = a_2^i \bar{a}_1^i, \\
J_3(i) &= \tilde{a}^i \sigma_3 \bar{a}^i/2 = (a_1^i \bar{a}_1^i - a_2^i \bar{a}_2^i)/2,
\end{aligned}
\tag{5.67}
$$

for $i = 1, 2$. It is clear that the two sets of operators

$$\{J_+(1), J_3(1), J_-(1)\} \tag{5.68}$$

and

$$\{J_+(2), J_3(2), J_-(2)\} \tag{5.69}$$

mutually commute,

$$[\mathbf{J}(1), \mathbf{J}(2)] = 0, \tag{5.70}$$

and define distinct quadratic invariants $\mathbf{J}^2(1)$ and $\mathbf{J}^2(2)$; that is,

$$\mathbf{J}^2(1) \neq \mathbf{J}^2(2). \tag{5.71}$$

The eigenvalues of these quadratic invariants are given by $\mathbf{J}^2(i) \to j_i(j_i + 1)$, where $2j_i$ is the eigenvalue of the Euler operator (the Jordan map of each unit operator): $\mathscr{E}(i) \to 2j_i$, the number of quanta for the ith boson pair. The remaining eigenvalues labeling this (uncoupled) basis are: $J_3(i) \to m_i$, the "magnetic" quantum numbers.

In terms of this notation, the basis vectors for this alternative realization are the monomials:

$$|j_1 m_1; j_2 m_2\rangle = P_{j_1 m_1}(\mathbf{a}^1) P_{j_2 m_2}(\mathbf{a}^2)|0\rangle, \tag{5.72}$$

where

$$P_{j_i m_i}(\mathbf{a}^i) = \frac{(a_1^i)^{j_i + m_i}(a_2^i)^{j_i - m_i}}{[(j_i + m_i)!(j_i - m_i)!]^{\frac{1}{2}}}. \tag{5.73}$$

Let us denote by $\mathscr{H}^{(j_1, j_2)}$ the space spanned by these $(2j_1 + 1) \times (2j_2 + 1)$ basis vectors above, enumerated as m_1 and m_2 run over their respective ranges: $m_1 = j_1, j_1 - 1, \ldots, -j_1$, and $m_2 = j_2, j_2 - 1, \ldots, -j_2$. The generators, Eqs. (5.68) and (5.69), have the standard action on this basis.

Now let us consider what these results mean. We have found two different realizations: the boson polynomials $(\det A)^{k-j} D^j(A)$ of Section 6, and the boson polynomials belonging to $\mathscr{H}^{(j_1, j_2)}$ above, each of which is a homogeneous polynomial over the same four boson operators, and each of which has four independent labels. The two sets must therefore be related.

To determine this relationship, let us note first that the operator $\mathbf{J}$ [for the polynomials over the matrix boson (A)] is the sum of the two angular momenta $\mathbf{J}(i)$:

$$\mathbf{J} = \mathbf{J}(1) + \mathbf{J}(2). \tag{5.74}$$

(This relation is the reason for calling the matrix boson realization the coupled basis.) Equation (5.74) implies that the m-eigenvalues are related by

$$m = m_1 + m_2. \tag{5.75}$$

Second, one notes that the "number of quanta" operator $\mathscr{E}$ [Eq. (5.60)] is similarly the sum of two "number of quanta" operators $\mathscr{E}(i)$:

$$\mathscr{E} = \mathscr{E}(1) + \mathscr{E}(2). \tag{5.76}$$

This implies that we have the eigenvalue relation

$$k = j_1 + j_2. \tag{5.77}$$

Alternatively, and equivalently, one may regard this result as stating that we are considering homogeneous boson polynomials of the same degree in both realizations.

Finally, one notes that the operator K_3 [Eq. (5.65)] for the polynomials over the matrix boson A is the *difference* of the two operators $\mathscr{E}(i)$, so that we have

$$K_3 = \tfrac{1}{2}[\mathscr{E}(1) - \mathscr{E}(2)], \tag{5.78}$$

and accordingly the eigenvalue relation:

$$m' = j_1 - j_2. \tag{5.79}$$

Let us summarize: We have found, in Eq. (5.72), an orthonormal set of $(2j_1 + 1) \times (2j_2 + 1)$ basis vectors, $\{|j_1m_1; j_2m_2\rangle\}$, for the space $\mathscr{H}^{(j_1,j_2)}$. We have also found a second realization, Eq. (5.63), for the same space using the orthonormal set of basis vectors:

$$|j_1j_2; jm\rangle = \left[\frac{2j+1}{(j_1+j_2-j)!(j_1+j_2+j+1)!}\right]^{\frac{1}{2}} (\det A)^{j_1+j_2-j} D^j_{m,j_1-j_2}(A)|0\rangle. \tag{5.80}$$

This latter basis is enumerated (for fixed j_1 and j_2) by j and m with the ranges $j = j_1 + j_2, j_1 + j_2 - 1, \ldots, |j_1 - j_2|$ and $m = j, j-1, \ldots, -j$.

Since the number of orthonormal vectors is the same in both sets, [using $\sum_j (2j+1) = (2j_1+1)(2j_2+1)$], we conclude that both sets span the same space $\mathscr{H}^{(j_1,j_2)}$. There is accordingly a unitary transformation linking the two sets of basis vectors. The transformation is, in fact, real-orthogonal (since the scalar products between vectors in the two basis sets are real).

We may conclude that *the transformation coefficients between bases are none other than the Wigner coefficients.* This follows, since, by Eq. (5.74), the transformation effects the angular momentum coupling, $\mathbf{J} = \mathbf{J}(1) + \mathbf{J}(2)$, and the action of all angular momenta is in standard form.

We have therefore shown that the transformation between bases:

$$|j_1j_2; jm\rangle = \sum_{m_1,m_2} C^{j_1j_2j}_{m_1m_2m} |j_1m_1; j_2m_2\rangle, \tag{5.81}$$

constitutes a definition of the Wigner coefficient—namely,

$$\begin{aligned} C^{j_1j_2j}_{m_1m_2m} &= \langle j_1m_1; j_2m_2 | j_1j_2; jm\rangle \\ &= \left[\frac{(2j+1)}{(j_1+j_2-j)!(j_1+j_2+j+1)!}\right]^{\frac{1}{2}} \\ &\quad \times \langle 0| P_{j_1m_1}(\bar{\mathbf{a}}^1) P_{j_2m_2}(\bar{\mathbf{a}}^2) (\det A)^{j_1+j_2-j} D^j_{m,j_1-j_2}(A)|0\rangle. \end{aligned} \tag{5.82}$$

The boson polynomial $D^j(A)$ is explicitly defined by Eqs. (5.44) and (5.53), the boson monomials $P_{j_i m_i}(\mathbf{a}^i)$ by Eq. (5.73).

This boson operator determination of the Wigner coefficients by means of Jordan mappings is the major result we wish to establish in this chapter. [The explicit evaluation of Eq. (5.82) is given below.] The geometric concept of rotations as turns, implemented by the boson calculus, has led us directly to the essential construct in the quantum theorem of angular momentum. On the basis of this result it appears justified to claim that boson operator techniques allow a unified and elegant presentation of the elements of that theory.

The relationship between boson polynomials, rotation matrices, and Wigner coefficients is developed further in Appendices B and C. In Appendix B we obtain a multiplication law for boson polynomials and prove a factorization lemma of considerable practical utility. The symmetry properties of these structures are explored in Appendix C; in particular, we obtain all 72 symmetries of the Wigner coefficient in a uniform way.

8. Wigner Coefficients as "Discretized" Rotation Matrices

We have obtained in the previous section an expression for the Wigner coefficient as a matrix element between boson polynomial bases. One purpose of the present section is to determine the explicit algebraic form for the coefficients, Eq. (5.87) below.

We shall then interpret our results in a novel way as an umbral operator (Rota [15–17]) effecting a transformation between the Jacobi polynomials and the Wigner coefficients. Expressed more suggestively (in the phraseology of physics), the Wigner coefficients are a discretized form of the rotation matrices.

Explicit algebraic form of the Wigner coefficients. We obtained in Section 7 a result for the Wigner coefficient that may equivalently be written as

$$C^{j_1 j_2 j}_{m_1 m_2 m} = \left[\frac{(2j+1)}{(j_1+j_2-j)!(j_1+j_2+j+1)!}\right]^{\frac{1}{2}}$$

$$\times \langle 0|(\bar{a}^{12}_{12})^{j_1+j_2-j} D^j_{m,j_1-j_2}(\bar{A})|j_1 m_1; j_2 m_2\rangle, \tag{5.83}$$

where the ket vector on the right has the form

$$|j_1 m_1; j_2 m_2\rangle = \frac{(a^1_1)^{j_1+m_1}(a^1_2)^{j_1-m_1}(a^2_1)^{j_2+m_2}(a^2_2)^{j_2-m_2}}{[(j_1+m_1)!(j_1-m_1)!(j_2+m_2)!(j_2-m_2)!]^{\frac{1}{2}}}|0\rangle, \tag{5.84}$$

and $D^j(\bar{A})$ denotes the rotation matrices (polynomials) over the (conjugate) matrix boson, $\bar{A}$.

It is useful at this point to introduce a more symmetric, "combinatorial," form for the rotation matrices, replacing the form derived earlier in Eq. (5.44). Let us introduce a 2×2 array, denoted by $\boxed{\alpha}$, of nonnegative integers $\{\alpha_i^j\}$ such that the row and column sums have the values indicated in the display below:

$$\begin{array}{|cc|l} \alpha_1^1 & \alpha_1^2 & j+m' \\ \alpha_2^1 & \alpha_2^2 & j-m' \\ \hline j+m & j-m & \end{array}$$

In terms of this notation, Eq. (5.44) takes the form

$$D^j_{m'm}(A) = [(j+m)!(j-m)!(j+m')!(j-m')]^{\frac{1}{2}}$$

$$\times \sum_{\boxed{\alpha}} \frac{(a_1^1)^{\alpha_1^1}(a_2^1)^{\alpha_2^1}(a_1^2)^{\alpha_1^2}(a_2^2)^{\alpha_2^2}}{\alpha_1^1!\alpha_2^1!\alpha_1^2!\alpha_2^2!}. \tag{5.85}$$

The summation in Eq. (5.85) is to be carried out over all arrays satisfying the constraints. Note that in this equation we have introduced the matrix boson $A = (a_i^j)$.

Let us return to the evaluation of the matrix element in Eq. (5.83). One first specializes Eq. (5.85) to the values $m' = m$ and $m = j_1 - j_2$, and introduces the conjugate matrix boson. Using the inner product of the boson calculus then yields a terminating series for the Wigner coefficient, which has the form

$$C^{j_1j_2j}_{m_1m_2m} = \delta_{(m_1+m_2,m)}(-1)^{j_1-m_1}\left[\frac{(2j+1)(j_1+j_2-j)!}{(j_1+j_2+j+1)!}\right]^{\frac{1}{2}}$$

$$\times [(j+m)!(j-m)!(j+j_1-j_2)!(j-j_1+j_2)!]^{\frac{1}{2}}$$

$$\times \sum_{\boxed{\alpha}} \frac{\left[\dfrac{(j_1+m_1)!}{(j_1+m_1-\alpha_1^1)!}\right]^{\frac{1}{2}}}{(\alpha_1^1)!} \times \frac{(-1)^{\alpha_2^1}\left[\dfrac{(j_1-m_1)!}{(j_1-m_1-\alpha_2^1)!}\right]^{\frac{1}{2}}}{(\alpha_2^1)!}$$

$$\times \frac{\left[\dfrac{(j_2+m_2)!}{(j_2+m_2-\alpha_1^2)!}\right]^{\frac{1}{2}}}{(\alpha_1^2)!} \times \frac{\left[\dfrac{(j_2-m_2)!}{(j_2-m_2-\alpha_2^2)!}\right]^{\frac{1}{2}}}{(\alpha_2^2)!}. \tag{5.86}$$

The result expressed by Eq. (5.86) may be put in the van der Waerden form upon eliminating all constraints among the α_i^j. One finds

$$C^{j_1 j_2 j}_{m_1 m_2 m} = \delta_{(m_1+m_2,m)}\left[\frac{(2j+1)(j_1+j_2-j)!(j+j_1-j_2)!(j+j_2-j_1)!}{(j_1+j_2+j+1)!}\right]^{\frac{1}{2}}$$

$$\times \sum_z (-1)^z \frac{[(j_1+m_1)!(j_1-m_1)!(j_2+m_2)!(j_2-m_2)!(j+m)!(j-m)]^{\frac{1}{2}}}{z!(j_1+j_2-j-z)!(j_1-m_1-z)!(j_2+m_2-z)!(j-j_2+m_1+z)!(j-j_1-m_2+z)!}. \tag{5.87}$$

There are many equivalent forms of this result as discussed in Chapter 3, Section 12. We also give in RWA the relationship between the Wigner coefficients and generalized hypergeometric series.

An interpretation in terms of the umbral calculus. Let us turn now to the task of interpreting this evaluation of the Wigner coefficient. If we look back to our original form for the Wigner coefficient, Eq. (5.83), we see that the essence of our evaluation, Eq. (5.86), is *to replace every boson operator term* $(\bar{a}^j_i)^{\alpha^j_i}$ *in the polynomial* $D^j(\bar{A})$ *by a number, the value of the inner product for this term.*

In other words, the evaluation is nothing other than a *linear functional* which maps the monomials of the basis into (real) numbers. This is precisely the foundation on which Rota [16, 17] has placed the umbral calculus, and our observation identifies the Wigner coefficient conceptually as an umbral operator acting on the rotation matrices.

The umbral calculus, or Blissard calculus, is the technique by which (in its simplest form) an element of a sequence of numbers $\{a_n\}$ is interpreted as a symbolic power: $a_n \leftrightarrow a^n$. In Rota's interpretation of the umbral notation one defines a linear functional on a space of polynomials such that $a_n = L(x^n)$. This is precisely the interpretation implied by our result for the Wigner coefficients.

The symbolic, or generalized, powers that we shall need are of the form

$$(\pm\sqrt{k})^s \rightarrow (\pm 1)^s\left[\frac{k!}{(k-s)!}\right]^{\frac{1}{2}}. \tag{5.88}$$

Using these symbolic powers to interpret the elements of the rotation matrix, we may express our result, Eq. (5.86), in the suggestive form:

$$C^{j_1 j_2 j}_{m_1 m_2 m} = \delta_{(m_1+m_2,m)}(-1)^{j_1-m_1}\left[\frac{(2j+1)(j_1+j_2-j)!}{(j_1+j_2+j+1)!}\right]^{\frac{1}{2}}$$

$$\times D^j_{m,j_1-j_2}\begin{pmatrix} \sqrt{j_1+m_1} & \sqrt{j_2+m_2} \\ -\sqrt{j_1-m_1} & \sqrt{j_2-m_2} \end{pmatrix}_{\text{symbolic powers}}. \tag{5.89}$$

This notation has the merit of emphasizing the relationship of the Wigner coefficient to a discretized rotation matrix.

This relationship was first pointed out in Ref. [1] (see Supplement III of Ref. [1], taken from a paper by Gel'fand and Tseitlin[1]), where the Wigner coefficients were interpreted in terms of the Jacobi functions using generalized powers. They simply *postulated* the appropriate generalized power law, using the known series for the two functions. The results above demonstrate that the origin of this *ad hoc* rule is to be found in the Jordan map to the matrix boson A combined with the fundamental fact that the boson operators come equipped with their own inner product structure.

The result expressed by Eq. (5.89) is also the source of a well-known asymptotic relationship between Wigner coefficients and representation functions. Thus, using the symmetry relations, Eqs. (3.180), we may rewrite Eq. (5.89) in the form

$$\begin{aligned} C^{j\,k\,j+\Delta}_{m,\mu,m+\mu} &= (-1)^{j-m}\left[\frac{(2j+2\Delta+1)}{(2k+1)}\right]^{\frac{1}{2}} C^{j+\Delta\,j\,k}_{m+\mu,-m,\mu} \\ &= (-1)^{\Delta-\mu}\left[\frac{(2j+2\Delta+1)(2j+\Delta-k)!}{(2j+\Delta+k+1)!}\right]^{\frac{1}{2}} \\ &\quad\times D^k_{\mu\Delta}\begin{pmatrix} \sqrt{j+m+\Delta+\mu} & \sqrt{j-m} \\ -\sqrt{j-m+\Delta-\mu} & \sqrt{j+m}\end{pmatrix}_{\text{symbolic powers}} \end{aligned} \tag{5.90}$$

In this result we now take k to have a fixed numerical value and let Δ, μ denote any allowed values in the range $k, k-1, \ldots, -k$. For sufficiently large j, the square-root factor in Eq. (5.90) may be approximated by $1/(2j)^k$. This factor may, in turn, be combined with the symbolic powers occurring in the expansion of the symbolic D-function in such a way that, for sufficiently large j, the symbolic powers are replaced by ordinary powers in

$$\cos\frac{\beta}{2} = \sqrt{\frac{j+m}{2j}}, \qquad \sin\frac{\beta}{2} = \sqrt{\frac{j-m}{2j}}. \tag{5.91}$$

Carrying out these details yields the asymptotic relation:

$$C^{j\,k\,j+\Delta}_{m,\mu,m+\mu} \sim (-1)^{\Delta-\mu} D^k_{\mu\Delta}\begin{pmatrix} \cos\frac{\beta}{2} & \sin\frac{\beta}{2} \\ -\sin\frac{\beta}{2} & \cos\frac{\beta}{2}\end{pmatrix} = d^k_{\mu\Delta}(\beta), \tag{5.92}$$

[1] The Gel'fand-Tseitlin result has been discussed by Vilenkin [17a] and interpreted in terms of an integral transform by Smorodinskiĭ [17b].

where for large j we may also write

$$\cos\beta = m/\sqrt{j(j+1)} \sim m/j. \tag{5.93}$$

In particular, one has

$$C^{jkj}_{m0m} \sim P_k(\cos\beta). \tag{5.94}$$

Racah coefficients have a similar asymptotic expression. This result follows directly from Eq. (3.280) of Chapter 3:

$$W^{abc}_{\rho\sigma\tau}(j) \sim C^{abc}_{\rho\sigma\tau} \tag{5.95}$$

for large j. Thus, using the asymptotic relation (5.92), we find

$$W^{j\,k\,j+\Delta}_{m,\mu,m+\mu}(J) \sim d^k_{\mu\Delta}(\beta) \tag{5.96}$$

for large j and large J. In conventional notation this result becomes

$$[(2J-2\mu+1)(2j+2\Delta+1)]^{\frac{1}{2}}W(J-m-\mu,j,J,k;J-\mu,j+\Delta) \sim d^k_{\mu\Delta}(\beta), \tag{5.97}$$

where β is defined by Eq. (5.93). In particular, setting $\Delta = \mu = 0$ and replacing m by $-m$, one has

$$(-1)^k[(2j+1)(2J+1)]^{\frac{1}{2}}W(j,k,J+m,J;jJ) \sim P_k(\cos\beta). \tag{5.98}$$

The angle β appearing in the asymptotic expansion (5.98) may be estimated more accurately than that given by Eq. (5.93) by *defining* $\cos\beta$ to be the result obtained from Eq. (5.98) for $k = 1$:

$$\cos\beta = \frac{(J+m)(J+m+1) - j(j+1) - J(J+1)}{2[j(j+1)J(J+1)]^{\frac{1}{2}}}. \tag{5.93$'$}$$

Taking the asymptotic limits, *first* $J \to \infty$ and then $j \to \infty$, yields the result given by Eq. (5.93), thereby demonstrating consistency (the order of the limits is determined by the sequence $W \to C \to D$, as in our derivation).

9. Appendices

A. Gel'fand Patterns, Young Frames, Weyl Patterns, and Generalizations of the Boson Polynomials

To introduce another notation for the quantum theory of angular momentum is certainly excessive, and quite unnecessary for the theory itself, but the new notation simplifies enormously the task of comprehending the relationship of the theory to that of the symmetric group and the generalizations to the unitary group $U(n)$. The essence of this notation is to recognize that the boson calculus deals inherently with integers – in fact, sets of integers – and we can use geometric constraints to put structure into these integer arrays.

The situation is, in fact, very similar to that for the symmetric groups, where the introduction of the Young frame defines the irrep ("irrep" denotes irreducible representation), and the *standard Young tableaux* (the Young frame filled in with the integers 1 to n in all ways satisfying the constraints described below) specify the state vectors carrying the irrep.

Although it is outside the scope of the present book to develop the details of the representation theory of the symmetric group S_n and the unitary group $U(n)$, it is useful to summarize briefly those aspects of that theory which relate to angular momentum. In this Appendix we present the relevant nomenclature and results for angular momentum theory, indicating the generalizations.

Weyl patterns.[1] The first concept required is that of a *Young frame*. A Young frame $Y_{[\lambda]}$ of *shape* $[\lambda] = [\lambda_1 \lambda_2 \cdots \lambda_n]$, where the λ_i are nonnegative integers satisfying $\lambda_1 \geqslant \lambda_2 \geqslant \cdots \geqslant \lambda_n$, is a diagram consisting of λ_1 boxes (nodes) in row 1, λ_2 boxes in row 2, ..., λ_n boxes in row n, arranged as illustrated below:

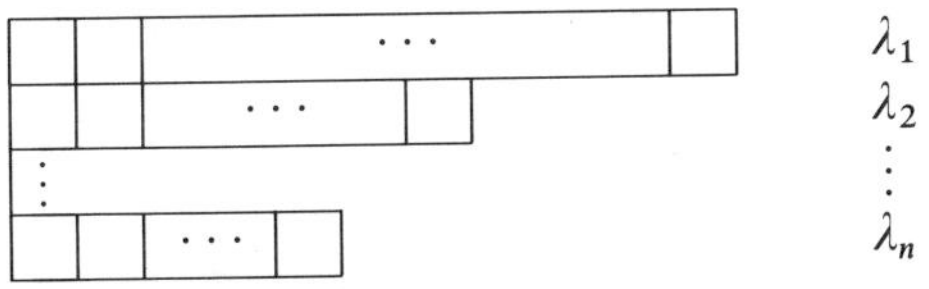

(A-1)

A *Weyl pattern* (or tableau) is a Young frame in which the boxes have been "filled in" with integers selected from $1, 2, \ldots, n$. A Weyl pattern is *standard* if the sequence of integers appearing in each row of $Y_{[\lambda]}$ is *nondecreasing* as read from left to right and the sequence of integers appearing in each column is *strictly increasing* as read from top to bottom. The *weight* or *content* (W) of a Weyl tableau $Y_{[\lambda]}$ is defined to be the row vector $(W) = (w_1, w_2, \ldots, w_n)$, where w_k equals the number of times integer k appears in the pattern. If $\lambda_1 + \lambda_2 + \cdots + \lambda_n = N$, then also $w_1 + w_2 + \cdots + w_n = N$. We shall call $[\lambda]$ a *partition* of N into n parts or, more often, a partition when N is unspecified. We generally count the zeros in determining the parts of a partition. For example, the partitions of 4 into 3 parts are [4 0 0], [3 1 0], [2 2 0], and [2 1 1]. When the number of parts is understood, one frequently omits the zeros (writing [4], [3 1], and [2 2] in the examples).

Example. The standard Weyl patterns corresponding to the Young frame

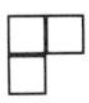 are

---

[1] Young's [18] interest was in invariant theory, utilizing the symmetric group, and he considered frames with n nodes filled in with integers 1 to n. To our knowledge Weyl [19] was the first to use Young frames filled in with repeated integers. We therefore refer to these latter tableaux as *Weyl tableaux*, reserving the term *Young tableaux* for the more restricted cases.

$$\begin{array}{|c|c|}\hline 1&1\\\hline 2\\\cline{1-1}\end{array}\qquad\begin{array}{|c|c|}\hline 1&2\\\hline 2\\\cline{1-1}\end{array}$$

$$\begin{array}{|c|c|}\hline 1&1\\\hline 3\\\cline{1-1}\end{array}\qquad\begin{array}{|c|c|}\hline 1&2\\\hline 3\\\cline{1-1}\end{array}\qquad\begin{array}{|c|c|}\hline 2&2\\\hline 3\\\cline{1-1}\end{array}$$

$$\begin{array}{|c|c|}\hline 1&3\\\hline 3\\\cline{1-1}\end{array}\qquad\begin{array}{|c|c|}\hline 2&3\\\hline 3\\\cline{1-1}\end{array}$$

$$\begin{array}{|c|c|}\hline 1&3\\\hline 2\\\cline{1-1}\end{array}\,. \tag{A-2}$$

Gel'fand patterns. An elegant geometric notation for codifying the constraints imposed on the entries of a Young pattern is provided by a *Gel'fand pattern*, which we now define.

A Gel'fand pattern[1] is a triangular array of n rows of integers, there being one entry in the first row, two entries in the second row, ..., and n entries in the nth row. The entries in each row $1, 2, \ldots, n-1$, are arranged so as to fall between the entries in the row above, as illustrated below:

$$\begin{pmatrix}[m]\\(m)\end{pmatrix} = \left(\begin{array}{ccccccc} m_{1n} & & m_{2n} & & \cdots & & m_{nn} \\ & \ddots & & \ddots & & \iddots & \\ & m_{13} & & m_{23} & & m_{33} & \\ & & m_{12} & & m_{22} & & \\ & & & m_{11} & & & \end{array}\right). \tag{A-3}$$

The integral entries m_{ij}, $i \leqslant j = 1, 2, \ldots, n$, in this array are required to satisfy the following rules:[2]

(*a*)

$$m_{1n} \geqslant m_{2n} \geqslant \cdots \geqslant m_{nn}. \tag{A-4}$$

(*b*) For each specified partition $[m_{1n} \cdots m_{nn}]$, the entries in the remaining rows $j = n-1, n-2, \ldots, 1$ may be any integers that satisfy the "betweenness conditions":[2]

$$m_{1j+1} \geqslant m_{1j} \geqslant m_{2j+1} \geqslant m_{2j} \geqslant m_{3j+1} \geqslant m_{3j} \geqslant \cdots \geqslant m_{j-1j} \geqslant m_{jj} \geqslant m_{j+1j+1}. \tag{A-5}$$

For example, for $n = 3$, and $[m_{13}m_{23}m_{33}] = [2\,1\,0]$, there are eight Gel'fand patterns as displayed below:

---

[1] These patterns were first introduced by Gel'fand and Tseitlin [20] to enumerate basis vectors in the carrier space of an irrep of the general linear group (see also Supplement I in Ref. [1]).

[2] The m_{in} may be *positive, zero, or negative* integers satisfying the inequalities (A-4). The inequalities given by Eq. (A-5) are expressions of the Weyl [21, p. 391] branching law for the general linear group.

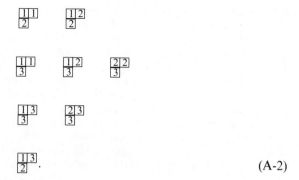

$$\begin{pmatrix} 2 & & 1 & & 0 \\ & 2 & & 1 & \\ & & 2 & & \end{pmatrix} \quad \begin{pmatrix} 2 & & 1 & & 0 \\ & 2 & & 1 & \\ & & 1 & & \end{pmatrix}$$

$$\begin{pmatrix} 2 & & 1 & & 0 \\ & 2 & & 0 & \\ & & 2 & & \end{pmatrix} \quad \begin{pmatrix} 2 & & 1 & & 0 \\ & 2 & & 0 & \\ & & 1 & & \end{pmatrix} \quad \begin{pmatrix} 2 & & 1 & & 0 \\ & 2 & & 0 & \\ & & 0 & & \end{pmatrix}$$

$$\begin{pmatrix} 2 & & 1 & & 0 \\ & 1 & & 0 & \\ & & 1 & & \end{pmatrix} \quad \begin{pmatrix} 2 & & 1 & & 0 \\ & 1 & & 0 & \\ & & 0 & & \end{pmatrix}$$

$$\begin{pmatrix} 2 & & 1 & & 0 \\ & 1 & & 1 & \\ & & 1 & & \end{pmatrix}. \tag{A-6}$$

Mapping between Gel'fand and Weyl patterns. There is a one-to-one correspondence between the set of Gel'fand patterns (m) having nth row $[m_{1n}m_{2n} \cdots m_{nn}]$ (with $m_{nn} \geqslant 0$) and the set of standard Weyl patterns of this shape.

The mapping between Gel'fand patterns and standard Weyl patterns is described as follows: The shape of the frame is $[m_{1n}m_{2n} \cdots m_{nn}]$, and the rows of the frame are filled in according to the following rules (read along the diagonals of the Gel'fand pattern):

$$\begin{array}{lll} \text{Row } 1\colon m_{11}\ 1\text{'s}, m_{12} - m_{11}\ 2\text{'s}, m_{13} - m_{12}\ 3\text{'s}, \ldots, & & m_{1n} - m_{1n-1} n\text{'s} \\ \text{Row } 2\colon \quad m_{22}\ 2\text{'s}, \quad m_{23} - m_{22}\ 3\text{'s}, \ldots, & & m_{2n} - m_{2n-1}\ n\text{'s} \\ \vdots & & \\ \text{Row } j\colon \quad m_{jj} j\text{'s}, \ m_{jj+1} - m_{jj}(j+1)\text{'s}, \ldots, m_{jn} - m_{jn-1}\ n\text{'s} & & \\ \vdots & & \\ \text{Row } n\colon & & m_{nn}\ n\text{'s}. \end{array} \tag{A-7}$$

Using the rule (A-7), we see that the set of Gel'fand patterns (A-6) is mapped to the set of Weyl patterns (A-2). Conversely, from each standard Weyl pattern (A-2), we construct in an obvious way the Gel'fand pattern in the set (A-6).

The *weight* or *content* of a Gel'fand pattern (m) is the row vector $(W) = (w_1, w_2, \ldots, w_n)$, where w_j is defined to be the sum of the entries in row j of (m) minus the sum of the entries in row $j-1$ $(w_1 \equiv m_{11})$:

$$w_j = \sum_{i=1}^{j} m_{ij} - \sum_{i=1}^{j-1} m_{ij-1}. \tag{A-8}$$

Clearly, this definition of weight coincides with that given earlier for a standard Weyl pattern.

The constraint in a standard Weyl pattern that each row (column) should comprise a set of nondecreasing (strictly increasing) nonnegative integers is realized in a Gel'fand pattern by the "geometric" rule that the integers (m_{ij}) satisfy the betweenness conditions.

Carrier spaces of the representations of the symmetric group. The important pattern results for the symmetric group S_n are as follows: (*a*) The set of irreps of S_n is in one-to-one correspondence with the set of partitions $\{[\lambda]\}$ of n into n parts; (*b*) the set of basis vectors of the carrier space of irrep $[\lambda]$ of S_n is in one-to-one correspondence with the set of standard Young patterns of shape $[\lambda]$ having weight $(W) = (1, 1, \ldots, 1)$ (the number of basis vectors or the number of standard patterns is then the dimension of the irrep). This result may, of course, also be expressed in terms of Gel'fand patterns. For example, the irreps of S_3 are enumerated by the partitions of 3 into 3 parts [3 0 0], [2 1 0], and [1 1 1]. The standard Young patterns of weight (1, 1, 1) having these shapes, respectively, are

$$\boxed{1}\boxed{2}\boxed{3}\,; \qquad \begin{array}{l}\boxed{1}\boxed{2}\\ \boxed{3}\end{array}, \qquad \begin{array}{l}\boxed{1}\boxed{3}\\ \boxed{2}\end{array}; \qquad \begin{array}{l}\boxed{1}\\ \boxed{2}\\ \boxed{3}\end{array}. \tag{A-9}$$

Thus, the irreps [3 0 0], [2 1 0], and [1 1 1] are of dimensions 1, 2, and 1, respectively. These same results are enumerated by the Gel'fand patterns:

$$\begin{pmatrix} 3 & & 0 & & 0 \\ & 2 & & 0 & \\ & & 1 & & \end{pmatrix}; \quad \begin{pmatrix} 2 & & 1 & & 0 \\ & 2 & & 0 & \\ & & 1 & & \end{pmatrix}, \quad \begin{pmatrix} 2 & & 1 & & 0 \\ & 1 & & 1 & \\ & & 1 & & \end{pmatrix}; \quad \begin{pmatrix} 1 & & 1 & & 1 \\ & 1 & & 1 & \\ & & 1 & & \end{pmatrix}. \tag{A-10}$$

The standard Young patterns for S_n are also often enumerated by another indexing scheme—the *Yamanouchi* [22] *symbol*:

$$(y) = (y_1, y_2, \ldots, y_n). \tag{A-11}$$

Here y_{n-j+1} is the positive integer equal to the row in which j appears in a given standard Young tableau of shape $[\lambda]$ and weight $(1, 1, \ldots, 1)$; y_{n-j+1} is also the position (counting from the left) in which 1 occurs in the set of differences

$$m_{1j} - m_{1j-1}, m_{2j} - m_{2j-1}, \ldots, m_{j-1j} - m_{j-1j-1}, m_{jj} \tag{A-12}$$

formed from the entries in the Gel'fand pattern corresponding to $[m_{1n} \cdots m_{nn}] = [\lambda_1 \cdots \lambda_n]$ and having weight $(1, 1, \ldots, 1)$. For example, the Yamanouchi symbols for the Young patterns (A-9) [and Gel'fand patterns (A-10)] are, respectively,

$$(1, 1, 1); \; (2, 1, 1), \; (1, 2, 1); \; (3, 2, 1). \tag{A-13}$$

The Yamanouchi symbols are, in fact, *lattice permutations*. A lattice permutation of n symbols that include the symbol a repeated α times, b repeated β times, $\ldots$, z repeated ζ times with $\alpha + \beta + \cdots + \zeta = n$ is defined to be any sequence (displayed, say, across the page) of these n symbols (a repeated

α times, b repeated β times, ...) such that the first p symbols in the sequence (counting from the left) contain at least as many a's as b's, at least as many b's as c's, ..., and at least as many y's as z's, where this property must hold for each $p = 1, 2, \ldots, n$. A lattice permutation put together in this way is said to be of the type

$$a^{\alpha} b^{\beta} \cdots z^{\zeta}. \tag{A-14}$$

The Yamanouchi symbols of the set of standard Young patterns of shape $[\lambda]$ are just the lattice permutations of the type

$$n^{\lambda_n} \cdots 2^{\lambda_2} 1^{\lambda_1}. \tag{A-15}$$

Carrier spaces of the representations of the rotation group. The important pattern results for the rotation group $[SU(2)]$ are as follows: (*a*) The set of irreps of the rotation group is in one-to-one correspondence with the set of partitions $[2j\ 0]$, $j = 0, \frac{1}{2}, 1, \ldots$; (*b*) the set of basis vectors of the carrier space of irrep $[2j\ 0]$ is in one-to-one correspondence with the set of Gel'fand patterns having the partition $[2j\ 0]$:

$$\begin{pmatrix} 2j & & 0 \\ & j+m & \end{pmatrix}, \qquad m = -j, -j+1, \ldots, j. \tag{A-16}$$

(Observe that the betweenness rule embodies in a natural way the fact that the projection quantum number m runs over the values $m = -j, \ldots, j$.)

The Weyl pattern corresponding to the Gel'fand pattern (A-16) is the one-row pattern

$$\begin{array}{|c|c|c|c|c|c|c|c|} \hline 1 & 1 & \cdots & 1 & 2 & 2 & \cdots & 2 \\ \hline \multicolumn{4}{|c|}{\longleftarrow j+m \longrightarrow} & \multicolumn{4}{c|}{\longleftarrow j-m \longrightarrow} \end{array}\ . \tag{A-17}$$

One sees at once that the boson operator form of the eigenket, Eq. (5.33), is a direct transcription of this pattern, replacing "1" by a_1, "2" by a_2, and normalizing.

The normalization factor is easily found for these boson states, but the $U(n)$ generalization is more interesting and represents one of the many pattern functions of unitary symmetry. This will be discussed below, using the concept of "entanglement" of bosons.

Consider next the standard Weyl pattern of two rows corresponding to the Gel'fand pattern

$$\begin{pmatrix} m_{12} & & m_{22} \\ & m_{11} & \end{pmatrix}, \tag{A-18}$$

where $m_{12} \geqslant m_{11} \geqslant m_{22}$:

$$\begin{array}{|c|c|c|c|c|c|c|c|c|c|c|c|}
\multicolumn{4}{c}{\longleftarrow m_{22} \longrightarrow} & \multicolumn{4}{c}{\leftarrow m_{11} - m_{22} \rightarrow} & \multicolumn{4}{c}{\leftarrow m_{12} - m_{11} \rightarrow} \\
\hline
1 & 1 & \cdots & 1 & 1 & 1 & \cdots & 1 & 2 & 2 & \cdots & 2 \\
\hline
2 & 2 & \cdots & 2 \\
\cline{1-4}
\end{array} \; . \tag{A-19}$$

The mapping from Weyl patterns to bosons is given by

$$\boxed{1} \rightarrow \begin{pmatrix} 1 & & 0 \\ & 1 & \end{pmatrix} \rightarrow a_1^1,$$

$$\boxed{2} \rightarrow \begin{pmatrix} 1 & & 0 \\ & 0 & \end{pmatrix} \rightarrow a_2^1,$$

$$\begin{array}{|c|}\hline 1 \\ \hline 2 \\ \hline \end{array} \rightarrow \begin{pmatrix} 1 & & 1 \\ & 1 & \end{pmatrix} \rightarrow \det \begin{pmatrix} a_1^1 & a_1^2 \\ a_2^1 & a_2^2 \end{pmatrix} \equiv a_{12}^{12}. \tag{A-20}$$

The Weyl pattern $\begin{array}{|c|}\hline 1 \\ \hline 2 \\ \hline \end{array}$ corresponds to antisymmetrized bosons made up of two independent bosons a_i^1 and a_i^2 $(i = 1, 2)$.

Using the correspondence (A-20), we obtain the following boson state vector, corresponding to the Gel'fand pattern (A-18) and the Weyl pattern (A-19):

$$\left| \begin{pmatrix} m_{12} & & m_{22} \\ & m_{11} & \end{pmatrix} \right\rangle = M^{-\frac{1}{2}} (a_{12}^{12})^{m_{22}} (a_1^1)^{m_{11} - m_{22}} (a_2^1)^{m_{12} - m_{11}} |0\rangle, \tag{A-21}$$

where the normalization factor is given by

$$M = \frac{(m_{12} + 1)!(m_{11} - m_{22})!(m_{12} - m_{11})!(m_{22})!}{(m_{12} - m_{22} + 1)!} . \tag{A-22}$$

The angular momentum labels for the states (A-21) are

$$j = \frac{m_{12} - m_{22}}{2},$$

$$m = m_{11} - \frac{m_{12} + m_{22}}{2} . \tag{A-23}$$

The $2m_{22}$ antisymmetric (paired) bosons are inert as far as angular momentum is concerned; that is, a_{12}^{12} is invariant under unitary unimodular transformations of the form given by Eq. (5.43). See Appendix E for the application of these results to the group $U(2)$.

The normalization factor M given by Eq. (A-22) shows a quite interesting structure ("entanglement"). To discuss this, one first requires the concepts of the Nakayama [23] "hook" and "hook length" (see Frame *et al.* [23a], Robinson [24]).

The definitions are as follows: (a) The (i,j)-hook of the Young frame $Y_{[\lambda]}$ consists of the box in row i ($i = 1, 2, \ldots, n$) and column j ($j = 1, 2, \ldots, \lambda_i$; for fixed i) together with the $\lambda_i - j$ boxes to the right (called the *arm* of the hook) and the $\lambda'_j - i$ boxes[1] below (called the *leg* of the hook); (b) the hook length h_{ij} of the (i,j)-hook is the number of boxes in the hook; that is,

$$h_{ij} = (\lambda_i - j) + (\lambda'_j - i) + 1. \qquad \text{(A-24)}$$

Example.

Young frame: . (A-25)

	hooks	hook lengths	
$(1,1)$		$h_{11} = 5$	
$(1,2)$		$h_{12} = 4$	(A-26)
$(1,3)$		$h_{13} = 2$	
$(1,4)$		$h_{14} = 1$	
$(2,1)$		$h_{21} = 2$	
$(2,2)$		$h_{22} = 1$	

The *hook graph* of $Y_{[\lambda]}$ is obtained by writing the hook length h_{ij} in the (i,j)-box.

Example. The hook graph of $Y_{[42]}$ is

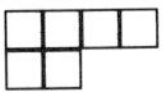

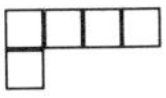

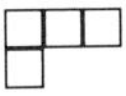

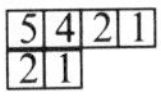

$$\boxed{\begin{array}{|c|c|c|c|} \hline 5 & 4 & 2 & 1 \\ \hline \end{array}\,\begin{array}{|c|c|} \hline 2 & 1 \\ \hline \end{array}} . \qquad \text{(A-27)}$$

The product of all the h_{ij} appearing in the hook graph of $Y_{[\lambda]}$ is denoted by $H^{[\lambda]}$.

The fundamental formulas involving hook lengths are (Robinson [24])

$$\dim[\lambda] = n!/H^{[\lambda]} \qquad \text{for } S_n, \qquad \text{(A-28)}$$

where $\dim[\lambda]$ denotes the dimension of irrep $[\lambda]$ of S_n; and

$$\text{Dim}[\lambda] = \prod_{i,j}(n + j - i)\Big/ H^{[\lambda]} \qquad \text{for } U(n), \qquad \text{(A-29)}$$

---

[1] The notation $Y_{[\lambda']}$ denotes the Young frame *conjugate* to $Y_{[\lambda]}$ and is obtained from $Y_{[\lambda]}$ by interchanging rows and columns.

where Dim[λ] denotes the dimension of irrep [λ] of $U(n)$, and the product (i,j) is over all boxes of the Young frame.

The concept of entanglement of a hook of a Young frame depends on the standard Weyl pattern assigned to the frame. The definition is as follows (Ciftan and Biedenharn [25]): The (i,j)-hook is *entangled* in a Weyl pattern if its arm contains at least one integer in common with the integers in the (i,j)-box and the leg.

Example. The (1, 1)-, (1, 2)-, and (2, 1)-hooks displayed in (A-26) are entangled in the Weyl pattern:

$$\begin{array}{|c|c|c|c|}\hline 1&1&1&2\\\hline 2&2\\\cline{1-2}\end{array}\,. \tag{A-30}$$

The *entangled hook graph* of a Weyl pattern is obtained by writing the hook length h_{ij} of each entangled hook in the (i,j)-box and 1 in each remaining box.

Example. The entangled hook graph of the Weyl pattern (A-30) is

$$\begin{array}{|c|c|c|c|}\hline 5&4&1&1\\\hline 2&1\\\cline{1-2}\end{array}\,. \tag{A-31}$$

The result for the normalization factor M [Eq. (A-22)] may now be stated: *The normalization factor M is the product of the entries in the entangled hook graph of the Weyl pattern* (A-19).

Double tableaux and the rotation matrices. A closer inspection [see (A-20)] of the basis vectors (A-21) reveals that the Weyl pattern (A-19) has been used to assign the *subscripts* to the bosons. One sees, in fact, that the superscript assignment originates from the Weyl pattern

$$\begin{array}{|c|c|c|c|c|c|c|c|}\multicolumn{4}{|c|}{\longleftarrow m_{22}\longrightarrow} & \multicolumn{4}{c|}{\leftarrow m_{12}-m_{22}\rightarrow}\\\hline 1&1&\cdots&1&1&1&\cdots&1\\\hline 2&2&\cdots&2\\\cline{1-4}\end{array} \tag{A-32}$$

corresponding to the maximal Gel'fand pattern (m_{11} chosen as large as possible)

$$\begin{pmatrix} m_{12} & & m_{22} \\ & m_{12} & \end{pmatrix}. \tag{A-33}$$

A more descriptive notation for the state vector (A-21) uses a *double Weyl pattern* or a *double Gel'fand pattern*:

$$\left|\left(\begin{array}{|c|c|c|c|c|c|c|c|c|}\hline 1&\cdots&1&1&\cdots&1&2&\cdots&2\\\hline 2&\cdots&2\\\cline{1-3}\end{array}\;\middle\|\;\begin{array}{|c|c|c|c|c|c|c|c|c|}\hline 1&\cdots&1&1&\cdots&1&1&\cdots&1\\\hline 2&\cdots&2\\\cline{1-3}\end{array}\right)\right\rangle =$$

$$= \left| \begin{pmatrix} & m_{12} & \\ m_{12} & & m_{22} \\ & m_{11} & \end{pmatrix} \right\rangle\!\!\Big\rangle \equiv M^{-\frac{1}{2}} (a^{12}_{12})^{m_{22}} (a^{1}_{1})^{m_{11}-m_{22}} (a^{1}_{2})^{m_{12}-m_{11}} |0\rangle, \tag{A-34}$$

where we observe that (*a*) the Young frames have the *same shape*; (*b*) by convention the second Gel'fand pattern (A-33) is inverted over the first one (A-18) in order to depict explicitly the shared labels $[m_{12}m_{22}]$ giving the common shape of the Young frame; (*c*) the mapping from the double Weyl pattern to bosons is obtained by pairing off the columns occurring in the *same positions* in the two Weyl patterns:

$$\left\{ \begin{smallmatrix}\boxed{1}\\ \boxed{2}\end{smallmatrix}, \begin{smallmatrix}\boxed{1}\\ \boxed{2}\end{smallmatrix} \right\} \to a^{12}_{12} \tag{A-35}$$

$$\{\boxed{i}, \boxed{j}\} \to a^{j}_{i}, \qquad i,j = 1,2. \tag{A-36}$$

In the patterns in (A-34) the column pair

$$\left\{ \begin{smallmatrix}\boxed{1}\\ \boxed{2}\end{smallmatrix}, \begin{smallmatrix}\boxed{1}\\ \boxed{2}\end{smallmatrix} \right\}$$

occurs m_{22} times, the column pair $\{\boxed{1}, \boxed{1}\}$ occurs $m_{11} - m_{22}$ times, and the column pair $\{\boxed{1}, \boxed{2}\}$ occurs $m_{12} - m_{11}$ times.

The significance of rewriting Eq. (A-21) in the form of Eq. (A-34) is that one now recognizes that the latter result generalizes: *The Weyl pattern in the second position* (the upper Gel'fand pattern) *may also be taken to be any standard pattern corresponding to the shape* $[m_{12}m_{22}]$. The mappings (A-35) and (A-36) then assign a definite state vector (or boson polynomial) to each pair of standard Weyl patterns of the same shape.

In terms of the Gel'fand pattern notation the polynomials that one reads off the corresponding double Weyl patterns, using the rules (A-35) and (A-36), are as follows:

$$P\begin{pmatrix} & m'_{11} & \\ m_{12} & & m_{22} \\ & m_{11} & \end{pmatrix}(A) \equiv (a^{12}_{12})^{m_{22}} (a^{1}_{1})^{m_{11}-m_{22}} (a^{1}_{2})^{m'_{11}-m_{11}} (a^{2}_{2})^{m_{12}-m'_{11}}, \tag{A-37}$$

if $m'_{11} \geqslant m_{11}$; and

$$P\begin{pmatrix} & m'_{11} & \\ m_{12} & & m_{22} \\ & m_{11} & \end{pmatrix}(A) \equiv (a^{12}_{12})^{m_{22}} (a^{1}_{1})^{m_{11}-m_{22}} (a^{2}_{1})^{m_{11}-m'_{11}} (a^{2}_{2})^{m_{12}-m_{11}}, \tag{A-38}$$

if $m'_{11} < m_{11}$.

The weight (W, W') (or content) of the double standard tableau polynomial (A-37) or (A-38) is defined to be

$$\begin{aligned}(W, W') &= (w_1, w_2, w'_1, w'_2)\\ &= (m_{11}, m_{12} + m_{22} - m_{11}, m'_{11}, m_{12} + m_{22} - m'_{11}). \quad \text{(A-39)}\end{aligned}$$

$[w_i(w'_i)$ is the number of occurrences of i in the left (right) Weyl pattern of a double standard tableau; $w_i(w'_i)$ is also the number of subscripts (superscripts) i appearing in the polynomials (A-37) and (A-38).$]$

The important properties of the double standard tableau polynomials include the following:

(*a*) The set of double standard tableau polynomials of weight (W, W') is a (linearly independent) basis of the vector space spanned by all monomials in the bosons $\{a_i^j\}$ that contain w_i occurrences of the subscript i and w'_j occurrences of the superscript j.

(*b*) The set of double standard tableau polynomials corresponding to all partitions $[m]$ of the nonnegative integer N is a basis of the vector space of homogeneous polynomials of degree N in the bosons $\{a_i^j\}$.

(*c*) The matrix $P^{[m]}(A)$ having element in row m_{11} $(m_{11} = m_{12}, \ldots, m_{22})$ and column m'_{11} $(m'_{11} = m_{12}, \ldots, m_{22})$ given by the boson polynomials (A-37) and (A-38) is an irreducible representation of the unitary group $U(2)$ when the matrix A is replaced by the unitary matrix $U \in U(2)$. (The representation is, in general, nonunitary.)

The boson polynomials occurring in the basis vectors, Eq. (5.80), are an orthonormalized version of the double standard tableau polynomials. For general double Gel'fand patterns these boson polynomials are defined by

$$B\begin{pmatrix} & m'_{11} & \\ m_{12} & & m_{22} \\ & m_{11} & \end{pmatrix}(A) \equiv (a_{12}^{12})^{m_{22}} D^j_{mm'}(A), \quad \text{(A-40)}$$

where

$$\begin{aligned} j &= (m_{12} - m_{22})/2, \qquad m = m_{11} - (m_{12} + m_{22})/2,\\ m' &= m'_{11} - (m_{12} + m_{22})/2. \end{aligned} \quad \text{(A-41)}$$

The corresponding orthonormalized basis vectors are

$$\left|\begin{pmatrix} & \mu_{11} & \\ m_{12} & & m_{22} \\ & m_{11} & \end{pmatrix}\right\rangle \equiv \left[\frac{(m_{12}+1)!(m_{22})!}{(m_{12}-m_{22}+1)}\right]^{-\frac{1}{2}} B\begin{pmatrix} & \mu_{11} & \\ m_{12} & & m_{22} \\ & m_{11} & \end{pmatrix}(A)|0\rangle, \quad \text{(A-42)}$$

$$\left\langle\begin{pmatrix} & \mu'_{11} & \\ m'_{12} & & m'_{22} \\ & m'_{11} & \end{pmatrix}\middle|\begin{pmatrix} & \mu_{11} & \\ m_{12} & & m_{22} \\ & m_{11} & \end{pmatrix}\right\rangle = \delta_{\mu'_{11}\mu_{11}} \prod_{i,j} \delta_{m'_{ij}, m_{ij}}. \quad \text{(A-43)}$$

For $\mu_{11} = m_{12}$ (or for $m_{11} = m_{12}$), the polynomials (A-40) coincide (up to a numerical factor) with those [(A-37) and (A-38)] read directly off the double standard Weyl patterns. However, the general polynomial (A-40) is not read directly off the associated double tableaux, but is generated from the $\mu_{11} = m_{12}$ polynomial by application (multiple commutation) of the lowering operator E^{21} [see Eq. (5.59)], thus assuring orthonormality of the corresponding basis vectors (A-42):

$$B\begin{pmatrix} & \mu_{11} & \\ m_{12} & & m_{22} \\ & m_{11} & \end{pmatrix}(A) = \left[\frac{(\mu_{11} - m_{22})!}{(m_{12} - m_{22})!(m_{12} - \mu_{11})!}\right]^{\frac{1}{2}}$$

$$\times \left[E^{21}, B\begin{pmatrix} & m_{12} & \\ m_{12} & & m_{22} \\ & m_{11} & \end{pmatrix}(A)\right]_{(m_{12}-\mu_{11})}$$

$$= \left[\frac{(\mu_{11} - m_{22})!}{(m_{12} - m_{22})!(m_{12} - \mu_{11})!}\right]^{\frac{1}{2}} \left[\frac{(m_{11} - m_{22})!}{(m_{12} - m_{22})!(m_{12} - m_{11})!}\right]^{\frac{1}{2}}$$

$$\times \left[E^{21}, \left[E_{21}, B\begin{pmatrix} & m_{12} & \\ m_{12} & & m_{22} \\ & m_{12} & \end{pmatrix}(A)\right]_{(m_{12}-m_{11})}\right]_{(m_{12}-\mu_{11})}, \tag{A-44}$$

in which

$$B\begin{pmatrix} & m_{12} & \\ m_{12} & & m_{22} \\ & m_{12} & \end{pmatrix}(A) = (a^{12}_{12})^{m_{22}}(a^{1}_{1})^{m_{12}-m_{22}}, \tag{A-45}$$

and the brackets denote multiple commutators.

The properties (*a*)–(*c*) above [following Eq. (A-39)] apply also to the orthonormalized boson polynomials,

$$B\begin{pmatrix} & \mu_{11} & \\ m_{12} & & m_{22} \\ & m_{11} & \end{pmatrix}(A),$$

except that in property (*c*) we now obtain the unitary irreps of $U(2)$ [see Appendix E].

The properties of the boson polynomials (A-40) are developed further in Appendices B–E. (Corresponding detailed relations for the double standard tableau polynomials do not appear in the literature.)

Generalizations. Consider the double standard Weyl tableau or pattern of shape $[m] = [\lambda] = [\lambda_1 \lambda_2 \cdots \lambda_n]$:

$$\left(\begin{array}{llll} i_{11} & i_{12} & \cdots & i_{1\lambda_1} \\ i_{21} & i_{22} & \cdots & i_{2\lambda_2} \\ \vdots & & & \\ i_{n1} & i_{n2} & \cdots & i_{n\lambda_n} \end{array}\right|\left.\begin{array}{llll} j_{11} & j_{12} & \cdots & j_{1\lambda_1} \\ j_{21} & j_{22} & \cdots & j_{2\lambda_2} \\ \vdots & & & \\ j_{n1} & j_{n2} & \cdots & j_{n\lambda_n} \end{array}\right). \tag{A-46}$$

The double standard Weyl tableaux of shape $[m]$ are in one-to-one correspondence with the double Gel'fand patterns

$$\begin{pmatrix} (m') \\ [m] \\ (m) \end{pmatrix}, \tag{A-47}$$

where the left Weyl pattern in (A-46) is constructed from the Gel'fand pattern

$$\begin{pmatrix} [m] \\ (m) \end{pmatrix} \tag{A-48}$$

according to the rule (A-7), and the right Weyl pattern in (A-46) is similarly constructed from the Gel'fand pattern

$$\begin{pmatrix} [m] \\ (m') \end{pmatrix}. \tag{A-49}$$

With each pair of columns in corresponding positions in the left and right patterns of the double standard Weyl pattern (A-46), we now associate a determinantal boson by the rule

$$\left\{ \begin{array}{|c|} \hline i_{1k} \\ \hline i_{2k} \\ \hline \vdots \\ \hline i_{\lambda'_k k} \\ \hline \end{array}, \begin{array}{|c|} \hline j_{1k} \\ \hline j_{2k} \\ \hline \vdots \\ \hline j_{\lambda'_k k} \\ \hline \end{array} \right\} \to a^{j_{1k} \cdots j_{\lambda'_k k}}_{i_{1k} \cdots i_{\lambda'_k k}}, \tag{A-50}$$

where

$$a^{j_1 \cdots j_k}_{i_1 \cdots i_k} \equiv \det \begin{pmatrix} a^{j_1}_{i_1} & \cdots & a^{j_k}_{i_1} \\ \vdots & & \vdots \\ a^{j_1}_{i_k} & \cdots & a^{j_k}_{i_k} \end{pmatrix}. \tag{A-51}$$

We next define a double standard Weyl pattern boson polynomial (equivalently, a double Gel'fand pattern boson polynomial) by forming the product of the determinantal bosons (A-50) over all columns $1, 2, \ldots, \lambda_1$ of the frame. Using the double Gel'fand patterns to denote the polynomials, we have

$$P\begin{pmatrix} (m') \\ [m] \\ (m) \end{pmatrix}(A) \equiv \prod_{k=1}^{\lambda_1} a^{j_{1k} \cdots j_{\lambda'_k k}}_{i_{1k} \cdots i_{\lambda'_k k}}. \tag{A-52}$$

We note two special cases of Eq. (A-52):

$$P\begin{pmatrix} (\max) \\ [m] \\ (\max) \end{pmatrix}(A) = \prod_{k=1}^{n} (a^{12\ldots k}_{12\ldots k})^{m_{kn} - m_{k+1\,n}}, \tag{A-53}$$

$$P\begin{pmatrix} (\text{max}) \\ [m] \\ (\text{semi-max}) \end{pmatrix}(A) = \prod_{k=1}^{n-1} (a^{12\ldots k}_{12\ldots k})^{m_{kn-1}-m_{k+1n}} \times \prod_{k=1}^{n} (a^{12\ldots k-1k}_{12\ldots k-1n})^{m_{kn}-m_{kn-1}}, \tag{A-54}$$

where $m_{ij} \equiv 0$ for $i > j$, $a^{12\ldots k-1k}_{12\ldots k-1n} \equiv a^k_n$ for $k = 1$, and special pattern notations have been introduced:

$$\begin{pmatrix} [m] \\ (\text{max}) \end{pmatrix} \equiv \begin{pmatrix} m_{1n} & & m_{2n} & \cdots & m_{n-1n} & & m_{nn} \\ & \ddots & & \ddots & & m_{n-1n} & \\ & & m_{1n} & & m_{2n} & & \\ & & & m_{1n} & & & \end{pmatrix}, \tag{A-55}$$

$$\begin{pmatrix} [m] \\ (\text{semi-max}) \end{pmatrix} \equiv \begin{pmatrix} m_{1n} & & m_{2n} & \cdots & m_{n-1n} & & m_{nn} \\ & m_{1n-1} & & m_{2n-1} & \cdots & m_{n-1n-1} & \\ & & & (\text{max}) & & & \end{pmatrix}. \tag{A-56}$$

The weight or content (W, W') of the double standard tableau (A-46) [and of the double Gel'fand pattern (A-47)] is defined to be

$$(W, W') = (w_1, \ldots, w_n, w'_1, \ldots, w'_n), \tag{A-57}$$

where (W) and (W') are, respectively, the weights of the left and right standard tableaux (lower and upper Gel'fand patterns). w_i and w'_i are also equal to the number of occurrences of index i among the subscripts and superscripts, respectively, of the boson polynomial (A-52). [We also say that the polynomial (A-52) has weight (W, W').]

The important properties (*a*) *and* (*b*) *noted earlier* (*p.* 240) *generalize to the double standard Weyl tableau polynomials*

$$P\begin{pmatrix} (m') \\ [m] \\ (m) \end{pmatrix}(A).$$

Property (*c*) *also generalizes to the general unitary group,*[1] *where the rows and columns of the matrix* $P^{[m]}(A)$ *are now to be enumerated by the* $U(n-1)$ *Gel'fand patterns* $((m), (m'))$.

A great deal of physics literature has been devoted to the study of the family of unitary groups, $U(n)$, through the use of the $n \times n$ boson as discussed in Section 5. The main emphasis has been on the construction of *orthonormal basis vectors* denoted in the double Gel'fand pattern notation by

$$\left|\begin{pmatrix} (m') \\ [m] \\ (m) \end{pmatrix}\right\rangle \equiv [\mathcal{M}([m])]^{-\frac{1}{2}} B\begin{pmatrix} (m') \\ [m] \\ (m) \end{pmatrix}(A)|0\rangle, \tag{A-58}$$

where

[1] Actually, the result is valid for the general linear group.

$$\mathscr{M}([m]) = \prod_{i=1}^{n} p_{in}! \Big/ \prod_{i<j} (p_{in} - p_{jn}), \tag{A-59}$$

in which

$$p_{in} \equiv m_{in} + n - i. \tag{A-60}$$

$\mathscr{M}([m])$ denotes the product of the entries in the entangled hook graph of the maximal Young pattern corresponding to the maximal Gel'fand pattern (A-55). $\mathscr{M}([m])$ is also called the *measure* of the Young frame of shape $[m]$.

The basis vectors (A-58) are *uniquely* characterized (up to phase) as being the *normalized simultaneous eigenvectors* of a set of (independent) mutually commuting Hermitian operators:

$$\mathscr{L}_k^{(\lambda)} \equiv \sum_{i_1 i_2 \cdots i_k}{}_1^{\lambda} \; E_{i_1 i_2} E_{i_2 i_3} \cdots E_{i_k i_1}, \tag{A-61}$$

$$\mathscr{U}_k^{(\lambda)} \equiv \sum_{j_1 j_2 \cdots j_k}{}_1^{\lambda} \; E^{j_1 j_2} E^{j_2 j_3} \cdots E^{j_k j_1}, \tag{A-62}$$

where $\lambda = 1, 2, \ldots, n$, and for each λ the range of k is $k = 1, 2, \ldots, \lambda$. Since it may be shown that

$$\mathscr{U}_k^{(n)} \equiv \mathscr{L}_k^{(n)}, \qquad k = 1, 2, \ldots, n, \tag{A-63}$$

there are n^2 distinct operators defined by Eqs. (A-61) and (A-62), and the eigenvalues of the simultaneous eigenvectors determine the n^2 entries appearing in the double Gel'fand pattern (A-47).

The boson polynomials

$$B\begin{pmatrix} (m') \\ [m] \\ (m) \end{pmatrix}(A) \tag{A-64}$$

occurring in Eq. (A-58) and the double tableau polynomials

$$P\begin{pmatrix} (m') \\ [m] \\ (m) \end{pmatrix}(A)$$

given by Eq. (A-52) span the same vector space. However, only for the patterns

$$\begin{pmatrix} (\text{max}) \\ [m] \\ (\text{max}) \end{pmatrix}, \quad \begin{pmatrix} (\text{max}) \\ [m] \\ (\text{semi-max}) \end{pmatrix}, \quad \begin{pmatrix} (\text{semi-max}) \\ [m] \\ (\text{max}) \end{pmatrix} \tag{A-65}$$

do the polynomials agree (up to a normalization factor). In particular, we note that

$$B\begin{pmatrix}(\max)\\ [m]\\ (\max)\end{pmatrix}(A) = P\begin{pmatrix}(\max)\\ [m]\\ (\max)\end{pmatrix}(A), \tag{A-66}$$

$$B\begin{pmatrix}(\max)\\ [m]\\ (\text{semi-max})\end{pmatrix}(A) = \left[\frac{M}{H^{[m]}}\right]^{-\frac{1}{2}} P\begin{pmatrix}(\max)\\ [m]\\ (\text{semi-max})\end{pmatrix}(A). \tag{A-67}$$

A significant property (Louck [14]) of the boson polynomials is the following: Under the mapping $A \to U \in U(n)$, the polynomials

$$B\begin{pmatrix}(m')\\ [m]\\ (m)\end{pmatrix}(U) \tag{A-68}$$

are the elements of the unitary matrix irrep $D^{[m]}(U)$ of the group $U(n)$.

There is a vast literature developing various properties of the boson polynomials (A-64) themselves, their relation to Wigner coefficients of $U(n)$, and their relation to state vectors of sharp angular momentum of n-particle systems, but we must refer to this literature for these details (see the *Remarks* at the end of this Appendix).

We conclude this Appendix by noting the boson interpretation of a classical mathematical result – the *Capelli identity* (Weyl [26, p. 41]) – and the application of this result in determining the measure factor (A-59). With a slight modification of the results given by Weyl, the Capelli identity may be written

$$\det\begin{pmatrix} E^{11}+n-1 & E^{12} & \cdots & E^{1n}\\ E^{21} & E^{22}+n-2 & \cdots & E^{2n}\\ \vdots & \vdots & & \vdots\\ E^{n1} & E^{n2} & \cdots & E^{nn}\end{pmatrix} = a^{12\ldots n}_{12\ldots n}\,\bar{a}^{12\ldots n}_{12\ldots n}, \tag{A-69}$$

where the factors in the expanded determinant are ordered such that any factor occurring in column i of the determinant lies to the left of any factor occurring in column j for $i < j$. For example, for $n = 2$, we have

$$\det\begin{pmatrix} E^{11}+1 & E^{12}\\ E^{21} & E^{22}\end{pmatrix} = (E^{11}+1)E^{22} - E^{21}E^{12} = a^{12}_{12}\bar{a}^{12}_{12}. \tag{A-70}$$

The E^{ij} are defined by Eq. (5.50).

Noting that

$$E^{ij}\left|\begin{pmatrix}(\max)\\ [m]\\ (\max)\end{pmatrix}\right\rangle = 0, \qquad i < j, \tag{A-71}$$

one finds, using the Capelli identity, the following result:

$$a_{1\,2\ldots n}^{1\,2\ldots n}\bar{a}_{1\,2\ldots n}^{1\,2\ldots n}\left|\begin{pmatrix}(\max)\\ [m]\\ (\max)\end{pmatrix}\right\rangle = \prod_{i=1}^{n} p_{in}\left|\begin{pmatrix}(\max)\\ [m]\\ (\max)\end{pmatrix}\right\rangle. \tag{A-72}$$

Observing that the $U(n)$ double maximal weight vector for irrep $[m_{1n}\cdots m_{kn}0\cdots 0]$ coincides with the $U(k)$ double maximal weight vector for irrep $[m_{1n}\cdots m_{kn}]$—that is,

$$\left|\begin{pmatrix}(\max)\\ [m_{1n}\cdots m_{kn}0\cdots 0]\\ (\max)\end{pmatrix}\right\rangle \equiv \left|\begin{pmatrix}(\max)\\ [m_{1n}\cdots m_{kn}]\\ (\max)\end{pmatrix}\right\rangle,$$

we see that the Capelli identity (A-69) applied to $n-k$ also yields

$$a_{1\,2\ldots k}^{1\,2\ldots k}\bar{a}_{1\,2\ldots k}^{1\,2\ldots k}\left|\begin{pmatrix}(\max)\\ [m_{1n}\cdots m_{kn}0\cdots 0]\\ (\max)\end{pmatrix}\right\rangle$$
$$= \prod_{i=1}^{k}(p_{in}+k-n)\left|\begin{pmatrix}(\max)\\ [m_{1n}\cdots m_{kn}0\cdots 0]\\ (\max)\end{pmatrix}\right\rangle.$$

This result implies (since the vectors are normalized)

$$a_{1\,2\ldots k}^{1\,2\ldots k}\left|\begin{pmatrix}(\max)\\ [m_{1n}\cdots m_{kn}0\cdots 0]\\ (\max)\end{pmatrix}\right\rangle$$
$$= \left[\prod_{i=1}^{k}(p_{in}+k-n+1)\right]^{\frac{1}{2}}\left|\begin{pmatrix}(\max)\\ [m_{1n}+1\cdots m_{kn}+1\,0\cdots 0]\\ (\max)\end{pmatrix}\right\rangle.$$

Iterating this result λ times, then setting $m_{kn}=0$ followed by setting $\lambda = m_{kn}$ and replacing m_{in} by $m_{in}-m_{kn}$ $(i=1,2,\ldots,k)$, we find

$$\left|\begin{pmatrix}(\max)\\ [m_{1n}\cdots m_{kn}0\cdots 0]\\ (\max)\end{pmatrix}\right\rangle$$
$$= \left[\prod_{i=1}^{k}\frac{(p_{in}-p_{kn})!}{(p_{in}+k-n)!}\right]^{\frac{1}{2}}(a_{1\,2\ldots k}^{1\,2\ldots k})^{m_{kn}}\left|\begin{pmatrix}(\max)\\ [m_{1n}-m_{kn}\cdots m_{k-1\,n}-m_{kn}0\cdots 0]\\ (\max)\end{pmatrix}\right\rangle. \tag{A-73}$$

Using this result repeatedly for $k=n$, then $k=n-1,\ldots,k=1$, and making the appropriate shifts in labels, we obtain the measure factor given by Eq. (A-59).

Remarks. (*a*) The concept of a standard tableau was introduced by Young [18] in his work on invariant theory. (*b*) Weyl [19] introduced (single) standard tableaux with repeated integers in his work on the representations of the classical groups [hence, the polynomials (A-52) corresponding to the right tableau being maximal in (A-46) are often called *Weyl basis vectors*]. (*c*) Single Gel'fand patterns were introduced by Gel'fand and Tseitlin [20] to denote the (abstract) orthonormal basis vectors in a Hilbert space carrying a unitary irrep of $U(n)$. (*d*) The mapping from Gel'fand patterns to standard Young patterns was given by Baird and Biedenharn [27]. (*e*) Brody *et al.* [28] and Louck [29] were the first to note explicitly the role of double Gel'fand pattern basis vectors (A-58) in spanning the space of homogeneous polynomials of degree N as $[m]$ runs over all partitions of N (these works may be considered as the natural outgrowth of the earlier work of Moshinsky and collaborators [30–33] and of Baird and Biedenharn [27]). Biedenharn *et al.* [34] gave an abstract proof that the orthonormal double Gel'fand pattern basis vectors possessed the vector space properties described above. (*f*) The double standard tableau polynomials (A-52) (in which the bosons a_i^j are replaced by transcendentals $\langle x_i | u_j \rangle$ adjoined to an arbitrary field) were introduced by Doubilet *et al.* [35]; see also Désarménien *et al.* [35a]. In these papers one finds discussed many applications of these forms to problems in mathematics. Additional papers relating to the boson calculus and the development of their properties include Refs. [36–52a].

B. Multiplication of Boson Polynomials and the Factorization Lemma

The boson polynomials occurring in Eq. (5.80), and defined by

$$B\begin{pmatrix} & k+m' & \\ k+j & & k-j \\ & k+m & \end{pmatrix}(A) \equiv (\det A)^{k-j} D^{j}_{mm'}(A) \tag{B-1}$$

are interesting for several reasons:

(*a*) The set of polynomials corresponding to $2k =$ nonnegative integer, $2j = 0, 1, \ldots, 2k$, and $m, m' = j, j-1, \ldots, -j$, constitutes a basis of the space of homogeneous polynomials of degree $2k$ in the a_i^j. The dimension of this space is

$$\sum_j (2j+1)^2 = \binom{2k+3}{3}.$$

(*b*) The polynomials in this set satisfy a multiplication rule that is easily deduced from the corresponding multiplication rule, Eq. (3.189), for the D-functions (rotation matrices). Using this result, we are able to generalize Eq. (3.190) for the Wigner coefficients in a way that can be extended to the general unitary group, $SU(n)$. We shall see that this result is the analog, in the boson calculus, of the integration over group space of the product of three D-functions [Eq. (3.190)].

(*c*) The boson polynomials are an orthogonalized version of the double tableau functions introduced by Doubilet *et al.* [35] (for the case of Young frames having two rows), and we see that a general investigation of the *multiplication properties* of double tableau functions leads to a study of Wigner coefficients.

(*d*) The symmetry properties of the Wigner coefficients can be deduced systematically from these polynomials.

We begin by noting the relation

$$\delta_{j'j}\varepsilon_{j_1j_2j}(\det A)^{j_1+j_2-j}D^j_{m'm}(A) = \sum_{\substack{m_1'm_2'\\ m_1m_2}} C^{j_1j_2j'}_{m_1'm_2'm'}\, C^{j_1j_2j}_{m_1m_2m}\, D^{j_1}_{m_1'm_1}(A)D^{j_2}_{m_2'm_2}(A), \tag{B-2}$$

which follows immediately from the corresponding relation, Eq. (3.187), for the D-functions and the fact that the polynomials

$$(\det A)^{j_1+j_2-j}D^j_{m'm}(A) \tag{B-3}$$

are a basis for the homogeneous polynomials in the a^j_i of degree $2(j_1+j_2)$. Multiplying Eq. (B-2) through by $(\det A)^{k_1-j_1}(\det A)^{k_2-j_2}$, we obtain the following general relation between boson polynomials (B-1):

$$\begin{aligned}
&\delta_{j'j}\varepsilon_{j_1j_2j} B\begin{pmatrix} & k_1+k_2+m' & \\ k_1+k_2+j & & k_1+k_2-j \\ & k_1+k_2+m & \end{pmatrix}(A) \\
&\quad = \sum_{\substack{m_1'm_2'\\ m_1m_2}} C^{j_1j_2j'}_{m_1'm_2'm'}\, C^{j_1j_2j}_{m_1m_2m}\, B\begin{pmatrix} & k_2+m_2' & \\ k_2+j_2 & & k_2-j_2 \\ & k_2+m_2 & \end{pmatrix}(A) \\
&\qquad \times B\begin{pmatrix} & k_1+m_1' & \\ k_1+j_1 & & k_1-j_1 \\ & k_1+m_1 & \end{pmatrix}(A).
\end{aligned} \tag{B-4}$$

The inversion of this relation, using the orthogonality of the Wigner coefficients, yields the *product law* for boson polynomials:

$$\begin{aligned}
&B\begin{pmatrix} & k_2+m_2' & \\ k_2+j_2 & & k_2-j_2 \\ & k_2+m_2 & \end{pmatrix}(A) \times B\begin{pmatrix} & k_1+m_1' & \\ k_1+j_1 & & k_1-j_1 \\ & k_1+m_1 & \end{pmatrix}(A) \\
&\quad = \sum_j C^{j_1j_2j}_{m_1',m_2',m_1'+m_2'}\, C^{j_1j_2j}_{m_1,m_2,m_1+m_2} \\
&\qquad \times B\begin{pmatrix} & k_1+k_2+m_1'+m_2' & \\ k_1+k_2+j & & k_1+k_2-j \\ & k_1+k_2+m_1+m_2 & \end{pmatrix}(A).
\end{aligned} \tag{B-5}$$

Let us now reinterpret this result in terms of the normalized *boson state vectors*:

$$\left| \begin{matrix} & k+m' & \\ k+j & & k-j \\ & k+m & \end{matrix} \right\rangle = \left[\frac{(2j+1)}{(k-j)!(k+j+1)!} \right]^{\frac{1}{2}} B \begin{pmatrix} & k+m' & \\ k+j & & k-j \\ & k+m & \end{pmatrix} (A)|0\rangle. \tag{B-6}$$

Operating with Eq. (B-5) on the vacuum ket $|0\rangle$ and using the orthonormality of the boson state vectors, we obtain the nonzero matrix elements of the boson polynomials between arbitrary state vectors:

$$\left\langle \begin{matrix} & k_1+k_2+m' & \\ k_1+k_2+j & & k_1+k_2-j \\ & k_1+k_2+m & \end{matrix} \right| B \begin{pmatrix} & k_2+m'_2 & \\ k_2+j_2 & & k_2-j_2 \\ & k_2+m_2 & \end{pmatrix} (A) \left| \begin{matrix} & k_1+m'_1 & \\ k_1+j_1 & & k_1-j_1 \\ & k_1+m_1 & \end{matrix} \right\rangle$$

$$= \left[\frac{(k_1+k_2-j)!(k_1+k_2+j+1)!(2j_1+1)}{(k_1-j_1)!(k_1+j_1+1)!(2j+1)} \right]^{\frac{1}{2}} C^{j_1j_2j}_{m_1'm_2'm'} C^{j_1j_2j}_{m_1m_2m}. \tag{B-7}$$

Equation (B-7) generalizes our original formula for the Wigner coefficients, Eq. (5.82). To show this, we first note from Eqs. (5.73), (5.85), and (B-1) that

$$B \begin{pmatrix} & 2j_1 & \\ 2j_1 & & 0 \\ & j_1+m_1 & \end{pmatrix} (A) = [(2j_1)!]^{\frac{1}{2}} P_{j_1m_1}(\mathbf{a}^1), \tag{B-8}$$

$$B \begin{pmatrix} & 0 & \\ 2j_2 & & 0 \\ & j_2+m_2 & \end{pmatrix} (A) = [(2j_2)!]^{\frac{1}{2}} P_{j_2m_2}(\mathbf{a}^2). \tag{B-9}$$

Thus, accounting for the normalization factors, Eq. (5.82) may be written

$$\left\langle \begin{matrix} & 2j_1 & \\ j_1+j_2+j & & j_1+j_2-j \\ & j_1+j_2+m & \end{matrix} \right| B \begin{pmatrix} & 0 & \\ 2j_2 & & 0 \\ & j_2+m_2 & \end{pmatrix} (A) \left| \begin{matrix} & 2j_1 & \\ 2j_1 & & 0 \\ & j_1+m_1 & \end{matrix} \right\rangle$$

$$= [(2j_2)!]^{\frac{1}{2}} C^{j_1j_2j}_{m_1m_2m}. \tag{B-10}$$

The left-hand side of this expression is the specialization of Eq. (B-7) obtained by choosing $k_1 = j_1$, $k_2 = j_2$, $m'_1 = j_1$, $m'_2 = -j_2$, and $m' = j_1 - j_2$. The right-hand side of Eq. (B-7) then reduces to

$$\left[\frac{(j_1+j_2-j)!(j_1+j_2+j+1)!}{(2j_1)!(2j+1)} \right]^{\frac{1}{2}} C^{j_1j_2j}_{j_1,-j_2,j_1-j_2} C^{j_1j_2j}_{m_1m_2m},$$

which must equal the right-hand side of Eq. (B-10). Thus, not only is Eq. (B-10) obtained as a special case of the general result, Eq. (B-7), but we obtain also the particular Wigner coefficient

$$C^{j_1 j_2 j}_{j_1, -j_2, j_1 - j_2} = \left[\frac{(2j+1)(2j_1)!(2j_2)!}{(j_1+j_2-j)!(j_1+j_2+j+1)!}\right]^{\frac{1}{2}} \tag{B-11}$$

without having to evaluate it directly.

We call the result expressed by Eq. (B-7) the *factorization lemma for bosons.*

Remarks. (*a*) The extension of Eqs. (B-7) and (B-10) to the general unitary group $SU(n)$ has been a principal tool in the determination of the Wigner coefficients of $SU(n)$ (Biedenharn and Louck, and collaborators [14, 27, 34, 40, 44, 46, 49, 51, 53–58]; Moshinsky and collaborators [28, 30–33, 41]).

(*b*) The results expressed by Eqs. (B-8) and (B-9) may also be obtained by projecting the boson variables with $e_1 = \left(\begin{smallmatrix}1 & 0\\ 0 & 0\end{smallmatrix}\right)$ and $e_2 = \left(\begin{smallmatrix}0 & 0\\ 0 & 1\end{smallmatrix}\right)$; that is, $Ae_1 = (\mathbf{a}^1\, \mathbf{0})$ and $Ae_2 = (\mathbf{0}\, \mathbf{a}^2)$:

$$B\begin{pmatrix} & k+m' & \\ k+j & & k-j \\ & k+m & \end{pmatrix}(Ae_1) = \delta_{kj}\delta_{jm'}B\begin{pmatrix} & 2j & \\ 2j & & 0 \\ & j+m & \end{pmatrix}(A),$$

$$B\begin{pmatrix} & k+m' & \\ k+j & & k-j \\ & k+m & \end{pmatrix}(Ae_2) = \delta_{kj}\delta_{j,-m'}B\begin{pmatrix} & 0 & \\ 2j & & 0 \\ & j+m & \end{pmatrix}(A).$$

C. Finite Symmetry Group of the Boson Polynomials and the Wigner Coefficients

In this Appendix we obtain a finite group of transformations of the boson polynomials—hence, the rotation matrices—and use these symmetries to derive the closely related symmetries of the Wigner coefficients.

A finite group of transformations of the rotation matrices and the boson polynomials. To develop these symmetry properties in their fullest generality (independently of parametrizations), we introduce a 2×2 matrix Z:

$$Z = \begin{pmatrix} a & b \\ c & d \end{pmatrix}. \tag{C-1}$$

The notation $F(Z) \equiv F(a, b, c, d)$ denotes an arbitrary polynomial (with coefficients in $\mathbb{C}$) defined on the indeterminates a, b, c, d.

Consider now the new polynomials $\mathscr{R}F$, $\mathscr{C}F$, and $\mathscr{T}F$ defined by

Row interchange: $$(\mathscr{R}F)\begin{pmatrix} a & b \\ c & d \end{pmatrix} = F\begin{pmatrix} c & d \\ a & b \end{pmatrix},$$

Column interchange: $$(\mathscr{C}F)\begin{pmatrix} a & b \\ c & d \end{pmatrix} = F\begin{pmatrix} b & a \\ d & c \end{pmatrix},$$

Transposition: $$(\mathscr{T}F)\begin{pmatrix} a & b \\ c & d \end{pmatrix} = F\begin{pmatrix} a & c \\ b & d \end{pmatrix}. \tag{C-2}$$

One should note very carefully the product rule for these operations; for example,

$$(\mathscr{C}\mathscr{T}F)\begin{pmatrix} a & b \\ c & d \end{pmatrix} = (\mathscr{T}F)\begin{pmatrix} b & a \\ d & c \end{pmatrix} = F\begin{pmatrix} b & d \\ a & c \end{pmatrix}.$$

The operators $\mathscr{R}$, $\mathscr{C}$, and $\mathscr{T}$ generate a group of order eight, the elements of this group being

$$\mathscr{H} = \{1, \mathscr{R}, \mathscr{C}, \mathscr{T}, \mathscr{R}\mathscr{C} = \mathscr{C}\mathscr{R}, \mathscr{T}\mathscr{R} = \mathscr{C}\mathscr{T}, \mathscr{T}\mathscr{C} = \mathscr{R}\mathscr{T}, \mathscr{R}\mathscr{C}\mathscr{T}\}, \tag{C-3}$$

where we note that

$$\begin{aligned} &\mathscr{R}^2 = \mathscr{C}^2 = \mathscr{T}^2 = 1, \\ &\mathscr{T}\mathscr{R}\mathscr{C} = \mathscr{T}\mathscr{C}\mathscr{R} = \mathscr{R}\mathscr{C}\mathscr{T} = \mathscr{C}\mathscr{R}\mathscr{T}, \\ &\mathscr{R}\mathscr{T}\mathscr{C} = \mathscr{C}\mathscr{T}\mathscr{R} = \mathscr{T}. \end{aligned} \tag{C-4}$$

[This group is clearly isomorphic to a subgroup of the symmetric group S_4; a realization of the isomorphism is $\mathscr{R} \to (13)(24)$, $\mathscr{C} \to (12)(34)$, $\mathscr{T} \to (1)(4)(23)$.]

The action of the above operators on the rotation matrices is determined directly from the definition, Eq. (5.85), to be[1]

$$\begin{aligned} \mathscr{R}D^j_{mm'} &= D^j_{-mm'}, \\ \mathscr{C}D^j_{mm'} &= D^j_{m,-m'}, \\ \mathscr{T}D^j_{mm'} &= D^j_{m'm}. \end{aligned} \tag{C-5}$$

In its action on the rotation matrices, the group of operators, $\mathscr{H}$, induces the eight transformations on the quantum labels (m, m') corresponding to any number of sign changes and transposition.

There is still another operation – complex conjugation – which is significant in the study of the rotation matrices. If U is unitary unimodular, it has the form

$$U = \begin{pmatrix} a & b \\ -b^* & a^* \end{pmatrix}, \qquad |a|^2 + |b|^2 = 1. \tag{C-6}$$

In this case

$$U^* = \begin{pmatrix} a^* & b^* \\ -b & a \end{pmatrix}. \tag{C-7}$$

[1] The symmetries of the rotation matrices, Eqs. (C-5) and (C-9), were first given in this form by Louck and Galbraith [59].

This result motivates us to introduce the operation $\mathscr{K}$, called *conjugation*, which we define by

$$(\mathscr{K}F)\begin{pmatrix} a & b \\ c & d \end{pmatrix} \equiv F\begin{pmatrix} d & -c \\ -b & a \end{pmatrix}. \tag{C-8}$$

Adjoining $\mathscr{K}$ to the generators $\mathscr{R}, \mathscr{C}, \mathscr{T}$, we obtain the group $\{\mathscr{H}, \mathscr{H}\mathscr{K}, \mathscr{H}\mathscr{K}\mathscr{R}, \mathscr{H}\mathscr{K}\mathscr{R}\mathscr{K}\}$ containing 32 elements (see Ref. [59]).

The action of $\mathscr{K}$ on $D^j_{mm'}$ is determined directly from Eq. (5.85) to be

$$\mathscr{K} D^j_{mm'} = (-1)^{m-m'} D^j_{-m,-m'}. \tag{C-9}$$

Note that this result is valid for arbitrary arguments (a, b, c, d) when using the general definition (C-8) in the form given by Eq. (5.85). If in addition (a, b, c, d) are the elements of a unimodular unitary matrix U, then one has

$$[D^j_{mm'}(U)]^* = D^j_{mm'}(U^*) = (-1)^{m-m'} D^j_{-m,-m'}(U). \tag{C-10}$$

Accounting for the transformation properties of the determinant that appears in the definition (B-1), we obtain the following transformations of the boson polynomials:

$$(\mathscr{R}B)\begin{pmatrix} & k+m' & \\ k+j & & k-j \\ & k+m & \end{pmatrix} = (-1)^{k-j} B\begin{pmatrix} & k-m' & \\ k+j & & k-j \\ & k+m & \end{pmatrix},$$

$$(\mathscr{C}B)\begin{pmatrix} & k+m' & \\ k+j & & k-j \\ & k+m & \end{pmatrix} = (-1)^{k-j} B\begin{pmatrix} & k+m' & \\ k+j & & k-j \\ & k-m & \end{pmatrix},$$

$$(\mathscr{T}B)\begin{pmatrix} & k+m' & \\ k+j & & k-j \\ & k+m & \end{pmatrix} = B\begin{pmatrix} & k+m & \\ k+j & & k-j \\ & k+m' & \end{pmatrix},$$

$$(\mathscr{K}B)\begin{pmatrix} & k+m' & \\ k+j & & k-j \\ & k+m & \end{pmatrix} = (-1)^{m-m'} B\begin{pmatrix} & k-m' & \\ k+j & & k-j \\ & k-m & \end{pmatrix}. \tag{C-11}$$

A finite group of symmetries of the Wigner coefficients. Discussions of the symmetry properties of the Wigner coefficients are given in the original work of Wigner [60] and Racah [61], the result being a symmetry group of 12 elements.

It came somewhat as a surprise to learn from Regge's work [62] that, in fact, the symmetry group is much larger—a 72-element group. An excellent exposition of Regge's observation is given by Bargmann [9]. (We discuss Regge's method in Appendix D.)

We shall present here a different approach to the Regge symmetries, demonstrating that their origin is, in fact, due to the basic $\mathscr{R}$, $\mathscr{C}$, $\mathscr{T}$, $\mathscr{K}$ symmetries of the rotation matrices discussed above.

It is convenient to introduce the notation

$$(F, F') \equiv \langle 0|F^*(\bar{A})F'(A)|0\rangle \tag{C-12}$$

to denote the scalar product[1] of two polynomials in the bosons a_i^j (with coefficients in $\mathbb{C}$). It then follows that[2]

$$(\mathscr{G}F, \mathscr{G}F') = (F, F') \tag{C-13}$$

for $\mathscr{G} = \mathscr{R}, \mathscr{C}, \mathscr{T}$, or $\mathscr{K}$; that is, the operators $\mathscr{R}$, $\mathscr{C}$, $\mathscr{T}$, and $\mathscr{K}$ are unitary operators with respect to the boson scalar product (C-12).

Consider now the Wigner coefficients given by the boson scalar product [see Eq. (5.83) and Appendix B]:

$$\begin{aligned} C^{j_1j_2j}_{m_1m_2m} &= \langle j_1j_2;jm|j_1m_1;j_2m_2\rangle \\ &= \left[\frac{2j+1}{(j_1+j_2-j)!(j_1+j_2+j+1)!(2j_1)!(2j_2)!}\right]^{\frac{1}{2}} \\ &\quad\times \langle 0|(\det \bar{A})^{j_1+j_2-j} D^j_{m,j_1-j_2}(\bar{A}) D^{j_2}_{m_2,-j_2}(A) D^{j_1}_{m_1j_1}(A)|0\rangle. \end{aligned} \tag{C-14}$$

The transformation properties of the various factors in the scalar product are listed as follows (for reasons discussed below, we momentarily set aside the discussion of $\mathscr{K}$):

$$\begin{aligned} \mathscr{R}D^j_{m,j_1-j_2} &= D^j_{-m,j_1-j_2} \\ \mathscr{R}(D^{j_2}_{m_2,-j_2}D^{j_1}_{m_1j_1}) &= D^{j_2}_{-m_2,-j_2}D^{j_1}_{-m_1,j_1}; \end{aligned} \tag{C-15}$$

$$\begin{aligned} \mathscr{C}D^j_{m,j_1-j_2} &= D^j_{m,j_2-j_1} \\ \mathscr{C}(D^{j_2}_{m_2,-j_2}D^{j_1}_{m_1j_1}) &= D^{j_1}_{m_1,-j_1}D^{j_2}_{m_2j_2}; \end{aligned} \tag{C-16}$$

$$\begin{aligned} \mathscr{T}D^j_{m,j_1-j_2} &= D^j_{j_1-j_2,m} \\ \mathscr{T}(D^{j_2}_{m_2,-j_2}D^{j_1}_{m_1j_1}) &= D^{j_2}_{-j_2,m_2}D^{j_1}_{j_1m_1} \\ &= D^{\frac{1}{2}(j_1+j_2-m_1-m_2)}_{\frac{1}{2}(j_1-j_2-m_1+m_2),-j_2} D^{\frac{1}{2}(j_1+j_2+m_1+m_2)}_{\frac{1}{2}(j_1-j_2+m_1-m_2),j_1}. \end{aligned} \tag{C-17}$$

---

[1] If F is the polynomial given by $F(A) = \sum_{(\alpha)} c(\alpha) \prod_{ij} (a_i^j)^{\alpha_i^j}$, then $F^*(\bar{A}) = \sum_{(\alpha)} c^*(\alpha) \prod_{ij} (\bar{a}_i^j)^{\alpha_i^j}$. Note, however, that the boson polynomials are real, that is, $c^*(\alpha) = c(\alpha)$.

[2] This unitary property is a consequence of the fact that the transformed bosons b_i^j and their conjugates satisfy the same commutation relations as the a_i^j. For example, the interchange of rows in the boson matrix A defines new bosons $b_1^1 = a_2^1$, $b_2^1 = a_1^1$, $b_1^2 = a_2^2$, $b_2^2 = a_1^2$, satisfying $[\bar{b}_i^j, b_k^l] = \delta^{jl}\delta_{ik}$.

(The last relation follows directly from the definition of $\mathscr{T}$ and the forms of the particular D-functions occurring in the product.)

Using the unitary property of $\mathscr{R}, \mathscr{C}$, and $\mathscr{T}$, the above relations, and the fact that $\mathscr{R}$ and $\mathscr{C}$ change the sign of the determinant $ad-bc$, while $\mathscr{T}$ leaves this determinant invariant, we immediately obtain the first three symmetry relations listed below (in order to have the complete list of the generators of the symmetries in one place, the symmetry property associated with $\mathscr{K}$ is included here and proved subsequently):

$$\mathscr{R}: C^{j_1 j_2 j}_{m_1 m_2 m} = (-1)^{j_1+j_2-j}\, C^{j_1 j_2 j}_{-m_1, -m_2, -m},$$

$$\mathscr{C}: C^{j_1 j_2 j}_{m_1 m_2 m} = (-1)^{j_1+j_2-j}\, C^{j_2 j_1 j}_{m_2 m_1 m},$$

$$\mathscr{T}: C^{j_1 j_2 j}_{m_1, m_2, m_1+m_2} = C^{\frac{1}{2}(j_1+j_2+m_1+m_2), \frac{1}{2}(j_1+j_2-m_1-m_2), j}_{\frac{1}{2}(j_1-j_2+m_1-m_2), \frac{1}{2}(j_1-j_2-m_1+m_2), j_1-j_2},$$

$$\mathscr{K}: C^{j_1 j_2 j}_{m_1 m_2 m} = (-1)^{j_2+m_2}[(2j+1)/(2j_1+1)]^{\frac{1}{2}}\, C^{j j_2 j_1}_{-m, m_2, -m_1}. \qquad \text{(C-18)}$$

The unitary operators $\mathscr{R}, \mathscr{C}, \mathscr{T}$ standing to the left induce the symmetries of the Wigner coefficients on the right. The group $\mathscr{H}$ of symmetries induces eight corresponding symmetries of the Wigner coefficients (counting the identity).

An even more remarkable fact is that these eight symmetries of the Wigner coefficients are precisely those of the discretized functions D^j_{m, j_1-j_2} [see Eq. (5.89)] under the action of the transformations of $\mathscr{H}$; that is, *these symmetries of the D-functions extend exactly to the discretized D-functions.*

Let us consider now the conjugation operator $\mathscr{K}$. We have already observed that $\mathscr{K}$ is unitary with respect to the boson scalar product. If we use the property (C-9) in the boson scalar product, the result that obtains is

$$C^{j_2 j_1 j}_{-m_2, -m_1, -m} = C^{j_1 j_2 j}_{m_1 m_2 m}, \qquad \text{(C-19)}$$

which is just the result of combining the first two symmetries in Eqs. (C-18). *The operation $\mathscr{K}$ does not lead to an independent relation between Wigner coefficients when implemented in the boson scalar product.*

There is, however, a new symmetry of the Wigner coefficients corresponding to the conjugation operation $\mathscr{K}$, but it is not realized in the boson polynomial structures, where a complex conjugation property analogous to Eq. (C-10) is not defined – that is, $[D^j_{m'm}(A)]^*$ is not defined for bosons.

To utilize the property (C-10), one uses a scalar product defined in terms of integration over the unit sphere in four-space (see Chapter 3, Section 9):

$$(f, g) = \int d\Omega [f(U)]^* g(U), \qquad \text{(C-20)}$$

where $U = U(\psi, \hat{n})$ [see Eqs. (2.27) and (2.28)] and the integration is carried out over all points of the unit sphere S^3.

[With respect to this scalar product the complex conjugation operator defined by $\mathscr{K}f = f^*$ is *anti-unitary*: $(\mathscr{K}f, \mathscr{K}g) = (f, g)^*$. The operator $\mathscr{T}$ is unitary with respect to the scalar product (C-20), but $(\mathscr{R}f, \mathscr{R}g)$ and $(\mathscr{C}f, \mathscr{C}g)$ are not defined since $(\mathscr{R}f)(U) = f(U')$ and $(\mathscr{C}f)(U) = f(U'')$, where for each

$U \in SU(2)$, one has $\det U' = \det U'' = -1$, so that $U', U'' \in U(2)$. A uniform derivation of the symmetries (C-18) may be given by taking the matrix Z in Eq. (C-1) to belong to the group $U(2)$ and using the scalar product defined by Eq. (C-24) below. In this case the operators $\mathscr{R}, \mathscr{C}, \mathscr{T}, \mathscr{K}$ constitute a group in which $\mathscr{R}$, $\mathscr{C}$, and $\mathscr{T}$ are unitary and $\mathscr{K}$ is anti-unitary (see Topic 1, Section 8, RWA for a discussion of the theory of corepresentations of groups containing both unitary and anti-unitary operators).]

In terms of the notation (C-20), the integral over three rotation matrices, Eq. (3.190), is expressed by

$$\frac{2\pi^2}{2j+1} C^{j_1 j_2 j}_{m_1 m_2 m} C^{j_1 j_2 j}_{m_1' m_2' m'} = (D^j_{mm'}, D^{j_1}_{m_1 m_1'} D^{j_2}_{m_2 m_2'}). \tag{C-21}$$

Using the complex conjugation property, Eq. (C-10), we obtain

$$\frac{(-1)^{j+m+j+m'}}{2j+1} C^{j_1 j_2 j}_{m_1 m_2 m} C^{j_1 j_2 j}_{m_1' m_2' m'}$$
$$= \frac{1}{2\pi^2} \int d\Omega\, D^j_{-m,-m'}(U) D^{j_2}_{m_2 m_2'}(U) D^{j_1}_{m_1 m_1'}(U). \tag{C-22}$$

Since there are $3! = 6$ arrangements of the factors in the integral, we see that there are correspondingly six identities relating the *product* of Wigner coefficients on the left. In these six relations, we can make choices of m_1', m_2', m' that lead to simple coefficients. In this way, we can obtain six relations between the Wigner coefficients corresponding to the six arrangements of

$$\begin{pmatrix} j_1 \\ m_1 \end{pmatrix} \begin{pmatrix} j_2 \\ m_2 \end{pmatrix} \begin{pmatrix} j \\ m \end{pmatrix}$$

with the appropriate sign changes on the projection quantum numbers. Of these six relations, only one is new in the sense that we can generate the remaining ones from it and the first three relations of (C-18). The new relation may be chosen as either the one that interchanges j and j_1 or the one that interchanges j and j_2.

Thus, for example, we obtain

$$\frac{(-1)^{j_1-m_1-j_2-m_2}}{2j+1} C^{j_1 j_2 j}_{m_1 m_2 m} C^{j_1 j_2 j}_{j_1, -j_2, j_1-j_2}$$
$$= \frac{(-1)^{j_1-m_1}}{2j_1+1} C^{j j_2 j_1}_{-m, m_2, -m_1} C^{j j_2 j_1}_{j_2-j_1, -j_2, -j_1}.$$

We use the explicit form of the Wigner coefficients to show

$$C^{j j_2 j}_{j_2-j_1, -j_2, -j_1} = \left[\frac{2j_1+1}{2j+1}\right]^{\frac{1}{2}} C^{j_1 j_2 j}_{j_1, -j_2, j_1-j_2},$$

thus obtaining

$$\mathscr{K}: C^{j_1 j_2 j}_{m_1 m_2 m} = (-1)^{j_2+m_2}\left[\frac{2j+1}{2j_1+1}\right]^{\frac{1}{2}} C^{j j_2 j_1}_{-m, m_2, -m_1}, \tag{C-23}$$

where the $\mathscr{K}$ standing to the left indicates that this symmetry has its origin in the complex conjugation properties of the D-functions.

Ignoring phase changes and dimensionality factors, the four symmetries (C-18) generate a group of order 72, as discussed in Chapter 3, Section 12.

Remark. We have used the boson scalar product in deriving the first three symmetry relations in Eqs. (C-18) and the integration over the unit sphere S^3 in deriving the last one. This lack of uniformity in the derivation can be removed by using the scalar product appropriate to functions $\{f\}$ defined on the elements of the unitary group $U(2)$ (unimodular restriction removed):

$$(f, g) = \int_0^{2\pi} d\chi \int d\Omega [f(U)]^* g(U). \tag{C-24}$$

[We discuss the representations of the group $U(2)$ in Appendix E.]

The generalization of the integral over three rotation matrices, Eq. (3.190), is given by

$$\int_0^{2\pi} d\chi \int d\Omega \left[D\begin{pmatrix} & \mu''_{11} & \\ m''_{12} & & m''_{22} \\ & m''_{11} & \end{pmatrix}(U)\right]^*$$

$$\times D\begin{pmatrix} & \mu'_{11} & \\ m'_{12} & & m'_{22} \\ & m'_{11} & \end{pmatrix}(U) D\begin{pmatrix} & \mu_{11} & \\ m_{12} & & m_{22} \\ & m_{11} & \end{pmatrix}(U)$$

$$= \frac{4\pi^3}{2j''+1} C^{jj'j''}_{mm'm''} C^{jj'j''}_{\mu\mu'\mu''}, \tag{C-25}$$

where

$$j = (m_{12} - m_{22})/2,$$

$$m = m_{11} - (m_{12} + m_{22})/2,$$

$$\mu = \mu_{11} - (m_{12} + m_{22})/2,$$

with corresponding definitions for (j', m', μ') and (j'', m'', μ'').

D. Generating Functions for Wigner and Racah Coefficients

Generating function for the Wigner coefficients. The derivation of generating functions for Wigner coefficients is based on the construction of invariants (a method used extensively by van der Waerden [63], Kramers [64], and Brinkman [65]). The key formula is Eq. (3.335) of Chapter 3.

Here we apply this invariant construction to the boson polynomials [see Eq. (5.33)],

$$P_{j_i m_i}(\mathbf{a}^i), \qquad i = 1, 2, 3, \tag{D-1}$$

where

$$\mathbf{a}^i = (a_1^i, a_2^i). \tag{D-2}$$

Observe that under the action of $\mathscr{T}_U$ [see Eq. (5.42)] the boson polynomials are tensor operators:

$$\begin{aligned}\mathscr{T}_U : P_{j_i m_i}(\mathbf{a}^i) &\to \mathscr{T}_U P_{j_i m_i}(\mathbf{a}^i) \mathscr{T}_{U^{-1}} \\ &= \sum_{m_i'} D^{j_i}_{m_i' m_i}(U) P_{j_i m_i'}(\mathbf{a}^i).\end{aligned} \tag{D-3}$$

Thus, the construction (3.335) is applicable.

The rotational invariant

$$\sum_{m_1 m_2 m_3} \begin{pmatrix} j_1 & j_2 & j_3 \\ m_1 & m_2 & m_3 \end{pmatrix} P_{j_1 m_1}(\mathbf{a}^1) P_{j_2 m_2}(\mathbf{a}^2) P_{j_3 m_3}(\mathbf{a}^3) \tag{D-4}$$

may be summed to a *monomial form*, which is defined on the three rotational invariants

$$a_{12}^{12}, a_{12}^{31}, a_{12}^{23}, \tag{D-5}$$

where

$$a_{12}^{ij} \equiv \det \begin{pmatrix} a_1^i & a_1^j \\ a_2^i & a_2^j \end{pmatrix}. \tag{D-6}$$

To prove this result, we use:

(*a*) A theorem of Weyl [26, p. 45], which states that every polynomial invariant of the form (D-4) is a *polynomial form* on the three invariants given by (D-5).

(*b*) The homogeneity properties of the form (D-4) – it is homogeneous of degree $2j_i$ in (a_1^i, a_2^i), and its total degree is $2(j_1 + j_2 + j_3)$.

These two properties suffice to show that the polynomial (D-4) may also be written

$$\#(a_{12}^{12})^{j_1+j_2-j_3}(a_{12}^{31})^{j_3+j_1-j_2}(a_{12}^{23})^{j_2+j_3-j_1}, \tag{D-7}$$

where # denotes a constant.

The constant, #, appearing in (D-7) may be evaluated by setting $a_2^1 = a_1^2 = 0$ in Eq. (D-4), which thereby reduces to the monomial term corresponding to $m_1 = j_1$, $m_2 = -j_2$, $m_3 = -j_1 + j_2$. Using the special 3-j coefficient for these m-quantum numbers [see Eq. (B-11)] yields the result for #.

We thus obtain

$$\sum_{m_1 m_2 m_3} \begin{pmatrix} j_1 & j_2 & j_3 \\ m_1 & m_2 & m_3 \end{pmatrix} P_{j_1 m_1}(\mathbf{a}^1) P_{j_2 m_2}(\mathbf{a}^2) P_{j_3 m_3}(\mathbf{a}^3)$$

$$= [(j_1 + j_2 + j_3 + 1)!]^{-\frac{1}{2}} \frac{(a_{12}^{12})^{j_1 + j_2 - j_3} (a_{12}^{31})^{j_3 + j_1 - j_2} (a_{12}^{23})^{j_2 + j_3 - j_1}}{[(j_1 + j_2 - j_3)!(j_3 + j_1 - j_2)!(j_2 + j_3 - j_1)!]^{\frac{1}{2}}}. \tag{D-8}$$

Regge [62] observed that one can multiply Eq. (D-8) by

$$\Phi_{j_1 j_2 j_3}(\mathbf{a}_3) \equiv \frac{(a_3^3)^{j_1 + j_2 - j_3} (a_3^2)^{j_3 + j_1 - j_2} (a_3^1)^{j_2 + j_3 - j_1}}{[(j_1 + j_2 - j_3)!(j_3 + j_1 - j_2)!(j_2 + j_3 - j_1)!]^{\frac{1}{2}}}, \tag{D-9}$$

sum over all $(j) = (j_1, j_2, j_3)$ such that the sum

$$j_1 + j_2 + j_3 = J \tag{D-10}$$

is fixed (at a nonnegative integral value J), and thus obtain $[(J+1)!]^{-\frac{1}{2}}(\det A)^J/J!$ from the right-hand side of (D-8). (One uses the trinomial expansion in obtaining this result.) Here A denotes the 3×3 matrix boson:

$$A \equiv \begin{pmatrix} a_1^1 & a_1^2 & a_1^3 \\ a_2^1 & a_2^2 & a_2^3 \\ a_3^1 & a_3^2 & a_3^3 \end{pmatrix}. \tag{D-11}$$

The result obtained above may be written more concisely as

$$\frac{(\det A)^J}{J!} = [(J+1)!]^{\frac{1}{2}} \sum_{\boxed{\alpha}} R\boxed{\alpha} \prod_{ij=1}^{3} \frac{(a_i^j)^{\alpha_i^j}}{[(\alpha_i^j)!]^{\frac{1}{2}}}, \tag{D-12}$$

where we have introduced the combinatorial 3×3 array [see Eq. (5.85)], denoted by $\boxed{\alpha}$, of nonnegative integers $\{\alpha_i^j\}$ such that the row and column sums[1] are J

$$\begin{array}{|ccc|c} \hline \alpha_1^1 & \alpha_1^2 & \alpha_1^3 & J \\ \alpha_2^1 & \alpha_2^2 & \alpha_2^3 & J \\ \alpha_3^1 & \alpha_3^2 & \alpha_3^3 & J. \\ \hline J & J & J & \end{array} \tag{D-13}$$

[1] The form (D-12) is a special case of a more general result, analogous to Eq. (5.85), giving the irreducible representations of the unitary group $U(3)$ (Louck [14], Holman [48], Louck and Biedenharn [51]).

The summation in Eq. (D-12) is to be carried out over all arrays satisfying the constraints. We have also used the notation $R\boxed{\alpha}$ to denote the coefficient appearing in the expansion (D-12). In terms of the (j_i, m_i) notation, we have

$$R\begin{array}{|ccc|} j_1+m_1 & j_2+m_2 & j_3+m_3 \\ j_1-m_1 & j_2-m_2 & j_3-m_3 \\ j_2+j_3-j_1 & j_3+j_1-j_2 & j_1+j_2-j_3 \end{array} = \begin{pmatrix} j_1 & j_2 & j_3 \\ m_1 & m_2 & m_3 \end{pmatrix}. \qquad \text{(D-14)}$$

Equation (D-12) (Regge's result) puts the determinantal symmetries in evidence: The operations on the matrix boson A of interchanging rows, columns, and transposing are transferred, in an obvious way, by the symmetry of the form $\prod_{ij}(a_i^j)^{\alpha_i^j}/[(\alpha_i^j)!]^{\frac{1}{2}}$, to the coefficients $R\boxed{\alpha}$ themselves.

Schwinger [8] also found the result,[1] Eq. (D-12), in the exponential form obtained by multiplying Eq. (D-12) by λ^J and summing all $J = 0, 1, \ldots$:

$$e^{\lambda \det A} = \sum_J [(J+1)!]^{\frac{1}{2}} \lambda^J \sum_{\boxed{\alpha}} R\boxed{\alpha} \prod_{ij} \frac{(a_i^j)^{\alpha_i^j}}{[(\alpha_i^j)!]^{\frac{1}{2}}}. \qquad \text{(D-15)}$$

The exponential function, $e^{\lambda \det A}$, is called the *generating function* for the 3-j coefficients.

Generating function for the Racah coefficients. A generating function for the set of Racah coefficients has been given by Schwinger [8]. We derive Schwinger's relation by appealing directly to the explicit form (3.292) for the Racah coefficients. For this purpose, it is convenient to rewrite Eq. (3.292) by introducing the (invertible) transformation from the parameters a, b, c, d, e, f and the summation index z to the nonnegative integer parameters $\alpha_1, \alpha_2, \alpha_3, \alpha_4, \beta_1, \beta_2, \beta_3$ given by

$$\begin{aligned} \alpha_1 &= z - a - b - e, \\ \alpha_2 &= z - c - d - e, \\ \alpha_3 &= z - a - c - f, \\ \alpha_4 &= z - b - d - f, \\ \beta_1 &= a + d + e + f - z, \\ \beta_2 &= b + c + e + f - z, \\ \beta_3 &= a + b + c + d - z. \end{aligned} \qquad \text{(D-16)}$$

These relations imply the equality of the two 4×3 matrices given by

[1] Schwinger failed, however, to point out that all the 72 symmetries of the 3-j coefficients are implied by Eq. (D-15).

$$\begin{pmatrix} \alpha_1+\beta_1 & \alpha_1+\beta_2 & \alpha_1+\beta_3 \\ \alpha_2+\beta_1 & \alpha_2+\beta_2 & \alpha_2+\beta_3 \\ \alpha_3+\beta_1 & \alpha_3+\beta_2 & \alpha_3+\beta_3 \\ \alpha_4+\beta_1 & \alpha_4+\beta_2 & \alpha_4+\beta_3 \end{pmatrix} = \begin{pmatrix} d+f-b & c+f-a & c+d-e \\ a+f-c & b+f-d & a+b-e \\ d+e-c & b+e-a & b+d-f \\ a+e-b & c+e-d & a+c-f \end{pmatrix}$$

$$\equiv K. \tag{D-17}$$

In terms of the parameters $(\alpha) = (\alpha_1, \alpha_2, \alpha_3, \alpha_4)$ and $(\beta) = (\beta_1, \beta_2, \beta_3)$, the 6-$j$ coefficient [see Eq. (3.292)] is expressed as

$$\begin{Bmatrix} a & b & e \\ d & c & f \end{Bmatrix} = \left[\frac{\prod_{ij} k_{ij}!}{\prod_i (k_i+1)!}\right]^{\frac{1}{2}} \sum_{\substack{(\alpha)(\beta) \\ \alpha_i+\beta_j=k_{ij}}} \frac{(-1)^{\sum\alpha_i+\sum\beta_j}(\sum\alpha_i + \sum\beta_j + 1)!}{\prod_{ij}\alpha_i!\,\beta_j!}, \tag{D-18}$$

where

(*a*) $\prod_{ij}$ denotes $\prod_{i=1}^{4} \prod_{j=1}^{3}$;
(*b*) $\sum \alpha_i + \sum \beta_j$ denotes $\sum_{i=1}^{4} \alpha_i + \sum_{j=1}^{3} \beta_j$;
(*c*) the summation is over all nonnegative integral values of the α_i and β_j such that $\alpha_i + \beta_j = k_{ij}$, where $K = (k_{ij})$ is the matrix on the right in (D-17);
(*d*) k_i is given by $k_i = (\sum \alpha_i + \sum \beta_j) - \alpha_i$, where we note that each k_i depends only on the elements of K, and, in fact, $k_1 = a + b + e$, $k_2 = c + d + e$, $k_3 = a + c + f$, $k_4 = b + d + f$.

Let us next consider a 4×3 matrix boson:

$$A \equiv \begin{pmatrix} a_1^1 & a_1^2 & a_1^3 \\ a_2^1 & a_2^2 & a_2^3 \\ a_3^1 & a_3^2 & a_3^3 \\ a_4^1 & a_4^2 & a_4^3 \end{pmatrix}. \tag{D-19}$$

We denote by R_i and C_j the product of the elements occurring in row i and column j of A, respectively:

$$R_i = a_i^1 a_i^2 a_i^3, \qquad i = 1, 2, 3, 4;$$

$$C_j = a_1^j a_2^j a_3^j a_4^j, \qquad j = 1, 2, 3. \tag{D-20}$$

We further define the form $S(A)$ by

$$S(A) \equiv R_1 + R_2 + R_3 + R_4 + C_1 + C_2 + C_3. \tag{D-21}$$

Now consider the expansion of $[1 + S(A)]^{-2}$. We first expand in the form

$$[1 + S(A)]^{-2} = \sum_{N=0}^{\infty} (-1)^N (N+1)[S(A)]^N. \tag{D-22}$$

The multinomial expansion of $[S(A)]^N$ yields

$$\frac{[S(A)]^N}{N!} = \sum_{\substack{(\alpha)\,(\beta)\\ \sum\alpha_i+\sum\beta_j=N}} \prod_{ij} \frac{R_i^{\alpha_i} S_j^{\beta_j}}{\alpha_i!\beta_j!} = \sum_{\substack{(\alpha)\,(\beta)\\ \sum\alpha_i+\sum\beta_j=N}} \prod_{ij} \frac{(a_i^j)^{\alpha_i+\beta_j}}{\alpha_i!\beta_j!}. \tag{D-23}$$

Using this result in Eq. (D-22), we obtain

$$[1 + S(A)]^{-2} = \sum_{(\alpha)(\beta)} (-1)^{\sum\alpha_i+\sum\beta_j}\left(\sum\alpha_i + \sum\beta_j + 1\right)! \prod_{ij} \frac{(a_i^j)^{\alpha_i+\beta_j}}{\alpha_i!\beta_j!}, \tag{D-24}$$

where the α_i and β_j range independently from 0 to ∞.

We next write the summation as

$$\sum_{(\alpha)(\beta)} = \sum_{(abcdef)} \sum_{\substack{(\alpha)\,(\beta)\\ \alpha_i+\beta_j=k_{ij}}}, \tag{D-25}$$

where $\sum_{(abcdef)}$ denotes the summation over all angular momenta quantum numbers a, b, c, d, e, f consistent with the triangle conditions in the definition of a Racah coefficient [see Eq. (3.264)]. Using the split summation (D-25) and the explicit form (D-18) of the 6-j coefficient, we find that Eq. (D-24) becomes

$$[1 + S(A)]^{-2} = \sum_{(abcdef)} \left(\prod_i (k_i + 1)!\right)^{\frac{1}{2}} \begin{Bmatrix} a & b & e \\ d & c & f \end{Bmatrix} \prod_{ij} \frac{(a_i^j)^{k_{ij}}}{(k_{ij}!)^{\frac{1}{2}}}, \tag{D-26}$$

where we may consider that $abcdef$ run independently from 0 to ∞ in consequence of the definition, Eq. (3.264).

Since the function $S(A)$ is invariant to the permutations of the rows and columns of the 4×3 matrix boson A, and since each such permutation of $\prod_{ij}(a_i^j)^{k_{ij}}$ is transferred to the corresponding permutation of the rows and columns of the 4×3 matrix K, it follows at once from the generating function (D-26) that the 6-j coefficients possess the symmetries induced on the $(abcdef)$ by the row and column permutations of K.

Remarks. (*a*) Schwinger [8] derived the generating function (D-26) without using the 6-j coefficients explicitly (see also Bargmann [9]). This formula was then used to derive the 6-j coefficients in the Racah symmetric form, Eq. (D-18). (*b*) The use of bosons in the above derivation is a purely formal device, and we could have used complex numbers (z_i^j). (*c*) Giovannini and Smith [66] have discussed certain algebraic structures related to magic squares of the type representing the 3-j and 6-j coefficients. This is discussed further in Topic 8, Section 5, RWA.

E. The Representations of the Group $U(2)$

The group $U(2)$ is the set of all 2×2 unitary matrices in which group multiplication is matrix multiplication. It is straightforward to obtain the irreps of $U(2)$ from the boson polynomials after making several observations.

An arbitrary unitary matrix $U \in U(2)$ can be *uniquely* written in the form[1]

$$U = e^{(i\chi/2)} U_0, \qquad 0 \leqslant \chi < 2\pi, \tag{E-1}$$

where $U_0 \in SU(2)$. Thus, χ is the unique angle on the unit circle such that

$$e^{i\chi} = \det U. \tag{E-2}$$

Now consider the boson polynomials given by Eq. (B-1), in which we set $A = U$:

$$D\begin{pmatrix} & \mu_{11} & \\ m_{12} & & m_{22} \\ & m_{11} & \end{pmatrix}(U) \equiv (\det U)^{m_{22}} D^{\frac{1}{2}(m_{12}-m_{22})}_{m_{11}-\frac{1}{2}(m_{12}+m_{22}),\mu_{11}-\frac{1}{2}(m_{12}+m_{22})}(U), \tag{E-3}$$

where we have introduced the Gel'fand notation (Appendix A). Let $D^{[m]}(U)$, $[m] = [m_{12}m_{22}]$, denote the unitary matrix that has the function (E-3) appearing in row[2] m_{11} and column μ_{11}. Then

$$U \to D^{[m]}(U) \tag{E-4}$$

is a representation of the group $U(2)$ by unitary matrices of dimension $m_{12} - m_{22} + 1$, this result being a consequence of the multiplication property, Eq. (5.61), for the boson polynomials.

In order that the representations be single-valued functions of the elements u_{ij} of U, we require that m_{12} and m_{22} be integers, *positive, zero, or negative*, that satisfy $m_{12} \geqslant m_{22}$.

The elements of the matrix, Eq. (E-4), are homogeneous functions of degree $m_{12} + m_{22}$ in the variables u_{ij}; if $m_{22} \geqslant 0$, they are *homogeneous polynomials*; if $m_{22} < 0$, they are ratios of homogeneous polynomials of degree $m_{12} - m_{22}$ to $(\det U)^{-m_{22}}$. That negative integers m_{22} must be considered in Eq. (E-4) is already indicated by the fact that the correspondence $U \to U^*$, each $U \in U(2)$, is itself a representation of $U(2)$. Note, however, that

$$U^* = (\det U)^{-1}\begin{pmatrix} u_{22} & -u_{21} \\ -u_{12} & u_{11} \end{pmatrix}. \tag{E-5}$$

Thus, to each ordered pair of integers, $[m_{12}m_{22}]$, such that $m_{12} \geqslant m_{22}$, there corresponds a unitary irrep of the group $U(2)$. [The fact that the representation

---

[1] If U_0 is parametrized in terms of ψ and $\hat{n}$ [see Eq. (2.27)], we may write $U = U(\chi, \psi, \hat{n}) = \exp(i\chi/2)U_0(\psi, \hat{n})$. Then the relation $U(2\pi + \chi, 2\pi - \psi, -\hat{n}) = U(\chi, \psi, \hat{n})$ shows that, when χ is allowed to run over all values $0 \leqslant \chi < 4\pi$, the set of unitary matrices is covered twice, once for $0 \leqslant \chi < 2\pi$, and once for $2\pi \leqslant \chi < 4\pi$. We therefore restrict χ to the interval $0 \leqslant \chi < 2\pi$, noting, however, that the periodicity in χ is 4π.

[2] Rows $1, 2, \ldots, m_{12} - m_{22} + 1$ of the matrix are labeled by $m_{11} = m_{12}, m_{12} - 1, \ldots, m_{22}$, respectively. A similar convention applies to the columns.

is irreducible follows immediately upon noting that it is irreducible when U belongs to $SU(2)$.]

Remark. We have obtained this result from an elementary argument, which is also the conclusion in technically more sophisticated treatments. In such treatments the starting point is usually Lie algebraic and hence topological. The Lie algebra of $U(2)$ admits a number of connected groups that have this Lie algebra. The universal covering group of the Lie algebra is the direct product group $G = R \times SU(2)$, where R is the additive group of real numbers.[1] The elements of G are the ordered pairs (χ, U_0), where $\chi \in R$ and $U_0 \in SU(2)$. Multiplication of elements is defined by $(\chi, U_0)(\chi', U_0') = (\chi + \chi', U_0 U_0')$. The mapping

$$(\chi, U_0) \to e^{i\chi/2} U_0 = U, \qquad \text{each } \chi \in R, \text{ each } U_0 \in SU(2) \tag{E-6}$$

is a homomorphism of $R \times SU(2)$ onto $U(2)$. The kernel of the homomorphism [the elements that map to the unit element σ_0 of $U(2)$] are determined by $e^{i\chi/2} U_0 = \sigma_0$. Since U_0 is unitary unimodular, we obtain $e^{i\chi} = 1$—that is, $\chi = 2k\pi$ and $U_0 = (-1)^k \sigma_0$, where $k = 0, \pm 1, \pm 2, \ldots$. The discrete group $Z \subset R \times SU(2)$ defined by

$$Z = \{(2k\pi, (-1)^k \sigma_0) : k = 0, \pm 1, \pm 2, \ldots\} \tag{E-7}$$

is thus an invariant subgroup of $R \times SU(2)$, and the factor group $R \times SU(2)/Z$ is isomorphic to $U(2)$. Notice that Z is isomorphic to the additive group of integers and is generated by $(2\pi, -\sigma_0)$. (For an excellent discussion of the points made here, see Michel [67].)

The next observation to be made is that the inequivalent, *unitary* irreps of $R \times SU(2)$ are given by $(\chi, U_0) \to e^{ir\chi/2} D^j(U_0)$, where r is an arbitrary real number. In particular, the generator $(2\pi, -\sigma_0)$ of Z is represented by $e^{ir\pi}(-1)^{2j} \mathbb{1}^{(j)}$, where $\mathbb{1}^{(j)}$ is the unit matrix of dimension $2j + 1$. But since $(2\pi, -\sigma_0) \to \sigma_0$ in the homomorphism $R \times SU(2) \to U(2)$, and σ_0 is represented by $\mathbb{1}^{(j)}$, we must have $(-1)^{2j+r} = 1$—that is, $2j + r =$ even integer. These constraints on j and r ($2j =$ integer $\geqslant 0$, $r =$ integer, $2j + r =$ even integer) are (uniquely) satisfied in terms of a pair $[m_{12} m_{22}]$ of ordered integers $m_{12} \geqslant m_{22}$ by defining $2j = m_{12} - m_{22}$, $r = m_{12} + m_{22}$. Thus, the inequivalent, unitary irreps of $U(2)$ are

$$e^{i(m_{12}+m_{22})\chi/2} D^{\frac{1}{2}(m_{12}-m_{22})}(U_0). \tag{E-8}$$

Choosing $[m_{12} m_{22}] = [1, 0]$, we obtain $e^{i\chi/2} U_0$, which is U itself. Using the homogeneity property $D^j(e^{i\chi/2} U_0) = e^{ij\chi/2} D^j(U_0)$, we obtain our result, Eq. (E-3).

The results of the general analysis of Michel thus agree with our elementary argument leading to Eq. (E-3). We have gone into this somewhat elaborate discussion because the points made have been a source of confusion, in practice.

---

[1] In this *Remark* we follow the notation of Michel [67] and depart from our standard notation where R denotes a rotation.

Let us note the following special results:

$$D^{[10]}(U) = U, \tag{E-9}$$

$$D\begin{pmatrix} 1 & \nu & 0 \\ & \mu & \end{pmatrix}(U^*) = (-1)^{\mu-\nu} D\begin{pmatrix} 0 & -\nu & -1 \\ & -\mu & \end{pmatrix}(U). \tag{E-10}$$

The second relation is a special case of the general relation

$$D\begin{pmatrix} & \mu_{11} & \\ m_{12} & & m_{22} \\ & m_{11} & \end{pmatrix}(U^*) = (-1)^{m_{11}-\mu_{11}} D\begin{pmatrix} & -\mu_{11} & \\ -m_{22} & & -m_{12} \\ & -m_{11} & \end{pmatrix}(U). \tag{E-11}$$

If we parametrize U by writing

$$U = e^{i\chi/2}(x_0\sigma_0 - i\mathbf{x}\cdot\boldsymbol{\sigma}), \tag{E-12}$$

where $(x_0, \mathbf{x})$ is a point on the unit sphere S^3, then the elements of the representation matrices satisfy the following orthogonality relations:[1]

$$\int_0^{2\pi} d\chi \int d\Omega \left[D\begin{pmatrix} & \mu'_{11} & \\ m'_{12} & & m'_{22} \\ & m'_{11} & \end{pmatrix}(U) \right]^* D\begin{pmatrix} & \mu_{11} & \\ m_{12} & & m_{22} \\ & m_{11} & \end{pmatrix}(U)$$

$$= \delta_{m'_{12}m_{12}} \delta_{m'_{22}m_{22}} \delta_{m'_{11}m_{11}} \delta_{\mu'_{11}\mu_{11}} \frac{4\pi^3}{m_{12} - m_{22} + 1}. \tag{E-13}$$

This result is a consequence of the definition Eq. (E-3), and the orthogonality relation, Eq. (3.315).

Finally, let us observe that we have obtained all the finite-dimensional, inequivalent irreps of the group $U(2)$.

References

1. I. M. Gel'fand, R. A. Minlos, and Z. Ya. Shapiro, *Representations of the Rotation and Lorentz Groups and Their Applications.* Pergamon, New York, 1963. Translated from the Russian by G. Cummins and T. Boddington.
2. P. A. M. Dirac, "La seconde quantification," *Ann. Inst. H. Poincaré* **11** (1949), 15–47.
3. S. Bergman, *The Kernel Function and Conformal Mapping.* Mathematical Surveys No. 5, Amer. Math. Soc., New York, 1950.
4. V. Bargmann, "On a Hilbert space of analytic functions and an associated integral transform," *Commun. Pure Appl. Math.* **14** (1961), 187–214.

[1] In actually carrying out this integral one finds the form $\int_0^{2\pi} d\chi\, \exp[i(m_{12} + m_{22} - m'_{12} - m'_{22})\chi/2] \times [SU(2)\text{ part}]$. At first glance the χ integral *seems* to require χ to range over 0 to 4π. However, the $SU(2)$ integration forces $m_{12} + m_{22} - m'_{12} - m'_{22}$ to be an *even* integer.

5. I. E. Segal, "Mathematical characterization of the physical vacuum for a linear Bose–Einstein field," *Illinois J. Math.* **6** (1962), 500–523.
6. J. R. Klauder, "Coherent and incoherent states of the radiation field," *Phys. Rev.* **131** (1963), 2766–2788.
7. J. R. Klauder and E. Sudarshan, *Fundamentals of Quantum Optics*. Benjamin, New York, 1968.
7a. N. Mukunda and E. C. G. Sudarshan, "New light on the optical equivalence theorem and a new type of discrete diagonal coherent state representation," *Pramāṇa* **10** (1978), 227–238.
7b. A. Grossman, "Geometry of real and complex canonical transformations in quantum mechanics," in *Group Theoretical Methods in Physics: Proceedings of the Sixth International Colloquium* (P. Kramer and A. Rieckers, eds.), pp. 162–179. Springer, Berlin, 1977.
7c. K. B. Wolf, "The Heisenberg-Weyl ring in quantum mechanics," in *Group Theory and its Applications* (E. M. Loebl, ed.), Vol. III, pp. 189–247. Academic Press, New York, 1975.
8. J. Schwinger, *On Angular Momentum*. U.S. Atomic Energy Commission Report NYO-3071, 1952, unpublished. Reprinted in *Quantum Theory of Angular Momentum* (L. C. Biedenharn and H. van Dam, eds.), pp. 229–279. Academic Press, New York, 1965.
9. V. Bargmann, "On the representations of the rotation group," *Rev. Mod. Phys.* **34** (1962), 829–845.
10. E. Schrödinger, "Der stetige Übergang von der Mikro- zur Makromechanik," *Naturwissenschaften* **14** (1926), 664–666.
11. J. M. Jauch, "On bras and kets. A commentary on Dirac's mathematical formalism of quantum mechanics," in *Aspects of Quantum Theory* (A. Salam and E. P. Wigner, eds.), pp. 137–167. Cambridge Univ. Press, London, 1972.
12. P. Jordan, "Der Zusammenhang der symmetrischen und linearen Gruppen und das Mehrkörperproblem," *Z. Physik* **94** (1935), 531–535.
12a. P. A. M. Dirac, "Unitary representations of the Lorentz group," *Proc. Roy. Soc.* (London) **A183** (1945), 284–295.
13. R. Jost, *The General Theory of Quantized Fields*. Amer. Math. Soc., Providence, R.I., 1965.
14. J. D. Louck, "Recent progress toward a theory of tensor operators in the unitary groups," *Amer. J. Phys.* **38** (1970), 3–42.
15. G.-C. Rota, *Finite Operator Calculus*. Academic Press, New York, 1975.
16. G.-C. Rota, "The number of partitions of a set," *Amer. Math. Monthly* **71** (1964), 498–504.
17. G.-C. Rota, D. Kahaner, and A. Odlyzko, "Finite operator calculus," *J. Math. Anal. Appl.* **42** (1973), 684–760.
17a. N. Ya. Vilenkin, *Special Functions and the Theory of Group Representations* (Amer. Math. Soc. Transl., Vol. 22). Amer. Math. Soc., Providence, R. I., 1968.
17b. Ya. A. Smorodinskiĭ, "The Clebsch-Gordan coefficients, the *d*-function of $SU(2)$, and their symmetry," *Sov. Phys. JETP* **48** (1978), 403–405.
18. A. Young, "Quantitative substitutional analysis," *PLMS* (1), I, **33** (1901), 97–146; II, **34** (1902), 361–397; *PLMS* (2), III, **28** (1928), 255–292; IV, **31** (1930), 253–272; V, **31** (1930), 273–288; VI, **34** (1932), 196–230; VII, **36** (1933), 304–368; VIII, **37** (1934), 441–495; IX, **54** (1952), 218–253.
19. H. Weyl, *The Structure and Representations of Continuous Groups*. Lectures at the Institute for Advanced Study, Princeton, N. J., 1934–1935, unpublished. Notes by R. Brauer.
20. I. M. Gel'fand and M. L. Tseitlin, "Matrix elements for the unitary groups," *Dokl. Akad. Nauk SSSR* (1950), 825–828.
21. H. Weyl, *Gruppentheorie und Quantenmechanik*, Hirzel, Leipzig, 1st ed., 1928; 2nd ed., 1931. Translated by H. P. Robertson as *The Theory of Groups and Quantum Mechanics*, Methuen, London, 1931; reissued by Dover, New York, 1949.

22. T. Yamanouchi, "On the calculation of atomic energy levels IV," *Proc. Phys.-Math. Soc. Japan* **18** (1936), 623–640; "On the construction of unitary irreducible representations of the symmetric group," *ibid.* **19** (1937), 436–450.
23. T. Nakayama, "On some modular properties of irreducible representations of S_n," *Jap. J. Math.* **17** (1940), I, 89–108; II, 411–423.
23a. J. S. Frame, G. de B. Robinson, and R. M. Thrall, "The hook graphs of the symmetric group," *Can. J. Math.* **6** (1954), 316–324.
24. G. de B. Robinson, *Representation Theory of the Symmetric Group*. Univ. of Toronto Press, Toronto, 1961.
25. M. Ciftan and L. C. Biedenharn, "Combinatorial structure of state vectors in U_n. I. Hook patterns for maximal and semimaximal states in U_n," *J. Math. Phys.* **10** (1969), 221–232.
26. H. Weyl, *The Classical Groups. Their Invariants and Representations.* Princeton Univ. Press, Princeton, N. J., 1946.
27. G. E. Baird and L. C. Biedenharn, "On the representations of semisimple Lie groups. II," *J. Math. Phys.* **4** (1963), 1449–1466; "III, The explicit conjugation operation for $SU(n)$," *ibid.* **5** (1964), 1723–1730; "IV, A canonical classification for tensor operators in SU_3," *ibid.* **5** (1965), 1730–1747.
28. T. A. Brody, M. Moshinsky, and I. Renero, "Recursion relations for the Wigner coefficients of unitary groups," *J. Math. Phys.* **6** (1965), 1540–1546.
29. J. D. Louck, "Group theory of harmonic oscillators in n-dimensional space," *J. Math. Phys.* **6** (1965), 1786–1804.
30. M. Moshinsky, "Wigner coefficients for the SU_3 group and some applications," *Rev. Mod. Phys.* **34** (1962), 813–828.
31. M. Moshinsky, "The harmonic oscillator and supermultiplet theory (I). The single particle picture," *Nucl. Phys.* **31** (1962), 384–405.
32. M. Moshinsky, "Bases for the irreducible representations of the unitary groups and some applications," *J. Math. Phys.* **4** (1963), 1128–1139.
33. J. G. Nagel and M. Moshinsky, "Operators that raise or lower the irreducible vector spaces of U_{n-1} contained in an irreducible vector space of U_n," *J. Math. Phys.* **6** (1965), 682–694.
34. L. C. Biedenharn, A. Giovannini, and J. D. Louck, "Canonical definition of Wigner operators in U_n," *J. Math. Phys.* **8** (1967), 691–700.
35. P. Doubilet, G.-C. Rota, and J. Stein, On the foundations of combinatorial theory: IX. Combinatorial methods in invariant theory, *Studies in Appl. Math.* **LIII**, No. 3 (1974), 185–216.
35a. J. Désarménien, J. P. S. Kung, and G.-C. Rota, "Invariant theory, Young bitableaux, and combinatorics," *Advan. in Math.* **27** (1978), 63–92.
36. J. M. Jauch and E. L. Hill, "On the problem of degeneracy in quantum mechanics," *Phys. Rev.* **57** (1940), 641–645.
37. G. A. Baker, "Degeneracy of the n-dimensional, isotropic, harmonic oscillator," *Phys. Rev.* **103** (1956), 1119–1120.
38. V. Bargmann and M. Moshinsky, "Group theory of harmonic oscillators (I). The collective modes," *Nucl. Phys.* **18** (1960), 697–712; "(II). The integrals of motion for the quadrupole–quadrupole interaction," *ibid.* **23** (1961), 177–199.
39. M. Moshinsky, "Gel'fand states and the irreducible representations of the symmetric group," *J. Math. Phys.* **7** (1966), 691–698.
40. L. C. Biedenharn and J. D. Louck, "A pattern calculus for tensor operators in the unitary groups," *Commun. Math. Phys.* **8** (1968), 80–131.
41. P. Kramer and M. Moshinsky, "Group theory of harmonic oscillators and nuclear structure," in *Group Theory and Its Applications* (E. M. Loebl, ed.), Vol. I, pp. 339–468. Academic Press, New York, 1968.

42. M. Ciftan, "On the combinatorial structure of state vectors in $U(n)$. II. The generalization of hypergeometric functions on $U(n)$ states," *J. Math. Phys.* **10** (1969), 1635–1646.
43. W. J. Holman, "Representation theory of $Sp(4)$ and $SO(5)$," *J. Math. Phys.* **10** (1969), 1710–1717.
44. W. J. Holman, "Tensor operators in Sp_4 and SO_5," *Nuovo Cimento* **66A** (1970), 619–643.
45. A. M. Bincer, "Interpretation of the symmetry of the Clebsch–Gordan coefficients discovered by Regge," *J. Math. Phys.* **11** (1970), 1835–1844.
46. J. D. Louck and L. C. Biedenharn, "Canonical unit adjoint tensor operators in $U(n)$," *J. Math. Phys.* **11** (1970), 2368–2414.
47. A. C. T. Wu, "Structure of the combinatorial generalization of hypergeometric functions for $SU(n)$ states, *J. Math. Phys.* **12** (1971), 437–440.
48. W. J. Holman, "On the general boson states of $U_n * U_n$ and $Sp_4 * Sp_4$," *Nuovo Cimento* **4A** (1971), 904–931.
49. W. J. Holman and L. C. Biedenharn, "The representations and tensor operators of the unitary groups $U(n)$," in *Group Theory and Its Applications* (E. M. Loebl, ed.), Vol. II, pp. 1–73. Academic Press, New York, 1971.
50. T. H. Seligman, "The Weyl basis of the unitary group $U(k)$," *J. Math. Phys.* **13** (1972), 876–879.
51. J. D. Louck and L. C. Biedenharn, "The structure of the canonical tensor operators in the unitary groups. III. Further developments of the boson polynomials and their implications," *J. Math. Phys.* **14** (1973), 1336–1357.
52. C. W. Patterson and W. G. Harter, "Canonical symmetrization for the unitary bases. I. Canonical Weyl bases," *J. Math. Phys.* **17** (1976), 1125–1136; "II. Boson and fermion bases," *ibid.* **17**, 1137–1142.
52a. Y. Fujiwara and H. Horiuchi, "Properties of double Gel'fand polynomials and their application to multiplicity-free problems," Memoirs of the Faculty of Science of Kyoto University (Japan), to be published in 1982.
53. J. D. Louck and L. C. Biedenharn, "Identity satisfied by the Racah coefficients of $U(n)$," *J. Math. Phys.* **12** (1971), 173–177.
54. E. Chacón, M. Ciftan, and L. C. Biedenharn, "On the evaluation of the multiplicity-free Wigner coefficients of $U(n)$," *J. Math. Phys.* **13** (1972), 577–589.
55. L. C. Biedenharn, J. D. Louck, E. Chacón, and M. Ciftan, "On the structure of the canonical tensor operators in the unitary groups. I. An extension of the pattern calculus rules and the canonical splitting in $U(3)$," *J. Math. Phys.* **13** (1972), 1957–1984.
56. L. C. Biedenharn and J. D. Louck, "On the structure of the canonical tensor operators in the unitary groups. II. The tensor operators in $U(3)$ characterized by maximal null space," *J. Math. Phys.* **13** (1972), 1985–2001.
57. J. D. Louck, M. A. Lohe, and L. C. Biedenharn, "Structure of the canonical $U(3)$ Racah functions and the $U(3){:}U(2)$ projective functions," *J. Math. Phys.* **16** (1975), 2408–2426.
58. M. A. Lohe, L. C. Biedenharn, and J. D. Louck, "Structural properties of the self-conjugate $SU(3)$ tensor operators," *J. Math. Phys.* **18** (1977), 1883–1891.
59. J. D. Louck and H. W. Galbraith, "Application of orthogonal and unitary group methods to the n-body problem," *Rev. Mod. Phys.* **44** (1972), 540–601.
60. E. P. Wigner, *Group Theory and Its Application to the Quantum Mechanics of Atomic Spectra*. Academic Press, New York, 1959. Translation from the 1931 German edition by J. J. Griffin.
61. G. Racah, "Theory of complex spectra. II," *Phys. Rev.* **62** (1942), 438–462.
62. T. Regge, "Symmetry properties of Clebsch–Gordan coefficients," *Nuovo Cimento* [10] **10** (1958), 544–545.
63. B. L. van der Waerden, *Die gruppentheoretische Methode in der Quantenmechanik*. Springer, Berlin, 1932.

64. H. A. Kramers, *Quantum Mechanics*. North-Holland, Amsterdam, 1957. Translation by D. ter Haar of Kramer's monograph published in the *Hand- und Jahrbuch der chemischen Physik*, 1937.

65. H. C. Brinkman, *Applications of Spinor Invariants in Atomic Physics*. North-Holland, Amsterdam, 1956.

66. A. Giovannini and D. A. Smith, "On algebraic structures associated with the 3-*j* and 6-*j* symbols," in *Spectroscopic and Group Theoretic Methods in Physics* (Racah Memorial Volume) (F. Bloch, S. G. Cohen, A. de-Shalit, S. Sambursky, and I. Talmi, eds.), pp. 89–97. Wiley (Interscience), New York, 1968.

67. L. Michel, "Invariance in quantum mechanics and group extension," in *Group Theoretical Concepts and Methods in Elementary Particle Physics* (F. Gürsey, ed.), pp. 135–200. Gordon and Breach, New York, 1962.

CHAPTER 6

Orbital Angular Momentum and Angular Functions on the Sphere

We have already discussed in Chapter 3, Section 2, the physical origin of the quantum theory of orbital angular momentum. Historically, the quantum angular momentum operators for a single particle were obtained from the classical angular momentum by the principle $\mathbf{p} \to -i\hbar\boldsymbol{\nabla}$, so that $\mathbf{L} = \mathbf{x} \times \mathbf{p}$ becomes the differential operator (in units of $\hbar$) $\mathscr{L} = -i\mathbf{x} \times \boldsymbol{\nabla}$. These operators may also be obtained by using the Jordan map [see Eq. (5.27)] of the Hermitian matrices H_i defined by $n_1H_1 + n_2H_2 + n_3H_3 = iN$ [see Eq. (2.6)]:

$$\mathscr{L}_i = \tilde{\mathbf{x}} H_i(\partial/\partial\mathbf{x}), \tag{6.1}$$

where $\mathbf{x} = \mathrm{col}(x_1, x_2, x_3)$, $\partial/\partial\mathbf{x} = \mathrm{col}(\partial/\partial x_1, \partial/\partial x_2, \partial/\partial x_3)$.

The relationship between the orbital angular momentum operators and spherical harmonics was developed in Sections 8 and 10 of Chapter 3 as an example of general methods in angular momentum theory (properties of rotation matrices, raising and lowering operator techniques, etc.). {We have given in this earlier discussion (*a*) explicit expressions for the spherical harmonics in terms of special functions [Eqs. (3.141)–(3.148)]; (*b*) the standard action of the operators (6.1) on the spherical harmonics [Eqs. (3.140)]; (*c*) the orthogonality property on the unit sphere, S^2 [Eq. (3.139)]; (*d*) the explicit and general form of the solid harmonics [Eq. (3.153)]; and (*e*) the Gaunt integral of three spherical harmonics [Eq. (3.192)].}

The theory of orbital angular momentum has, on its own, an important role in physical theories (as distinguished from the general theory of angular momentum), and we develop various aspects of this under the present topic.

ENCYCLOPEDIA OF MATHEMATICS and Its Applications, Gian-Carlo Rota (ed.).
Vol. 8: L. C. Biedenharn and J. D. Louck, Angular Momentum in
Quantum Physics: Theory and Application

Remark. It is still to be emphasized, however, that orbital rotational symmetry [$SO(3)$] can always be subsumed under the more general notion of general [$SU(2)$] rotational symmetry. This is the geometric content of the relation between orthogonal and unitary matrices expressed by Eqs. (2.15)–(2.17): *Each* 2×2 *unitary matrix defines a proper orthogonal transformation of the points of Euclidean three-space*. This result becomes particularly significant for unifying the treatment of composite systems (see Notes 5 and 8 of Chapter 3) in which kinematically independent parts may separately have either orbital rotational symmetry [$SO(3)$] or general rotational symmetry [$SU(2)$]. We may always consider that the invariance group of the (noninteracting) parts is the direct product group $SU(2) \times SU(2) \times \cdots$ with elements $(U_1, U_2, \ldots)$. The rotation group of the composite whole is then represented uniformly as the diagonal subgroup of the direct product group having elements $(U, U, \ldots)$.

1. Rotational Symmetry of a Simple Physical System

Consider the quantum mechanical description of one of the simplest of physical systems: a single, structureless, nonrelativistic particle of mass m moving in three-space in a force field derivable from a central potential. The Hamiltonian operator for this particle is

$$H(\mathbf{x}, \mathbf{p}) = \frac{\mathbf{p} \cdot \mathbf{p}}{2m} + V(r), \tag{6.2}$$

where $\mathbf{p} = -i\boldsymbol{\nabla}$ and $V(r)$ is the potential energy, which depends on the position $\mathbf{x}$ of the particle through $r = (x_1^2 + x_2^2 + x_3^2)^{\frac{1}{2}}$.

From the point of view of angular momentum theory, the most important property possessed by the operator (6.2) is expressed by

$$\mathscr{T}_R H \mathscr{T}_{R^{-1}} = H \tag{6.3}$$

for each *proper, orthogonal matrix R*. Here $\mathscr{T}_R$ is the *orbital rotation operator, which is defined by its action on a state* $\Psi(\mathbf{x})$ of the physical system described by the Hamiltonian (6.2):[1]

$$\begin{aligned}(\mathscr{T}_R \Psi)(\mathbf{x}) &= e^{-i\phi\hat{n} \cdot \mathscr{L}} \Psi(\mathbf{x}) \\ &= \Psi(R^{-1}\mathbf{x}),\end{aligned} \tag{6.4}$$

[1] We require the set $\{\Psi(\mathbf{x})\}$ of state functions to be analytic so that repeated application of $\mathscr{L}$ to a function again gives a function in the set, and also that each function in the set equal its Taylor series so that $\exp(-i\phi\hat{n} \cdot \mathscr{L})\Psi(\mathbf{x})$ actually equals $\Psi(R^{-1}\mathbf{x})$.

where $\mathbf{x}' = R^{-1}\mathbf{x}$ denotes the column matrix obtained from $\mathbf{x} = \text{col}(x_1, x_2, x_3)$ by matrix multiplication with R^{-1}.

The importance of the invariance property, Eq. (6.3), follows from the transformation properties of the Schrödinger equation for the physical system:

$$(H\Psi)(\mathbf{x}, t) = i(\partial/\partial t)\Psi(\mathbf{x}, t), \tag{6.5}$$

$$[(\mathscr{T}_R H \mathscr{T}_{R^{-1}})(\mathscr{T}_R \psi)](\mathbf{x}, t) = i(\partial/\partial t)(\mathscr{T}_R \Psi)(\mathbf{x}, t). \tag{6.6}$$

Since $\mathscr{T}_R H \mathscr{T}_{R^{-1}} = H$, Eq. (6.6) expresses the important result: *If Ψ is a possible state of the physical system described by H, then so is $\Psi' = \mathscr{T}_R \Psi$.*

An important special case of this general result applies to eigenstates: If Ψ_E is an eigenstate of energy E of the Hamiltonian H—that is, $H\Psi_R = E\Psi_E$—then $\Psi'_E = \mathscr{T}_R \Psi_E$ is also an eigenstate of energy E.

The space $\mathscr{H}_E$ spanned by the eigenstates of given energy E of a Hamiltonian of type (6.2) is invariant under the action of the group of operators $\{\mathscr{T}_R : R \in SO(3)\}$. This space is therefore a carrier space of a representation of the orthogonal group $SO(3)$. The space $\mathscr{H}_E$ must therefore split into a direct sum of spaces, each of which carries an irreducible representation of $SO(3)$. It is in the determination of these subspaces that the theory of *spherical harmonics* makes its appearance into the quantum theory of simple physical systems.[1]

From the point of view of operator algebra, the invariance property (6.3) is equivalent to the commutation property,

$$[\mathscr{L}_i, H(\mathbf{x}, \mathbf{p})] = 0; \tag{6.7}$$

for the construction of state vectors the problem is to diagonalize simultaneously the mutually commuting (Hermitian) operators: $H, \mathscr{L}^2$, and $\mathscr{L}_3$.

2. Scalar Product of State Vectors

The inner product of two state vectors for a physical system with Hamiltonian (6.2) is (in the Dirac bra–ket notation)

$$\langle \Psi | \Psi' \rangle \equiv \int d\mathbf{x} \langle \mathbf{x} | \Psi \rangle^* \langle \mathbf{x} | \Psi' \rangle, \tag{6.8}$$

where the integration extends over all of $\mathbb{R}^3$. Furthermore, because the

[1] One is thus using group theory as a tool in implementing the actual construction of state vectors, in contrast to the more general role (Chapter 1) of group theory in determining the form of the laws of nature (Houtappel *et al.* [1]).

potential is central, the value $\langle \mathbf{x}|\Psi\rangle$ of each state vector $|\Psi\rangle$ has the form

$$\langle \mathbf{x}|\Psi\rangle = f(r)\Phi(\hat{x}), \tag{6.9}$$

where $r = \|\mathbf{x}\| \equiv (x_1^2 + x_2^2 + x_3^2)^{\frac{1}{2}}$, and $\hat{x}$ is the unit vector $\hat{x} = \mathbf{x}/r$. The inner product, Eq. (6.8), thus factorizes into a product:

$$\langle \Psi|\Psi'\rangle = (\Phi, \Phi')\int_0^\infty r^2\, dr f^*(r) f'(r), \tag{6.10}$$

where

$$(\Phi, \Phi') \equiv \int dS_{\hat{x}} \Phi^*(\hat{x})\Phi'(\hat{x}) \tag{6.11}$$

in which $dS_{\hat{x}}$ designates the surface element on the unit sphere at $\hat{x}$ and the integration extends over the unit sphere.

Noting that (Φ, Φ') itself satisfies all the properties required of an inner product, it follows that the set of continuous, single-valued functions, which are square-integrable over the unit sphere, constitutes a Hilbert space $\mathscr{H}(S)$ (S designates sphere).

3. Unitarity of the Orbital Rotation Operator

In the general theory of angular momentum (Chapter 3), one assumes (on physical grounds) that the rotation operator $\exp(-i\psi\hat{n}\cdot\mathbf{J})$ is unitary—that is, the total angular momentum operator is Hermitian. In the orbital angular momentum realization of the theory, one can demonstrate this result explicitly. *The operators $\mathscr{T}_R$, $R \in SO(3)$, are unitary on the space $\mathscr{H}(S)$:*

$$(\mathscr{T}_R\Phi, \mathscr{T}_R\Phi') = (\Phi, \Phi') \tag{6.12}$$

for all $\Phi, \Phi' \in \mathscr{H}(S)$.

Proof. One has

$$(\mathscr{T}_R\Phi, \mathscr{T}_R\Phi') = \int dS_{\hat{x}}\Phi^*(R^{-1}\hat{x})\Phi'(R^{-1}\hat{x}) = \int dS_{R\hat{x}}\Phi^*(\hat{x})\Phi'(\hat{x}) = (\Phi, \Phi'),$$

where we have used $dS_{\hat{x}} = dS_{R\hat{x}}$, since all points on the sphere are equivalent under rotations. ■

The Hermitian property of the orbital angular momentum operators is expressed by[1]

$$(L_i\Phi, \Phi') = (\Phi, L_i\Phi'), \qquad \text{all} \quad \Phi, \Phi' \in \mathscr{H}(S), \tag{6.13}$$

and follows from the unitary property (6.12) by choosing $R = R_i = R(\phi, \hat{e}_i)$, differentiating with respect to ϕ, and setting $\phi = 0$. (This procedure entails differentiating under the integral; single-valuedness, and continuity are important.)

4. A (Dense) Subspace of $\mathscr{H}(S)$

The set $\mathscr{H}_l$ of polynomial functions $\{P_l\}$, which are homogeneous of degree l—that is,

$$\mathscr{H}_l = \{P_l : P_l(\lambda \mathbf{x}) = \lambda^l P_l(\mathbf{x})\}, \tag{6.14}$$

is clearly an invariant subspace of $\mathscr{H}(S)$ with respect to the group of orbital rotations (hence, with respect to each of the operators L_i). Furthermore, if the polynomials are also solutions to Laplace's equation, $\nabla^2 P_l(\mathbf{x}) = 0$, it follows immediately from

$$\mathscr{L}^2 = -r^2\nabla^2 + (\mathbf{x}\cdot\nabla)^2 + \mathbf{x}\cdot\nabla \tag{6.15}$$

and Euler's theorem on homogeneous functions, $(\mathbf{x}\cdot\nabla)P_l(\mathbf{x}) = lP_l(\mathbf{x})$, that

$$\mathscr{L}^2 P_l(\mathbf{x}) = l(l+1)P_l(\mathbf{x}), \qquad l = 0, 1, 2, \ldots. \tag{6.16}$$

In quantum mechanics the basis of $\mathscr{H}_l$ is, by convention, fixed (up to overall phase) by the standard action of the operators $\mathscr{L}_i$ [see Eqs. (3.140) and (3.154)]. The solution to

$$\mathscr{L}_+\mathscr{Y}_{ll}(\mathbf{x}) = 0, \qquad \mathscr{L}_3\mathscr{Y}_{ll}(\mathbf{x}) = l\mathscr{Y}_{ll}(\mathbf{x}) \tag{6.17}$$

on the space $\mathscr{H}_l$ is *uniquely* determined to be

$$\mathscr{Y}_{ll}(\mathbf{x}) = \frac{1}{2^l l!}\left[\frac{(2l+1)!}{4\pi}\right]^{\frac{1}{2}}(-x_1 - ix_2)^l, \tag{6.18}$$

---

[1] The operator L_i acts on functions, whereas the differential operator $\mathscr{L}_i$ acts on values of a function: Thus, $(L_i f)(\mathbf{x}) = \mathscr{L}_i(f(\mathbf{x}))$ is a correct usage, as is $L_i f = f'$, but $\mathscr{L}_i f$ is not. Ignoring this seemingly pedantic distinction can lead to errors (a cautionary example is given by Wigner [1a, p. 106]). It is a common practice, unfortunately, not to distinguish L_i and $\mathscr{L}_i$ by the notation (as done here).

where $\mathscr{Y}_{ll}$ has been normalized on the unit sphere (and a phase has been chosen to conform to the Condon and Shortley convention). Thus, the space $\mathscr{H}_l$ contains but one highest-weight vector. It follows that the orthonormal basis of $\mathscr{H}_l$ is obtained by the standard lowering procedure [see Eq. (3.151)], thus leading to the solid harmonics as given explicitly by Eq. (3.153).

The direct sum space defined by

$$\sum_{l=0}^{\infty} \oplus \mathscr{H}_l \tag{6.19}$$

is then a dense subspace of $\mathscr{H}(S)$.

5. Only Integral Values of l Can Occur in the Quantization of Spatial (Orbital) Angular Momentum

It is often suggested that the criterion of single-valuedness of wave functions *must* be invoked in the theory of orbital angular momentum in order to eliminate from consideration the half-integral values of j, which are allowed in the general theory of angular momentum. This point did not arise in the presentation given above, since we picked the space $\mathscr{H}_l$ (always of odd dimension $2l + 1$) at the beginning, with l an integer. The fact is, however, that the general theory of angular momentum implies, without any additional assumption about single-valuedness, the nonexistence of a Hilbert space $\mathscr{H}_j$ (of even dimension $2j + 1$) of functions $\{f(\mathbf{x})\}$ with respect to which the orbital angular momentum operators $\mathscr{L} = -i\mathbf{x} \times \nabla$ are Hermitian.

The proof is by contradiction. Assume the existence of a Hilbert space $\mathscr{H}$ of dimension $2n$ ($n = 1, 2, \ldots$) such that the hypotheses of the general theory of angular momentum[1] are valid in the orbital angular momentum realization of $\mathbf{J}$—that is, $\mathbf{J} = \mathscr{L} = -i\mathbf{x} \times \nabla$. Then the construction of the angular momentum multiplets given in Chapter 3, Section 3, must be valid. In particular, it must be possible to span $\mathscr{H}$ by the set of basis vectors generated by repeated application of the lowering operator $\mathscr{L}_-$ to the set of linearly independent solutions to the differential equations

$$\mathscr{L}_+ f_{jj}(\mathbf{x}) = 0, \qquad \mathscr{L}_3 f_{jj}(\mathbf{x}) = j f_{jj}(\mathbf{x}). \tag{6.20}$$

These equations possess the unique solution

$$f_{jj}(\mathbf{x}) = c_j(-x_1 - ix_2)^j, \tag{6.21}$$

[1] The hypotheses of the general theory are as follows: (*a*) There exists a Hilbert space $\mathscr{H}$ on which the angular momentum $\mathbf{J}$ is represented by a linear Hermitian operator, $\mathbf{J}: \mathscr{H} \to \mathscr{H}$; and (*b*) $\mathbf{J}$ obeys the commutation rule $\mathbf{J} \times \mathbf{J} = i\mathbf{J}$ (see Chapter 3, Section 3).

where $j = (2n - 1)/2$, and c_j is a constant determined by normalizing $|f_{jj}(\mathbf{x})|^2$ over the unit sphere (observe that $f_{jj}(\mathbf{x})$ is square-integrable over the unit sphere). Consider next the result of repeated application of $\mathscr{L}_-$ to $f_{jj}(\mathbf{x})$. Then, by assumption, $f_p(\mathbf{x}) \equiv \mathscr{L}_-^{p-1} f_{jj}(\mathbf{x}) \in \mathscr{H}$ for each $p = 1, 2, 3, \ldots$, and furthermore $f_1(\mathbf{x}), f_2(\mathbf{x}), \ldots, f_{2n}(\mathbf{x}), f_{2n+1}(\mathbf{x}), \ldots$ are perpendicular (Hermitian property of $\mathscr{L}$). It follows then (from the multiplet construction) that one must have

$$\mathscr{L}_-^{2j+1} f_{jj}(\mathbf{x}) \equiv 0. \tag{6.22}$$

But this property *fails* to obtain for $2j + 1 =$ even integer, as we now show directly.

Define $u = -x_1 - ix_2$, $z = x_3$, $v = x_1 - ix_2$. Then we must show that

$$f(u, v, z) \equiv \left(v \frac{\partial}{\partial z} + 2z \frac{\partial}{\partial u} \right)^{2n} u^{n-\frac{1}{2}} \not\equiv 0$$

for $n = 1, 2, \ldots$. It suffices to demonstrate that $f(u, v, 0) \neq 0$. But

$$f(u, v, 0) = \frac{(2n)!}{n!} v^n \left(\frac{\partial}{\partial u} \right)^n u^{n-\frac{1}{2}}$$

is clearly not identically zero. ■

We are forced to conclude that the assumed Hilbert space $\mathscr{H}$ does not exist. Only integral values l of orbital angular momentum can occur; the Hilbert space $\mathscr{H}_l$ $(l = 0, 1, \ldots)$ of polynomials comprise all possible irrep spaces for the group $SO(3)$ of rotations of the space $\mathbb{R}^3$.

(The preceding method of proof does not reveal what assumptions in the general theory of angular momentum are violated for $l =$ half-integer. Closer examination of this point shows that it is the Hermitian property of the operators L_i that fails. The literature on this subject is quite extensive and is surveyed in the Note to this Chapter.)

6. Transformations of the Solid Harmonics under Orbital Rotation

The explicit form that the abstract result, Eq. (3.34), takes for the solid (or spherical) harmonics is

$$(\mathscr{T}_R \mathscr{Y}_{lm})(\mathbf{x}) = \mathscr{Y}_{lm}(R^{-1}\mathbf{x}) = \sum_{m'} \mathscr{D}^l_{m'm}(R) \mathscr{Y}_{lm'}(\mathbf{x}). \tag{6.23}$$

Thus, the elements of the rotation matrix are given by

$$\mathscr{D}^l_{m'm}(R) = (\mathscr{Y}_{lm'}, \mathscr{T}_R \mathscr{Y}_{lm}); \tag{6.24}$$

also, the correspondence

$$R \to \mathscr{D}^l(R) \tag{6.25}$$

is a representation of $SO(3)$ by unitary matrices. (The unitary property of the matrix is a consequence of the orbital rotation operator $\mathscr{T}_R$ being unitary; the multiplication property

$$\mathscr{D}^l(R')\mathscr{D}^l(R) = \mathscr{D}^l(R'R) \tag{6.26}$$

follows similarly from $\mathscr{T}_{R'}\mathscr{T}_R = \mathscr{T}_{R'R}$.)

The relation of the rotation matrices $\mathscr{D}^l(R)$ to the $D^j(U)$ found earlier (see Chapter 3) is determined by comparing Eq. (6.4), defining $\mathscr{T}_R$, with the abstract rotation operator defined by Eq. (3.34). One sees that

$$\mathscr{T}_{R(\phi,\hat{n})} \equiv \mathscr{U}(\phi,\hat{n}) \qquad \text{on } \mathscr{H}_l. \tag{6.27}$$

Hence, it follows from Eqs. (3.34) and (6.23) that

$$\mathscr{D}^l(R(\phi,\hat{n})) = D^l(\phi,\hat{n}). \tag{6.28}$$

More generally, we find

$$\mathscr{D}^l(R) = D^l(U)/(\det U)^l \tag{6.29}$$

for R and U related by Eq. (2.16).

Remarks. (*a*) Throughout this chapter, we emphasize the properties of the solid harmonics as opposed to those of the spherical harmonics. The relation between these functions is given by $\mathscr{Y}_{lm}(\mathbf{x}) = r^l Y_{lm}(\theta\phi)$ for $\mathbf{x} = (r \sin\theta\cos\phi,$ $\sin\theta\sin\phi,\ \cos\theta)$, where $Y_{lm}(\theta\phi)$ are spherical harmonics. There are several advantages to using solid harmonics: (1) many relations between the solid harmonics are easily comprehended as relations between polynomials (in three variables) – see, for example, Eqs. (6.23), (6.56), and Sections 15–19; (2) relations between spherical harmonics are easily obtained from those between solid harmonics by parametrizing the points $\mathbf{x} \in \mathbb{R}^3$ in terms of spherical coordinates; and (3) relations between solid harmonics, including those with gradient and curl operations, are easily given in terms of any parametrization (curvilinear coordinates) of the points $\mathbf{x} \in \mathbb{R}^3$.

(*b*) The group representation law (6.26) includes as special cases the transformation property (6.23) for the spherical harmonics as well as the addition theorem, Eq. (6.137): Equation (6.23) is obtained from $\mathscr{D}^l_{0,m}(R'R) = \sum_{m'} \mathscr{D}^l_{m'm}(R)\mathscr{D}^l_{0,m'}(R')$ and Eqs. (6.181) by setting $\boldsymbol{\eta} = \mathbf{x} = \mathrm{col}(R'_{31}, R'_{32}, R'_{33})$; the further specialization of this result to $m = 0$ then yields Eq. (6.137) upon setting $\mathbf{y} = (R_{13}, R_{23}, R_{33})$ and using property (6.182). These results have been discussed in the context of the group representation law for $SU(2)$ in Chapter 3, Sections 7 and 8. The relation between group representations and special functions has been developed by Wigner [1b] (see also Talman [1c], Vilenkin [1d], Miller [1e], Koornwinder [1f]).

7. The Elements of the Rotation Matrix $\mathscr{D}^l(R)$ Are Homogeneous Polynomials

We have previously pointed out [Eq. (3.86)] that the elements of the matrix $D^j(U)$ $(j = 0, \frac{1}{2}, 1, \ldots)$ are homogeneous polynomial forms of degree $2j$ in the elements u_{ij} of U. A similar result is true for $\mathscr{D}^l(R)$. To establish this result we use the orthogonality property $R^{-1} = \tilde{R}$ to rewrite Eq. (6.23) in the form

$$\mathscr{Y}_{lm}(\tilde{R}\mathbf{x}) = \sum_{m'} \mathscr{D}^l_{m'm}(R)\mathscr{Y}_{lm'}(\mathbf{x}). \tag{6.30}$$

As a relation between *polynomials*, Eq. (6.30) is valid for (*a*) $(x_1, x_2, x_3) = (z_1, z_2, z_3) = \mathbf{z}$, an arbitrary triple of complex variables (or indeterminates); and (*b*) R an arbitrary *nonsingular complex matrix* satisfying $\tilde{R}R = \alpha I$ $(\alpha \in \mathbb{C})$.

Proof. The function $\mathscr{Y}_{lm}(\mathbf{z})$ satisfies the complex Laplace equation (replace $\partial/\partial x_i$ by $\partial/\partial z_i$), and the general validity of Eq. (6.30) depends only on having $\mathscr{Y}_{lm}(\tilde{R}\mathbf{z})$ satisfy the complex Laplace equation; this requires condition (*b*).■

As a particular example of the validity of Eq. (6.30) under this extended domain, we may replace R by λR and use the homogeneity property of the solid harmonics,

$$\mathscr{Y}_{lm}(\lambda\mathbf{x}) = \lambda^l \mathscr{Y}_{lm}(\mathbf{x}), \tag{6.31}$$

thus proving

$$\mathscr{D}^l(\lambda R) = \lambda^l \mathscr{D}^l(R). \tag{6.32}$$

Since Eqs. (6.24), (6.23), and (6.30) imply that $\mathscr{D}^l_{m'm}(R)$ is a polynomial in the R_{ij}, we conclude from Eq. (6.32) that *the elements of the matrix $\mathscr{D}^l(R)$ are homogeneous polynomials of degree l in the elements R_{ij} of R.*

Remarks. (*a*) The result obtained above, extending the domain over which Eq. (6.30) is valid, shows that the matrices $\mathscr{D}^l(R)$ are irreps of the group of nonsingular complex matrices satisfying $\tilde{R}R = \alpha I$. In particular, they are irreps of the group of real orthogonal matrices. (*b*) In Sections 19 and 21 we develop the $\mathscr{D}^l_{mm'}(R)$ as polynomial forms in the $\{R_{ij}\}$ and note the relationship to the irreps of $SU(2)$.

8. The Energy Eigenvalue Equation

Let us return now to the development of the implications of the preceding results (for spherical harmonics) in solving the energy eigenvalue problem: $H\Psi_E = E\Psi_E$. Since (well-defined) functions of $r = (x_1^2 + x_2^2 + x_3^2)^{\frac{1}{2}}$ are the only rotational invariants that can be formed from the coordinates (x_1, x_2, x_3) of a single point, the orbital angular momentum analysis above shows that each energy eigenspace (the space spanned by the linearly independent solutions to $H\Psi_E = E\Psi_E$) must split into orbital angular momentum multiplets,

$$\{\Psi_{Elm}(\mathbf{x}) \equiv F_{El}(r)\mathscr{Y}_{lm}(\mathbf{x})\colon m = l, l-1, \ldots, -l\}, \tag{6.33}$$

where $l \in \{0, 1, 2, \ldots\}$ (the rotational invariant functions $F_{El}(r)$ cannot depend on m). Thus, the general solution to the simultaneous eigenvalue problem for $\mathscr{L}_3, \mathscr{L}^2$, and H is

$$\begin{aligned} \mathscr{L}_3\Psi_{Elm}(\mathbf{x}) &= m\Psi_{Elm}(\mathbf{x}), \\ \mathscr{L}^2\Psi_{Elm}(\mathbf{x}) &= l(l+1)\Psi_{Elm}(\mathbf{x}), \\ H(\mathbf{x}, \mathbf{p})\Psi_{Elm}(\mathbf{x}) &= E\Psi_{Elm}(\mathbf{x}), \end{aligned} \tag{6.34}$$

in which the energy eigenvalue E is independent of m.

It is convenient to use Eq. (6.15) (which expresses $\mathscr{L}^2$ in terms of the rotational invariants $\mathbf{x} \cdot \nabla, r^2$, and $\nabla \cdot \nabla = \nabla^2$) to derive the radial eigenvalue equation. {This is particularly simple in consequence of the derivation property $(\mathbf{x} \cdot \nabla)[f(\mathbf{x})g(\mathbf{x})] = [(\mathbf{x} \cdot \nabla)f(\mathbf{x})]g(\mathbf{x}) + f(\mathbf{x})[(\mathbf{x} \cdot \nabla)g(\mathbf{x})]$ and $(\mathbf{x} \cdot \nabla)F(r) = r\,(d/dr)\,F(r)$.} Thus, one obtains

$$r^2\nabla^2[F(r)\mathscr{Y}_{lm}(\mathbf{x})] = \left[r\frac{d}{dr}\left(r\frac{d}{dr} + 2l + 1\right)F(r)\right]\mathscr{Y}_{lm}(\mathbf{x}).$$

Using this result in Eq. (6.34) yields

$$\left[\frac{-1}{2mr^2}\left(r\frac{d}{dr}\right)\left(r\frac{d}{dr} + 2l + 1\right) + V(r)\right]F_{El}(r) = EF_{El}(r). \tag{6.35}$$

It is customary in quantum mechanics to introduce explicitly the spherical polar coordinates defined by

$$x_1 \pm ix_2 = r \sin\theta\, e^{\pm i\phi}, \qquad x_3 = r \cos\phi \tag{6.36}$$

$(0 \leqslant \phi < 2\pi, 0 \leqslant \theta \leqslant \pi)$, so that

$$\mathscr{Y}_{lm}(\mathbf{x}) = r^l Y_{lm}(\theta\phi). \tag{6.37}$$

One then finds that the rotational invariant function

$$R_{El}(r) = r^l F_{El}(r) \tag{6.38}$$

satisfies

$$\left[-\frac{1}{2m}\left(\frac{d^2}{dr^2} + \frac{2}{r}\frac{d}{dr} - \frac{l(l+1)}{r^2}\right) + V(r)\right] R_{El}(r) = ER_{El}(r). \tag{6.39}$$

The principal consequence of the rotational invariance of a physical system described by a Hamiltonian of the form (6.2) is as follows: *The wave functions of sharp orbital angular momentum must be of the product form* (6.33), *where the radial functions and energy eigenvalues satisfy the radial differential equation* (6.35); see also (6.39).

Spherical harmonics (solid harmonics restricted to the unit sphere) are the simplest example of a class of functions, called *tensor spherical harmonics*, which we shall refer to (generically) as *angular functions on the unit sphere*. Let us next motivate the introduction of this more general class of functions into physics.

9. Tensor Spherical Harmonics

Spherical harmonics are of general interest in quantum physics because they characterize the states of sharp orbital angular momentum for every orbital rotationally symmetric one-particle system. Tensor spherical harmonics play a similar role for "two-part" composite systems, where the first part has orbital rotational symmetry [$SO(3)$] and the second (independent) part has general rotational symmetry [$SU(2)$]. Since any mutual interaction between the parts must possess the rotational invariance of the composite whole [the diagonal subgroup of $SU(2) \times SU(2)$ – see the Remark at the beginning of this chapter], one seeks to extract the essential total angular momentum characteristics of the system by coupling the orbital angular momentum states of the first part with the angular momentum states of the second part. However, we wish to be noncommittal (in order to be general) about the details of the second part of the system. We therefore designate its angular momentum states in general abstract form. This provides us with the model we seek for abstracting the concept of tensor spherical harmonics.

Consider a physical system that possesses an orbital angular momentum **L** as well as a second (commuting) angular momentum **S**.

We denote by $\mathscr{H}_l$ the $(2l + 1)$-dimensional vector space spanned by the orthonormal basis

$$\{\mathscr{Y}_{l\mu} : \mu = l, l-1, \ldots, -l\}. \tag{6.40}$$

The orbital angular momentum operators (L_1, L_2, L_3) have (by definition) the standard action on this basis (see footnote p. 273):

$$L_\pm \mathscr{Y}_{l\mu} = [(l \mp \mu)(l \pm \mu + 1)]^{\frac{1}{2}} \mathscr{Y}_{l,\mu \pm 1},$$

$$L_3 \mathscr{Y}_{l\mu} = \mu \mathscr{Y}_{l\mu}. \tag{6.41}$$

We denote by $\mathscr{H}'_s$, $s \in \{0, \frac{1}{2}, 1, \ldots\}$, the $(2s + 1)$-dimensional vector space spanned by the orthonormal basis[1]

$$\{\xi_\nu : \nu = s, s-1, \ldots, -s\}. \tag{6.42}$$

The angular momentum operators (S_1, S_2, S_3) have (by definition) the standard action on this basis:

$$S_\pm \xi_\nu = [(s \mp \nu)(s \pm \nu + 1)]^{\frac{1}{2}} \xi_{\nu \pm 1},$$

$$S_3 \xi_\nu = \nu \xi_\nu. \tag{6.43}$$

The space $\mathscr{H}_l$ possesses the inner product (6.11). We assume also that $\mathscr{H}'_s$ is equipped with its own inner product, which we denote by $(\xi', \xi)'$ for $\xi', \xi \in \mathscr{H}'_s$. Thus,

$$(\xi_{\nu'}, \xi_\nu)' = \delta_{\nu'\nu}. \tag{6.44}$$

The *tensor solid harmonics* are functions belonging to the tensor product space $\mathscr{H}_l \otimes \mathscr{H}'_s$ (see Note 5, Chapter 3) and are defined by

$$\mathscr{Y}^{(ls)jm} \equiv \sum_\nu C^{l\,s\,j}_{m-\nu,\nu,m} \mathscr{Y}_{l,m-\nu} \otimes \xi_\nu. \tag{6.45}$$

[1] We regard s as arbitrary, but fixed, and suppress it in the notation.

The "value" of this function at the point $\mathbf{x} = (x_1, x_2, x_3)$ is denoted by

$$\mathcal{Y}^{(ls)jm}(\mathbf{x}) \equiv \sum_{\mu,\nu} C^{lsj}_{\mu\nu m}\mathcal{Y}_{l\mu}(\mathbf{x}) \otimes \xi_\nu. \tag{6.46}$$

Remark. The basis vectors ξ_ν and the angular momentum operators (S_1, S_2, S_3) are deliberately left in abstract and general form so as to admit a variety of interpretations of definition (6.45). In developing the properties of the tensor solid harmonics, one focuses on the solid harmonics in the definition (6.45), *keeping the* ξ_ν *fixed* and utilizing only the properties implied by Eqs. (6.43) and (6.44). We have also retained the general tensor product notation of Note 5, Chapter 3, to avoid confusion of general results with particular realizations. No uniform notation for the tensor harmonics has been adopted in the literature. We have chosen a notation that shows explicitly the angular momentum coupling.

Let us summarize some of the properties of the tensor solid harmonics that follow directly from the fact that they are a special case of coupled angular momentum basis vectors. The total angular momentum $\mathbf{J}$ has the definition

$$\mathbf{J} \equiv \mathbf{L} \otimes \mathbb{1}' + \mathbb{1} \otimes \mathbf{S}, \tag{6.47}$$

where $\mathbb{1}$ and $\mathbb{1}'$ are the unit operators on the vector spaces $\mathcal{H}_l$ and $\mathcal{H}'_s$, respectively. Principal results are as follows:

(*a*) *Vector space properties.* The set of vectors

$$\{\mathcal{Y}^{(ls)jm}: m = j, j-1, \ldots, -j;\, j = l+s, l+s-1, \ldots, |l-s|\} \tag{6.48}$$

is an orthonormal basis of the space $\mathcal{H}_l \otimes \mathcal{H}'_s$ (see Note 5, Chapter 3, for the definition of inner product). Using the notation $\langle\,,\rangle$ to designate the inner product on the space $\mathcal{H}_l \otimes \mathcal{H}'_s$, we have

$$\langle \mathcal{Y}^{(l's)j'm'}, \mathcal{Y}^{(ls)jm}\rangle \equiv \sum_{\nu\nu'} C^{l'sj'}_{m'-\nu',\nu',m'} C^{lsj}_{m-\nu,\nu,m} (\mathcal{Y}_{l',m'-\nu'}, \mathcal{Y}_{l,m-\nu})(\xi_{\nu'}, \xi_\nu)'$$

$$= \delta_{j'j}\delta_{l'l}\delta_{m'm}. \tag{6.49}$$

(*b*) *Simultaneous eigenvector properties.* The tensor solid harmonics are harmonic, and they solve the simultaneous eigenvector problem shown below:

$$(\boldsymbol{\nabla}^2 \otimes \mathbb{1}')\mathcal{Y}^{(ls)jm} = 0,$$

$$\mathbf{J}^2\mathscr{Y}^{(ls)jm} = j(j+1)\mathscr{Y}^{(ls)jm},$$

$$J_3\mathscr{Y}^{(ls)jm} = m\mathscr{Y}^{(ls)jm},$$

$$(\mathbf{L}^2 \otimes \mathbb{1}')\mathscr{Y}^{(ls)jm} = l(l+1)\mathscr{Y}^{(ls)jm},$$

$$(\mathbb{1} \otimes \mathbf{S}^2)\mathscr{Y}^{(ls)jm} = s(s+1)\mathscr{Y}^{(ls)jm}, \tag{6.50}$$

where we note that

$$\mathbf{J}^2 = \mathbf{L}^2 \otimes \mathbb{1}' + \mathbb{1} \otimes \mathbf{S}^2 + 2\sum_i L_i \otimes S_i. \tag{6.51}$$

It is also correct that the operators $J_\pm$ have the standard action on the tensor solid harmonics:

$$J_\pm\mathscr{Y}^{(ls)jm} = [(j \mp m)(j \pm m + 1)]^{\frac{1}{2}}\mathscr{Y}^{(ls)j,m\pm 1}. \tag{6.52}$$

(*c*) *Transformation property under rotations.* The action of the unitary rotation operator $\mathscr{U} = \exp(-i\psi\hat{n}\cdot\mathbf{J})$ is given by

$$\mathscr{U}\mathscr{Y}^{(ls)jm} = \sum_{m'} D^j_{m'm}(\psi, \hat{n})\mathscr{Y}^{(ls)jm'}. \tag{6.53}$$

(*d*) *Derivation property.* If the operators S_i possess the derivation property [see Chapter 3, Eq. (3.217)], the solid tensor harmonics are also irreducible tensors with respect to $\mathbf{J}$. In fact, in the notation of Eq. (3.232), one has

$$\mathscr{Y}^{(ls)jm} = [\mathscr{Y}^l \times \boldsymbol{\xi}^s]^j_m, \tag{6.54}$$

where $\mathscr{Y}^l$ denotes the set of solid harmonics (6.40) and $\boldsymbol{\xi}^s$ denotes the set of basis vectors (6.42).

Remarks. (*a*) The tensor *spherical* harmonics are obtained from the tensor solid harmonics by restricting the point $\mathbf{x}$ to the surface of the unit sphere. (*b*) A relation between ordinary solid harmonics implies a relation between tensor solid harmonics (because the ξ_ν are fixed); the irreducible tensor property [Chapter 3, Eq. (3.206)] signifies that the whole apparatus of irreducible tensor theory is applicable to tensor solid harmonics.

It is customary [see, however, Eq. (6.59) for vector spherical harmonics given below] to select the vectors ξ_ν in the definition (6.45) to be the column matrices defined by

$$\xi_\nu = \mathrm{col}(0 \cdots 010 \cdots 0), \qquad 1 \text{ in position } s - \nu + 1, \tag{6.55}$$

where $\nu = s, s-1, \ldots, -s$. The operators S_i then become the standard angular momentum matrices $S_i^{(s)}$ (see Section 4, Chapter 3).

The tensor product $\mathscr{Y}_{lm} \otimes \xi_\nu$ now becomes juxtaposition of a function and a column matrix: $\mathscr{Y}_{lm}\xi_\nu$. The operators $\mathbb{1}$ and $\mathbb{1}'$ become the $(2s+1) \times (2s+1)$ unit matrix and 1, respectively, so that, for example, $\mathbf{L}^2 \otimes \mathbb{1}' = \mathbf{L}^2$ and $\mathbb{1} \otimes S_i = S_i^{(s)} =$ matrix: juxtaposition of operators and state vectors has a well-defined and standard meaning.

Using this interpretation, the action of the rotation operator $\mathscr{U}$ corresponding to a rotation R of $\mathbb{R}^3$ may be given the specific form:

$$(\mathscr{U}\mathscr{Y}^{(ls)jm})(\mathbf{x}) = D^s(U)[\mathscr{Y}^{(ls)jm}(R^{-1}\mathbf{x})] = \sum_{m'} D^j_{m'm}(U)\mathscr{Y}^{(ls)jm'}(\mathbf{x}), \tag{6.56}$$

where $\pm U \to R$ in the homomorphism of $SU(2)$ onto $SO(3)$.

10. Spinor Spherical Harmonics

The spinor harmonics correspond to the case $s = \frac{1}{2}$ in Eq. (6.45) and are given in the matrix form of Eqs. (6.55) and (6.56) by[1,2]

$$\mathscr{Y}^{(j-\frac{1}{2},\frac{1}{2})jm} = \begin{pmatrix} \sqrt{\dfrac{j+m}{2j}}\,\mathscr{Y}_{j-\frac{1}{2},m-\frac{1}{2}} \\ \sqrt{\dfrac{j-m}{2j}}\,\mathscr{Y}_{j-\frac{1}{2},m+\frac{1}{2}} \end{pmatrix},$$

$$\mathscr{Y}^{(j+\frac{1}{2},\frac{1}{2})jm} = \begin{pmatrix} -\sqrt{\dfrac{j-m+1}{2j+2}}\,\mathscr{Y}_{j+\frac{1}{2},m-\frac{1}{2}} \\ \sqrt{\dfrac{j+m+1}{2j+2}}\,\mathscr{Y}_{j+\frac{1}{2},m+\frac{1}{2}} \end{pmatrix}. \tag{6.57}$$

The quantum number j is half-integral in these functions, which are sometimes called the *Pauli central field spinors* because of Pauli's use of them in solving the hydrogen atom problem with spin-orbit coupling (see Chapter 7, Section 4f).

---

[1] Since it is the total angular momentum that is conserved in physical problems, one usually lists the functions for fixed j.

[2] Further discussions of spinor spherical harmonics can be found in Edmonds [2] and Rose [3, 4]. A recent treatment that also discusses tensor spherical harmonics has been given by Campbell [4a].

11. Vector Spherical Harmonics

The vector spherical harmonics correspond to $s = 1$ in Eq. (6.45) and are given in the matrix form of Eqs. (6.55) and (6.56) by

$$\mathcal{Y}^{(j-1,1)jm} = \begin{pmatrix} \sqrt{\dfrac{(j+m-1)(j+m)}{2j(2j-1)}}\,\mathcal{Y}_{j-1,m-1} \\ \sqrt{\dfrac{(j-m)(j+m)}{j(2j-1)}}\,\mathcal{Y}_{j-1,m} \\ \sqrt{\dfrac{(j-m-1)(j-m)}{2j(2j-1)}}\,\mathcal{Y}_{j-1,m+1} \end{pmatrix},$$

$$\mathcal{Y}^{(j1)jm} = \begin{pmatrix} -\sqrt{\dfrac{(j+m)(j-m+1)}{2j(j+1)}}\,\mathcal{Y}_{j,m-1} \\ \dfrac{m}{\sqrt{j(j+1)}}\,\mathcal{Y}_{j,m} \\ \sqrt{\dfrac{(j-m)(j+m+1)}{2j(j+1)}}\,\mathcal{Y}_{j,m+1} \end{pmatrix},$$

$$\mathcal{Y}^{(j+1,1)jm} = \begin{pmatrix} \sqrt{\dfrac{(j+1-m)(j+2-m)}{2(j+1)(2j+3)}}\,\mathcal{Y}_{j+1,m-1} \\ -\sqrt{\dfrac{(j+1-m)(j+1+m)}{(j+1)(2j+3)}}\,\mathcal{Y}_{j+1,m} \\ \sqrt{\dfrac{(j+m+2)(j+m+1)}{2(j+1)(2j+3)}}\,\mathcal{Y}_{j+1,m+1} \end{pmatrix}. \tag{6.58}$$

(The quantum number j is integral in these results.)

The vector spherical harmonics have an important role in expressing the multipole expansion of the electromagnetic interaction in quantum physics [see Chapter 7, Section 6]. Accordingly, the theory of vector spherical harmonics has been developed and utilized by various authors (Hansen [5], Corben and Schwinger [6], Berestetski [7], Blatt and Weisskopf [8], Biedenharn and Rose [9], see also Ref. [10]).

A purely vectorial viewpoint may be adoped, based on the realization of spin-1 angular momentum operators presented in detail in Appendix D to Chapter 3.

The interpretation of Eq. (6.45) is that

$$\mathscr{Y}^{(l1)jm} = \sum_{\mu} C^{l\,1\,j}_{m-\mu,\mu,m} \mathscr{Y}_{l,m-\mu} \hat{\xi}_{\mu} \tag{6.59}$$

defines a *vector function* in the usual sense of vector calculus. Let us recall that the basis vectors are given by

$$\begin{aligned}\hat{\xi}_{+1} &= -(\hat{e}_1 + i\hat{e}_2)/\sqrt{2},\\ \hat{\xi}_0 &= \hat{e}_3,\\ \hat{\xi}_{-1} &= (\hat{e}_1 - i\hat{e}_2)/\sqrt{2}.\end{aligned} \tag{6.60}$$

Furthermore, under the action of $S_k = \Gamma_k \equiv i\hat{e}_k\,\mathbf{x}$, one has the standard action of spin-1 angular momentum operators (see Appendix D, Chapter 3):

$$\begin{aligned}S_{\pm}\hat{\xi}_{\mu} &= [(1 \mp \mu)(2 \pm \mu)]^{\frac{1}{2}}\hat{\xi}_{\mu \pm 1},\\ S_3\hat{\xi}_{\mu} &= \mu\hat{\xi}_{\mu},\\ \mathbf{S}^2\hat{\xi}_{\mu} &= 2\hat{\xi}_{\mu}.\end{aligned} \tag{6.61}$$

Keeping in mind that differential operators are to act on the components of these vector functions and that the vector operations $\cdot$ and $\mathbf{x}$ act on the basis vectors, $\hat{\xi}_{\mu}$, one obtains a fully consistent realization of the properties of the general tensor $\otimes$ product. We may write Eqs. (6.50) as

$$\begin{aligned}\nabla^2\mathscr{Y}^{(l1)jm} &= 0,\\ \mathbf{J}^2\mathscr{Y}^{(l1)jm} &= j(j+1)\mathscr{Y}^{(l1)jm},\\ J_3\mathscr{Y}^{(l1)jm} &= m\mathscr{Y}^{(l1)jm},\\ \mathbf{L}^2\mathscr{Y}^{(l1)jm} &= l(l+1)\mathscr{Y}^{(l1)jm},\\ \mathbf{S}^2\mathscr{Y}^{(l1)jm} &= 2\mathscr{Y}^{(l1)jm}.\end{aligned} \tag{6.62}$$

The result implied by Eq. (6.51) is quite interesting. Since

$$\sum_k L_k \otimes S_k = \sum_k L_k S_k = i\left(\sum_k L_k \hat{e}_k\right)\mathbf{x} = i\mathbf{L}\,\mathbf{x}, \tag{6.63}$$

one obtains (Corben and Schwinger [6]):

$$2i\mathbf{L}\times\mathcal{Y}^{(l1)jm} = [j(j+1) - l(l+1) - 2]\mathcal{Y}^{(l1)jm}. \tag{6.64}$$

Thus, *the vector spherical harmonics are eigenvectors of the operator* $2i\mathbf{L}\times$. Since j may assume only the values $l+1$, l, and $l-1$ ($l \geqslant 1$), the eigenvalues of the operator $2i\mathbf{L}\times$ are correspondingly $2l$, -2, and $-2l-2$. (For $l = 0$ one obtains the single eigenvector $\mathcal{Y}^{(01)1m}$ with eigenvalue 0, since, by definition, the other two vector harmonics are zero.)

The structure suggested by Eq. (6.64) is that of an underlying operator algebra. To support this suggestion, let us observe that the equations expressing the standard action of the orbital angular momentum operators on the solid harmonics become the following single concise statement in terms of the vector solid harmonics:

$$\mathcal{Y}^{(l1)lm} = [l(l+1)]^{-\frac{1}{2}}\mathbf{L}\mathcal{Y}_{lm}, \qquad l \neq 0. \tag{6.65}$$

Using this result in both sides of Eq. (6.64) for $j = l$, we obtain (after canceling factors)

$$(\mathbf{L}\times\mathbf{L})\mathcal{Y}_{lm} = i\mathbf{L}\mathcal{Y}_{lm}; \tag{6.66}$$

that is, $\mathbf{L}\times\mathbf{L} = i\mathbf{L}$. Thus, for $j = l$, the result expressed by Eq. (6.64) is just a restatement of the fundamental commutation relations of angular momentum and the standard angular momentum action on state vectors. Noticing that $\mathbf{L}$ is an eigenvector of the operator $\Gamma_{\mathbf{L}} \equiv \mathbf{L}\times$ – that is,

$$\Gamma_{\mathbf{L}}\mathbf{L} = i\mathbf{L}, \tag{6.67}$$

one is led to attempt to interpret the $j = l \pm 1$ cases of Eq. (6.64) in terms of other (possible) operator eigenvalues of $\Gamma_{\mathbf{L}}$, which we now discuss.

12. Algebraic Aspects of Vector Spherical Harmonics

As might be expected, it is the theory of vector operators (with respect to $\mathbf{L}$) that underlies the properties of the vector spherical harmonics noted above. Let us recall that a vector operator $\mathbf{A}$ (with respect to $\mathbf{L}$) satisfies, by definition, the commutation rules [see Eqs. (3.221) and (3.222)]:

$$\mathbf{L}\times\mathbf{A} + \mathbf{A}\times\mathbf{L} = 2i\mathbf{A}. \tag{6.68}$$

Consider now some consequences of this relation, which are derived directly by standard vector algebra techniques, keeping in mind that the order of operators is important[1] – that is, by definition, we have

$$(\mathbf{A}\times\mathbf{B})_i \equiv A_jB_k - A_kB_j, \qquad i, j, k \text{ cyclic.}$$

We obtain

$$(\mathbf{A}\times\mathbf{L})\times\mathbf{L} = i\mathbf{A}\times\mathbf{L} - \mathbf{A}\mathbf{L}^2 + (\mathbf{A}\cdot\mathbf{L})\mathbf{L}, \tag{6.69}$$

$$[(\mathbf{A}\times\mathbf{L})\times\mathbf{L}]\times\mathbf{L} = i(\mathbf{A}\times\mathbf{L})\times\mathbf{L} - (\mathbf{A}\times\mathbf{L})\mathbf{L}^2 + i(\mathbf{A}\cdot\mathbf{L})\mathbf{L}. \tag{6.70}$$

We have deliberately written these relations in a form in which the angular momentum operators stand to the *right*. This form is significant because it allows us to utilize the known action of the angular momentum operators when we let these operator identities act on $\mathscr{Y}_{lm}$. (A form that places the angular momentum operators to the left would be appropriate for a dual space theory.) For these same reasons, it is also more convenient to define a **x**-product operator $C_{\mathbf{L}}$ by

$$\mathbf{A}C_{\mathbf{L}} \equiv \mathbf{A}\times\mathbf{L}. \tag{6.71}$$

[This seemingly awkward "left-handed action" has definite merits, as we demonstrate below in Eqs. (6.74) and (6.75).]

Using the notation (6.71), we may rewrite Eqs. (6.69) and (6.70) as

$$\mathbf{A}C_{\mathbf{L}}^2 = i\mathbf{A}C_{\mathbf{L}} - \mathbf{A}\mathbf{L}^2 + (\mathbf{A}\cdot\mathbf{L})\mathbf{L}, \tag{6.72}$$

$$\mathbf{A}C_{\mathbf{L}}^3 = i\mathbf{A}C_{\mathbf{L}}^2 - (\mathbf{A}C_{\mathbf{L}})\mathbf{L}^2 + i(\mathbf{A}\cdot\mathbf{L})\mathbf{L}, \tag{6.73}$$

where repeated application of $C_{\mathbf{L}}$ to a vector $\mathbf{A}$ is defined by $\mathbf{A}C_{\mathbf{L}}^n \equiv (\mathbf{A}C_{\mathbf{L}}^{n-1})C_{\mathbf{L}}$, $n = 2, 3, \ldots$. Eliminating the term $(\mathbf{A}\cdot\mathbf{L})\mathbf{L}$ between these two equations, we obtain the *characteristic equation* (Cayley–Hamilton theorem) for the operator $C_{\mathbf{L}}$:

$$\mathbf{A}[C_{\mathbf{L}}^3 - 2iC_{\mathbf{L}}^2 + C_{\mathbf{L}}(\mathbf{L}^2 - 1) - i\mathbf{L}^2] = \mathbf{0}. \tag{6.74}$$

Operating on $\mathscr{Y}_{lm}$ with this vector operator, we obtain

$$[\mathbf{A}(C_{\mathbf{L}} + il)(C_{\mathbf{L}} - i)(C_{\mathbf{L}} - il - i)]\mathscr{Y}_{lm} = 0. \tag{6.75}$$

Accordingly, *the eigenvalues of the operator* $C_{\mathbf{L}}$ *are* $-il$, i, *and* $i(l+1)$; *since* $\mathbf{A}$ *is arbitrary, these are the only possible eigenvalues.*

[1] The extension of the ordinary formulas of vector analysis to vectors having noncommuting components has been given by Shortley and Kimball [11] (see also Condon and Shortley [12]).

Equation (6.74) may be expressed as the operator identity (Cayley–Hamilton theorem for the operator $C_{\mathbf{L}}$):

$$\prod_{\delta=-1}^{+1} (C_{\mathbf{L}} - \Omega_{\delta}) = 0, \tag{6.76}$$

which is valid when applied to an arbitrary vector operator. In this result, the Ω_{δ} are invariant operators defined by

$$\Omega_{\delta} = i - i\delta\left(\frac{D+\delta}{2}\right), \qquad \delta = +1, 0, -1, \tag{6.77}$$

in which $D = \sqrt{4\mathbf{L}^2 + 1}$ is the (positive definite) invariant dimension operator (eigenvalues $2l+1$, $l = 0, 1, \ldots$).

Consider now the vector operators defined by

$$\mathbf{A}^{(\delta)} = \mathbf{A}\prod_{\substack{\lambda \\ \lambda\neq\delta}} (C_{\mathbf{L}} - \Omega_{\lambda}). \tag{6.78}$$

The implication of the Cayley–Hamilton theorem is the eigenvalue equation

$$\mathbf{A}^{(\delta)}C_{\mathbf{L}} = \mathbf{A}^{(\delta)}\Omega_{\delta}, \tag{6.79}$$

or, equivalently,

$$\mathbf{L}\times\mathbf{A}^{(\delta)} = \mathbf{A}^{(\delta)}(2i - \Omega_{\delta}). \tag{6.80}$$

Each of the vector operators $\mathbf{A}^{(\delta)}$ *is an eigenvector of the operator* $\mathbf{L}\times$, *this result being valid for an arbitrary vector operator.*

In order to interpret these results in relation to Eq. (6.64), we use the following identities:

$$[\mathbf{L}^2, \mathbf{A}] = -2i\mathbf{L}\times\mathbf{A} - 2\mathbf{A},$$

$$[L_3, \mathbf{A}] = -i\hat{e}_3\times\mathbf{A} = -S_3\mathbf{A}. \tag{6.81}$$

Applying these relations to the solid harmonic $\mathscr{Y}_{lm}$ and employing $\mathbf{A}^{(\delta)}$ in place of $\mathbf{A}$ yields

$$\mathbf{L}^2(\mathbf{A}^{(\delta)}\mathscr{Y}_{lm}) = (l+\delta)(l+\delta+1)(\mathbf{A}^{(\delta)}\mathscr{Y}_{lm}),$$

$$J_3(\mathbf{A}^{(\delta)}\mathscr{Y}_{lm}) = m(\mathbf{A}^{(\delta)}\mathscr{Y}_{lm}). \tag{6.82}$$

These two results imply that $\mathbf{A}^{(\delta)}\mathscr{Y}_{lm}$ is proportional to $\boldsymbol{\mathscr{Y}}^{(l+\delta,1)lm}$.

We now state the fundamental result for vector operators: *Let* $\mathbf{A}$ *be an arbitrary vector operator and define the vector operators* $\mathbf{A}^{(\delta)}$ *by*

$$\mathbf{A}^{(\delta)} \equiv \mathbf{A} \prod_{\substack{\lambda \\ \lambda \neq \delta}} (C_{\mathbf{L}} - \Omega_\lambda). \tag{6.83}$$

Then

$$\mathbf{A}^{(\delta)} \mathscr{Y}_{lm} = a_{l\delta}(l + \delta \| \mathbf{A} \| l) \boldsymbol{\mathscr{Y}}^{(l+\delta,1)lm}, \tag{6.84}$$

where $(l + \delta \| \mathbf{A} \| l)$ *denotes the reduced matrix element of the vector operator* $\mathbf{A}$ *and the coefficients* $a_{l\delta}$ *are defined by*

$$a_{l\delta} = \begin{cases} (l+1)[(2l+1)(2l+3)]^{\frac{1}{2}}, & \delta = +1 \\ l(l+1), & \delta = 0 \\ l[(2l-1)(2l+1)]^{\frac{1}{2}}, & \delta = -1. \end{cases} \tag{6.85}$$

Proof. It remains only to verify the correctness of the proportionality factor in Eq. (6.84). We do this by comparing Eq. (6.84) with the Wigner–Eckart theorem for vector operators. This yields

$$a_{l\delta}(l + \delta \| \mathbf{A} \| l) = (-1)^\delta \left(\frac{2l + 2\delta + 1}{2l + 1} \right)^{\frac{1}{2}} (l + \delta \| \mathbf{A}^{(\delta)} \| l). \tag{6.86}$$

However, Eq. (6.78) relates the operator $\mathbf{A}$ to the $\mathbf{A}^{(\delta)}$; hence, the reduced matrix elements are related. The basic result required is

$$\begin{aligned} \mathbf{A} \mathscr{Y}_{lm} = &- [(l+1)(2l+1)]^{-1} \mathbf{A}^{(+1)} \mathscr{Y}_{lm} + [l(l+1)]^{-1} \mathbf{A}^{(0)} \mathscr{Y}_{lm} \\ &- [l(2l+1)]^{-1} \mathbf{A}^{(-1)} \mathscr{Y}_{lm} \end{aligned} \tag{6.87}$$

for $l \neq 0$. (For $l = 0$ one retains only the first term in this result. Note that $\mathscr{Y}_{00}$ belongs to the null spaces of the vector operators $\mathbf{A}^{(0)}$ and $\mathbf{A}^{(-1)}$ acting in the space of the spherical harmonics.) Taking matrix elements of Eq. (6.87), we establish the full validity of Eq. (6.84). ■

One must be careful to note that the "reduced matrix elements" occurring in Eq. (6.84) are the quantities that appear in the expansion

$$A_\mu \mathscr{Y}_{lm} = \sum_{l'} (l' \| \mathbf{A} \| l) C^{l\,1\,l'}_{m,\mu,m+\mu} \mathscr{Y}_{l',m+\mu}, \tag{6.88}$$

and, if the vector operator A_μ has a radial dependence, the coefficients $(l'\|\mathbf{A}\|l)$ will, in general, be radial functions. We have introduced the parentheses in place of Dirac bra–kets to denote such functions.

The result expressed by Eq. (6.84) finds many applications, and it is useful to write it out fully in terms of vector notation [expand Eq. (6.83) and use Eq. (6.72)]:

$$\begin{aligned}
\mathbf{A}^{(+1)}\mathscr{Y}_{lm} &= (l+1)[(2l+1)(2l+3)]^{\frac{1}{2}}(l+1\|\mathbf{A}\|l)\boldsymbol{\mathscr{Y}}^{(l+1,1)lm} \\
&= [(\mathbf{A}\cdot\mathbf{L})\mathbf{L} - (l+1)^2\mathbf{A} - i(l+1)(\mathbf{A}\times\mathbf{L})]\mathscr{Y}_{lm}, \\
\mathbf{A}^{(0)}\mathscr{Y}_{lm} &= l(l+1)(l\|\mathbf{A}\|l)\boldsymbol{\mathscr{Y}}^{(l1)lm} \\
&= (\mathbf{A}\cdot\mathbf{L})\mathbf{L}\mathscr{Y}_{lm}, \\
\mathbf{A}^{(-1)}\mathscr{Y}_{lm} &= l[(2l-1)(2l+1)]^{\frac{1}{2}}(l-1\|\mathbf{A}\|l)\boldsymbol{\mathscr{Y}}^{(l-1,1)lm} \\
&= [(\mathbf{A}\cdot\mathbf{L})\mathbf{L} - l^2\mathbf{A} + il(\mathbf{A}\times\mathbf{L})]\mathscr{Y}_{lm}.
\end{aligned} \tag{6.89}$$

It is also useful to display these relations in inverted form:

$$\begin{aligned}
\mathbf{A}\mathscr{Y}_{lm} &= -\left(\frac{2l+3}{2l+1}\right)^{\frac{1}{2}}(l+1\|\mathbf{A}\|l)\boldsymbol{\mathscr{Y}}^{(l+1,1)lm} \\
&\quad + (l\|\mathbf{A}\|l)\boldsymbol{\mathscr{Y}}^{(l1)lm} - \left(\frac{2l-1}{2l+1}\right)^{\frac{1}{2}}(l-1\|\mathbf{A}\|l)\boldsymbol{\mathscr{Y}}^{(l-1,1)lm}, \\
(\mathbf{A}\cdot\mathbf{L})\mathbf{L}\mathscr{Y}_{lm} &= l(l+1)(l\|\mathbf{A}\|l)\boldsymbol{\mathscr{Y}}^{(l1)lm}, \\
i(\mathbf{A}\times\mathbf{L})\mathscr{Y}_{lm} &= -l\left(\frac{2l+3}{2l+1}\right)^{\frac{1}{2}}(l+1\|\mathbf{A}\|l)\boldsymbol{\mathscr{Y}}^{(l+1,1)lm} \\
&\quad - (l\|\mathbf{A}\|l)\boldsymbol{\mathscr{Y}}^{(l1)lm} + (l+1)\left(\frac{2l-1}{2l+1}\right)^{\frac{1}{2}}(l-1\|\mathbf{A}\|l)\boldsymbol{\mathscr{Y}}^{(l-1,1)lm}.
\end{aligned} \tag{6.90}$$

Remarks. (*a*) The algebraic results given by Eqs. (6.68)–(6.90) depend only on the fact that **A** is a vector operator with respect to **L**. Hence, they apply for **L** a generic angular momentum (integers and half-integers allowed), it only being necessary to replace the spherical harmonics by general basis vectors $\{|lm\rangle\}$, $l = 0, \frac{1}{2}, 1, \ldots$.

(*b*) The substitution of Eq. (6.84) into the expansion of **A** given by Eq. (6.87) yields the Wigner–Eckart theorem for vector operators. The algebraic method thus provides an alternative method of proof of the Wigner–Eckart theorem for this special case.

(*c*) The essential features of the algebra of vector operators are contained in Eqs. (6.68)–(6.80). *The basic result that has been obtained is the eigenvalue equation* [Eq. (6.79)]:

$$\mathbf{A}^{(\delta)} C_{\mathbf{L}} = \mathbf{A}^{(\delta)} \Omega_{\delta}. \tag{6.91}$$

Clearly, the procedure has been one of solving, by the method of principal idempotents, a characteristic eigenvalue problem over the space of vector operators in which the scalars are rotational invariants. It is the existence of a Cayley–Hamilton theorem that keeps the problem finite. It is the eigenvalue properties ($l > 0$) that assure that a vector operator **A**, such that **A**, **L**, and $\mathbf{L} \times \mathbf{A}$ are linearly independent (in the numerical sense), will define three vector operators, $\mathbf{A}^{(+1)}$, $\mathbf{A}^{(0)}$, and $\mathbf{A}^{(-1)}$, which span the space of vector operators. One has simply projected out of a generic vector operator the "components" in the direction of the three vector ($J = 1$) Wigner operators having $\Delta = +1, 0, -1$ (see Chapter 3, Section 21). What one has accomplished by the construction given by Eqs. (6.89) is to define *physical vector operators* that up to an invariant multiplicative factor have the action of the three corresponding vector Wigner operators on states.

(*d*) The procedure given above generalizes to arbitrary tensors of rank k and to arbitrary angular momenta **J**. This generalization will be seen from a single remark: Define the operator Λ (see Note 9, Chapter 3) to be commutation with $\mathbf{L}^2$ (from the right):

$$\mathbf{A}\Lambda \equiv [\mathbf{A}, \mathbf{L}^2]. \tag{6.92}$$

Then, the first of Eqs. (6.81) expresses the identity

$$\Lambda = -2iC_{\mathbf{L}} - 2; \tag{6.93}$$

that is,

$$C_{\mathbf{L}} = \frac{i}{2}(\Lambda + 2). \tag{6.94}$$

Thus, the entire algebraic development above, using the operation **x**—which is particular to three-space—can be replaced by the operation of commutation with $\mathbf{L}^2$, which has significance for general tensor operators. We sketch this general theory in RWA.

13. Summary of Properties of Vector Solid Harmonics

The properties of the vector solid harmonics are summarized in this section after we have discussed the use of Eqs. (6.89) and (6.90) in obtaining the results listed.

It is a common practice to use the vector **x** in Eqs. (6.89) to give an alternative [to Eq. (6.59)] definition of the vector solid harmonics. We have followed this practice in tabulating the results (*a*) in Eqs. (6.105).

The operator form of Eqs. (6.89) also has useful applications for $\mathbf{A} = \mathbf{x}$, and we note here these results before discussing the additional properties of the vector solid harmonics given below in (*a*)–(*h*). For this purpose, we have used the reduced matrix elements of **x** given by Eqs. (6.100). Choosing $\mathbf{A} = \mathbf{x}$ in Eq. (6.89), we obtain

$$\mathcal{Y}^{(l+\delta,1)lm} = \mathbf{R}^{(\delta)}\mathcal{Y}_{lm}, \tag{6.95}$$

where $\mathbf{R}^{(\delta)}$ ($\delta = +1, 0, -1$) are the vector operators given by

$$\mathbf{R}^{(+1)} = -\left[\mathbf{x}\left(\frac{D+1}{2}\right) + i\mathbf{x}\times\mathbf{L}\right]\left(\frac{D(D+1)}{2}\right)^{-\frac{1}{2}},$$

$$\mathbf{R}^{(0)} = \mathbf{L}(\mathbf{L}^2)^{-\frac{1}{2}} = \mathbf{L}\left[\frac{(D-1)(D+1)}{2}\right]^{-\frac{1}{2}},$$

$$r^{-2}\mathbf{R}^{(-1)} = -\left[-\mathbf{x}\left(\frac{D-1}{2}\right) + i\mathbf{x}\times\mathbf{L}\right]\left(\frac{[(D-1)D]}{2}\right)^{-\frac{1}{2}}, \tag{6.96}$$

in which D is the dimension operator [the middle result comes directly from Eqs. (6.84) and (6.85)].

One may verify by direct computation that these operators obey the scalar product[1] relation

$$\mathbf{R}^{(\delta)\dagger}\cdot\mathbf{R}^{(\delta')} = \delta_{\delta\delta'}r^{\delta+\delta'}. \tag{6.97}$$

Let us recall also that these vector operators solve the eigenvalue relation, Eq. (6.80), which it is convenient (for a subsequent application) to write in the form

$$i\mathbf{L}\times\mathbf{R}^{(\delta)} + \mathbf{R}^{(\delta)} = -\mathbf{R}^{(\delta)}\left(\frac{D+\delta}{2}\right)\delta. \tag{6.98}$$

---

[1] Technically, one must pay careful attention to the null space of the operator $\mathbf{R}^{(\delta)}$ in formulating this result as well as other relations obtained in this discussion. Some relations will be invalid for $l = 0$.

The results (*a*)–(*h*) below may be obtained from the general results, Eqs. (6.89) and (6.90), applied to the vector operators $\mathbf{r}$, $\mathbf{L}$, ∇F, and $\nabla \times \mathbf{R}^{(\delta)}F$, where F is an arbitrary differentiable function of r. [The juxtaposition of an operator $\mathbf{A}$ such as $\mathbf{A} = \nabla$ or $\mathbf{A} = \nabla \times \mathbf{R}^{(\delta)}$ with a function F denotes "operator" multiplication in the usual sense in that the action of $\mathbf{A}F$ on an arbitrary wave function Ψ is defined by $\mathbf{A}F\Psi \equiv \mathbf{A}(F\Psi)$.] Alternatively, they may be derived by application of the appropriate vector operation to the defining relations (*a*).

Several reduced matrix elements are required to implement Eqs. (6.89) and (6.90) for the special vector operators introduced above.

The reduced matrix elements of $\mathbf{x}$ are found from Eq. (3.437), using

$$(l'\|\mathbf{x}\|l) = \sqrt{\frac{4\pi}{3}}\,\langle l'\|\mathcal{Y}_1\|l\rangle. \tag{6.99}$$

Thus, we obtain

$$(l+1\|\mathbf{x}\|l) = \left(\frac{l+1}{2l+3}\right)^{\frac{1}{2}},$$

$$(l\|\mathbf{x}\|l) = 0,$$

$$(l-1\|\mathbf{x}\|l) = -\left(\frac{l}{2l-1}\right)^{\frac{1}{2}} r^2. \tag{6.100}$$

The reduced matrix elements of the vector operator ∇F are obtained from those of $\mathbf{x}$ by the formula

$$(l'\|\nabla F\|l) = \left\{\frac{dF}{dr} + \frac{[(l+1)(l+2) - l'(l'+1)]F}{2r}\right\}\frac{(l'\|\mathbf{x}\|l)}{r}. \tag{6.101}$$

The verification of this result may be given by using the operator identity

$$[\mathbf{L}^2, \mathbf{x}] = 2\mathbf{x} + 2\mathbf{x}(\mathbf{x}\cdot\nabla) - 2r^2\nabla. \tag{6.102}$$

One may obtain the curl equations below by using the fact that $\nabla \times \mathbf{R}^{(\delta)}F$ is a vector operator, or, more readily, simply by applying the curl operator directly to Eqs. (6.105) (multiplied by F), using the identities[1]

$$\nabla \times \mathbf{x} = -i\mathbf{L},$$

[1] Equations (6.103) and (6.104) are operator identities when applied to differentiable functions.

$$\nabla \times (\mathbf{x} \times \mathbf{L}) = -(\mathbf{x} \cdot \nabla)\mathbf{L} - 2\mathbf{L}. \tag{6.103}$$

Similarly, one obtains the divergence equations below by applying $\nabla \cdot$ directly to Eqs. (6.105) (multiplied by F), using

$$\nabla \cdot \mathbf{L} = 0,$$

$$\nabla \cdot \mathbf{x} = \mathbf{x} \cdot \nabla + 3,$$

$$\nabla \cdot (\mathbf{x} \times \mathbf{L}) = -i\mathbf{L}^2. \tag{6.104}$$

We now summarize the properties we have established for the vector solid harmonics:

(*a*) Defining equations:

$$\boldsymbol{\mathcal{Y}}^{(l+1,1)lm} = -[(l+1)(2l+1)]^{-\frac{1}{2}}[(l+1)\mathbf{x} + i\mathbf{x} \times \mathbf{L}]\mathcal{Y}_{lm}$$

$$\boldsymbol{\mathcal{Y}}^{(l1)lm} = [l(l+1)]^{-\frac{1}{2}}\mathbf{L}\mathcal{Y}_{lm}$$

$$r^2\boldsymbol{\mathcal{Y}}^{(l-1,1)lm} = -[l(2l+1)]^{-\frac{1}{2}}[-l\mathbf{x} + i\mathbf{x} \times \mathbf{L}]\mathcal{Y}_{lm} \tag{6.105}$$

(*b*) Eigenvalue properties:

$$\mathbf{J}^2\boldsymbol{\mathcal{Y}}^{(l1)jm} = j(j+1)\boldsymbol{\mathcal{Y}}^{(l1)jm}$$

$$\mathbf{L}^2\boldsymbol{\mathcal{Y}}^{(l1)jm} = l(l+1)\boldsymbol{\mathcal{Y}}^{(l1)jm}$$

$$\mathbf{S}^2\boldsymbol{\mathcal{Y}}^{(l1)jm} = 2\boldsymbol{\mathcal{Y}}^{(l1)jm}$$

$$J_3\boldsymbol{\mathcal{Y}}^{(l1)jm} = m\boldsymbol{\mathcal{Y}}^{(l1)jm}$$

$$\nabla^2\boldsymbol{\mathcal{Y}}^{(l1)jm} = 0$$

$$2i\mathbf{L} \times \boldsymbol{\mathcal{Y}}^{(l1)jm} = [j(j+1) - l(l+1) - 2]\boldsymbol{\mathcal{Y}}^{(l1)jm}. \tag{6.106}$$

(*c*) Orthogonality:

$$\int dS_{\hat{x}}\boldsymbol{\mathcal{Y}}^{(l'1)j'm'*}(\mathbf{x}) \cdot \boldsymbol{\mathcal{Y}}^{(l1)jm}(\mathbf{x}) = \delta_{l'l}\delta_{j'j}\delta_{m'm}r^{l'+l} \tag{6.107}$$

(*d*) Complex conjugation:

$$\boldsymbol{\mathcal{Y}}^{(l1)jm*} = (-1)^{l+1-j}(-1)^m\boldsymbol{\mathcal{Y}}^{(l1)j,-m} \tag{6.108}$$

(*e*) Vector and gradient formulas:

$$\mathbf{x}\mathcal{Y}_{lm} = -\left[\frac{l+1}{2l+1}\right]^{\frac{1}{2}}\mathcal{Y}^{(l+1,1)lm} + \left[\frac{l}{2l+1}\right]^{\frac{1}{2}} r^2\mathcal{Y}^{(l-1,1)lm}$$

$$[(l+1)\nabla + i\nabla\times\mathbf{L}](F\mathcal{Y}_{lm}) = -[(l+1)(2l+1)]^{\frac{1}{2}}\left(\frac{1}{r}\frac{dF}{dr}\right)\mathcal{Y}^{(l+1,1)lm}$$

$$[-l\nabla + i\nabla\times\mathbf{L}](F\mathcal{Y}_{lm}) = -[l(2l+1)]^{\frac{1}{2}}\left[r\frac{dF}{dr} + (2l+1)F\right]\mathcal{Y}^{(l-1,1)lm}$$

$$\nabla(F\mathcal{Y}_{lm}) = -\left[\frac{l+1}{2l+1}\right]^{\frac{1}{2}}\left(\frac{1}{r}\frac{dF}{dr}\right)\mathcal{Y}^{(l+1,1)lm} + \left[\frac{l}{2l+1}\right]^{\frac{1}{2}}\left(r\frac{dF}{dr} + (2l+1)F\right)\mathcal{Y}^{(l-1,1)lm}$$

$$i\nabla\times\mathbf{L}(F\mathcal{Y}_{lm}) = -l\left(\frac{l+1}{2l+1}\right)^{\frac{1}{2}}\left(\frac{1}{r}\frac{dF}{dr}\right)\mathcal{Y}^{(l+1,1)lm} - (l+1)\left(\frac{l}{2l+1}\right)^{\frac{1}{2}}\left(r\frac{dF}{dr} + (2l+1)F\right)\mathcal{Y}^{(l-1,1)lm} \tag{6.109}$$

(*f*) Curl equations:

$$i\nabla\times(F\mathcal{Y}^{(l+1,1)lm}) = -\left[\frac{l}{2l+1}\right]^{\frac{1}{2}}\left(r\frac{dF}{dr} + (2l+3)F\right)\mathcal{Y}^{(l1)lm}$$

$$i\nabla\times(F\mathcal{Y}^{(l1)lm}) = -\left[\frac{l}{2l+1}\right]^{\frac{1}{2}}\left(\frac{1}{r}\frac{dF}{dr}\right)\mathcal{Y}^{(l+1,1)lm} - \left[\frac{l+1}{2l+1}\right]^{\frac{1}{2}}\left(r\frac{dF}{dr} + (2l+1)F\right)\mathcal{Y}^{(l-1,1)lm}$$

$$i\nabla\times(F\mathcal{Y}^{(l-1,1)lm}) = -\left[\frac{l+1}{2l+1}\right]^{\frac{1}{2}}\left(\frac{1}{r}\frac{dF}{dr}\right)\mathcal{Y}^{(l1)lm} \tag{6.110}$$

(*g*) Divergence equations:

$$\nabla \cdot (F\mathcal{Y}^{(l+1,1)lm}) = -\left[\frac{l+1}{2l+1}\right]^{\frac{1}{2}}\left(r\frac{dF}{dr} + (2l+3)F\right)\mathcal{Y}_{lm}$$

$$\nabla \cdot (F\mathcal{Y}^{(l1)lm}) = 0,$$

$$\nabla \cdot (F\mathcal{Y}^{(l-1,1)lm}) = \left[\frac{l}{2l+1}\right]^{\frac{1}{2}}\left(\frac{1}{r}\frac{dF}{dr}\right)\mathcal{Y}_{lm} \tag{6.111}$$

(*h*) Parity property:

$$\mathcal{Y}^{(l+\delta,1)lm}(-\mathbf{x}) = (-1)^{l+\delta}\mathcal{Y}^{(l+\delta,1)lm}(\mathbf{x}) \tag{6.112}$$

Remarks. (*a*) We have chosen to write Eqs. (6.105)–(6.112) in terms of the vector solid harmonics, since the results are then easily transformed to any curvilinear coordinate system. However, one easily converts our results to vector spherical harmonics by making the substitution

$$\mathcal{Y}^{(l+\delta,1)lm}(\mathbf{x}) = r^{l+\delta}\mathcal{Y}^{(l+\delta,1)lm}(\hat{x}). \tag{6.113}$$

One then customarily (Edmonds [2]) also substitutes

$$F = r^{-l}G \qquad \text{or} \qquad F = r^{-l-\delta}G \tag{6.114}$$

in the left-hand sides of Eqs. (6.109)–(6.111) and re-expresses the right-hand radial functions in terms of G.

(*b*) It is amusing to note that the only vector solid harmonic for $l = 0$ [see Eq. (6.105)] is $\mathcal{Y}^{(11)00}(\mathbf{x}) \propto \mathbf{x}$. Thus, the vector $\mathbf{x}$ is a rotational invariant! This is a consistent, but somewhat unusual viewpoint (see Fano and Racah [12a, p. 5]).

We conclude this summary by noting two more important results: the dot and cross product expansions for two vector solid harmonics. The proof of these relations are elegant applications of the Wigner–Eckart theorem and the general expression, Eq. (3.248), for reduced matrix elements. We state the results first and give the proofs below.

$$\mathcal{Y}^{(l'1)j'm'} \cdot \mathcal{Y}^{(l1)jm}$$

$$= \sum_{l''} r^{l+l'-l''}\left[\frac{(2j+1)(2j'+1)(2l+1)(2l'+1)}{4\pi(2l''+1)}\right]^{\frac{1}{2}} \times$$

$$\times (-1)^{l+j'+l''} C^{ll'l''}_{000} C^{jj'l''}_{m,m',m+m'} \begin{Bmatrix} l' & j' & 1 \\ j & l & l'' \end{Bmatrix} \mathscr{Y}_{l'',m+m'}, \qquad (6.115)$$

$\boldsymbol{\mathscr{Y}}^{(l'1)j'm'} \times \boldsymbol{\mathscr{Y}}^{(l1)jm}$

$$= (-i\sqrt{2}) \sum_{l''j''} r^{l+l'-l''} \left[\frac{(2j+1)(2j'+1)(3)(2l+1)(2l'+1)}{4\pi}\right]^{\frac{1}{2}}$$

$$\times C^{ll'l''}_{000} C^{jj'j''}_{m,m',m+m'} \begin{Bmatrix} l & 1 & j \\ l' & 1 & j' \\ l'' & 1 & j'' \end{Bmatrix} \boldsymbol{\mathscr{Y}}^{(l''1)j'',m+m'}. \qquad (6.116)$$

Proof. Consider the proof of the cross product relation first. It is only necessary to rewrite the scalar product of vector spherical harmonics, Eq. (6.107), in standard bra–ket notation, using the spin-1 realization, $S_\mu = i\hat{\xi}_\mu \times$, to prove the following relation:

$$\int dS_{\hat{x}}\, \boldsymbol{\mathscr{Y}}^{(l''1)j''m''*}(\hat{x}) \cdot (\boldsymbol{\mathscr{Y}}^{(l'1)j'm'}(\hat{x}) \times \boldsymbol{\mathscr{Y}}^{(l1)jm}(\hat{x}))$$

$$= \langle (l''1)j''m''| - i[\mathbf{Y}_{l'} \times \mathbf{S}]^{j'}_{m'} |(l1)jm\rangle,$$

where

$$|(l1)jm\rangle = \sum_{\mu} C^{l1j}_{m-\mu,\mu,m} Y_{l,m-\mu} \otimes \hat{\xi}_\mu,$$

$$[\mathbf{Y}_{l'} \times \mathbf{S}]^{j'}_{m'} = \sum_{\mu'} C^{l'1j'}_{m'-\mu',\mu',m'} Y_{l',m'-\mu'} \otimes S_{\mu'}.$$

Thus, the Wigner–Eckart theorem and the reduced matrix element expression, Eq. (3.248), are directly applicable. The reduced matrix elements of the solid harmonics are given by Eq. (3.437) and the nonvanishing reduced matrix element of $\mathbf{S}$ is $\langle 1\|\mathbf{S}\|1\rangle = \sqrt{2} = [s(s+1)]^{\frac{1}{2}}$ $(s = 1)$. Using these results in Eq. (3.248) yields the cross product expansion, Eq. (6.116).

The dot product expansion has a similar interpretation. Thus, we find

$$\int dS_{\hat{x}}\, \mathscr{Y}^*_{l''m''}(\hat{x}) \boldsymbol{\mathscr{Y}}^{(l'1)j'm'}(\hat{x}) \cdot \boldsymbol{\mathscr{Y}}^{(l1)jm}(\hat{x}) = \langle (l''0)l''m''|[\mathbf{Y}_{l'} \times \mathbf{t}]^{j'}_{m'}|(l1)jm\rangle,$$

where

$$|(l''0)l''m''\rangle = Y_{l''m''} \otimes |00\rangle,$$

in which $|00\rangle$ denotes a spin-0 eigenvector, and $\mathbf{t}$ denotes a rank-1 tensor operator that has an action on the spin-1 eigenvector $\hat{\xi}_\nu$ ($\nu = +1, 0, -1$) given by

$$t_\mu \hat{\xi}_\nu = \hat{\xi}_\mu \cdot \hat{\xi}_\nu |00\rangle = -\sqrt{3}C^{110}_{\nu\mu 0}|00\rangle$$

(see Chapter 3, Section 21). The relevant reduced matrix element is thus $\langle 0\|\mathbf{t}\|1\rangle = -\sqrt{3}$. Using the Wigner–Eckart theorem and Eq. (3.248) for the reduced matrix elements, we obtain the dot product expansion, Eq. (6.115). ■

Remarks. The vector operator $\mathbf{W}^{(0)}$ and $\mathbf{W}^{(-1)}$ with components $(i/\sqrt{2})\hat{\xi}_\mu \times$ and $-t_\mu/\sqrt{3}\ \cdot$, respectively, are realizations of the Wigner operators (see Chapter 3, Section 21):

$$\left\langle \begin{matrix} & 1 & \\ 2 & & 0 \\ & 1+\mu & \end{matrix} \right\rangle \quad \text{and} \quad \left\langle \begin{matrix} & 0 & \\ 2 & & 0 \\ & 1+\mu & \end{matrix} \right\rangle,$$

where the only "states" are those realized by $|00\rangle = 1$ and $|1\mu\rangle = \hat{\xi}_\mu$ (dot and cross products of a vector with a scalar are, by definition, zero). Thus, in this elementary case the structural feature that distinguishes between these two Wigner operators is the definition of a *geometric action* (dot and cross product) on the set of vectors of $\mathbb{R}^3$. Observe also that the multiplication of the operators $W^{(0)}_\mu$ and $W^{(-1)}_\nu$ ($\mu, \nu = 1, 2, 3$) closes:

$$\begin{aligned} W^{(0)}_\mu W^{(0)}_\nu &= C^{111}_{\nu,\mu,\mu+\nu} W^{(0)}_{\mu+\nu}, \\ W^{(0)}_\mu W^{(-1)}_\nu &= 0, \\ W^{(-1)}_\mu W^{(0)}_\nu &= C^{111}_{\nu,\mu,\mu+\nu} W^{(-1)}_{\mu+\nu}, \\ W^{(-1)}_\mu W^{(-1)}_\nu &= 0. \end{aligned} \tag{6.117}$$

We demonstrate in RWA (Note 2, p. 41) that this closure property is general.

14. Decomposition Theorem for Vector Functions Defined on the Sphere

Algebraic methods may be used to derive the so-called *angular momentum Helmholtz theorem* (Lomont and Moses [13], Moses [14]).

The key algebraic relations for establishing this theorem are the eigenvalue properties, Eqs. (6.98), rewritten as

$$\mathbf{R}^{(+1)} = i\mathbf{L}\,\mathbf{x}\,\mathbf{S}^{(+1)} + \mathbf{S}^{(+1)},$$

$$\mathbf{R}^{(0)} = i\mathbf{L}\,\mathbf{x}\,\mathbf{S}^{(0)},$$

$$\mathbf{R}^{(-1)} = i\mathbf{L}\,\mathbf{x}\,\mathbf{S}^{(-1)} + \mathbf{S}^{(-1)}, \tag{6.118}$$

where we have defined[1]

$$\mathbf{S}^{(+1)} \equiv -\,\mathbf{R}^{(+1)}\left(\frac{D+1}{2}\right)^{-1},$$

$$\mathbf{S}^{(0)} \equiv -\,\mathbf{R}^{(0)},$$

$$\mathbf{S}^{(-1)} \equiv \mathbf{R}^{(-1)}\left(\frac{D-1}{2}\right)^{-1}. \tag{6.119}$$

Using the definitions, Eqs. (6.96), one also obtains the following dot and cross product relations satisfied by the vector operators $\mathbf{R}^{(\delta)}$ [see Eqs. (6.96)–(6.98)]:

$$\mathbf{L}\cdot\mathbf{R}^{(\delta)} = 0 \qquad \text{for } \delta = \pm 1,$$

$$\mathbf{L}\cdot\mathbf{R}^{(0)} = (\mathbf{L}^2)^{\frac{1}{2}} \qquad \text{for } \delta = 0,$$

$$i\mathbf{L}\,\mathbf{x}\,\mathbf{R}^{(\delta)} = -\,\mathbf{R}^{(\delta)}\left[\delta\left(\frac{D+\delta}{2}\right)+1\right], \qquad \delta = +1, 0, -1. \tag{6.120}$$

Consider now any vector function $\mathbf{F}$ that can be represented by the expansion

[1] We have divided Eqs. (6.98) by the appropriate dimension factors and introduced the new vector operators $\mathbf{S}^{(\delta)}$ in anticipation of using these latter operators in the definition of "vector potentials" in Eq. (6.129).

$$\mathbf{F} = \sum_{lm\delta} f_{lm}^{(\delta)} \mathcal{Y}^{(l+\delta,1)lm}, \tag{6.121}$$

where $f_{lm}^{(\delta)}$ is a function of r only. Using the defining property, Eq. (6.95), of the vector solid harmonics, we may rewrite Eq. (6.121) as

$$\mathbf{F} = \mathbf{F}^{(+1)} + \mathbf{F}^{(0)} + \mathbf{F}^{(-1)}, \tag{6.122}$$

where[1]

$$\mathbf{F}^{(\delta)} \equiv (\mathbf{R}^{(\delta)} f^{(\delta)}), \tag{6.123}$$

$$f^{(\delta)} \equiv \sum_{lm} f_{lm}^{(\delta)} \mathcal{Y}_{lm}. \tag{6.124}$$

We may conclude immediately from the orthogonality relations, Eqs. (6.97), for the vector operators $\mathbf{R}^{(\delta)}$ that the vector functions $\mathbf{F}^{(\delta)}$ satisfy the orthogonality relations:

$$\int dS_{\hat{x}}\, F^{(\delta)*}(\mathbf{x}) \cdot F^{(\delta')}(\mathbf{x}) = \delta_{\delta\delta'} r^{\delta+\delta'} \sum_{lm} r^{2l} f_{lm}^{(\delta)*}(r) f_{lm}^{(\delta)}(r). \tag{6.125}$$

The vector functions, $\mathbf{F}^{(\delta)}$, also possess dot and cross product properties with $\mathbf{L}$ as a direct consequence of the corresponding properties of the vector operators $\mathbf{R}^{(\delta)}$ [see Eqs. (6.120)]:

(*a*) Dot product relations:

$$(\mathbf{L} \cdot \mathbf{F}^{(\delta)}) = 0, \qquad \delta = \pm 1,$$

$$(\mathbf{L} \cdot \mathbf{F}^{(0)}) = (\mathbf{L}^2)^{\frac{1}{2}} f^{(0)} = \sum_{lm} [l(l+1)]^{\frac{1}{2}} f_{lm}^{(0)} \mathcal{Y}_{lm}; \tag{6.126}$$

(*b*) Cross product relations:

$$i(\mathbf{L} \times \mathbf{F}^{(+1)}) = -\mathbf{R}^{(+1)} \left(\frac{D+3}{2}\right) f^{(+1)}$$

$$= -\sum_{lm} (l+2) f_{lm}^{(+1)} \mathcal{Y}^{(l+1,1)lm},$$

[1] When there is a possibility of an ambiguous interpretation, we use parentheses around an operator $\mathbf{A}$ and a function f in juxtaposition, $(\mathbf{A}f)$, to designate the vector function obtained by the action of $\mathbf{A}$ *on* f.

$$i(\mathbf{L}\times\mathbf{F}^{(0)}) = -\mathbf{F}^{(0)},$$

$$i(\mathbf{L}\times\mathbf{F}^{(-1)}) = \mathbf{R}^{(-1)}\left(\frac{D-3}{2}\right)f^{(-1)} = \sum_{lm}(l-1)f_{lm}^{(-1)}\boldsymbol{\mathscr{Y}}^{(l-1,1)lm}. \quad (6.127)$$

More interesting, however, are the results obtained from the "inverted" form of the eigenvalue equations [Eqs. (6.118)]:

$$\mathbf{F}^{(+1)} = i(\mathbf{L}\times\mathscr{A}^{(+1)}) + \mathscr{A}^{(+1)},$$

$$\mathbf{F}^{(0)} = i(\mathbf{L}\times\mathscr{A}^{(0)}),$$

$$\mathbf{F}^{(-1)} = i(\mathbf{L}\times\mathscr{A}^{(-1)}) + \mathscr{A}^{(-1)}, \quad (6.128)$$

where

$$\mathscr{A}^{(\delta)} \equiv (\mathbf{S}^{(\delta)}f^{(\delta)}). \quad (6.129)$$

These equations establish the existence of "vector potentials," $\mathscr{A}^{(\delta)}$, from which the vector functions $\mathbf{F}^{(\delta)}$ may be derived.

The explicit results obtained from Eqs. (6.129), (6.119), and (6.95) are

$$\mathscr{A}^{(+1)} = -\sum_{lm}(l+1)^{-1}f_{lm}^{(+1)}\boldsymbol{\mathscr{Y}}^{(l+1,1)lm},$$

$$\mathscr{A}^{(0)} = -\sum_{lm}f_{lm}^{(0)}\boldsymbol{\mathscr{Y}}^{(l1)lm},$$

$$\mathscr{A}^{(-1)} = \sum_{\substack{lm\\ l>0}}l^{-1}f_{lm}^{(-1)}\boldsymbol{\mathscr{Y}}^{(l-1,1)lm}. \quad (6.130)$$

Since $\mathscr{A}^{(0)}$ has the form $\mathscr{A}^{(0)} = -i\mathbf{L}\Phi$, where

$$\Phi = -i(\mathbf{L}^2)^{-\frac{1}{2}}f^{(0)} = -i\sum_{\substack{lm\\ l>0}}[l(l+1)]^{-\frac{1}{2}}f_{lm}^{(0)}\mathscr{Y}_{lm}, \quad (6.131)$$

one also obtains $\mathbf{F}^{(0)}$ from the "scalar potential" Φ:

$$\mathbf{F}^{(0)} = i(\mathbf{L}\Phi). \quad (6.132)$$

The same $\mathbf{F}^{(\delta)}$ are obtained if we add an arbitrary constant to Φ and an arbitrary vector function of the form $\mathbf{L}\psi^{(\pm 1)}$ to $\mathscr{A}^{(\pm 1)}$.

The angular momentum Helmholtz theorem asserts: *A vector function* $\mathbf{F}$ *can be decomposed into two components* $\mathbf{F}_1$ *and* $\mathbf{F}_2$,

$$\mathbf{F} = \mathbf{F}_1 + \mathbf{F}_2, \tag{6.133}$$

such that

$$i(\mathbf{L} \times \mathbf{F}_1) = -\mathbf{F}_1, \qquad (\mathbf{L} \cdot \mathbf{F}_2) = 0;$$

$$\mathbf{F}_1 = i(\mathbf{L}\Phi),$$

$$\mathbf{F}_2 = i(\mathbf{L} \times \mathscr{A}) + \mathscr{A}, \tag{6.134}$$

where Φ *and* $\mathscr{A}$ *are "potentials" defined in Eqs.* (6.131) *and* (6.130), *respectively.*

Proof. Choose $\mathbf{F}_1 = \mathbf{F}^{(0)}$, $\mathbf{F}_2 = \mathbf{F}^{(+1)} + \mathbf{F}^{(-1)}$, and $\mathscr{A} = \mathscr{A}^{(+1)} + \mathscr{A}^{(-1)}$ in the results above. ∎

Remark. As shown above an arbitrary (well-behaved) vector function can always be decomposed into *three orthogonal parts*, this result being a consequence of the existence of *three vector Wigner operators*. Each part may be derived from a potential as we have demonstrated.

15. Rotationally Invariant Spherical Functions of Two Vectors

We have discussed general methods of constructing rotational invariants in Chapter 3, Section 20. Here we examine the structure of the orbital rotational invariant $I_l(\mathbf{x}, \mathbf{y})$ defined by

$$I_l(\mathbf{x}, \mathbf{y}) \equiv \frac{4\pi}{2l+1} \sum_m (-1)^m \mathscr{Y}_{lm}(\mathbf{x}) \mathscr{Y}_{l,-m}(\mathbf{y}). \tag{6.135}$$

Although one generally considers this invariant only for real vectors $\mathbf{x}$ and $\mathbf{y}$, it is, in fact, an invariant under *complex orthogonal transformations* [see the comments below Eq. (6.30)]. Accordingly, it is convenient to regard $\mathbf{x}$ and $\mathbf{y}$ as complex vectors in Eq. (6.135). [$\mathscr{Y}_{l,-m}(\mathbf{y})$ for $\mathbf{y}$ complex is obtained directly from the defining relation, Eq. (3.153).]

The relations presented in this section are consequences of the fact that one can also express $I_l(\mathbf{x}, \mathbf{y})$ as a polynomial form in the basic rotational invariants

$$\mathbf{x} \cdot \mathbf{x}, \ \mathbf{y} \cdot \mathbf{y}, \text{ and } \mathbf{x} \cdot \mathbf{y}.$$

The explicit form is

$$I_l(\mathbf{x}, \mathbf{y}) = \frac{1}{2^l} \sum_k (-1)^k \binom{l}{k} \binom{2l-2k}{l} (\mathbf{x} \cdot \mathbf{y})^{l-2k} (\mathbf{x} \cdot \mathbf{x})^k (\mathbf{y} \cdot \mathbf{y})^k. \quad (6.136)$$

Equations (6.135) and (6.136) generalize some of the well-known properties of the solid harmonics; it is appropriate to note these results before turning to the proof of Eq. (6.136).

Example 1. Choose $\mathbf{x}$ and $\mathbf{y}$ to be real unit vectors $\mathbf{x} = \hat{x}$ and $\mathbf{y} = \hat{y}$. Then Eqs. (6.135) and (6.136) yield[1]

$$P_l(\cos\theta) = \frac{4\pi}{2l+1} \sum_m (-1)^m \mathscr{Y}_{l,-m}(\hat{y}) \mathscr{Y}_{lm}(\hat{x}), \quad (6.137)$$

where $\cos\theta = \hat{x} \cdot \hat{y}$ and the P_l are the Legendre polynomials; that is,

$$I_l(\hat{x}, \hat{y}) = P_l(\cos\theta) = \frac{1}{2^l} \sum_k (-1)^k \binom{l}{k} \binom{2l-2k}{l} (\cos\theta)^{l-2k}. \quad (6.138)$$

Example 2. Choose $\mathbf{y} = (0, 0, 1)$ and use

$$\mathscr{Y}_{lm}(0, 0, 1) = \delta_{0,m} \left[\frac{2l+1}{4\pi}\right]^{\frac{1}{2}}$$

to obtain

$$\left[\frac{4\pi}{2l+1}\right]^{\frac{1}{2}} \mathscr{Y}_{l0}(\mathbf{x}) = \frac{1}{2^l} \sum_k (-1)^k \binom{l}{k} \binom{2l-2k}{l} x_3^{l-2k} (\mathbf{x} \cdot \mathbf{x})^k,$$

and, in particular,

$$\left[\frac{4\pi}{2l+1}\right]^{\frac{1}{2}} \mathscr{Y}_{l0}(\hat{x}) = P_l(\cos\theta) \quad (6.139)$$

for all vectors $\hat{x} = (\sin\theta\cos\phi, \sin\theta\,\sin\phi, \cos\theta)$.

---

[1] This relation is known as the *addition theorem for spherical harmonics*; it is usually written in terms of the spherical harmonics corresponding to the solid harmonics in which the unit vectors $\hat{x}$ and $\hat{y}$ are parametrized by spherical polar coordinates (see the *Remarks*, p. 276).

Example 3. Choose $\mathbf{y} = \mathbf{x}$ to obtain

$$(\mathbf{x} \cdot \mathbf{x})^l = \frac{4\pi}{2l+1} \sum_m (-1)^m \mathscr{Y}_{l,-m}(\mathbf{x}) \mathscr{Y}_{lm}(\mathbf{x}) \tag{6.140}$$

by using

$$\frac{1}{2^l} \sum_k (-1)^k \binom{l}{k} \binom{2l-2k}{l} = 1.$$

Example 4. Choose $\mathbf{y} \perp \mathbf{x}$ to obtain

$$\frac{4\pi}{2l+1} \sum_m (-1)^m \mathscr{Y}_{l,-m}(\mathbf{y}) \mathscr{Y}_{lm}(\mathbf{x})$$

$$= \begin{cases} 0 & \text{for } l \text{ odd} \\ (-1)^{l/2} \dfrac{1}{2^l} \dbinom{l}{l/2} (\mathbf{x} \cdot \mathbf{x})^{l/2} (\mathbf{y} \cdot \mathbf{y})^{l/2} & \text{for } l \text{ even.} \end{cases} \tag{6.141}$$

Let us now give the proof of expression (6.136) for $I_l(\mathbf{x}, \mathbf{y})$.

Proof. The only invariants under (complex) rotations that can be built from two vectors $\mathbf{x}$ and $\mathbf{y}$ are functions of $\mathbf{x} \cdot \mathbf{x}$, $\mathbf{x} \cdot \mathbf{y}$, $\mathbf{y} \cdot \mathbf{y}$ (Weyl [15]). Furthermore, since $I_l(\mathbf{x}, \mathbf{y})$ is clearly a polynomial that is homogeneous of degree l in $\mathbf{x}$ and $\mathbf{y}$ separately, and is symmetric in the interchange of $\mathbf{x}$ and $\mathbf{y}$, the most general form of $I_l(\mathbf{x}, \mathbf{y})$ is given by

$$I_l(\mathbf{x}, \mathbf{y}) = \sum_k a_k (\mathbf{x} \cdot \mathbf{y})^{l-2k} (\mathbf{x} \cdot \mathbf{x})^k (\mathbf{y} \cdot \mathbf{y})^k,$$

where the summation is over $k = 0, 1, \ldots, l/2$ or $(l-1)/2$. Finally, $I_l(\mathbf{x}, \mathbf{y})$ must satisfy Laplace's equation in the variables x_1, x_2, x_3 (also in y_1, y_2, y_3), even for the complex case.

For nonnegative integers n and m, one verifies by direct application of $\nabla^2 = \sum_i \partial^2/\partial x_i^2$ that

$$\nabla^2 (\mathbf{x} \cdot \mathbf{y})^n (\mathbf{x} \cdot \mathbf{x})^m = 2m(2m + 2n + 1)(\mathbf{x} \cdot \mathbf{y})^n (\mathbf{x} \cdot \mathbf{x})^{m-1}$$
$$+ n(n-1)(\mathbf{x} \cdot \mathbf{y})^{n-2} (\mathbf{x} \cdot \mathbf{x})^m (\mathbf{y} \cdot \mathbf{y}).$$

The property

$$\nabla^2 I_l(\mathbf{x}, \mathbf{y}) = 0$$

now yields a two-term recursion relation for the coefficients a_k:

$$a_{k-1}(l - 2k + 2)(l - 2k + 1) + a_k(2k)(2l - 2k + 1) = 0.$$

The general solution to this recursion relation is

$$a_k = (-1)^k \binom{l}{k}\binom{2l - 2k}{l} A_l,$$

where A_l is arbitrary. The value of A_l is found to be $1/2^l$ by evaluating $I_l(\mathbf{x}, \mathbf{y})$ at the point $\mathbf{x} = \mathbf{y} = (0, 0, 1)$. ■

The Cartan map, Eq. (2.7), when applied to rotational invariants leads to further useful formulas. We discuss this technique now, since it will be useful for the theory of rotational invariants in several vectors.

16. Applications of the Cartan Map to Spherical Functions

It is generally recognized that the theory of orbital angular momentum can be obtained from the theory of $SU(2)$ rotations by restriction of the latter. {Explicitly, this may be accomplished by the methods discussed in Chapter 3 [see Eq. (3.138) and the discussion below Eq. (3.106)].} *It is less well-known that this procedure may be reversed*; that is, we can obtain the theory of $SU(2)$ rotations from that of orbital rotations *by the use of the Cartan map*, extending the results then to include half-integral angular momenta.

Let us make this result explicit. It is convenient to formulate the Cartan map [see Eq. (2.7)] in terms of bosons:

$$x_1 + ix_2 = -2(a_1)^2, \qquad x_1 - ix_2 = 2(a_2)^2, \qquad x_3 = 2a_1a_2. \tag{6.142}$$

The key relation is

$$\mathcal{Y}_{lm}(\mathbf{x}) = \frac{(2l)!}{l!}\left[\frac{2l + 1}{4\pi}\right]^{\frac{1}{2}} P_{lm}(a_1, a_2), \tag{6.143}$$

where P_{lm} is the boson polynomial given by Eq. (5.33). This result is proved by direct substitution of the map (6.142) into Eq. (3.153), using the identity

$$\sum_k \frac{2^{l+m-2k}}{k!(k - m)!(l + m - 2k)!} = \frac{(2l)!}{l!(l + m)!(l - m)!}.$$

Similarly, if we choose $\mathbf{y}$ to be the (conjugate) Cartan map of the bosons (a_1, a_2) given by

$$y_1 + iy_2 = 2(a_2)^2, \qquad y_1 - iy_2 = -2(a_1)^2, \qquad y_3 = 2a_1a_2, \tag{6.144}$$

we obtain

$$(-1)^m \mathscr{Y}_{l,-m}(\mathbf{y}) = \frac{(2l)!}{l!}\left[\frac{2l+1}{4\pi}\right]^{\frac{1}{2}} P_{lm}(a_1, a_2). \tag{6.145}$$

One can also define the Cartan map of an operator $\mathscr{O}$, which has a well-defined action on the solid harmonics, to an operator Λ, which has a well-defined action on the boson state vectors, by the rule

$$\mathscr{O} \overset{\text{Cartan map}}{\rightarrow} \Lambda \tag{6.146}$$

if

$$F_{lm}(\mathbf{x}) = (\mathscr{O}\mathscr{Y}_{lm})(\mathbf{x}) \overset{\text{Cartan map}}{\rightarrow} f_{lm}(\mathbf{a}) = (\#)\Lambda P_{lm}(\mathbf{a})|0\rangle. \tag{6.147}$$

In general, because of the different normalization rules, it will be necessary to remove a numerical factor, #, before the map (6.146) becomes explicit.

Thus, relations between spherical functions can be mapped into relations between boson polynomials. While the relations thus obtained between boson polynomials are thereby proved to be valid only for integral values of l, the relations are, in fact, valid for half-integral values as well. We shall demonstrate this latter fact by examples.

Example 1. The multiplication law, Eq. (3.436), maps to the relation

$$P_{k\mu}(\mathbf{a})P_{lm}(\mathbf{a}) = \left[\frac{(k+l+m+\mu)!(k+l-m-\mu)!}{(k+\mu)!(k-\mu)!(l+m)!(l-m)!}\right]^{\frac{1}{2}} P_{k+l,m+\mu}(\mathbf{a}). \tag{6.148}$$

Example 2. The generators $\mathbf{L}$ of orbital rotations are mapped to the generators $\mathbf{J}$ [see Eqs. (5.34)] of unitary rotations by the map (6.147).

Example 3. Choose $\mathbf{y}$ in Eqs. (6.135) and (6.136) to be the (conjugate) Cartan map given by Eq. (6.144) to obtain

$$\frac{(\mathbf{y}\cdot\mathbf{x})^l}{2^l l!} = \left[\frac{4\pi}{2l+1}\right]^{\frac{1}{2}} \sum_m P_{lm}(\mathbf{a})\mathscr{Y}_{lm}(\mathbf{x}), \qquad l \text{ integral} \tag{6.149}$$

for

$$y_1 + iy_2 = 2(a_2)^2, \qquad y_1 - iy_2 = -2(a_1)^2, \qquad y_3 = 2a_1a_2.$$

[The expression $(\mathbf{y}\cdot\mathbf{x})^l$ is sometimes called a *generating function* for the solid harmonics in consequence of the expansion (6.149).]

Example 4. Replace (a_1, a_2) by $(a_1^1, a_2^1) = \mathbf{a}^1$ in Eq. (6.149) and choose $\mathbf{x}$ to be the Cartan map (6.142) of $(a_1^2, a_2^2) = \mathbf{a}^2$ to obtain

$$(a_{1\,2}^{1\,2})^{2l}/(2l)! = \sum_m (-1)^{l-m} P_{lm}(\mathbf{a}^1)P_{l,-m}(\mathbf{a}^2), \tag{6.150}$$

where we note that

$$\mathbf{y}\cdot\mathbf{x} \to -2(a_{1\,2}^{1\,2})^2, \tag{6.151}$$

in which $a_{1\,2}^{1\,2} = \det A$ [see Eq. (5.54)].

17. Rotationally Invariant Spherical Functions in Several Vectors

The important structure theorem for rotational invariants is a special case of a general theorem proved by Weyl [15, p. 53]: *Every even invariant depending on n vectors* $\mathbf{x}^1, \mathbf{x}^2, \ldots, \mathbf{x}^n$ *in the space* $\mathbb{R}^3$ *is expressible in terms of the* n^2 *scalar products* $\mathbf{x}^\alpha\cdot\mathbf{x}^\beta$. *Every odd invariant is a sum of terms*

$$[\mathbf{x}^\alpha\cdot(\mathbf{x}^\beta\times\mathbf{x}^\gamma)]I(\mathbf{x}^1, \ldots, \mathbf{x}^n), \tag{6.152}$$

where $\mathbf{x}^\alpha, \mathbf{x}^\beta, \mathbf{x}^\gamma$ *are selected from* $\mathbf{x}^1, \ldots, \mathbf{x}^n$ *and* $I(\mathbf{x}^1, \ldots, \mathbf{x}^n)$ *is an even invariant.*

The invariant (see Ref. [16]) defined by

$$\begin{aligned} I_{(l_1l_2l_3)}(\mathbf{x}^1, \mathbf{x}^2, \mathbf{x}^3) \equiv {} & \frac{(4\pi)^{\frac{3}{2}}}{[(2l_1+1)(2l_2+1)(2l_3+1)]^{\frac{1}{2}}} \sum_{m_1m_2m_3} \begin{pmatrix} l_1 & l_2 & l_3 \\ m_1 & m_2 & m_3 \end{pmatrix} \\ & \times \mathscr{Y}_{l_1m_1}(\mathbf{x}^1)\mathscr{Y}_{l_2m_2}(\mathbf{x}^2)\mathscr{Y}_{l_3m_3}(\mathbf{x}^3) \end{aligned} \tag{6.153}$$

plays an important role in the theory of angular correlations in reactions (Chapter 7, Section 8). It is an even (odd) invariant for even (odd) $l_1 + l_2 + l_3$. Notice that

$$I_{(l_1l_2l_3)}(\mathbf{x}^1, \mathbf{x}^2, \mathbf{0}) = \delta_{l_1l_2}\delta_{l_30}(-1)^{l_1}I_{l_1}(\mathbf{x}^1, \mathbf{x}^2)/(2l_1+1)^{\frac{1}{2}}.$$

The properties of the invariant spherical functions in two vectors are thus subsumed under those of the invariant spherical functions in three vectors.

Applying the Weyl theorem stated above, we find that it must also be possible to express $I_{(l)}$ in the form:

$$I_{(l_1 l_2 l_3)}(\mathbf{x}^1, \mathbf{x}^2, \mathbf{x}^3) = [\mathbf{x}^1 \cdot (\mathbf{x}^2 \times \mathbf{x}^3)]^{\varepsilon} I^{(\varepsilon)}_{(l_1 l_2 l_3)}(\mathbf{x}^1, \mathbf{x}^2, \mathbf{x}^3), \tag{6.154}$$

where

$$\varepsilon = \begin{cases} 0 & \text{for } l_1 + l_2 + l_3 \text{ even,} \\ 1 & \text{for } l_1 + l_2 + l_3 \text{ odd.} \end{cases}$$

In this result $I^{(\varepsilon)}_{(l)}$ is an even invariant, which may be written in the form:

$$I^{(\varepsilon)}_{(l_1 l_2 l_3)}(\mathbf{x}^1, \mathbf{x}^2, \mathbf{x}^3) = \sum_{(k)} A_{\varepsilon} \begin{bmatrix} l_1 & l_2 & l_3 \\ k_1 & k_2 & k_3 \end{bmatrix} \prod_{\alpha \leqslant \beta} (\mathbf{x}^{\alpha} \cdot \mathbf{x}^{\beta})^{k_{\alpha\beta}}, \tag{6.155}$$

where the $A_{\varepsilon}[:\,:\,:]$ denote numerical coefficients, and the indices $k_{\alpha\beta}$ are related to the indices $(k) = (k_1, k_2, k_3)$ and $(l) = (l_1, l_2, l_3)$ by

$$k_{\alpha\alpha} = (l_{\alpha} - k_{\alpha} - \varepsilon)/2, \qquad \alpha = 1, 2, 3,$$

$$k_{\beta\alpha} \equiv k_{\alpha\beta} = (k_{\alpha} + k_{\beta} - k_{\gamma})/2, \qquad (\alpha, \beta, \gamma) \text{ cyclic in } 1, 2, 3.$$

Moreover, the $k_{\alpha}\,(\alpha = 1, 2, 3)$ are nonnegative integers such that

(*a*) $k_1 + k_2 + k_3$ is even and (k_1, k_2, k_3) satisfy the triangle conditions;

(*b*) $k_{\alpha} = l_{\alpha} - \varepsilon, l_{\alpha} - \varepsilon - 2, \ldots,$ 1 or 0. (6.156)

The summation on (k) is over all such triples.

The permutational symmetry of the 3-*j* coefficients in Eq. (6.153) implies a corresponding symmetry of the functions $I_{(l_1 l_2 l_3)}(\mathbf{x}^1, \mathbf{x}^2, \mathbf{x}^3)$ under *simultaneous* permutations of $(l_1 l_2 l_3)$ and $(\mathbf{x}^1 \mathbf{x}^2 \mathbf{x}^3)$: $I_{(l_1 l_2 l_3)}(\mathbf{x}^1, \mathbf{x}^2, \mathbf{x}^3)$ is invariant under even permutations and multiplied by $(-1)^{l_1 + l_2 + l_3}$ under odd permutations.

This symmetry implies a symmetry of the coefficients in Eq. (6.155): *The coefficient*

$$A_{\varepsilon} \begin{bmatrix} l_1 & l_2 & l_3 \\ k_1 & k_2 & k_3 \end{bmatrix} \tag{6.157}$$

is invariant under all permutations of its columns.

Unfortunately, the expression for the general coefficient (6.157) has not been given in the literature, and one has had to work out these invariant polynomials from the definition, Eq. (6.153).

We give here a general relation (product law), which the invariant polynomials

$$I_{(l)}(\mathbf{x}) = I_{(l_1 l_2 l_3)}(\mathbf{x}^1, \mathbf{x}^2, \mathbf{x}^3) \tag{6.118}$$

satisfy, and a special, but useful, case of these invariants expanded in terms of the scalar products $\mathbf{x}^\alpha \cdot \mathbf{x}^\beta$:

(*a*) Product law:

$$I_{(l)}(\mathbf{x})I_{(k)}(\mathbf{x}) = \sum_{(j)} \left[\prod_{\alpha=1}^{3} (-1)^{j_\alpha}(2j_\alpha + 1) \begin{pmatrix} l_\alpha & k_\alpha & j_\alpha \\ 0 & 0 & 0 \end{pmatrix} (\mathbf{x}^\alpha \cdot \mathbf{x}^\alpha)^{(l_\alpha + k_\alpha - j_\alpha)/2} \right]$$

$$\times \begin{Bmatrix} l_1 & l_2 & l_3 \\ k_1 & k_2 & k_3 \\ j_1 & j_2 & j_3 \end{Bmatrix} I_{(j)}(\mathbf{x}). \tag{6.159}$$

(*b*) Coplanar vectors:

$$I_{(l)}(\mathbf{x}^1, \mathbf{x}^2, \alpha\mathbf{x}^1 + \beta\mathbf{x}^2)$$

$$= \sum_{kl} \left[\frac{(2l_3 + 1)!}{(2l_3 - 2k)!(2k)!} \right]^{\frac{1}{2}} \alpha^{l_3 - k}\beta^k$$

$$\times (-1)^{l_1 + l_3 + k}(2l + 1) \begin{pmatrix} l_3 - k & l_1 & l \\ 0 & 0 & 0 \end{pmatrix} \begin{pmatrix} k & l_2 & l \\ 0 & 0 & 0 \end{pmatrix} \begin{Bmatrix} l_3 - k & l_3 & k \\ l_2 & l & l_1 \end{Bmatrix}$$

$$\times (\mathbf{x}^1 \cdot \mathbf{x}^1)^{(l_1 + l_3 - l - k)/2}(\mathbf{x}^2 \cdot \mathbf{x}^2)^{(l_2 + k - l)/2} \mathcal{I}_l(\mathbf{x}^1, \mathbf{x}^2). \tag{6.160}$$

The derivation of Eq. (6.159) is a straightforward application of the product law, Eq. (3.436), for the solid harmonics and a standard reduction of a multiple summation over Wigner coefficients to a 9-*j* symbol, using Eq. (3.316).

The proof of Eq. (6.160) requires a new result, the *vectorial addition theorem* for the solid harmonics, which is given below [Eq. (6.176)]. Using this addition theorem, the proof of Eq. (6.160) is again an application of the product law for solid harmonics and a reduction of a summation over Wigner coefficients to a Racah coefficient using Eq. (3.267).

An interesting application of the Cartan map can be made to Eq. (6.153). We put

$$x_1^\alpha + ix_2^\alpha = -2(a_1^\alpha)^2, \qquad x_1^\alpha - ix_2^\alpha = 2(a_2^\alpha)^2, \qquad x_3^\alpha = 2a_1^\alpha a_2^\alpha \tag{6.161}$$

noting that

$$\mathbf{x}^{\alpha} \cdot \mathbf{x}^{\beta} = -2(a^{\alpha\beta}_{12})^2,$$

$$\mathbf{x}^1 \cdot (\mathbf{x}^2 \times \mathbf{x}^3) = -4ia^{12}_{12}a^{23}_{12}a^{31}_{12}. \tag{6.162}$$

Making this map in Eq. (6.153), we obtain [see Eqs. (6.143) and (D-8) in Appendix D, Chapter 5]

$$I_{(l)}(\mathbf{x}) \to [(l_1 + l_2 + l_3 + 1)!]^{-\frac{1}{2}} \prod_{\alpha=1}^{3} \frac{(2l_\alpha)!}{(l_\alpha)!} \prod_{\substack{\alpha\beta\gamma \\ \text{cyclic}}} \frac{(a^{\alpha\beta}_{12})^{l_\alpha + l_\beta - l_\gamma}}{[(l_\alpha + l_\beta - l_\gamma)!]^{\frac{1}{2}}}. \tag{6.163}$$

On the other hand, this same Cartan map applied to the form given by Eqs. (6.154) and (6.155) yields

$$I_{(l)}(\mathbf{x}) \to (-i)^{\varepsilon}(-2)^{(l_1 + l_2 + l_3 + \varepsilon)/2}$$

$$\times A_\varepsilon \begin{bmatrix} l_1 & l_2 & l_3 \\ l_1 - \varepsilon & l_2 - \varepsilon & l_3 - \varepsilon \end{bmatrix} \prod_{\substack{\alpha\beta\gamma \\ \text{cyclic}}} (a^{\alpha\beta}_{12})^{l_\alpha + l_\beta - l_\gamma}.$$

Comparing this last result with Eq. (6.163), we obtain the special coefficient:

$$A_\varepsilon \begin{bmatrix} l_1 & l_2 & l_3 \\ l_1 - \varepsilon & l_2 - \varepsilon & l_3 - \varepsilon \end{bmatrix}$$

$$= \frac{(i)^{\varepsilon}}{(-2)^L[(2L - \varepsilon + 1)!]^{\frac{1}{2}}} \prod_{\alpha=1}^{3} \frac{(2l_\alpha)!}{l_\alpha!} \prod_{\substack{\alpha\beta\gamma \\ \text{cyclic}}} \frac{1}{[(l_\alpha + l_\beta - l_\gamma)!]^{\frac{1}{2}}}, \tag{6.164}$$

where $L \equiv (l_1 + l_2 + l_3 + \varepsilon)/2$ and $\varepsilon = 0, 1$.

Finally, applying the Cartan map to the product law (6.159), we obtain the value of the following 9-j coefficient:

$$\begin{Bmatrix} l_1 & l_2 & l_3 \\ k_1 & k_2 & k_3 \\ l_1 + k_1 & l_2 + k_2 & l_3 + k_3 \end{Bmatrix}$$

$$= \left[\frac{[1 + \sum_\alpha (l_\alpha + k_\alpha)]!}{(1 + \sum_\alpha l_\alpha)!(1 + \sum_\alpha k_\alpha)!}\right]^{\frac{1}{2}} \times \prod_{\alpha=1}^{3} \left[\frac{(2l_\alpha)!(2k_\alpha)!}{(2l_\alpha + 2k_\alpha + 1)!}\right]^{\frac{1}{2}} \times$$

$$\times \prod_{\substack{\alpha\beta\gamma \\ \text{cyclic}}} \left[\frac{(l_\alpha + k_\alpha + l_\beta + k_\beta - l_\gamma - k_\gamma)!}{(l_\alpha + l_\beta - l_\gamma)!(k_\alpha + k_\beta - k_\gamma)!} \right]^{\frac{1}{2}}. \qquad (6.165)$$

(This result is then valid for half-integers as well.)

18. Relationship of Solid Harmonics to Potential Theory

The form of the solid harmonics given by Eq. (3.153) is adapted to quantum mechanics where one is led to consider the simultaneous diagonalization of the observables $\mathscr{L}^2$ and $\mathscr{L}_3$. Traditionally, however, spherical harmonics had their origin in *potential theory* and the study of *harmonic functions* (solutions to Laplace's equation). We indicate here briefly the connection between the solid harmonics developed above and the classical results of potential theory (Maxwell [17]).

One begins with the observation that $1/r$ satisfies Laplace's equation:

$$\nabla^2 \left(\frac{1}{r}\right) = 0, \qquad r \neq 0. \qquad (6.166)$$

Since the derivatives $\partial/\partial x_i$ commute with ∇^2, it follows then that the functions defined by

$$F_{(\lambda)}(\mathbf{x}) \equiv \left(\frac{\partial}{\partial x_1}\right)^{\lambda_1} \left(\frac{\partial}{\partial x_2}\right)^{\lambda_2} \left(\frac{\partial}{\partial x_3}\right)^{\lambda_3} \left(\frac{1}{r}\right) \qquad (6.167)$$

are also harmonic for each triple $(\lambda) = (\lambda_1, \lambda_2, \lambda_3)$ of nonnegative integers.

$F_{(\lambda)}(\mathbf{x})$ is clearly homogeneous of degree $-\lambda_1 - \lambda_2 - \lambda_3 - 1$. Using $\partial r^{-1}/\partial x_i = -x_i/r^3$, one sees, however, that $F_{(\lambda)}(\mathbf{x})$ has the form

$$F_{(\lambda)}(\mathbf{x}) = P_{(\lambda)}(\mathbf{x})/r^{2\lambda_1 + 2\lambda_2 + 2\lambda_3 + 1}, \qquad (6.168)$$

where $P_{(\lambda)}(\mathbf{x})$ is a homogeneous polynomial of degree $l = \lambda_1 + \lambda_2 + \lambda_3$. The relation (6.15) between $\mathscr{L}^2, r^2\nabla^2$, and $\mathbf{x} \cdot \nabla$, and the fact that $\mathscr{L}^2$ commutes with functions of r, allow us to conclude:

(*a*) $P_{(\lambda)}(\mathbf{x})$ is an eigenfunction of $\mathscr{L}^2$ with eigenvalue $l(l + 1)$;
(*b*) $P_{(\lambda)}(\mathbf{x})$ is harmonic.

The number of homogeneous harmonic polynomials of degree l defined by Eqs. (6.167) and (6.168) is $(l + 1)(l + 2)/2$. Since the dimension of the space $\mathscr{H}_l$ is $2l + 1$, the polynomials $P_{(\lambda)}(\mathbf{x})$, $\lambda_1 + \lambda_2 + \lambda_3 = l$, are *dependent* $(l > 1)$.

We can, however, use the idea behind Eq. (6.167) to obtain an interesting relationship for the solid harmonics, $\mathscr{Y}_{lm}(\mathbf{x})$.

The solid harmonics $\mathscr{Y}_{lm}(\mathbf{x})$ may be obtained as a sum of derivatives of $1/r$. The appropiate operator action is given by[1]

$$\mathscr{Y}_{lm}(\mathbf{\nabla})\left(\frac{1}{r}\right) = \frac{(-1)^l(2l)!}{2^l l!}\frac{\mathscr{Y}_{lm}(\mathbf{x})}{r^{2l+1}}, \tag{6.169}$$

where $\mathscr{Y}_{lm}(\mathbf{\nabla})$ denotes the differential operator obtained from the solid harmonics, Eq. (3.153), by replacing x_i by $\partial/\partial x_i$.

Proof. The proof of Eq. (6.169) may be given by noticing that $\mathscr{Y}_{lm}(\mathbf{\nabla})$ is a tensor operator with respect to $\mathscr{L}$; hence,

$$[\mathscr{L}_3, \mathscr{Y}_{lm}(\mathbf{\nabla})] = m\mathscr{Y}_{lm}(\mathbf{\nabla}). \tag{6.170}$$

Accordingly, the function $\mathscr{Y}_{lm}(\mathbf{\nabla})(1/r)$ is a simultaneous eigenfunction of $\mathscr{L}^2$ and $\mathscr{L}_3$ with eigenvalues $l(l+1)$ and m, respectively. The polynomial in the numerator on the right-hand side of Eq. (6.169) must therefore be proportional to $\mathscr{Y}_{lm}(\mathbf{x})$. The proportionality factor must be independent of m, and a simple recursive calculation (for $m = l$) establishes the value of this constant. ■

A principal result in potential theory is the expansion of the potential $V(\mathbf{x})$ at a point $\mathbf{x}$ due to a particle (unit mass or charge) located at point $\mathbf{y}$ into an absolutely convergent series about $\mathbf{y}$ (see Fig. 6.1).

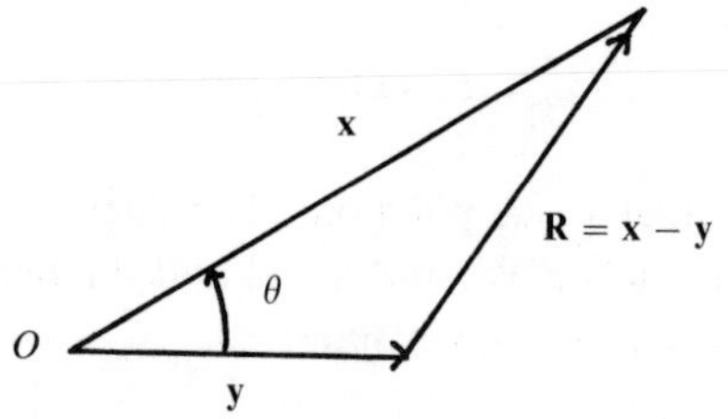

Figure 6.1.

[1] Observe, however, that this relationship requires knowledge of the solid harmonics themselves.

The potential is $1/R$ and has the expansion [for $(\mathbf{x}\cdot\mathbf{x})^{\frac{1}{2}} \geqslant (\mathbf{y}\cdot\mathbf{y})^{\frac{1}{2}}$]

$$1/R = H_0(\mathbf{x},\mathbf{y}) + H_1(\mathbf{x},\mathbf{y}) + \cdots, \tag{6.171}$$

where $H_l(\mathbf{x},\mathbf{y})$ is harmonic and homogeneous of degree l in $\mathbf{y}$ and harmonic in $\mathbf{x}$. Indeed, the expansion (6.171) is the Taylor series expansion for an analytic function given by

$$\frac{1}{R} = e^{-\mathbf{y}\cdot\mathbf{\nabla}}\left(\frac{1}{r}\right) = \sum_l \frac{(-1)l}{l!}(\mathbf{y}\cdot\mathbf{\nabla})^l\left(\frac{1}{r}\right). \tag{6.172}$$

Thus, we find

$$H_l(\mathbf{x},\mathbf{y}) = \frac{(-1)^l}{l!}(\mathbf{y}\cdot\mathbf{\nabla})^l\left(\frac{1}{r}\right). \tag{6.173}$$

Clearly, $H_l(\mathbf{x},\mathbf{y})$ is harmonic in $\mathbf{x}$ (since $\mathbf{y}\cdot\mathbf{\nabla}$ commutes with $\mathbf{\nabla}^2$), and, furthermore, it is a rotational invariant.

We shall now prove that

$$H_l(\mathbf{x},\mathbf{y}) = I_l(\mathbf{x},\mathbf{y})/r^{2l+1}, \tag{6.174}$$

where $I_l(\mathbf{x},\mathbf{y})$ is the invariant polynomial defined by Eq. (6.135). [Hence, $H_l(\mathbf{x},\mathbf{y})$ is also harmonic in $\mathbf{y}$.]

Proof. Replace $\mathbf{x}$ by $\mathbf{\nabla}$ in Eq. (6.135) and operate on $1/r$, using Eqs. (6.136) and (6.169), noting that $\mathbf{\nabla}^2(1/r) = 0$. ■

Using the homogeneity properties of $I_l(\mathbf{x},\mathbf{y})$ in $\mathbf{x}$ and $\mathbf{y}$, we now obtain the classical result [see Eqs. (6.137) and (6.138)]:

$$\frac{1}{R} = \sum_l P_l(\cos\theta)\frac{s^l}{r^{l+1}}, \qquad s \leqslant r \tag{6.175}$$

where r and s denote the lengths of $\mathbf{x}$ and $\mathbf{y}$, respectively, and θ is the angle between $\mathbf{x}$ and $\mathbf{y}$ (see Fig. 6.1). For $s \geqslant r$, the expansion still has the form (6.171), where we replace r by s in Eq. (6.174) and exchange r and s in Eq. (6.175).

19. The Orbital Rotation Matrices as Forms

We have proved previously [Eq. (6.32)] that the elements of the orbital rotation matrix $\mathscr{D}^l(R)$ may be written as homogeneous polynomial forms in the elements R_{ij} of R. These forms are interesting for several reasons: (*a*) They illustrate in a particularly vivid manner the distinctions between orbital rotations and unitary [$SU(2)$] rotations (as discussed below); and (*b*) they are irreps (when extended) of the group of complex orthogonal matrices.

We sketch here a method for obtaining these forms directly from Eq. (6.30). For this purpose we require the *vectorial addition theorem for the solid harmonics*:[1]

$$\mathcal{Y}_{lm}(\mathbf{z}+\mathbf{z}') = \sum_{k\mu}\left[\frac{4\pi(2l+1)!}{(2l-2k+1)!(2k+1)!}\right]^{\frac{1}{2}} C^{l-k\,k\,l}_{m-\mu,\mu,m} \times \mathcal{Y}_{l-k,m-\mu}(\mathbf{z})\mathcal{Y}_{k\mu}(\mathbf{z}'), \tag{6.176}$$

where we note that

$$C^{l-k\,k\,l}_{m-\mu,\mu,m} = \left[\binom{l+m}{k+\mu}\binom{l-m}{k-\mu}\Big/\binom{2l}{2k}\right]^{\frac{1}{2}}. \tag{6.177}$$

This result (as a relation between polynomials) is valid for arbitrary complex vectors $\mathbf{z}$ and $\mathbf{z}'$. {Equation (6.176) may be proved by direct substitution of the definition [Eq. (3.153)] of the solid harmonics into the right-hand side, followed by carrying out three internal summations.}

We now carry out the determination of $\mathcal{D}^l_{m'm}(R)$ in four steps:

(*a*) Set $\mathbf{x} = (-1, i, z)$ in Eq. (6.30) noting from Eq. (3.153) that

$$\mathcal{Y}_{lm}(-1, i, z) = \begin{cases} \left[\dfrac{(2l+1)(l+m)!}{4\pi(l-m)!}\right]^{\frac{1}{2}} \dfrac{z^{l-m}}{m!}, & m \geqslant 0 \\ 0, & m < 0. \end{cases}$$

The transformed point $\tilde{R}\mathbf{x}$ corresponding to $\mathbf{x}$ is then

$$\mathbf{x}' \equiv \tilde{R}\mathbf{x} = \boldsymbol{\zeta} + z\boldsymbol{\eta},$$

where

$$\zeta_j = -R_{1j} + iR_{2j}, \qquad \eta_j = R_{3j}.$$

[1] This relation for the solid harmonics, in which the argument vectors in (complex) three-space are added, is distinct from the relation given by the "addition theorem for spherical harmonics," Eq. (6.137), which refers to "addition of arc lengths" on the unit sphere (spherical trigonometry); hence, the designation "vectorial addition theorem" for relation (6.176).

(*b*) Expand $\mathscr{Y}_{lm}(\boldsymbol{\zeta} + z\boldsymbol{\eta})$, using the addition theorem, Eq. (6.176), with $\mathbf{z} = \boldsymbol{\zeta}$ and $\mathbf{z}' = z\boldsymbol{\eta}$.

(*c*) Equate powers of z in the two sides of Eq. (6.30) resulting from steps (*a*) and (*b*).

(*d*) Repeat the procedure, using the point $\mathbf{x} = (1, i, z)$.

The results of these calculations are

$$\mathscr{D}^l_{l-k,m}(R) = 4\pi \left[\frac{k!(l-k)!(l-k)!}{(2l-2k+1)(2k+1)(2l-k)!}\right]^{\frac{1}{2}}$$

$$\times \sum_\mu \left[\binom{l+m}{k+\mu}\binom{l-m}{k-\mu}\right]^{\frac{1}{2}} \mathscr{Y}_{l-k,m-\mu}(\boldsymbol{\zeta})\mathscr{Y}_{k\mu}(\boldsymbol{\eta}), \quad (6.178)$$

$$\mathscr{D}^l_{k-l,m}(R) = 4\pi \left[\frac{k!(l-k)!(l-k)!}{(2l-2k+1)(2k+1)(2l-k)!}\right]^{\frac{1}{2}}$$

$$\times \sum_\mu \left[\binom{l+m}{k+\mu}\binom{l-m}{k-\mu}\right]^{\frac{1}{2}} \mathscr{Y}_{l-k,m-\mu}(\boldsymbol{\xi})\mathscr{Y}_{k\mu}(\boldsymbol{\eta}), \quad (6.179)$$

where $k = 0, 1, \ldots, l$ and

$$\xi_j \equiv R_{1j} + iR_{2j}, \qquad \eta_j \equiv R_{3j}, \qquad \zeta_j \equiv -R_{1j} + iR_{2j}. \quad (6.180)$$

We have thus obtained explicitly the elements of the unitary matrices, $\mathscr{D}^l(R)$, $l = 0, 1, \ldots,$ *representing the group of proper orthogonal matrices, independently of any particular parametrization.* The $\mathscr{D}_{m'm}(R)$ are expressed as sums over products of pairs of solid harmonics whose arguments are complex numbers formed from the R_{ij}.

Interesting particular cases of Eqs. (6.178) and (6.179) are

$$\mathscr{D}^l_{lm}(R) = l! \left[\frac{4\pi}{(2l+1)!}\right]^{\frac{1}{2}} \mathscr{Y}_{lm}(\boldsymbol{\zeta}),$$

$$\mathscr{D}^l_{-l,m}(R) = l! \left[\frac{4\pi}{(2l+1)!}\right]^{\frac{1}{2}} \mathscr{Y}_{lm}(\boldsymbol{\xi}),$$

$$\mathscr{D}^l_{0,m}(R) = \left[\frac{4\pi}{2l+1}\right]^{\frac{1}{2}} \mathscr{Y}_{lm}(\boldsymbol{\eta}). \quad (6.181)$$

For each $R \in 0(3)$, one may also verify the relations:

$$[\mathscr{D}^l_{m'm}(R)]^* = (-1)^{m'-m}\mathscr{D}^l_{-m',-m}(R) = \mathscr{D}^l_{mm'}(\tilde{R}). \quad (6.182)$$

20. The Orbital Rotation Matrices Are Equivalent to Real Orthogonal Matrices

Consider the unitary change of basis of the space $\mathscr{H}_l$ [(see Eq. (6.14)] given by

$$\bar{\mathscr{Y}}_{lm} = -\frac{1}{\sqrt{2}}[\mathscr{Y}_{lm} + (-1)^m \mathscr{Y}_{l,-m}],$$

$$\bar{\mathscr{Y}}_{l0} = \mathscr{Y}_{l0},$$

$$\bar{\mathscr{Y}}_{l,-m} = \frac{i}{\sqrt{2}}[\mathscr{Y}_{lm} - (-1)^m \mathscr{Y}_{l,-m}], \tag{6.183}$$

where $m = l, l-1, \ldots, 1$ in these equations. In consequence of the property $\mathscr{Y}^*_{lm} = (-1)^m \mathscr{Y}_{l,-m}$, these functions are real:

$$\bar{\mathscr{Y}}^*_{lm} = \bar{\mathscr{Y}}_{lm}. \tag{6.184}$$

Hence, under the transformations $\mathscr{T}_R$ defined by Eq. (6.4), these basis functions will yield real rotation matrices:

$$(\mathscr{T}_R \bar{\mathscr{Y}}_{lm})(\mathbf{x}) = \bar{\mathscr{Y}}_{lm}(R^{-1}\mathbf{x}) = \sum_{m'} \bar{\mathscr{D}}^l_{m'm}(R) \bar{\mathscr{Y}}_{lm'}(\mathbf{x}). \tag{6.185}$$

The relation between the real rotation matrices $\bar{\mathscr{D}}^l(R)$ and the unitary rotation matrices $\mathscr{D}^l(R)$ is given by

$$\bar{\mathscr{D}}^l(R) = A^l \mathscr{D}^l(R) A^{l\dagger}, \tag{6.186}$$

where A^l is the $(2l+1) \times (2l+1)$ unitary matrix defined by the transformation (6.183):

$$\bar{\mathscr{Y}}_{lm} = \sum_{m'} A^{l*}_{mm'} \mathscr{Y}_{lm'}. \tag{6.187}$$

21. The "Doubled-Valued Representations" of the Proper Orthogonal Group $SO(3)$

We have shown earlier that the orbital angular momentum operator $\mathscr{L}$ can only possess states of integer values $l = 0, 1, \ldots$ of angular momenta. Correspondingly, $\mathscr{L}$ generates rotation matrices $\mathscr{D}^l(R)$ having only these integer values of l and odd-dimensional $(2l + 1)$ irreps of $SO(3)$, the group of proper orthogonal matrices. The question arises as to whether there exist *irreducible* unitary matrices of dimensions 2, 4, ..., which represent the group of proper orthogonal matrices.

That no even dimensional *irreps* of $SO(3)$ exist may be proved by showing that the even dimensional irreps of $SU(2)$ are not irreps of $SO(3)$.

To address this question, let us consider the relationship (2.16) between proper orthogonal matrices and unitary unimodular matrices. For each R, there exist two U's, which we now denote by $+U(R)$ and $-U(R)$, that satisfy Eq. (2.16); they are given explicitly by Eqs. (2.23) and (2.21). Either solution, $\pm U(R)$, when substituted back into Eq. (2.16), gives R—that is, $R(\pm U(R)) = R$. Thus, putting $U = \pm U(R)$ in the explicit form for $D^l(U)$ given by Eq. (3.86), we obtain, for integral l,

$$D^l(\pm U(R)) = D^l(U(R)) = \mathscr{D}^l(R), \tag{6.188}$$

where $\mathscr{D}^l(R)$ is given by Eqs. (6.178) and (6.179).

This result expresses the fact that, despite the occurrence of the square root factor in Eq. (2.23), it is always possible, for integral l, to express the functions $D^l_{mm'}(U(R))$ as homogeneous polynomials of degree l in the R_{ij}—that is, in the form $\mathscr{D}^l_{mm'}(R)$. Equivalently, for integral l the elements of $D^l(U)$ can always be expressed as polynomials in the elements $D^1_{\mu\mu'}(U)$ of $D^1(U)$ (these matrix elements are quadratic in the elements u_{ij} of U).

Let us consider now whether or not it is possible to use this same procedure to *define* $\mathscr{D}^j(R)$ for j half-integral.

In order to be specific, it is convenient to use the parametrization $(x_0, \mathbf{x})$ of U: $U(x_0, \mathbf{x}) = x_0\sigma_0 - i\mathbf{x}\cdot\boldsymbol{\sigma}$, $x_0^2 + \mathbf{x}\cdot\mathbf{x} = 1$; that is, $(x_0, \mathbf{x})$ is a point on the surface of the unit sphere, S^3. For prescribed R, there are two diametrically opposite points on the surface of S^3 that yield R. The first solution point may be picked *uniquely* by specifying that the first nonzero component of

$$[x_0(R), x_1(R), x_2(R), x_3(R)], \qquad \text{each } R \in SO(3), \tag{6.189}$$

shall be positive. For convenience of expression, we shall say that the point (6.189) lies on the *upper* hemisphere of S^3. The second solution point (diametrically opposite) then lies on the *lower* hemisphere and is the point

$$[-x_0(R), -x_1(R), -x_2(R), -x_3(R)]. \tag{6.190}$$

The unitary unimodular matrix $U(R)$, defined by the point (6.189) is

$$U(R) = x_0(R)\sigma_0 - i\mathbf{x}(R)\cdot\boldsymbol{\sigma}, \tag{6.191}$$

while the unitary unimodular matrix defined by the point (6.190) is $-U(R)$.

Corresponding to the two solution points, Eqs. (6.189) and (6.190), we may define two sets of matrices:

$$\{\mathscr{D}^j(R) \equiv D^j(U(R)) : R \in SO(3)\},$$

$$\{\mathscr{D}'^j(R) \equiv D^j(-U(R)) = -D^j(U(R)) : R \in SO(3)\}, \qquad (6.192)$$

where j is half-integral. [Since the elements of $D^j(U(x_0, \mathbf{x}))$ are continuous functions (they are polynomials) on S^3, the two solutions $\mathscr{D}^j(R)$ and $\mathscr{D}'^j(R)$ must join smoothly.]

The question now is whether either of the two sets of matrices of Eq. (6.192) is a representation of the group of proper orthogonal matrices – that is, whether or not they multiply according to the group law.

Using the matrix defined by the first of Eqs. (6.192), we find

$$\mathscr{D}^j(R)\mathscr{D}^j(R') = D^j(U(R))D^j(U(R')) = D^j(U(R)U(R')). \qquad (6.193)$$

By definition, $U(R)$ and $U(R')$ are parametrized by points belonging to the upper hemisphere – that is, they are given by Eq. (6.191).

Consider now the point of S^3 defined by the product $U(R)U(R')$:

$$\zeta_0(R, R') \equiv x_0(R)x_0(R') - \mathbf{x}(R) \cdot \mathbf{x}(R'),$$

$$\boldsymbol{\zeta}(R, R') \equiv x_0(R)\mathbf{x}(R') + x_0(R')\mathbf{x}(R) + \mathbf{x}(R) \times \mathbf{x}(R'). \qquad (6.194)$$

The point $(\zeta_0, \boldsymbol{\zeta})$ will sometimes (for certain pairs R, R') belong to the upper hemisphere, sometimes to the lower.

If we put $R'' \equiv RR'$ and again take $U(R'')$ to be the matrix having the (unique) parameters $(x_0(R''), \mathbf{x}(R''))$ belonging to the upper hemisphere, then

$$(x_0(R''), \mathbf{x}(R'')) = S(R, R')(\zeta_0, \boldsymbol{\zeta}), \qquad (6.195)$$

where $S(R, R')$ denotes the *sign* of the first nonvanishing component of $(\zeta_0, \boldsymbol{\zeta})$ as determined from Eqs. (6.194). Thus, *by selecting parametrization points always to belong to the upper hemisphere, we find that the multiplication rule is*

$$U(R)U(R') = S(R, R')U(R''). \qquad (6.196)$$

Corresponding to this multiplication of $SU(2)$ matrices, we have the multiplication rule for the rotation matrix:

$$\mathscr{D}^j(R)\mathscr{D}^j(R') = [S(R, R')]^{2j}\mathscr{D}^j(R'') \qquad (6.197)$$

for $RR' = R''$.

The result of the above analysis is as follows: The matrices $\mathcal{D}^j(R)$, j half-integral, are *not* a matrix representation of the group of proper orthogonal matrices in the strict definition of a representation, because the factor $S(R, R')$ takes on the values $+1$ or -1 (in a systematic fashion). [The second set of matrices defined by Eqs. (6.192) also possesses this feature.] These matrices are said to be a *representation up to a* ($\pm$) *factor, or a projective representation.*

Remarks. The distinction between the true representations $\mathcal{D}^j(R)$, j integral, and the $\mathcal{D}^j(R)$, j half-integral, almost disappears in either parametrization $R(\phi, \hat{n})$ or $R(\alpha\beta\gamma)$ of R—the functions have the same formal appearance, j merely appearing as an integer or half-integer. The essential difference in these functions is underscored when they are considered as functions of the R_{ij}: In one case (j integral), we are dealing with homogeneous polynomials; in the other case (j half-integral), we are dealing with functions that necessarily contain the irrational factor $[1 + \operatorname{tr} R]^{\frac{1}{2}}$. Indeed, the "double-valued property" has its origin in the solution of "quadratic-like" equations.

22. Note

We proved in Section 5 of this chapter that there exists no Hilbert space of even dimension of square-integrable functions (of $\mathbf{x}$) over the unit sphere on which the action of the orbital angular momentum operator $\mathbf{L} = \mathbf{x} \times \mathbf{p}$ is Hermitian and irreducible. The method of proof (Ref. [18]) made use of the fact that the condition $L_-^{2j+1} \mathcal{Y}_{jj} = 0$ (j half-integral) cannot be satisfied by the Hermitian action of $\mathbf{L}$ on a (finite) Hilbert space. The purpose of this Note[1] is to survey the literature that has considered the problem of half-integral orbital angular momentum l, a topic of perennial interest to physicists.

The question of whether half-integral values of (l, m) are allowed in the eigenvalues $l(l+1)$ of $\mathbf{L}^2$ and m of L_3 was of concern early in quantum mechanics and may be subsumed under the more general question of whether the wave functions of physical systems are necessarily single-valued.[2] (Many textbooks invoke single-valuedness in order to exclude half-integral orbital angular momenta l—this is an unnecessary assumption, as we have seen.) Eddington [20] considered the possibility of using multivalued wave functions, and Schrödinger [21] showed that (under very general assumptions) there were only two alternatives: Ψ is either a single-valued function of position, or it is a double-valued function,[3] the two values at a given point differing only by a sign.

It is generally agreed that the *probability density* $|\Psi(\mathbf{x})|^2$ should be single-valued. Doubts concerning the single-valued condition for $\Psi(\mathbf{x})$ itself have

---

[1] We are indebted to Dr. M. Lohe for assistance in preparing this survey.

[2] Merzbacher [19] has given a nice survey of the early literature dealing with the single-valuedness problem and the orbital angular momentum problem.

[3] We agree strongly with Dieudonné [21a, p. 198] that the concept of a "multivalued function" is meaningless, but this unfortunate language is common in the physics literature.

been expressed most concisely by Blatt and Weisskopf [8, footnotes on pp. 783 and 787].

Pauli [22, p. 126] was very much concerned with the single-valuedness problem, in general. He also examined [23] the double-valued eigensolutions of the orbital angular momentum problem in some detail (see also Merzbacher [19]). It is this latter problem to which this Note is addressed.

Pauli recognized that the repeated application of the ladder operators $L_\pm$ ultimately generate from a particular eigensolution $\phi_{jm}(\mathbf{x})$ (j, m half-integral) singular functions with respect to which $\mathbf{L}$ fails to be Hermitian (Merzbacher [19] gives a detailed discussion of Pauli's analysis). Pauli's results again demonstrate that the assumptions made (see footnote p. 274) in the general (j, m)-multiplet construction are violated for $\mathbf{L} = \mathbf{x} \times \mathbf{p}$ and l half-integral.

Buchdahl [24], using a three-dimensional harmonic oscillator model (and only the canonical commutation relations between $\mathbf{x}$ and $\mathbf{p}$), concludes that only integral l can occur. This result is, however, irrelevant to the general problem – it shows that only integral l can arise when one splits the Hilbert space of harmonic oscillator states into invariant and irreducible subspaces with respect to $\mathbf{L} = \mathbf{x} \times \mathbf{p}$.

Whippman [25] rederives results given by Pauli [23] by constructing solutions to the differential equations $\mathbf{L}^2 Y_{lm} = l(l+1) Y_{lm}$, $L_3 Y_{lm} = m Y_{lm}$ for both integral and half-integral l ($m = l, l-1, \ldots, -l$). These solutions are characterized by

$$Y_{lm} = (\#) L_-^{l-m} Y_{ll}, \qquad Y_{l,-m} = (\#) L_+^{l-m} Y_{ll}^*, \tag{6.198}$$

where $Y_{ll} = a_l (\sin\theta)^l e^{il\phi}$, $\#$ denotes a nonvanishing number, and, for l half-integral, one has that m is nonnegative: $m = l-1, l-2, \ldots, \frac{1}{2}$.

The difficulties here arise because the solutions $Y_{l,\frac{1}{2}}$ and $Y_{l,-\frac{1}{2}}$ are no longer related by the ladder operators:

$$\begin{aligned} L_- Y_{l,\frac{1}{2}} &\neq (\#) Y_{l,-\frac{1}{2}}, \\ L_+ Y_{l,-\frac{1}{2}} &\neq (\#) Y_{l,\frac{1}{2}}. \end{aligned} \tag{6.199}$$

As recognized by Pauli [23], and reemphasized by Merzbacher [19], these relations imply the key result: *For l half-integral,* $\mathbf{L}$ *fails to be a Hermitian operator on states generated by repeated application of* $\mathbf{L}$ *to eigenstates of* $\mathbf{L}^2$ *and* L_3. This property alone is sufficient to reject half-integral values of l (within the conventional framework of quantum mechanics). The arguments given by Whippman excluding half-integral l because of undesirable or unphysical properties of the functions $L_- Y_{l,\frac{1}{2}}$ and $L_+ Y_{l,-\frac{1}{2}}$ are subordinate to the failure of the Hermitian property.

Green [26] uses the factorization of $\mathbf{L}^2$ given by $\mathbf{L}^2 = (\boldsymbol{\sigma} \cdot \mathbf{L})(\boldsymbol{\sigma} \cdot \mathbf{L} + 1)$, the Dirac identity, $(\mathbf{a} \cdot \boldsymbol{\sigma})(\mathbf{b} \cdot \boldsymbol{\sigma}) = (\mathbf{a} \cdot \mathbf{b})\sigma_0 + i(\mathbf{a} \times \mathbf{b}) \cdot \boldsymbol{\sigma}$, and $\mathbf{r} \cdot \mathbf{L} = 0$ to prove[1] that $(\boldsymbol{\sigma} \cdot \mathbf{r})(\boldsymbol{\sigma} \cdot \mathbf{L} - l)\psi$ and $(\boldsymbol{\sigma} \cdot \mathbf{r})(\boldsymbol{\sigma} \cdot \mathbf{L} + l + 1)\psi$ are eigenvectors of $\mathbf{L}^2$ with eigenvalues $l(l-1)$ and $(l+1)(l+2)$, respectively, if ψ is an eigenvector with eigenvalue $l(l+1)$. Using the Hermitian, semipositive definite property of $\mathbf{L}^2$,

---

[1] See the discussion of Dirac's κ quantum number in Chapter 7, Section 4f.

he then concludes that $l = \frac{1}{2}$ cannot occur, since $\frac{1}{2}(\frac{1}{2} - 1) < 0$. The key point, however, is the use of the Hermiticity of **L** in establishing $l(l + 1) \geqslant 0$.

Rorschach [26a] constructs raising and lowering operators [see Eqs. (6.96)] that shift l by one unit; in effect, this adjoins the Hermitian operators **x** and **p** to $\mathbf{L} = \mathbf{x} \times \mathbf{p}$ and suffices to obtain an orbital $SO(4)$ Lie algebra in which only integer l are admitted (see the discussion below).

The work of Pandres [27] and Pandres and Jacobson [28] attempts to include half-integral l in the definition of spherical harmonics by introducing a concept of a "nonzero function" that is a "representative" of the zero vector in a Hilbert space. This concept has not been given sufficient mathematical basis to be considered as providing an alternative to standard angular momentum theory [or to the usual theory of the irreps of $SU(2)$].

Another method [18, 26a, 29–31] for showing that only integral values of orbital angular momentum can occur considers the Lie algebra $\mathbf{L} \times \mathbf{L} = i\mathbf{L}$ to be a subalgebra of the Lie algebra of the orthogonal group $SO(4)$. Thus, introducing the vector operator (with respect to **L**) defined by $\mathbf{K} = (K_1, K_2, K_3)$, $K_j = -i(x_j\,\partial/\partial x_4 - x_4\,\partial/\partial x_j)$, one finds

$$\begin{aligned} \mathbf{L} \times \mathbf{L} &= i\mathbf{L}, \\ \mathbf{L} \times \mathbf{K} &= -\mathbf{K} \times \mathbf{L} = i\mathbf{K}, \\ \mathbf{K} \times \mathbf{K} &= i\mathbf{L}. \end{aligned} \tag{6.200}$$

Defining $\mathbf{J}_1 = (\mathbf{L} + \mathbf{K})/2$ and $\mathbf{J}_2 = (\mathbf{L} - \mathbf{K})/2$, one obtains the Lie algebra (6.200) in the $SU(2) \times SU(2)$ form

$$\mathbf{J}_1 \times \mathbf{J}_1 = i\mathbf{J}_1, \qquad \mathbf{J}_2 \times \mathbf{J}_2 = i\mathbf{J}_2, \tag{6.201}$$

where each component of $\mathbf{J}_1$ commutes with each component of $\mathbf{J}_2$. Using now the Wigner coupled angular momentum basis $|(j_1 j_2)lm\rangle$ corresponding to the angular momentum addition $\mathbf{L} = \mathbf{J}_1 + \mathbf{J}_2$, one finds that the abstract algebra (6.201) (Hermitian operators acting on a Hilbert space) admits, *in general*, the l-values

$$l = |j_1 - j_2|, \ldots, j_1 + j_2, \tag{6.202}$$

where j_i $(i = 1, 2)$ may be integral or half-integral; hence, l may be integral or half-integral. The half-integral values are excluded in the realization of the algebra obtained from **L** and **K** in consequence of the algebraic identity

$$\mathbf{L} \cdot \mathbf{K} = \mathbf{J}_1^2 - \mathbf{J}_2^2 = 0, \tag{6.203}$$

which requires $j_1 = j_2$; hence, $l = 0, 1, \ldots, 2j_1$.

The structure of the orbital rotation group within the framework of the general tensor and spinor representations of the orthogonal group $SO(n)$ can also be determined through the boson calculus using the concept of a traceless boson operator (Lohe and Hurst [32]; Lohe [33]).

The best reference for a detailed survey of the literature on this subject is a critical review by van Winter [34]. This paper continues the review by Merzbacher, but inadvertently omits Ref. [26a].

References

1. R. M. F. Houtappel, H. van Dam, and E. P. Wigner, "The conceptual basis and use of the geometric invariance principles," *Rev. Mod. Phys.* **37** (1965), 595–632.

1a. E. P. Wigner, *Group Theory and Its Application to the Quantum Mechanics of Atomic Spectra.* Academic Press, New York, 1959. Translation by J. J. Griffin of the 1931 German edition.

1b. E. P. Wigner, *Application of Group Theory to the Special Functions of Mathematical Physics.* Lecture notes, unpublished, Princeton University, Princeton, N.J., 1955.

1c. J. D. Talman, *Special Functions: A Group Theoretic Approach.* Benjamin, New York, 1968.

1d. N. Vilenkin, *Special Functions and the Theory of Group Representations* (transl. from the Russian; Amer. Math. Soc. Transl., Vol. 22). Amer. Math. Soc., Providence, R. I., 1968.

1e. W. Miller, Jr., *Symmetry and Separation of Variables. Encyclopedia of Mathematics and its Applications.* Vol. 4, Addison-Wesley Publ. Co., Reading, Massachusetts, 1977.

1f. T. Koornwinder, "The addition formula of Jacobi polynomials and spherical harmonics," in *Lie Algebras: Applications and Computational Methods* (B. Kolman, ed.), pp. 68–78. SIAM, Philadelphia, Penn., 1973.

2. A. R. Edmonds, *Angular Momentum in Quantum Mechanics.* Princeton Univ. Press, Princeton, N. J., 1957.

3. M. E. Rose, *Elementary Theory of Angular Momentum.* Wiley, New York, 1957.

4. M. E. Rose, *Relativistic Electron Theory.* Wiley, New York, 1961.

4a. W. B. Campbell, "Tensor and spinor spherical harmonics ${}_sY_{lm}(\theta, \phi)$," *J. Math. Phys.* **12** (1971), 1763–1770.

5. W. W. Hansen, "New type of expansion in radiation problems," *Phys. Rev.* **47** (1935), 139–143.

6. H. C. Corben and J. Schwinger, "The electromagnetic properties of mesotrons," *Phys. Rev.* **58** (1940), 953–968.

7. V. B. Berestetski, *J. Expt. Theoret. Phys. (USSR)* **18** (1948), 1057–1069; *ibid.*, 1070–1080 (in Russian).

8. J. M. Blatt and V. F. Weisskopf, *Theoretical Nuclear Physics.* Wiley, New York, 1952.

9. L. C. Biedenharn and M. E. Rose, "Theory of angular correlations of nuclear radiations," *Rev. Mod. Phys.* **25** (1953), 729–777.

10. L. C. Biedenharn, *Notes on Multipole Fields.* Lecture notes at Yale University, New Haven, Conn., 1952, unpublished.

11. G. H. Shortley and G. E. Kimball, "Analysis of non-commuting vectors with application to quantum mechanics and vector calculus," *Proc. Natl. Acad. Sci. U.S.* **20** (1934), 82–84.

12. E. U. Condon and G. H. Shortley, *The Theory of Atomic Spectra.* Cambridge Univ. Press, New York, 1935.

12a. U. Fano and G. Racah, *Irreducible Tensorial Sets.* Academic Press, New York, 1959.

13. J. S. Lomont and H. E. Moses, "An angular momentum Helmholtz theorem," *Commun. Pure Appl. Math.* **14** (1961), 69–76.

14. H. E. Moses, "A simple proof of the angular momentum Helmholtz theorem and the relation of the theorem to the decomposition of solenoidal vectors into poloidal and toroidal components," *J. Math. Phys.* **17** (1976), 1821–1823.

15. H. Weyl, *The Classical Groups: Their Invariants and Representations.* Princeton Univ. Press, Princeton, N. J., 1946.

16. L. C. Biedenharn, "Angular correlations in nuclear spectroscopy," in *Nuclear Spectroscopy*: Part B (F. Ajzenberg-Selove, ed.), pp. 732–810. Academic Press, New York, 1960.

17. J. C. Maxwell, *A Treatise on Electricity and Magnetism,* Vol. I, 3rd ed. Oxford Univ. Press, London, 1892, reprinted 1904 (see Chapter IX).

18. J. D. Louck, "Special nature of orbital angular momentum," *Amer. J. Phys.* **31** (1963), 378–383.
19. E. Merzbacher, "Single valuedness of wave functions," *Amer. J. Phys.* **30** (1962), 237–247.
20. A. S. Eddington, *Relativity Theory of Protons and Electrons.* Cambridge Univ. Press, New York, 1936.
21. E. Schrödinger, "Die Mehrdeutigkeit der Wellenfunktion," *Ann. Physik* **32** (1938), 49–55.
21a. J. Dieudonné, *Foundations of Modern Analysis.* Academic Press, New York, 1969.
22. W. Pauli, "Allgemeine Prinzipien der Wellenmechanik," in *Handbuch der Physik* (H. Geiger and K. Scheel, eds.), Vol. 24, Part 1, pp. 83–272. Springer Berlin, 1933. Later published in *Encyclopedia of Physics* (S. Flügge, ed.), Vol. 5, Part 1. Springer, Berlin, 1958 (see pp. 45, 46).
23. W. Pauli, "Über ein Kriterium für Ein- oder Zweiwertigkeit der Eigenfunktionen in der Wellenmechanik," *Helv. Phys. Acta* **12** (1939), 147–168.
24. H. A. Buchdahl, "Remark concerning the eigenvalues of orbital angular momentum," *Amer. J. Phys.* **30** (1962), 829–831.
25. M. L. Whippman, "Orbital angular momentum in quantum mechanics," *Amer. J. Phys.* **34** (1966), 656–659.
26. H. S. Green, *Matrix Mechanics in Quantum Mechanics.* Barnes and Noble, New York, 1968.
26a. H. E. Rorschach, "Single-valued wave functions and orbital angular momentum," *Bull. Amer. Phys. Soc.* **7** (1962), 121.
27. D. Pandres, "Schrödinger basis for spinor representations of the three-dimensional rotation group," *J. Math. Phys.* **6** (1965), 1098–1102.
28. D. Pandres and D. A. Jacobson, "Scalar product for harmonic functions of the group $SU(2)$," *J. Math. Phys.* **9** (1968), 1401–1403.
29. J. D. Louck, *Theory of Angular Momentum in N-Dimensional Space.* Los Alamos Scientific Laboratory Report LA-2451, 1960, unpublished.
30. L. C. Biedenharn, "Wigner coefficients for the R_4 group and some applications," *J. Math. Phys.* **2** (1961), 433–441.
31. C. E. Wulfman, "Dynamical groups in atomic and molecular physics," in *Group Theory and Its Applications* (E. M. Loebl, ed.), Vol. II, pp. 145–197. Academic Press, New York, 1971.
32. M. A. Lohe and C. A. Hurst, "The boson calculus for the orthogonal and symplectic groups," *J. Math. Phys.* **12** (1971), 1882–1889.
33. M. A. Lohe, "Spinor representations of the orthogonal groups," *J. Math. Phys.* **14** (1973), 1959–1964.
34. C. van Winter, "Orbital angular momentum and group representations," *Ann. of Phys.* (N.Y.) **47** (1968), 232–274.

CHAPTER 7

Some Applications to Physical Problems

1. Introductory Remarks

The concepts and techniques of angular momentum theory are by now all but ubiquitous in present-day quantum physics; there is scarcely a single issue of a physics journal that does not in some article directly use angular momentum constructs. It would accordingly be rather fatuous, if not indeed quite impossible, to attempt any definitive (or even very detailed) survey of applications.

The purpose of this chapter is therefore rather more modest; we shall focus attention on several typical applications of the theory, in widely different contexts, with an emphasis toward illustrating the basic principles. In discussing any particular field of application, however, we have attempted not to be extensive, but instead have chosen to present a novel, or, at least, unfamiliar, aspect of the subject in some detail. (The discussion in Section 6g – of a result due to Casimir – provides a good example of our intentions.)

2. Basic Principles Underlying the Applications

Before turning to specific examples, it will be useful to review briefly the principles underlying the applications. We have seen that a symmetry in quantum physics is a transformation[1] on state vectors (active view – vectors in

[1] This usage of "active" versus "passive" transformations in Hilbert space and operator space is taken from Houtappel *et al.* [1]; it accords with the definition of active and passive transformations in $\mathbb{R}^3$ given in Chapter 2, Section 2, when $\mathbb{R}^3$ is viewed as a linear space.

ENCYCLOPEDIA OF MATHEMATICS and Its Applications, Gian-Carlo Rota (ed.).
Vol. 8: L. C. Biedenharn and J. D. Louck, Angular Momentum in
Quantum Physics: Theory and Application

Hilbert space are transformed and operators are kept fixed) and on observables (passive view – vectors in Hilbert space are kept fixed and operators are transformed) such that all probabilities are preserved (Chapter 3, Section 14). Using the Wigner theorem, we found that this led to either a unitary, or an antiunitary, realization of the symmetry by operators.

The physical problems most amenable to solution by invariance techniques are characterized by belonging to linear manifolds that have a finite basis; we have mentioned (in Chapter 1) that this accounts for the enhanced importance of symmetry in quantum physics as opposed to classical physics (where such manifolds do not occur).

It follows that in applications where the overall symmetry structure is the Poincaré ($\mathscr{P}$) group or the Galilei ($\mathscr{G}$) group that the most important case is one in which the states are restricted[1] such that the spatial components of the momentum are zero (that is, the states are invariant under spatial displacements) and the energy has a definite value E.

In effect, one has restricted the general, $\mathscr{P}$ or $\mathscr{G}$, invariance to the rotation subgroup, augmented (if valid) by the various possible reflections. This is, for example, the point of view of spectroscopy, where one is concerned only with the *internal* energy of a given system and not with its kinetic energy as a whole.

Aside from the rather uninteresting case where there are no states of zero momentum and fixed energy, there are two fundamentally different cases to be distinguished:[2] (*a*) The states singled out belong to the discrete spectrum of the Hamiltonian; or (*b*) the states singled out belong to the continuous spectrum of the Hamiltonian.

Both cases are important. For case (*a*), one recognizes that this is the problem of spectroscopy (molecular, atomic, nuclear, and/or hadronic). The application of symmetry techniques to spectroscopy has been, and continues to be, one of the major successes in quantum physics. As Wigner [1] has expressed it: "It is hardly an exaggeration to say that all qualitative rules of spectroscopy, including the intensity ratios of multiplet spectra, can be derived from invariance principles – some of them exact, others approximate." Examples taken from atomic spectroscopy are discussed in Sections 3–5; from nuclear spectroscopy in Section 9; and from molecular spectroscopy in Section 10.

---

[1] The states so restricted form a linear manifold but, strictly speaking, do not lie in the original Hilbert space (since this space contains no states with sharply defined momentum). Using a limiting process, a new Hilbert space whose elements are ideal elements (limiting cases) can, however, be defined (see Houtappel *et al.* [1], von Neumann [2], Mackey [3]).

[2] In principle, one might have discrete states embedded in the continuum. An example of this occurs in the perturbation approach to autoionization in the helium atom (Fano [4, and references therein]; Friedrichs [5]).

For case (*b*) – that is, states lying in the continuous spectrum – one recognizes that this implies that the system is unbound and can separate into two or more components. There are accordingly two types of phenomena involving continuous spectra to which symmetry principles, and hence angular momentum techniques, apply most advantageously. The first of these concerns the angular distribution of disintegration products and, even more important, the correlations between the direction of motion of successively emitted particles in the disintegration. The second type of process in which invariance principles have proved most useful concerns the angular distribution of products resulting from collisions – that is, the angular distributions in scattering and reaction processes. We shall discuss, in Section 8, examples of both types of application. The possible measurements in these two types of phenomena are not limited to observing directions of motion only, but may also include measurements of the polarization state of one or more of the particles involved (having nonzero spin). Angular momentum techniques are of particular importance in polarization measurements, especially in the relativistic regime; this is discussed in Sections 6 and 7.

The central importance of angular momentum theory in both the relativistic and nonrelativistic domains should be clear even from this brief review (see Houtappel *et al.* [1, p. 628]).

References

1. R. M. F. Houtappel, H. van Dam, and E. P. Wigner, "The conceptual basis and use of the geometric invariance principles," *Rev. Mod. Phys.* **37** (1965), 595–632.
2. J. von Neumann, "On rings of operators. Reduction theory," *Ann. Math.* **50** (1949), 401–485.
3. G. W. Mackey, "Induced representations of locally compact groups. II. The Frobenius reciprocity theorem," *Ann. Math.* **58** (1953), 193–221.
4. U. Fano, "Effects of configuration interaction on intensities and phase shifts," *Phys. Rev.* **124** (1961), 1866–1878.
5. K. O. Friedrichs, *Perturbation of Spectra in Hilbert Space*. Lectures in Applied Mathematics, Vol. III, Amer. Math. Soc., Providence, R.I., 1965.

3. The Zeeman Effect

a. Background. The Zeeman effect is, broadly speaking, the effect of a magnetic field on the emission (or absorption) of light. Such an effect was believed by Faraday to exist, and was unsuccessfully looked for in 1845 and afterward, but it was not found experimentally until 1896, by Zeeman, and confirmed (and extended) by Preston (1896–1899). In the next twenty-five years the Zeeman effect (and its modifications for strong fields, the Paschen–Back and Back–Goudsmit effects) was *the* principal source of information in atomic spectroscopy.

The explication of these Zeeman spectroscopic data was the key problem whose solution led to the founding of quantum mechanics. The Lorentz theory of electrons (1895) could easily explain the "normal" Zeeman triplet, but it was quite inadequate for the much more frequent "anomalous" Zeeman patterns.

Discussion of the Zeeman effect is, at present, a standard topic in physics textbooks. Our purpose in repeating this discussion is to emphasize the essential contributions to current understanding made by angular momentum theory as well as to trace the origins of angular momentum theory itself as inductive, *empirical* constructs from the spectroscopic data.

b. The normal Zeeman effect. From the viewpoint of symmetry techniques, one may categorize the Zeeman effect as *the breaking of rotational symmetry by distinguishing a special direction*, the direction of the magnetic field **B**. Taking an atom to be, schematically, a fixed nucleus attracting an electron (via a central force), the problem is to discuss the electron's motion as changed by the presence of the magnetic field.

This problem was solved classically by Larmor, leading to the Larmor theorem: *The effect of a (weak) magnetic field on the motion of a particle of mass m_0 and charge e (confined to a finite region of space) is to superimpose on the original motion a rotation with circular frequency*

$$\boldsymbol{\omega}_L = -\left(\frac{e}{2m_0c}\right)\mathbf{B} \tag{7.3.1}$$

around an axis along the field direction.

This classical result easily explained the (Larmor) shifted frequencies in the normal Zeeman pattern, and even the finer details such as polarization. We forego these details, since the quantal treatment is at once simpler and more general.

c. Quantal treatment. The quantal treatment (Newtonian relativity) starts from a classical Hamiltonian. In the absence of the magnetic field, one has the Hamiltonian

$$H_0 = \frac{\mathbf{p}^2}{2m_0} + V(r) \tag{7.3.2}$$

for an electron moving in an effective central potential $V(r)$.

To introduce the magnetic field one uses the classically defined substitution for a charge e,

$$\mathbf{p} \to \mathbf{p} - e\mathbf{A}/c,$$

where the vector potential $\mathbf{A}$ is given by $\mathbf{A} = \frac{1}{2}(\mathbf{B} \times \mathbf{x})$ in which $\mathbf{B}$ denotes a constant magnetic field.

Neglecting the quadratic terms in the vector potential ("weak field approximation"), one finds the Hamiltonian

$$H = H_0 - \left(\frac{e\hbar}{2m_0c}\right)\mathbf{B} \cdot \mathbf{L}, \tag{7.3.3}$$

where $\mathbf{L}$ is the (dimensionless) orbital angular momentum operator

$$\mathbf{L} = -i\mathbf{x} \times \nabla. \tag{7.3.4}$$

The eigenvalues of H are easily found by using the eigenfunctions of the original rotationally invariant Hamiltonian H_0 (see Chapter 6, Sections 1–8). One sees, in fact, that l is still a good quantum number, since $[\mathbf{L}^2, H] = 0$. Thus, choosing the z-axis parallel to $\mathbf{B}$ so that $\mathbf{B} = B\hat{e}_3$ and $\mathbf{B} \cdot \mathbf{L} = BL_3$, one obtains the eigenvalues $E(n, l, m)$ of H in terms of the eigenvalues $E_0(n, l)$ of H_0:

$$E(n, l, m) = E_0(n, l) - \left(\frac{e\hbar B}{2m_0c}\right)m. \tag{7.3.5}$$

One sees that angular momentum techniques have completely determined the changes in the energy levels. Note further that the magnetic field has removed the (spatial) degeneracies of the various levels. From the viewpoint of group theory, one has restricted the symmetry group $SO(3)$ to the subgroup $SO(2)$ by distinguishing the axial direction $\mathbf{B}$. This physical symmetry breaking did not affect the angular momentum commutation relations (which were used in obtaining the eigenvalues of the Hamiltonian). This illustrates a simple, but very important, point: *Commutation rules are kinematic constructs, independent of dynamics.*

Angular momentum techniques also solve the second part of the problem: the polarization and angular distribution of the emitted radiation. In Section 6 we discuss systematically electromagnetic multipole radiation; for the present purpose we need only recall that, for atomic spectra, $r_{\text{atom}}/(\lambda_{\text{radiation}}) \approx 10^{-2}$–$10^{-3}$, so that only electric dipole radiation enters. Thus (see Section 6), the radiative part of the Hamiltonian that induces optical transitions is

$$H_{\text{rad}} = \mathbf{j} \cdot \mathbf{A}_{\text{rad}} = \frac{e\hbar}{m_0c}(\hat{\varepsilon} \cdot \mathbf{p})A_{\text{rad}}\,. \tag{7.3.6}$$

($\hat{\varepsilon}$ is the linear polarization of the emitted light.) This result leads to the well-known formula (see Bethe and Salpeter [1, Chapter IV]) for the radiation intensity (per atom) for a transition from the upper energy state $E(n', l', m')$ to the lower state $E(n, l, m)$:

$$I = \frac{4e^2\omega^4}{3c^3}|\langle n', l', m'|\mathbf{x}|nlm\rangle|^2, \tag{7.3.7}$$

where $\hbar\omega = E(n', l', m') - E(n, l, m)$.

The matrix element in the intensity formula involves the position vector $\mathbf{x}$, and we may use the Wigner–Eckart theorem to conclude that

$$\Delta l = \pm 1; \qquad \Delta m = \pm 1, 0. \tag{7.3.8}$$

[See Eqs. (6.100).]

Electric dipole transitions are possible only between states of different parity, since the vector $\mathbf{x}$ changes sign under inversion.

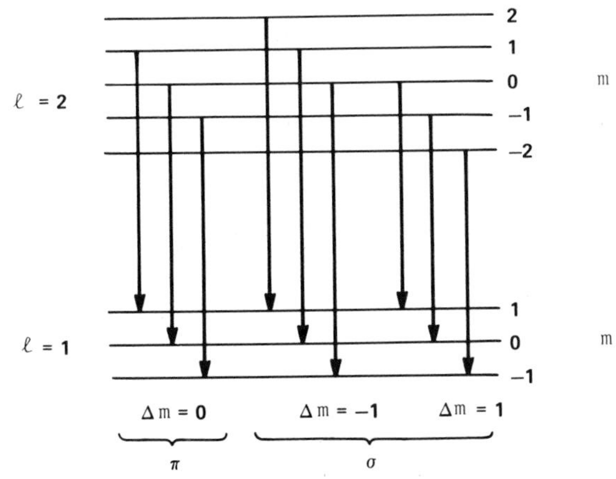

Figure 7.1. Allowed electric dipole transitions between two levels with $l = 2$ and $l = 1$.

Let us, for definiteness, consider an $l = 2$ level decaying radiatively to an $l = 1$ level. Thus, $\Delta l = 1$, and for $\Delta m = \pm 1, 0$ there are nine allowed electric dipole transitions. We have the energy-level diagram shown in Fig. 7.1. The emitted radiation consists, therefore, of three frequencies (normal triplet): $\omega_0 + \omega_L, \omega_0, \omega_0 - \omega_L$. The relative intensities (I_+, I_0, I_{-1}) of these three lines are obtained from Eq. (7.3.7). For the line $\omega_0 + k\omega_L$, we find

$$I_k = \sum_{m'm} (C^{l+1\,1\,l}_{m', -k, m})^2 = \frac{(2l+1)}{3}. \tag{7.3.9}$$

Thus, the three lines have the same total intensity (integrated over all directions). (This assumes that the populations of the excited states are uniform.)

To determine the various angular distributions, we may proceed in this way. The radiation of frequency $\omega_+ = \omega_0 + \omega_L$, if observed along the direction of the magnetic field, would carry an intrinsic angular momentum of magnitude $S = 1$, and projection $M_L = 1$ [$S = 1$, since the radiation is dipole; $M_S = 1$, since M_S of radiation is the Δm of levels]; hence, it is right circularly polarized.

The angular distribution of right circularly[1] polarized radiation of frequency $\omega = \omega_0 + \omega_L$ can be found directly from the following angular momentum considerations: (*a*) The radiation carries an angular momentum of magnitude $S = 1$; (*b*) along the direction of motion **k**, the radiation carries one unit of internal angular momentum (right circular polarization); and (*c*) for **k** along the direction of the field **B**, the angular momentum component is $M_S = 1$. These facts imply that the probability for observing the radiation, along an arbitrary direction **k**, is proportional to the square of a rotation matrix:

$$\text{Probability} \propto |\mathscr{D}^1_{1,1}(R^k)|^2\,,$$

where R^k is an orthogonal matrix that rotates the direction $\hat{e}_3$ into the direction $\hat{k} = \mathbf{k}/k$. One finds

$$|\mathscr{D}^1_{1,1}(R^k)|^2 = \sum_S (C^{1\,1\,S}_{1,-1,0})^2\, d^S_{0,0}(\cos\theta)$$

$$= \tfrac{1}{3} + \tfrac{1}{2}P_1(\cos\theta) + \tfrac{1}{6}P_2(\cos\theta), \qquad (7.3.10)$$

where $\cos\theta = \hat{e}_3 \cdot \hat{k}$.

These considerations show that along the direction of the magnetic field there are only two components (of equal intensity): right circular (clockwise), frequency $\omega = \omega_0 + \omega_L$, and left circular, frequency $\omega = \omega_0 - \omega_L$. *Note that this association fixes the sign of the electron charge* (for opposite charge the left–right relation reverses); the normal Zeeman effect gave the first proof that the radiating charges in the atom were negative. (The data also yield a value for e/m_0.)

Perpendicular to the field direction one finds three components, all linearly polarized; the unshifted components ($\Delta m = 0$), from Eq. (7.3.6), are polarized along **B**; the shifted components, perpendicular to **B**.

[1] This means that the radiation detector is assumed (ideally) to measure only right circularly polarized radiation of frequency $\omega_0 + \omega_L$.

We conclude that the normal Zeeman effect can be completely discussed directly from angular momentum considerations.

d. The anomalous Zeeman effect. The anomalous Zeeman effect is entirely a matter of properly including the spin of the electron. This insight was not won quickly or cheaply. It took the efforts of many (Sommerfeld, Pauli, Dirac), but above all it was Landé who made the significant breakthrough. [See Eq. (7.3.19).]

It is commonly said that spin (intrinsic angular momentum) is a relativistic effect[1]; this is incorrect, as the discussion in Chapter 1 of Galilean relativity showed (see Chapter 1, Note 3). It is easy, in fact, to put spin properly into the Galilean Hamiltonian H_0. Using the identity $(\boldsymbol{\sigma} \cdot \mathbf{p})^2 = \mathbf{p}^2$, where $\{\sigma_i\}$ are the Pauli matrices, one writes the Hamilton H_0 in the equivalent form

$$H_0 = \frac{1}{2m_0} (\boldsymbol{\sigma} \cdot \mathbf{p})^2 + V(r). \tag{7.3.11}$$

This new form becomes significant when one introduces an external magnetic field by the gauge invariance prescription: $\mathbf{p} \to \mathbf{p} - e\mathbf{A}/c$. Using the Dirac rule, $(\boldsymbol{\sigma} \cdot \mathbf{A})(\boldsymbol{\sigma} \cdot \mathbf{B}) = \mathbf{A} \cdot \mathbf{B} + i\boldsymbol{\sigma} \cdot (\mathbf{A} \times \mathbf{B})$, one finds

$$H = H_0 - \left(\frac{e\hbar}{2m_0c}\right)\mathbf{B} \cdot \mathbf{L} - \left(\frac{e\hbar}{2m_0c}\right)\boldsymbol{\sigma} \cdot \mathbf{B}. \tag{7.3.12}$$

Since the spin angular momentum is $\mathbf{S} = \frac{1}{2}\boldsymbol{\sigma}$, we see that the interaction is now

$$H' = -\frac{e\hbar}{2m_0c}(\mathbf{L} + 2\mathbf{S}) \cdot \mathbf{B} = -\frac{e\hbar}{2m_0c}(\mathbf{J} + \mathbf{S}) \cdot \mathbf{B}, \tag{7.3.13}$$

where $\mathbf{J} = \mathbf{L} + \mathbf{S}$.

Defining the g-factor as the numerical ratio of magnetic moment to angular momentum, one sees from Eq. (7.3.13) that

$$g_{\text{orbital}} = 1 \qquad \text{and} \qquad g_{\text{spin}} = 2 \tag{7.3.14}$$

(in units of Bohr magnetons $-e\hbar/2m_0c$).

Observing next that H' commutes with $\mathbf{L}^2$ and $\mathbf{S}^2$, one sees that the appropriate basis for calculating the energy correction due to H' is the coupled

[1] This view is based on the fact that Dirac's relativistic electron equation automatically predicted spin-$\frac{1}{2}$ for the electron and a g-factor of 2.

basis:

$$|n(l\tfrac{1}{2})jm\rangle = \sum_{\mu\nu} C^{l\frac{1}{2}j}_{\mu\nu m}|nl\mu\rangle \otimes |\tfrac{1}{2}\nu\rangle$$

$$= R_{nl}\,\mathscr{Y}^{(l\frac{1}{2})jm}. \tag{7.3.15}$$

[See Eqs. (6.33)–(6.39) and (6.57).]

Furthermore, in first-order perturbation theory, one requires only the diagonal matrix elements in j. Since **S** is a vector operator with respect to **J**, it follows from the Wigner–Eckart theorem that [see Eq. (3.246), Chapter 3]

$$\langle n(l\tfrac{1}{2})jm'|S_\mu|n(l\tfrac{1}{2})jm\rangle = (-1)^{j+1+l+\frac{1}{2}}[2(2j+1)]^{\frac{1}{2}}\langle\tfrac{1}{2}\|\mathbf{S}\|\tfrac{1}{2}\rangle\begin{Bmatrix} j & l & \frac{1}{2} \\ \frac{1}{2} & 1 & j \end{Bmatrix} C^{j1j}_{m\mu m'}$$

$$= [j(j+1)]^{-1}\langle n(l\tfrac{1}{2})jm'|(\mathbf{S}\cdot\mathbf{J})J_\mu|n(l\tfrac{1}{2})jm\rangle. \tag{7.3.16}$$

We have obtained the right-hand side of this result directly from Eqs. (6.87) and (6.89), Chapter 6, by replacing **A** by **S**, l by j, and $\mathscr{Y}_{lm}$ by $|jm\rangle$.

Insofar as first-order perturbation theory is concerned, one may replace the operator H' by its *restriction* to the $(2j+1)$-dimensional subspace spanned by the orthonormal vectors (7.3.15) for $m = j, j-1, \ldots, -j$:

$$H'_{\text{restriction}} = -\frac{e\hbar}{2m_0c}\left(1 + \frac{\mathbf{S}\cdot\mathbf{J}}{\mathbf{J}^2}\right)\mathbf{J}\cdot\mathbf{B}. \tag{7.3.17}$$

(This operator then has the same matrix elements diagonal in j as does H'.)

Since the scalar product $\mathbf{S}\cdot\mathbf{J}$ may be interpreted, from the addition of angular momentum, to be

$$\mathbf{S}\cdot\mathbf{J} = (\mathbf{J}^2 + \mathbf{S}^2 - \mathbf{L}^2)/2, \tag{7.3.18}$$

we obtain the general g-factor (Landé formula) for the basis (7.3.15):

$$g = \left\langle n(l\tfrac{1}{2})jm\left|1 + \frac{\mathbf{S}\cdot\mathbf{J}}{\mathbf{J}^2}\right|n(l\tfrac{1}{2})jm\right\rangle$$

$$= 1 + \frac{j(j+1) + s(s+1) - l(l+1)}{j(j+1)}, \tag{7.3.19}$$

in which, for the case considered here, one has $s = \frac{1}{2}$. The corresponding energy levels are [see Eq. (7.3.5)]

$$E[n(l\tfrac{1}{2})jm] = E_0(n, l) - \left(\frac{e\hbar B}{2m_0c}\right)gm. \tag{7.3.20}$$

Note that, from angular momentum theory, *the Landé factor*, $g - 1$, *is a Racah coefficient*, to within invariant factors.

It is clear from these results that the level shifts are no longer integer multiples of the Larmor energy $\hbar\omega_L$. Hence, the characteristic triplet pattern of the normal Zeeman effect is lost. It is equally clear that the frequency, polarization, and intensity of each component can be calculated from the relevant Wigner coefficients and rotation matrices, using the same techniques as before. To pursue this further would carry us into relatively uninteresting detail.

It is essential to remark, however, that the interaction Eq. (7.3.17) is not quite the same as that of Eq. (7.3.13), since Eq. (7.3.17) is valid only for matrix elements diagonal in j. The Wigner–Eckart theorem applies to the vector operator $\mathbf{J} + \mathbf{S}$ in Eq. (7.3.13) and may be used to calculate all matrix elements of the interaction (7.3.13). The diagonalization of the resulting Hamiltonian matrix is more complicated for the anomalous Zeeman case, and results are simple only in various limits.

e. Relation to the development of angular momentum theory. The unraveling of the complexities of the anomalous Zeeman patterns was a monumental task. The Bohr quantum theory had introduced (1913) the heuristic idea of quantizing the orbital angular momentum in the hydrogen atom (see the Note at the end of this section). The electron orbits were still viewed as classical orbits, with the electron *moving in a plane perpendicular to the angular momentum direction*. Sommerfeld and Wilson generalized (1916) the Bohr quantization rule to the phase integrals of the Hamilton–Jacobi theory. More to the point for our interests, Sommerfeld (in 1918) introduced the concept of the "spatial quantization" of angular momentum; this is the heuristic picture that symbolized the quantized angular momentum vector $\mathbf{L}$ as having discrete orientations (the beginnings of the vector model). (We discuss this further in RWA.)

Attention shifted then to the two-electron problem, the helium atom, and (such is the force of unexamined prior ideas – in this case the analog to the planetary orbits around the sun) the two electrons were always taken *in a common plane*. (See Yourgrau and van der Merwe [2, p. 2]). Landé, with Born, was the first to break this habit, and it was Landé (1919) who first considered the addition of quantized angular momenta. (He invented the now-familiar vector model for this.)

But it was only in 1923, after lengthy *numerical* calculations, that Landé arrived at his result for the g-factor. His replacement of $\mathbf{L}^2$ by $l(l + 1)$ (and similarly for $\mathbf{S}^2$ and $\mathbf{J}^2$) was *entirely empirical*, as was his assignment of $g_{\text{spin}} = 2$.

The explanation of these curious, experimentally determined facts was only found later in the quantum theory of angular momentum as developed by Born and Jordan and by Wigner.

It is amusing to note that Sommerfeld's "spatial quantization" already implied the result $\mathbf{J}^2 \to j(j+1)$. To see this, note that the model shows that $-j \leqslant m \leqslant j$, by steps of unity. Taking $J_3 \to m$ as known, we seek to calculate $\mathbf{J}^2 = J_1^2 + J_2^2 + J_3^2$. If we average over the multiplet, then

$$\langle \mathbf{J}^2 \rangle = \frac{1}{2j+1} \sum_m (\langle J_1^2 \rangle + \langle J_2^2 \rangle + \langle J_3^2 \rangle). \tag{7.3.21}$$

Using $\langle J_1^2 \rangle = \langle J_2^2 \rangle = \langle J_3^2 \rangle = m^2$, since the average can prefer no direction (an "ergodic"-like assumption), one sees that

$$\langle \mathbf{J}^2 \rangle = \frac{3}{2j+1} \sum_m m^2 = \frac{3}{2j+1} \left[\frac{j(j+1)(2j+1)}{3} \right]. \tag{7.3.22}$$

Thus, we obtain $\mathbf{J}^2 \to j(j+1)$ as sought[1]. This argument, as noted earlier (Chapter 3, Section 3), is due to Feynman and can be easily generalized to higher unitary groups.

f. Concluding remarks. The Zeeman effect, as we have discussed it above, is very crude by the standards of spectroscopy of today. The techniques of optical pumping, double resonance (Brossel–Bitter), Mössbauer spectroscopy, and above all high-intensity laser sources have achieved a resolution that is qualitatively far beyond the beginnings we have discussed. Yet the principles of exploiting angular momentum techniques are unchanged, and the complications are essentially ones of detail, with the physics limited only by the ingenuity of the experimenter.

The fact that the g-factor of the electron is changed slightly ($\approx 10^{-4}$) by quantization effects (of both the electromagnetic field and the electron–positron field) is well-known to physicists. It is worth remarking that present-day experiments can readily show (via the Hanle effect) qualitative, and large, changes in the observed g-factor due to the electromagnetic environment. A very nice discussion of this point is given by Kastler [3].

---

[1] Let us note that a physical explanation of the "extra" term j in $j(j+1)$ is to be found in the uncertainty principle. As discussed in RWA, states ψ of the form $J_+\psi = 0$ (alternatively $J_-\psi = 0$) have minimal uncertainty properties. For such a state, J_3 is sharp and assumes a maximal eigenvalue: $J_3 \to j$. The fluctuations in J_1 and J_2 are then minimal: $(\Delta J_1^2) = (\Delta J_2^2) = (j/2)$. It is these fluctuations in J_1 and J_2 that contribute the "extra" term.

g. Note. An engaging account of the beginnings of the quantum theory has been given by Armin Hermann [4]. According to this account, the first suggestion that rotational motion must be quantized was by W. Nernst in 1911 (see [4, Ref. 28, p. 144]). Niels Bjerrum applied this idea to band spectra in 1912 (see [4, Ref. 29, p. 144]) but got unsatisfactory results, since the *rotational energy*, and not the *angular momentum*, was taken to be quantized.

Surprisingly, Niels Bohr in his classic paper [5] quantized the energy of the hydrogen atom in units of *one-half* of Planck's constant, and the number one-half was in fact *empirical*, and chosen as the best numerical fit to Rydberg's formula! However, in the latter part of this paper Bohr gave a "simple explanation" of his results, explicitly enunciating the rule quantizing the (orbital) angular momentum (for each electron) in units of $h/2\pi$. (Hermann [4] does not make this last point clear.)

References

1. H. A. Bethe and E. E. Salpeter, *Quantum Mechanics of One- and Two-Electron Atoms.* Springer, Berlin, 1957.
2. W. Yourgrau and A. van der Merwe (eds.), *Perspectives in Quantum Theory* (Essays in honor of Alfred Landé). MIT Press, Cambridge, Mass., 1971.
3. A. Kastler, "How the Landé factor of an atom can be changed by putting the atom in a radiofrequency bath," *ibid.*, pp. 71–77.
4. A. Hermann, *The Genesis of the Quantum Theory.* MIT Press, Cambridge, Mass., 1971. Translation of *Frühgeschichte der Quantentheorie* (1899–1913), Physik Verlag, Mosbach/ Baden, 1969.
5. N. Bohr, "On the constitution of atoms and molecules," *Phil. Mag.* **26** (1913), 1–25.

4. The Nonrelativistic Hydrogen Atom

The hydrogen atom problem presents an excellent example for the application of general angular momentum techniques. It is from this particular viewpoint that we discuss the problem.

Pauli [1] first determined the energy spectrum of the hydrogen atom (thus deriving the Balmer formula), using the algebraic methods of the new matrix mechanics. This verification of the concepts of matrix mechanics was important, but the significance (for symmetry techniques) was not fully appreciated, since the later solution to the same problem using Schrödinger's wave mechanics [2] was more accessible to physicists, and this latter technique predominated.

Pauli's methods were fundamental from another viewpoint. He recognized the role of the Runge–Lenz observable[1,2] $\mathbf{A}_0$, defined by[3]

$$\mathbf{A}_0 \equiv \frac{\mathbf{x}}{r} + (2Zm_0e^2)^{-1}(\mathbf{L}\times\mathbf{p} - \mathbf{p}\times\mathbf{L}), \tag{7.4.1}$$

in forming a closed matrix algebra with the angular momentum observable $\mathbf{L} = \mathbf{x}\times\mathbf{p}$ [see Eqs. (7.4.8) and (7.4.9) below].

As we shall see below, the nonrelativistic[4] hydrogen atom problem can be completely solved by using the properties of $\mathbf{L}$ and $\mathbf{A}_0$ (we shall discuss both the spinless and the spin-$\frac{1}{2}$ cases).

The contents of the present section include (*a*) the identification of the Lie algebras defined by $\mathbf{L}$ and $\mathbf{A}_0$; (*b*) a detailed development of the representations of the algebra (in the finite case) carried by the *energy eigenstates* in the two bases corresponding to the separation of the Schrödinger equation in parabolic and spherical coordinates; (*c*) a detailed discussion of the relationship between four-space spherical harmonics and the momentum space representation of the states of the hydrogen atom; and (*d*) an elegant solution to the hydrogen atom with spin (Pauli particle) obtained by using Dirac's kappa quantum number.

a. Algebraic aspects. The Hamiltonian for an electron (charge e, mass m_0) moving in the Coulomb field of a fixed nucleus (charge $-Ze$) is given by

$$H = (2m_0)^{-1}\mathbf{p}^2 - Ze^2/r, \tag{7.4.2}$$

where

$$r = (x_1^2 + x_2^2 + x_3^2)^{\frac{1}{2}} \qquad \text{and} \qquad \mathbf{p}^2 = p_1^2 + p_2^2 + p_3^2. \tag{7.4.3}$$

It is convenient to use dimensionless variables, choosing as the unit of length a/Z, where $a = \hbar^2/m_0e^2$ is the radius of the first Bohr orbit. Measuring $\mathbf{L}$ in units of $\hbar$, momentum $\mathbf{p}$ in units of $\hbar Z/a$, and energy H in units of $Z^2e^2/2a$, we

---

[1] The quantization of the observable $\mathbf{A}_0$ was also one of the first examples of the quantization of a classical observable going beyond the fundamental rules for coordinates, linear momenta, and angular momenta; one must replace a product AB of classical observables by the symmetrized product $(AB + BA)/2$.

[2] Excellent surveys on the historical aspects of the Runge–Lenz vector have been given by McIntosh [3] and by Goldstein [4]. Variants of the Runge–Lenz vector are discussed by Kobayashi [4a].

[3] We use the symbol $\mathbf{A}_0$ for the physical Runge–Lenze operator to distinguish it from the normalized form $\mathbf{A}$ introduced in Eq. (7.4.12).

[4] This is an essential limitation, since relativistic (symmetry-breaking) effects can be important. An excellent survey of the relativistic Coulomb problem is given in the review by McIntosh [3].

find for the three operators of interest the dimensionless forms

$$\mathbf{L} = \mathbf{x} \times \mathbf{p},$$

$$\mathbf{A}_0 = \frac{\mathbf{x}}{r} + \frac{1}{2}(\mathbf{L} \times \mathbf{p} - \mathbf{p} \times \mathbf{L}),$$

$$H = \mathbf{p}^2 - \frac{2}{r}. \tag{7.4.4}$$

Notice that the Runge–Lenz–Pauli operator, $\mathbf{A}_0$, is a vector operator with respect to $\mathbf{L}$:

$$[L_i, A_{0j}] = ie_{ijk}A_{0k}. \tag{7.4.5}$$

For the purpose of taking matrix elements, it is also useful to write $\mathbf{A}_0$ in the form

$$\mathbf{A}_0 = \frac{\mathbf{x}}{r} + \frac{i}{2}[\mathbf{L}^2, \mathbf{p}]. \tag{7.4.6}$$

The operator $\mathbf{A}_0^2 \equiv \mathbf{A}_0 \cdot \mathbf{A}_0$ is related to the angular momentum operator $\mathbf{L}^2$ and the Hamiltonian H by the relation

$$\mathbf{A}_0 \cdot \mathbf{A}_0 = H(\mathbf{L}^2 + 1) + 1. \tag{7.4.7}$$

[This relation is the operator version of the classical orbit relation (Kepler's law of conic sections).]

Since $\mathbf{A}_0 \cdot \mathbf{A}_0$ is an invariant under rotations (commutes with $\mathbf{L}$), one sees from Eq. (7.4.7) that the simultaneous diagonalization of $\mathbf{A}_0 \cdot \mathbf{A}_0$ and $\mathbf{L}^2$ will solve the eigenvalue problem for H also. One is thus motivated to calculate the commutators of $\mathbf{A}_0$ with H and with itself.

The properties of the commutator algebra that one obtains for H, $\mathbf{L}$, and $\mathbf{A}_0$ are the following:

$$\mathbf{L} \times \mathbf{L} = i\mathbf{L},$$

$$[L_i, A_{0j}] = ie_{ijk}A_{0k}; \tag{7.4.8}$$

$$\mathbf{A}_0 \times \mathbf{A}_0 = -i\mathbf{L}H,$$

$$[H, \mathbf{L}] = [H, \mathbf{A}_0] = \mathbf{0}. \tag{7.4.9}$$

To relations (7.4.7)–(7.4.9), one also adds

$$\mathbf{L} \cdot \mathbf{A}_0 = 0. \tag{7.4.10}$$

There are two significant features that one should note about the commutator algebra (7.4.9): (*a*) The Hamiltonian H enters into the commutation relations; (*b*) the vector operator $\mathbf{A}_0$ defined by Eq. (7.4.4) *fails* to possess the derivation property [see Eq. (3.217), Chapter 3] (the implications of this are discussed subsequently).

Consider the consequences of (*a*) above. Although H is Hermitian, it is not positive definite, and it may, accordingly, possess both positive, zero, and negative eigenvalues:

$$H\Psi_E = E\Psi_E. \tag{7.4.11}$$

The structure of the commutation relations (7.4.9) is thus seen to depend on the energy eigenspace on which the operator relations (7.4.9) act. Thus, in interpreting Eqs. (7.4.9) in terms of the usual notions of Lie algebras, *we must assume that the vector space of state vectors on which these equations are allowed to act are energy eigenspaces.*

Subject to these conditions, we can now normalize the operator $\mathbf{A}_0$:

$$\mathbf{A} = \begin{cases} \mathbf{A}_0(-H)^{-\frac{1}{2}} & \text{for } E < 0, \\ \mathbf{A}_0 & \text{for } E = 0, \\ \mathbf{A}_0(H)^{-\frac{1}{2}} & \text{for } E > 0. \end{cases} \tag{7.4.12}$$

Equations (7.4.8) and (7.4.9) can now be put in the Lie algebra form

$$\begin{aligned} \mathbf{L} \times \mathbf{L} &= i\mathbf{L}, \\ [L_i, A_j] &= ie_{ijk}A_k, \\ \mathbf{A} \times \mathbf{A} &= \varepsilon i\mathbf{L}, \end{aligned} \tag{7.4.13}$$

where $\varepsilon = 1, 0, -1$ corresponding to $E < 0, E = 0, E > 0$, respectively. Under the renorming of $\mathbf{A}_0$, Eqs. (7.4.7) and (7.4.10) become

$$\begin{aligned} (\mathbf{L}^2 + \varepsilon\mathbf{A}^2 + 1)H &= -1, \qquad \varepsilon = \pm 1 \\ H(\mathbf{L}^2 + 1) + 1 &= \mathbf{A}^2, \qquad \varepsilon = 0 \\ \mathbf{L} \cdot \mathbf{A} &= 0. \end{aligned} \tag{7.4.14}$$

Let us next discuss the three Lie algebras (7.4.13) corresponding to $\varepsilon = +1, 0, -1$:

(1) For $\varepsilon = +1$ ($E < 0$, spherical geometry) the commutation relations (7.4.13) are in the standard $SO(4) \supset SO(3)$ form of the Lie algebra of the group of 4×4 proper orthogonal matrices, $SO(4)$. Fock [5] demonstrated that $SO(4)$ is the symmetry group of the Kepler problem by writing the Schrödinger equation as an integral differential equation in momentum space and relating the hydrogen atom wave functions in momentum space to the spherical harmonics in four-space. Bargmann [6] showed that the infinitesimal generators of Fock's transformations were the angular momentum **L** and the Runge–Lenz vector **A**. (We discuss the momentum representation from the point of view of the algebra of **L** and **A** below.)

Equations (7.4.13) may be put into the familiar form of "addition of two angular momenta" by a change of basis of the algebra:

$$\mathbf{J}(1) \equiv \tfrac{1}{2}(\mathbf{L} + \mathbf{A}),$$
$$\mathbf{J}(2) \equiv \tfrac{1}{2}(\mathbf{L} - \mathbf{A}). \qquad (7.4.15)$$

One then obtains the familiar $SU(2) \times SU(2)$ form of the commutation relations:

$$\mathbf{J}(a) \times \mathbf{J}(a) = i\mathbf{J}(a), \qquad a = 1, 2$$
$$[\mathbf{J}(1), \mathbf{J}(2)] = 0. \qquad (7.4.16)$$

However, because of the relation $\mathbf{L} \cdot \mathbf{A} = 0$, the invariant operators $\mathbf{J}^2(1)$ and $\mathbf{J}^2(2)$ are identically equal:

$$\mathbf{J}^2(1) = \mathbf{J}^2(2) = (\mathbf{L}^2 + \mathbf{A}^2)/4. \qquad (7.4.17)$$

We may now utilize the results of angular momentum coupling theory to obtain the possible eigenvalues E of H for the case $E < 0$. Since the operators $\mathbf{J}(1)$ and $\mathbf{J}(2)$ are Hermitian, the eigenvalues of $\mathbf{J}^2(1)$ and $\mathbf{J}^2(2)$ are $j_1(j_1 + 1)$ and $j_2(j_2 + 1)$, respectively, and the constraint (7.4.17) admits only the solution $j = j_1 = j_2$. Thus, the eigenvalues of H for $E < 0$ are obtained from Eq. (7.4.14) to be of the form

$$E = -1/(2j + 1)^2, \qquad (7.4.18)$$

where $j \in \{0, \tfrac{1}{2}, 1, \ldots\}$ (which of these j-values are actually realized still must be determined).

The results obtained above in Eqs. (7.4.15)–(7.4.18) correspond to the *bound states* of the hydrogen atom; the explicit energy eigenstates are discussed below. The integer $2j + 1$ in Eq. (7.4.18) is identified in Eq. (7.4.22) below to be the *principal quantum number n* of the bound state spectrum.

(2) For $\varepsilon = -1$ ($E > 0$, hyperbolic geometry) the commutation relations (7.4.13) are in the standard $SO(3, 1) \supset SO(3)$ form of the Lie algebra of the Lorentz group. In this case (see p. 358) the eigenvalues and eigenfunctions can be obtained by replacing the principal quantum number n $(= 2j + 1)$ in the discrete case, (1) above, by $\pm i\eta$, where η is an arbitrary real number $(0 < \eta < \infty)$. In this way one obtains, after proper normalization of the radial eigenfunctions (see Landau and Lifshitz [7], Biedenharn and Brussaard [8]), all the wave functions corresponding to sharp angular momentum. It is important to realize in this case that to each value of the energy $E > 0$ there corresponds an infinity of states, a multiplet for each value $l = 0, 1, 2, \ldots$ of the orbital angular momentum.

(3) For $\varepsilon = 0$ ($E = 0$, parabolic geometry) Eqs. (7.4.13) show that the components of the Runge–Lenz vector $\mathbf{A} = \mathbf{A}_0$ commute, and that $\mathbf{A}_0$ is a vector operator with respect to $\mathbf{L}$. (The group is the Euclidean group in three dimensions.) One can obtain the radial eigenfunctions by taking the limit $\eta \to \infty$ of the hyperbolic radial functions. One still finds, for $E = 0$, an infinity of states corresponding to orbital angular momenta $l = 0, 1, 2, \ldots$ (see Refs. [7] and [8]).

It is beyond the scope of the present volume to deal in depth with these *continuum Coulomb states*, and we refer the reader to Ref. [8] and literature citations therein.

b. Properties of the bound states of the hydrogen atom.[1] We next adapt the results (Chapter 3) of the coupling of two angular momenta to the description of the bound states of the hydrogen atom. Since $\mathbf{J}(1)$ and $\mathbf{J}(2)$ act in the same function space, one now interprets the tensor product of kets $|j_1 m_1\rangle \otimes |j_2 m_2\rangle$ as a function with values

$$\Phi_{j_1 m_1;\, j_2 m_2}(\mathbf{x}) \equiv \langle \mathbf{x} | j_1 m_1 \rangle \langle \mathbf{x} | j_2 m_2 \rangle. \tag{7.4.19}$$

The angular momentum addition represented by Eqs. (7.4.15) is

$$\mathbf{L} = \mathbf{J}(1) + \mathbf{J}(2),$$

$$\mathbf{A} = \mathbf{J}(1) - \mathbf{J}(2). \tag{7.4.20}$$

[1] The treatment given here has been adapted from the work of Meadors [9].

Thus, the states Ψ_{Elm} (see Chapter 6, Sections 1–8) of sharp orbital angular momentum are the "coupled states" of the angular momenta **J**(1) and **J**(2).

Since we wish to focus attention on the properties of the operators **L** and **A** instead of **J**(1) and **J**(2), we introduce the notations

$$m = m_1 + m_2,$$
$$q = m_1 - m_2, \tag{7.4.21}$$

so that m and q are, respectively, eigenvalues of L_3 and A_3. Accounting for $j_1 = j_2 = j$ and defining n by

$$n \equiv 2j + 1, \tag{7.4.22}$$

we also introduce the notation

$$\Phi_{nqm}(\mathbf{x}) \equiv \Phi_{(n-1)/2,\,(m+q)/2;\,(n-1)/2,\,(m-q)/2}(\mathbf{x}). \tag{7.4.23}$$

Using Eqs. (7.4.20) and the standard action of $\mathbf{J}(a)$ on the states (7.4.19), we find (after accounting for the notational changes above) the following actions for the operators **L** and **A** on the states (7.4.23):

$$\begin{aligned} L_+\Phi_{nqm} &= \tfrac{1}{2}[(n-m-q-1)(n+m+q+1)]^{\frac{1}{2}}\,\Phi_{n,q+1,m+1} \\ &\quad + \tfrac{1}{2}[(n-m+q-1)(n+m-q+1)]^{\frac{1}{2}}\,\Phi_{n,q-1,m+1}, \\ L_-\Phi_{nqm} &= \tfrac{1}{2}[(n+m+q-1)(n-m-q+1)]^{\frac{1}{2}}\,\Phi_{n,q-1,m-1} \\ &\quad + \tfrac{1}{2}[(n+m-q-1)(n-m+q+1)]^{\frac{1}{2}}\,\Phi_{n,q+1,m-1}; \end{aligned} \tag{7.4.24}$$

$$\begin{aligned} A_+\Phi_{nqm} &= \tfrac{1}{2}[(n-m-q-1)(n+m+q+1)]^{\frac{1}{2}}\,\Phi_{n,q+1,m+1} \\ &\quad - \tfrac{1}{2}[(n-m+q-1)(n+m-q+1)]^{\frac{1}{2}}\,\Phi_{n,q-1,m+1}, \\ A_-\Phi_{nqm} &= \tfrac{1}{2}[(n+m+q-1)(n-m-q+1)]^{\frac{1}{2}}\,\Phi_{n,q-1,m-1} \\ &\quad - \tfrac{1}{2}[(n+m-q-1)(n-m+q+1)]^{\frac{1}{2}}\,\Phi_{n,q+1,m-1}. \end{aligned} \tag{7.4.25}$$

One also has the eigenvalue equations

$$L_3\Phi_{nqm} = m\Phi_{nqm},$$
$$A_3\Phi_{nqm} = q\Phi_{nqm},$$

$$(\mathbf{L}^2 + \mathbf{A}^2 + 1)\Phi_{nqm} = n^2\Phi_{nqm},$$

$$H\Phi_{nqm} = (-1/n^2)\Phi_{nqm}. \tag{7.4.26}$$

For each integer $n \geqslant 1$, one finds from Eqs. (7.4.21) and (7.4.22) that the ranges of m and q in Eqs. (7.4.24)–(7.4.26) are

$$m = n-1, n-2, \ldots, -(n-1),$$

$$q = n - |m| - 1, n - |m| - 3, \ldots, -n + |m| + 1; \tag{7.4.27}$$

or, alternatively,

$$q = n-1, n-2, \ldots, -(n-1),$$

$$m = n - |q| - 1, n - |q| - 3, \ldots, -n + |q| + 1. \tag{7.4.28}$$

Thus, for each n that is allowed in Eqs. (7.4.24)–(7.4.26), the degeneracy of the energy level

$$E_n = -1/n^2 \tag{7.4.29}$$

is

$$\sum_q (2n-1) = n^2. \tag{7.4.30}$$

All n^2 states $\{\Phi_{nqm}\}$ corresponding to the values of m and q given by Eqs. (7.4.27) must exist if $\Phi_{n,0,n-1}$ does. The function $\Phi_{n,0,n-1}$ itself is determined to within normalization by the equations

$$L_+\Phi_{n,0,n-1} = 0, \qquad A_+\Phi_{n,0,n-1} = 0,$$

$$L_3\Phi_{n,0,n-1} = (n-1)\Phi_{n,0,n-1}, \qquad A_3\Phi_{n,0,n-1} = 0. \tag{7.4.31}$$

The unique normalized solution[1] (see Chapter 6, Section 2) to these equations is

[1] An arbitrary phase factor (which could depend on n) has been chosen so that the sign of the radial function

$$R_{n,n-1}(r) = \left(\frac{2}{n}\right)^{n+\frac{1}{2}} \frac{r^{n-1}}{\sqrt{(2n)!}} e^{-r/n}$$

agrees with the general result, Eq. (7.4.43), derived below. An additional factor $(Z/a)^{3/2}$ will also multiply Eq. (7.4.32) if the normalization is made to unity in physical space instead of in dimensionless coordinate space.

$$\Phi_{n,0,n-1}(\mathbf{x}) = \left(\frac{2}{n}\right)^{n+\frac{1}{2}} \frac{e^{-r/n}}{\sqrt{(2n)!}} \mathscr{Y}_{n-1,n-1}(\mathbf{x}). \tag{7.4.32}$$

This solution is valid for each $n = 1, 2, 3, \ldots$. Thus, all positive integral n are admitted in the energy eigenvalues, Eq. (7.4.29). [Technically, for consistency of the results taken over from angular momentum theory, one must verify that the Runge–Lenz–Pauli vector $\mathbf{A}$ is, in fact, a Hermitian operator with respect to the inner product given by Eq. (6.8) in Chapter 6 when restricted to functions $\Phi(\mathbf{x})$ belonging to the finite-dimensional Hilbert space spanned by the n^2 functions $\{\Phi_{nqm}: n \text{ specified}\}$.]

The Hermitian property of the three operators L_3, A_3, and $\mathbf{L}^2 + \mathbf{A}^2$ (equivalently, H) then implies that the functions Φ_{nlm} are orthogonal (orthonormal when normalized):

$$\int d\mathbf{x}\, \Phi^*_{n'q'm'}(\mathbf{x})\Phi_{nqm}(\mathbf{x}) = \delta_{n'n}\delta_{q'q}\delta_{m'm}. \tag{7.4.33}$$

The preceding results, Eqs. (7.4.19)–(7.4.33), are realizations of abstract angular momentum theory in the uncoupled basis $|j_1m_1\rangle \otimes |j_2m_2\rangle$. The Wigner coupled states are, according to Eq. (7.4.20), the more familiar states of sharp orbital angular momentum (Parks [10]):

$$\Psi_{nlm} \equiv \sum_q C^{(n-1)/2\ \ (n-1)/2\ \ l}_{(m+q)/2,\ (m-q)/2,\ m}\, \Phi_{nqm}, \tag{7.4.34}$$

$$\mathbf{L}^2\Psi_{nlm} = l(l+1)\Psi_{nlm},$$

$$L_3\Psi_{nlm} = m\Psi_{nlm},$$

$$H\Psi_{nlm} = -\frac{1}{n^2}\Psi_{nlm}, \tag{7.4.35}$$

where now, for each $n = 1, 2, \ldots$, the ranges of l and m are

$$l = 0, 1, 2, \ldots, n-1,$$

$$m = l, l-1, \ldots, -l. \tag{7.4.36}$$

The action of $L_\pm$ on the states ψ_{nlm} is the standard one:

$$L_\pm\Psi_{nlm} = [(l \mp m)(l \pm m + 1)]^{\frac{1}{2}}\, \Psi_{n,l,m\pm 1}. \tag{7.4.37}$$

For completeness, we also give the action of the Runge–Lenz–Pauli operator on the coupled basis (7.4.34):[1]

$$A_{\pm}\Psi_{nlm} = \mp\left[\frac{(n+l+1)(n-l-1)(l\pm m+1)(l\pm m+2)}{(2l+1)(2l+3)}\right]^{\frac{1}{2}}\Psi_{n,l+1,m\pm 1}$$

$$\pm\left[\frac{(n+l)(n-l)(l\mp m-1)(l\mp m)}{(2l-1)(2l+1)}\right]^{\frac{1}{2}}\Psi_{n,l-1,m\pm 1},$$

$$A_3\Psi_{nlm} = \left[\frac{(n+l)(n-l)(l+m)(l-m)}{(2l-1)(2l+1)}\right]^{\frac{1}{2}}\Psi_{n,l-1,m}$$

$$+\left[\frac{(n+l+1)(n-l-1)(l+m+1)(l-m+1)}{(2l+1)(2l+3)}\right]^{\frac{1}{2}}\Psi_{n,l+1,m}. \tag{7.4.38}$$

These latter results are just applications (to the finite vector space of functions having energy eigenvalue E_n) of the Wigner–Eckart theorem, where the reduced matrix elements are (see Table 8 in Appendix of Tables)

$$\langle n, l+1\|\mathbf{A}\|nl\rangle = \left[\frac{(l+1)(n+l+1)(n-l-1)}{(2l+3)}\right]^{\frac{1}{2}},$$

$$\langle nl\|\mathbf{A}\|nl\rangle = 0,$$

$$\langle n, l-1\|\mathbf{A}\|nl\rangle = -\left[\frac{l(n+l)(n-l)}{(2l-1)}\right]^{\frac{1}{2}}. \tag{7.4.39}$$

(Recall that these reduced matrix elements refer to the spherical components $A_{+1} \equiv -A_+/\sqrt{2}$, $A_0 \equiv A_3$, $A_{-1} \equiv A_-/\sqrt{2}$.)

The results expressed by Eqs. (7.4.37) and (7.4.38) may be written in very concise form using the Pauli spin matrices and central field spinors [see Eqs. (7.4.94)].

c. Explicit hydrogen atom wave functions. One may use Eqs. (7.4.24), (7.4.25), (7.4.37), and (7.4.38) in various ways to generate the explicit forms of the functions Φ_{nqm} and Ψ_{nlm}. For example, one such procedure would be to

[1] The symmetry of the coefficients in the second of Eqs. (7.4.38) under the interchange of n and m is quite unexpected because n and m have different physical interpretations. The explanation of this symmetry is quite recent and is taken up in a general discussion of radial integrals in RWA.

construct suitable shift operators for generating all states from [see Eq. (7.4.32)]

$$\Psi_{n,n-1,n-1}(\mathbf{x}) = \Phi_{n,0,n-1}(\mathbf{x}). \tag{7.4.40}$$

This procedure would serve to demonstrate that the $SO(4)$ algebra associated with **L** and **A** fully determines (up to phase and normalization conventions) the solution to the hydrogen atom bound state problem. We give on p. 357 an explicit procedure for determining the functions Ψ_{nlm}. Here we note the results for the functions Ψ_{nlm} and Φ_{nqm}.

We know from the discussion in Chapter 6, Section 8, that the Ψ_{nlm} wave functions must have the form

$$\Psi_{nlm}(\mathbf{x}) = \left(\frac{2}{n}\right)^l F_{nl}(r)\,\mathcal{Y}_{lm}(\mathbf{x}) = R_{nl}(r)\,Y_{lm}(\theta\phi), \tag{7.4.41}$$

where

$$R_{nl}(r) = \left(\frac{2r}{n}\right)^l F_{nl}(r). \tag{7.4.42}$$

The explicit form of the radial function F_{nl} is [see Eqs. (7.4.104)–(7.4.108)]

$$\begin{aligned} F_{nl}(r) &= (-1)^{n-l-1}\frac{2}{n^2}\left[\frac{(n-l-1)!}{(n+l)!}\right]^{\frac{1}{2}} e^{-r/n} L_{n-l-1}^{2l+1}\left(\frac{2r}{n}\right) \\ &= \frac{2(-1)^{n-l-1}}{n^2(2l+1)!}\left[\frac{(n+l)!}{(n-l-1)!}\right]^{\frac{1}{2}} e^{-r/n}\,{}_1F_1\left(-n+l+1;2l+2;\frac{2r}{n}\right). \end{aligned} \tag{7.4.43}$$

In this result the definition of *associated Laguerre polynomials* is[1]

$$L_k^\alpha(\xi) \equiv \sum_{s=0}^{k} \binom{k+\alpha}{k-s}\frac{(-\xi)^s}{s!} = \binom{k+\alpha}{k}\,{}_1F_1(-k;\alpha+1;\xi) \tag{7.4.44}$$

---

[1] An alternative definition for $\alpha = 0, 1, 2, \ldots$ is often used in the literature:

$$\bar{L}_{k+\alpha}^{\alpha}(\xi) \equiv (-1)^\alpha (k+\alpha)!\,L_k^\alpha(\xi) = \frac{d^\alpha}{d\xi^\alpha}\bar{L}_{k+\alpha}(\xi),$$

where $\bar{L}_k$ is a Laguerre polynomial.

for $k = 0, 1, 2, \ldots$ and *arbitrary parameter* α, where ${}_1F_1(a;b;\xi)$ denotes the confluent hypergeometric function.

In Eqs. (7.4.41)–(7.4.43), r is the dimensionless radial distance $r = ZR/a$, where R is physical distance. The normalization is such that

$$\int_0^\infty r^2\, dr[R_{nl}(r)]^2 = 1.$$

The functions $\Phi_{nqm}(\mathbf{x})$ are given by[1]

$$\Phi_{nqm}(\mathbf{x}) = \frac{(-1)^{(n+m+q-1)/2}}{n^2\sqrt{\pi}} \left[\frac{\left(\frac{n-m+q-1}{2}\right)!\left(\frac{n-m-q-1}{2}\right)!}{\left(\frac{n+m+q-1}{2}\right)!\left(\frac{n+m-q-1}{2}\right)!} \right]^{\frac{1}{2}}$$

$$\times \left(\frac{x_1 + ix_2}{n}\right)^m e^{-r/n} L^m_{(n-m+q-1)/2}\left(\frac{r + x_3}{n}\right)$$

$$\times L^m_{(n-m-q-1)/2}\left(\frac{r - x_3}{n}\right), \tag{7.4.45}$$

$$\Phi_{n,q,-m}(\mathbf{x}) = (-1)^m \Phi^*_{nqm}(\mathbf{x}), \tag{7.4.46}$$

where $m \geqslant 0$ in these results, and we regard r as $r = (x_1^2 + x_2^2 + x_3^2)^{\frac{1}{2}}$.

[We have purposely written Eq. (7.4.45) in Cartesian form so that the transformation (7.4.34) may be regarded as being between functions defined on points $\mathbf{x}$ of $\mathbb{R}^3$.]

The eigenfunctions (7.4.41) diagonalize $H, \mathbf{L}^2$, and L_3 and become a product of functions $R(r)\Theta(\theta)\Phi(\phi)$, when expressed in terms of spherical polar coordinates.

The eigenfunctions (7.4.45) diagonalize H, L_3, and A_3 and become a product of functions $F_1(\xi)F_2(\eta)\Phi(\phi)$, when expressed in terms of parabolic coordinates:

$$x_1 = \sqrt{\xi\eta}\cos\phi, \qquad x_2 = \sqrt{\xi\eta}\sin\phi, \qquad x_3 = (\xi - \eta)/2,$$

$$0 \leqslant \xi < \infty, \qquad 0 \leqslant \eta < \infty, \qquad 0 \leqslant \phi < 2\pi. \tag{7.4.47}$$

---

[1] Once the functions Ψ_{nlm} have been determined, the functions Φ_{nqm} are uniquely defined by the inverse of Eq. (7.4.34) (and conversely). It is nontrivial to verify that the explicit functions given by Eqs. (7.4.45) and (7.4.46) are those defined by Eq. (7.4.34). This may be accomplished by using Eqs. (7.4.24) and (7.4.25) to generate Φ_{nqm} from $\Phi_{n,0,n-1}$, as previously mentioned.

(Kalnins *et al.* [11] have carried out a thorough investigation of the relationship between the separation of variables for the hydrogen atom problem and the existence of different complete sets of mutually commuting observables; we refer the reader to their paper for a full account. See also Miller [11a] for a systematic treatment of the relation between symmetry and separation of variables.)

d. Momentum space representation. In Note 4 to Chapter 3, we developed the theory of the rotation matrices from the viewpoint of quaternions and spherical harmonics in four-space. Comparing Eqs. (7.4.15) and (7.4.16) satisfied by the generators (**L**, **A**) of the hydrogen atom $SO(4)$ group with Eqs. (3.387), (3.389), and (3.391) satisfied by the generators ($\mathscr{J}$, $\mathscr{K}$) of the $SO(4)$ (rigid rotator) group of the rotation matrices, we see that these two structures are identical under the correspondences:

$$\begin{aligned} &\mathbf{L} \to -i(\mathbf{x} \times \nabla), \\ &\mathbf{A} \to -i\left(\mathbf{x}\frac{\partial}{\partial x_0} - x_0\nabla\right), \\ &\mathbf{J}(1) \to \mathscr{K}, \qquad \mathbf{J}(2) \to \mathscr{J}. \end{aligned} \tag{7.4.48}$$

(We have introduced $p_0 = -i\partial/\partial x_0$ and $\mathbf{p} = -i\nabla$ into the results of Note 4 in order that p_0 and $\mathbf{p}$ may denote momentum–space *coordinates* here.)

Although the orbital angular momentum operator is identical in each realization, the Runge–Lenz–Pauli vector **A** cannot be taken *equal* to the corresponding operator in Eqs. (7.4.48), since one operator possesses the derivation property and the other, **A**, does not (see Ref. [12]). It is for this reason that the functions [see Eqs. (7.4.21) and (7.4.22), and Note 4, Chapter 3][1]

$$\Psi^j_{m'm}(x_0, \mathbf{x}) \equiv (-1)^{j+m'}\sqrt{\frac{2j+1}{2\pi^2}}\, D^j_{-m',m}(x_0, \mathbf{x}) \tag{7.4.49}$$

cannot be identified directly with the hydrogen atom wave functions in coordinate space (given in the last section).

The relationship of the harmonic functions (7.4.49) to the hydrogen atom problem is found by representing the point $(x_0, \mathbf{x})$ on the sphere S^3 in terms of the momentum **p** projected onto the sphere S^3 (Fock [5], Judd [13], Wulfman [14])[2],

---

[1] $\mathscr{J}$ and $\mathscr{K}$ (see Chapter 3, Note 4) have the standard action on the functions $\Psi^j_{m'm}$.

[2] Our discussion here is closely related to that given by Judd [13] and Wulfman [14].

$$\mathbf{x} = 2p_0\mathbf{p}/(p_0^2 + \mathbf{p}^2),$$

$$x_0 = (p_0^2 - \mathbf{p}^2)/(p_0^2 + \mathbf{p}^2), \tag{7.4.50}$$

where $\mathbf{p}^2 = p_1^2 + p_2^2 + p_3^2$, and p_0 is a real (numerical) parameter (to be later identified with $\sqrt{-E}$). We shall see below that the transformation (7.4.50) of the operator $-i(\mathbf{x}\,\partial/\partial x_0 - x_0\boldsymbol{\nabla})$ allows us to introduce a new operator $\mathscr{B}$ that is related in a simple way to the momentum space representation, $\mathscr{A}$, of the Runge–Lenz–Pauli vector [see Eqs. (7.4.58) and (7.4.62)]. It is this relation that leads directly to the occurrence of the four-space spherical harmonics in the hydrogen atom wave functions. [The relations $1 + x_0 = 2p_0^2/(p_0^2 + \mathbf{p}^2)$, $\mathbf{p} = p_0\mathbf{x}/(1 + x_0)$, are useful in carrying out the calculations below.]

Using Eqs. (7.4.50), we find

$$\boldsymbol{\nabla} = \frac{p_0^2 + \mathbf{p}^2}{2p_0}\boldsymbol{\nabla}_p,$$

$$\frac{\partial}{\partial x_0} = -\frac{p_0^2 + \mathbf{p}^2}{2p_0^2}(\mathbf{p}\cdot\boldsymbol{\nabla}_p), \tag{7.4.51}$$

where

$$\boldsymbol{\nabla}_p = \left(\frac{\partial}{\partial p_1}, \frac{\partial}{\partial p_2}, \frac{\partial}{\partial p_3}\right). \tag{7.4.52}$$

Substituting these results into Eq. (7.4.48), one obtains the transformation

$$-i\left(\mathbf{x}\frac{\partial}{\partial x_0} - x_0\boldsymbol{\nabla}\right) = ip_0^{-1}\left[\mathbf{p}(\mathbf{p}\cdot\boldsymbol{\nabla}_p) + \frac{(p_0^2 - \mathbf{p}^2)}{2}\boldsymbol{\nabla}_p\right] \equiv \mathscr{B}. \tag{7.4.53}$$

[Observe that $\mathscr{B}$ *does* possess the derivation property when acting on product functions, say, $\Phi(\mathbf{p})\Psi(\mathbf{p})$.]

In order to interpret this result, Eq. (7.4.53), in relation to the Runge–Lenz–Pauli operator $\mathbf{A}$, we rewrite $\mathbf{A}$ [using the basic commutator $[x_i, p_j] = i\delta_{ij}$ and the expansion $\mathbf{L}\times\mathbf{p} = -\mathbf{x}\mathbf{p}^2 + (\mathbf{x}\cdot\mathbf{p})\mathbf{p}$] in the form

$$\mathbf{A} = \left[\mathbf{p}(\mathbf{p}\cdot\mathbf{x}) + 3i\mathbf{p} + \mathbf{x}\left(\frac{1}{r} - \mathbf{p}^2\right)\right]\Big/\sqrt{-H}. \tag{7.4.54}$$

Introducing $\mathbf{x} = i\boldsymbol{\nabla}_p$ into the definition of $\mathscr{B}$, and employing the commutator

$$\left[\frac{(p_0^2 - \mathbf{p}^2)}{2}, \nabla_p\right] = \mathbf{p}, \tag{7.4.55}$$

we bring the operator $\mathscr{B}$ to the form

$$\mathscr{B} = p_0^{-1}[\mathbf{p}(\mathbf{p} \cdot \mathbf{x}) + i\mathbf{p} + \tfrac{1}{2}\mathbf{x}(p_0^2 - \mathbf{p}^2)]. \tag{7.4.56}$$

Finally, letting $\Phi'_n(\mathbf{p})$ denote an eigenfunction of H in the momentum representation,

$$H\Phi'_n = -\frac{1}{n^2}\Phi'_n = -p_0^2\Phi'_n, \qquad p_0 \equiv 1/n, \tag{7.4.57}$$

we now obtain the desired relationship:

$$(\mathscr{B} + 2in\mathbf{p})\Phi'_n(\mathbf{p}) = \mathscr{A}\Phi'_n(\mathbf{p}), \tag{7.4.58}$$

where $\mathscr{A}$ denotes the momentum representation of the Runge–Lenz–Pauli operator [*set* $\mathbf{x} = i\nabla_p$ *in Eq.* (7.4.54)].

We need one more result to connect Eq. (7.4.58) to the theory of harmonic functions on S^3. Let us put

$$\Phi'_n(\mathbf{p}) = F(\mathbf{p})\Lambda_n(\mathbf{p}). \tag{7.4.59}$$

Now, *using the derivation property of* $\mathscr{B}$, we find

$$(\mathscr{B} + 2ip_0^{-1}\mathbf{p})\,\Phi'_n(\mathbf{p}) = [(\mathscr{B} + 2in\mathbf{p})F(\mathbf{p})]\,\Phi'_n(\mathbf{p}) + F(\mathbf{p})(\mathscr{B}\Lambda_n(\mathbf{p})). \tag{7.4.60}$$

Choosing $F(\mathbf{p})$ to satisfy the first-order differential equation

$$(\mathscr{B} + 2ip_0^{-1}\mathbf{p})\,F(\mathbf{p}) = \mathbf{0}, \tag{7.4.61}$$

one can effectively dispose of the lack of the derivation property for $\mathscr{A}$ through the relation

$$\mathscr{A}(F(\mathbf{p})\Lambda_n(\mathbf{p})) = F(\mathbf{p})(\mathscr{B}\Lambda_n(\mathbf{p})). \tag{7.4.62}$$

This relation is the key result. It shows that if we choose the functions $\Lambda_n(\mathbf{p})$ such that $\mathscr{L} = -i\mathbf{p} \times \nabla_p$ and $\mathscr{B}$ have the standard action on these functions [relation of the forms (7.4.24) and (7.4.25) or (7.4.37) and (7.4.38)], then $\mathscr{L} = -i\mathbf{p} \times \nabla_p$ and $\mathscr{A}$ will have this same standard action on the functions $F(\mathbf{p})\Lambda_n(\mathbf{p})$ for which $[\mathscr{L}, F(\mathbf{p})] = \mathbf{0}$.

Up to an arbitrary multiplicative numerical factor, the solution to Eq. (7.4.61) is the rotational invariant function:

$$F(\mathbf{p}) = (p_0^2 + \mathbf{p}^2)^{-2}. \tag{7.4.63}$$

The results obtained above may now be summarized in the following way: Since $\mathscr{L}$ and $\mathscr{B}$ have the standard action on the functions (7.4.49) with $(x_0, \mathbf{x})$ given by the transformation (7.4.50), the angular momentum operator $\mathscr{L}$ and the Runge–Lenz–Pauli operator $\mathscr{A}$ have, in the momentum–space representation, the standard action [Eqs. (7.4.24)–(7.4.26)] on the momentum–space functions defined by [see Eq. (7.4.49)]:

$$\Phi'_{nqm}(\mathbf{p}) = \frac{4n^{3/2}}{[1 + (np)^2]^2} \Psi^{(n-1)/2}_{(m+q)/2,\,(m-q)/2}(x_0, \mathbf{x}), \tag{7.4.64}$$

where x_0, $\mathbf{x}$ are to be expressed in terms of momentum–space coordinates by Eqs. (7.4.50) $[p_0 = 1/n,\ p = (\mathbf{p} \cdot \mathbf{p})^{\frac{1}{2}}]$.

The rotation-matrix functions in Eq. (7.4.64) are those given by Eq. (7.4.49) with the normalization factor $[(2j + 1)/2\pi^2]^{\frac{1}{2}}$ and phase included. The quantum numbers $m' = m_1$ [eigenvalue of $J_3(1) = \mathscr{K}_3$] and $m = m_2$ [eigenvalue of $J_3(2) = \mathscr{J}_3$] are identified as $(m + q)/2$ and $(m - q)/2$, respectively. The normalization is such that

$$\int d\mathbf{p}\, \Phi'^{*}_{n'q'm'}(\mathbf{p})\Phi'_{nqm}(\mathbf{p}) = \delta_{n'n}\delta_{q'q}\delta_{m'm}. \tag{7.4.65}$$

The phase of $\Phi'_{nqm}(\mathbf{p})$ has been chosen such that $\Psi'_{nlm}(\mathbf{p})$ defined by Eq. (7.4.67) below is the Fourier transform of $\Psi_{nlm}(\mathbf{x})$:

$$\Psi'_{nlm}(\mathbf{p}) = \frac{1}{(2\pi)^{3/2}} \int e^{-i\mathbf{x}\cdot\mathbf{p}}\, \Psi_{nlm}(\mathbf{x})\, d\mathbf{x}. \tag{7.4.66}$$

The function $\Phi'_{nqm}(\mathbf{p})$ is then also the Fourier transform of $\Phi_{nqm}(\mathbf{x})$. No direct use of the Fourier transform is required in the algebraic techniques employed here, but an overall phase (depending only on n) has been fixed by Eq. (7.4.66) for consistency of results obtained by different methods.

The momentum–space states of sharp angular momentum are obtained from the Φ'_{nqm} through the Wigner coupling given by Eq. (7.4.34):

$$\Psi'_{nlm}(\mathbf{p}) = \sum_q C^{(n-1)/2\ (n-1)/2\ l}_{(m+q)/2,\,(m-q)/2,\,m}\, \Phi'_{nqm}(\mathbf{p}). \tag{7.4.67}$$

These momentum–space state vectors, Ψ'_{nlm}, then satisfy Eqs. (7.4.35)–(7.4.38), with **L** and **A** now replaced by their momentum–space counterparts, $\mathscr{L}$ and $\mathscr{A}$. It must also be possible to express the functions $\{\Psi'_{nlm}\}$ in the form of Eq. (7.4.41), which contains the solid harmonics (now evaluated in momentum space):

$$\Psi'_{nlm}(\mathbf{p}) = R'_{nl}(p)\,\mathscr{Y}_{lm}(\mathbf{p}), \tag{7.4.68}$$

where $R'_n(p)$ denotes a radial momentum–space function depending only on the invariant $p = (\mathbf{p}\cdot\mathbf{p})^{\frac{1}{2}}$ and p_0. We show below that the explicit form of the radial momentum–space function[1] is {Podolsky and Pauling [15, Eq. (28)]}:

$$\mathscr{R}'_{nl}(p) = (-1)^{n-1}(2i)^l l!\left[\frac{2n(n-l-1)!}{(n+l)!}\right]^{\frac{1}{2}}\left[\frac{2n}{1+(np)^2}\right]^{l+2} C^{(l+1)}_{n-l-1}(\cos\chi), \tag{7.4.69}$$

where

$$\cos\chi \equiv \frac{p_0^2 - \mathbf{p}^2}{p_0^2 + \mathbf{p}^2} = \frac{1-(np)^2}{1+(np)^2}, \tag{7.4.70}$$

and $C^{(\alpha)}_n(x)$ denotes a Gegenbauer polynomial [see Eq. (7.4.75) below].

To obtain this result, one substitutes the following definitions in Eq. (7.4.72) below:

$$\begin{aligned} x_0 &= \cos\chi = (p_0^2 - \mathbf{p}^2)/(p_0^2 + \mathbf{p}^2),\\ \mathbf{x} &= 2p_0\mathbf{p}/(p_0^2 + \mathbf{p}^2) = (\mathbf{p}/p)\sin\chi,\\ r &= \sin\chi = 2p_0 p/(p_0^2 + \mathbf{p}^2). \end{aligned} \tag{7.4.71}$$

One then uses Eqs. (7.4.64) and (7.4.67) and the relation of $H_{(n-1)/2,\,l}$ in Eq. (7.4.73) to the Gegenbauer polynomials given by Eq. (7.4.74) to obtain the final result, Eq. (7.4.69).

e. Relationship between rotation matrices and hyperspherical harmonics.[2] We have seen above that a relationship must hold between the rotation matrices and the hyperspherical harmonics. As follows from the above discussion (and

---

[1] Equation (28) of Ref. [15] is repeated in the abstract with the important phase factor $(-1)(-i)^l$ missing.

[2] The term "hyperspherical harmonic," as used here, designates a solution to Laplace's equation in four-space that is labeled by the irreps of $O(4) \supset O(3)$ [see the right-hand side of Eq. (7.4.72)].

as is already evident from Note 4, Chapter 3), the form of this relationship must be

$$\sum_{m'm} C^{j\,j\,L}_{m'm\,M}\Psi^{j}_{m'm}(x_0,\mathbf{x}) = H_{jL}(x_0, r)\mathscr{Y}_{LM}(\mathbf{x}), \tag{7.4.72}$$

where $H_{jL}(x_0, r)$ denotes a homogeneous polynomial of degree $2j - L$ in x_0 and $\mathbf{x}$ $[r = (\mathbf{x}\cdot\mathbf{x})^{\frac{1}{2}}]$. (The relation between Wigner D-functions and hyperspherical harmonics has been noted by Sharp [16] and Talman [17].)

Let us first state the form of the polynomial (rotational) invariant H_{jL} and relate it to the Jacobi and Gegenbauer polynomials. We find

$$H_{jL}(x_0, r) = (-1)^{2j}(2i)^L[2(2j+1)(2j-L)!(2j+L+1)!/\pi]^{\frac{1}{2}}$$
$$\times \sum_s \frac{(-1)^s(L+s)!\,x_0^{2j-2s-L}\,r^{2s}}{(2L+2s+1)!(2j-2s-L)!s!}, \tag{7.4.73}$$

where the sum is over all integral s such that the factorials are nonnegative. The relation of H_{jL} to special functions is

$$H_{jL}(\cos\chi, \sin\chi) = (-1)^{2j}(2i)^L\,L!\left[\frac{2(2j+1)(2j-L)!}{\pi(2j+L+1)!}\right]^{\frac{1}{2}} C^{(L+1)}_{2j-L}(\cos\chi), \tag{7.4.74}$$

where $C_n^{(\alpha)}(x)$ denotes a Gegenbauer polynomial.

The Gegenbauer polynomials for $n = 0, 1, 2, \ldots$ may be defined in terms of the Jacobi polynomials [see Eq. (3.67), Chapter 3] or the hypergeometric functions as

$$C_n^{(\alpha)}(x) = \frac{(2\alpha)_n}{(\alpha+\frac{1}{2})_n} P_n^{(\alpha-\frac{1}{2},\alpha-\frac{1}{2})}(x)$$
$$= \frac{(2\alpha)_n}{n!}\,{}_2F_1\!\left(-n, n+2\alpha; \alpha+\frac{1}{2}; \frac{1-x}{2}\right), \tag{7.4.75}$$

where $(a)_n = a(a+1)\cdots(a+n-1)$ is Pochhammer's notation for a rising factorial. The generating function for the Gegenbauer polynomials is

$$(1 - 2tx + t^2)^{-\alpha} = \sum_{n=0}^{\infty} C_n^{(\alpha)}(x)t^n. \tag{7.4.76}$$

Expression (7.4.73) for the invariant polynomials H_{jL} may be derived by setting $M = L$ in Eq. (7.4.72) and evaluating at the special point $\mathbf{x} = (0, x_2, 0)$ [see Eq. (3.89), Chapter 3]. This procedure yields an expression for H_{jL} that involves a summation over m' and m that includes a boundary ($M = L$) Wigner coefficient. When the value of the Wigner coefficient is explicitly substituted, one finds that the summation can be carried out using Eq. (3.163), Chapter 3, thus yielding the form (7.4.73).

f. Pauli particle (hydrogen atom with spin). It is an interesting fact that it is easier to discuss the motion in a Coulomb field of a Pauli particle (nonrelativistic spin-$\frac{1}{2}$ particle with kinematically independent spin) than the motion of a spinless particle (see Biedenharn and Brussaard [8, p. 62]). This simplicity results from using the spin operator $\boldsymbol{\sigma}$, together with the two vector operators $\mathbf{L}$ and $\mathbf{A}$, to construct invariant linear operators that enable one to factorize both the angular momentum operator (7.4.80) and the radial equation (7.4.100). Let us sketch briefly how this comes about.

The energy eigenstates are now of the form

$$|nlm\rangle \otimes |\tfrac{1}{2}\mu\rangle, \tag{7.4.77}$$

or, equivalently, they may be taken to be

$$R_{nl}\,\mathbf{Y}^{(l\frac{1}{2})jm}, \tag{7.4.78}$$

where the $\mathbf{Y}^{(l\frac{1}{2})jm}$ are the Pauli central field spinors constructed in Chapter 6, Section 10.

Consider now the introduction of Dirac's quantum number κ. This number is the eigenvalue of the scalar operator $\mathscr{K}$ defined by

$$\mathscr{K} \equiv -(\boldsymbol{\sigma}\cdot\mathbf{L} + 1) = \mathbf{L}^2 - \mathbf{J}^2 - \tfrac{1}{4}. \tag{7.4.79}$$

Using this definition and the Dirac rule [see line above Eq. (7.3.12)], one finds that the orbital angular momentum $\mathbf{L}$ and the total angular momentum $\mathbf{J} = \mathbf{L} + \frac{1}{2}\boldsymbol{\sigma}$ are each expressible as a quadratic form in $\mathscr{K}$:

$$\mathbf{L}^2 = \mathscr{K}(\mathscr{K} + 1), \qquad \mathbf{J}^2 = \mathscr{K}^2 - \tfrac{1}{4}. \tag{7.4.80}$$

One next finds that the eigenvalue κ of $\mathscr{K}$ may be any integer, *zero excluded*:

$$\kappa = \ldots, -2, -1, 1, 2, \ldots. \tag{7.4.81}$$

For a prescribed value of κ the quantum numbers l and j are then obtained as

$$l = l(\kappa) = |\kappa| + \tfrac{1}{2}(-1 + \operatorname{sgn} \kappa), \qquad j = j(\kappa) = |\kappa| - \tfrac{1}{2}, \tag{7.4.82}$$

where sgn κ denotes the sign of κ.

The Pauli central field spinors may now be fully enumerated by a notation that employs only κ and m:

$$\chi_m^\kappa \equiv \mathbf{Y}^{[l(\kappa)\frac{1}{2}]j(\kappa)m}, \tag{7.4.83}$$

where for each prescribed integer κ from the set (7.4.81) the quantum number m takes on the values

$$m = |\kappa| - \tfrac{1}{2}, |\kappa| - \tfrac{3}{2}, \ldots, -|\kappa| + \tfrac{1}{2}. \tag{7.4.84}$$

The operator $\mathscr{K}$ possesses the important property of *anticommuting* with the operator $\boldsymbol{\sigma} \cdot \mathbf{v}$, where $\mathbf{v}$ is any vector operator with respect to $\mathbf{L}$ that is also perpendicular to $\mathbf{L}$. Thus,

$$\mathscr{K}(\boldsymbol{\sigma} \cdot \mathbf{v}) = -(\boldsymbol{\sigma} \cdot \mathbf{v})\mathscr{K} = \frac{i}{2}\, \boldsymbol{\sigma} \cdot (\mathbf{L} \times \mathbf{v} - \mathbf{v} \times \mathbf{L}) \tag{7.4.85}$$

for

$$\mathbf{v} \cdot \mathbf{L} = \mathbf{L} \cdot \mathbf{v} = 0,$$

$$\mathbf{L} \times \mathbf{v} + \mathbf{v} \times \mathbf{L} = 2i\mathbf{v}. \tag{7.4.86}$$

[The result, Eq. (7.4.85), is proved by a straightforward application of the Dirac rule.] Important special cases of Eq. (7.4.85) include $\mathbf{v} = \hat{r} \equiv \mathbf{x}/r$ (unit radial vector), $\mathbf{v} = \mathbf{p}$ (linear momentum), and $\mathbf{v} = \mathbf{A}$ (Runge–Lenz vector).

The importance of the anticommutation property is the fact that the spinor

$$(\boldsymbol{\sigma} \cdot \mathbf{v})\chi_m^\kappa \tag{7.4.87}$$

is an eigenfunction of $\mathscr{K}$ with eigenvalue $-\kappa$. In particular, one finds

$$(\boldsymbol{\sigma} \cdot \hat{r})\chi_m^\kappa = -\chi_m^{-\kappa}, \tag{7.4.88}$$

in consequence of $(\boldsymbol{\sigma} \cdot \hat{r})^2 = 1$ and the phase conventions implied by the definition (7.4.83). Similarly, using the definition

$$\phi_{E\kappa m} \equiv R_{E,l(\kappa)}\chi_m^\kappa, \tag{7.4.89}$$

and the fact that $\boldsymbol{\sigma}$ and $\mathbf{A}$ commute with H, one finds that

$$(\boldsymbol{\sigma}\cdot\mathbf{A})\phi_{E\kappa m} = \alpha(E,\kappa)\phi_{E,-\kappa,m}, \tag{7.4.90}$$

where $\alpha(E,\kappa)$ is a constant. Upon using the results [see Eqs. (7.4.14)]

$$[\mathscr{K}^2 + \varepsilon(\boldsymbol{\sigma}\cdot\mathbf{A})^2]H = -1, \qquad \varepsilon = \pm 1$$

$$(\boldsymbol{\sigma}\cdot\mathbf{A})^2 = (\mathscr{K}^2 + \mathscr{K} + 1)H + 1, \qquad \varepsilon = 0 \tag{7.4.91}$$

the constant in Eq. (7.4.90) is found to satisfy

$$\alpha(E,\kappa)\alpha(E,-\kappa) = \varepsilon(-E^{-1} - \kappa^2), \qquad \varepsilon = \pm 1$$

$$\alpha(0,\kappa)\alpha(0,-\kappa) = 1, \qquad \varepsilon = 0. \tag{7.4.92}$$

In particular for bound states, $-E^{-1} = n^2$, and one finds [choosing a phase to obtain agreement with Eqs. (7.4.38)] that

$$\alpha(E,\kappa) = -[n^2 - \kappa^2]^{\frac{1}{2}}. \tag{7.4.93}$$

Quite remarkably, the complicated results[1] given by Eqs. (7.4.37) and (7.4.38) are now succinctly stated as

$$\mathscr{K}\phi_{n\kappa m} = \kappa\phi_{n\kappa m},$$

$$(\boldsymbol{\sigma}\cdot\mathbf{A})\phi_{n\kappa m} = -(n^2 - \kappa^2)^{\frac{1}{2}}\,\phi_{n,-\kappa,m},$$

$$\phi_{n\kappa m} \equiv R_{n,l(\kappa)}\chi^{\kappa}_{m}. \tag{7.4.94}$$

For arbitrary energy E, it is convenient to define the real parameter η by

$$\eta = \begin{cases} n = 1, 2, \ldots \text{ for } E < 0, & \varepsilon = +1 \\ 1 \text{ for } E = 0, & \varepsilon = 0 \\ \text{arbitrary positive number for } E > 0, & \varepsilon = -1. \end{cases} \tag{7.4.95}$$

One then obtains the generalization of Eqs. (7.4.37) *and* (7.4.38) *for all states of the hydrogen atom* ($\varepsilon = +1, 0, -1$):

---

[1] The use of the Dirac $\mathscr{K}$ operator can greatly simplify angular momentum calculations for half-integral spin systems. We summarize in an Appendix (Section h) several identities for the 3-j and 6-j symbols that have been derived in this way. These identities are used later in the applications (see Chapter 7, Sections 5 and 9).

$$(\boldsymbol{\sigma}\cdot\mathbf{A})\boldsymbol{\phi}_{\eta\kappa m} = -(\eta^2-\varepsilon\kappa^2)^{\frac{1}{2}}\boldsymbol{\phi}_{\eta,-\kappa,m},$$

$$\mathscr{K}\boldsymbol{\phi}_{\eta\kappa m} = \kappa\boldsymbol{\phi}_{\eta\kappa m}, \tag{7.4.96}$$

where

$$\boldsymbol{\phi}_{\eta\kappa m} \equiv R_{\eta,l(\kappa)}\boldsymbol{\chi}_m^{\kappa}.$$

[A phase has been chosen to agree with that in Eqs. (7.4.94) for the bound states.] Using $\mathbf{A} = \eta\mathbf{A}_0$, we may also write

$$(\boldsymbol{\sigma}\cdot\mathbf{A}_0)\boldsymbol{\phi}_{\eta\kappa m} = -\left(1-\varepsilon\frac{\kappa^2}{\eta^2}\right)^{\frac{1}{2}}\boldsymbol{\phi}_{\eta,-\kappa,m}. \tag{7.4.97}$$

Upon employing Dirac's radial decomposition,

$$\boldsymbol{\sigma}\cdot\mathbf{p} = (\boldsymbol{\sigma}\cdot\hat{r})[(\hat{r}\cdot\mathbf{p}) - ir^{-1}(\mathscr{K}+1)], \tag{7.4.98}$$

we can write the operator $\boldsymbol{\sigma}\cdot\mathbf{A}_0$ in the form [see Eqs. (7.4.4) and (7.4.85) $(\mathbf{v}=\mathbf{p})$]

$$\boldsymbol{\sigma}\cdot\mathbf{A}_0 = \boldsymbol{\sigma}\cdot\hat{r} - i\mathscr{K}(\boldsymbol{\sigma}\cdot\mathbf{p})$$

$$= (\boldsymbol{\sigma}\cdot\hat{r})\{1-\mathscr{K}[i\hat{r}\cdot\mathbf{p} + r^{-1}(\mathscr{K}+1)]\}. \tag{7.4.99}$$

This result, when combined with Eqs. (7.4.88) and (7.4.97), leads directly to the radial differential equation in the form

$$\left[1-\kappa\left(\frac{d}{dr}+\frac{\kappa+1}{r}\right)\right]R_{\eta,l(\kappa)}(r) = \left(1-\varepsilon\frac{\kappa^2}{\eta^2}\right)^{\frac{1}{2}}R_{\eta,l(-\kappa)}(r). \tag{7.4.100}$$

This equation defines a raising (κ negative) and a lowering (κ positive) operator on the values of l. *It is a concise and complete expression of the recurrence relations for the nonrelativistic Coulomb radial functions*:[1]

$$\left[1+(l+1)\left(\frac{d}{dr}-\frac{l}{r}\right)\right]R_{\eta l}(r) = \left[1-\varepsilon\frac{(l+1)^2}{\eta^2}\right]^{\frac{1}{2}}R_{\eta,l+1}(r),$$

$$\left[1-l\left(\frac{d}{dr}+\frac{(l+1)}{r}\right)\right]R_{\eta l}(r) = \left[1-\varepsilon\frac{l^2}{\eta^2}\right]^{\frac{1}{2}}R_{\eta,l-1}(r). \tag{7.4.101}$$

[1] For $\kappa > 0$, one has $l = l(\kappa) = \kappa$, and $l(-\kappa) = \kappa - 1 = l - 1$, while for $\kappa < 0$, one has $l = l(-\kappa) = -\kappa - 1$ and $l(-\kappa) = -\kappa = l + 1$.

By combining the raising and lowering operators we can construct the second-order differential equation for the radial functions [Eq. (6.39), Chapter 6, with $V(r) = -2/r$ and $E = -\varepsilon/\eta^2$].

The results expressed by Eqs. (7.4.101) are just those obtained from the *factorization method* introduced by Schrödinger [18] and developed further by Infeld and Hull [19]. *This factorization has occurred here as a natural consequence of the Dirac radial decomposition and the properties of the operators* $\mathscr{K}$ *and* $\boldsymbol{\sigma} \cdot \mathbf{A}$.

Let us turn now to the case of the discrete states where η is replaced by the principal quantum number n and $\varepsilon = +1$. Defining raising and lowering operators by

$$\Omega_+(l) \equiv 1 + (l+1)\left(\frac{d}{dr} - \frac{l}{r}\right),$$

$$\Omega_-(l) \equiv 1 - l\left(\frac{d}{dr} + \frac{l+1}{r}\right), \tag{7.4.102}$$

we find that $R_{n,n-1}$ must satisfy the first-order differential equation

$$\Omega_+(n-1)R_{n,n-1}(r) = 0. \tag{7.4.103}$$

The normalized solution to this equation is given by the radial part of Eq. (7.4.32) (see footnote p. 342). The remaining radial functions are then generated, *with appropriate normalization and relative phase*, by repeated application of the lowering operator:

$$R_{nl}(r) = \prod_{s=1}^{n-l-1}\left[\frac{n^2}{n^2-(n-s)^2}\right]^{\frac{1}{2}} \times \Omega_-(l+1)\Omega_-(l+2)\cdots\Omega_-(n-1)R_{n,n-1}(r) \tag{7.4.104}$$

for $l = n-2, n-3, \ldots, 0$.

One then finds from this result that $R_{nl}(r)$ has the form

$$R_{nl}(r) = (-1)^{n-l-1}\frac{2}{n^2}\left[\frac{(n-l-1)!}{(n+l)!}\right]^{\frac{1}{2}}\left(\frac{2r}{n}\right)^l P_{nl}\left(\frac{2r}{n}\right)e^{-r/n}, \tag{7.4.105}$$

where $P_{nl}(\xi)$ is the polynomial that satisfies

$$(n-l)(n+l)P_{n,l-1}(\xi) = 2l\xi\,[dP_{nl}(\xi)/d\xi] - (n+l)\xi P_{nl}(\xi) + 2l(2l+1)P_{nl}(\xi), \tag{7.4.106}$$

and

$$P_{n,n-1}(\xi) = 1. \tag{7.4.107}$$

These relations serve to identify (uniquely) the polynomials in question as the associated Laguerre polynomials:

$$P_{nl}(\xi) = L_{n-l-1}^{2l+1}(\xi). \tag{7.4.108}$$

Observe that, except for an overall phase factor (depending only on n), there is no choice of phase factor in Eq. (7.4.105), this phase being already determined by the choices made in defining the central field spinors [Eq. (7.4.88)] and in Eq. (7.4.93).

The radial wave functions for the continuum case ($E > 0$) can be obtained by the device of replacing n by $-i\eta$ in the general result for the discrete case [observe that this correctly gives the raising and lowering operators for the continuum states Eqs. (7.4.101)]. In this way, one obtains, after properly normalizing one of the newly obtained radial functions, all the wave functions for a charged particle moving in the attractive Coulomb field of a fixed charge.[1] The results are given by

$$\langle \mathbf{x}|\eta lm\rangle = (kr)^{-1}F_{\eta l}(kr)\,Y_{lm}(\theta\phi), \tag{7.4.109}$$

with $k = \eta^{-1} = \sqrt{E}$, and

$$F_{\eta l}(kr) = C_l(-\eta)(kr)^{l+1}e^{-ikr}\,{}_1F_1(l+1+i\eta;2l+2;2ikr),$$

$$C_l(-\eta) = \frac{2^l e^{\pi\eta/2}|\Gamma(l+1-i\eta)|}{\Gamma(2l+2)}. \tag{7.4.110}$$

These functions are so chosen that in the limit of large kr one obtains

$$F_{\eta l}(kr) \sim \sin\left[kr - \frac{l\pi}{2} + \eta ln(2kr) + \sigma_l(-\eta)\right] \tag{7.4.111}$$

with the Coulomb phase shift defined by

$$\sigma_l(-\eta) = \arg\Gamma(l+1-i\eta). \tag{7.4.112}$$

---

[1] The properties of the Coulomb wave functions for the repulsive field are developed extensively in an article by Hull and Breit [20]. The notations $C_l(\eta)$ and $\sigma_l(\eta)$ for the normalization constant and phase shift, respectively, have been chosen to agree with that of this standard reference work for $\eta > 0$.

The radial wave functions for the zero energy case [$\varepsilon = 0$ in Eqs. (7.4.101)] can be obtained from the radial functions above ($E > 0$) by using the limit relation

$$\lim_{\eta\to\infty} {}_1F_1(l + 1 + i\eta; 2l + 2; i\eta^{-1}r) = (2l + 1)!(2r)^{-l-\frac{1}{2}} J_{2l+1}(\sqrt{8r}). \tag{7.4.113}$$

One obtains the zero energy radial function as

$$R_{0l}(r) = \sqrt{\frac{2}{r}}\, J_{2l+1}(\sqrt{8r}), \tag{7.4.114}$$

where

$$R_{0l}(r) \equiv \lim_{\eta\to\infty} \eta^{\frac{1}{2}} R_{\eta l}(r), \tag{7.4.115}$$

and for large r,

$$R_{0l}(r) \sim \left(\frac{2}{\pi^2 r^3}\right)^{\frac{1}{4}} \sin(\sqrt{8r} - l\pi - \tfrac{1}{4}\pi). \tag{7.4.116}$$

g. Remarks. (*a*) The algebraic methods given in this section, developing various properties of the hydrogen atom, have emphasized the role of angular momentum techniques. The presentation has been adapted from the work of Biedenharn and Brussaard [8], Meadors [9], Judd [13], Wulfman [14], and Bacry [21]. Numerous papers have been written developing the subject from various viewpoints.[1] Extensive bibliographies may be found in the review papers by McIntosh [3], Bander and Itzykson [22], Györgyi [22a], and in the monograph by Englefield [23]. The generalization of the $O(4)$ Coulomb symmetry to the $O(4, 2)$ symmetry group is discussed in the monograph by Barut [23a]; references to the literature are given there.

(*b*) The coordinate and function transformation given by

$$r = \rho^2/2\sqrt{-E}, \qquad E < 0,$$

$$F(\rho) \equiv \frac{\rho}{(2\sqrt{-E})^{\frac{1}{2}}} R\left(\frac{\rho^2}{2\sqrt{-E}}\right) \tag{7.4.117}$$

[1] Ibragimov and Anderson [21a] have treated the problem from the novel viewpoint of Lie-Bäcklund tangent transformations.

may be used to write the radial differential equation for the hydrogen atom in the form of the radial differential equation for the two-dimensional isotropic harmonic oscillator. This relation may be used to obtain the energy eigenvalues and eigenfunctions of the hydrogen atom from those of the oscillator, and conversely. This result was pointed out by Schrödinger [18], was used in Ref. [24] in developing radial integrals, and has subsequently been rediscovered in various contexts by other authors [25, 26].

(*c*) In the *Stark effect* (hydrogen atom in a constant external electric field) the hydrogen atom Hamiltonian H of Eq. (7.4.4) is perturbed by the energy H_1 owing to the potential energy of the electron in the field $\mathbf{E} = E\hat{e}_3$:

$$H_1 = \lambda x_3. \tag{7.4.118}$$

In first-order formal perturbation theory[1] (weak electric field), the unperturbed solution to the hydrogen atom in the form given by Eqs. (7.4.26) is adapted to the perturbation H_1: To see this, one expresses (Meadors [9], Becker and Bleuler [28]) x_3 in the form given by

$$x_3 = \tfrac{3}{2}(\mathbf{L}^2 + \mathbf{A}^2 + 1)^{\frac{1}{2}} A_3 - i[S, H], \tag{7.4.119}$$

where

$$S \equiv \tfrac{1}{8}[T(\mathbf{L}^2 + \mathbf{A}^2 + 1) + (\text{Hermitian conjugate})],$$

$$T \equiv -2ix_3 - 2r^2 p_3 + x_3(\mathbf{x} \cdot \mathbf{p}).$$

Since the commutator term contributes nothing in first-order, one finds, using the representation given by Eqs. (7.4.26), that the degenerate energy level $E_n = -1/n^2$ is split to first-order into the $2n - 1$ sublevels

$$E_n + \tfrac{3}{2}\lambda n q, \tag{7.4.120}$$

where $q = n - 1, n - 2, \ldots, -(n - 1)$.

(*d*) Products of hydrogen atom wave functions in coordinate space do not satisfy a simple product law (addition theorem) as one might expect in consequence of the $SU(2) \times SU(2)$ Lie algebra structure. This may be attributed to the lack of the derivation property for the Runge–Lenz vector $\mathbf{A}$ and the fact the $SU(2) \times SU(2)$ Lie algebra structure is realized only on bound energy eigenspaces.

[1] In formal perturbation theory the existence of a perturbation series is not examined; in fact, the perturbation series does not exist for the present problem. See the review by Killingbeck [27] (and references therein), where modern work in perturbation theory is surveyed.

(*e*) The extension of the relation (7.4.34) between two bases of a bound energy eigenspace (uncoupled and coupled bases in the language of angular momentum theory) to continuum states is accomplished by analytic continuation of the Wigner coefficients. We discuss in RWA how the Wigner coefficients $C^{jJj+\Delta}_{m,M,m+M}$ may be defined for all complex j. Thus, up to phases and normalizations, one will find a relationship of the form

$$\langle \mathbf{x}|\eta q m\rangle = \sum_{l=|m|}^{\infty} (\text{phase})\sqrt{2l+1}\; C^{(-i\eta-1)/2\; l\;\; (-i\eta-1)/2}_{(q-m)/2,\;\; m,\; (q+m)/2} \langle \mathbf{x}|\eta l m\rangle. \quad (7.4.121)$$

It is of interest to note that the Coulomb phase shifts (7.4.112) are obtainable in this way (see Ref. [8, p. 107] and Zwanziger [29]).

h. Appendix. Special identities for 3*n*-*j symbols.* A number of useful identities between 3*n*-*j* symbols may be derived by specializing general relations between these coefficients to spin-$\frac{1}{2}$ systems and writing the spin-$\frac{1}{2}$ Wigner coefficients in terms of Dirac's kappa quantum number:

$$(2l+1)^{\frac{1}{2}}\, C^{l\frac{1}{2}j}_{0mm} = |\kappa|^{\frac{1}{2}}[-S(\kappa)]^{m+\frac{1}{2}}, \quad (\text{A-1})$$

where $S(\kappa)$ denotes the sign of κ. [Equation (A-1) may be verified directly from the spin-$\frac{1}{2}$ Wigner coefficients tabulated in Table 1 in the Appendix of Tables.] Young [30] has used this technique to derive the following two relations in which we have $a + a' + c$ even, $b = a \pm \frac{1}{2}$, $b' = a' \pm \frac{1}{2}$:

$$\begin{pmatrix} b & b' & c \\ \frac{1}{2} & -\frac{1}{2} & 0 \end{pmatrix} = -[(2a+1)(2a'+1)]^{\frac{1}{2}} \begin{pmatrix} a & a' & c \\ 0 & 0 & 0 \end{pmatrix} \begin{Bmatrix} a & a' & c \\ b' & b & \frac{1}{2} \end{Bmatrix}, \quad (\text{A-2})$$

$$\begin{pmatrix} b & b' & c \\ \frac{1}{2} & \frac{1}{2} & -1 \end{pmatrix} = [6(2a+1)(2b'+1)(2c+1)]^{\frac{1}{2}} \begin{pmatrix} a & a' & c \\ 0 & 0 & 0 \end{pmatrix} \begin{Bmatrix} a & a' & c \\ \frac{1}{2} & \frac{1}{2} & 1 \\ b & b' & c \end{Bmatrix}. \quad (\text{A-3})$$

Equation (A-3) may be brought to a form similar to Eq. (A-2) by using the following relation[1] between 9-*j* symbols and 6-*j* symbols (see Varshalovich *et al.* [32, p. 306]):

$$\begin{Bmatrix} a & a' & c \\ d & d & 1 \\ b & b' & c \end{Bmatrix} = 2\frac{(a-a')(a+a'+1)-(b-b')(b+b'+1)}{[(2c)(2c+1)(2c+2)(2d)(2d+1)(2d+2)]^{\frac{1}{2}}}$$

$$\times (-1)^{a+c+b'+d+1} \begin{Bmatrix} a & a' & c \\ b' & b & d \end{Bmatrix}. \quad (\text{A-4})$$

[1] This relation is stated incorrectly in de-Shalit and Talmi [31, p. 520] and in Rotenberg *et al.* [33, p. 24].

Specializing Eq. (A-4) to $d = \frac{1}{2}$, we may write Eq. (A-3) in the following form, where again we have $a + a' + c$ even, $b = a \pm \frac{1}{2}$, $b' = a' \pm \frac{1}{2}$:

$$\begin{pmatrix} b & b' & c \\ \frac{1}{2} & \frac{1}{2} & -1 \end{pmatrix} = (-1)^{a+c+b'+\frac{3}{2}}[(a-a')(a+a'+1)-(b-b')(b+b'+1)]$$
$$\times \left[\frac{(2a+1)(2a'+1)}{c(c+1)}\right]^{\frac{1}{2}} \begin{pmatrix} a & a' & c \\ 0 & 0 & 0 \end{pmatrix} \begin{Bmatrix} a & a' & c \\ b' & b & \frac{1}{2} \end{Bmatrix}. \tag{A-5}$$

The following special relations for 3-j symbols were also noted by Young [30] [the spin-$\frac{1}{2}$ identities are special cases of Eq. (A-5)]:
c even:

$$\begin{pmatrix} a & a & c \\ 1 & -1 & 0 \end{pmatrix} = \frac{c(c+1)-2a(a+1)}{2a(a+1)} \begin{pmatrix} a & a & c \\ 0 & 0 & 0 \end{pmatrix},$$

$$\begin{pmatrix} a+\frac{1}{2} & a-\frac{1}{2} & c \\ \frac{1}{2} & \frac{1}{2} & -1 \end{pmatrix} = \frac{2a+1}{2[a(a+1)]^{\frac{1}{2}}} \begin{pmatrix} a & a & c \\ 0 & 0 & 0 \end{pmatrix};$$

c odd:

$$\begin{pmatrix} a & a & c \\ 1 & -1 & 0 \end{pmatrix} = -\frac{[c(c+1)(2a+2-c)(2a+1-c)]^{\frac{1}{2}}}{2a(a+1)} \begin{pmatrix} a+1 & a & c \\ 0 & 0 & 0 \end{pmatrix},$$

$$\begin{pmatrix} a-\frac{1}{2} & a-\frac{1}{2} & c \\ \frac{1}{2} & \frac{1}{2} & -1 \end{pmatrix} = \begin{pmatrix} a & a-1 & c \\ 0 & 0 & 0 \end{pmatrix},$$

$$\begin{pmatrix} a+\frac{1}{2} & a-\frac{1}{2} & c \\ \frac{1}{2} & \frac{1}{2} & -1 \end{pmatrix} = -\left[\frac{(2a+2+c)(2a+1-c)}{4a(a+1)c(c+1)}\right]^{\frac{1}{2}} \begin{pmatrix} a & a+1 & c \\ 0 & 0 & 0 \end{pmatrix}.$$

References

1. W. Pauli, "Über das Wasserstoffspektrum vom Standpunkt der neuen Quantenmechanik," *Z. Physik* **36** (1926), 336–363.
2. E. Schrödinger, "Quantisierung als Eigenwertproblem. (Erste Mitteilung)," *Ann. Physik* **79** (1926), 361–376; (Zweite Mitteilung), *ibid.*, 489–525.
3. H. V. McIntosh, "Symmetry and degeneracy," in *Group Theory and Its Applications* (E. M. Loebl, ed.), Vol. II, pp. 75–144. Academic Press, New York, 1971.
4. H. Goldstein, "Prehistory of the 'Runge–Lenz' vector," *Amer. J. Phys.* **43** (1975), 737–738; "More on the prehistory of the Laplace or Runge–Lenz vector," *ibid.* **44** (1976), 1123–1124.

4a. K. Kobayashi, "A derivation of the Pauli–Lenz vector and its variants," *J. Phys. A*: *Math. Gen.* **13** (1980), 425–430.

5. V. Fock, "Zur Theorie des Wasserstoffatoms," *Z. Physik* **98** (1935), 145–154.
6. V. Bargmann, "Zur Theorie des Wasserstoffatoms," *Z. Physik* **99** (1936), 169–188.
7. L. D. Landau and E. M. Lifshitz, *Quantum Mechanics. Nonrelativistic Theory.* Addison-Wesley, Reading, Mass., 1958. Translated from the Russian by J. B. Sykes and J. S. Bell.
8. L. C. Biedenharn and P. J. Brussaard, *Coulomb Excitation.* Clarendon, Oxford, 1965.

9. J. G. Meadors, *The Stark Effect in Hydrogen.* Thesis, Auburn University, Auburn, Ala., 1961, unpublished.
10. D. Parks, "Relation between the parabolic and spherical eigenfunctions of hydrogen," *Z. Physik* **159** (1960), 155–157.
11. E. G. Kalnins, W. Miller, and P. Winternitz, "The group $O(4)$, separation of variables and the hydrogen atom," *SIAM J. Appl. Math.* **30** (1976), 630–664.
11a. W. Miller, Jr., *Symmetry and Separation of Variables. Encyclopedia of Mathematics and Its Applications.* (G.-C. Rota, ed.), Vol. 4, Addison-Wesley, Reading, Mass., 1977.
12. D. Aebersold and L. C. Biedenharn, "A cautionary remark on applying symmetry techniques to the Coulomb problem," *Phys. Rev.* **A15** (1977), 441–443.
13. B. R. Judd, *Operator Techniques in Atomic Spectroscopy.* McGraw-Hill, New York, 1963.
14. C. E. Wulfman, "Dynamical groups in atomic and molecular physics," in *Group Theory and Its Applications* (E. M. Loebl, ed.), Vol II, pp. 145–197. Academic Press, New York, 1971.
15. B. Podolsky and L. Pauling, "The momentum distribution in hydrogenlike atoms," *Phys. Rev.* **34** (1929), 109–116.
16. R. T. Sharp, "Notes on $O(4)$ representations," *J. Math. Phys.* **47** (1968), 359–365.
17. J. D. Talman, *Special Functions, A Group Theoretic Approach.* Benjamin, New York, 1968.
18. E. Schrödinger, "A method of determining quantum-mechanical eigenvalues and eigenfunctions," *Proc. Roy. Irish Acad.* **46A** (1940), 9–15; "Further studies on solving eigenvalue problems by factorization," *ibid.* **46A** (1941), 183–205; "The factorization of the hypergeometric equation," *ibid.* **47A** (1941), 53–54.
19. L. Infeld and T. E. Hull, "The factorization method," *Rev. Mod. Phys.* **23** (1951), 21–68.
20. M. H. Hull, Jr., and G. Breit, "Coulomb wave functions," in *Encyclopedia of Physics* (S. Flügge, ed.), Vol. 41/1, pp. 408–465. Springer, Berlin, 1959.
21. H. Bacry, "The de Sitter group $L_{4,1}$ and the bound states of the hydrogen atom," *Nuovo Cimento* **41A** (1966), 222–234.
21a. N. H. Ibragimov and R. L. Anderson, "Lie-Bäcklund tangent transformations," *J. Math. Anal. Appl.* **59** (1977), 145–162.
22. M. Bander and C. Itzykson, "Group theory and the hydrogen atom (I)," *Rev. Mod. Phys.* **38** (1966), 330–345; (II), *ibid.*, 346–358.
22a. G. Györgyi, "Kepler's equation, Fock variables, Bacry's generators and Dirac brackets," *Nuovo Cimento* **53A** (1968), 717–736.
23. M. J. Englefield, *Group Theory and the Coulomb Problem.* Wiley (Interscience), New York, 1972.
23a. A. O. Barut, *Dynamical Groups and Generalized Symmetries in Quantum Theory.* University of Canterbury, Christchurch, N. Z., 1972.
24. J. D. Louck, "Generalized orbital angular momentum and the n-fold degenerate quantum-mechanical oscillator III. Radial integrals," *J. Mol. Spectrosc.* **4** (1960), 334–341.
25. D. Bergmann and Y. Frishman, "A relation between the hydrogen atom and multidimensional harmonic oscillators," *J. Math. Phys.* **6** (1965), 1855–1856.
26. F. Ravndal and T. Toyoda, "A new approach to the $SU(2) \times SU(2)$ symmetry of the Kepler motion," *Nucl. Phys.* **B3** (1967), 312–322.
27. J. Killingbeck, "Quantum-mechanical perturbation theory," *Rept. Prog. Phys.* **40** (1977), 963–1031.
28. H. G. Becker and K. Bleuler, "$O(4)$-Symmetrie und Stark-Effect des H-Atoms," *Z. Naturforsch.* **31A** (1976), 517–523.
29. D. Zwanziger, "Algebraic calculation of nonrelativistic Coulomb phase shifts," *J. Math. Phys.* **8** (1967), 1858–1860; "Exactly soluble nonrelativistic model of particles with both electric and magnetic charges," *Phys. Rev.* **176** (1968), 1480–1488.
30. R. C. Young, "Conversion electron angular correlations: general K-shell formulation and threshold limit," *Phys. Rev.* **115** (1959), 577–585.
31. A. de-Shalit and I. Talmi, *Nuclear Shell Theory.* Academic Press, New York, 1963.

32. D. A. Varshalovich, A. N. Moskalev, and V. K. Khersonskii, *Quantum Theory of Angular Momentum*. Nauka, Leningrad, 1975 (in Russian).
33. M. Rotenberg, R. Bivins, N. Metropolis, and J. K. Wooten, Jr., *The 3-j and 6-j Symbols*. M.I.T. Press, Cambridge, 1959.

5. Atomic Spectroscopy

a. Introduction. Modern quantum theory was founded to overcome the conceptual difficulties with the Bohr–Sommerfeld theory of quantized orbits of electrons in atoms and the attendant emission of light by the sudden jump of an electron from one orbit to another. There already existed a large accumulation of observational data (Fowler [1], Paschen and Götze [2]) on the line spectra due to atoms, and empirical rules for interpreting the data (Back and Landé [3], Van Vleck [4]). (Textbooks that reference those early developments include those by Ruark and Urey [5], Pauling and Goudsmit [6], White [7], and Sommerfeld [8]. Slater [9] presents the developments in atomic spectra from a historical perspective.) The remarkable successes of quantum theory in explaining these results culminated in the publication in 1935 of the classic work[1] *The Theory of Atomic Spectra* by Condon and Shortley [10].

In the short period of less than a decade, following Heisenberg's [11] pioneering paper, an understanding of atomic spectra had been reached that is optimistically explained in the opening paragraphs in the Preface to *The Theory of Atomic Spectra* and that reflected the general feeling of the time that the interpretation of atomic spectra "seems to us to be in a fairly closed and highly satisfactory state."

The theory of groups was not part of the ordinary mathematical equipment of physicists at that time, and Condon and Shortley chose to develop the quantum theory of vector addition and its applications to atomic spectra from an algebraic viewpoint, which minimized any direct reliance on symmetry (group-theoretic) concepts.

Wigner's book [12], using a group-theoretic approach to the interpretation of atomic spectra, had appeared earlier (as had Weyl's book [13] and the lengthy review article by Eckart [14]). Wigner originally was of the opinion (see the Preface to the English translation of his book) that the most important results in his book were the explanation of Laporte's rule (parity concept) and the quantum theory of vector addition of angular momentum, but he later came to agree with the opinion expressed to him by M. von Laue that "the

---

[1] This work has recently been extensively revised and updated by Condon and Odabaşi [10a]. Racah's (group theoretical) techniques have been incorporated into the revision, but not the Gel'fand (unitary group) representation discussed in Section 5h.

recognition that almost all rules of spectroscopy follow from the symmetry of the problem is the most remarkable result."

It was Racah [15, 16] more than anyone else who developed during the period of 1942–1949 a combination of algebraic and group-theoretic methods for dealing with the specific problems of the atomic and nuclear spectroscopy of many-particle systems, thus changing greatly the mathematical techniques utilized in the interpretation of spectra.

The mathematical methods introduced into (atomic) spectroscopy by Racah include (*a*) the angular momentum techniques given in Chapter 3; (*b*) the concept of fractional parentage coefficients for implementing the Pauli principle for equivalent electrons; and (*c*) the use of larger groups containing the rotation group as a subgroup for replacing the undefined set of labels (α) appearing (see p. 32) in the state vector $|(\alpha)jm\rangle$ by a more nearly complete set of mathematically meaningful labels, even though they may not always be *good quantum numbers* [labels associated with the irreps of an invariance (or non-invariance) group of the Hamiltonian].

We cannot, in one section, survey completely the properties of atomic systems. Our aim is to illustrate, through selected examples, how the apparatus of tensor operator theory is used to bring structure into a problem that logically had already been solved by the more conventional methods introduced by Slater [17] (the basis of the method given by Condon and Shortley). We shall also indicate the role of more general group-theoretic methods in spectroscopy. We address only that part of the theory dealing with energy levels and the corresponding state vectors, referring to the literature for the theory of the radiative process giving rise to spectral lines themselves.[1]

b. The approximate Hamiltonian for many-electron atoms. The approximate Hamiltonian for a collection of N identical mass points (electrons), each characterized by a mass $m_0 > 0$, a spin $\mathbf{S}$ $(s = \frac{1}{2})$, and a charge $e < 0$, and moving in the Coulomb field of a fixed charge $-Ze$ (the nucleus), is

$$H = \sum_{\alpha=1}^{N} \left(\frac{\mathbf{p}_\alpha \cdot \mathbf{p}_\alpha}{2m_0} - \frac{Ze^2}{r_\alpha} \right) + H_1 + H_2, \tag{7.5.1}$$

where

$$H_1 = \sum_{\beta>\alpha=1}^{N} \frac{e^2}{r_{\alpha\beta}}, \tag{7.5.2}$$

[1] Our presentation is based, in part, on Condon and Shortley's book, Wigner's book, the papers by Racah, and the more recent textbooks by Slater [9], Judd [18], Sobel'man [19], and de-Shalit and Talmi [20].

$$H_2 = \sum_{\alpha=1}^{N} \xi(r_\alpha)\mathbf{L}_\alpha \cdot \mathbf{S}_\alpha. \tag{7.5.3}$$

Here the αth electron has position vector $\mathbf{x}_\alpha$ relative to the nucleus (regarded as fixed in an inertial frame), linear momentum $\mathbf{p}_\alpha$, orbital angular momentum $\mathbf{L}_\alpha$, and spin $\mathbf{S}_\alpha$. The notations $r_\alpha \equiv (\mathbf{x}_\alpha \cdot \mathbf{x}_\alpha)^{\frac{1}{2}}$ and $r_{\alpha\beta} \equiv [(\mathbf{x}_\alpha - \mathbf{x}_\beta) \cdot (\mathbf{x}_\alpha - \mathbf{x}_\beta)]^{\frac{1}{2}}$ denote, respectively, the distance from the nucleus to the αth electron and the mutual distance between the αth and βth electrons.

The first term in the Hamiltonian (7.5.1) is the kinetic energy plus the Coulomb potential energy for N electrons interacting only with the central nucleus; H_1 is the mutual Coulomb potential energy between the electrons; and H_2 is a spin–orbit interaction term whose form is based on the Pauli theory of the spin–orbit interaction in the hydrogen atom [see Eqs. (7.5.46) and (7.5.53) below].

The Hamiltonian (7.5.1) is approximate in that it does not include terms accounting for (*a*) the motion of the nucleus due to its finite mass; (*b*) relativistic motions of the electrons; (*c*) the magnetic hyperfine interaction (interaction of the magnetic moment of the nucleus with the magnetic field at the nucleus produced by the electrons); and (*d*) the magnetic interaction between the spins of the electrons. (This list is not exhaustive.) The effects due to these phenomena may be treated successfully for many atoms as perturbations; the initial focus is therefore on finding the energy eigenstates and eigenfunctions of the Hamiltonian (7.5.1).

The exact solution to $H\Psi = E\Psi$ for more than one electron has thus far been impossible, although for two electrons (helium) Hylleraas [21] and others [22, 23] have obtained approximate solutions giving good agreement with experiment. For the many-electron problem, the starting point of most calculations is the *central-field approximation* due to Slater [17].

c. The central-field model. A systematic theory of the spectra of atoms, yielding good quantitative agreement with experiment, can be built by starting with the assumption that each electron moves in a spherically symmetric potential $V(r)$, which is produced by the nucleus and all the other electrons.[1] The question as to the best assumptions for $V(r)$ are beyond the scope of the present survey; we refer the reader to the extensive literature on this subject (Slater [9]).

[1] Methods for determining explicit forms for $V(r)$ include (*a*) the Thomas–Fermi statistical model (Thomas [24], Fermi [25]); (*b*) the Hartree self-consistent field method (Hartree [26]); and (*c*) the Hartree–Fock method (Fock [27]) (see also Slater [9]). One must not underestimate the difficult problem of obtaining good "effective central potentials," but this problem is not central to the angular momentum techniques to be illustrated in this section. We note, however, that the distribution of angular momentum in the Thomas-Fermi atom is often incorrectly associated with the shell structure. The correct result was given by Jensen and Luttinger [27a].

For our purposes it is sufficient to assume that the potential in the single-particle Hamiltonian

$$H(\alpha) \equiv \frac{\mathbf{p}_\alpha \cdot \mathbf{p}_\alpha}{2m_0} + V(r_\alpha) \tag{7.5.4}$$

is known and that one has solved the single-particle Schrödinger equation,

$$H(\alpha)\Psi_E(\alpha) = E\Psi_E(\alpha), \tag{7.5.5}$$

both for the set of energy eigenvalues $\{E\}$ and for the set of energy eigenstates $\{\Psi_E(\alpha)\}$.

We have discussed in detail the form of the solution to Eq. (7.5.5) in Chapter 6, Sections 1–8. The state vectors are fully characterized as simultaneous wave functions of the Hamiltonian $H(\alpha)$, the square $\mathbf{L}_\alpha \cdot \mathbf{L}_\alpha$ of the orbital angular momentum $\mathbf{L}_\alpha = \mathbf{x}_\alpha \times \mathbf{p}_\alpha$ of the αth electron, and the z-component $L_{\alpha 3}$ of $\mathbf{L}_\alpha$:

$$\begin{aligned} H(\alpha)\Psi_{n_\alpha l_\alpha m_\alpha} &= E_{n_\alpha l_\alpha}\Psi_{n_\alpha l_\alpha m_\alpha}, \\ (\mathbf{L}_\alpha \cdot \mathbf{L}_\alpha)\Psi_{n_\alpha l_\alpha m_\alpha} &= l_\alpha(l_\alpha + 1)\hbar^2\Psi_{n_\alpha l_\alpha m_\alpha}, \\ L_{\alpha 3}\Psi_{n_\alpha l_\alpha m_\alpha} &= m_\alpha\hbar\Psi_{n_\alpha l_\alpha m_\alpha}, \end{aligned} \tag{7.5.6}$$

where the wave function has the form

$$\Psi_{n_\alpha l_\alpha m}(\mathbf{x}_\alpha) = R_{n_\alpha l_\alpha}(r_\alpha)Y_{l_\alpha m_\alpha}(\theta_\alpha \phi_\alpha). \tag{7.5.7}$$

The radial function $R_{n_\alpha l_\alpha}$ itself satisfies the radial differential equation given by Eq. (6.39), Chapter 6. In general, the $E_{n_\alpha l_\alpha}$ will depend on the angular momentum quantum number l_α as well as on a principal quantum number n_α.

For a specified energy eigenvalue $E_{n_\alpha l_\alpha}$, one cannot say for a general potential what the allowed values of n_α and l_α will be. It is generally assumed for the purpose of nomenclature that the radial functions are *hydrogen-like* and that the range of n_α and l_α have a well-defined meaning. The spherical symmetry ensures that $l_\alpha \in \{0, 1, 2, \ldots\}$ and that for each allowed value of l_α one has $m_\alpha = l_\alpha, l_\alpha - 1, \ldots, -l_\alpha$. Each energy level $E_{n_\alpha l_\alpha}$ is therefore $(2l_\alpha + 1)$-fold degenerate, owing to the orbital rotational symmetry of the Hamiltonian.

The single-particle Hamiltonian $H(\alpha)$ does not depend explicitly on the spin of the electron. Each of the state vectors

$$|n_\alpha l_\alpha m_\alpha \mu_\alpha\rangle \equiv |n_\alpha l_\alpha m_\alpha\rangle \otimes |\tfrac{1}{2}\mu_\alpha\rangle \tag{7.5.8}$$

corresponding to the two spin orientations $\mu_\alpha = \pm\frac{1}{2}$ is also an eigenstate of $H(\alpha)$ with energy eigenvalue $E_{n_\alpha l_\alpha}$. [In Eq. (7.5.8) we employ the notation $\langle \mathbf{x}_\alpha | n_\alpha l_\alpha m_\alpha \rangle = \Psi_{n_\alpha l_\alpha m_\alpha}(\mathbf{x}_\alpha)$.] Accordingly, each energy level $E_{n_\alpha l_\alpha}$ is $2(2l_\alpha + 1)$-fold degenerate.

We next proceed to relate the single-particle Hamiltonians to the approximate Hamiltonian of the last section. Defining the N-particle Hamiltonian H_0 and the potential term V_1 by

$$H_0 \equiv \sum_{\alpha=1}^{N} H(\alpha), \tag{7.5.9}$$

$$V_1 \equiv -\sum_{\alpha=1}^{N} \left[\frac{Ze^2}{r_\alpha} + V(r_\alpha)\right], \tag{7.5.10}$$

we find that the Hamiltonian H defined by Eq. (7.5.1) is given by

$$H = H_0 + H', \tag{7.5.11}$$

$$H' = V_1 + H_1 + H_2. \tag{7.5.12}$$

Finally, we note that it is the practice to relate the function $\xi(r)$ appearing in the definition of H_2, Eq. (7.5.3), to the potential $V(r)$ in the same way as it occurs in the hydrogen atom problem:

$$\xi(r) = \frac{\hbar^2}{2m_0^2c^2r}\left(\frac{dV}{dr}\right). \tag{7.5.13}$$

The physical motivation for the central-field model is the expectation that the main contribution to an energy level E will come from H_0, which itself describes the composite system obtained by taking N replicas of the single-particle system consisting of a single electron moving in the central field $V(r)$. The idea is then to treat all other interactions occurring in the actual physical system as perturbations. This program has, in fact, had considerable success in the development of the systematics of atomic spectra, including an explanation of the regularities in the chemical properties of the elements in the periodic table (see Slater [9]).

The relative importance of different types of interactions varies a great deal from element to element, and although it is this phenomenon that helps one to understand the origin of the different spectroscopic and chemical behavior of the elements, it also means that no single perturbation scheme will be valid for all elements of the periodic table. Much of atomic spectroscopy is concerned with identifying the important interactions for specific elements and developing appropriate approximate methods for calculating energy corrections due to these interactions.

Consider next the energy eigenspaces of the Hamiltonian H_0 given by Eq. (7.5.9). Since we are dealing with N kinematically independent systems, the appropriate vector space is the tensor product space composed from the state vectors of the N individual electrons. Thus, each of the vectors

$$|n_1 l_1 m_1 \mu_1\rangle \otimes |n_2 l_2 m_2 \mu_2\rangle \otimes \cdots \otimes |n_N l_N m_N \mu_N\rangle \tag{7.5.14}$$

corresponding to $m_\alpha = l_\alpha, l_\alpha - 1, \ldots, -l_\alpha$ and $\mu_\alpha = \frac{1}{2}, -\frac{1}{2}$ has the energy eigenvalue

$$E_0 = \sum_\alpha E_{n_\alpha l_\alpha}. \tag{7.5.15}$$

Notice, however, that each of the vectors

$$|n_{\alpha_1} l_{\alpha_1} m_{\alpha_1} \mu_{\alpha_1}\rangle \otimes |n_{\alpha_2} l_{\alpha_2} m_{\alpha_2} \mu_{\alpha_2}\rangle \otimes \cdots \otimes |n_{\alpha_N} l_{\alpha_N} m_{\alpha_N} \mu_{\alpha_N}\rangle, \tag{7.5.16}$$

where $(\alpha_1 \alpha_2 \cdots \alpha_N)$ is a permutation of $(12 \cdots N)$, is also an energy eigenstate having eigenvalue E_0, where again we may have $m_\alpha = l_\alpha, l_\alpha - 1, \ldots, -l_\alpha$ and $\mu_\alpha = \frac{1}{2}, -\frac{1}{2}$.

The state vectors (7.5.14) and (7.5.16) do not, however, correspond to *physical states* of a noninteracting N-electron atom. One must account for the *Pauli exclusion principle* [two or more electrons cannot have the same assignment of the four quantum numbers $(nlm\mu)$]. The Pauli exclusion principle is introduced into the theory by the requirement due to Dirac that the state vectors of the system must be *antisymmetric* in the interchange of all pairs of electron states. We are thus led to the Slater states defined by[1]

$$|\{(n_1 l_1 m_1 \mu_1); \cdots; (n_N l_N m_N \mu_N)\}\rangle$$

$$\equiv \frac{1}{\sqrt{N!}} \sum_P (-1)^\sigma P(|n_1 l_1 m_1 \mu_1\rangle \otimes \cdots \otimes |n_N l_N m_N \mu_N\rangle), \tag{7.5.17}$$

where the summation extends over all $N!$ permutations $\{P\}$ of the subscripts $(12 \cdots N)$, and $(-1)^\sigma$ is the signature of P ($\sigma = 0$ for P an even permutation and 1 for P an odd permutation).

---

[1] The braces enclosing a set of quantum numbers will be used to designate that the state is antisymmetric under the exchange of any pair of 4-tuples of quantum numbers.

The nonzero vectors among the state vectors defined by Eq. (7.5.17) are still orthonormal and span the vector space, $\mathscr{V}_{E_0}^A$, of antisymmetrized products of N single-electron state vectors of energy E_0. If the angular momentum states of the individual electrons are all distinct, $l_1 \neq l_2 \neq \cdots \neq l_N$, then the dimension of the space $\mathscr{V}_{E_0}^A$ is

$$\dim \mathscr{V}_{E_0}^A = \prod_{\alpha=1}^{N} 2(2l_\alpha + 1). \tag{7.5.18}$$

If, however, two or more (nl)-pairs in the sequence

$$(n_1 l_1)(n_2 l_2) \cdots (n_N l_N) \tag{7.5.19}$$

are equal, then some of the basis vectors defined by Eq. (7.5.17) will be zero or dependent, and, correspondingly,

$$\dim \mathscr{V}_{E_0}^A < \prod_{\alpha=1}^{N} 2(2l_\alpha + 1). \tag{7.5.20}$$

In the central-field approximation the state of an N-electron atom is specified by the set of quantum numbers $\{(n_\alpha l_\alpha m_\alpha \mu_\alpha)\colon \alpha = 1, 2, \ldots, N\}$, the corresponding energy $E_0 = \sum_\alpha E_{n_\alpha l_\alpha}$, and the antisymmetric Slater state vectors $|\{(n_1 l_1 m_1 \mu_1) \cdots (n_N l_N m_N \mu_N)\}\rangle$.

Under the permutation P of the sets of electron quantum numbers among themselves given by

$$P\colon (n_1 l_1 m_1 \mu_1) \cdots (n_N l_N m_N \mu_N) \to (n_{\alpha_1} l_{\alpha_1} m_{\alpha_1} \mu_{\alpha_1}) \cdots (n_{\alpha_N} l_{\alpha_N} m_{\alpha_N} \mu_{\alpha_N}) \tag{7.5.21}$$

the state vector (7.5.17) is mapped into itself (even permutation) or into its negative (odd permutation). It is therefore impossible to speak of a definite electron as possessing a particular set of quantum numbers. The implication of this result is that the state vectors (7.5.17) are no longer simultaneous eigenvectors of the complete set of commuting operators

$$\{H(\alpha), \mathbf{L}_\alpha \cdot \mathbf{L}_\alpha, L_{\alpha 3}, \mathbf{S}_\alpha \cdot \mathbf{S}_\alpha, S_{\alpha 3}\colon \alpha = 1, 2, \ldots, N\}. \tag{7.5.22}$$

[The notation (7.5.17) for a state vector must accordingly be used cautiously, since the labels do not correspond to eigenvalues of the corresponding operators!]

The Hamiltonian $H_0 = \sum_\alpha H(\alpha)$ is, of course, diagonal on the space $\mathscr{V}_{E_0}^A$, as is any other totally symmetric operator constructed from the operators in the set (7.5.22). In particular, the z-components

$$L_3 = \sum_\alpha L_{\alpha 3}, \qquad S_3 = \sum_\alpha S_{\alpha 3} \tag{7.5.23}$$

of the total orbital angular momentum and the total spin are diagonal with corresponding eigenvalues given by

$$L_3 \to M_L \equiv \sum_\alpha m_\alpha, \qquad S_3 \to M_S \equiv \sum_\alpha \mu_\alpha. \tag{7.5.24}$$

The total orbital angular momentum operator **L** and the total spin operator **S** defined by

$$\mathbf{L} = \sum_\alpha \mathbf{L}_\alpha, \qquad \mathbf{S} = \sum_\alpha \mathbf{S}_\alpha \tag{7.5.25}$$

may also be seen to map the space $\mathscr{V}^A_{E_0}$ into itself. This result follows from the commutation properties

$$[H, \mathbf{L}] = \mathbf{0}, \qquad [H, \mathbf{S}] = \mathbf{0}, \tag{7.5.26}$$

$$[P, \mathbf{L}] = \mathbf{0}, \qquad [P, \mathbf{S}] = \mathbf{0}, \qquad [P, H] = 0, \tag{7.5.27}$$

for each $P \in S_N$. It must be possible then to split the space $\mathscr{V}^A_{E_0}$ into a direct sum of spaces, each of which is characterized by sharp total orbital angular momentum quantum numbers (LM_L) and sharp total spin quantum numbers (SM_S). Equivalently, it must be possible to span the space $\mathscr{V}^A_{E_0}$ by sets of basis vectors of the type

$$\left\{ |(\gamma)LM_L; SM_S\rangle : \begin{matrix} M_L = L, L-1, \ldots, -L \\ M_S = S, S-1, \ldots, -S \end{matrix} \right\}, \tag{7.5.28}$$

where **L** and **S** have the standard action on these basis vectors. Here (γ) denotes an unspecified set of labels, which may depend on the method used to decompose the space $\mathscr{V}^A_{E_0}$ into the basic orbital and spin multiplets.

The ranges and multiple occurrences of L and S required to enumerate a basis (7.5.28) of $\mathscr{V}^A_{E_0}$ must be compatible with the reduction (Clebsch–Gordan series) of the direct products

$$[l_1] \otimes [l_2] \otimes \cdots \otimes [l_N],$$

$$[\tfrac{1}{2}] \otimes [\tfrac{1}{2}] \otimes \cdots \otimes [\tfrac{1}{2}], \tag{7.5.29}$$

but the general result is unknown {here $[j]$ denotes an irrep of $SU(2)$}.

The importance of reducing the space $\mathscr{V}^A_{E_0}$ into carrier spaces of the irreps of the orbital rotation group (generators **L**) and the spin group (generators **S**), or of finding other methods of generating a basis (7.5.28) of $\mathscr{V}^A_{E_0}$, is further accentuated by the observation that $V_1 + H_1$ commutes with **L** and **S** as well as with the permutations $\{P\}$. This result implies that the matrix elements of $V_1 + H_1$ on the basis (7.5.28) will be diagonal in the quantum numbers $(LM_L)(SM_S)$; that is, the only nonvanishing matrix elements will be of the form

$$\langle(\gamma')LM_L; SM_S|V_1 + H_1|(\gamma)LM_L; SM_S\rangle. \tag{7.5.30}$$

Thus, for atoms for which the spin-orbit interaction can be ignored in comparison with the electrostatic interaction, the construction of a basis of $\mathscr{V}^A_{E_0}$ of sharp orbital angular momentum (LM_L) and sharp spin (SM_S) is an important step toward the solution to the energy-level problem. Indeed, for certain cases where there are no multiple occurrences of (LS) [so that $(\gamma') = (\gamma)$], and the levels of H_0 are sufficiently separated, the construction of the basis (7.5.28) solves the first-order perturbation problem, since the corrections to E_0 are then just

$$E_1(LS) \equiv \langle(\gamma)LM_L; SM_S|V_1 + H_1|(\gamma)LM_L; SM_L\rangle. \tag{7.5.31}$$

This result is further simplified when one notes that the matrix elements (7.5.31) will be independent of M_L and M_S in consequence of the invariance of $V_1 + H_1$ separately under orbital rotations and $SU(2)$ spin rotations. Thus a degeneracy of $(2L + 1)(2S + 1)$ will remain in the levels $E_0 + E_1(LS)$ resulting from the electrostatic perturbation $V_1 + H_1$.

Consider next how one might construct a basis of $\mathscr{V}^A_{E_0}$ having sharp quantum numbers $(LM_L)(SM_S)$. For simplicity let us assume that no multiple occurrences of the pair (LS) arise. Because the Slater states (7.5.17) have sharp M_L and M_S, we can construct (uniquely, up to phases) a basis of $\mathscr{V}^A_{E_0}$ by using the method of highest weights given in Chapter 3, Section 12, but now generalized to accommodate two angular momenta, **L** and **S**. The method is best illustrated through examples.

Example 1. $N = 2$, $n_1 = n_2 = n \geqslant 2$, $l_1 = l_2 = 1$.

The space $\mathscr{V}^A_{E_0}$ is spanned by fifteen Slater vectors as given by Eq. (7.5.17) and having the (M_L, M_S)-pairs of projection quantum numbers

$$\begin{matrix} (2,0) & (1,1) & 2(1,0) & (1,-1) & (0,1) & 3(0,0) \\ (-2,0) & (-1,-1) & 2(-1,0) & (-1,1) & (0,-1), & \end{matrix} \tag{7.5.32}$$

where the notation $k(M_L, M_S)$ designates k occurrences of (M_L, M_S). We order these pairs by the rules $(M_L, M_S) < (M'_L, M'_S)$ if $M_L < M'_L$, and $(M_L, M_S) < (M_L, M'_S)$ if $M_S < M'_S$.

We now proceed to find the $|LM_L; SM_S\rangle$ multiplets by the method of highest weights: The vector $|\{(n\,1\,1\,\tfrac{1}{2});\ (n\,1\,1\,-\tfrac{1}{2})\}\rangle$ corresponding to the highest (M_L, M_S)-pair, $(2, 0)$, is the highest-weight vector

$$|22;00\rangle \equiv |\{(n\,1\,1\,-\tfrac{1}{2});\ (n\,1\,1\,-\tfrac{1}{2})\}\rangle$$

of the LS-multiplet $|2M_L;00\rangle$ ($M_L = 2, 1, 0, -1, -2$). {We apply the lowering operator L_- repeatedly [see Eq. (3.162), Chapter 3] to $|22;00\rangle$ to generate explicitly the full multiplet.} We next delete the (M_L, M_S)-pairs $(2, 0)$, $(1, 0)$, $(0, 0)$, $(-1, 0)$, $(-2, 0)$ from the set (7.5.32) and are left with

$$\begin{array}{cccc} (1,1) & (1,0) & (1,-1) & (0,1)\quad 2(0,0) \\ (-1,-1) & (-1,0) & (-1,1) & (0,-1). \end{array} \tag{7.5.33}$$

The vector $|\{(n\,1\,1\,\tfrac{1}{2});\ (n\,1\,0\,\tfrac{1}{2})\}\rangle$ corresponding to the highest (M_L, M_S)-pair, $(1, 1)$, is the highest-weight vector $|11;11\rangle \equiv |\{(n\,1\,1\,\tfrac{1}{2});\ (n\,1\,0\,\tfrac{1}{2})\}\rangle$ of the LS-multiplet $|1M_L;1M_S\rangle$ ($M_L, M_S = 1, 0, -1$), which is generated by the independent applications of L_- and S_- to $|11;11\rangle$ according to Eq. (3.162), Chapter 3. We now delete all (M_L, M_S)-pairs for M_L, $M_S = 1, 0, -1$ from the set (7.5.33) and are left with

$$(0, 0).$$

There are three Slater states corresponding to $(0, 0)$ – namely,

$$|m\rangle \equiv |\{(n\,1\,m\,\tfrac{1}{2});(n\,1\,-m\,-\tfrac{1}{2})\}\rangle, \qquad m = 1, 0, -1.$$

There is a unique normalized vector of the form

$$|00;00\rangle = \sum_m \alpha_m |m\rangle$$

that is perpendicular to the $|20;00\rangle$ and $|10;10\rangle$ vectors, and this vector, $|00;00\rangle$, is the last vector needed to span the fifteen-dimensional space $\mathscr{V}^A_{E_0}$ with the LS-multiplets in the set $\{|2M_L;00\rangle$ ($M_L = 2, 1, 0, -1, -2$); $|1M_L;1M_S\rangle$ ($M_L, M_S = 1, 0, -1$); $|00;00\rangle\}$.

Example 2. $N = 2$, $n_1 = n_2 = n \geqslant 3$, $l_1 = 0$, $l_2 = 2$.

The space $\mathscr{V}^A_{E_0}$ is spanned by the twenty Slater state vectors as given by Eq. (7.5.17) and having the (M_L, M_S)-pairs of projection quantum numbers:

$$\begin{array}{l}(2,1)\quad 2(2,0)\ (2,-1)\qquad (1,1)\quad 2(1,0)\ (1,-1)\quad (0,1)\ 2(0,0)\\ (-2,-1)\ 2(-2,0)\ (-2,1)\ (-1,-1)\ 2(-1,0)\ (-1,1)\ (0,-1).\end{array} \tag{7.5.34}$$

The vector $|\{(n\,0\,0\,\tfrac{1}{2});\ (n\,2\,2\,\tfrac{1}{2})\rangle$ corresponding to the highest (M_L, M_S)-pair, (2, 1), is the highest-weight vector $|22;11\rangle \equiv |\{(n\,0\,0\,\tfrac{1}{2});\ (n\,2\,2\,\tfrac{1}{2})\}\rangle$ of the LS-multiplet $|2M_L;1M_S\rangle$ generated by the independent and repeated applications of L_- and S_- to $|22;11\rangle$. Removing the fifteen (M_L, M_S)-pairs corresponding to $M_L = 2, 1, 0, -1, -2$; $M_S = 1, 0, -1$, from the set (7.5.34) leaves

$$(2,0)\quad (1,0)\quad (0,0)\quad (-1,0)\quad (-2,0). \tag{7.5.35}$$

There are two Slater state vectors corresponding to (2, 0):

$$\begin{aligned}|+\rangle &= |\{(n\,0\,0\,\tfrac{1}{2});(n\,2\,2\,-\tfrac{1}{2})\}\rangle,\\ |-\rangle &= |\{(n\,0\,0\,-\tfrac{1}{2});(n\,2\,2\,\tfrac{1}{2})\}\rangle.\end{aligned} \tag{7.5.36}$$

There is a unique normalized vector of the form

$$|22;00\rangle = a_+|+\rangle + a_-|-\rangle \tag{7.5.37}$$

that is perpendicular to the vector $|22;10\rangle$ belonging to the LS-multiplet $\{|2M_L;1M_S\rangle\}$. This vector, $|22;00\rangle$, is the highest-weight vector of the LS-multiplet $|2M_L;00\rangle$ generated by the repeated application of L_- to $|22;00\rangle$.

The procedure outlined above will produce, except for phase choices, a unique basis of $\mathscr{V}^A_{E_0}$ of the form

$$\left\{|LM_L;SM_S\rangle\colon \begin{array}{l}M_L = L,\ldots,-L;\ L = L', L'',\ldots\\ M_S = S,\ldots,-S;\ S = S', S'',\ldots\end{array}\right\} \tag{7.5.38}$$

in the case of no multiplicity of the pair (LS).

In principle, the method of highest weights is still applicable even when multiple values of a given pair, (LS), occurs. One must now introduce additional indices (γ) to distinguish between the various (perpendicular) highest-weight vectors having the same L and S:

$$\{|(\gamma)LL;SS\rangle\colon (\gamma) = (\gamma'), (\gamma''), \ldots\}. \tag{7.5.39}$$

In general the existence of k occurrences of (LS) in the reduction of the space $\mathscr{V}^A_{E_0}$ into LS-multiplets will require that one diagonalize the $k \times k$ matrix $\mathbf{H}'_1$

with entries

$$(\mathbf{H}'_1)_{(\gamma')(\gamma)} \equiv \langle(\gamma')LS; LS|V_1 + H_1|(\gamma)LS; LS\rangle \qquad (7.5.40)$$

in order to obtain the first-order energy splittings (for elements in which the electrostatic interaction dominates).

(In simple cases one may avoid this last step in so far as the calculation of energy eigenvalues are concerned by using Slater's *diagonal sum rule*. This rule is just the statement of the invariance of the trace of a matrix A under the similarity transformation $A \to SAS^{-1}$. In particular, since the LS-multiplet basis is related to the Slater basis (7.5.17) by a real, orthogonal transformation, the sum of all the first-order energy corrections is equal to the sum of the diagonal matrix elements of $V_1 + H_1$ calculated in the Slater basis.)

Given the central potential, the main problem in the calculation of energy levels of atoms (ions) is to determine a basis for the space $\mathscr{V}^A_{E_0}$ (the vector space of antisymmetrized states of energy E_0) that is optimal (in some sense) for the calculation of the splittings due to the electrostatic interaction, the spin–orbit interaction, etc. [We have indicated above two bases of this space: Slater's determinantal basis (7.5.17) and the LS-multiplet basis. The jj-coupling scheme (see below) offers yet another method for spanning $\mathscr{V}^A_{E_0}$. Mathematically, there are, of course, an infinity of such bases.] Although seemingly mathematically trivial (since the problem has been reduced to the study of the properties of a finite-dimensional vector space), the investigations of the energy-level problem have been the source of the invention (principally by Racah and Wigner) of new mathematical structures (the tensor operator concept) having applications across all areas of spectroscopy (molecular, atomic, nuclear, and hadronic).

Slater's methods are direct and conceptually simple (they were introduced in 1929 before the Wigner coefficients and tensor operators), but difficult to implement generally. Racah recognized that there must be recurrent structures in Slater's methods, since, for example, the method of highest weights must, in fact, be tantamount to solving the coupling problem for N angular momenta. But the solution to the latter problem is geometric and is the same for all physical systems. Thus, certain systematics could be brought to the problem that could be tabulated once and for all. It was this insight by Racah that brought about the development and introduction of new mathematical techniques for spectroscopy.

There are three aspects to Racah's program: (*a*) the introduction of general methods for spanning the space $\mathscr{V}^A_{E_0}$ itself; (*b*) the classifications of interactions as invariant tensors (thus bringing in the apparatus of the Wigner–Eckart theorem); and (*c*) the use of more general groups as a means for understanding the occurrence of the labels (γ) in state vectors.

We give selected examples of these methods after introducing a short vocabulary of spectroscopic terminology.

d. A short vocabulary of spectroscopy terminology

Electron configuration. A definite sequence of quantum numbers

$$(n_1 l_1)(n_2 l_2) \cdots (n_N l_N) \tag{7.5.41}$$

that determines the energy E_0 is said to specify an electron configuration (the sequence gives the N energy levels among which the N electrons are distributed). One often refers to the sequence itself as the electron configuration. It is customary to use hydrogenic values for (nl) $(n = 1, 2, \ldots;$ $l = 0, 1, \ldots, n-1)$ and to denote specific values of l by Latin letters as shown below:

$$\begin{array}{l} l\colon 0\ 1\ 2\ 3\ 4\ 5\ 6 \ldots \\ \quad s\ p\ d\ f\ g\ h\ i \ldots . \end{array} \tag{7.5.42}$$

(The first four letters denote characteristics of early spectra referred to as *s*harp, *p*rincipal, *d*iffuse, and *f*undamental.) A sequence (nl) repeated r times is denoted by $(nl)^r \equiv nl^r$. For example, the 57 electrons of the Nd^{3+} ion $(Z = 60)$ are distributed in the ground energy state in the electron configuration

$$1s^2\ 2s^2\ 2p^6\ 3s^2\ 3p^6\ 3d^{10}\ 4s^2\ 4p^6\ 4d^{10}\ 5s^2\ 5p^6\ 4f^3. \tag{7.5.43}$$

The sequence is ordered by ascending values of the energy E_{nl}, which, as a rule, obey $E_{nl} > E_{n'l'}$ for $n + l > n' + l'$ ("Madelung's rule").

Closed shell. The Pauli principle allows only one electron to be in the state $(nlm\mu)$. For given (nl) there can be no more than $2(2l + 1)$ electrons having energy E_{nl}. We say that the quantum numbers (nl) specify a *shell* and that this shell is *closed* when electrons occupy all possible $2(2l + 1)$ states. Thus, $(nl)^{2(2l+1)}$ denotes a closed shell. There are eleven closed shells in the configuration (7.5.43), and the $4f$-shell, which contains fourteen possible electron states, is *less than half-filled*, with three electrons.

Equivalent electrons. In the sense of the Pauli exclusion principle, all electrons are equivalent. The term *equivalent electrons* is, however, often used to refer to electrons in the same nl-shell.

Level. Used generically, "level" refers to any energy eigenvalue. In atomic spectroscopy "level" means only one thing: the collection of $2J + 1$ states belonging to an energy level with specified J, the total angular momentum of the electrons in the atom.

LS-coupling. This refers to a coupling scheme that couples the individual electron orbital angular momenta to obtain the total orbital angular momentum, and separately couples the individual electron spins to obtain the total spin.

LS-multiplet or term. In *LS*-coupling, the word *term* refers to the set of states

$$\{|(\gamma)LM_L; SM_S\rangle : M_L = L, \ldots, -L; S = M_S, \ldots, -M_S\}. \quad (7.5.44)$$

It is customary to denote specific values of L by the capital letters in the scheme (7.5.44)

$$\begin{array}{lccccccc} L: & 0 & 1 & 2 & 3 & 4 & 5 & 6 \\ & S & P & D & F & G & H & I. \end{array}$$

The set of states (7.5.44), or term, is then denoted by ^{2S+1}L. For example, the term ^5D denotes the (5)(5) = 25 states for which $L = 2$ and $S = 2$.

jj-coupling. This refers to a coupling scheme in which one first couples the individual electron orbital angular momenta and individual spin to obtain $\mathbf{J}_\alpha = \mathbf{L}_\alpha + \mathbf{S}_\alpha$, followed by a coupling scheme to obtain the total angular momentum: $\mathbf{J} = \sum_\alpha \mathbf{J}_\alpha$. This scheme is more appropriate for the description of the states of elements in which the spin–orbit interaction dominates the electrostatic repulsion interaction.

Hund's rule. Hund's rule is an empirical rule stating that the electrostatic interaction splits the energy E_0 into a series of levels such that the level with highest S and highest L (for this S) has the lowest energy. This rule works for the ground terms of all atoms and ions so far analyzed experimentally, but there are exceptions for excited configurations.

e. Closed shells. The number of physical (antisymmetric) states for n equivalent electrons[1] is given by the binomial coefficient

$$\binom{4l+2}{n}, \qquad n = 1, 2, \ldots, 4l + 2. \quad (7.5.45)$$

This configuration is denoted by l^n ($l = s, p, d, f, \ldots$) in which one drops the principal quantum number. In the single state of the closed-shell configuration, l^{4l+2}, a negative value of m_α and μ_α occurs for every positive value, so that $M_L = M_S = 0$, and hence, also, $L = S = 0$. The (zeroth-approximation) state

[1] At the risk of some confusion we use the symbol n to denote both the number of equivalent electrons in a shell and the principal quantum number of a single-electron state. We often suppress the principal quantum number to avoid ambiguity.

of a closed-shell configuration is thus both an orbital–rotational and a spin–rotational invariant.

The rotational invariance of the states of closed shells gives such atoms (He, Ne, Ar, Kr, Xe) a stable and inert quality, which often allows one to consider neighboring atoms in the periodic table as having spectra that arise solely from the transitions of electrons outside closed shells. Since the effective field due to electrons in closed shells is centrally symmetric ($L = S = 0$), this is a useful simplification, allowing one, in effect, to replace the N-electron problem by an n-electron problem. For example, the spectra of the alkali metals (Li, Na, K, Rb, Fr) may be interpreted as arising from a single (valence) electron in a centrally symmetric field (at great distances the field is effectively that of a Coulomb field due to a nucleus of charge $-e$, since the electrons in the closed shells screen the nucleus). One must, of course, pay attention to the fact that quantum numbers $(n_\alpha l_\alpha)$ of the single electron cannot assume the values already used up by electrons in closed shells.

The simplifications mentioned above may not be applicable when levels corresponding to different shells are comparable in energy (for example, the $3d$ and $4s$ shells in Cu). Nonetheless, *the motivation for developing the theory of one electron states, two-electron states, three-electron states, ... is considerably enhanced, since the results have implications for many-electron atoms as well.* Since these model-type problems serve very well to illustrate angular momentum techniques, we shall emphasize here this aspect of atomic spectroscopy, leaving the more detailed considerations of their applications to specific atoms to the textbooks on the subject.

f. The one-electron problem with spin-orbit coupling. For a single electron the Hamiltonian (7.5.1) reduces to

$$H = \frac{\mathbf{p}^2}{2m_0} - \frac{Ze^2}{r} + \xi(r)\mathbf{L} \cdot \mathbf{S}. \qquad (7.5.46)$$

It is customary to treat the spin-orbit interaction as a perturbation on the energy levels,

$$E_n = -Z^2e^2/2an^2, \qquad (7.5.47)$$

of the hydrogen atom without spin interaction (see Section 4 of this chapter). The zeroth-order states are given by the Pauli central-field spinors [see Eq. (6.57), Chapter 6]:

$$\boldsymbol{\Psi}_{nljm} \equiv R_{nl}\,\mathbf{Y}^{(l\frac{1}{2})jm}. \qquad (7.5.48)$$

In obtaining this result, we have utilized the fact that the total angular momentum $\mathbf{J} = \mathbf{L} + \mathbf{S}$ commutes separately with the zeroth-order Hamiltonian and the spin–orbit interaction. Hence, the spinor states (7.5.48) diagonalize the $\mathbf{L} \cdot \mathbf{S}$ part of the spin–orbit term [see Eqs. (6.50) and (6.51), Chapter 6]. The R_{nl} are the standard hydrogen radial functions as discussed in Section 4.

The energy levels are given in first-order perturbation theory by

$$E_{nlj} = E_n + \tfrac{1}{2}\zeta_{nl}[j(j+1) - l(l+1) - \tfrac{3}{4}], \tag{7.5.49}$$

where we recall that for a given orbital angular momentum state l the total angular momentum j may assume the two possible values $l + \frac{1}{2}$ and $l - \frac{1}{2}$ $(l > 0)$ $(j = \frac{1}{2}$ only for $l = 0)$ leading to[1]

$$E_{nlj} = \begin{cases} E_n + \tfrac{1}{2}\zeta_{nl} l, & \text{for } j = l + \tfrac{1}{2} \\ E_n - \tfrac{1}{2}\zeta_{nl}(l+1), & \text{for } j = l - \tfrac{1}{2}; \end{cases} \tag{7.5.50}$$

$$E_{n0\frac{1}{2}} = E_n + \frac{(\alpha Z)^4 m_0 c^2}{2n^3}, \qquad \text{for } l = 0. \tag{7.5.51}$$

In these results ζ_{nl} is given by the radial integral

$$\zeta_{nl} \equiv \int_0^\infty r^2\, dr\, \xi(r)[R_{nl}(r)]^2, \tag{7.5.52}$$

which, when evaluated for $\xi(r)$ given by [see Eq. (7.5.13)]

$$\zeta(r) = Ze^2\hbar^2/2m_0^2c^2r^3, \tag{7.5.53}$$

yields

$$\zeta_{nl} = \frac{(\alpha Z)^4\, m_0 c^2}{n^3 l(l+1)(2l+1)}, \qquad l > 0. \tag{7.5.54}$$

In Eqs. (7.5.51) and (7.5.54), $\alpha = e^2/\hbar c$ denotes the *fine structure constant*. (The

[1] The spin displacement of an $l = 0$ level is indeterminate, since the radial integral for ζ_{n0} is infinite. An approximate relativistic calculation (Condon and Shortley [10, p. 130]) shows that the contribution coming from the spin–orbit interaction itself is zero, but that a compensating term that is nonzero only for $l = 0$ states exactly restores the term obtained by first combining $\frac{1}{2}\zeta_{nl} l$, canceling l, and then setting $l = 0$. Equations (7.5.50) and (7.5.51) do not, however, include relativistic mass effects.

evaluation and properties of Coulomb radial integrals are developed systematically in RWA using angular momentum techniques.)

The two levels (7.5.50) into which each configuration (nl) $(l > 0)$ is split are together called a *doublet term.*

The preceding calculation may, of course, be carried out for any central field and, in particular, for Coulomb-like potentials representing the interaction of a single valence electron with the nucleus and closed-shell electrons.

The relativistic effects due to the variation of mass with velocity may also be included in a first-order perturbation treatment. The approximate interaction may be obtained from the Dirac equation; it is found to be

$$H'_1 = -\mathbf{p}^4/8m_0^3c^2. \tag{7.5.55}$$

In an energy eigenstate E_n, we may write the momentum as

$$\mathbf{p}^2 = 2m_0\left(E_n + \frac{Ze^2}{r}\right) \tag{7.5.56}$$

so that the first-order correction due to H'_1 is obtained by taking diagonal matrix elements of

$$-\frac{1}{2m_0c^2}\left[(E_n)^2 + 2E_n\left(\frac{Ze^2}{r}\right) + \left(\frac{Ze^2}{r}\right)^2\right]. \tag{7.5.57}$$

The result of this calculation is

$$E'_{nl} = -(\alpha Z)^4 m_0c^2\left(\frac{1}{2l+1} - \frac{3}{8n}\right)\frac{1}{n^3}. \tag{7.5.58}$$

Combining this energy correction with the energy given by Eq. (7.5.50), we obtain the following system of energy levels for the hydrogen atom, including relativistic effects to first-order in perturbation theory:

$$E_{nj} = E_n\left[1 + \frac{(\alpha Z)^2}{n^2}\left(\frac{n}{j+\frac{1}{2}} - \frac{3}{4}\right)\right]. \tag{7.5.59}$$

It is significant that the final form (7.5.59) of the energy levels, which includes both mass variation and spin–orbit interactions, is independent of the two orbital angular momentum states $l = j \pm \frac{1}{2}$, since the exact solution to Dirac's relativistic Hamiltonian yields energy levels that are independent of l. To the approximation considered, the levels given by Eq. (7.5.59) agree with the Dirac relativistic theory.

g. Two-electron configurations. We now turn to the illustration of Racah's methods in spectroscopy. The application of tensor operator methods requires that the interactions be expressed in terms of the rotational invariants associated with kth-rank irreducible tensors.

The expression of the electrostatic interaction H_1 in terms of rotational invariants constructed from kth-rank tensors is found by first expanding $1/r_{\alpha\beta}$ in terms of the Legendre polynomials [Eq. (6.175), Chapter 6] and then substituting the invariant form given by Eq. (6.137), Chapter 6, for the Legendre polynomials. The result is

$$H_1 = e^2 \sum_{\beta>\alpha=1}^{N} \frac{1}{r_{\alpha\beta}} = \sum_{\beta>\alpha=1}^{N} \sum_{k=0}^{\infty} v_k(r_\alpha, r_\beta) P_k(\cos\theta_{\alpha\beta}), \tag{7.5.60}$$

where

$$P_k(\cos\theta_{\alpha\beta}) = \frac{4\pi(-1)^k}{(2k+1)^{\frac{1}{2}}} [\mathscr{Y}^k(\hat{x}_\alpha) \times \mathscr{Y}^k(\hat{x}_\beta)]_0^0, \tag{7.5.61}$$

$$v_k(r, s) = \begin{cases} e^2 r^k / s^{k+1}, & r \leqslant s \\ e^2 s^k / r^{k+1}, & r \geqslant s. \end{cases} \tag{7.5.62}$$

States of sharp total angular momentum quantum numbers (LM) are constructed for two particles by the Wigner coupling of the single-particle states:

$$|(n_1 n_2)(l_1 l_2) L M_L\rangle \equiv \sum_{m_1 m_2} C^{l_1 l_2 L}_{m_1 m_2 M_L} |n_1 l_1 m_1\rangle \otimes |n_2 l_2 m_2\rangle. \tag{7.5.63}$$

These states possess the following symmetry under the exchange P_{12} of the quantum numbers of the single-electron states:

$$\begin{aligned} P_{12}|(n_1 n_1)(l_1 l_2) L M_L\rangle &\equiv \sum_{m_1 m_2} C^{l_1 l_2 L}_{m_1 m_2 m_L} |n_2 l_2 m_2\rangle \otimes |n_1 l_1 m_1\rangle \\ &= (-1)^{l_1 + l_2 - L} |(n_2 n_1)(l_2 l_1) L M_L\rangle, \end{aligned} \tag{7.5.64}$$

where we have used the symmetry of the Wigner coefficient under the exchange of $(l_1 m_1)$ and $(l_2 m_2)$ [see Eq. (3.180)].

The next step in obtaining antisymmetrized states of sharp orbital and spin angular momenta is to couple the orbital angular momentum states (7.5.63) with the spin states of the two electrons (the explicit construction of sharp spin states for N spin-$\frac{1}{2}$ particles is given in the Appendix to this section). In terms of the double tableau notation (see Chapter 5, Appendix A), the spin singlet and triplet are denoted by

$$S = 0:\ \left|\left(\begin{array}{|c|}\hline 1\\\hline 2\\\hline\end{array}\ \middle|\ \begin{array}{|c|}\hline 1\\\hline 2\\\hline\end{array}\right)\right\rangle = |00\rangle,$$

$$S = 1:\ |(\begin{array}{|c|c|}\hline a & b\\\hline\end{array}\mid\begin{array}{|c|c|}\hline 1 & 2\\\hline\end{array})\rangle = |1M_S\rangle,$$

$$\begin{array}{|c|c|}\hline a & b\\\hline\end{array} = \quad \begin{array}{|c|c|}\hline 1 & 1\\\hline\end{array}, \qquad \begin{array}{|c|c|}\hline 1 & 2\\\hline\end{array}, \qquad \begin{array}{|c|c|}\hline 2 & 2\\\hline\end{array} \tag{7.5.65}$$
$$(M_S = 1),\ (M_S = 0),\ (M_S = -1)$$

These spin states have the following symmetries under the exchange P_{12} of the quantum numbers of the single-electron spin states:

$$P_{12}\left|\left(\begin{array}{|c|}\hline 1\\\hline 2\\\hline\end{array}\ \middle|\ \begin{array}{|c|}\hline 1\\\hline 2\\\hline\end{array}\right)\right\rangle = -\left|\left(\begin{array}{|c|}\hline 1\\\hline 2\\\hline\end{array}\ \middle|\ \begin{array}{|c|}\hline 1\\\hline 2\\\hline\end{array}\right)\right\rangle,$$

$$P_{12}|(\begin{array}{|c|c|}\hline a & b\\\hline\end{array}\mid\begin{array}{|c|c|}\hline 1 & 2\\\hline\end{array})\rangle = |(\begin{array}{|c|c|}\hline a & b\\\hline\end{array}\mid\begin{array}{|c|c|}\hline 1 & 2\\\hline\end{array})\rangle. \tag{7.5.66}$$

In the case of two particles, these spin states are also just the Wigner coupling of the single-electron spin states:

$$|SM_S\rangle = |(\tfrac{1}{2}\tfrac{1}{2})SM_S\rangle = \sum_{\mu\nu} C^{\frac{1}{2}\frac{1}{2}S}_{\mu\nu M_S}|\tfrac{1}{2}\mu\rangle \otimes |\tfrac{1}{2}\nu\rangle. \tag{7.5.67}$$

The exchange symmetry (7.5.66) then follows from the Wigner coefficient symmetry.

Combining the spatial states (7.5.63) with the spin states, we obtain the (tensor) product states, which are antisymmetric under the action of the exchange operator P_{12}:

Singlet states:

$$|\{(n_1n_2)(l_1l_2)LM_L;00\}\rangle$$

$$\equiv \frac{1}{\sqrt{2}}(|(n_1n_2)(l_1l_2)LM_L\rangle + (-1)^{l_1+l_2-L}|(n_2n_1)(l_2l_1)LM_L\rangle)$$

$$\otimes\left|\left(\begin{array}{|c|}\hline 1\\\hline 2\\\hline\end{array}\ \middle|\ \begin{array}{|c|}\hline 1\\\hline 2\\\hline\end{array}\right)\right\rangle; \tag{7.5.68}$$

Triplet states:

$$|\{(n_1n_2)(l_1l_2)LM_L;1M_S\}\rangle$$
$$\equiv \frac{1}{\sqrt{2}}(|(n_1n_2)(l_1l_2)LM_L\rangle - (-1)^{l_1+l_2-L}|(n_2n_1)(l_2l_1)LM_L\rangle)$$
$$\otimes|(\boxed{a}\boxed{b}\,|\,\boxed{1}\boxed{2})\rangle. \tag{7.5.69}$$

For $l_1 \neq l_2$, the number of vectors enumerated by $L = |l_1 - l_2|, \ldots, l_1 + l_2$; $M_L = L, \ldots, -L$; and $\boxed{a}\boxed{b} = \boxed{1}\boxed{1}, \boxed{1}\boxed{2}, \boxed{2}\boxed{2}$ is $4(2l_1 + 1)(2l_2 + 1)$. These vectors are orthonormal and antisymmetric. These vectors therefore constitute a basis of the space $\mathscr{V}^A_{E_0}$.

We have noted that the basis states (7.5.68) and (7.5.69) have been constructed to be antisymmetric under the action of the exchange operator P_{12} expressed by Eqs. (7.5.64) and (7.5.66). In addition, these states possess the important symmetry property

$$|\{(n_2n_1)(l_2l_1)LM_L;SM_S\}\rangle$$
$$= (-1)^{l_1+l_2+L+S}|\{(n_1n_2)(l_1l_2)LM_L;SM_S\}\rangle. \tag{7.5.70}$$

For the configuration nl^2 it is not necessary to form the linear combination of spatial functions given in Eqs. (7.5.68) and (7.5.69), since the states $|(nl^2)LM_L\rangle$ [$(n_1l_1) = (n_2l_2) = (nl)$ in Eq. (7.5.63)] are already symmetric (L even) or antisymmetric (L odd) under the action of P_{12}:

$$P_{12}|(nl^2)LM_L\rangle = (-1)^L|(nl^2)LM_L\rangle. \tag{7.5.71}$$

The basis of the space $\mathscr{V}^A_{E_0}$ is now given by the $(2l + 1)(4l + 1)$ orthonormal vectors:

Singlet states:

$$|\{(nl^2)LM_L;00\}\rangle \equiv |(nl^2)LM_L\rangle \otimes \left|\left(\begin{smallmatrix}\boxed{1}\\\boxed{2}\end{smallmatrix}\,\middle|\,\begin{smallmatrix}\boxed{1}\\\boxed{2}\end{smallmatrix}\right)\right\rangle, \tag{7.5.72}$$

where $L = 0, 2, \ldots, 2l$; $M = L, \ldots, -L$.

Triplet states:

$$|\{(nl^2)LM_L;1M_S\}\rangle \equiv |(nl^2)LM_L\rangle \otimes |(\boxed{a}\boxed{b}\,|\,\boxed{1}\boxed{2})\rangle, \tag{7.5.73}$$

where $L = 1, 3, \ldots, 2l - 1$; $M = L, \ldots, -L$; $M_S = 1, 0, -1$.

Observe that the symmetry given by Eq. (7.5.70) is also valid for these states, $|\{(nl^2)LM_L;\, SM_S\}\rangle$.

The construction given above (Racah [15]) thus solves in full generality the problem of constructing a basis of the space of antisymmetrized products of single-particle states of sharp total orbital angular momentum (LM_L) and sharp total spin (SM_S) for two electrons. [The Slater procedure would, if implemented in this generality, lead to the same basis, since there are no multiple occurrences of (LS).]

The next step in a first-order perturbation treatment is to calculate the matrix elements

$$\langle\{(n_1n_2)(l_1l_2)L'M'_L;\, S'M'_S\}|H_1|\{(n_1n_2)(l_1l_2)LM_L;\, SM_S\}\rangle$$

of the electrostatic interaction given by Eq. (7.5.60). Since **L** and **S** commute with H_1, these matrix elements are zero unless $(L'M'_L) = (LM_L)$ and $(S'M'_S) = (SM_S)$. Furthermore, the invariance of H_1 under permutations and the antisymmetry of the basis states lead to the following simplification:

$$\begin{aligned}
\Delta E_{l_1l_2LS} &\equiv \langle\{(n_1n_2)(l_1l_2)LM_L;\, SM_S\}|H_1|\{(n_1n_2)(l_1l_2)LM_L;\, SM_S\}\rangle \\
&= \langle(n_1n_2)(l_1l_2)LM_L|H_1|(n_1n_2)(l_1l_2)LM_L\rangle \\
&\quad + (-1)^{l_1+l_2+L+S}\langle(n_1n_2)(l_1l_2)LM_L|H_1|(n_2n_1)(l_2l_1)LM_L\rangle \\
&= \sum_k F^k(n_1l_1;\, n_2l_2)f_k(l_1l_2L) \\
&\quad + (-1)^{l_1+l_2+L+S}\sum_k G^k(n_1l_1;\, n_2l_2)g_k(l_1l_2L),
\end{aligned} \tag{7.5.74}$$

where F^k, G^k, f_k, and g_k have the following definitions: F^k and G^k are the Slater integrals defined by

$$F^k(n_1l_1;\, n_2l_2) = \iint dr_1\, dr_2 (r_1)^2(r_2)^2 v_k(r_1, r_2)[R_{n_1l_1}(r_1)R_{n_2l_2}(r_2)]^2, \tag{7.5.75}$$

$$G^k(n_1l_1;\, n_2l_2) = \iint dr_1\, dr_2 (r_1)^2(r_2)^2 v_k(r_1, r_2)R_{n_1l_1}(r_1)R_{n_2l_2}(r_2)R_{n_1l_1}(r_2)R_{n_2l_2}(r_1); \tag{7.5.76}$$

and f_k and g_k are the purely geometric (angular momentum) factors (the same for all central fields) given by

$$f_k(l_1l_2L) = \langle (l_1l_2)LM_L|P_k(\cos\theta_{12})|(l_1l_2)LM_L\rangle, \tag{7.5.77}$$

$$g_k(l_1l_2L) = \langle (l_1l_2)LM_L|P_k(\cos\theta_{12})|(l_2l_1)LM_L\rangle, \tag{7.5.78}$$

in which $|(l_1l_2)LM_L\rangle$ denotes the coupled angular momentum state

$$|(l_1l_2)LM_L\rangle = \sum_{m_1m_2} C^{l_1l_2L}_{m_1m_2M_L}\mathscr{Y}_{l_1m_1}\otimes\mathscr{Y}_{l_2m_2}. \tag{7.5.79}$$

The summation over k in the first term of Eq. (7.5.74) is over $k = 0, 2, \ldots, 2\min(l_1, l_2)$; in the second term it is over $k = |l_1 - l_2|, \ldots, l_1 + l_2$ [see Eqs. (7.5.81) and (7.5.82) below].

The evaluation of the geometric factors f_k and g_k is carried out by using the general matrix element, Eq. (3.260), Chapter 3, the relation (7.5.61), and the reduced matrix elements of the spherical harmonics, Eq. (3.437), Chapter 3. We find

$$\begin{aligned}&\langle (l'_1l'_2)L'M'_L|P_k(\cos\theta_{12})|(l_1l_2)LM_L\rangle\\ &= \delta_{L'L}\delta_{M'_LM_L}(-1)^{l_2+l'_2+L}[(2l_1+1)(2l_2+1)(2l'_1+1)(2l'_2+1)]^{\frac{1}{2}}\\ &\times\begin{pmatrix} l_1 & k & l'_1\\ 0 & 0 & 0\end{pmatrix}\begin{pmatrix} l_2 & k & l'_2\\ 0 & 0 & 0\end{pmatrix}\begin{Bmatrix} l_1 & l_2 & L\\ l'_2 & l'_1 & k\end{Bmatrix}.\end{aligned} \tag{7.5.80}$$

This result yields the following expressions for f_k and g_k:

$$f_k(l_1l_2L) = (-1)^L(2l_1+1)(2l_2+1)\begin{pmatrix} l_1 & k & l_1\\ 0 & 0 & 0\end{pmatrix}\begin{pmatrix} l_2 & k & l_2\\ 0 & 0 & 0\end{pmatrix}\begin{Bmatrix} l_1 & l_2 & L\\ l_2 & l_1 & k\end{Bmatrix}, \tag{7.5.81}$$

$$g_k(l_1l_2L) = (-1)^{k+L}(2l_1+1)(2l_2+1)\begin{pmatrix} l_1 & k & l_2\\ 0 & 0 & 0\end{pmatrix}^2\begin{Bmatrix} l_1 & l_2 & L\\ l_1 & l_2 & k\end{Bmatrix}. \tag{7.5.82}$$

The modifications of the energy correction (7.5.74) in the case $(n_1l_1) = (n_2l_2) = (nl)$ are carried out directly by using the basis vectors (7.5.72) and (7.5.73). The result is

$$\begin{aligned}\Delta E_{(l^2)L} &= \langle\{(nl)^2LM_L;SM_S\}|H_1|\{(nl)^2LM_L;SM_S\}\rangle\\ &= \sum_{k\,\text{even}} F^k(nl;nl)f_k(llL).\end{aligned} \tag{7.5.83}$$

Finally, it should be noted that the energy corrections (7.5.74) and (7.5.83) are invariant to the change of basis from states of sharp orbital angular momentum (LM_L) and sharp spin angular momentum (SM_S) to the basis states of sharp total angular momentum (JM_J):[1]

$$|\{(n_1n_2)[(l_1l_2)_L(\tfrac{1}{2}\tfrac{1}{2})_S]_{JM_J}\}\rangle$$
$$\equiv \sum_{M_LM_S} C^{L\,S\,J}_{M_LM_SM_J}|\{(n_1n_2)(l_1l_2)LM_L; SM_S\}\rangle. \tag{7.5.84}$$

The antisymmetry of the basis vectors is also preserved by this transformation:

$$P_{12}|\{(n_1n_2)[(l_1l_2)_L(\tfrac{1}{2}\tfrac{1}{2})_S]_{JM_J}\}\rangle = -|\{(n_1n_2)[(l_1l_2)_L(\tfrac{1}{2}\tfrac{1}{2})_S]_{JM_J}\}\rangle. \tag{7.5.85}$$

These states also satisfy the symmetry relation

$$|\{(n_2n_1)[(l_2l_1)_L(\tfrac{1}{2}\tfrac{1}{2})_S]_{JM_J}\}\rangle$$
$$= (-1)^{l_1+l_2+L+S}|\{(n_1n_2)[(l_1l_2)_L(\tfrac{1}{2}\tfrac{1}{2})_S]_{JM_J}\}\rangle. \tag{7.5.86}$$

The coupling scheme represented by Eq. (7.5.84) is known as the *LS*-coupling scheme. It corresponds to the coupling of four angular momenta given by

$$\mathbf{J} = (\mathbf{L}_1 + \mathbf{L}_2) + (\mathbf{S}_1 + \mathbf{S}_2). \tag{7.5.87}$$

Equivalently, this scheme corresponds to the labeled tree (see RWA) given by Fig. 7.2.

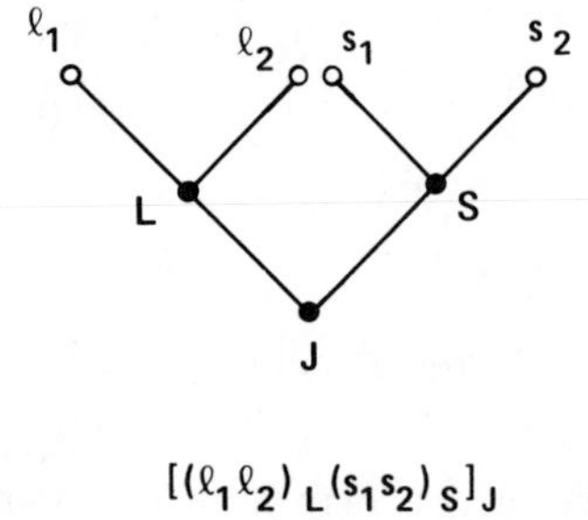

Figure 7.2. Tree diagram for the LS-coupling scheme: $\mathbf{L} = \mathbf{L}_1 + \mathbf{L}_2$, $\mathbf{S} = \mathbf{S}_1 + \mathbf{S}_2$, $\mathbf{J} = \mathbf{L} + \mathbf{S}$.

---

[1] The notation $(j_1j_2)_j$ designates the coupling of two angular momenta j_1 and j_2 to the total angular momentum j corresponding to $\mathbf{J} = \mathbf{J}(1) + \mathbf{J}(2)$. This (unconventional) notation is suggested naturally from the relationship of the coupling theory of angular momenta to the theory of "Cayley trees" and the "problem of parentheses." This subject is developed in detail in RWA.

The LS-coupling scheme is appropriate to perturbations that are scalars with respect to both the total orbital angular momentum $\mathbf{L}$ and the total spin angular momentum $\mathbf{S}$, since the matrices of such perturbations are then diagonal in L and S. The energy correction in this case always retains the $(2L + 1)(2S + 1)$ multiplicity associated with arbitrary spatial and spin–space orientations of the atom.

Consider next the effect of including the spin-orbit interaction

$$H_2 \equiv \xi(r_1)\mathbf{L}_1 \cdot \mathbf{S}_1 + \xi(r_2)\mathbf{L}_2 \cdot \mathbf{S}_2. \tag{7.5.88}$$

The total orbital angular momentum operator $\mathbf{L}$ and the total spin operator $\mathbf{S}$ do not now separately commute with H_2, although the total angular momentum $\mathbf{J}$ does. Thus, in order to determine the effect (in first-order perturbation theory) of H_2 by using the LS-coupling scheme (7.5.84), we would need to diagonalize the matrix with elements:

$$\langle\{(n_1n_2)[(l_1l_2)_{L'}(\tfrac{1}{2}\tfrac{1}{2})_{S'}]_{JM_J}\}|H_2|\{(n_1n_2)[(l_1l_2)_L(\tfrac{1}{2}\tfrac{1}{2})_S]_{JM_J}\}\rangle. \tag{7.5.89}$$

We shall not consider the problem in this generality but note a simplified case.

If we ignore the contributions of the off-diagonal elements in (7.5.89), then for configurations of equivalent particles $[(n_1l_1) = (n_2l_2) = (nl)]$, we find that the energy shift due to the spin–orbit interaction [diagonal matrix element in (7.5.89)] takes a particularly simple form:

$$\Delta E_{(l^2)LSJ} = \tfrac{1}{2}\zeta_{nl}[J(J + 1) - L(L + 1) - S(S + 1)]. \tag{7.5.90}$$

[This result may be obtained by observing that the radial integration (for the case being considered) replaces H_2 by

$$\zeta_{nl}(\mathbf{L}_1 \cdot \mathbf{S}_1 + \mathbf{L}_2 \cdot \mathbf{S}_2) = \zeta_{nl}(\mathbf{L} \cdot \mathbf{S} - \mathbf{L}_1 \cdot \mathbf{S}_2 - \mathbf{L}_2 \cdot \mathbf{S}_1),$$

and $\mathbf{L}_1 \cdot \mathbf{S}_2 + \mathbf{L}_2 \cdot \mathbf{S}_1$ has diagonal elements equal to zero in the basis (7.5.84).]

Let us summarize the results for two-particle configurations in LS-coupling: The zeroth-order energy E_0 of two electrons in the $(n_1l_1)(n_2l_2)$-configuration is

$$E_{n_1l_1} + E_{n_2l_2}, \qquad \left\{\begin{matrix} & \text{degeneracy} & \\ 4(2l_1 + 1)(2l_2 + 1), & & l_1 \neq l_2 \\ (2l + 1)(4l + 1), & & l_1 = l_2 = l \end{matrix}\right\}; \tag{7.5.91}$$

the electrostatic repulsion interaction splits this level into a number of LS-multiplets:

$$E_{n_1l_1} + E_{n_2l_2} + \Delta E_{l_1l_2LS}, \qquad \left\{\begin{matrix}\text{degeneracy} \\ (2L + 1)(2S + 1)\end{matrix}\right\}. \tag{7.5.92}$$

The energy term (7.5.92) is independent of the total angular momentum $J = |L - S|, \ldots, L + S$. The introduction of the spin–orbit interaction splits the different J states within an LS-multiplet. For the case of equivalent electrons, this splitting is given by

$$2E_{nl} + \Delta E_{(l^2)LS} + \Delta E_{(l^2)LSJ}, \qquad \begin{Bmatrix} \text{degeneracy} \\ 2J+1 \end{Bmatrix}, \tag{7.5.93}$$

provided one can ignore off-diagonal terms coming from the spin–orbit interaction. The $(2J + 1)$-fold degeneracy that remains in the state (7.5.93) is due to the arbitrary orientation of the total angular momentum $\mathbf{J}$.

The description of two-particle configurations given above is appropriate to many atomic spectra where the spin-orbit correction is very small compared with the splitting between two LS-multiplets.

The interval between two adjacent levels in an LS-multiplet that has been split by the spin–orbit interaction between two equivalent electrons is obtained from Eq. (7.5.90):

$$\Delta E_{(l^2)L,S,J+1} - \Delta E_{(l^2)LSJ} = \zeta_{nl}(J + 1). \tag{7.5.94}$$

This relationship was discovered empirically by Landé [28] and is known as the *Landé interval rule*.

Let us next discuss briefly the situation for two-particle configurations in which the spin–orbit interaction (7.5.88) dominates the electron repulsion interaction. It is more appropriate then to span the space $\mathscr{V}^A_{E_0}$ using the basis that corresponds to the coupling scheme indicated by writing

$$\mathbf{J} = (\mathbf{L}_1 + \mathbf{S}_1) + (\mathbf{L}_2 + \mathbf{S}_2). \tag{7.5.95}$$

This scheme corresponds to the labeled tree (Fig. 7.3):

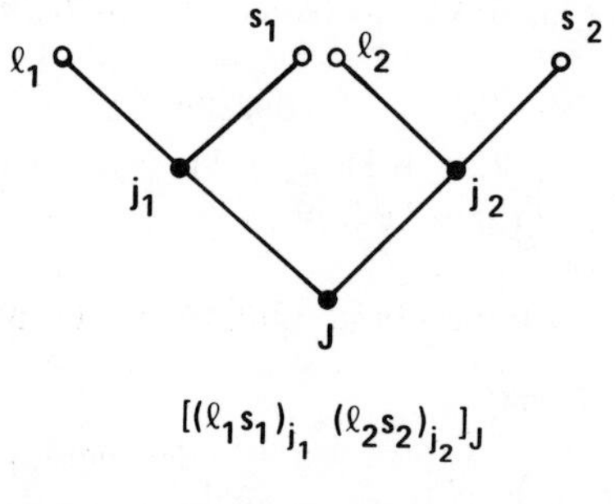

Figure 7.3. Tree diagram for the jj-coupling scheme: $\mathbf{J}_1 = \mathbf{L}_1 + \mathbf{S}_1$, $\mathbf{J}_2 = \mathbf{L}_2 + \mathbf{S}_2$, $\mathbf{J} = \mathbf{J}_1 + \mathbf{J}_2$.

This scheme is known as *jj*-coupling. It is the natural coupling scheme for calculating matrix elements of the spin–orbit interaction, since $\mathbf{L}_1 \cdot \mathbf{L}_1$, $\mathbf{L}_2 \cdot \mathbf{L}_2$, $\mathbf{S}_1 \cdot \mathbf{S}_1$, $\mathbf{S}_2 \cdot \mathbf{S}_2$, $\mathbf{J} \cdot \mathbf{J}$, and J_3 all commute with the operator H_2.

It is sometimes useful to transform between the *LS*- and *jj*-coupling schemes. The transformation coefficient between the *LS*-coupling scheme and the *jj*-coupling scheme is the 9-*j* coefficient (see Chapter 5, RWA):

$$\langle[(l_1 l_2)_L(s_1 s_2)_S]_J|[(l_1 s_1)_{j_1}(l_2 s_2)_{j_2}]_J\rangle$$

$$= [(2L+1)(2S+1)(2j_1+1)(2j_2+1)]^{\frac{1}{2}} \begin{Bmatrix} l_1 & l_2 & L \\ s_1 & s_2 & S \\ j_1 & j_2 & J \end{Bmatrix}, \tag{7.5.96}$$

where for the case at hand we have $s_1 = s_2 = \frac{1}{2}$ and $S = 0, 1$.

Thus, the *jj*-coupling scheme corresponds to the orthogonal change of basis of the space $\mathscr{V}^A_{E_0}$ given by

$$|\{(n_1 n_2)[(l_1 \tfrac{1}{2})_{j_1}(l_2 \tfrac{1}{2})_{j_2}]_{JM_J}\}\rangle$$

$$= \sum_{LS} [(2L+1)(2S+1)(2j_1+1)(2j_2+1)]^{\frac{1}{2}} \begin{Bmatrix} l_1 & l_2 & L \\ \frac{1}{2} & \frac{1}{2} & S \\ j_1 & j_2 & J \end{Bmatrix}$$

$$\times\, |\{(n_1 n_2)[(l_1 l_2)_L(\tfrac{1}{2}\tfrac{1}{2})_S]_{JM_J}\}\rangle. \tag{7.5.97}$$

Alternatively, these states may be expressed as

$$|\{(n_1 n_2)[(l_1 \tfrac{1}{2})_{j_1}(l_2 \tfrac{1}{2})_{j_2}]_{JM_J}\}\rangle$$

$$= \frac{1}{\sqrt{2}}\big(|(n_1 n_2)[(l_1 \tfrac{1}{2})_{j_1}(l_2 \tfrac{1}{2})_{j_2}]_{JM_J}\rangle$$

$$- (-1)^{j_1+j_2-J}|(n_2 n_1)[(l_2 \tfrac{1}{2})_{j_2}(l_1 \tfrac{1}{2})_{j_1}]_{JM_J}\rangle\big), \tag{7.5.98}$$

where

$$|(n_1 n_2)[(l_1 \tfrac{1}{2})_{j_1}(l_2 \tfrac{1}{2})_{j_2}]_{JM_J}\rangle$$

$$\equiv \sum_{\substack{\mu_1 \nu_1 m_1 \\ \mu_2 \nu_2 m_2}} C^{l_1 \frac{1}{2} j_1}_{\mu_1 \nu_1 m_1} C^{l_2 \frac{1}{2} j_2}_{\mu_2 \nu_2 m_2} C^{j_1 j_2 J}_{m_1 m_2 M_J} |n_1 l_1 \mu_1 \nu_1\rangle \otimes |n_2 l_2 \mu_2 \nu_2\rangle. \tag{7.5.99}$$

The states (7.5.97) are antisymmetric under the action of P_{12}:

$$P_{12}|\{(n_1n_2)[(l_1\tfrac{1}{2})_{j_1}(l_2\tfrac{1}{2})_{j_2}]_{JM_J}\}\rangle = -|\{(n_1n_2)[(l_1\tfrac{1}{2})_{j_1}(l_2\tfrac{1}{2})_{j_2}]_{JM_J}\}\rangle. \qquad (7.5.100)$$

They also satisfy the symmetry relation

$$|\{(n_2n_1)[(l_2\tfrac{1}{2})_{j_2}(l_1\tfrac{1}{2})_{j_1}]_{JM_J}\}\rangle$$
$$= (-1)^{j_1+j_2-J+1}|\{(n_1n_2)[(l_1\tfrac{1}{2})_{j_1}(l_2\tfrac{1}{2})_{j_2}]_{JM_J}\}\rangle. \qquad (7.5.101)$$

For the states $(n_1l_1j_1) = (n_2l_2j_2) = (nlj)$, the antisymmetric states are the coupled states themselves,

$$|\{(nn)[(l\tfrac{1}{2})_j(l\tfrac{1}{2})_j]_{JM_J}\}\rangle \equiv |(nn)[(l\tfrac{1}{2})_j(l\tfrac{1}{2})_j]_{JM_J}\rangle, \qquad (7.5.102)$$

provided J is an *even integer*. (The odd J states are symmetric under the action of P_{12}.) Observe that Eq. (7.5.97) remains valid for special case (7.5.102).

The matrix elements of the spin–orbit interaction are diagonal in all angular momentum quantum numbers appearing in the jj-coupled state vector given by Eq. (7.5.97). Accordingly, the energy correction due to spin–orbit interaction is given by

$$\Delta E_{l_1l_2j_1j_2} \equiv \langle\{(n_1n_2)[(l_1\tfrac{1}{2})_{j_1}(l_2\tfrac{1}{2})_{j_2}]_{JM_J}\}|H_2|\{(n_1n_2)[(l_1\tfrac{1}{2})_{j_1}(l_2\tfrac{1}{2})_{j_2}]_{JM_J}\}\rangle$$
$$= \tfrac{1}{2}\zeta_{n_1l_1}[j_1(j_1+1) - l_1(l_1+1) - \tfrac{3}{4}]$$
$$+ \tfrac{1}{2}\zeta_{n_2l_2}[j_2(j_2+1) - l_2(l_2+1) - \tfrac{3}{4}], \qquad (7.5.103)$$

where $j_1 = l_1 \pm \frac{1}{2}$, $j_2 = l_2 \pm \frac{1}{2}$, and $J = |j_1 - j_2|, \ldots, j_1 + j_2$.

Each $(n_1l_1)(n_2l_2)$-configuration is thus split into four *jj-coupling configurations* $(l_1j_1l_2j_2)$ corresponding to the four combinations of $j_1 = l_1 \pm \frac{1}{2}$, $j_2 = l_2 \pm \frac{1}{2}$ (for l_1 or l_2 zero, there are two jj-coupling configurations; for $l_1 = l_2 = 0$, there is only one configuration). Each jj-coupling configuration $(l_1j_1l_2j_2)$ has the $(2j_1+1)(2j_2+1)$-fold degeneracy corresponding to the allowed values of the total angular momentum quantum numbers (JM): $J = |j_1 - j_2|, \ldots, j_1 + j_2$; $M = J, \ldots, -J$. The electrostatic interaction H_1 removes the degeneracy in J, and, to the extent that the off-diagonal matrix elements of H_1 can be ignored, the first-order energy correction due to H_1 is given by

$$\Delta E_{l_1j_1l_2j_2\mathbf{J}}$$
$$\equiv \langle\{(n_1n_2)[(l_1\tfrac{1}{2})_{j_1}(l_2\tfrac{1}{2})_{j_2}]_{JM_J}\}|H_1|\{(n_1n_2)[(l_1\tfrac{1}{2})_{j_1}(l_2\tfrac{1}{2})_{j_2}]_{JM_J}\}\rangle =$$

$$= \sum_{LS} (2j_1 + 1)(2j_2 + 1)(2L + 1)(2S + 1) \begin{Bmatrix} l_1 & l_2 & L \\ \frac{1}{2} & \frac{1}{2} & S \\ j_1 & j_2 & J \end{Bmatrix}^2 \Delta E_{l_1 l_2 LS}$$

$$= \sum_k F^k(n_1 l_1; n_2 l_2) f_k(l_1 j_1 l_2 j_2 J)$$

$$- (-1)^{j_1 + j_2 - J} \sum_k G^k(n_1 l_1; n_2 l_2) g_k(l_1 j_1 l_2 j_2 J), \qquad (7.5.104)$$

where

$$f_k(l_1 j_1 l_2 j_2 J)$$

$$= \langle [(l_1 \tfrac{1}{2})_{j_1} (l_2 \tfrac{1}{2})_{j_2}]_{JM_J} | P_k(\cos\theta_{12}) | [(l_1 \tfrac{1}{2})_{j_1} (l_2 \tfrac{1}{2})_{j_2}]_{JM_J} \rangle$$

$$= \sum_{LS} (2L + 1)(2S + 1)(2j_1 + 1)(2j_2 + 1) \begin{Bmatrix} l_1 & l_2 & L \\ \frac{1}{2} & \frac{1}{2} & S \\ j_1 & j_2 & J \end{Bmatrix}^2 f_k(l_1 l_2 L), \qquad (7.5.105)$$

$$g_k(l_1 j_1 l_2 j_2 J)$$

$$= \langle [(l_1 \tfrac{1}{2})_{j_1} (l_2 \tfrac{1}{2})_{j_2}]_{JM_J} | P_k(\cos\theta_{12}) | [(l_2 \tfrac{1}{2})_{j_2} (l_1 \tfrac{1}{2})_{j_1}]_{JM_J} \rangle$$

$$= \sum_{LS} (2L + 1)(2S + 1)(2j_1 + 1)(2j_2 + 1) \begin{Bmatrix} l_1 & l_2 & L \\ \frac{1}{2} & \frac{1}{2} & S \\ j_1 & j_2 & J \end{Bmatrix} \begin{Bmatrix} l_2 & l_1 & L \\ \frac{1}{2} & \frac{1}{2} & S \\ j_2 & j_1 & J \end{Bmatrix}$$

$$\times\, g_k(l_1 l_2 L). \qquad (7.5.106)$$

[The two forms of these results have been obtained by (*a*) using the state vector given by Eq. (7.5.98) directly in the left-hand side of Eq. (7.5.104), and (*b*) substituting $\Delta E_{l_1 l_2 LS}$ as given by Eq. (7.5.74) into Eq. (7.5.104).]

The evaluation of the matrix elements of the Legendre polynomials in the coupled angular momentum basis $|[(l_1 \tfrac{1}{2})_{j_1} (l_2 \tfrac{1}{2})_{j_2}]_{JM_J}\rangle$ follows the derivation of Eq. (7.5.80). Thus, Eqs. (3.259) and (3.260), Chapter 3, yield

$$\langle [(l'_1 \tfrac{1}{2})_{j'_1} (l'_2 \tfrac{1}{2})_{j'_2}]_{J'M'_J} | P_k(\cos\theta_{12}) | [(l_1 \tfrac{1}{2})_{j_1} (l_2 \tfrac{1}{2})_{j_2}]_{JM_J} \rangle$$

$$= \delta_{J'J} \delta_{M'_J M_J} (-1)^{j'_2 + j_1 + J} [(2j'_1 + 1)(2j'_2 + 1)]^{\frac{1}{2}} \begin{Bmatrix} j_1 & j_2 & J \\ j'_2 & j'_1 & k \end{Bmatrix} \times$$

$$\times \frac{4\pi}{2k+1} \langle (l'_1 \tfrac{1}{2}) j'_1 \| \mathscr{Y}^k(1) \| (l_1 \tfrac{1}{2}) j_1 \rangle$$

$$\times \langle (l'_2 \tfrac{1}{2}) j'_2 \| \mathscr{Y}^k(2) \| (l_2 \tfrac{1}{2}) j_2 \rangle, \tag{7.5.107}$$

where the reduced matrix elements are found from Eqs. (3.246) and (3.437), Chapter 3, to be

$$\langle (l' \tfrac{1}{2}) j' \| \mathscr{Y}^k \| (l \tfrac{1}{2}) j \rangle$$

$$= (-1)^{l'+k+j+\frac{1}{2}} \left[\frac{(2l+1)(2l'+1)(2k+1)(2j+1)}{4\pi} \right]^{\frac{1}{2}}$$

$$\times \begin{pmatrix} l & k & l' \\ 0 & 0 & 0 \end{pmatrix} \begin{Bmatrix} j' & \frac{1}{2} & l' \\ l & k & j \end{Bmatrix}$$

$$= \begin{cases} (-1)^{j'+\frac{1}{2}} \left[\dfrac{(2k+1)(2j+1)}{4\pi} \right]^{\frac{1}{2}} \begin{pmatrix} j & k & j' \\ \frac{1}{2} & 0 & -\frac{1}{2} \end{pmatrix}, & l + l' + k \text{ even} \\ 0, \quad l + l' + k \text{ odd}. \end{cases} \tag{7.5.108}$$

[We have used Eq. (A-2) in the appendix to Section 4 in obtaining the right-hand side in this result.] Thus, we find

$$f_k(l_1 j_1 l_2 j_2 J) = 0, \qquad \text{for } k \text{ odd},$$

$$f_k(l_1 j_1 l_2 j_2 J) = (-1)^{J-1}(2j_1+1)(2j_2+1) \begin{pmatrix} j_1 & k & j_1 \\ \frac{1}{2} & 0 & -\frac{1}{2} \end{pmatrix} \begin{pmatrix} j_2 & k & j_2 \\ \frac{1}{2} & 0 & -\frac{1}{2} \end{pmatrix}$$

$$\times \begin{Bmatrix} j_1 & j_2 & J \\ j_2 & j_1 & k \end{Bmatrix}, \qquad \text{for } k \text{ even};$$

$$g_k(l_1 j_1 l_2 j_2 J) = 0 \qquad \text{for } l_1 + l_2 + k \text{ odd},$$

$$g_k(l_1 j_1 l_2 j_2 J) = (-1)^{j_1+j_2-J}(2j_1+1)(2j_2+1) \begin{pmatrix} j_1 & k & j_2 \\ \frac{1}{2} & 0 & -\frac{1}{2} \end{pmatrix}^2$$

$$\times \begin{Bmatrix} j_1 & j_2 & J \\ j_1 & j_2 & k \end{Bmatrix}, \qquad \text{for } l_1 + l_2 + k \text{ even}. \tag{7.5.109}$$

The modification of Eq. (7.5.104) for *equivalent electrons in a jj-configuration*—that is, $(n_1 l_1 j_2) = (n_2 l_2 j_2) = (nlj)$—is

$$\Delta E_{(lj)^2J} = \sum_k F^k(nl;nl) f_k[(lj)^2 J], \tag{7.5.110}$$

where we recall that J must be an even integer in consequence of the Pauli principle.

Let us summarize our results for the *jj*-coupling scheme. The two-electron levels in the $(n_1l_1)(n_2l_2)$-configuration in the *jj*-coupling scheme are given by

$$E_{n_1l_1} + E_{n_2l_2} + \Delta E_{l_1l_2j_1j_2} + \Delta E_{l_1j_1l_2j_2J}, \tag{7.5.111}$$

where it has been assumed that (*a*) the spin–orbit interaction dominates the electrostatic interaction, and (*b*) contributions to the energy coming from the off-diagonal matrix elements of the electrostatic interaction may be ignored.

For discussions of the applicability of the two-electron levels obtained above, in either the *LS*-coupling approximation or the *jj*-coupling approximation [Eqs. (7.5.92), (7.5.93), and (7.5.111)], to the systems of levels of atoms having two electrons outside closed shells, we refer to the literature on atomic spectroscopy (see footnote, p. 365).

In the case of two-electron configurations the techniques of coupling of angular momenta, augmented with the treatment of interactions in terms of tensor operator invariants, provides a conceptually complete method for spanning the energy eigenspace $\mathscr{V}^A_{E_0}$ of antisymmetrized states with basis vectors possessing sharp total angular momentum **J**. The lack of one general solution or scheme of levels may be attributed to the varying significance of different perturbations in different atoms rather than to shortcomings of the general method. Different coupling schemes or mixing of the various schemes may be required to approximate the levels of two-electron configurations of particular atoms, but the conceptual problem of constructing states of sharp angular momentum obeying the Pauli principle may be considered as solved.

In the case of three-electron configurations one encounters new problems in constructing antisymmetrized states of sharp total angular momentum. In illustrating these difficulties, we shall confine our discussion to equivalent electron configurations l^n (see footnote p. 377).

h. Equivalent electron configurations. Let us denote the single-electron states by[1]

$$|l - m + 1, \tfrac{3}{2} - \mu\rangle \equiv |nlm\rangle \otimes |\tfrac{1}{2}, \mu\rangle \tag{7.5.112}$$

[1] We have suppressed n, l, and $s = \frac{1}{2}$ in this notation and replaced the pair of projection quantum numbers (m, μ) by $(l - m + 1, \frac{3}{2} - \mu)$. We have chosen this (unconventional) notation in anticipation of mapping the single-particle states to boson states [see Eq. (7.5.121)]. This latter indexing (of bosons) is convenient for utilizing results from general unitary group theory, which will be the principal theme of the remainder of this section.

for $m = l, l-1, \ldots, -l$ and $\mu = \frac{1}{2}, -\frac{1}{2}$. We shall also order the pairs of integers in the set

$$\{(l - m + 1, \tfrac{3}{2} - \mu) \colon m = l, l-1, \ldots, -l; \mu = \tfrac{1}{2}, -\tfrac{1}{2}\} \tag{7.5.113}$$

by the rule

$$\begin{aligned} &(1,1) < (2,1) < \cdots < (2l+1, 1) \\ &< (1,2) < (2,2) < \cdots < (2l+1, 2). \end{aligned} \tag{7.5.114}$$

In terms of this notation a Slater state for the configuration l^n may be denoted in the standard Young tableau notation by

$$\left| \begin{array}{|c|} \hline (k_1, \sigma_1) \\ \hline (k_2, \sigma_2) \\ \hline \vdots \\ \hline (k_n, \sigma_n) \\ \hline \end{array} \right\rangle \equiv \frac{1}{\sqrt{n!}} \sum_P (-1)^\sigma P(|k_1, \sigma_1\rangle \otimes \cdots \otimes |k_n, \sigma_n\rangle), \tag{7.5.115}$$

where $(k_1, \sigma_1)(k_2, \sigma_2) \cdots (k_n, \sigma_n)$ is any sequence of n pairs, selected from the set (7.5.114), that satisfies

$$(k_1, \sigma_1) < (k_2, \sigma_2) < \cdots < (k_n, \sigma_n). \tag{7.5.116}$$

The sum in Eq. (7.5.115) is over all $n!$ permutations of the string of n pairs $(k_1, \sigma_1)(k_2, \sigma_2) \cdots (k_n, \sigma_n)$.

Introducing the notation

$$\zeta_\alpha = (\mathbf{x}_\alpha, \xi_\alpha) \tag{7.5.117}$$

to denote the position and spin coordinates of the αth electron, we find that the value of the Slater state at the point $(\zeta_1, \zeta_2, \ldots, \zeta_n)$ is given by the Slater determinant:

$$\left\langle \begin{array}{|c|} \hline \zeta_1 \\ \hline \zeta_2 \\ \hline \vdots \\ \hline \zeta_n \\ \hline \end{array} \middle| \begin{array}{|c|} \hline (k_1, \sigma_1) \\ \hline (k_2, \sigma_2) \\ \hline \vdots \\ \hline (k_n, \sigma_n) \\ \hline \end{array} \right\rangle \equiv \frac{1}{\sqrt{n!}} \sum_P (-1)^\sigma P(\langle \zeta_1 | k_1, \sigma_1\rangle \cdots \langle \zeta_n | k_n, \sigma_n\rangle), \tag{7.5.118}$$

where

$$\langle \zeta_\alpha | l - m + 1, \tfrac{3}{2} - \mu\rangle \equiv \langle \mathbf{x}_\alpha | nlm\rangle \langle \xi_\alpha | \tfrac{1}{2}, \mu\rangle.$$

The energy eigenspace $\mathscr{V}^A(l^n) \equiv \mathscr{V}^A_{E_0}$ of the l^n-configuration is the vector space of dimension

$$\dim \mathscr{V}^A(l^n) = \binom{4l+2}{n} \qquad (7.5.119)$$

that is spanned by the set of orthonormal vectors given by Eq. (7.5.115) corresponding to the set of standard Young tableaux of shape (1, 1, ..., 1), which are filled in with the $(4l + 2)$ ordered pairs (7.5.114).

We shall consider here only the *LS*-coupling scheme aspects of l^n-configurations. *The basic problem is to introduce a new basis into the space* $\mathscr{V}^A(l^n)$ *such that the total orbital angular momentum* $\mathbf{L} = \sum_{\alpha=1}^{n} \mathbf{L}_\alpha$ *and the total spin* $\mathbf{S} = \sum_{\alpha=1}^{n} \mathbf{S}_\alpha$ *have the standard action on the new basis vectors.* [The matrix elements of electrostatic interaction H_1 will then be diagonal in (LM_L) and (SM_S).]

A new element of complexity now enters the problem: For $n = 2$ the coupling of angular momenta indicated by $\mathbf{L} = \mathbf{L}_1 + \mathbf{L}_2$ and $\mathbf{S} = \mathbf{S}_1 + \mathbf{S}_2$, followed by antisymmetrization, led uniquely to the desired basis of $\mathscr{V}^A(l^n)$. Recognizing that the labeling of the electrons in any ordering of $1, 2, \ldots, n$ will be irrelevant because of the permutational symmetry of H_1, we are still confronted with the fact that there are many distinct coupling schemes[1] for the orbital as well as the spin angular momenta. This result signifies that there exist several antisymmetrized *LS*-coupled bases of the space $\mathscr{V}^A(l^n)$ for $n > 2$. Thus, there is no unique coupling scheme leading to a solution to the problem.

Let us summarize the classical methods that have been applied to the l^n-configuration problem:

(*a*) The earliest methods of Slater [17] and Gray and Wills [29] attempt to construct states of sharp total orbital and spin angular momenta by applying the method of highest weights directly to the basis vectors (7.5.115) as described earlier. [We shall not discuss this method further—its merit is that the antisymmetry is maintained from the outset, but the construction of the angular momentum states is difficult.]

(*b*) Racah [16, papers II and III] introduced a method that combines angular momentum coupling and antisymmetrization recursively. More specifically, one assumes that antisymmetrized states of sharp angular momenta $(L'M'_L)$ and $(S'M'_S)$ have been constructed for the configuration l^{n-1}, $(\mathbf{L}' = \sum_{\alpha=1}^{n-1} \mathbf{L}_\alpha,\ \mathbf{S}' = \sum_{\alpha=1}^{n-1} \mathbf{S}_\alpha)$. One then couples in one more electron

[1] The number of distinct coupling schemes for n angular momenta is $(2n-3)!! = (1)(3)(5)\cdots(2n-3)$, where schemes differing only by a phase in the coupled state vector are counted as equivalent. This counting problem (and its relationship to Cayley trees) is developed in RWA.

according to the scheme $\mathbf{L} = \mathbf{L}' + \mathbf{L}_n$, $\mathbf{S} = \mathbf{S}' + \mathbf{S}_n$, and then seeks the linear combinations of these coupled states that are antisymmetric with respect to pairwise interchanges of the n electrons. The coefficients of this linear combination are called *coefficients of fractional parentage*.

(c) Racah's second method [15-IV, 16] utilizes more general group-theoretic techniques for interpreting the meaning and properties of fractional parentage coefficients.

We now describe in more detail Racah's second method (c), since it provides the natural framework for discussing the structures of the earlier technique (b).

We shall not, however, follow the traditional way of describing Racah's ideas. Greater generality and insight into the group-theoretic structures involved can be gained by employing a *vector space of boson states* that is isomorphic to the space $\mathscr{V}^A(l^n)$ (vector space isomorphism).

Let us introduce two sets of independent boson operators denoted by

$$\{a_k^\alpha : \alpha = 1, 2, \ldots, n; k = 1, 2, \ldots, 2l + 1\},$$

$$\{b_\sigma^\alpha : \alpha = 1, 2, \ldots, n; \sigma = 1, 2\}, \tag{7.5.120}$$

where these boson operators obey the standard commutation relations (see Chapter 5, Section 5).

We next map the physical state vectors appearing in Eq. (7.5.115) into boson state vectors using the rule

$$|k_1, \sigma_1\rangle \otimes |k_2, \sigma_2\rangle \otimes \cdots \otimes |k_n, \sigma_n\rangle \to a_{k_1}^1 b_{\sigma_1}^1 a_{k_2}^2 b_{\sigma_2}^2 \cdots a_{k_n}^n b_{\sigma_n}^n |0\rangle \tag{7.5.121}$$

Using this map on the Slater state vectors (7.5.115), we obtain the following boson state vectors:

$$\left| \begin{array}{|c|} \hline (k_1, \sigma_1) \\ \hline (k_2, \sigma_2) \\ \hline \vdots \\ \hline (k_n, \sigma_n) \\ \hline \end{array} \right\rangle = \frac{1}{\sqrt{n!}} \mathscr{P}_{(k_1, \sigma_1)(k_2, \sigma_2) \cdots (k_n, \sigma_n)}(A, B)|0\rangle, \tag{7.5.122}$$

where

$$\mathscr{P}_{(k_1, \sigma_1)(k_2, \sigma_2) \cdots (k_n, \sigma_n)}(A, B)$$

$$\equiv \det \begin{bmatrix} a_{k_1}^1 b_{\sigma_1}^1 & a_{k_1}^2 b_{\sigma_1}^2 & \cdots & a_{k_1}^n b_{\sigma_1}^n \\ a_{k_2}^1 b_{\sigma_2}^1 & a_{k_2}^2 b_{\sigma_2}^2 & \cdots & a_{k_2}^n b_{\sigma_2}^n \\ \vdots & \vdots & & \vdots \\ a_{k_n}^1 b_{\sigma_n}^1 & a_{k_n}^2 b_{\sigma_n}^2 & \cdots & a_{k_n}^n b_{\sigma_n}^n \end{bmatrix}, \tag{7.5.123}$$

in which A denotes the $(2l+1) \times n$ matrix with elements (a_k^α), B denotes the $2 \times n$ matrix with elements (b_σ^α), and the notation $\mathscr{P}(A, B)$ means that $\mathscr{P}$ is a polynomial defined on the elements (a_k^α) of A and the elements (b_σ^α) of B.

The set of boson state vectors (7.5.122) corresponding to all standard Young tableaux of one column [ordering given by (7.5.114)] is an orthonormal basis of a vector space of dimension $\binom{4l+2}{n}$, which we shall denote by $\mathscr{B}^A(l^n)$.

Using now the Jordan map (Chapter 5, Section 3), we may introduce the (boson) operators $\mathscr{L}_\alpha = (\mathscr{L}_{\alpha 1}, \mathscr{L}_{\alpha 2}, \mathscr{L}_{\alpha 3})$ and $\mathscr{S}_\alpha = (\mathscr{S}_{\alpha 1}, \mathscr{S}_{\alpha 2}, \mathscr{S}_{\alpha 3})$ defined by

$$\mathscr{L}_{\alpha i} = \tilde{a}^\alpha J_i^{(l)} \bar{a}^\alpha,$$

$$\mathscr{S}_{\alpha i} = \tfrac{1}{2} \tilde{b}^\alpha \sigma_i \bar{b}^\alpha, \tag{7.5.124}$$

where $a^\alpha = \text{col}(a_1^\alpha, a_2^\alpha, \ldots, a_{2l+1}^\alpha)$ and $b^\alpha = \text{col}(b_1^\alpha, b_2^\alpha)$. The operators $\mathscr{L}_\alpha$ and $\mathscr{S}_\alpha$ then have the same action, respectively, on the space $\mathscr{B}^A(l^n)$ as do the orbital angular momentum $\mathbf{L}_\alpha$ and the spin $\mathbf{S}_\alpha$ on the space $\mathscr{V}^A(l^n)$. In this manner the problem of finding states of sharp $(LM_L)(SM_S)$ in the space $\mathscr{V}^A(l^n)$ has been transformed into an identical problem in the space $\mathscr{B}^A(l^n)$, now using the realizations of the total angular momentum $\mathscr{L}$ and the total spin $\mathscr{S}$ given by $\mathscr{L} = \sum_\alpha \mathscr{L}_\alpha$ and $\mathscr{S}_\alpha = \sum_\alpha \mathscr{S}_\alpha$. Once the latter problem is solved, one may transform the results back to the space $\mathscr{V}^A(l^n)$ by using the inverse map to that given by Eq. (7.5.121).

Our reason for making the map to the boson realization $\mathscr{B}^A(l^n)$ of the vector space $\mathscr{V}^A(l^n)$ is to make accessible for the determination of the properties of these spaces the wealth of known properties of boson state vectors as discussed in Appendices A and B to Chapter 5 (and in the cited literature). Greater generality and simplicity are gained by making this step: It is now meaningful to consider the following four types of unitary transformations of the bosons (left and right translations):

$$A \to \tilde{W}A, \qquad B \to \tilde{U}B,$$

$$A \to AV, \qquad B \to BV', \tag{7.5.125}$$

where W, U, V, and V' are unitary matrices of dimensions $2l+1$, 2, n, and n, respectively.

The transformations $A \to \tilde{W}A$ and $B \to \tilde{U}B$ correspond, respectively, to unitary transformations of the set of spatial one-particle states $\{|nlm\rangle : m = l, l-1, \ldots, -l\}$ and the set of single-particle spin states $\{|\tfrac{1}{2}\mu\rangle : \mu = \tfrac{1}{2}, -\tfrac{1}{2}\}$. The space $\mathscr{B}^A(l^n)$ is invariant under these transformations. In general, either of the transformations $A \rightarrow AV$ and $B \rightarrow BV'$ maps a vector of $\mathscr{B}^A(l^n)$ out of the space. [For the space $\mathscr{V}^A(l^n)$ such transformations would correspond to unitary transformations on the *order* of the single-particle states in the tensor product space and would have no physical significance except for the symmetric subgroup $S_n \subset U(n)$ transformations.] Notice, however, that *all the transformations* (7.5.125) *are mappings of the larger space* $\mathscr{B}(l^n) \supset \mathscr{B}^A(l^n)$ *of homogeneous polynomials of degree n* in the a_k^α *and in the* b_σ^α *into itself.*

We may now use the known properties of the space $\mathscr{B}(l^n)$ to determine explicitly a new basis of the subspace $\mathscr{B}^A(l^n)$ with the following properties: (*a*) The basis is split fully into sub-bases such that each sub-basis spans a subspace that transforms irreducible under unitary transformations $U(2l+1) \supset SO(3)$, where $SO(3)$ is the rotation group generated by the total orbital angular momentum $\mathscr{L}$; and (*b*) each basis vector has sharp total spin (SM_S).

One is thus able to carry out fully a first step in Racah's program—namely, that of classifying l^n-configurations by their irreducible transformation properties under the group $U(2l+1)$ and under the spin group $U(2)$. The much more difficult problem of reducing the irreps of $U(2l+1)$ into irreps of the total orbital angular momentum group $SO(3)$ will be discussed subsequently.

Let us now describe in detail this new basis of $\mathscr{B}^A(l^n)$, using the double tableau and Gel'fand pattern notations introduced in Appendices A and B of Chapter 5. This description is itself a rather lengthy task and is carried out in ten steps, (*a*)–(*j*) below. Following this description we shall sketch a proof of the main result, Eq. (7.5.126) below. (A more detailed proof would not serve our purpose of illustrating angular momentum techniques, since it relies heavily on results found in detail only in the literature cited in Chapter 5.)

A basis of the space $\mathscr{B}^A(l^n)$ *is given by a set of orthonormal vectors labeled by the following double standard tableau notation*:

$$|(\tilde{T}_{U_{2l+1}}|T_{U_2})\rangle \equiv \sum_{T_{S_n}} C_{T_{S_n}} |(\tilde{T}_{U_{2l+1}}|\tilde{T}_{S_n})\rangle \otimes |(T_{U_2}|T_{S_n})\rangle. \qquad (7.5.126)$$

The notations in this expression have the following definitions and meanings:

(*a*) T denotes a Young frame of shape

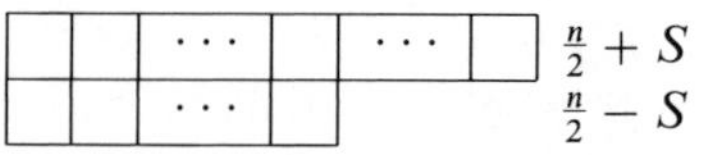

(7.5.127)

(*b*) $\tilde{T}$ denotes the dual or conjugate frame of shape

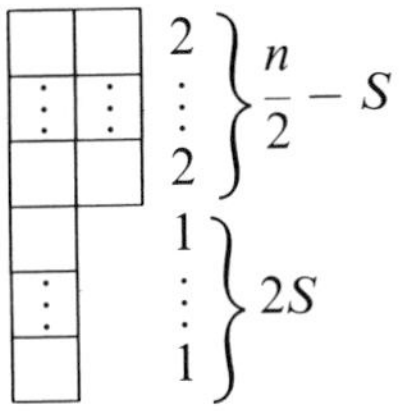

(7.5.128)

(*c*) T_{S_n} denotes a standard Young tableau for the symmetric group S_n [the frame (7.5.127) filled in lexically with the integers $1, 2, \ldots, n$].

(*d*) T_{U_2} denotes a standard Weyl tableau for the unitary group $U(2)$ [the frame (7.5.127) filled in lexically with the integers 1, 2].

(*e*) $\tilde{T}_{S_n}$ denotes the standard Young tableau conjugate to T_{S_n} [the frame (7.5.128) filled in lexically with $1, 2, \ldots, n$].

(*f*) $\tilde{T}_{U_{2l+1}}$ denotes a standard Weyl tableau for the unitary group $U(2l+1)$ [the frame (7.5.128) filled in lexically with the integers $1, 2, \ldots, 2l+1$].

(*g*) The basis vectors $|(T_{U_2}|T_{S_n})\rangle$ are the boson realization of the explicit spin states (constructed in Appendix A of this section) given by

$$
\begin{aligned}
|(T_{U_2}|T_{S_n})\rangle &= |(y_1 y_2 \cdots y_n); SM_S\rangle \\
&= \sum_{\sigma_1 \cdots \sigma_n = 1}^{2} \left\langle \left(\begin{matrix} \frac{n}{2}+S & & \frac{n}{2}-S \\ & \frac{n}{2}+M_S & \end{matrix} \right) \middle| \left\langle \begin{matrix} & y_1 & \\ 1 & & 0 \\ & \sigma_1 & \end{matrix} \right\rangle \cdots \left\langle \begin{matrix} & y_n & \\ 1 & & 0 \\ & \sigma_n & \end{matrix} \right\rangle \middle| \left(\begin{matrix} 0 & & 0 \\ & 0 & \end{matrix} \right) \right\rangle \\
&\quad \times b^1_{\sigma_1} b^2_{\sigma_2} \cdots b^n_{\sigma_n} |0\rangle,
\end{aligned}
\tag{7.5.129}
$$

where $(y_1 y_2 \cdots y_n)$ is the Yamanouchi symbol of the standard Young tableau T_{S_n}.

(*h*) The basis vectors $|(\tilde{T}_{U_{2l+1}}|\tilde{T}_{S_n})\rangle$ are boson state vectors in the bosons (a^α_k). [They are the boson realization of the tensor product space constructed from the spatial states $\{|nlm\rangle\}$ alone.] These boson state vectors are obtained as special cases of the general boson state vectors given by Eq. (A-58) (in Chapter 5, Appendix A): For $n \leqslant 2l+1$, the double Gel'fand patterns are $U(2l+1)$ patterns of the type

$$
\begin{matrix} \text{row } n - \\ \\ \text{row } 2l+1 - \end{matrix}
\left(\begin{matrix} & & (\mu) & & \\ & [2 \cdots 2\ 1 & \cdots & 1\ 0 \cdots 0] & \\ & / \quad // & & // \quad / & \\ [2 \cdots 2\ 1 & \cdots & 1\ 0 \cdots 0 & \cdots & 0] \\ & & (m) & & \end{matrix} \right)
\tag{7.5.130}
$$

in which each of the rows $n, n+1, \ldots, 2l+1$ in the upper pattern contains: 2's [$(n/2)$-S in number] and 1's [$2S$ in number] with the rest of each row filled in with 0's. The lower pattern (m) is the lexical $U(2l+1)$ Gel'fand pattern for the partition $[2^{(n/2)-S}\,1^{2S}]$ of n corresponding to the standard tableau $\tilde{T}_{U2l+1}$. The upper pattern (μ) is the lexical S_n Gel'fand pattern for the partition $[2^{(n/2)-S}\ 1^{2S}]$ of n corresponding to the standard tableau $\tilde{T}_{S_n}$.

For $n \geqslant 2l+1$ the double Gel'fand patterns are $U(n)$ patterns of the type

$$\begin{array}{r} \text{row } n - \\ \\ \text{row } 2l+1 - \end{array}\left(\begin{array}{c} (\mu) \\ [2\ \cdots\ 2\ 1\ \cdots\ 1\ 0\ \cdots\ 0\ \cdots\ 0] \\ \diagdown \quad \diagdown\diagdown \quad \diagdown\diagdown \quad \diagdown \\ [2\ \cdots\ 2\ 1\ \cdots\ 1\ 0\ \cdots\ 0] \\ (m) \end{array}\right) \tag{7.5.131}$$

in which each of the rows $2l+1, 2l+2, \ldots, n$ in the lower pattern contains $(n/2)-S$ 2's and $2S$ 1's. The patterns (m) and (μ) have the same significance as described above.

The explicit basis vectors are

$$|(\tilde{T}_{U2l+1}|\tilde{T}_{S_n})\rangle$$

$$= \sum_{k_1\cdots k_n=1}^{2l+1} \left\langle \begin{pmatrix} [m] \\ (m) \end{pmatrix} \middle| \left\langle \begin{matrix} \gamma_1 \\ [1\ \ \dot{0}] \\ k_1 \end{matrix} \right\rangle \left\langle \begin{matrix} \gamma_2 \\ [1\ \ \dot{0}] \\ k_2 \end{matrix} \right\rangle \cdots \left\langle \begin{matrix} \gamma_n \\ [1\ \ \dot{0}] \\ k_n \end{matrix} \right\rangle \middle| (0) \right\rangle$$

$$\times\, a^1_{k_1}\, a^2_{k_2} \cdots a^n_{k_n}|0\rangle, \tag{7.5.132}$$

where $(\gamma_1\gamma_2\cdots\gamma_n)$ is the Yamanouchi symbol of the standard Young tableau $\tilde{T}_{S_n}$. For $n \leqslant 2l+1$, the Gel'fand pattern $\binom{[m]}{(m)}$,

$$[m] = [2^{(n/2)-S}\ 1^{2S}\ 0^{2l+1-(n/2)-S}],$$

is the lower $U(2l+1)$ pattern of the double pattern (7.5.130); for $n \geqslant 2l+1$, the irrep labels $[m]$ in $\binom{[m]}{(m)}$ are

$$[m] = [2^{(n/2)-S}\ 1^{2S}\ 0^{(n/2)-S}],$$

and the pattern is the *full* lower $U(n)$ pattern in (7.5.131), including rows $2l+1$ to n. The notation

$$\left\langle \begin{matrix} \gamma \\ [1\ \ \dot{0}] \\ k \end{matrix} \right\rangle \tag{7.5.133}$$

designates a fundamental Wigner operator of the unitary group $U(r)$, $r = \max(n, 2l+1)$, where k and γ ($k, \gamma = 1, 2, \ldots, r$) signify that the lower and upper patterns are those having weights $W(k) = [0\cdots010\cdots0]$ (1 in position k) and $W(\gamma) = [0\cdots010\cdots0]$ (1 in position γ), respectively. The matrix elements of the fundamental Wigner operators are completely known for all $U(n)$ (see Ref. [30]) and may be obtained, in fact, from some simple rules known as the "pattern calculus rules" described in Ref. [31] and also in RWA. Thus, Eq. (7.5.132) is a fully explicit expression for the basis vectors $|(\tilde{T}_{U_{2l+1}}|\tilde{T}_{S_n})\rangle$.

(*i*) The summation in Eq. (7.5.126) is carried out over all standard Young tableaux of shape T {hence, there are $\dim[(n/2) + S, (n/2) - S]$ terms in the summation [see Eq. (A-18) in the Appendix at the end of this section]}. The coefficients $C_{T_{S_n}}$ in this expression are the Clebsch–Gordan coefficients that give the antisymmetric irrep component in the direct product representation $\tilde{T} \times T$ of S_n. These coefficients are given explicitly by (see, for example, Hamermesh [32, p. 266])

$$C_{T_{S_n}} = \left\langle \begin{array}{|c|}\hline \\ \hline \vdots \\ \hline \\ \hline \end{array} \middle| (\tilde{T}_{S_n}|T_{S_n}) \right\rangle$$

$$= \sigma(T_{S_n}) \Big/ \left(\dim\left[\frac{n}{2} + S \quad \frac{n}{2} - S \right] \right)^{\frac{1}{2}}$$

$$= \sigma(T_{S_n}) \Big/ \left[\frac{2S+1}{\frac{n}{2} + S + 1} \binom{n}{\frac{n}{2} - S} \right]^{\frac{1}{2}}, \qquad (7.5.134)$$

where $\sigma(T_{S_n})$ is the signature of the tableau T_{S_n}. [If the entries in the standard tableau T_{S_n} are $i_1 i_2 \cdots i_n$ as read left to right across row one, followed by row two, then the signature $\sigma(T_{S_n})$ of the tableau is defined to be the signature of the permutation $\begin{pmatrix} 1 & \cdots & n \\ i_1 & \cdots & i_n \end{pmatrix}$.]

(*j*) The basis (7.5.126) of the space $\mathscr{B}^A(l^n)$ is enumerated in the following manner: For each Young frame T (fixed spin S), the tableaux $\tilde{T}_{U_{2l+1}}$ and T_{U_2} run over all standard tableaux. [This is equivalent to letting (m) run over all lexical Gel'fand patterns and M_S over $S, S-1, \ldots, -S$.] The spin S is then allowed to range over all values

$$S = \frac{n}{2}, \frac{n}{2} - 1, \ldots, \frac{1}{2} \text{ or } 0, \qquad (7.5.135)$$

corresponding to the different Young frames T. For the consistency of the dimensionalities of the vector spaces occurring in the product of spaces in Eq. (7.5.126) with that of $\mathscr{B}^A(l^n)$, one has the relation

$$\sum_S (2S+1)\dim[2^{(n/2)-S}\,1^{2S}] = \binom{4l+2}{n}, \tag{7.5.136}$$

where the dimension of irrep $[2^{(n/2)-S}\ 1^{2S}]$ of $U(2l+1)$ is given by

$$\dim[2^{(n/2)-S}\,1^{2S}] = \frac{2S+1}{\dfrac{n}{2}+S+1}\binom{2l+1}{\dfrac{n}{2}+S}\binom{2l+2}{\dfrac{n}{2}-S}. \tag{7.5.137}$$

Remarks. (*a*) The result expressed by Eq. (7.5.126) solves completely the problem of spanning the space $\mathscr{B}^A(l^n)$ [hence, $\mathscr{V}^A(l^n)$] by basis vectors that (*i*) are antisymmetric in the pairwise exchange of the electrons; (*ii*) have sharp spin quantum numbers (SM_S); and (*iii*) are a basis of a carrier space of irrep $[2^{(n/2)-S}\ 1^{2S}]$ of $U(2l+1)$. (*b*) Racah recognized the tableau structure of the basis vectors $|(\tilde{T}_{U_{2l+1}}|T_{U_2})\rangle$ corresponding to the irreps of $U(2l+1)\times U(2)$, but he did not obtain the explicit decomposition of these vectors into a summation over spatial and spin parts, although he knew this general decomposition law from the work of Weyl [13]. (*c*) By using projection operators, Goddard [33] has also constructed the basis vectors $|(\tilde{T}_{U_{2l+1}}|T_{U_2})\rangle$ in a form that is equivalent to the boson state vector relation

$$|(\tilde{T}_{U_{2l+1}}|T_{U_2})\rangle = \sum_{(k,\sigma)} \langle (k,\sigma)|(\tilde{T}_{U_{2l+1}}|T_{U_2})\rangle |(k,\sigma)\rangle, \tag{7.5.138}$$

where $(k,\sigma) = \text{col}[(k_1,\sigma_1)\cdots(k_n,\sigma_n)]$ denotes the 1-column tableau in Eq. (7.5.122), $|(k,\sigma)\rangle$ the state vector, and the summation is over all standard tableaux. Goddard's contribution was in giving procedures for the computation of the coefficients

$$\langle (k,\sigma)|(\tilde{T}_{U_{2l+1}}|T_{U_2})\rangle. \tag{7.5.139}$$

{These coefficients are the subduction coefficients for the reduction of the antisymmetric irrep $[(k_1,\sigma_1)(k_2,\sigma_2)\cdots(k_n,\sigma_n)]$ of $U(4l+2)$ into irreps of the subgroup $U(2l+1)\times U(2) \subset U(4l+2)$. The coefficients (7.5.139) are here obtained by evaluating the boson scalar product indicated by the notation:

$$\langle (k,\sigma)|(\tilde{T}_{U2l+1}|T_{U2})\rangle$$

$$= \left(\dim\begin{bmatrix}\frac{n}{2}+S & \frac{n}{2}-S\end{bmatrix}\right)^{-\frac{1}{2}}$$

$$\times \sum_{T_{S_n}} \sigma(T_{S_n})$$

$$\times \sum_P (-1)^\sigma \left\langle \begin{pmatrix} \frac{n}{2}+S & & \frac{n}{2}-S \\ & \frac{n}{2}+M_S & \end{pmatrix} \middle| P\left(\left\langle \begin{matrix} & y_1 & \\ 1 & & 0 \\ & \sigma_1 & \end{matrix} \right\rangle \cdots \left\langle \begin{matrix} & y_n & \\ 1 & & 0 \\ & \sigma_n & \end{matrix} \right\rangle \right) \middle| \begin{pmatrix} 0 & & 0 \\ & 0 & \end{pmatrix} \right\rangle$$

$$\times \left\langle \begin{pmatrix} [m] \\ (m) \end{pmatrix} \middle| P\left(\left\langle \begin{matrix} & \gamma_1 & \\ [1 & & \dot{0}] \\ & k_1 & \end{matrix} \right\rangle \cdots \left\langle \begin{matrix} & \gamma_n & \\ [1 & & \dot{0}] \\ & k_n & \end{matrix} \right\rangle \right) \middle| (0) \right\rangle, \tag{7.5.140}$$

where the sum P is over all $n!$ permutations of the n pairs in the sequence $(k_1,\sigma_1)\cdots(k_n,\sigma_n)$. We also recall that $(y_1 y_2 \cdots y_n)$ is the Yamanouchi symbol of the tableau T_{S_n}; $(\gamma_1\gamma_2\cdots\gamma_n)$ is that of the conjugate tableau $\tilde{T}_{S_n}$; and $\binom{[m]}{(m)}$ is the Gel'fand pattern occurring in Eq. (7.5.132). (*d*) Unitary group and related tableau methods have found extensive applications recently in many-body physics, and results equivalent to those given in this section have been obtained by other authors, using different methods. See, for example, Kramer [34–36], Kramer and Seligman [37], Seligman [38], Matsen [39], Matsen *et al.* [40], Klein and Junker [41], Junker and Klein [42], Patterson and Harter [43], and Paldus [44, 45]. These methods lead to alternative forms for the matrix elements of the string of Wigner operators occurring in our results (see Ref. [46] for a summary of the use of the general boson polynomials in obtaining particular basis vectors of interest in physical applications).

The result, Eq. (7.5.126), *is of fundamental importance in the study of l^n-configurations from the viewpoint of unitary symmetry.* However, this result does not solve the problem of constructing states of sharp orbital angular momentum (LM_L). This requires reducing the irrep $[2^{(n/2)-S}\,1^{2S}]$ of $U(2l+1)$ into irreps of the orthogonal subgroup $SO(3)$ corresponding to orbital rotations of the n electrons. The construction of the carrier spaces of these irreps will, however, preserve the symmetric group (S_n) structure of the right-hand side of Eq. (7.5.126) and of the coefficients (7.5.140), since this construction will only involve summations over the standard tableaux $\tilde{T}_{U2l+1}$ [of the Gel'fand patterns $\binom{[m]}{(m)}$ in Eq. (7.5.140)]. [A similar result is true for the reduction of irrep $[2^{(n/2)-S}\,1^{2S}]$ of $U(2l+1)$ into the irreps of any subgroup.]

Let us now sketch the proof of Eq. (7.5.126).

Proof. We first state four general results (*i*)–(*iv*): (*i*) *The space* $\mathscr{B}(l^n)$ *is the carrier space of a unitary representation of the direct product group*

$$U(2l+1) \times U(n) \times U(2) \times U(n). \tag{7.5.141}$$

Indeed, letting $W \in U(2l+1)$, $V \in U(n)$, $U \in U(2)$, and $V' \in U(n)$, we find that the set of unitary operators [see Eqs. (5.57) and (5.42) of Chapter 5] defined by

$$\mathscr{T}(W, V, U, V')P(A, B)|0\rangle \equiv P(\tilde{W}AV, UBV')|0\rangle \tag{7.5.142}$$

is a unitary representation of $U(2l+1) \times U(n) \times U(2) \times U(n)$.

(*ii*) *The space* $\mathscr{B}^A(l^n)$ *is the carrier space of the direct product group*

$$U(2l+1) \times U(2) \times S_n, \tag{7.5.143}$$

which is a subgroup of the direct product group (7.5.141), where S_n is the diagonal subgroup of $S_n \times S_n \subset U(n) \times U(n)$. This result is obtained by choosing $V = V' = \Gamma(P)$ in Eq. (7.4.142), where $\Gamma(P)$ is the $n \times n$ unitary permutation matrix defined by

$$\Gamma(P) = [e_{\alpha 1} e_{\alpha 2} \cdots e_{\alpha n}] \tag{7.5.144}$$

for

$$P = \begin{pmatrix} 1 & 2 & \cdots & n \\ \alpha_1 & \alpha_2 & \cdots & \alpha_n \end{pmatrix}. \tag{7.5.145}$$

The symbol e_α denotes the column matrix with elements $\delta_{\alpha i}$, $i = 1, 2, \ldots, n$.

(*iii*) *The antisymmetric irrep* $[1^n]$ *of* S_n *occurs in the direct product* $[\nu] \otimes [\mu]$ *of two irreps of* S_n *if and only if* $[\nu]$ *is conjugate to* $[\mu]$, *and then* $[1^n]$ *occurs exactly once.* (This is a well-known result of Weyl [13].)

(*iv*) The subgroup of $U(2l+1) \times U(n)$ transformations $\{(W, V, \mathbb{1}_2, \mathbb{1}_n)\}$ ($\mathbb{1}_k$ = unit $k \times k$ matrix) is carried by the $(2l+1) \times n$ boson matrix A. It is known [see Chapter 5, Appendix A] that the boson polynomial vectors denoted in the double Gel'fand pattern notation by (we consider only $n \leqslant 2l+1$)

$$\left|\begin{pmatrix} & (\mu) & \\ [m_{1,2l+1} & \cdots & m_{n,2l+1}] \\ / & & / \\ [m_{1,2l+1} & \cdots & m_{n,2l+1}\; 0 \cdots 0] \\ & (m) & \end{pmatrix}\right\rangle \tag{7.5.146}$$

are a basis for all homogeneous polynomial boson state vectors in the (a_k^α) of degree n. (The basis is enumerated by letting $[m_{1,2l+1} \cdots m_{n,2l+1}\ 0 \cdots 0]$ run over all partitions of n, and (μ) and (m) over all allowed Gel'fand patterns.)

Similarly, the subgroup of $U(2) \times U(n)$ transformations $\{(\mathbb{1}_{2l+1}, \mathbb{1}_n, U, V')\}$ is carried by the $2 \times n$ boson matrix B, and the boson polynomial vectors

$$\left| \begin{pmatrix} & (\mu') & \\ [(n/2)+S & (n/2)-S & 0 \cdots 0] \\ (n/2)+S & (n/2)-S & \\ & (n/2)+M_S & \end{pmatrix} \right\rangle \tag{7.5.147}$$

are a basis for all homogeneous polynomial boson state vectors in the (b_σ^α) of degree n. [The basis is enumerated by letting $S = (n/2), (n/2) - 1, \ldots, \frac{1}{2}$ or 0; $M_S = S, \ldots, -S$; while (μ') runs over all allowed Gel'fand patterns.]

Using properties (*i*)–(*iv*), we now complete the proof of Eq. (7.5.126).

The space $\mathscr{B}(l^n)$ is spanned by the set of orthonormal vectors obtained by forming the tensor product of the basis (7.5.146) with the basis (7.5.147). The space $\mathscr{B}(l^n)$ is thus the carrier space [under the action of the operators defined by Eq. (7.5.142)] of the representation

$$\sum \oplus [m_{1,2l+1} \cdots m_{n,2l+1}\, 0 \cdots 0] \otimes [m_{1,2l+1} \cdots m_{n,2l+1}]$$
$$\otimes \left[\frac{n}{2}+S \quad \frac{n}{2}-S\right] \otimes \left[\frac{n}{2}+S \quad \frac{n}{2}-S \quad 0 \cdots 0\right] \tag{7.5.148}$$

of $[U(2l+1) \times U(n)] \times [U(2) \times U(n)]$, where the summation in this expression is over all partitions $[m_{1,2l+1} \cdots m_{n,2l+1}]$ of n and over all $S = \frac{n}{2}, \frac{n}{2} - 1, \ldots, \frac{1}{2}$ or 0.

The subspace $\mathscr{B}'(l^n)$ of $\mathscr{B}(l^n)$ spanned by the product vectors obtained by restricting the Gel'fand patterns (μ) and (μ') in Eqs. (7.5.146) and (7.5.147), respectively, to the set of patterns having weight $(1, 1, \ldots, 1)$ is the carrier space of the representation (7.5.148) of the subgroup

$$[U(2l+1) \times S_n] \times [U(2) \times S_n]. \tag{7.5.149}$$

Using property (*iii*) above, we next observe that the antisymmetric representation $[1^n]$ of S_n [diagonal subgroup of $S_n \times S_n$ in (7.5.149)] is carried by those subspaces of $\mathscr{B}'(l^n)$ for which the irrep $[m_{1,2l+1} \cdots m_{n,2l+1}]$ of the S_n in the direct product $U(2l+1) \times S_n$ is conjugate to the irrep $[\frac{n}{2}+S\ \frac{n}{2}-S\ 0 \cdots 0]$ of the S_n in $U(2) \times S_n$. Selecting the basis vectors satisfying this condition from the set of basis vectors of $\mathscr{B}'(l^n)$ and using the coefficients (7.5.134) to form the linear combinations of basis vectors transforming as the $[1^n]$ irrep of S_n under the action of $\mathscr{T}(\mathbb{1}_{2l+1}, \Gamma(P), \mathbb{1}_2, \Gamma(P))$ ($\mathbb{1}_k$ = identity matrix of dimension k), we obtain the subspace $\mathscr{B}''(l) \subset \mathscr{B}'(l) \subset \mathscr{B}(l^n)$ with the basis

$$\left\{\begin{array}{l} |(\tilde{T}_{U_{2l+1}}|T_{U_2})\rangle\colon\ \tilde{T}_{U_{2l+1}} \text{ is a standard Weyl tableau for } U_{2l+1} \text{ of shape } \tilde{T}; \\ \qquad T_{U_2} \text{ is a standard Weyl tableau for } U_2 \text{ of shape } T; \\ \qquad S = (n/2), (n/2) - 1, \ldots, \tfrac{1}{2} \text{ or } 0. \end{array}\right\} \tag{7.5.150}$$

The basis (7.5.150) is (*a*) orthonormal; (*b*) of dimension $\binom{4l+2}{n}$; (*c*) homogeneous of degree n in the (a_k^α) and in the (b_σ^α); and (*d*) antisymmetric under transpositions of superscript labels $(1, 2, \ldots, n)$. Hence, $\mathscr{B}''(l^n) = \mathscr{B}^A(l^n)$. ■

i. Operator structures in l^n-configurations. We have noted earlier that the construction of the basis vectors (7.5.126) does not solve the problem of constructing states of sharp total orbital angular momentum (LM_L). This latter problem may be described in two useful and related ways: (*a*) It may be viewed as the global group-theoretic problem of reducing the irrep $[2^{(n/2)-S}\ 1^{2S}]$ of $U(2l+1)$ into irreps of the orbital angular momentum subgroup $SO(3) \subset U(2l+1)$; and (*b*) it may be viewed as the Lie algebraic problem of determining all the linear combinations of the basis vectors (7.5.132) on which the total orbital angular momentum operators (more precisely, their boson operator analogs) have the standard multiplet action. In either approach to the problem, it is essential to identify correctly the $SO(3)$ subgroup of $U(2l+1)$ that is generated by the total angular momentum operators. [The group $SO(3)$ may be embedded in $U(2l+1)$ in various ways that do not correspond to the physical orbital angular momentum.]

Let us first examine the problem from a global viewpoint. Under the action of the operators $\mathscr{T}(U, \mathbb{1}_n, \mathbb{1}_2, \mathbb{1}_n)$, each $U \in U(2l+1)$ [see Eq. (7.5.142)], the basis vectors (7.5.132) are transformed linearly and irreducibly among themselves for each specified tableau $\tilde{T}_{S_n}$. The coefficients in this linear transformation yield the irrep of $U(2l+1)$ denoted by

$$D^{[2^{(n/2)-S}\ 1^{2S}]}(U), \qquad U \in U(2l+1) \tag{7.5.151}$$

[unitary matrix of the dimension given by Eq. (7.5.137)]. The underlying transformation of the bosons themselves is given by

$$a_k^\alpha \to \sum_{j=1}^{2l+1} u_{jk}\, a_j^\alpha \tag{7.5.152}$$

corresponding to the transformation of single-particle states (see footnote, p. 393):

$$|nlm\rangle \to \sum_{m'} u_{l-m'+1,l-m+1}|nlm'\rangle. \tag{7.5.153}$$

Under a rotation of ϕ about direction $\hat{n}$, the single-particle states undergo the transformation

$$|nlm\rangle \to \sum_{m'=l}^{-l} (U(\phi,\hat{n}))_{l-m'+1,l-m+1}|nlm'\rangle, \tag{7.5.154}$$

where $U(\phi,\hat{n})$ is the $(2l+1)\times(2l+1)$ unitary rotation matrix

$$U(\phi,\hat{n}) = e^{-i\phi\hat{n}\cdot\mathbf{J}^{(l)}}. \tag{7.5.155}$$

(Rows and columns are indexed by $j, k = 1, 2, \ldots, 2l+1$.) Thus, the transformation in the boson space corresponding to a rotation in physical space is

$$a_k^\alpha \to \sum_{j=1}^{2l+1} (U(\phi,\hat{n}))_{jk} a_j^\alpha. \tag{7.5.156}$$

This result shows that the space $\mathscr{B}^A(l^n)$ carries the representation

$$D^{[2^{(n/2)-S}1^{2S}]}(U(\phi,\hat{n})) \tag{7.5.157}$$

of the physical rotation group (group generated by the total orbital angular operators).

The representation (7.5.157) of the rotation group is, in general, reducible. *The problem of finding states of sharp orbital angular momenta is as follows: Reduce the representation* (7.5.157) *into its irreducible components* $D^L(\phi,\hat{n})$, $L = 0, 1, 2, \ldots, L_{\max}$.

The generators of the representation (7.5.157) of $SO(3)$ are the $(2l+1)\times(2l+1)$ Hermitian angular momentum matrices $J_i^{(l)}$ themselves. Using now the Jordan map, Eq. (5.46), Chapter 5, we find the boson realization of the orbital angular momentum operators $\mathscr{L}_i$ $(i = 1, 2, 3)$:

$$\mathscr{L}_i = \sum_{m,m'=l}^{-l} (J_i^{(l)})_{mm'} E_{l-m+1,l-m'+1}, \tag{7.5.158}$$

where

$$E_{jk} = \sum_{\alpha=1}^{n} a_j^\alpha \bar{a}_k^\alpha, \qquad j,k = 1, 2, \ldots, 2l+1. \tag{7.5.159}$$

More specifically, one has the relations

$$\mathscr{L}_+ = \sum_{m=l}^{-l} [(l-m)(l+m+1)]^{\frac{1}{2}} E_{l-m,l-m+1},$$

$$\mathscr{L}_- = \sum_{m=l}^{-l} [(l-m)(l+m+1)]^{\frac{1}{2}} E_{l-m+1,l-m},$$

$$\mathscr{L}_3 = \sum_{m=l}^{-l} mE_{l-m+1,l-m+1}. \tag{7.5.160}$$

Phrased in terms of the Lie algebra, *the problem of finding states of sharp orbital angular momentum is as follows: Determine all linear combinations of the vectors* $|(\tilde{T}_{U_{2l+1}}|\tilde{T}_{S_n})\rangle$ [Eq. (7.5.132)] *on which the operators* $\mathscr{L}_\pm$, $\mathscr{L}_3$ *have the standard action* (method of highest weights).

The general solution to the problem posed (in two ways) above has not yet appeared in the literature, although for small l-values (multiplicity 1 or multiplicity 2) the solution has been given in many places.

The difficulty in solving the orbital angular momentum problem is due to the multiple occurrence of a given value L of the total orbital angular momentum: The operators $\mathscr{L}_\pm$, $\mathscr{L}_3$ do not distinguish between subspaces of $\mathscr{B}^A(l^n)$ that are characterized by the same L. Racah [15, 16] attempted to resolve this difficulty by introducing other subgroups of $U(2l+1)$, which contain the rotation group as a subgroup. For a discussion of this method, we require further elaboration of the Jordan map technique used above.

The result to be made more explicit is the following: *Each finite-dimensional inner product space* $\mathscr{V}$ *as well as the vector space of all linear mappings of* $\mathscr{V}$ *into* $\mathscr{V}$ *may be realized in terms of bosons.* Let us prove this result in terms of an arbitrary inner product space $\mathscr{V}$ of dimension $\dim \mathscr{V} = d$.

Proof. Let ψ_i, $i = 1, 2, \ldots, d$, denote an orthonormal basis of $\mathscr{V}$. Then $\mathscr{V}$ is mapped to a vector space $\mathscr{B}$ ($\dim \mathscr{B} = d$) of boson states by mapping the basis of $\mathscr{V}$ by the rule

$$\psi_i \to a_i|0\rangle. \tag{7.5.161}$$

An arbitrary vector $\psi = \sum_i \lambda_i \psi_i \in \mathscr{V}$ is then mapped to the vector $|\psi\rangle = \sum_i \lambda_i a_i|0\rangle \in \mathscr{B}$ ($\lambda_i \in \mathbb{C}$).

Each linear operator $\mathscr{T}$ defined on $\mathscr{V}$ by

$$\mathscr{T}\psi_i = \sum_j \mathscr{T}_{ji}\psi_j \tag{7.5.162}$$

is now mapped via the Jordan map to the linear operator

$$\mathscr{L}_{\mathscr{T}} = \sum_{ij} \mathscr{T}_{ij} a_i \bar{a}_j. \tag{7.5.163}$$

This operator has the action on $\mathscr{B}$ given by

$$\mathscr{L}_{\mathscr{T}}(a_i|0\rangle) = \sum_j \mathscr{T}_{ji}(a_j|0\rangle). \tag{7.5.164}$$

More generally, if $\mathscr{T}\psi = \psi'$, then $\mathscr{L}_{\mathscr{T}}|\psi\rangle = |\psi'\rangle$. We note again that the correspondence

$$\mathscr{T} \to \mathscr{L}_{\mathscr{T}} \tag{7.5.165}$$

is linear and preserves commutation:

$$\mathscr{L}_{\alpha\mathscr{T}+\beta\mathscr{T}'} = \alpha\mathscr{L}_{\mathscr{T}} + \beta\mathscr{L}_{\mathscr{T}'},$$

$$[\mathscr{L}_{\mathscr{T}}, \mathscr{L}_{\mathscr{T}'}] = \mathscr{L}_{[\mathscr{T},\mathscr{T}']}. \qquad \blacksquare \tag{7.5.166}$$

Let us now apply the preceding results to the tensor product space with the basis

$$|(l^n)(m)\rangle \equiv |lm_1\rangle \otimes |lm_2\rangle \otimes \cdots \otimes |lm_n\rangle, \tag{7.5.167}$$

where $(m) = (m_1, m_2, \ldots, m_n)$, and each $m_\alpha = l, l-1, \ldots, -l$. (The principal quantum number has been suppressed in the notation for basis vectors.) The map (7.5.161) of this basis vector to a boson basis vector is given by

$$|(l^n)(m)\rangle \to a^1_{l-m_1+1} a^2_{l-m_2+1} \cdots a^n_{l-m_n+1}|0\rangle. \tag{7.5.168}$$

The boson operator corresponding to the operator $\mathscr{T}$ defined by

$$\mathscr{T}|(l^n)(m)\rangle = \sum_{(m')} \langle (l^n)(m')|\mathscr{T}|(l^n)(m)\rangle |(l^n)(m')\rangle \tag{7.5.169}$$

is then

$$\mathscr{L}_{\mathscr{T}} = \sum_{(m)(m')} \langle (l^n)(m')|\mathscr{T}|(l^n)(m)\rangle \prod_{\alpha=1}^{n} a^\alpha_{l-m'_\alpha+1} \bar{a}^\alpha_{l-m_\alpha+1}. \tag{7.5.170}$$

Let us summarize next the special cases of Eqs. (7.5.169) and (7.5.170) corresponding to one- and two-body operators.

One body-operator:

$$\mathscr{T}_1 = T \otimes \mathbb{1} \otimes \cdots \otimes \mathbb{1} + \mathbb{1} \otimes T \otimes \cdots \otimes \mathbb{1} + \cdots$$

$$+ \mathbb{1} \otimes \mathbb{1} \otimes \cdots \otimes \mathbb{1} \otimes T, \qquad n \text{ terms},$$

$$T|lm\rangle = \sum_{m'} \langle lm'|T|lm\rangle |lm'\rangle,$$

$$\langle (l^n)(m')|\mathscr{T}_1|(l^n)(m)\rangle = \sum_{\alpha=1}^{n} \left(\prod_{\beta \neq \alpha} \delta_{m'_\beta m_\beta} \right) \langle lm'_\alpha|T|lm_\alpha\rangle,$$

$$\mathscr{L}_{\mathscr{T}_1} = \sum_{mm'} \langle lm'|T|lm\rangle\, E_{l-m'+1,\,l-m+1}. \tag{7.5.171}$$

Here $E_{kk'}$ is defined by Eq. (7.5.159), and we have used the fact that $\sum_k a_k^\alpha \bar{a}_k^\alpha$ has eigenvalue 1 on the boson state vector (7.5.168) for each $\alpha = 1, \ldots, n$.

Two-body operator:

$$\mathscr{T}_2 = T_1 \otimes T_2 \otimes \mathbb{1} \otimes \cdots \otimes \mathbb{1} + T_1 \otimes \mathbb{1} \otimes T_2 \otimes \mathbb{1} \otimes \cdots \otimes \mathbb{1} + \cdots$$

$$+ \mathbb{1} \otimes \cdots \otimes \mathbb{1} \otimes T_1 \otimes T_2, \qquad \binom{n}{2} \text{ terms},$$

$$T_s|lm\rangle = \sum_{m'} \langle lm'|T_s|lm\rangle |lm'\rangle, \qquad s = 1, 2,$$

$$\langle (l^n)(m')|\mathscr{T}_2|(l^n)(m)\rangle = \sum_{\beta > \alpha = 1}^{n} \left(\prod_{\gamma \neq \alpha, \beta} \delta_{m'_\gamma m_\gamma} \right) \langle lm'_\alpha|T_1|lm_\alpha\rangle \langle lm'_\beta|T_2|lm_\beta\rangle,$$

$$\mathscr{L}_{\mathscr{T}_2} = \sum_{\beta > \alpha = 1}^{n} \sum_{mm'm''m'''} \langle lm'''|T_1|lm''\rangle \langle lm'|T_2|lm\rangle$$

$$\times E^{\alpha\alpha}_{l-m'''+1,\,l-m''+1}\, E^{\beta\beta}_{l-m'+1,\,l-m+1}, \tag{7.5.172}$$

where

$$E^{\alpha\alpha}_{kk'} = a_k^\alpha \bar{a}_{k'}^\alpha.$$

(These results have an obvious extension to a k-body operator.)

Several one-body operators of special interest are the following:

(*i*) Choose T to be the ket–bra operator $T = |lm'\rangle\langle lm|$. Then the one-body operator $\mathscr{T}_1$ has the boson operator map

$$\mathscr{L}_{\mathscr{T}_1} = E_{l-m'+1,l-m+1}. \tag{7.5.173}$$

(*ii*) Choose T to be the Wigner operator $\left\langle \begin{matrix} & 0 & \\ a & & -a \\ & \lambda & \end{matrix} \right\rangle$ defined by

$$\left\langle \begin{matrix} & 0 & \\ a & & -a \\ & \lambda & \end{matrix} \right\rangle |lm\rangle = C^{l\,a\,l}_{m,\lambda,m+\lambda}|l, m+\lambda\rangle. \tag{7.5.174}$$

Then the one-body operator $\mathscr{T}_1$ has the boson operator map

$$\mathscr{L}_{\mathscr{T}_1} = \sum_m C^{l\,a\,l}_{m,\lambda,m+\lambda} E_{l-m-\lambda+1,l-m+1}. \tag{7.5.175}$$

These latter operators are important for understanding the role of angular momentum in l^n-configurations, and we turn to a discussion of their properties before giving an example of a two-body operator.

The operator given by Eq. (7.5.175) is defined for all $a = 0, 1, 2, \ldots, 2l$, and $\lambda = a, a-1, \ldots, -a$. It is convenient to give a slight redefinition of these operators:

$$V^a_\lambda \equiv \sqrt{(2a+1)/(2l+1)} \sum_{mm'} C^{l\,a\,l}_{m\lambda m'} E_{l-m'+1,l-m+1}. \tag{7.5.176}$$

The coefficients

$$W_{a\lambda,mm'} = \sqrt{(2a+1)/(2l+1)}\, C^{l\,a\,l}_{m\lambda m'} \tag{7.5.177}$$

in this redefinition are then the elements of a $(2l+1)^2$-dimensional real orthogonal matrix where the rows are enumerated by the index pair $(a\lambda)$ $(\lambda = a, a-1, \ldots, -a;\ a = 0, 1, \ldots, 2l)$, and columns by the index pair (mm') $(m = l, l-1, \ldots, -l;\ m' = l, l-1, \ldots, -l)$.

The inversion of Eq. (7.5.176) is thus given by

$$E_{l-m'+1,l-m+1} = \sum_{a\lambda} \sqrt{(2a+1)/(2l+1)}\, C^{l\,a\,l}_{m\lambda m'} V^a_\lambda. \tag{7.5.178}$$

Equations (7.5.176) and (7.5.178) show that the sets of operators

$$\{E_{jk}: j, k = 1, 2, \ldots, 2l + 1\} \tag{7.5.179}$$

and

$$\{V^a_\lambda: \lambda = a, a - 1, \ldots, -a; a = 0, 1, \ldots, 2l\} \tag{7.5.180}$$

are simply *different bases* of the Lie algebra of $U(2l + 1)$.

The physical significance of the basis set (7.5.180) is found by verifying the following relations:

$$\begin{aligned} V^1_{+1} &= -[3/(2l)(l+1)(2l+1)]^{\frac{1}{2}} \mathscr{L}_+, \\ V^1_0 &= [3/l(l+1)(2l+1)]^{\frac{1}{2}} \mathscr{L}_3, \\ V^1_{-1} &= [3/(2l)(l+1)(2l+1)]^{\frac{1}{2}} \mathscr{L}_-; \end{aligned} \tag{7.5.181}$$

$$\begin{aligned} [\mathscr{L}_\pm, V^a_\lambda] &= [(a \mp \lambda)(a \pm \lambda + 1)]^{\frac{1}{2}} V^a_{\lambda \pm 1}, \\ [\mathscr{L}_3, V^a_\lambda] &= \lambda V^a_\lambda. \end{aligned} \tag{7.5.182}$$

Thus, *the basis* $\{V^a_\lambda\}$ *of the Lie algebra of* $U(2l + 1)$ *corresponds to the complete classification of the generators of* $U(2l + 1)$ *as irreducible tensor operators with respect to the rotation subgroup generated by the total orbital angular momentum.*

The basis $\{V^a_\lambda\}$ of the Lie algebra of $U(2l + 1)$ was introduced (explicitly) by Racah[1] [15, 16] for the purpose of keeping visible the total angular momentum properties of the unitary group $U(2l + 1)$ and its subgroups.

Using the commutation relations for the *Weyl basis* $\{E_{jk}\}$, namely,

$$[E_{jk}, E_{j'k'}] = \delta_{kj'} E_{jk'} - \delta_{jk'} E_{j'k}, \tag{7.5.183}$$

one finds the commutation relations for the *Racah basis* $\{V^k_\mu\}$:

$$[V^a_\lambda, V^b_\mu] = \sum_{c\nu} A^{abc}_{\lambda\mu\nu} V^c_\nu, \tag{7.5.184}$$

where the structure constants $A^{abc}_{\lambda\mu\nu}$ are defined by

$$A^{abc}_{\lambda\mu\nu} \equiv [(-1)^{a+b-c} - 1] C^{abc}_{\lambda\mu\nu} [(2a+1)(2b+1)]^{\frac{1}{2}} W(abll; cl). \tag{7.5.185}$$

---

[1] Racah's tensor operator $\mathbf{U}^{(a)}$ is related to $\mathbf{V}^a$ by $\mathbf{U}^{(a)} = \mathbf{V}^a/\sqrt{2a+1}$.

[The derivation of the structure constants in this form utilizes Eq. (7.5.178) as well as Eq. (3.267) of Chapter 3 and the symmetries of the Wigner coefficients.]

The results obtained above solve completely the problem of embedding the physical orbital rotation group $SO(3)$ in the unitary group $U(2l+1)$:

$$U(2l+1) \supset SO(3). \tag{7.5.186}$$

However, since an irrep of $SO(3)$ will generally occur with repetition in the reduction of an irrep of $U(2l+1)$ into a direct sum of irreps of $SO(3)$, the construction of states of sharp orbital angular momentum is still unsolved. Additional labels for enumerating the subspaces of $\mathscr{B}^A(l^n)$ that are irreducible under the action of $\mathscr{L}$ and that have the same angular momentum label L may be introduced if one can find a group G such that

$$U(2l+1) \supset G \supset SO(3). \tag{7.5.187}$$

The irrep spaces of G will then be invariant under the action of $\mathscr{L}$, and the irrep labels of G will provide additional quantum numbers.

One of the advantages of using the Racah basis $\{\mathbf{V}^a: a = 0, 1, \ldots, 2l\}$ of the Lie algebra of $U(2l+1)$ is that one can identify immediately a basis of the Lie algebra of the proper orthogonal group $SO(2l+1)$. The set of operators

$$\{\mathbf{V}^a: a = 1, 3, \ldots, 2l-1\} \tag{7.5.188}$$

is seen from Eqs. (7.5.184) and (7.5.185) to form a closed subalgebra, which contains $\mathscr{L}$. Furthermore, each basis element in the set (7.5.188) may be shown directly from Eq. (7.5.176) to commute with the positive definite quadratic form

$$\sum_m (-1)^m a^\alpha_{l+m+1} a^\alpha_{l-m+1} \tag{7.5.189}$$

for each $\alpha = 1, 2, \ldots, n$. This result shows that the basis (7.5.188) is a set of generators of the orthogonal group of transformations, $SO(2l+1)$, that leaves the quadratic form (7.5.189) invariant.

The reduction of the irrep

$$[2^{(n/2)-S}\, 1^{2S}\, 0^{2l+1-(n/2)-S}] \tag{7.5.190}$$

of $U(2l+1)$ into irreps of $SO(2l+1)$ provides only one additional label: the seniority quantum number v (Racah [15, 16]). Let us next discuss the group-theoretic origin of this result.[1]

[1] Racah introduced the seniority quantum number using the concept of an "exchange operator," noting, however, the relationship to the irreps of $SO(2l+1)$. This approach is discussed in Section 9 on nuclear physics.

The irreps of the group $SO(2l+1)$ are labeled by a partition of the form

$$[j_1 j_2 \cdots j_l], \tag{7.5.191}$$

where $j_1, j_2, \ldots, j_l$ are integers satisfying

$$j_1 \geqslant j_2 \geqslant \cdots \geqslant j_l \geqslant 0. \tag{7.5.192}$$

The only irreps of $SO(2l+1)$ that occur in the irrep (7.5.190) of $U(2l+1)$ are of the form (Hamermesh [32], Wybourne [47])

$$[2^p\, 1^q\, 0^{l-p-q}], \tag{7.5.193}$$

where

$$p = v/2 - S,$$
$$q = \min(2S,\ 2l + 1 - v). \tag{7.5.194}$$

In this result, v is an integer, which may assume the values

$$v = 2S, 2S + 2, \ldots, n. \tag{7.5.195}$$

Thus, one additional label, v, enumerates all irreps of $SO(2l+1)$ occurring in the special irrep (7.5.190) of $U(2l+1)$, since there is no multiplicity, as we now show.

There is no multiplicity in the reduction of an irrep of $U(2l+1)$ of the type $[2^a\, 1^b\, 0^{2l+1-a-b}]$ into irreps of $SO(2l+1)$. A proof of this result may be given in the following manner. The dimension of irrep $[2^p\, 1^q\, 0^{l-p-q}]$ of $SO(2l+1)$ is

$$d_{2l+1}(p,q) = \frac{(q+1)(2l-2p-q+2)}{(2l+2)(2l+3)} \binom{2l+3}{p} \binom{2l+3}{p+q+1}. \tag{7.5.196}$$

When this dimension formula is summed over all (p, q) pairs given by

$$(p,\, q) = \big(p,\, \min(b, 2l+1-b-2p)\big), \qquad p = 0, 1, \ldots, a,$$

one obtains the dimension of irrep $[2^a\, 1^b\, 0^{2l+1-a-b}]$ of $U(2l+1)$ [see Eq. (7.5.137)]:

$$\sum_{(p,q)} d_{2l+1}(p,q) = \frac{b+1}{a+b+1} \binom{2l+2}{a} \binom{2l+1}{a+b}. \quad \blacksquare \tag{7.5.197}$$

The explicit branching of irreps of $U(2l+1)$ into irreps of $SO(2l+1)$ may be found for $l = 1, 2, \ldots, 6$ in Wybourne [47].

An alternative procedure (Racah [15]) for introducing the quantum number v is to diagonalize the quadratic Casimir operator[1] of the group $SO(2l+1)$:

$$\mathbf{C}^2 \equiv 2 \sum_{a \text{ odd}} \sum_{\lambda} (-1)^{\lambda} V_{\lambda}^{a} V_{-\lambda}^{a}. \tag{7.5.198}$$

The form of the eigenvalue of $\mathbf{C}^2$ on a general irrep space of $SO(2l+1)$ labeled by $[j_1 j_2 \cdots j_l]$ is

$$j_1(j_1 + 2l - 1) + j_2(j_2 + 2l - 3) + \cdots + j_l(j_l + 1).$$

Using the special irreps of the form (7.5.193), one thus finds that the diagonalization of $\mathbf{C}^2$ on a carrier space of the irrep (7.5.190) of $U(2l+1)$ will yield the eigenvalues λ_v,

$$\lambda_v \equiv v\left(2l + 2 - \frac{v}{2}\right) - 2S(S+1), \tag{7.5.199}$$

corresponding to the values (7.5.195) of v.

Let us next apply the results of the group–subgroup reduction, $U(2l+1) \supset SO(2l+1)$, to the space $\mathscr{B}^A(l^n)$ [basis given by Eq. (7.5.126)]. Since the spin part of the basis has already been determined, we may confine our attention to the subspace $\mathscr{B}^A(l^n, S)$ with basis

$$\left\{ |\tilde{T}_{U_{2l+1}}| T_{U_2}\rangle \colon T \text{ is a standard Weyl tableau of shape } [2^{(n/2)-S}\, 1^{2S}];\ n,\ S,\ M_S \text{ prescribed} \right\}. \tag{7.5.200}$$

Thus, one is actually concerned only with the reduction of the irrep space of $U(2l+1)$ labeled by $[2^{(n/2)-S}\ 1^{2S}]$ into irrep spaces of $SO(2l+1)$, as mentioned earlier.

The analysis above shows that one may split the space $\mathscr{B}^A(l^n, S)$ uniquely into carrier spaces $\mathscr{B}^A(l^n, S, v)$ of irreps of $SO(2l+1)$ of the type (7.5.193) that are labeled by the seniority v:

$$\mathscr{B}^A(l^n, S) = \sum_{v} \oplus\, \mathscr{B}^A(l^n, S, v). \tag{7.5.201}$$

The summation in this direct sum is over $v = 2S, 2S+2, \ldots, n$. Each subspace $\mathscr{B}^A(l^n, S, v)$ is invariant under the action of the total angular momentum

---

[1] There are l mutually commuting independent polynomial forms in the generators of $SO(2l+1)$ of degrees $2, 4, \ldots, 2l$ that commute with all elements of the algebra. When applied to special irrep spaces of the type (7.5.193), this set of l operators becomes dependent on the quadratic (Casimir) and quartic operators.

operator $\mathscr{L}$ and irreducible under the action of the set of generators $\{\mathbf{V}^a: a = 1, 3, \ldots, 2l-1\}$ of $SO(2l+1)$. The branching of irreps of $U(2l+1)$ into irreps of $SO(2l+1)$ corresponding to Eq. (7.5.201) is given in Table 7.1 for $l = 1$ and $l = 2$.

Table 7.1. *The Branching of* $[2^{\frac{n}{2}-S}1^{2S}0^{2l+1-\frac{n}{2}-S}]$ *Irreps of* $U(2l+1)$ *into Irreps of* $SO(2l+1)$ *Labeled by Seniority for* $l=1$ *and* 2.

l^n	$SU(2l+1)$	$SO(2l+1)$	(S, v)
p^1	$[100] =$	$[1]$	$(\frac{1}{2}, 1)$
p^2	$[110] =$	$[1]$	$(1, 2)$
	$[200] =$	$[0] \oplus [2]$	$(0,0) + (0,2)$
p^3	$[111] =$	$[1]$	$(\frac{3}{2}, 3)$
d^1	$[10000] =$	$[1]$	$(\frac{1}{2}, 1)$
d^2	$[11000] =$	$[11]$	$(1, 2)$
	$[20000] =$	$[00] \oplus [20]$	$(0,0) + (0,2)$
d^3	$[11100] =$	$[11]$	$(\frac{3}{2}, 3)$
	$[21000] =$	$[10] \oplus [21]$	$(\frac{1}{2}, 1) + (\frac{1}{2}, 3)$
d^4	$[11110] =$	$[10]$	$(2, 4)$
	$[21100] =$	$[11] \oplus [21]$	$(1,2) + (1,4)$
	$[22000] =$	$[00] \oplus [20] \oplus [22]$	$(0,0) + (0,2) + (0,4)$
d^5	$[11111] =$	$[00]$	$(\frac{5}{2}, 5)$
	$[21110] =$	$[11] \oplus [20]$	$(\frac{3}{2}, 3) + (\frac{3}{2}, 5)$
	$[22100] =$	$[10] \oplus [21] \oplus [22]$	$(\frac{1}{2}, 1) + (\frac{1}{2}, 3) + (\frac{1}{2}, 5)$

Let us now combine the results of d^n-configurations given in Table 7.1 with the branching of irreps of $SO(5)$ into irreps $[L]$ of $SO(3)$:

$$
\begin{aligned}
[00] &= [0] \\
[10] &= [2] \\
[11] &= [1] \oplus [3] \\
[20] &= [2] \oplus [4] \\
[21] &= [1] \oplus [2] \oplus [3] \oplus [4] \oplus [5] \\
[22] &= [0] \oplus [2] \oplus [3] \oplus [4] \oplus [6].
\end{aligned}
\tag{7.5.202}
$$

In these reductions $[L]$ occurs without repetition, and each space $\mathscr{B}^A(l^n, S, v)$ therefore splits uniquely into irrep spaces $\mathscr{B}^A(l^n, v, L, S)$ under the action of $\mathscr{L}$:

$$
\mathscr{B}^A(l^n, v, S) = \sum_L \oplus\, \mathscr{B}^A(l^n, v, L, S). \tag{7.5.203}
$$

The seniority quantum number v thus provides the extra label required to classify completely all states of sharp angular momentum in d^n-configurations (v is meaningful, but superfluous, for p^n-configurations).

The complete group-theoretic classification of states described above for p^n- and d^n-configurations does not usually solve the problem of constructing states of sharp angular momentum for a physical system. This is because the Casimir operator $\mathbf{C}^2$ [see Eq. (7.5.198)] does not usually commute with the effective perturbation operator [the physical perturbation restricted to the space $\mathscr{B}^A(l^n)$] that is required for approximating the physical energy levels. For example, while the electrostatic repulsion operator $\mathscr{E}_R$ [see Eq. (7.5.217) below] necessarily commutes with $\mathscr{L}^2$ and $\mathscr{L}_3$, it does not commute with $\mathbf{C}^2$. This result does not, however, nullify altogether the results obtained by the group-theoretic analysis, as may be made clear by an example we now give.

Consider the space $\mathscr{B}^A(d^5, \frac{1}{2})$ corresponding to the $U(5)$ irrep labels [22100]. This space contains three orthonormal vectors having $L = M_L = 2$ (highest-weight vectors) and $v = 1, 3, 5$ [see Table 7.1 and Eqs. (7.5.202)]. Let us denote these three vectors by the notation $|v, L\rangle$:

$$|1, 2\rangle,\ |3, 2\rangle,\ |5, 2\rangle. \tag{7.5.204}$$

Since the electrostatic repulsion operator $\mathscr{E}_R$ commutes with the total angular momentum $\mathscr{L}$, the three-dimensional space spanned by the vectors (7.5.204) is invariant under the action of $\mathscr{E}_R$. Thus, one needs only to diagonalize a 3×3 matrix to calculate the first-order energy corrections to the zeroth-order energy of the state having $n = 5$, $l = 2$, and $S = \frac{1}{2}$.

This example illustrates still another important point: From the viewpoint of diagonalizing a perturbation operator that commutes with $\mathscr{L}$, any scheme that classifies the vectors belonging to the space of highest-weight vectors for a specified L would serve just as well as a group-theoretic scheme.

It is useful to note that, from the viewpoint of highest weights of $SO(3)$, the group-theoretic classification, $U(2l+1) \supset SO(2l+1)$, corresponds to the determination of the set of vectors $|(\gamma)v, L\rangle \in \mathscr{B}^A(l^n, S)$ such that

$$\mathscr{L}_+|(\gamma)v, L\rangle = 0, \qquad \mathscr{L}_3|(\gamma)v, L\rangle = L|(\gamma)v, L\rangle,$$

$$\mathbf{C}^2|(\gamma)v, L\rangle = \lambda_v|(\gamma)v, L\rangle, \tag{7.5.205}$$

where the eigenvalue λ_v is given by Eq. (7.5.199). Thus, for d^n-configurations all solutions to Eqs. (7.5.205) are labeled by the eigenvalues of the operators themselves – that is, by v and L. For $l \geqslant 3$, additional labels (γ) are required to enumerate the solutions to Eqs. (7.5.205), as we now discuss.

The group $SO(2l+1)$ is not sufficient for the classification of all angular momentum states in l^n-configurations for $l \geqslant 3$ as noted above: The irreps of type (7.5.193) of $SO(2l+1)$ branch into more than one irrep of $SO(3)$ (see Refs. [18] and [47]). In a group-theoretic approach one attempts to resolve this problem by finding yet another group G, this time in the chain of subgroups

$$U(2l+1) \supset SO(2l+1) \supset G \supset SO(3). \tag{7.5.206}$$

It is quite remarkable that for f^n-configurations ($l = 3$) such a group G exists, as was first pointed out by Racah [15]. Again, it is the Racah basis of the Lie algebra of $U(2l+1)$ that makes this fact transparent when combined with the observation that the Racah coefficient $W(5533;33)$ vanishes:[1]

$$W(5533;33) = 0. \tag{7.5.207}$$

Thus, setting $l = 3$ in Eq. (7.5.184), we find that the set of tensor operators of ranks 1 and 5 (fourteen operators in all) forms a closed subalgebra of the Lie algebra of $SO(7)$. Racah has identified this algebra to be that of the exceptional Lie group G_2. Thus, a basis of the Lie algebra of G_2 is

$$\{V^1_\lambda, V^5_\mu : \lambda = +1, 0, -1;\ \mu = +5, +4, \ldots, -5\}. \tag{7.5.208}$$

The irreps of G_2 are labeled by a pair of integers $u = (u_1 u_2)$, $u_1 \geqslant u_2 \geqslant 0$, and the quadratic Casimir operator for G_2 is

$$\mathbf{G}^2 \equiv \tfrac{1}{4} \sum_{a=1,5} \sum_{\lambda} (-1)^\lambda V^a_\lambda V^a_{-\lambda}. \tag{7.5.209}$$

(Recall that $l = 3$ in this result.) The eigenvalue λ_u of $\mathbf{G}^2$ on an irrep space labeled by $u = (u_1 u_2)$ is

$$\lambda_u = (u_1^2 + u_2^2 + u_1 + u_2 + 5u_1 + 4u_2)/12. \tag{7.5.210}$$

(See Racah [15, 16], Judd [18].)

One may now proceed, using the irreps that appear in the group chain

$$U(7) \supset SO(7) \supset G_2 \supset SO(3), \tag{7.5.211}$$

[1] This Racah coefficient vanishes despite the fact that the triangle conditions and all symmetries are satisfied. No general explanation of this phenomenon exists, although here the special vanishing may be attributed to the existence of the embedding $SO(7) \supset G_2 \supset SO(3)$. Such zeros are discussed in more detail in RWA. When expressed in terms of the generating function for the 6-j coefficients (see Appendix D, Chapter 5), these vanishings imply that certain monomials in the bosons that might otherwise be expected to be present will be missing in the expansion. This same phenomenon happens for the 3-j coefficients (see RWA).

to split the space $\mathscr{B}^A(f^n, S, v)$ into irrep spaces of G_2. The procedure is completely analogous to that illustrated above for d^n-configurations. The only irreps of G_2 that appear in the group reduction of f^n-configurations in the chain (7.5.211) are (00), (10), (11), (20), (21), (22), (30), (31), and (40) (see Wybourne [47]).

The space of highest-weight vectors of $SO(3)$ is spanned by the set of vectors $|(\gamma)u, v, L\rangle \in \mathscr{B}^A(f^3, S)$ that satisfy

$$\mathscr{L}_+|(\gamma)u, v, L\rangle = 0, \qquad \mathscr{L}_3|(\gamma)u, v, L\rangle = L|(\gamma)u, v, L\rangle,$$

$$\mathbf{C}^2|(\gamma)u, v, L\rangle = \lambda_v|(\gamma)u, v, L\rangle, \qquad \mathbf{G}^2|(\gamma)u, v, L\rangle = \lambda_u|(\gamma)u, v, L\rangle. \tag{7.5.212}$$

The labels u, v, and L do not, however, now enumerate all the solutions to Eqs. (7.5.212) [the angular momenta $L = 3, 5, 6, 7$ occur with multiplicity 2 in the reduction of irrep (31) of G_2; and $L = 4, 6, 8$ occur with multiplicity 2 in the reduction of irrep (40)]. Thus, the group chain (7.5.211) fails to yield a complete classification of the angular momentum states. (This failure is not very serious, however, since only multiplicity-2 states occur.)

The explicit construction of bases possessing the group–subgroup properties introduced above is a sizable task and will be discussed only in a general way in the Remarks below.

The preceding examples illustrate quite nicely how angular momentum techniques lead directly to the tensor operator classification of the generators of Lie groups that contain the angular momentum subgroup, and how such groups are used in physical problems. We have emphasized, however, that even a complete classification of all angular momentum states in l^n-configurations would, in general, not correspond to the construction of states that diagonalize, for example, the effective electrostatic repulsion operator. To make this result more explicit, let us now consider how the electrostatic repulsion interaction, Eq. (7.5.60), is mapped to a boson operator $\mathscr{E}_R$ acting in the space $\mathscr{B}^A(l^n)$.

Let us rewrite Eq. (7.5.60) as

$$H_1 = \sum_{\beta > \alpha = 1}^{n} \sum_{a=0}^{\infty} v_a(r_\alpha, r_\beta) \frac{4\pi}{2a+1} \sum_{\lambda} (-1)^\lambda Y_{a\lambda}(\hat{x}_\alpha) Y_{a,-\lambda}(\hat{x}_\beta). \tag{7.5.213}$$

We may interpret H_1 as a sum of two-body operators and apply Eqs. (7.5.172), or we may appeal directly to Eqs. (7.5.169) and (7.5.170). We follow the latter course, beginning with the relation[1]

[1] To avoid confusing the number of equivalent electrons, n, with the principal quantum number, we have denoted the latter by n' below.

$$\langle (l^n)(m')|H_1|(l^n)(m)\rangle = \sum_{\beta>\alpha=1}^{n} \left(\prod_{\gamma\neq\alpha,\beta} \delta_{m'_\gamma m_\gamma} \right) \sum_{a=0}^{2l} F^a(n'l;n'l)\,(C^{lal}_{000})^2$$

$$\times \sum_{\lambda} (-1)^\lambda C^{lal}_{m_\alpha \lambda m'_\alpha} C^{lal}_{m_\beta, -\lambda, m'_\beta}. \qquad (7.5.214)$$

[The Gaunt integral of three spherical harmonics, Eq. (3.192), Chapter 3, has been used in obtaining this result.] Substituting this result into Eq. (7.5.170) yields

$$\mathscr{E}_R = \sum_{a=0}^{2l} F^a(n'l;n'l)(C^{lal}_{000})^2 \sum_{\beta>\alpha=1}^{n} \sum_{\lambda} (-1)^\lambda \left(\sum_{\nu\nu'} C^{lal}_{\nu\lambda\nu'} E^{\alpha\alpha}_{l-\nu'+1,l-\nu+1} \right)$$

$$\times \left(\sum_{mm'} C^{lal}_{m,-\lambda,m'} E^{\beta\beta}_{l-m'+1,l-m+1} \right). \qquad (7.5.215)$$

[See the remark following Eq. (7.5.171).] Equation (7.5.215) may be simplified further by using the symmetry of the terms in the summation under the interchange of α and β, the definition of V^a_λ given by Eq. (7.5.176), and the operator identity [on states (7.5.168)] given by

$$E^{\alpha\alpha}_{jk} E^{\alpha\alpha}_{j'k'} = \delta_{kj'} E^{\alpha\alpha}_{jk'}. \qquad (7.5.216)$$

Carrying out this step, we obtain the following form for the electrostatic repulsion operator $\mathscr{E}_R$:

$$\mathscr{E}_R = \frac{1}{2} \sum_a F^a(n'l;n'l) \frac{2l+1}{2a+1} (C^{lal}_{000})^2$$

$$\times \left[\sum_\lambda (-1)^\lambda V^a_\lambda V^a_{-\lambda} - \frac{2a+1}{2l+1} \sum_k E_{kk} \right]. \qquad (7.5.217)$$

The operator $\mathscr{E}_R$ is the restriction of the electrostatic repulsion energy to the space $\mathscr{B}^A(l^n)$. It is this Hermitian operator, $\mathscr{E}_R$, that must be diagonalized to obtain the energy corrections due to the electrostatic repulsion. Notice that $\mathscr{E}_R$ is written in a form that makes its invariance under the $SO(3)$ group generated by $\mathscr{L}$ obvious ($\sum_k E_{kk}$ is the number operator and may be replaced by n). Since the operator $\mathscr{E}_R$ is, by construction, a polynomial form in the generators $\{E_{jk}\}$ of $U(2l+1)$, it also commutes with each of the $(2l+1)$ mutually commuting Hermitian operators $\mathscr{L}^{(2l+1)}_k$ ($k = 1, 2, \ldots, 2l+1$), whose simultaneous eigenvectors define a general irrep space $[m_{1,2l+1}, m_{2,2l+1}, \ldots, m_{2l+1,2l+1}]$ of $U(2l+1)$ [see Eq. (A-61), Appendix A, Chapter 5]. [Only two of these operators will be independent when acting on special states of the type given by Eq. (7.5.168).] The operator $\mathscr{E}_R$ does not commute with the Casimir operator of $SO(2l+1)$, nor with the Casimir operator of G_2 in the special case $l = 3$.

We remark again that the problem of diagonalizing $\mathscr{E}_R$ on the space $\mathscr{B}^A(l^n)$ may be approached in numerous ways. In a "worst" approach, one may simply diagonalize the $\binom{4l+2}{n}$-dimensional matrix representing $\mathscr{E}_R$ on some basis of $\mathscr{B}^A(l^n)$. In a "best" approach, one seeks principles whereby the space of highest-weight vectors for each prescribed L may be found. The group–subgroup procedure discussed above is but one such approach to the "multiplicity problem for highest weights."

As noted above, there are many other techniques for addressing the "multiplicity problem" in the literature. We cite the following references as having special relevance, although not all the techniques discussed therein have been developed for atomic spectroscopy applications: Shudeman [48], Armstrong and Judd [49], Bargmann and Moshinsky [50], Moshinsky and Devi [51], Devi [52], Wybourne [47, 53], Judd *et al.* [54], Chacón *et al.* [55], Kramer *et al.* [55a], and Patera and Sharp [55b]. (For a review of some of these techniques, see Ref. [56].)

Remarks. (*a*) This concludes our discussion of atomic spectroscopy, although the omission of many important topics is obvious (coefficients of fractional parentage, *jj* and other coupling schemes, etc.). Angular momentum techniques are equally important for these topics, and we refer to the textbook literature for these developments (Slater [9], Judd [18, 57], Sobel'man [19], de-Shalit and Talmi [20], Wybourne [47], Cowan [58], Weissbluth [59]). (*b*) We have not discussed how various techniques are implemented for the calculation of energy levels and the comparison with actual spectra. This latter problem is one of enormous proportions, depending on conceptual structures as well as detailed computational procedures. Furthermore, new methods of computation (based on the unitary group) have also been advocated (Harter and Patterson [60–62], Paldus [44, 45], Shavitt [63], Matsen [64], Drake *et al.* [65], Drake and Schlesinger [66]). To discuss these details beyond the introduction given in this section would carry us further into current research in atomic spectroscopy, as opposed to angular momentum techniques, which is our purpose in the present volume.

j. Appendix: Explicit spin states. In this Appendix we give the construction of the states of sharp total spin (SM_S) that occur in the coupling of n spin-$\frac{1}{2}$ angular momenta. There is an extensive literature dealing with the problem, using a variety of methods. The projection operator technique (Löwdin [67]) has, perhaps, been the most popular. We base our method on the concept of a Wigner operator introduced in Chapter 3, Section 21. The structural simplicity of the principal result, Eq. (A.23) below, is an excellent example of the scope of angular momentum methods when combined with results from the theory of the symmetric group. For other approaches to the problem, we refer to the literature (Goddard [33], Kramer [34], Seligman [38], Patterson and Harter [43], Löwdin [67], Pauncz [68], Rotenberg [69], Murty and Sarma [70], and references cited therein.)

Let us recall briefly the definition of the action of a permutation $P \in S_n$ on the tensor product space $\mathscr{H}(j_1 j_2 \cdots j_n)$ corresponding to n kinematically independent angular momenta and having the basis

$$\{|j_1 m_1\rangle \otimes |j_2 m_2\rangle \otimes \cdots \otimes |j_n m_n\rangle : \text{each } m_i = j_i, j_i - 1, \ldots, -j_i\}. \quad \text{(A-1)}$$

For P given by

$$P = \begin{pmatrix} 1 & 2 & \cdots & n \\ i_1 & i_2 & \cdots & i_n \end{pmatrix}, \quad \text{(A-2)}$$

one has[1]

$$P: |j_1 m_1\rangle \otimes \cdots \otimes |j_n m_n\rangle \to |j_{i_1} m_{i_1}\rangle \otimes \cdots \otimes |j_{i_n} m_{i_n}\rangle. \quad \text{(A-3)}$$

In general, the action of a permutation is to map the space $\mathscr{H}(j_1 j_2 \cdots j_n)$ to the space $\mathscr{H}(j_{i_1} j_{i_2} \cdots j_{i_n})$.

Henceforth, we shall particularize to the case of equal angular momenta, $k = j_1 = j_2 = \cdots = j_n$, so that each $P \in S_n$ maps the $(2k+1)^n$-dimensional vector space $\mathscr{V}_k^{(n)} \equiv \mathscr{H}(kk \cdots k)$ spanned by the basis vectors

$$\{|km_1\rangle \otimes \cdots \otimes |km_n\rangle : \text{ each } m_i = k, k-1, \ldots, -k\} \quad \text{(A-4)}$$

onto itself. Indeed, the matrix $D(P)$ obtained from the action (A-3) of P on the basis (A-4) is a $(2k+1)^n$-dimensional matrix whose columns (rows) are a permutation of the columns (rows) of the unit matrix of dimension $(2k+1)^n$. Furthermore, the trace of $D(P)$ is given by

$$\operatorname{tr} D(P) = \text{number of particle labels invariant under } P. \quad \text{(A-5)}$$

The representation $P \to D(P)$ of S_n is, in general, reducible.

Consider next the total angular momentum $\mathbf{J} = \sum_{\alpha=1}^{n} \mathbf{J}(\alpha)$, recalling that the space $\mathscr{V}_k^{(n)}$ is also the carrier space of the representation

$$U \to \underbrace{D^k(U) \otimes D^k(U) \otimes \cdots \otimes D^k(U)}_{n \text{ times}} = D(U) \quad \text{(A-6)}$$

of $SU(2)$. Since $[P, \mathbf{J}] = \mathbf{0}$, each $P \in S_n$, it follows that P also commutes with the

[1] Weyl's book [13] (see, particularly, Chapter V) is the standard work developing the consequences of the Pauli principle for the classification of terms in atomic spectra by using general techniques from symmetric group theory (group algebra, Young tableaux, Young symmetry operators).

rotation operator $\mathscr{U} = \exp[-i\psi\hat{n}\cdot\mathbf{J}]$. Accordingly, the matrices $D(P)$ and $D(U)$ commute for each $P \in S_n$ and each $U \in SU(2)$.

The space $\mathscr{V}_k^{(n)}$ is thus the carrier space of the representation

$$D(P)D(U) = D(U)D(P) \tag{A-7}$$

of the direct product group $SU(2) \times S_n$. Utilizing the properties of the irreps of S_n introduced in Chapter 5, Appendix A, we are able to conclude that the space $\mathscr{V}_k^{(n)}$ possesses an *orthonormal basis*, which may be labeled by

$$|(\alpha);[\lambda](y);jm\rangle, \tag{A-8}$$

where $[\lambda]$ is a partition of n into n parts, (y) is a Yamanouchi symbol, jm are the $SU(2)$ angular momentum quantum numbers, and (α) denotes some unspecified set of additional labels required to complete the basis of $\mathscr{V}_k^{(n)}$. The basis vectors (A-8) are characterized by their transformation properties under $\mathscr{U}$ (see Chapter 3, Section 6) and P:

$$\mathscr{U}|(\alpha);[\lambda](y);jm\rangle = \sum_{m'} D^j_{m'm}(U)|(\alpha)[\lambda](y);jm'\rangle$$

$$P|(\alpha);[\lambda](y);jm\rangle = \sum_{(y')} D^{[\lambda]}_{(y')(y)}(P)|(\alpha);[\lambda](y');jm\rangle, \tag{A-9}$$

where

$$P \to D^{[\lambda]}(P) \tag{A-10}$$

denotes a (unitary) irrep of S_n.

It is clear that the set of basis vectors

$$\left\{ |(\alpha);[\lambda](y);jm\rangle : \begin{array}{l} m = j, j-1, \ldots, -j; \\ (y) \text{ runs over the Yamanouchi} \\ \text{symbols of the standard Young} \\ \text{tableau of shape } [\lambda] \end{array} \right\} \tag{A-11}$$

spans a carrier space of irrep

$$D^j(U) \otimes D^{[\lambda]}(P) \tag{A-12}$$

of the direct product group $SU(2) \times S_n$.

Unfortunately, for arbitrary n and k, the construction of the basis vectors (A-8) has never been given (fully explicitly), principally because of unsolved problems relating to the additional labels (α), which are required to specify a basis of the space $\mathscr{V}_k^{(n)}$, and which imply a multiplicity of occurrence of the irrep (A-12) of $SU(2) \times S_n$ in the representation (A-7). However, *for the case* $k = \frac{1}{2}$, it may be proved, using standard character formulas (Weyl [13]), that *the representation* (A-7) *contains no multiply occurring irreps of* $SU(2) \times S_n$. It is this case to which we now direct our attention (see also Hamermesh [32], Landau and Lifshitz [71]).

For the case of the coupling of n spin-$\frac{1}{2}$ angular momenta, the indices (α) are not required in the notation (A-8) for the basis vectors. Furthermore, the allowed partitions $[\lambda]$ of n into n parts now assume a very special form, related to j itself (Weyl [13, p. 370]):

$$[\lambda] = \left[\frac{n}{2}+j \quad \frac{n}{2}-j \quad 0\cdots 0\right], \tag{A-13}$$

which corresponds to the two-rowed Young frame

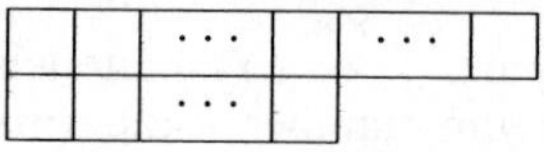

$\frac{n}{2}+j$ boxes
$\frac{n}{2}-j$ boxes. (A-14)

Finally, the values that j may assume are those in the set

$$\left\{\frac{n}{2}, \frac{n}{2}-1, \ldots, \frac{1}{2} \text{ or } 0\right\}, \tag{A-15}$$

each value exactly once. The occurrence of only these j-values may be derived by elementary means by using the $SU(2)$ Clebsch–Gordan series applied repeatedly to n pairwise couplings of $k = \frac{1}{2}$.

The space $\mathscr{V}_{\frac{1}{2}}^{(n)}$ thus splits into the direct sum

$$\mathscr{V}_{\frac{1}{2}}^{(n)} = \sum_j \oplus \mathscr{V}\left[\frac{n}{2}+j \quad \frac{n}{2}-j\right], \tag{A-16}$$

where the sum on j is over those values in the set (A-15). Here $\mathscr{V}[\frac{n}{2}+j\ \frac{n}{2}-j]$ denotes the carrier space of irrep

$$D^j(U) \otimes D^{[\frac{n}{2}+j\ \frac{n}{2}-j]}(P) \tag{A-17}$$

of $SU(2) \times S_n$. The dimension of the irrep $[\lambda] = [\frac{n}{2}+j\ \frac{n}{2}-j]$ of S_n is given by

$$\dim\left[\frac{n}{2}+j \quad \frac{n}{2}-j\right] = \frac{2(2j+1)}{2j+2+n}\binom{n}{\frac{n}{2}-j}, \tag{A-18}$$

so that the dimension of the space $\mathscr{V}[\frac{n}{2}+j\ \frac{n}{2}-j]$ is

$$\dim \mathscr{V}\left[\frac{n}{2}+j \quad \frac{n}{2}-j\right] = (2j+1)\dim\left[\frac{n}{2}+j \quad \frac{n}{2}-j\right]. \tag{A-19}$$

Equation (A-16) thus implies the identity

$$\sum_j (2j+1)\dim\begin{bmatrix}\frac{n}{2}+j & \frac{n}{2}-j\end{bmatrix} = 2^n. \tag{A-20}$$

We may introduce various notations for the orthonormal basis vectors of the space $\mathscr{V}[\frac{n}{2}+j\ \frac{n}{2}-j]$, which reflect in various degrees the group-theoretic structure of the basis. The most succinct notation is that given by (A-8), where the extra (α) is no longer required and we may drop $[\lambda]$ as well, since n is implicit anyway and j already occurs. Although we shall use this notation below (for typographical reasons), there are several other notations that exhibit more clearly the underlying group structures. Indeed, the fact that the shape of the frame is related to j through (A-14) has a deep and significant origin that we shall now indicate.

The Young frame (A-14) plays a double role. We have seen in Appendix A, Chapter 5, that, when it is filled in with $1, 2, \ldots, n$ in the standard way, we obtain Young tableaux that may be used to enumerate the basis vectors in the carrier space for an irrep of S_n. We have also seen that the frame may be filled in with 1's and 2's to yield a set of standard Weyl tableaux that may be used to enumerate basis vectors in a carrier space of an irrep of $U(2)$.

The interesting point is that the *shape of the Young frame is the same* for the S_n irreps and the $SU(2)$ irreps occurring in the direct product (A-12).

An appropriate notation for basis vectors, which displays explicitly the structural feature noted above, is

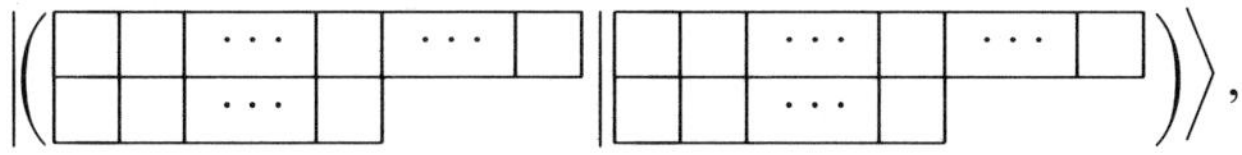

(A-21)

where the two frames are of the same shape $[\frac{n}{2}+j\ \frac{n}{2}-j]$. We then fill in the frame to the right with $1, 2, \ldots, n$ to obtain a standard Young tableau of S_n; similarly, we fill in the frame to the left with 1's and 2's to obtain a standard Weyl tableau for $SU(2)$. In this manner, one enumerates a basis for the space $\mathscr{V}[\frac{n}{2}+j\ \frac{n}{2}-j]$.

The double Gel'fand pattern (Appendix A, Chapter 5) notation for the state vector corresponding to (A-21) is

$$\left|\left(\begin{array}{ccccccc} & & & (m') & & & \\ \frac{n}{2}+j & \frac{n}{2}-j & & 0 & & \cdots & 0 \\ & \ddots & & \ddots & & \ddots & \\ & \frac{n}{2}+j & & \frac{n}{2}-j & & 0 & \\ & & \frac{n}{2}+j & & \frac{n}{2}-j & & \\ & & & \frac{n}{2}+m & & & \end{array}\right)\right\rangle \tag{A-22}$$

in which (m') is an array with weight $(1, 1, \ldots, 1)$; that is, the entries in row j must add to j (and, of course, satisfy betweenness).

The notation (A-22), although very explicit in showing group–subgroup structures, is highly redundant. We have displayed it because it illustrates nicely how special results may be obtained from the very general results of general unitary group theory. Here the relevant subgroup structure is $SU(2) * S_n \subset U(n) * U(n)$.

As indicated above, one may obtain the explicit spin states by specializing known results from general unitary group representation theory (see Ref. [72]). Using this procedure here would, however, make the result needlessly inaccessible. Since the final result itself is easily understood in terms of the matrix elements of the fundamental $SU(2)$ Wigner operators, $\left\langle \begin{matrix} & i & \\ 1 & & 0 \\ & k & \end{matrix} \right\rangle$, $i, k = 1, 2$, we shall simply state the result and prove it directly, using properties of the fundamental Wigner operators themselves (see Chapter 3, Section 21).

The orthonormalized coupled spin-j states for n particles of spin-$\frac{1}{2}$ are given by

$$|(i_1 i_2 \cdots i_n); jm\rangle$$

$$= \sum_{k_1 k_2 \cdots k_n} \left\langle \begin{matrix} 2j & 0 \\ j+m & \end{matrix} \right| \left\langle \begin{matrix} & i_i & \\ 1 & & 0 \\ & k_n & \end{matrix} \right\rangle \left\langle \begin{matrix} & i_2 & \\ 1 & & 0 \\ & k_{n-1} & \end{matrix} \right\rangle \cdots \left\langle \begin{matrix} & i_n & \\ 1 & & 0 \\ & k_1 & \end{matrix} \right\rangle \left| \begin{matrix} 0 & 0 \\ & 0 \end{matrix} \right\rangle$$

$$\times \left| \begin{matrix} 1 & 0 \\ & k_1 \end{matrix} \right\rangle \otimes \left| \begin{matrix} 1 & 0 \\ & k_2 \end{matrix} \right\rangle \otimes \cdots \otimes \left| \begin{matrix} 1 & 0 \\ & k_n \end{matrix} \right\rangle, \tag{A-23}$$

where the notations have the following meanings:

(i) Each index i_s and k_s ($s = 1, 2, \ldots, n$) may assume the value 0 or 1.

(ii)
$$\left| \begin{matrix} 1 & 0 \\ & 1 \end{matrix} \right\rangle = \left| \frac{1}{2}, \frac{1}{2} \right\rangle, \quad \left| \begin{matrix} 1 & 0 \\ & 0 \end{matrix} \right\rangle = \left| \frac{1}{2}, -\frac{1}{2} \right\rangle. \tag{A-24}$$

(iii) The sum $k_1 k_2 \cdots k_n$ is over all sets containing $(n/2) + m$ 1's and $(n/2) - m$ 0's, but since the coefficient will automatically vanish when these constraints are violated, we may ignore these conditions.

Observe that each i_s in $(i_1 i_2 \cdots i_n)$ may assume the value 0 or 1. However, one may verify, using the shift action of a Wigner operator on a state vector [see Eq. (3.341), Chapter 3] that

$$\left\langle \begin{matrix} & i_1 & \\ 1 & & 0 \\ & \cdot & \end{matrix} \right\rangle \left\langle \begin{matrix} & i_2 & \\ 1 & & 0 \\ & \cdot & \end{matrix} \right\rangle \cdots \left\langle \begin{matrix} & i_n & \\ 1 & & 0 \\ & \cdot & \end{matrix} \right\rangle \left| \begin{matrix} 0 & 0 \\ & 0 \end{matrix} \right\rangle = 0 \tag{A-25}$$

unless $(i_1 i_2 \cdots i_n)$ is a *lattice permutation* of type

$$0^{\frac{n}{2}-j} 1^{\frac{n}{2}+j}. \tag{A-26}$$

Equivalently, $\left|\begin{matrix}0 & 0\\ & 0\end{matrix}\right\rangle$ belongs to the null space of the string of fundamental Wigner operators occurring in (A-25) unless $(i_1 i_2 \cdots i_n)$ is a lattice permutation of type (A-26). Thus, the vectors defined by Eq. (A-23) are zero unless $(i_1 i_2 \cdots i_n)$ is a lattice permutation of type (A-26), and these permutations are just the Yamanouchi symbols (change the 0's to 2's) of S_n for the Young frame $[\frac{n}{2}+j\ \frac{n}{2}-j]$. In terms of *operator patterns* of strings of fundamental Wigner coefficients and null space, we thus find a very natural interpretation of lattice permutations and Yamanouchi symbols [this result generalizes to any Young frame (see Ref. [72])]. Let us now give the proof of Eq. (A-23).

Proof. The fact that the basis vectors (A-23) comprise all the states of sharp angular momentum may be given simply by demonstrating the orthonormality of the vectors:

$$\langle (i'_1 i'_2 \cdots i'_n); j'm' \,|(i_1 i_2 \cdots i_n); jm\rangle = \delta_{j'j}\,\delta_{m'm} \prod_{s=1}^{n} \delta_{i'_s i_s}, \tag{A-27}$$

where $(i'_1 i'_2 \cdots i'_n)$ and $(i_1 i_2 \cdots i_n)$ are lattice permutations of type $0^{\frac{n}{2}-j'}\, 1^{\frac{n}{2}+j'}$ and $0^{\frac{n}{2}-j}\, 1^{\frac{n}{2}+j}$, respectively. Since we already know that the counting over m, over j, and over lattice permutations yields 2^n vectors, we shall, by demonstrating perpendicularity, have established that the vectors constitute a basis of the 2^n-dimensional space $\mathscr{V}^{(n)}_{\frac{1}{2}}$.

Using the orthonormality of the product of fundamental state vectors appearing in the right-hand side of Eq. (A-23), we find

$$\langle i'_1 i'_2 \cdots i'_n); j'm' |(i_1 i_2 \cdots i_n); jm\rangle$$

$$= \sum_{k_1 \cdots k_n} \left\langle \begin{matrix} 2j' & 0 \\ j'+m' & \end{matrix} \right| \left\langle \begin{matrix} & i'_1 & \\ 1 & & 0 \\ & k_n & \end{matrix} \right\rangle \cdots \left\langle \begin{matrix} & i'_n & \\ 1 & & 0 \\ & k_1 & \end{matrix} \right\rangle \left| \begin{matrix} 0 & 0 \\ & 0 \end{matrix} \right\rangle$$

$$\times \left\langle \begin{matrix} 2j & 0 \\ j+m & \end{matrix} \right| \left\langle \begin{matrix} & i_1 & \\ 1 & & 0 \\ & k_n & \end{matrix} \right\rangle \cdots \left\langle \begin{matrix} & i_n & \\ 1 & & 0 \\ & k_1 & \end{matrix} \right\rangle \left| \begin{matrix} 0 & 0 \\ & 0 \end{matrix} \right\rangle$$

$$= \left\langle \begin{matrix} 2j' & 0 \\ j'+m' & \end{matrix} \right| \sum_{k_1 \cdots k_n} \left\langle \begin{matrix} & i'_1 & \\ 1 & & 0 \\ & k_n & \end{matrix} \right\rangle \cdots \left\langle \begin{matrix} & i'_n & \\ 1 & & 0 \\ & k_1 & \end{matrix} \right\rangle \left\langle \begin{matrix} & i'_n & \\ 1 & & 0 \\ & k_1 & \end{matrix} \right\rangle^{\dagger} \cdots \left\langle \begin{matrix} & i_1 & \\ 1 & & 0 \\ & k_n & \end{matrix} \right\rangle^{\dagger} \left| \begin{matrix} 2j & 0 \\ j+m & \end{matrix} \right\rangle$$

$$= \delta_{j'j}\,\delta_{m'm}\ (\text{null space factor}) \prod_{s=1}^{n} \delta_{i'_s i_s},$$

where we have used the orthonormality of the spin-$\frac{1}{2}$ Wigner operators given by Eq. (3.343), Chapter 3. Finally, one sees that, when $(i'_1 i'_2 \cdots i'_n)$ and $(i_1 i_2 \cdots i_n)$ are lattice permutations, the null space factor is unity. ■

One may verify the following useful properties of Eq. (A-23) for the basis vectors: (*i*) The action of the components J_i ($i = 1, 2, 3$) of the total angular momentum on these basis vectors, which is defined as taking place on the vectors in the tensor product space, may be replaced by the action of J_i on the final state vector $\left|\begin{matrix} 2j & & 0 \\ & j+m & \end{matrix}\right\rangle$ appearing in the matrix element; and (*ii*) the action (A-3) of the permutation P: $1 \to i_1, 2 \to i_2, \ldots, n \to i_n$ on the indices $1, 2, \ldots, n$ appearing as subscripts on the angular momenta j_i may be replaced by the permutation $k_1 \to k_{i_1}, k_2 \to k_{i_2}, \ldots, k_n \to k_{i_n}$ applied to the $k_1, k_2, \ldots, k_n$ *appearing in the fundamental Wigner operators only.*

These two results allow us to conclude that (*i*) the action of J_3, $J_\pm$ on the basis vectors is the standard action, and under the unitary transformation $\mathscr{U}$ we have the standard transformation

$$\mathscr{U}|(i_1 i_2 \cdots i_n); jm\rangle = \sum_{m'} D^j_{m'm}(U)|(i_1 i_2 \cdots i_n); jm'\rangle; \tag{A-28}$$

and (*ii*) the matrix representation of P: $1 \to i_1, 2 \to i_2, \ldots, n \to i_n$ on the basis (A-22) is given by

$$\begin{aligned} &D^{[\frac{n}{2}+j\ \frac{n}{2}-j]}_{(i_1' i_2' \cdots i_n'),(i_1 i_2 \cdots i_n)}(P) \\ &= \langle (i_1' i_2' \cdots i_n'); jm|P|(i_1 i_2 \cdots i_n); jm\rangle \\ &= \left\langle \begin{matrix} 2j & & 0 \\ & j+m & \end{matrix}\right| \sum_{k_1 \cdots k_n} \left\langle \begin{matrix} & i_1' & \\ 1 & & 0 \\ & k_{i_n} & \end{matrix}\right\rangle \cdots \left\langle \begin{matrix} & i_n' & \\ 1 & & 0 \\ & k_{i_1} & \end{matrix}\right\rangle \\ &\times \left\langle \begin{matrix} & i_1 & \\ 1 & & 0 \\ & k_1 & \end{matrix}\right\rangle^\dagger \cdots \left\langle \begin{matrix} & i_n & \\ 1 & & 0 \\ & k_n & \end{matrix}\right\rangle^\dagger \left|\begin{matrix} 2j & & 0 \\ & j+m & \end{matrix}\right\rangle. \end{aligned} \tag{A-29}$$

This latter result is an explicit expression for the elements of the matrix representation of S_n carried by the space $\mathscr{V}\,[\frac{n}{2}+j\ \frac{n}{2}-j]$, given in terms of the matrix elements of products of fundamental Wigner coefficients (this result generalizes to all irreps of S_n; see Refs. [73] and [74]).

Remarks. (*a*) The rules of the pattern calculus (Ref. [31]) provide the tool for evaluating the matrix elements that appear in Eqs. (A-23) and (A-29). (*b*) The boson realization, Eq. (7.5.129), of the spin states (A-23) is obtained by making the mapping

$$|\tfrac{1}{2}, \tfrac{3}{2} - \sigma_1\rangle \otimes \cdots \otimes |\tfrac{1}{2}, \tfrac{3}{2} - \sigma_n\rangle \to b^1_{\sigma_1} b^2_{\sigma_2} \cdots b^n_{\sigma_n}|0\rangle, \qquad \sigma_i = 1, 2. \tag{A-30}$$

References

1. A. Fowler, *Report on Series in Line Spectra.* Fleetway, London, 1922.
2. F. Paschen and R. Götze, *Seriengesetze der Linienspektren.* Springer, Berlin, 1922.
3. E. Back and A. Landé, *Zeemaneffekt und Multiplettstruktur der Spektrallinien.* Springer, Berlin, 1925.

4. J. H. Van Vleck, *Quantum Principles of Line Spectra*. National Research Council Bulletin 54, Washington, 1926.
5. A. E. Ruark and H. C. Urey, *Atoms, Molecules, and Quanta*. McGraw-Hill, New York, 1930.
6. L. Pauling and S. Goudsmit, *The Structure of Line Spectra*. McGraw-Hill, New York, 1930.
7. H. E. White, *Introduction to Atomic Spectra*. McGraw-Hill, New York, 1934.
8. A. Sommerfeld, *Atombau und Spektrallinien* II, Vieweg, Braunschweig, 1939.
9. J. C. Slater, *Quantum Theory of Atomic Spectra*, Vols. I, II. McGraw-Hill, New York, 1960.
10. E. U. Condon and G. H. Shortley, *The Theory of Atomic Spectra*. Cambridge Univ. Press, London, first printed 1935; reprinted 1953.
10a. E. U. Condon and H. Odabaşi, *Atomic Spectra*. Cambridge Univ. Press, New York, 1980.
11. W. Heisenberg, "Über quantentheoretische Umdeutung kinematischer und mechanischer Beziehungen," *Z. Physik* **33** (1925), 879–893.
12. E. P. Wigner, *Group Theory and Its Application to the Quantum Mechanics of Atomic Spectra*. Academic Press, New York, 1959. Translation by J. J. Griffin of the 1931 German edition.
13. H. Weyl, *Gruppentheorie und Quantenmechanik*. Hirzel, Leipzig, 1st ed., 1928; 2nd ed., 1931. Translated as *The Theory of Groups and Quantum Mechanics*, by H. P. Robertson, Methuen, London, 1931; reissued by Dover, New York, 1949.
14. C. Eckart, "The application of group theory to the quantum dynamics of monatomic systems," *Rev. Mod. Phys.* **2** (1930), 305–380.
15. G. Racah, "Theory of complex spectra I," *Phys. Rev.* **61** (1942), 186–197; II, *ibid.* **62** (1942), 438–462; III, *ibid.* **63** (1943), 368–382; IV, *ibid.* **76** (1949), 1352–1365.
16. G. Racah, *Group Theory and Spectroscopy*. Lectures at the Institute for Advanced Study, Princeton, N. J., 1951; published in *Ergeb. Exakt. Naturw.* **37** (1965), 28–84.
17. J. C. Slater, "The theory of complex spectra," *Phys. Rev.* **34** (1929), 1293–1322.
18. B. R. Judd, *Operator Techniques in Atomic Spectroscopy*. McGraw-Hill, New York, 1963.
19. I. I. Sobel'man, *Introduction to the Theory of Atomic Spectra*. Pergamon, New York, 1972; Russian edition, 1963.
20. A. de-Shalit and I. Talmi, *Nuclear Shell Theory*. Academic Press, New York, 1963.
21. E. A. Hylleraas, "Über den Grundzustand des Heliumatoms," *Z. Physik* **48** (1928), 469–494; "Neue Berechnung der Energie des Heliums im Grundzustande, sowie des tiefsten Terms von Ortho-Helium," *ibid.* **54** (1929), 347–366; "Über den Grundterm der Zweielektronenprobleme von H^-, He, Li^+, Be^{++} usw.," *ibid.* **65** (1930), 209–225.
22. T. Kinoshita, "Ground state of the helium atom," *Phys. Rev.* **105** (1957), 1490–1502.
23. C. L. Pekeris, "Ground state of two-electron atoms," *Phys. Rev.* **112** (1958), 1649–1658.
24. L. H. Thomas, "The calculation of atomic fields," *Proc. Cambridge Phil. Soc.* **23** (1927), 542–548.
25. E. Fermi, "Eine statistische Methode zur Bestimmung einiger Eigenschaften des Atoms und ihrer Anwendung auf die Theorie des periodischen Systems der Elemente," *Z. Physik* **48** (1928), 73–79.
26. D. R. Hartree, "The wave mechanics of an atom with a non-Coulomb central field. Part I. Theory and methods," *Proc. Cambridge Phil. Soc.* **24** (1928), 89–110; "Part II. Some results and discussion," *ibid.*, 111–132.
27. V. Fock, "Näherungsmethode zur Lösung des quantenmechanischen Mehrkörperproblems," *Z. Physik* **61** (1930), 126–148; ""Selfconsistent field" mit Austausch für Natrium," *ibid.* **62** (1930), 795–805.
27a. J. H. D. Jensen and J. M. Luttinger, "Angular momentum distributions in the Thomas–Fermi model," *Phys. Rev.* **86** (1952), 907–910.
28. A. Landé, "Termstruktur und Zeemaneffekt der Multipletts," *Z. Physik* **15** (1923), 189–205.
29. N. M. Gray and L. A. Wills, "Note on the calculation of zero order eigenfunctions," *Phys. Rev.* **38** (1931), 248–254.

30. G. E. Baird and L. C. Biedenharn, "On the representations of the semi-simple Lie groups," *J. Math. Phys.* **4** (1963), 1449–1466.
31. L. C. Biedenharn and J. D. Louck, "A pattern calculus for tensor operators in the unitary groups," *Commun. Math. Phys.* **8** (1968), 89–131.
32. M. Hamermesh, *Group Theory and Its Applications to Physical Problems.* Addison-Wesley, Reading, Mass., 1962.
33. W. A. Goddard, "Improved quantum theory of many-electron systems. I. Construction of eigenfunctions of S^2 which satisfy Pauli's principle," *Phys. Rev.* **157** (1967), 73–80;" II. The basic method," *ibid.*, 81–93.
34. P. Kramer, "Recoupling coefficients of the symmetric group for shell and cluster model configurations," *Z. Physik* **216** (1968), 68–83.
35. P. Kramer, "Irreducible representations of the semidirect-product group $K_n = A_n : S_n$ and the harmonic-oscillator shell model," *J. Math. Phys.* **4** (1968), 639–649.
36. P. Kramer, "Finite representations of the unitary group and their applications in many body physics," in *Group Theoretical Methods in Physics: Proceedings of the Fifth International Colloquium* (R. T. Sharp and B. Kolman, eds.), pp. 173–179. Academic Press, New York, 1977.
37. P. Kramer and T. H. Seligman, "Studies in the nuclear cluster model. I. Matrix elements in the supermultiplet scheme," *Nucl. Phys.* **A123** (1969), 161–172; "II. Two-cluster configurations," **A136** (1969), 545–563; "III. The *k*-cluster configuration," **A186** (1972), 49–64.
38. T. H. Seligman, *Double Coset Decompositions of Finite Groups and the Many Body Problem*, Burg Monographs in Science (D. Clement, ed.). Burg, Basel, 1975.
39. F. A. Matsen, "Spin-free quantum chemistry," *Advan. Quantum Chem.* **1** (1964), 59–114; "XVIII. The unitary group formulation of the many-electron problem," *Int. J. Quantum Chem.* **10** (1976), 525–544.
40. F. A. Matsen, D. J. Klein, and D. C. Foyt, "Spin-free quantum chemistry. X. The effective spin Hamiltonian," *J. Phys. Chem.* **75** (1971), 1866–1877.
41. D. J. Klein and B. R. Junker, "Spin-free computation of matrix elements. I. Group-theoretical computation of Pauling numbers," *J. Chem. Phys.* **54** (1971), 4290–4296.
42. B. R. Junker and D. J. Klein, "Spin-free computation of matrix elements. II. Simplifications due to invariance groups," *J. Chem. Phys.* **55** (1971), 5532–5542.
43. C. W. Patterson and W. G. Harter, "Canonical symmetrization for unitary bases. I. Canonical Weyl bases," *J. Math. Phys.* **17** (1976), 1125–1136; "II. Boson and fermion bases," *ibid.*, **17** (1976), 1137–1142.
44. J. Paldus, "Many-electron correlation problem. A group theoretical approach," in *Theoretical Chemistry: Advances and Perspectives* (H. Eyring and D. J. Henderson, eds.), Vol. 2, pp. 131–290. Academic Press, New York, 1976.
45. J. Paldus, "Group theoretical approach to the configuration interaction and perturbation theory calculations for atomic and molecular systems," *J. Chem. Phys.* **61** (1974), 5321–5330; "Unitary-group approach to the many-electron correlation problem: Relation of Gelfand and Weyl tableau formulations," *Phys. Rev.* **A14** (1976), 1620–1625.
46. J. D. Louck, "Application of the boson polynomials of $U(n)$ to physical problems," in *Group Theoretical Methods in Physics: Proceedings of the Seventh International Colloquium* (W. Beiglböck, A. Bohm, and E. Takasugi, eds.), pp. 39–50. Springer, Berlin, 1979.
47. B. G. Wybourne, *Symmetry Principles and Atomic Spectroscopy.* Wiley (Interscience), New York, 1970.
48. C. L. B. Shudeman, "Equivalent electrons and their spectroscopic terms," *J. Franklin Inst.* **224** (1937), 501–518.
49. L. Armstrong, Jr., and B. R. Judd, "Quasi-particles in atomic shell theory," *Proc. Roy. Soc. London* **A315** (1970), 27–37.

50. V. Bargmann and M. Moshinsky, "Group theory of harmonic oscillators (II). The integrals of motion for the quadrupole-quadrupole interaction," *Nucl. Phys.* **23** (1961), 177–199.
51. M. Moshinsky and V. Syamala Devi, "General approach to fractional parentage coefficients," *J. Math. Phys.* **10** (1969), 455–466.
52. V. Syamala Devi, "Bases for irreducible representations of the unitary group in the symplectic group chain," *J. Math. Phys.* **11** (1970), 162–168.
53. B. G. Wybourne, "Hermite's reciprocity law and the angular-momentum states of equivalent particle configurations," *J. Math. Phys.* **10** (1969), 467–471.
54. B. R. Judd, W. Miller, Jr., J. Patera, and P. Winternitz, "Complete sets of commuting operators and $O(3)$ scalars in the enveloping algebra of $SU(3)$," *J. Math. Phys.* **15** (1974), 1787–1799.
55. E. Chacón, M. Moshinsky, and R. T. Sharp, "$U(5) \supset O(5) \supset O(3)$ and the exact solution for the problem of quadrupole vibrations of the nucleus," *J. Math. Phys.* **17** (1976), 668–676.
55a. P. Kramer, G. John, and D. Schenzle, *Group Theory and the Interaction of Composite Nucleon Systems.* Vieweg, Braunschweig, 1980.
55b. J. Patera and R. T. Sharp, "Generating functions for plethysms of finite and continuous groups," *J. Phys. A: Math. Gen.* **13** (1980), 397–416.
56. J. D. Louck, "The state vector labelling problem: A review of structural principles," in *The Proceedings of the International Symposium on Mathematical Physics*, Vol. 1, pp. 121–154. Mexico City, 1976.
57. B. R. Judd, *Second Quantization and Atomic Spectroscopy.* Johns Hopkins Press, Baltimore, 1967.
58. R. D. Cowan, *The Theory of Atomic Structure and Spectra*, Univ. of California Press, Berkeley, 1981.
59. M. Weissbluth, *Atoms and Molecules.* Academic Press, New York, 1978.
60. W. G. Harter, "Alternative basis for the theory of complex spectra. I," *Phys. Rev.* **A8** (1973), 2819–2827.
61. W. G. Harter and C. W. Patterson, "Alternative basis for the theory of complex spectra. II," *Phys. Rev.* **13** (1976), 1067–1082.
62. W. G. Harter and C. W. Patterson, *A Unitary Calculus for Electronic Orbitals.* Springer, Berlin, 1976.
63. I. Shavitt, "Group theoretical concepts for the unitary group approach to the many-electron correlation problem" *Int. J. Quantum Chem.* **S11** (1977), 131–148.
64. F. A. Matsen, "The unitary group and the many-body problem," *Advan. Quantum Chem.* **11** (1978), 223–250.
65. J. Drake, G. W. F. Drake, and M. Schlesinger, "Spin-orbit parameters by the Gelfand–Harter method – a test calculation," *Phys. Rev.* **A15** (1977), 807–809.
66. G. W. F. Drake and M. Schlesinger, "Vector-coupling approach to orbital and spin-dependent tableau matrix elements in the theory of complex spectra," *Phys. Rev.* **A15** (1977), 1990–1999.
67. P.-O. Löwdin, "Quantum theory of many-particle systems. III. Extension of the Hartree–Fock scheme to include degenerate systems and correlation effects," *Phys. Rev.* **91** (1955), 1509–1520.
68. R. Pauncz, *Alternant Molecular Orbital Method.* Saunders, Philadelphia, 1967.
69. A. Rotenberg, "Calculation of exact eigenfunctions of spin and orbital angular momentum using the projection operator method," *J. Chem. Phys.* **39** (1963), 512–517.
70. J. S. Murty and C. R. Sarma, "A method for the construction of orthogonal spin eigenfunctions," *Int. J. Quantum Chem.* **9** (1975), 1097–1107.
71. L. D. Landau and E. M. Lifshitz, *Quantum Mechanics. Nonrelativistic Theory.* Addison-Wesley, Reading, Mass. 1958. Translated from the Russian by J. B. Sykes and J. S. Bell.

72. J. D. Louck and L. C. Biedenharn, "The permutation group and the coupling of n spin-$\frac{1}{2}$ angular momenta," in *The Permutation Group and Its Applications in Physics and Chemistry* (J. Hinze, ed.), pp. 121–147. *Lecture Notes in Chemistry* **12**, Springer, Berlin, 1979.
73. J. D. Louck and L. C. Biedenharn, "The structure of the canonical tensor operators in the unitary groups. III. Further developments of the boson polynomials and their implications," *J. Math. Phys.* **14** (1973), 1336–1357.
74. L. C. Biedenharn and J. D. Louck, "Representations of the symmetric group as special cases of the boson polynomials in $U(n)$," in *The Permutation Group and Its Applications in Physics and Chemistry* (J. Hinze, ed.), pp. 148–163. *Lecture Notes in Chemistry* **12**, Springer, Berlin, 1979.

6. Electromagnetic Processes

a. Preliminary remarks. The purpose of the present section is to illustrate the application of angular momentum techniques to electromagnetic phenomena; our aim is, as before, directed toward illustrating concepts by discussing explicit examples. This poses something of a problem in considering electromagnetic processes; the quantal theory of electromagnetism (quantum electrodynamics, QED) is one of the triumphs of theoretical physics, but this very fact means that any synoptic discussion will miss the mark. Accordingly, we shall limit our discussion to a consideration only of first-order, one-quantum processes for which the field-theoretic aspects are inessential (aside from justifying the treatment to be used for spontaneous and induced emission). This limitation is of little consequence for applications in nuclear and hadronic processes; but the limitation can be serious for the specialized discipline of nonlinear optics (laser phenomena).

Angular momentum techniques are of essential importance in the treatment of first-order (one-quantum) electromagnetic processes. Such a treatment is based on a quantal interpretation of Maxwell's equations, which considers the electromagnetic field not so much as field operators (which create and destroy photons in a field-theoretic treatment) but as heuristic analogs to wave functions (probability amplitudes) for the photon (see Ferretti [1]). Maxwell's equations are Poincaré covariant,[1] and invariant (so far as is presently known: Brodsky [4]) to both time reversal and space reflection (parity). These equations identify the photon to be a massless particle of unit intrinsic spin ($S = 1$). For zero-mass particles the helicity (intrinsic spin projection along the direction of motion: $\mathbf{S} \cdot \hat{P}$) is a Poincaré invariant with just two values, $\pm S$ (for spin $S > 0$ and assuming reflection invariance).

It follows that angular momentum constructs for the free electromagnetic field belong to an absolutely valid symmetry; accordingly, the photon field may itself be expanded in sharp angular momentum states. These states are the

[1] Maxwell's equations without sources are covariant under a larger group, the 15-parameter conformal group (Cunningham [2], Bateman [3]).

multipole fields for the Maxwell equation; it is in the analysis of the multipole fields that angular momentum techniques make their principal contribution.

These consequences of Poincaré group symmetry apply strictly to the free photon field, but it is a further consequence of the Poincaré symmetry (for a massless spin-1 field) that the potentials (see below) are necessarily gauge-invariant, and if there is an interaction with a current, then, to be consistent, the current must be conserved (van Dam [5]). This implies the form of the interaction between the photon field and the sources.

The method of multipoles assumes particular importance in quantum physics.[1] The reason for this importance is that the sources of the electromagnetic field (atomic, nuclear, and hadronic systems) have quantized angular momentum states; the nonvanishing multipole moments [see Eq. (7.6.19), below] for these systems are accordingly *finite in number*. Considerations on the *size* of the nonvanishing moments further limits, in practice, the multipole expansion for the electromagnetic fields, so that the method of multipoles becomes one of decisive utility. (Let us note that angular momentum techniques also enter into the treatment of the polarization properties of the photon field; this is discussed in Section *l*, below.)

b. Multipole radiation.[2] Maxwell's equations for the electromagnetic field (**E**, **B**) in free space—but with sources—take the form[3]

$$\nabla \times \mathbf{E} + \frac{\partial \mathbf{B}}{\partial t} = \mathbf{0}, \qquad \nabla \times \mathbf{B} - \frac{1}{c^2}\frac{\partial \mathbf{E}}{\partial t} = \mu_0 \mathbf{j},$$

$$\nabla \cdot \mathbf{B} = 0, \qquad \nabla \cdot \mathbf{E} = \rho/\varepsilon_0. \tag{7.6.1}$$

These equations imply the continuity equation

$$\nabla \cdot \mathbf{j} + \frac{\partial \rho}{\partial t} = 0 \tag{7.6.2}$$

for the charge and current sources (which we shall interpret shortly as quantum mechanical expectation values).

A first integral of the two source-free equations above introduces the vector potential **A** and the scalar potential ϕ (which together form a Lorentz four-

---

[1] The concept of multipoles, and multipole moments, belongs also to classical physics. Maxwell, in fact, developed an extensive theory of multipole moments (but not based directly upon angular momentum considerations).

[2] See Note 1.

[3] We use rationalized M.K.S. units. For brevity, we let **E** denote $\mathbf{E}(\mathbf{x}, t)$ and $\mathbf{B} = \mathbf{B}(\mathbf{x}, t)$ where the meaning is clear.

vector):

$$\mathbf{B} \equiv \nabla \times \mathbf{A},$$

$$\mathbf{E} \equiv -\frac{\partial \mathbf{A}}{\partial t} - \nabla \phi. \tag{7.6.3}$$

Introducing these relations into the Maxwell equations yields the familiar results

$$\left(\nabla^2 - \frac{1}{c^2}\frac{\partial^2}{\partial t^2}\right)\mathbf{A} = -\mu_0 \mathbf{j},$$

$$\left(\nabla^2 - \frac{1}{c^2}\frac{\partial^2}{\partial t^2}\right)\phi = -\rho/\varepsilon_0, \tag{7.6.4}$$

where we have imposed the Lorentz gauge condition,

$$\nabla \cdot \mathbf{A} + \frac{1}{c^2}\frac{\partial \phi}{\partial t} = 0, \tag{7.6.5}$$

to simplify the form of the equations.

For applications in quantum mechanics (using first-order perturbation) the sources are considered to be in sharp energy states before and after any transition. Accordingly, it is useful to make a Fourier analysis of the equations corresponding to harmonic sources [$\omega = (E_{\text{initial}} - E_{\text{final}})/\hbar$; the wave number k denotes ω/c]. Let us note explicitly that the time variation is taken to be $e^{-i\omega t}$ for positive frequency ω.

This step replaces the electromagnetic potentials ($\mathbf{A}$, ϕ) by their Fourier transforms: $\mathbf{A}(\mathbf{x}, \omega)$ and $\phi(\mathbf{x}, \omega)$; similarly, the sources become $\rho(\mathbf{x}, \omega)$ and $\mathbf{j}(\mathbf{x}, \omega)$. Equations (7.6.4) then take the form of the inhomogeneous scalar and vector Helmholtz equations (Morse [6, 7]):

$$(\nabla^2 + k^2)\mathbf{A}(\mathbf{x}, \omega) = -\mu_0 \mathbf{j}(\mathbf{x}, \omega),$$

$$(\nabla^2 + k^2)\phi(\mathbf{x}, \omega) = -\rho(\mathbf{x}, \omega)/\varepsilon_0. \tag{7.6.6}$$

The solution to these equations, corresponding to the radiation boundary condition (outgoing waves for $r \to \infty$), is, in the standard way, achieved by Green's functions:

$$\mathbf{A}(\mathbf{x}, \omega) = \mu_0 \int d^3\mathbf{x}' \, \mathbf{j}(\mathbf{x}', \omega) \cdot \underset{\sim}{\mathbf{G}}_\omega(\mathbf{x}, \mathbf{x}'),$$

$$\phi(\mathbf{x}, \omega) = \varepsilon_0^{-1} \int d^3\mathbf{x}' \rho(\mathbf{x}', \omega) G_\omega(\mathbf{x}, \mathbf{x}'), \qquad (7.6.7)$$

where the scalar and dyadic Green functions are explicitly given by

$$G_\omega(\mathbf{x}, \mathbf{x}') = e^{ikR}/(4\pi R),$$

$$\underset{\sim}{\mathbf{G}}_\omega(\mathbf{x}, \mathbf{x}') = \underset{\sim}{\mathbf{1}} G_\omega(\mathbf{x}, \mathbf{x}'), \qquad \text{with } R \equiv \|\mathbf{x} - \mathbf{x}'\|. \qquad (7.6.8)$$

To achieve the desired multipole expansion for the potentials $\mathbf{A}$ and ϕ it suffices to obtain a multipole expansion of the corresponding Green function. For the scalar case this introduces the spherical harmonics, and the result is well-known:

$$G_\omega(\mathbf{x}, \mathbf{x}') = ik \sum_{lm} j_l(kr_<) h_l^{\text{out}}(kr_>) Y_{lm}(\hat{x}) Y_{lm}^*(\hat{x}'). \qquad (7.6.9)$$

We use here the standard notation (Morse [6, 7]) for the regular spherical Bessel functions, j_l, and for the outgoing spherical Hankel function, h_l^{out}. [The notation $r_<$ denotes the lesser of the two radii (r, r') and $r_>$ the greater. The notation $\hat{x}$ denotes the unit vector in the direction of $\mathbf{x}$; that is, $\hat{x} = \mathbf{x}/r$.]

The analogous step for the dyadic Green function is achieved by means of the vector spherical harmonics (Chapter 6, Sections 9–13). This yields[1]

$$\underset{\sim}{\mathbf{G}}_\omega(\mathbf{x}, \mathbf{x}') = ik \sum_{jlm} j_l(kr_<) h_l^{\text{out}}(kr_>) \mathbf{Y}_{ljm}(\hat{x}) \mathbf{Y}_{ljm}^*(\hat{x}'). \qquad (7.6.10)$$

This result, although correct, is not in the most convenient form for applications. The reason is that the physically useful multipole solutions are based upon particular divergence and curl properties (which can mix orbital angular momenta, unlike the vector harmonics, each of which involves a single orbital angular momentum). The desired form for the multipole solutions is that of the Hansen solutions (Hansen [8]), which we now summarize.

c. The Hansen multipole fields. The magnetic multipole fields, $\mathbf{M}_{lm}(\mathbf{x})$, are defined by

$$\mathbf{M}_{lm}(\mathbf{x}) \equiv f_l(kr) \mathbf{Y}_{llm}(\hat{x}), \qquad (7.6.11)$$

[1] Since the only tensor harmonics that occur in this section are vector spherical harmonics, it is convenient to simplify the notation introduced in Chapter 6, Section 9, and define $\mathbf{Y}_{ljm} = \mathbf{Y}^{(l1)jm}$.

where $f_l(kr)$ is a spherical Bessel function (regular or irregular). [These magnetic multipole fields are also called transverse electric (TE) fields, since the radial electric field vanishes.]

The electric multipole fields, $\mathbf{N}_{lm}(\mathbf{x})$, [or transverse magnetic (TM) fields] are defined by

$$\begin{aligned}\mathbf{N}_{lm}(\mathbf{x}) &\equiv -ik^{-1}\nabla \times \mathbf{M}_{lm}(\mathbf{x})\\ &= [(l+1)/(2l+1)]^{\frac{1}{2}} f_{l-1}(kr)\mathbf{Y}_{l-1,l,m}(\hat{x})\\ &\quad - [l/(2l+1)]^{\frac{1}{2}} f_{l+1}(kr)\mathbf{Y}_{l+1,l,m}(\hat{x}).\end{aligned} \tag{7.6.12}$$

In obtaining this last result we have used Eq. (6.110), Chapter 6, for the curl operation as well as the properties of the spherical Bessel functions given by

$$\frac{df_l}{dx} - \frac{l}{x} f_l = -f_{l+1}, \qquad \frac{df_l}{dx} + \frac{l+1}{x} f_l = f_{l-1}. \tag{7.6.13}$$

The longitudinal multipole fields, $\mathbf{L}_{lm}(\mathbf{x})$, which do not occur in free space, since they are eliminated by the helicity constraint $\mathbf{S} \cdot \mathbf{P} \neq 0$, are necessary to complete the vector multipole fields. These fields are given by [see Eqs. (6.109)–(6.111), Chapter 6]

$$\begin{aligned}\mathbf{L}_{lm}(\mathbf{x}) &\equiv -ik^{-1}\nabla(f_l(kr)\, Y_{lm}(\hat{x}))\\ &= -i[l/(2l+1)]^{\frac{1}{2}} f_{l-1}(kr)\, \mathbf{Y}_{l-1,l,m}(\hat{x})\\ &\quad - i[(l+1)/(2l+1)]^{\frac{1}{2}} f_{l+1}(kr)\, \mathbf{Y}_{l+1,l,m}(\hat{x}).\end{aligned} \tag{7.6.14}$$

It follows from the definitions that the vector multipole fields **L**, **M**, and **N** have the divergence and curl properties expressed by [see Eqs. (6.105), Chapter 6]

$$\begin{aligned}&\nabla \cdot \mathbf{L}_{lm} = ikf_l Y_{lm}, && \nabla \times \mathbf{L}_{lm} = \mathbf{0},\\ &\nabla \cdot \mathbf{M}_{lm} = 0, && \nabla \times \mathbf{M}_{lm} = ik\mathbf{N}_{lm},\\ &\nabla \cdot \mathbf{N}_{lm} = 0, && \nabla \times \mathbf{N}_{lm} = -ik\mathbf{M}_{lm}.\end{aligned} \tag{7.6.15}$$

It is helpful (say, for recognizing the vectorial properties of these multipole fields) to express these fields in the more explicit form:

$$\mathbf{M}_{lm}(\mathbf{x}) = [l(l+1)]^{-\frac{1}{2}} f_l(kr) \mathbf{L} Y_{lm}(\hat{x}),$$

$$\mathbf{N}_{lm}(\mathbf{x}) = \frac{[l(l+1)]^{-\frac{1}{2}}}{(kr)} \left\{ l(l+1)\hat{x} f_l(kr) - (i\hat{x}\times\mathbf{L}) \frac{d[(kr)f_l(kr)]}{d(kr)} \right\} Y_{lm}(\hat{x}),$$

$$\mathbf{L}_{lm}(\mathbf{x}) = \left[-i\hat{x}\frac{df_l(kr)}{d(kr)} - (kr)^{-1} f_l(kr)(\hat{x}\times\mathbf{L}) \right] Y_{lm}(\hat{x}). \qquad (7.6.16)$$

The dyadic Green function can now be expressed in terms of the Hansen multipole fields:

$$\underset{\sim}{\mathbf{G}}_\omega(\mathbf{x}, \mathbf{x}') = ik \sum_{lm} \{ {}^{(\mathrm{st})}\mathbf{M}^*_{lm}(\mathbf{x}_<)\, {}^{(\mathrm{out})}\mathbf{M}_{lm}(\mathbf{x}_>)$$

$$+ {}^{(\mathrm{st})}\mathbf{N}^*_{lm}(\mathbf{x}_<)\, {}^{(\mathrm{out})}\mathbf{N}_{lm}(\mathbf{x}_>) + {}^{(\mathrm{st})}\mathbf{L}^*_{lm}(\mathbf{x}_<)\, {}^{(\mathrm{out})}\mathbf{L}_{lm}(\mathbf{x}_>)\}. \qquad (7.6.17)$$

The notations (st) and (out) denote standing waves (regular Bessel function) versus outgoing waves (Hankel function), and $\mathbf{x}_>$ denotes the vector of greater length of the pair $\mathbf{x}, \mathbf{x}'$ and $\mathbf{x}_<$ the smaller.

A formal solution for the multipole potentials radiated by arbitrary harmonic charge and current sources results from introducing Eqs. (7.6.17) and (7.6.9) into the integrals given by Eq. (7.6.7).

d. Classical multipole moments. Let us assume that the sources are limited to a finite spatial extent. Then the radiated multipole fields can be completely described, *outside the sources*, by means of the multipole moments. For the potentials we obtain from Eqs. (7.6.7), (7.6.9), and (7.6.17) the results (outside the sources):

$$\phi(\mathbf{x}, \omega) = (ik/\varepsilon_0) \sum_{lm} p_{lm} h_l^{(\mathrm{out})}(kr) Y_{lm}(\hat{x}),$$

$$\mathbf{A}(\mathbf{x}, \omega) = (ik\mu_0) \sum_{lm} [m_{lm}\, {}^{(\mathrm{out})}\mathbf{M}_{lm}(\mathbf{x}) + n_{lm}\, {}^{(\mathrm{out})}\mathbf{N}_{lm}(\mathbf{x}) + l_{lm}\, {}^{(\mathrm{out})}\mathbf{L}_{lm}(\mathbf{x})], \quad (7.6.18)$$

where we have introduced the *multipole moments* $l_{lm}, m_{lm}, n_{lm}, p_{lm}$ (longitudinal, magnetic, electric, scalar moments, respectively) defined by integrals over the sources:

$$m_{lm} \equiv \int d^3x\, \mathbf{j}(\mathbf{x}, \omega) \cdot {}^{(\mathrm{st})}\mathbf{M}^*_{lm}(\mathbf{x}),$$

$$n_{lm} \equiv \int d^3\mathbf{x}\, \mathbf{j}(\mathbf{x}, \omega) \cdot {}^{(\mathrm{st})}\mathbf{N}^*_{lm}(\mathbf{x}),$$

$$l_{lm} \equiv \int d^3\mathbf{x}\, \mathbf{j}(\mathbf{x}, \omega) \cdot {}^{(\mathrm{st})}\mathbf{L}^*_{lm}(\mathbf{x}),$$

$$p_{lm} \equiv \int d^3\mathbf{x}\, \rho(\mathbf{x}, \omega) j_l(kr) Y^*_{lm}(\hat{x}). \tag{7.6.19}$$

These four types of multipole moments are not independent, but are restricted by the Lorentz gauge condition, Eq. (7.6.5). This results effectively in eliminating the longitudinal moments, since one finds that

$$p_{lm} - (1/c) l_{lm} = 0, \tag{7.6.20}$$

so that only the scalar moments p_{lm} need be retained.

e. Reduction of the electric multipole moments. The electric multipole moments, n_{lm}, defined in Eq. (7.6.19) are not, at first glance, very simply related to the usual (approximate) electric moments defined by an integral over the charge density. The difference stems from the fact that the electric multipole moments, as defined above, are *exact* and take into account both charge and current contributions, as well as retardation effects. To illustrate this explicitly requires considerable technical manipulation, the result of which is to exhibit the electric multipole moment in the equivalent form

$$\begin{aligned} n_{lm} = {} & ic[l(l+1)]^{-\frac{1}{2}} \int d^3\mathbf{x}\, \rho(\mathbf{x}, \omega) Y^*_{lm}(\hat{x}) \frac{d}{dr}[r j_l(kr)] \\ & - k[(l+1)]^{-\frac{1}{2}} \int d^3\mathbf{x}\, \mathbf{x} \cdot \mathbf{j}(\mathbf{x}, \omega) j_l(kr) Y^*_{lm}(\hat{x}). \end{aligned} \tag{7.6.21}$$

The general electric multipole moment is thus seen to consist of two terms: one dependent explicitly upon the oscillating charge density, and the other dependent on the *radial* current density. (The *transverse* current density gives rise to magnetic multipoles.) Only in the long wavelength approximation ($\omega \to 0$) is the electric multipole moment proportional to the usual static moment. In this approximation the radial current contribution is of higher order, and one finds

$$n_{lm} \simeq ic \left(\frac{l+1}{l} \right)^{\frac{1}{2}} [(2l+1)!!]^{-1} k^l \int d^3\mathbf{x}\, \rho r^l Y^*_{lm}(\hat{x}). \tag{7.6.22}$$

The integral appearing in Eq. (7.6.22) is that of the usual (static) electric multipole moment.

f. The radiated multipole fields. The radiated electromagnetic fields can be found from Eq. (7.6.18) for the potentials, using Eqs. (7.6.3) and the divergence and curl properties of the Hansen solutions Eqs. (7.6.15).

For the magnetic field one finds

$$\mathbf{B}(\mathbf{x}, \omega) = -(\mu_0 k^2) \sum_{lm} \{m_{lm} {}^{(\text{out})}\mathbf{N}_{lm}(\mathbf{x}) - n_{lm} {}^{(\text{out})}\mathbf{M}_{lm}(\mathbf{x})\}, \qquad (7.6.23)$$

and for the electric field

$$\mathbf{E}(\mathbf{x}, \omega) = -(k^2 c \mu_0) \sum_{lm} \{n_{lm} {}^{(\text{out})}\mathbf{N}_{lm}(\mathbf{x}) + m_{lm} {}^{(\text{out})}\mathbf{M}_{lm}(\mathbf{x})\}. \qquad (7.6.24)$$

Remarks. (*a*). Outside the sources the electric and magnetic fields can be expressed as a sum of magnetic and electric multipole waves with amplitudes given by the corresponding multipole moments, m_{lm} and n_{lm}. (*Inside* the sources such a simple result does not obtain, for the amplitudes of the various multipole waves vary with position, and, moreover, the longitudinal waves enter in an essential way.)

(*b*) The multipole waves for **E**, **B** that occur outside the sources are of two types, distinguished by the labels "electric multipoles" and "magnetic multipoles." The electric multipoles n_{lm} correspond to transverse magnetic (TM) waves – since $\mathbf{Y}_{llm}$, and hence $\mathbf{M}_{lm}$, is perpendicular to the radius vector $\mathbf{x}$ – and the magnetic multipoles correspond to transverse electric (TE) waves.

(*c*) In a region free of sources, Maxwell's equations possess a symmetry called *duality*: $\mathbf{E} \to c\mathbf{B}$, $c\mathbf{B} \to -\mathbf{E}$; this transformation interchanges electric and magnetic multipoles, $m_{lm} \to n_{lm}$, $n_{lm} \to -m_{lm}$.

g. A curious property of the multipole expansion (*Casimir* [9]). Maxwell's equations in free space do not contain any physical parameter of the dimension of a length. It follows that electromagnetic multipole radiation of given type (electric versus magnetic) and multipole character (lm) – and of given strength – can be generated by sources confined inside a sphere of *arbitrary* radius. (This follows from the above multipole series for **E**, **B** by choosing the charge-current density to have the desired rotational symmetry, so that a given multipole is obtained; the strength of the source can be arbitrarily scaled.)

These considerations lead to several remarkable – and paradoxical appearing! – results, all of which may be deduced from a theorem stated by Casimir [9].

Let $\mathbf{j}(\mathbf{x}, \omega)$ *and* $\rho(\mathbf{x}, \omega)$ *be charge-current density distributions, varying with time as* $\exp(-i\omega t)$, *and vanishing outside a sphere of radius R. Then it is always possible to find another charge-current density* $(\mathbf{j}_1, \rho_1)$ *vanishing outside a radius* $R_1 < R$ *such that the radiation field outside R is identical to that produced by the original sources.*

(The proof is essentially contained in the observation of the previous paragraph using the multipole expansion. The possibility of a converse to the Casimir theorem has been examined by Bosco [9a].)

To give an extreme example applying this theorem, let us quote Casimir [9]: "Suppose an elephant in a spherical cage is illuminated only by coherent-light sources inside the cage. Then the spectators outside the cage cannot be sure that there really is an elephant: the cage might be empty but for a peculiar charge-current density at the center of the cage."

The importance of this theorem can be understood from the fact that is implies that an arbitrarily precise angular directivity can be produced by an arbitrarily small source. This becomes more paradoxical in appearance if one recalls that physicists are in the habit of using the uncertainty principle to estimate the directivity of an antenna by taking the angular deviation, $\Delta\phi$, to be related to the typical dimension of the antenna, D, by

$$\Delta\phi \approx \lambda/2\pi D = (kD)^{-1}. \qquad (7.6.25)$$

This naive use of the uncertainty principle is incorrect, as the theorem shows.

To obtain such high directivity in a small-size antenna ("supergain antenna," Boukamp and de Bruijn [10]) requires a large number of multipole fields of similar order of magnitude. In atomic, nuclear, and hadronic systems the actual multipole moments have very different orders of magnitude so that these considerations are not directly relevant for such domains.

h. The radiated power. To determine the radiated flux of energy, momentum, and angular momentum, one calculates the (time-averaged) Poynting vector, given by $\mathbf{S} = (1/2\mu_0)\,\mathrm{Re}(\mathbf{E} \times \mathbf{B}^*)$, using the asymptotic form of the electromagnetic fields. This is an extremely complicated result, as it must be, since it describes the most general possible angular distribution of radiation from an arbitrary finite source. Although the use of angular momentum techniques is capable of simplifying this general result considerably, the answer is still too complicated to justify including here.

The determination of the radiated power, however, is quite simple. One finds for this[1]

[1] The factor $[\mu_0/\varepsilon_0]^{\frac{1}{2}}$, which appears in M.K.S. units, has the dimension of ohms, and the approximate numerical value $120\pi \simeq 377$. The multipole moments have the dimension ec = (charge) $\times$ (velocity) so that the result checks dimensionally.

$$P \equiv \text{radiated power} \equiv r^2 \int d\omega\, \hat{x}\cdot \mathbf{S}$$

$$= \tfrac{1}{2}k^2 (\mu_0/\varepsilon_0)^{\frac{1}{2}} \sum_{lm} \{|m_{lm}|^2 + |n_{lm}|^2\}. \tag{7.6.26}$$

i. Angular momentum flux. The significance of the multipole solutions becomes clearer if we examine the problem of the angular momentum emitted by a radiating source. The angular momentum flux density is given by

$$\mathscr{J} = (1/2)\sqrt{\varepsilon_0/\mu_0}\, \mathrm{Re}\,[\mathbf{x}\,\mathbf{x}\,(\mathbf{E}\,\mathbf{x}\,\mathbf{B}^*)]. \tag{7.6.27}$$

Asymptotically (that is, for large r) the leading term in the Poynting vector, $\mathbf{E}\,\mathbf{x}\,\mathbf{B}^*$, is *radial*; for the *angular* momentum flow this contributes nothing. One must therefore carry the asymptotic forms of $\mathbf{E}$ and $\mathbf{B}$ to one higher order. Using a radial and tangential decomposition of the Hansen solutions, and the asymptotic forms for the spherical Bessel functions, one finds for the angular momentum flux density

$$\mathscr{J} \sim (k\mu_0/2r^2) \sum_{l'm'lm} \mathrm{Re}\{\cdots\}, \tag{7.6.28}$$

where

$$\begin{aligned}\{\cdots\} = {}& m_{lm}m^*_{l'm'}[l'(l'+1)]^{\frac{1}{2}} i^{l'-l}\, Y^*_{l'm'}\mathbf{Y}_{llm} \\ &+ n_{lm}m^*_{l'm'}[l'(l'+1)]^{\frac{1}{2}} i^{l'-l}\, Y^*_{l'm'}\mathbf{x}\,\mathbf{x}\,\mathbf{Y}_{llm} \\ &- n_{lm}m^*_{l'm'}[l(l+1)]^{\frac{1}{2}} i^{l'-l}\, Y_{lm}\mathbf{x}\,\mathbf{x}\,\mathbf{Y}^*_{l'l'm'} \\ &+ n_{lm}n^*_{l'm'}[l(l+1)]^{\frac{1}{2}} i^{l'-l}\, Y_{lm}\mathbf{Y}^*_{l'l'm'}\,.\end{aligned} \tag{7.6.29}$$

The angular momentum (about a given axis) emitted per unit time is obtained from the above expression by integrating the appropriate component of $\mathscr{J}$ over a large sphere. For the 3-component we find

$$d\mathscr{J}_3/dt = (k\mu_0/2)\sum_{lm} m(|m_{lm}|^2 + |n_{lm}|^2). \tag{7.6.30}$$

To interpret this result, let us consider a pure multipole component, say m_{lm}, and compare the radiated angular momentum (3-component) to the radiated power, P, given by Eq. (7.6.26). We find

$$\frac{1}{P}\frac{d\mathscr{J}_3}{dt} = \frac{m}{\omega}. \tag{7.6.31}$$

One can interpret this result by saying that – for a pure multipole – *if* the power is radiated in units (photons) of energy $\hbar\omega$, then the angular momentum is also transferred in units of $m\hbar$. It is in this way that Maxwell's equations, which do not contain Planck's constant, are nonetheless compatible with the quantum mechanics of photons. (The agreement necessarily fails for order $\hbar^2$. See Note 2.)

Remarks. (*a*) The angular momentum flux density for monochromatic photon fields has been the source of much error and confusion, even paradox, in the literature. There is no difficulty if one calculates, as above, with multipole fields. If, however, plane waves of sharp helicity are used, then special care has to be exercised to handle properly the surface integrations to avoid an erroneous result of zero angular momentum flux. (A careful discussion, with extensive literature citations, is given in Jauch and Rohrlich [11].)

(*b*) It should be noted that the multipole fields for a given value of the total angular momentum and type (electric versus magnetic) do not have (for both **E** and **B**) a definite value of the (orbital) angular momentum l. It is, indeed, *not* possible to separate the total angular momentum of the photon field into an "orbital" and a "spin" part (this would contradict gauge invariance); the best that can be done is to define the helicity operator $\mathbf{S} \cdot \hat{P}$, which is an observable (Beth [12]).

j. A vectorial analog to the Rayleigh expansion. For applications of multipole radiation to angular correlations (Section 8) it is essential to obtain the vectorial analog to the Rayleigh expansion of scalar plane waves into spherical waves. This is easily accomplished directly from the dyadic Green function [developed in Eq. (7.6.17)] by taking the limit as the source point $\mathbf{x}'$ goes to infinity.

Alternatively, we can take the Rayleigh plane wave result,

$$e^{i\mathbf{k}\cdot\mathbf{x}} = 4\pi \sum_{lm} i^l j_l(kr) Y_{lm}(\hat{x}) Y^*_{lm}(\hat{k}), \tag{7.6.32}$$

and use the Wigner coefficients to couple in the spin-1 basis functions [see Section 11, Chapter 6].

By either method, one obtains the following result for a circularly polarized plane wave propagating in the direction **k**:

$$e^{i\mathbf{k}\cdot\mathbf{x}} \xi'_p = \sum_{l=1}^{\infty} \sum_m i^{l-1} [2\pi(2l+1)]^{\frac{1}{2}} D^l_{mp}(\hat{k}) [{}^{(\mathrm{st})}\mathbf{N}_{lm}(\mathbf{x}) - ip\, {}^{(\mathrm{st})}\mathbf{M}_{lm}(\mathbf{x})], \tag{7.6.33}$$

where $p = +1$ and -1, respectively, for right- and left-circular polarization, and $\hat{\xi}'_p$ denotes a spherical basis vector defined in terms of a frame $(\hat{e}'_1, \hat{e}'_2, \hat{e}'_3)$ whose 3-axis is parallel to **k**. In this result, $D^l(\hat{k})$ denotes the rotation matrix defined by $D^l(\hat{k}) \equiv \mathscr{D}^l(R(\hat{k})) = D^l(\alpha\beta 0)$, where $R(\hat{k})$ is an orthogonal matrix that rotates the 3-axis, $\hat{e}_3 = \mathrm{col}(0, 0, 1)$, into the direction $\hat{k} = \mathrm{col}(\cos\alpha \sin\beta, \sin\alpha \sin\beta, \cos\beta) = \mathbf{k}/k$ – that is, $\hat{k} = R(\hat{k})\hat{e}_3$. [See Chapter 6, Eqs. (6.60), for the definition of the spherical vectors $\hat{\xi}'_p$, and Eq. (6.188) for the relationship between $\mathscr{D}^l(R)$ and $D^l(U)$.]

k. An illustrative example. To demonstrate the utility of angular momentum techniques let us reconsider briefly the old problem of the scattering of electromagnetic waves from a perfectly conducting sphere. Consider the incident plane waves to be circularly polarized. We seek to calculate the force on the sphere per unit time.

The incident electromagnetic wave, assumed to be propagating along the 3-axis, has for its **E** vector the form

$$\mathbf{E}_{\mathrm{inc}} = E_0 \sum_l [2\pi(2l+1)]^{\frac{1}{2}} i^{l-1} [{}^{(\mathrm{st})}\mathbf{N}_{lp} - ip\, {}^{(\mathrm{st})}\mathbf{M}_{lp}]. \tag{7.6.34}$$

The complete **E** vector consists of the incident field (above) plus outgoing scattered waves; this latter is determined from the Maxwell equation using the boundary condition that, for a perfect conductor, the tangential electric field vanishes at the surface of the sphere.

A convenient formulation for the complete **E** vector utilizes the phase shifts α_l and β_l for scattered outgoing waves:

$$\mathbf{E}_{\mathrm{total}} = E_0 \sum_l i^{l-1} [2\pi(2l+1)]^{\frac{1}{2}} \{\cdots\},$$

$$\{\cdots\} \equiv {}^{(\mathrm{st})}\mathbf{N}_{lp} + ie^{i\beta_l} \sin\beta_l \, {}^{(\mathrm{out})}\mathbf{N}_{lp} - ip\, {}^{(\mathrm{st})}\mathbf{M}_{lp} + e^{i\alpha_l} \sin\alpha_l \, {}^{(\mathrm{out})}\mathbf{M}_{lp}. \tag{7.6.35}$$

Using the boundary condition, the phase shifts are found to be

$$\tan\alpha_l = j_l(ka)/n_l(ka), \tag{7.6.36}$$

$$\tan\beta_l = \left[\frac{(d/dr)(rj_l)}{(d/dr)(rn_l)}\right]_{r=a}. \tag{7.6.37}$$

These equations completely determine the scattered wave. From Eq. (7.6.26) one finds that the scattered power is

$$P_{\mathrm{scatt}} = (\pi E_0^2/k^2)\sqrt{\varepsilon_0/\mu_0} \sum_l (2l+1)[\sin^2\alpha_l + \sin^2\beta_l]. \tag{7.6.38}$$

The incident energy flux is given by $\frac{1}{2}E_0^2\sqrt{\varepsilon_0/\mu_0}$. The ratio of the scattered power to the incident flux defines the *cross section* σ – that is, that area of the beam that the scattering in effect removes. The cross section σ is found to be

$$\sigma = (2\pi/k^2)\sum_l (2l + 1)(\sin^2 \alpha_l + \sin^2 \beta_l). \tag{7.6.39}$$

For $\lambda \gg a$ the phase shifts α_l and β_l take the limiting forms

$$\alpha_l \simeq -(2l + 1)\frac{(ka)^{2l+1}}{[(2l + 1)!!]^2},$$

$$\beta_l \simeq \frac{l + 1}{l}(2l + 1)\frac{(ka)^{2l+1}}{[(2l + 1)!!]^2}. \tag{7.6.40}$$

For this limit, the cross section takes the approximate value

$$\sigma \simeq \frac{10}{3}(ka)^4(\pi a^2). \tag{7.6.41}$$

This result agrees with the general result that in the long-wavelength limit one obtains Rayleigh scattering, $\sigma \propto (\lambda)^4$.

Let us turn next to the problem of determining the force on the sphere. This can be obtained from the Maxwell stress tensor evaluated over the surface of the sphere. Alternatively, we may evaluate the stress tensor over the sphere at infinity, since the volume forces cancel in the region between the two spheres.

The stress tensor is

$$T_{ij} = E_i D_j + H_i B_j - \tfrac{1}{2}\delta_{ij}(\mathbf{E} \cdot \mathbf{D} + \mathbf{B} \cdot \mathbf{H}), \tag{7.6.42}$$

so that the force on a spherical surface $r^2\, d\omega$ is

$$d\mathbf{F} = (r^2\, d\omega)\{\varepsilon_0 \mathbf{E}(\hat{x}\cdot\mathbf{E}) + \mu_0 \mathbf{H}(\hat{x}\cdot\mathbf{H}) - \tfrac{1}{2}(\varepsilon_0 \mathbf{E}^2 + \mu_0 \mathbf{H}^2)\hat{x}\}, \tag{7.6.43}$$

where $\hat{x} = \mathbf{x}/r$.

Now for $r \to \infty$, the lowest-order terms in both $\mathbf{E}$ and $\mathbf{H}$ are *tangential*, so that the effective force is just

$$d\mathbf{F} \simeq (r^2\, d\omega)(-\tfrac{1}{2})(\varepsilon_0 \mathbf{E}^2 + \mu_0 \mathbf{H}^2)\hat{x}. \tag{7.6.44}$$

One further simplification may be made: By symmetry, only the 3-component of the total force is nonvanishing.

The 3-component of the resultant force is thus given by the expression

$$F_3 = (-\tfrac{1}{2}) \int r^2\, d\omega \cos\theta\, [\varepsilon_0 \mathbf{E}^2 + \mu_0 \mathbf{H}^2]. \tag{7.6.45}$$

The evaluation of the integral is greatly facilitated by using angular momentum techniques (see Section 13, Chapter 6) for the vector spherical harmonics. The final result reduces to a remarkably simple form:

$$F_3 = (\pi\varepsilon_0 E_0^2 k^{-2}) \sum_{l=1}^{\infty} \left\{ \frac{2l+1}{l(l+1)} \sin^2(\alpha_l - \beta_l) + \frac{l(l+2)}{l+1} [\sin^2(\beta_l - \beta_{l+1}) + \sin^2(\alpha_l - \alpha_{l+1})] \right\}. \tag{7.6.46}$$

In the long-wavelength limit one finds that

$$F_3 \simeq (7\pi/3)\,\varepsilon_0 E_0^2 k^4 a^6 = \left(\frac{1}{2c} E_0^2 \sqrt{\frac{\varepsilon_0}{\mu_0}}\right)\left(\frac{10\pi}{3} k^4 a^6\right)\left(\frac{7}{5}\right)$$

$$= (\text{momentum flux incident}) \times (\text{cross section}) \times \tfrac{7}{5}. \tag{7.6.47}$$

The factor $\frac{7}{5}$ shows that the net force is slightly bigger than the force resulting from totally absorbing the momentum flux incident on an area the size of the cross section.

The problem of the force of a plane electromagnetic wave on a perfectly conducting sphere is a very old one, the lowest-order term (the result given above) having originally been given by Schwarzschild in 1901, and a more complete discussion by Debye in 1909. Let us consider the problem a bit further in order to exhibit the utility of angular momentum techniques, in a way not available in the literature on this problem.

A dimensionless quantity can be obtained by dividing the force F_3 by the product of the incident energy density multiplied by the cross-sectional area of the sphere, πa^2. Calling this quantity Γ, we find that

$$\Gamma \equiv F_3/(\tfrac{1}{2}\varepsilon_0 E_0^2)(\pi a^2), \tag{7.6.48}$$

$$\Gamma = 2(ka)^{-2} \sum_{l=1}^{\infty} \left\{ \frac{2l+1}{l(l+1)} \sin^2(\alpha_l - \beta_l) + \frac{l(l+2)}{l+1} [\sin^2(\alpha_l - \alpha_{l+1}) + \sin^2(\beta_l - \beta_{l+1})] \right\}. \tag{7.6.49}$$

In the following, let $z = ka$.

Consider the term $\sin^2(\alpha_l - \beta_l)$. Using the definition of the phase shifts α_l, β_l in Eqs. (7.6.36) and (7.6.37), and the properties of the spherical Bessel functions, one gets a simpler result:

$$\sin^2(\alpha_l - \beta_l) = [z^2(j_l^2 + n_l^2)]^{-1}\{[(zj_l)']^2 + [(zn_l)']^2\}^{-1}, \qquad (7.6.50)$$

where the prime denotes differentiation with respect to z.

The simplicity of this result lies in the fact that the quantities have a more direct physical significance. The term $[z^2(j_l^2 + n_l^2)]^{-1} \equiv A_l^{-1}$ is the *penetrability* of the angular momentum barrier. This type of quantity—barrier penetrability—has received a great deal of attention in quantum mechanics.

The second quantity, $B_l \equiv [(zn_l)']^2 + [(zj_l)']^2$, is not directly a barrier penetration, but is closely related, as shown by the following formula:

$$B_l = \left[1 - \frac{l(l+1)}{z^2}\right]A_l + \frac{1}{2}A_l''. \qquad (7.6.51)$$

It remains only to investigate the other terms in the series for Γ. It can be shown that

$$\sin^2(\alpha_l - \alpha_{l+1}) = A_l^{-1}A_{l+1}^{-1}, \qquad (7.6.52)$$

$$\sin^2(\beta_l - \beta_{l+1}) = \left[1 - \frac{(l+1)^2}{z^2}\right]^2 B_l^{-1}B_{l+1}^{-1}. \qquad (7.6.52)$$

The series for Γ is expressed in its simplest form upon substitution of these more physical quantities. The explicit result is

$$\Gamma = 2z^{-2}\sum_{l=1}^{\infty}\left\{\frac{2l+1}{l(l+1)}A_l^{-1}B_l^{-1} + \frac{l(l+2)}{l+1}A_l^{-1}A_{l+1}^{-1}\right.$$

$$\left. + \frac{l(l+2)}{l+1}\left[1 - \left(\frac{l+1}{z}\right)^2\right]^2 B_l^{-1}B_{l+1}^{-1}\right\}. \qquad (7.6.53)$$

The quantities A_l (the reciprocal of the penetrability) are *polynomials* in z^{-2}. It can be shown that they are given by the result

$$A_l(z) = {}_3F_0(\tfrac{1}{2}, l+1, -l; -z^{-2}). \qquad (7.6.54)$$

The generalized hypergeometric function ${}_3F_0$ in this result is defined by its series

$$_3F_0(a,b,c;x) = 1 + \frac{abc}{(1)}x + \frac{a(a+1)b(b+1)c(c+1)}{(1)(2)}x^2 + \cdots$$

For general values of a, b, c, this series does not converge, but if one of the parameters a, b, c is a *negative integer* (the case at hand), the series terminates and the function is a polynomial.

Utilizing these results, it is a straightforward matter to get a series expansion for Γ in the variable z.

$$\Gamma = \frac{14}{3}z^4\left(1 - \frac{1}{21}z^2 + \cdots\right). \tag{7.6.55}$$

Actually the series converges *very poorly* and is *not* useful for calculation.

It would appear from this that the multipole expansion is well-adapted to the long-wavelength limit, but ill-adapted away from this limit. Off-hand it would seem hopeless even to consider the multipole expansion for the *short-wavelength* limit. We will now show that – properly interpreted – the multipole expansion really works even here. To discuss this we must understand something about the penetrability formulas.

If we utilize the A_l and B_l themselves in the original series, it is quickly found that this multipole series converges markedly better. (A useful criterion is that one needs l-values up through the value $l \simeq z$. Thus for $z = 1$, using $l = 1$ only, yields an answer for Γ accurate to much better than 1%.)

For large z and $l \ll z$, it is clear that $A_l \to 1$, which agrees with physical intuition that the barrier becomes ineffective in this limit. The critical region, however, is the region where $l \simeq z$ and $z \to \infty$. This is the transition region, and the penetrability decreases rapidly. For $l \gg z$, the penetrability drops exponentially to zero. Thus, *the multipole series is always convergent*, effectively being cut off in the vicinity of $l \simeq z$.

For $z \to \infty$, $l < z$, and $l \simeq z$ we can obtain the asymptotic formula,

$$A_l(z) \sim (1 - l^2/z^2)^{-\frac{1}{2}}. \tag{7.6.56}$$

With this result we find for B_l the asymptotic formula

$$B_l \sim (1 - l^2/z^2)^{\frac{1}{2}}. \tag{7.6.57}$$

Consider now the expression for Γ. The sum will involve a great many l's up to $l \simeq z$ with $z \to \infty$. The sum goes over into an integral under such conditions, and we find, upon introducing the asymptotic formulas for A_l and B_l,

$$\Gamma \sim \frac{2}{z^2}\int_0^{l\simeq z} dl\,[2l^{-1} + 2l(1 - l^2/z^2)]. \tag{7.6.58}$$

The first term is clearly small compared with the second. If we introduce the variable $l/z \equiv u$, we find

$$\Gamma \sim 4\int_0^1 u\,du\,(1 - u^2) = 1. \tag{7.6.59}$$

Thus, for large z, Γ *becomes unity in the short-wavelength limit*, and we have obtained this result using the multipole series.

Such a result must be obtainable by more elementary methods, and, in fact, this is easily done, since the short-wavelength limit implies the validity of *ray optics*. The force along the 3-axis due to the rays striking an annular ring with angle θ is

$$\begin{aligned} dF_3 &= (\tfrac{1}{2}\varepsilon_0 E_0^2)(2\pi a^2 \sin\theta\, d\theta \cos\theta)(2\cos^2\theta) \\ &= \text{(incident light pressure)}\,\text{(projected area)}\,\text{(reflection factor)}. \end{aligned}$$

Thus, one finds

$$\begin{aligned} F_3 &= (\tfrac{1}{2}\varepsilon_0 E_0^2)(4\pi a^2)\int_0^{\pi/2} \cos^3\theta \sin\theta\, d\theta \\ &= (\tfrac{1}{2}\varepsilon_0 E_0^2)\pi a^2; \end{aligned}$$

that is,

$$\Gamma = \frac{F_3}{\frac{1}{2}\varepsilon_0 E_0^2(\pi a^2)} = 1.$$

To summarize: We have obtained from the multipole expansion of the dimensionless force parameter, Γ, *the correct limiting behavior in both the long- and short-wavelength limits*. [The behavior in the vicinity of the maximum (the Debye problem) can also be handled by multipole expansion.]

Our purpose, however, is illustrative: *The multipole expansion is a valid technique at all wavelengths* and is a practical technique if penetrability is properly taken into account (see Note 3).

l. The density matrix for photon angular correlation measurements. The angular distribution of the electromagnetic flux radiated by an arbitrary charge-current source density can be calculated from the explicit multipole fields **E** and **B**, Eqs. (7.6.23) and (7.6.24), using the Poynting vector, as indicated in our discussion of the radiated power. In order to interpret this result in quantum mechanical applications, an essential change in viewpoint is necessary. The sources are, first of all, considered as expectation values (probability amplitudes) corresponding to the atomic, nuclear, or hadronic transition. This provides an interpretation for the magnetic and electric multipole moments, m_{lm} and n_{lm}, respectively, as parameters characterizing important physical properties of the quantal sources. (As discussed in RWA, time reversal symmetry in quantum mechanics imposes important phase conditions on the multipole moments.)

The more important change in viewpoint concerns the photon interpretation of the electromagnetic flux. The angular distribution of the radiated flux at (θ, ϕ) is defined to be the probability for observing a photon of definite four-momentum, but having random polarization, unless a spin determination is simultaneously performed (that is, a measurement of the photon's polarization; see below). It is important to emphasize that the multipolarity of the photon is not measured directly; that is to say, the angular momentum labels (l, m) are not local properties but require integration over a surface enclosing the sources. The quantal sources do indeed emit photons in (a finite number of) multipolar states, but the photons are detected asymptotically as plane wave states.

We can conclude from this description of the measurement process that the information to be obtained depends not only on the multipole moment parameters but the (matrix) transformation between the multipole states and the observed (maximal measurement) helicity plane wave states. This transformation between the two basis systems (multipolar basis versus plane wave basis) is precisely the content of the vectorial analog to the Rayleigh expansion and is a structure determined entirely by angular momentum considerations.

With these concepts in mind let us now give the relevant details.

Consider first the spin-polarization measurements. As discussed in detail in Section 7, one defines a coordinate frame – a tetrad – based on the momentum four-vector of the photon. With respect to this tetrad, the most general pure state (coherent state) for the spin consists of a linear combination over the two (helicity basis) states: $\hat{\xi}_{+1} = |+1\rangle$ and $\hat{\xi}_{-1} = |-1\rangle$. This situation is *formally* equivalent to the situation for the spin states of a spin-$\frac{1}{2}$ particle[1] and leads to the polarization density matrix:

$$\rho_{\text{Pol}} = \tfrac{1}{2}(\mathbb{1}_2 + \boldsymbol{\sigma} \cdot \mathbf{w}), \tag{7.6.60}$$

[1] Note that the physical angular momentum is nonetheless in units of $\hbar$ (and *not* $\frac{1}{2}\hbar$).

where $\mathbb{1}_2 = \sigma_0$ denotes the 2×2 unit matrix. The vector[1] $\mathbf{w}$ defines the Stokes parameters for the photon (Stokes [13]); the magnitude $0 \leqslant w \leqslant 1$ defines the degree of polarization, and the direction (θ, ϕ) (the Poincaré sphere; see Ref. [14]) defines the type of polarization: $\theta = 0$ corresponds to left circular,[2] and $\theta = \pi$ to right circular; $\theta = \pi/2$ corresponds to linear polarization (with ϕ the azimuth); elliptic polarization is the generic pure state.

(It is interesting to realize that these parameters predated quantum mechanics by nearly 75 years and were reintroduced relatively late (Mueller [15], Jones [16], McMaster [16a].)

Now let us turn to the transformation matrix.[3] Using the fact that the spin-$\frac{1}{2}$ polarization (pure) states are transitive under rotations, we can describe the generic spin state vector by a (two-parameter) rotation:

$$\xi(\hat{w}) \equiv \sum_{\mu} D^{\frac{1}{2}}_{\mu\frac{1}{2}}(\hat{w})\xi_{2\mu},$$

where $\hat{w} = \mathbf{w}/w$ and $D^{\frac{1}{2}}(\hat{w}) = D^{\frac{1}{2}}(\phi\theta 0)$ for $\hat{w} = (\cos\phi \sin\theta, \sin\phi\sin\theta, \cos\theta)$. Using this spin state in the analog to the Rayleigh expansion, Eq. (7.6.33), we obtain

$$\sum_{\mu} D^{\frac{1}{2}}_{\mu\frac{1}{2}}(\hat{w})e^{ikx_3}\xi_{2\mu}$$

$$= \sum_{l\mu} i^{l-1}[2\pi(2l+1)]^{\frac{1}{2}} D^{\frac{1}{2}}_{\mu\frac{1}{2}}(\hat{w})[\mathbf{N}_{l,2\mu}(\mathbf{x}) + i(-2\mu)\mathbf{M}_{l,2\mu}(\mathbf{x})]. \qquad (7.6.61)$$

This result is particularized to a reference frame having the 3-axis oriented along the direction of propagation of the photon. Under a rotation of the coordinate frame, the spin direction and propagation direction rotate rigidly, so that in a general frame one finds

$$\left[\sum_{\mu} D^{\frac{1}{2}}_{\mu\frac{1}{2}}(\hat{w})e^{ikx_3}\right]_{\text{rotated}} = \sum_{lm\mu} i^{l-1}[2\pi(2l+1)]^{\frac{1}{2}} D^{\frac{1}{2}}_{\mu\frac{1}{2}}(\hat{w}) D^{l}_{m,2\mu}(\hat{k})$$

$$\times [\mathbf{N}_{lm}(\mathbf{x}) + i(-2\mu)\mathbf{M}_{lm}(\mathbf{x})]. \qquad (7.6.62)$$

[See Eq. (7.6.33) for the definition of the special rotation matrix $D^l(\hat{k})$.]

[1] The polarization vector is often denoted in this context by $\mathbf{P}$. We have chosen to use the unusual symbol $\mathbf{w}$ for two reasons: to avoid confusion with the momentum operator $\mathbf{P}$ and (more importantly) to recognize that *the four-vector* $\mathbf{W}$ *is used to denote the relativistic polarization vector operator*, and it is logical to use $\mathbf{w}$ to denote the numerical polarization three-vector.

[2] Conventionally defined by looking "at the photon," opposite to its direction of motion.

[3] This discussion makes essential use of concepts in Section 7 on the density matrix.

The electromagnetic multipoles that appear in Eq. (7.6.62) are (standing-wave) potentials normalized uniformly for each (lm) component to $(2\pi\hbar k)^{-1}$ quanta/sec. The overall normalization is less significant than the fact that the various multipoles are normalized uniformly, so that the $\{\mathbf{N}_{lm}, \mathbf{M}_{lm}\}$ correspond to basis states for a maximal measurement. In order to give the density matrix formulation corresponding to the state (7.6.62), let us rewrite this result in the form

$$|\hat{w}, \hat{k}\rangle = \sum_{\pi lm} A_{\pi lm}(\hat{w}, \hat{k})|\pi lm\rangle, \tag{7.6.63}$$

where

$$\langle \mathbf{x}|\pi lm\rangle \equiv \begin{cases} \mathbf{N}_{lm}(\mathbf{x}) & \text{for } \pi = (\text{electric}) \\ \mathbf{M}_{lm}(\mathbf{x}) & \text{for } \pi = (\text{magnetic}), \end{cases} \tag{7.6.64}$$

$$A_{\pi lm}(\hat{w}, \hat{k}) = i^{l-1}[2\pi(2l+1)]^{\frac{1}{2}} \sum_{\mu} D^{\frac{1}{2}}_{\mu\frac{1}{2}}(\hat{w}) D^{l}_{m,2\mu}(\hat{k})$$

$$\times \begin{cases} 1 & \text{for } \pi = (\text{electric}) \\ -2i\mu & \text{for } \pi = (\text{magnetic}). \end{cases} \tag{7.6.65}$$

One finds for the density matrix of the radiation (denoted ρ_{rad}):

$$\rho_{\text{rad}} = |\hat{w}, \hat{k}\rangle\langle \hat{w}, \hat{k}|.$$

Using the transformation coefficients $A_{\pi lm}(\hat{w}, \hat{k})$ of Eq. (7.6.65), we find:

$$\langle \pi' l' m'|\rho_{\text{rad}}|\pi lm\rangle = A^{*}_{\pi' l' m'}(\hat{w}, \hat{k}) A_{\pi lm}(\hat{w}, \hat{k}). \tag{7.6.66}$$

Using the Wigner coefficients (as will be discussed in detail in Section 7), one finds the tensor parameters of the radiation, denoted by $R^{k,\kappa}_{\pi' l';\pi l}(\hat{w}, \hat{k})$, to be

$$R^{k,\kappa}_{\pi' l';\pi l}(\hat{w}, \hat{k}) = \sum_{mm'} \langle \pi' l' m'|\rho_{\text{rad}}|\pi lm\rangle\, C^{ll'k}_{mm'\kappa}. \tag{7.6.67}$$

It is convenient to split the tensor parameters into three terms corresponding to the three terms in the photon polarization density matrix. One finds:

Polarization-independent term (see Note 4):

$$R^{k,\kappa}_{\pi' l';\pi l;\,\text{unpol}}(\hat{w}, \hat{k}) = (-)^{\kappa + l' + 1}[(2l+1)(2l'+1)]^{\frac{1}{2}}$$

$$\times\, C^{ll'k}_{1,-1,0} D^{k}_{\kappa,0}(\hat{k}), \tag{7.6.68}$$

where k is restricted to be an *even* integer when parity is conserved.

"Circular" polarization term:

$$R^{k,\kappa}_{\pi'l';\pi l;\mathrm{I}}(\hat{w},\hat{k}) = (-1)^{\kappa+l'+1} w\cos\theta[(2l+1)(2l'+1)]^{\frac{1}{2}} \times C^{ll'k}_{1,-1,0} D^{k}_{\kappa,0}(\hat{k}), \tag{7.6.69}$$

where k is restricted to be an *odd* integer when parity is conserved. (w and θ are Stokes parameters.)

"Linear" polarization term:

$$R^{k,\kappa}_{\pi'l';\pi l;\mathrm{II}}(\hat{w},\hat{k}) = \pm\tfrac{1}{2}(-1)^{\kappa+1} w\sin\theta[(2l+1)(2l'+1)]^{\frac{1}{2}} \times C^{ll'k}_{112}[e^{i\phi}D^{k}_{\kappa,2}(\hat{k}) + e^{-i\phi}D^{k}_{\kappa,-2}(\hat{k})], \tag{7.6.70}$$

where k is again restricted to be an even integer for parity and time reversal invariance. (w, θ, ϕ are again the Stokes parameters.) The sign ($\pm$) is determined by the parity of the multipole.

Remarks. (*a*) These results are not really as complicated as they may at first appear. It is clear that the Wigner coefficients that enter in Eqs. (7.6.68)–(7.6.70) simply express angular momentum conservation and the fact that the photon necessarily has spin ± 1 along its direction of motion.

(*b*) The polarization terms depend linearly on the degree of polarization. The fact that k is odd, in Eq. (7.6.69), expresses the fact that the circular polarization components (like helicity) reverse under reflection, in contrast to the linear polarization components, Eq. (7.6.70), which do not (k = even). However, the linear polarization components do change sign when the parity of the radiation changes (this expresses the fact that under duality the Stokes parameter ϕ goes to $\phi + \pi$).

(*c*) These results for the tensor parameters of the radiation are used extensively in nuclear spectroscopy involving gamma ray angular correlations; a comprehensive account is given in Ref. [16b].

m. Notes. 1. As best as we can establish, Mie [17] in 1908 was the first to determine the complete set of multipole solutions to the free-space Maxwell equations. It was, however, a technological advance that led to the rediscovery and widespread use of multipole waves—namely, Kallmann's invention (c. 1943) of (NaI) gamma-ray detectors. These detectors made possible for the first time angular correlation measurements of nuclear gamma-ray cascades (see Section 8 of this chapter); multipole fields were then developed as the natural technique for these angular correlation studies. See, for example, Franz [18], Blatt and Weisskopf [19], Rose and Biedenharn [20], Rose [21] and French and Shimamoto [22]. Our discussion is taken from unpublished lectures given at Yale University in 1952. The textbook of Jackson [23] contains an especially good treatment of multipole radiation. Corben and Schwinger [24] were, to our knowledge, the first to define the vector spherical harmonics in terms of the Wigner coefficients. An alternative formulation of electromagnetism in terms of multipolar Debye potentials has been given by Gray [25], among others.

2. A classical electromagnetic multipole (L, M), oscillating with circular frequency ω, radiates angular momentum as well as energy; as discussed in Section i, the ratio $\hbar M/\hbar\omega$ is independent of Planck's constant and accords with a quantal interpretation. However, if one calculates the *square* of the angular momentum radiated and compares this with the square of the radiated energy, the result is M^2/ω^2 and *not* the expected (quantal) value $L(L + 1)/\omega^2$ as befits a "particle" in an (L, M) state. This puzzle was explained by Morette-de Witt and Jensen [26], who showed (using quantized electromagnetic fields) that the correct quantal value for the square is

$$\{N^2M^2 + N[L(L + 1) - M^2]\}/\omega^2,$$

where N is the number of quanta in the multipole mode (L, M). This resolution of the puzzle shows that for $N = 1$ (one quantum) one indeed gets the expected single-particle quantum mechanical answer: $L(L + 1)/\omega^2$, but for large numbers of quanta (classical field limit, $N \gg 1$) the classical value M^2/ω^2 obtains. One sees, moreover, that the N quanta coherently add their 3-components of angular momentum (to give N^2M^2 in the ratio) but that the uncertainty principle requires the perpendicular components (1- and 2-components) of the angular momenta to add *randomly*, contributing the factor proportional to N in the ratio.

3. An alternative approach to the short-wavelength limit was developed by Sommerfeld [27], using analytic continuation into complex angular momenta. These methods are of particular importance for both heavy-ion nuclear reactions and high-energy physics ("Regge poles"), and are discussed in more detail in RWA.

4. It is of interest to note explicitly that the use of the density matrix formalism makes it very easy to treat "unpolarized" electromagnetic waves, but in so doing conceals a problem of the greatest theoretical importance. Every solution of the Maxwell equations, which propagates spatially as a plane wave, is *necessarily* completely polarized transversally; every additive superposition of two completely polarized solutions yields another completely polarized solution. An unpolarized wave *cannot* be a solution of the Maxwell equations! Thus, the concept of an unpolarized wave goes *beyond* Maxwell electrodynamics and involves quantal considerations (implicitly introduced through the density matrix formalism) in an essential way (Hepp and Jensen [28]). (Jensen points out that in 1821 Fresnel had already noted that the existence of unpolarized waves was a serious problem for classical optics.)

References

1. B. Ferretti, "Propagation of signals and particles," in *Old and New Problems in Elementary Particles* (G. Puppi, ed.), pp. 108–119. Academic Press, New York, 1968.

2. E. Cunningham, "The principle of relativity in electrodynamics and an extension thereof," *Proc. London Math. Soc., Series 2*, **8** (1910), 77–98.
3. H. Bateman, "The conformal transformations of a space of four dimensions and their applications to geometric optics," *Proc. London Math. Soc., Series 2*, **7** (1909), 70–89; see also *Electrical and Optical Wave-Motion*, reprinting by Dover, New York, 1955.
4. S. J. Brodsky, "Quantum electrodynamics theory: Its relation to precision low energy experiments," in *Precision Measurements and Experimental Constants* (D. N. Langenberg and B. N. Taylor, eds.), pp. 297–307. U.S. National Bureau of Standards Special Publication 343, Washington, 1971.
5. H. van Dam, *Gravitational Theory*, Lectures at the University of Nijmegen, Nijmegen, The Netherlands, 1975.
6. P. M. Morse, *Vibration and Sound*. McGraw-Hill, New York, 1948.
7. P. M. Morse and H. Feshbach, *Methods of Theoretical Physics*, Vol. I, pp. 621ff. McGraw-Hill, New York, 1953.
8. W. W. Hansen, "A new type of expansion in radiation problems," *Phys. Rev.* **47** (1935), 139–143.
9. H. B. G. Casimir, "On supergain antennae," in *Old and New Problems in Elementary Particles* (G. Puppi, ed.), pp. 73–79. Academic Press, New York, 1968.

9a. B. Bosco and M. T. Sacchi, "On the inversion problem in classical electrodynamics and the Casimir theorem," *Ann. Phys.* (N.Y.), to be published, Oct., 1981.

10. C. J. Bouwkamp and N. G. de Bruijn, *The Problems of Optimum Antenna Current Distribution*, *Phillips Research Reports* **1** (1946), 135–158.
11. J. M. Jauch and F. Rohrlich, *The Theory of Photons and Electrons*. Addison-Wesley, Reading, Mass., 1955 (see footnote p. 34).
12. R. A. Beth, "Mechanical detection and measurement of the angular momentum of light," *Phys. Rev.* **50** (1936), 115–125.
13. G. G. Stokes, "On the composition and resolution of streams of polarized light from different sources," *Trans. Cambridge Phil. Soc.* **9** (1852), 399–416. This formalism was used by S. Chandrasekhar in his elegant monograph, *Radiative Transfer*, Clarendon, Oxford, 1950.
14. H. Poincaré, *Theorie Mathematique de la Lumière*, Vol. 2, p. 92. G. Carré, Paris, 1892.
15. H. Mueller, *Memorandum on the Polarization Optics of the Photoelastic Shutter*. Report 2, OSRD Project OEMsr-576-15, 1943.
16. R. C. Jones, "A new calculus for the treatment of optical systems, V. A more general formulation and description of another calculus," *J. Opt. Soc. Amer.* **37** (1947), 107–110.

16a. W. M. McMaster, "Matrix representation of polarization," *Rev. Mod. Phys.* **33** (1961), 8–28.

16b. W. D. Hamilton (ed.), *The Electromagnetic Interaction in Nuclear Spectroscopy*. North-Holland, Amsterdam, 1975.

17. G. Mie, "Beiträge zur Optik trüber Medien, speziell kolloidaler Metallösungen," *Ann. Physik* **25** (1908), 377–445.
18. W. Franz, "Multipolstrahlung als Eigenwertproblem," *Z. Physik* **127** (1950), 363–370.
19. J. M. Blatt and V. F. Weisskopf, *Theoretical Nuclear Physics*. Wiley, New York, 1952.
20. M. E. Rose and L. C. Biedenharn, "Angular correlation of nuclear radiations," *Rev. Mod. Phys.* **25** (1953), 729–777.
21. M. E. Rose, *Multipole Fields*. Wiley, New York, 1955.
22. J. B. French and Y. Shimamoto, "Theory of multipole radiation," *Phys. Rev.* **91** (1953), 898–899.
23. J. D. Jackson, *Classical Electrodynamics*. Wiley, New York, 1962.
24. H. C. Corben and J. Schwinger, "The electromagnetic properties of mesotrons," *Phys. Rev.* **58** (1940), 953–968; see Appendix.

25. C. G. Gray, "Multipole expansions of fields using Debye potentials," *Amer. J. Phys.* **46** (1978), 169–179.
26. C. Morette-deWitt and J. H. D. Jensen, "Ueber den Drehimpuls der Multipolstrahlung," *Z. Naturforsch.* **8a** (1953), 267–270.
27. A. Sommerfeld, *Partial Differential Equations in Physics*. Academic Press, New York, 1949.
28. H. Hepp and H. Jensen, "Klassische Feldtheorie der polarisierten Kathodenstrahlung und ihre Quantelung," *Sitzber. Heidelberg. Akad. Wiss. Math.-Naturw. Kl.* 4 Abh. (1971), 89–122.

7. Angular Momentum Techniques in the Density Matrix Formulation of Quantum Mechanics

a. Preliminaries. The density matrix formulation of quantum mechanics (von Neumann [1]) is a presentation alternative to the usual formulation in terms of vectors in Hilbert space. Instead of representing the physical state ψ by the Hilbert space vector $|\psi\rangle$, one now represents the state ψ by the operator ρ_ψ – called the "density matrix of the state ψ" – which is given by

$$\rho_\psi \equiv |\psi\rangle\langle\psi|/\langle\psi|\psi\rangle. \tag{7.7.1}$$

By construction, the operator ρ_ψ is *idempotent*:

$$\rho_\psi\rho_\psi = \rho_\psi^2 = \rho_\psi. \tag{7.7.2}$$

To recover the standard formulation of quantum mechanics, one notes that the *probability amplitude* for the measurement of the operator A in the state ψ – that is, $\langle\psi|A|\psi\rangle$ – can be written as

$$\text{Probability amplitude} \equiv \langle\psi|A|\psi\rangle = \text{tr}(A\rho_\psi). \tag{7.7.3}$$

More abstractly, one may introduce an orthonormal basis into Hilbert space, indexed by the denumerable set $\{i = 1, 2, 3, \ldots\}$ and regard the associated orthogonal idempotents, e_i, as a complete set of commuting operators whose eigenvalues are 1 or 0.

The probability that a system prepared in the state ψ will be measured in the state η is expressed by

$$\text{Probability} = \text{tr}(\rho_\psi\rho_\eta), \tag{7.7.4}$$

which, upon introducing a basis (that is, $|\psi\rangle = \sum_i \psi_i|i\rangle, \psi_i \in \mathbb{C}$), becomes

$$\text{tr}(\rho_\psi\rho_\eta) = \left|\sum_i \psi_i^*\eta_i\right|^2. \tag{7.7.5}$$

Remark. The result expressed by Eq. (7.7.5) is the *fundamental content of quantum mechanics*, since it demonstrates "the (nonclassical) interference of probabilities" as opposed to the classical formulation of probability, which would replace Eq. (7.7.5) by

$$\begin{aligned}\text{Classical probability} &= \sum_i (\text{probability that } \psi \text{ is in state } i)\\ &\qquad \times (\text{probability that } \eta \text{ is in state } i)\\ &= \sum_i |\psi_i|^2\,|\eta_i|^2 = \sum_i |\psi_i^* \eta_i|^2. \end{aligned} \tag{7.7.6}$$

This aspect of quantum mechanics has been strongly emphasized by Feynman [2]; that Eq. (7.7.5) is essentially the only way to generalize Eq. (7.7.6) is the content of fundamental work by Jordan and von Neumann [3].

The motivation for the introduction of the density matrix concept is the desire to treat, in a natural way, quantal systems for which one has less than maximal information. (Unpolarized light, for example, is the incoherent superposition of two independent polarizations of equal probability. There is no vector in Hilbert space that corresponds to this system.)

Density matrices are easily adapted to this situation: One defines a general density matrix to be

$$\rho = \sum_i p_i e_i \tag{7.7.7}$$

with $0 \leqslant p_i \leqslant 1$ and $\sum_i p_i = 1$. The e_i are a basis of orthogonal idempotents: $e_i e_j = \delta_{ij} e_j$, $\operatorname{tr} e_i = 1$, $\sum_i e_i = \mathbb{1}$.

The condition that a general density matrix correspond to a *pure state* – that is, a vector in Hilbert space – is that it be idempotent; otherwise, the density matrix represents a *mixed* (or incoherent) *state*.

(From a mathematical viewpoint, the concept of a density matrix is equivalent to the concept of a positive linear functional. From this viewpoint, the pure states are the extremal points of the convex hull.)

The density matrix formulation has an advantage in that it makes clear that quantum mechanics does not actually deal with vectors in Hilbert space, but rather with "rays." That is, from Eq. (7.7.1), one sees that ρ_ψ is invariant to a multiplication of the vector $|\psi\rangle$ by a nonzero constant. This observation is the starting point of the geometric approach to quantum mechanics, which treats the subject as a *projective geometry* (Jauch [4]).

The density matrix formulation has the disadvantage that it obscures the fact that vectors in Hilbert space form a linear manifold.

The transition from the Hilbert space formulation to the density matrix formulation of quantum mechanics is given, in essence, by Eq. (7.7.1). It is of interest to enquire as to how one could go in the other direction. Given an idempotent ρ, how can one obtain the corresponding vector $|\psi\rangle$? The answer[1] is that the ket vector $|\psi\rangle$ corresponds to the *right ideal* generated by ρ, with the left ideal corresponding to the bra vector $\langle\psi|$.

Let us complete these preliminary remarks by noting the form that the Schrödinger equation assumes in the density matrix formulation. One finds [using Eq. (7.7.1) and linearity] that

$$i\hbar\frac{d\rho}{dt} = [H, \rho] \equiv H\rho - \rho H. \tag{7.7.8}$$

(Note that the sign is *opposite* to that in the Heisenberg equation of motion for operators.[2])

b. Statistical tensors. The application of angular momentum techniques to the density matrix concept is based on the fact that angular momentum is an exact symmetry, and hence the vector $|\psi\rangle$, and ρ_ψ as well, must admit an expansion in an angular momentum adapted basis. Such an expansion is actually always valid, but the fact that the angular momentum is a symmetry implies that there is a common (coherent) time dependence for the states in a given angular momentum block $(-j \leqslant m \leqslant j)$. Using the basis $\{|(\alpha)jm\rangle\}$ (Chapter 3) for the vector $|\psi\rangle$ implies that the density matrix has the form

$$\rho_\psi = (\langle(\alpha')j'm'|\rho|(\alpha)jm\rangle), \tag{7.7.9}$$

and we extend this form to all ρ (mixed states) by linearity, Eq. (7.7.7).

To proceed further, we use the fact that the Wigner operators are an orthonormal operator basis for all maps between the angular momentum subspaces that enter into the description of ρ_ψ in an angular momentum basis. (See Section 21 and Note 12 of Chapter 3.) It is convenient here to introduce the

---

[1] One can carry these considerations further to conclude that *the density matrix formulation is actually more general than*, and not equivalent to, *the Hilbert space formulation*. If the algebraic structure imposed on the operators is nonassociative, and not embeddable in an associative algebra, then there may be no ideals and hence no wave functions. This happens for the Jordan algebra of Hermitian 3 × 3 matrices over the Cayley numbers (M_3^8), (Jordan *et al.* [5]). This example is of much current interest as a possible model for the quark (Gürsey [6], Horwitz and Biedenharn [7]).

[2] To verify the consistency, consider the time-independent operator $\mathcal{O}$ (Schrödinger picture). Then, from Eq. (7.7.8) the expectation value $\langle\mathcal{O}\rangle = \mathrm{tr}(\rho\mathcal{O})$ has the time dependence given by $i\hbar(d/dt)\langle\mathcal{O}\rangle = \mathrm{tr}([H,\rho]\mathcal{O}) = \mathrm{tr}(\rho[\mathcal{O},H])$, which agrees with $i\hbar(d\mathcal{O}/dt) = [\mathcal{O},H]$ in the Heisenberg picture.

operators defined by

$$T_{\kappa}^{k,\Delta} \equiv (2k+1)^{\frac{1}{2}} D^{-\frac{1}{2}} \left\langle \begin{matrix} & k+\Delta & \\ 2k & & 0 \\ & k+\kappa & \end{matrix} \right\rangle, \tag{7.7.10}$$

where D is the dimension operator that is defined on a generic angular momentum state by $D|jm\rangle = (2j+1)|jm\rangle$. [This slight modification of a Wigner operator has the effect of removing the dimension factors from the trace expression, Eq. (3.462), Chapter 3.] Thus, one has

$$\operatorname{tr}(T_{\kappa'}^{k',\Delta'} T_{\kappa}^{k,\Delta\dagger}) = \delta_{k'k}\delta_{\Delta'\Delta}\delta_{\kappa'\kappa}. \tag{7.7.11}$$

Accordingly, we obtain an operator decomposition of the general density matrix operator (see Note 12 of Chapter 3):[1]

$$\rho(\alpha', \alpha) = \sum_{k\kappa\Delta} q_{\kappa}^{k,\Delta}(\alpha', \alpha) T_{\kappa}^{k,\Delta},$$

$$q_{\kappa}^{k,\Delta}(\alpha', \alpha, j) = \operatorname{tr}(\rho(\alpha', \alpha) T_{\kappa}^{k,\Delta\dagger}). \tag{7.7.12}$$

The expansion parameters $q_{\kappa}^{k,\Delta}(\alpha', \alpha, j)$ in Eq. (7.7.12), which are essentially reduced matrix elements [see Eq. (3.460), Chapter 3], are called tensor parameters or sometimes statistical tensors (Fano [8]).

The utility of this approach is most in evidence when only a few angular momentum states are involved. Let us illustrate the content by examples with a single angular momentum.

Example 1. Orientation of a spin-$\frac{1}{2}$ particle (rest frame).[2]

The description of the spin is given in terms of the two states $|\frac{1}{2}, \frac{1}{2}\rangle$ and $|\frac{1}{2}, -\frac{1}{2}\rangle$ [no additional labels (α) are required in Eq. (7.7.9)]. In this case there are four Wigner operators mapping $\mathscr{H}_{\frac{1}{2}}$ into itself: the identity operator and the three angular momentum operators J_i themselves. The density matrix operator is thus represented by a 2×2 Hermitian matrix on the basis $\{|\frac{1}{2}m\rangle : m = \pm\frac{1}{2}\}$. Thus, the most general matrix form of the operator (7.7.9) is

$$\rho = \tfrac{1}{2}(\mathbb{1}_2 + \boldsymbol{\sigma} \cdot \mathbf{w}), \tag{7.7.13}$$

[1] We use the same notation, ρ, for the density matrix and for the operator ρ whose matrix elements define the density matrix.

[2] The more general case (relativistic) is discussed in Section d.

where $\mathbb{1}_2 = \sigma_0$ denotes the 2×2 unit matrix, and $\mathbf{w} = (w_1, w_2, w_3)$, $w_i = \text{tr}(\rho\sigma_i)$, denotes the polarization. Using the Dirac rule,

$$(\boldsymbol{\sigma} \cdot \mathbf{A})(\boldsymbol{\sigma} \cdot \mathbf{B}) = \mathbf{A} \cdot \mathbf{B} + i\boldsymbol{\sigma} \cdot (\mathbf{A} \times \mathbf{B}),$$

one can evaluate the product ρ^2 to find

$$\rho^2 = \frac{1}{2}\left[\left(\frac{1 + \mathbf{w} \cdot \mathbf{w}}{2}\right)\mathbb{1}_2 + \boldsymbol{\sigma} \cdot \mathbf{w}\right]. \tag{7.7.14}$$

Since $\text{tr}(\rho - \rho^2)$ is necessarily nonnegative, we see that the system is in a pure state, $\rho^2 = \rho$, if and only if $\mathbf{w} \cdot \mathbf{w} = 1$. For an unpolarized state, one has $\mathbf{w} = \mathbf{0}$. *Thus, the set of all polarization states of a spin-$\frac{1}{2}$ particle (in its rest frame) may be represented geometrically by the points of the surface and interior of the unit sphere in three-space* (Poincaré sphere). Orthogonal pure states are antipodal points of the surface. The rotational symmetry with which we began is expressed by the fact that the *set* of polarization states is spherically symmetric and all states have the same energy. [Note that, although the particle is a *spinor* (the ket vectors reverse sign under a rotation by 2π), this property is lost in the operator ρ, which consists of a (pseudo) vector operator $\mathbf{J} = \frac{1}{2}\boldsymbol{\sigma}$ and the scalar (unit) operator only.]

The physical significance of the vector $\mathbf{w}$ is implied by Eq. (7.7.13): *The polarization vector for a pure state is that spatial direction along which the angular momentum has projection* $+\frac{1}{2}$.

The representation of the density matrix given by Eq. (7.7.13) leads to a simple treatment of the Larmor precession of a spin-$\frac{1}{2}$ particle in a uniform magnetic field, $\mathbf{B}$. [The existence of the field *destroys the rotational symmetry*; the remaining symmetry is that of the subgroup, $SO(2)$, consisting of rotations around the magnetic field direction.] The magnetic moment $\mathbf{M}$ is defined quantum mechanically by the operator

$$\mathbf{M} = g\mu\mathbf{J} = \tfrac{1}{2}g\mu\boldsymbol{\sigma}, \tag{7.7.15}$$

where g is the gyromagnetic ratio (a dimensionless constant), and μ denotes the appropriate dimensional factor, $e\hbar/2mc$, with e and m the charge and mass of the system. Hence, the Hamiltonian H for the system is given by

$$H = -\mathbf{M} \cdot \mathbf{B} = -\tfrac{1}{2}g\mu\boldsymbol{\sigma} \cdot \mathbf{B}. \tag{7.7.16}$$

The Schrödinger equation, Eq. (7.7.8), for this Hamiltonian yields the equation of motion for the density matrix:

$$i\hbar\frac{d\rho}{dt} = i\hbar\left(\frac{1}{2}\frac{d\mathbf{w}}{dt}\cdot\boldsymbol{\sigma}\right) = [H,\rho] = -\frac{i}{2}g\mu\boldsymbol{\sigma}\cdot\mathbf{B}\times\mathbf{w}, \qquad (7.7.17)$$

and, hence, we obtain from the quantum mechanical formulation precisely *the classical equation of motion* for the polarization:[1]

$$\frac{d\mathbf{w}}{dt} = \gamma\mathbf{w}\times\mathbf{B}, \qquad (7.7.18)$$

where $\gamma = g\mu/\hbar$ is independent of $\hbar$.

The motion of the polarization vector $\mathbf{w}$ is a uniform precession about the direction of the magnetic field $\mathbf{B}$ with $\mathbf{w}\cdot\mathbf{B}$ constant. Let us introduce a new reference frame rotating with the angular velocity $\boldsymbol{\omega}$. If $\partial/\partial t$ represents differentiation with respect to this rotating frame, then we have

$$\frac{d\mathbf{w}}{dt} = \frac{\partial\mathbf{w}}{\partial t} + \boldsymbol{\omega}\times\mathbf{w}.$$

Thus, Eq. (7.7.18) becomes

$$\frac{\partial\mathbf{w}}{\partial t} = \gamma\mathbf{w}\times(\mathbf{B} + \gamma^{-1}\boldsymbol{\omega}) \equiv \mathbf{w}\times\mathbf{B}_{\text{eff}}. \qquad (7.7.18')$$

Here $\mathbf{B}_{\text{eff}}$ is the effective magnetic field in the rotating frame. Choosing for $\boldsymbol{\omega}$ the particular value[2] $\boldsymbol{\omega} = -\gamma\mathbf{B}$, one finds that $\mathbf{w}$ is *constant* in this particular rotating frame. This procedure is valid quantum mechanically as well as classically, and is valid for a general spin-j particle.

The use of rotating frames is of great utility in discussing molecular beam (magnetic resonance) and nuclear induction experiments. By proper choice of rotating frame, more general (experimentally important) magnetic field configurations can effectively be reduced to statics (see Rabi *et al.* [11]). (The motion of a magnetic moment in a time varying magnetic field is discussed in Section h.)

Example 2. Spin-j particle (rest frame).

For a spin-j particle the basis states are the angular momentum states $\{|jm\rangle : m = j, j-1, \ldots, -j\}$ spanning the space $\mathscr{H}_j$. [No α appears in Eq.

---

[1] The treatment of spin precession by ordinary quantum mechanical methods is very complicated (Bloch and Rabi [9]). In his paper on nuclear induction Bloch [10] pointed out that Eq. (7.7.18) can be established without solving the Schrödinger equation, since the quantum mechanical mean value of any quantity follows in its time dependence exactly the classical equation of motion (Ehrenfest theorem). (See also Rabi *et al.* [11].)

[2] The frequency $\omega = \gamma B$ is the Larmor frequency.

(7.7.9).] In this case, there are $(2j+1)^2$ operators mapping $\mathscr{H}_j$ into itself, and these may be taken to be the Wigner operators $T^{k,0}_{\kappa}$ ($k = 0, 1, \ldots, 2j$; $\kappa = k, k-1, \ldots, -k$) having the shift parameter $\Delta = 0$. Thus, letting T^k_κ denote the $(2j+1) \times (2j+1)$ *matrix* with elements given by $\langle jm'|T^{k,0}_\kappa|jm\rangle = \left(\frac{2k+1}{2j+1}\right)^{\frac{1}{2}} C^{jkj}_{m\kappa m'}$, we find that the most general form of the density matrix is

$$\rho = \sum_{k=0}^{\infty} \sum_{\kappa} q^k_\kappa T^k_\kappa,$$

$$q^k_\kappa = \operatorname{tr}(\rho T^{k\dagger}_\kappa). \tag{7.7.19}$$

The parameters q^k_κ are called *multipole tensor parameters*. Since ρ is Hermitian with unit trace, there are, accordingly, $(2j+1)^2 - 1$ free parameters. Since $T^0_0 = (2j+1)^{-\frac{1}{2}} \times$ (unit matrix) and the T^k_μ $(k \neq 1)$ are traceless, one finds $q^0_0 = (2j+1)^{-\frac{1}{2}}$, using $\operatorname{tr}\rho = 1$.

To achieve some physical insight into the meaning of this structure, let us consider an arbitrary Hamiltonian acting on the space $\mathscr{H}_j$, and let H denote the Hermitian matrix representing this Hamiltonian. Such a Hamiltonian would no longer necessarily be rotationally invariant, and the tensor parameters q^k_κ (for $k \neq 0$) would become functions of the time.

The Schrödinger equation, Eq. (7.7.8), upon introducing Eq. (7.7.19) for ρ, takes the form of a set of linear equations in the parameters q^k_κ. One finds

$$i\hbar \frac{dq^k_\kappa}{dt} = \sum_{k'\kappa'} \operatorname{tr}(T^{k\dagger}_\kappa [H, T^{k'}_{\kappa'}]) q^{k'}_{\kappa'}$$

$$= \sum_{k'\kappa'} \operatorname{tr}(H[T^{k'}_{\kappa'}, T^{k\dagger}_\kappa]) q^{k'}_{\kappa'}. \tag{7.7.20}$$

To make the structure more perspicuous, let us write it in matrix form, defining $\mathbf{q} = \operatorname{col}(q^0_0, q^1_{+1}, q^1_0, q^1_{-1}, \ldots)$. Then the Schrödinger equation, as given by Eq. (7.7.20), takes the form

$$i\hbar \frac{d\mathbf{q}}{dt} = \Omega \mathbf{q}, \tag{7.7.21}$$

where Ω is a Hermitian matrix with rows and columns indexed by the pairs $(k\kappa;\, k'\kappa')$. Specifically, one has

$$\Omega_{k\kappa;k'\kappa'} = \operatorname{tr}(H[T^{k'}_{\kappa'}, T^{k\dagger}_\kappa]). \tag{7.7.22}$$

To gain further insight into the structure, let us note that the Wigner matrices T^k_κ have definite algebraic properties. In fact, the product in Eq. (7.7.22) is reducible into a sum by using the symmetry property $T^{k\dagger}_\kappa = (-1)^\kappa T^k_{-\kappa}$ and the product property [see Eq. (3.350), Chapter 3]:

$$T^k_\kappa T^{k'}_{\kappa'} = \sum_{k''\kappa''} (-1)^{2j+k''}[(2k+1)(2k'+1)]^{\frac{1}{2}} C^{kk'k''}_{\kappa\kappa'\kappa''} \begin{Bmatrix} k & k' & k'' \\ j & j & j \end{Bmatrix} T^{k''}_{\kappa''}. \tag{7.7.23}$$

The commutator is thus given by

$$[T^{k'}_{\kappa'}, T^k_\kappa] = \sum_{k''\kappa''} (k'\kappa'; k\kappa)^{k''\kappa''} T^{k''}_{\kappa''}, \tag{7.7.24}$$

where the *structure constants* are defined by

$$\begin{aligned}(k'\kappa'; k\kappa)^{k''\kappa''} \equiv (-1)^{2j+k''}[(-1)^{k+k'-k''} - 1] \\ \times [(2k+1)(2k'+1)]^{\frac{1}{2}} C^{kk'k''}_{\kappa\kappa'\kappa''} \begin{Bmatrix} k & k' & k'' \\ j & j & j \end{Bmatrix}.\end{aligned} \tag{7.7.25}$$

[These are the same structure constants that are denoted by $A^{k'kk''}_{\kappa'\kappa\kappa''}$ in Section 5, Eq. (7.5.185).]

The structure constants (7.7.25), when normalized conventionally, are precisely the structure constants of $[(2j+1)^2 - 1]$ generators $\{T^k_\kappa\}$ of the Lie group $SU(2j+1)$. The general motion in the tensor parameter space $\mathbf{q}$ induced by a generic Hamiltonian is accordingly a unitary "rotation" (see Ref. [12]).

The matrix Ω, which induces this transformation, can now be expressed in terms of the generators $\{T^k_\kappa\}$ by using the structure constants in Eq. (7.7.25):

$$\Omega_{k\kappa,k'\kappa'} = \sum_{k''\kappa''} (-1)^\kappa (k'\kappa'; k, -\kappa)^{k''\kappa''} \operatorname{tr}(HT^{k''}_{\kappa''}). \tag{7.7.26}$$

The generators that enter into Ω thus depend on the expansion of H in the operator basis $\{T^{k''}_{\kappa''}\}$.

It is of interest to note that the symmetry properties of the Wigner coefficients $C^{kk'k''}_{\kappa\kappa'\kappa''}$ appearing in Eq. (7.7.25) for the structure constants show quite directly that the subset of generators $\{T^k_\kappa : k = \text{odd}\}$ close to form a subgroup. For $j =$ integer, this is the rotation subgroup $SO(2j+1)$; for $j =$ half-integer, this is the symplectic subgroup, $Sp(2j)$. In this way, one achieves by means of the angular momentum functions a uniform presentation of the generators and structure constants of the unitary, orthogonal, and symplectic Lie groups (Biedenharn [12], Racah [13]).

This general approach to the quantum mechanics of particles having angular momentum j is very useful in constructing models in atomic and nuclear spectroscopy, as we discuss in Sections 5 and 9.

The interaction Hamiltonian for a uniform magnetic field is a great simplification of the above general result. For this special case, the Hamiltonian takes the form

$$H = -g\mu \mathbf{J} \cdot \mathbf{B}, \tag{7.7.27}$$

where g and μ, as before, are the gyromagnetic ratio and appropriate magneton, respectively. Because the operator $\mathbf{J}$ classifies the tensor operators $T_\kappa^{k,\Delta}$ as multipole operators, it leaves the k index unchanged. Accordingly, the commutator $[H, T_{\kappa'}^{k'}]$ in Eq. (7.7.20) also leaves this index invariant. It follows that the vector multipole moment for spin-j, $\mathbf{q}^{(1)} = \{q_\kappa^1\}$, obeys the same equation of motion, Eq. (7.7.17), as for the special case $j = \frac{1}{2}$, as stated earlier.

c. A geometric characterization of the density matrices for pure states of spin-j. A geometric characterization can be of considerable help in developing an intuitive understanding of the set of pure (rest frame) states of a particle of spin-j.

For the set of pure states of $j = \frac{1}{2}$, a geometric picture is easily found – this is the content of Example 1 above – but let us proceed more formally, in a way that can be generalized. The pure states for spin-$\frac{1}{2}$ are described by two-component ket vectors; it follows that two complex numbers label the pure states. The density matrix, Eq. (7.7.1), is, however, invariant to multiplying the ket vector by an arbitrary *nonzero* complex number; this reduces the space $\mathbb{C}^2$ (the space of pairs of complex numbers) to the *projective space* $\hat{\mathbb{C}}^2$. The standard model of this space is the spherical surface S^2, the Riemann sphere.

Hence, we obtain, in this way, a familiar result: *The pure states for $j = \frac{1}{2}$ may be characterized by the points of the sphere S^2.*

Let us now generalize this result. For the pure states of spin-j the ket vectors have $(2j + 1)$ components; the corresponding space is therefore $\mathbb{C}^{2j+1}$. The invariance of the density matrix to multiplication of the ket vectors by a nonzero complex number defines an equivalence relation on the ket vectors: The ket vector $(a_1, a_2, \ldots, a_{2j+1})$ is equivalent to the ket vector $(b_1, b_2, \ldots, b_{2j+1})$ if $a_i = \lambda b_i$, where λ is a nonzero complex number. The resulting space of inequivalent pure states is the set of points of the projective space denoted by $\hat{\mathbb{C}}^{2j+1}$.

Majorana [14] gave a geometric realization of the space $\hat{\mathbb{C}}^{2j+1}$ *as a constellation of $2j$ points on the sphere S^2*. (The points in a constellation need not all be distinct.)

Proof. The most economical proof (Majorana [14]) is to observe that an angular momentum j can always be written as the totally symmetric composition of $2j$ kinematically independent spin-$\frac{1}{2}$ angular momenta, each of which is characterized by a point of S^2.

A more informative variant of this proof (it is, in fact, equivalent[1]) has been given by Bacry [15]. Consider the set P_{2j} of nonzero homogeneous polynomials of degree $2j$ in two complex variables. To each polynomial in P_{2j} we may associate an element of $\mathbb{C}^{2j+1}$ by the bijective (onto) map:

$$(a_1, a_2, \ldots, a_{2j+1}) \leftrightarrow a_1 x^{2j} + a_2 x^{2j-1} y + \cdots + a_{2j+1} y^{2j}.$$

The polynomials P_{2j} allow – by the fundamental theorem of algebra – a factorization into a product of $2j$ homogeneous linear factors. Thus, we may write each element of P_{2j} as the product of $2j$ elements of P_1. But this factorization is not unique: (*a*) we may permute the factors, and (*b*) we may multiply any two factors by λ and λ^{-1}, respectively. To obviate (*a*), we consider only the totally symmetric factorizations; to obviate (*b*), we introduce the same equivalence relation on P_{2j} that led to $\hat{\mathbb{C}}^{2j+1}$, and consider the projective space $\hat{P}_{2j}$. It follows, then, from the theorem on the roots of polynomials that $\hat{P}_{2j}$ is isomorphic to the totally symmetrized product $(P_1)^{2j}$. From the mapping $\hat{\mathbb{C}}^{2j+1} \leftrightarrow \hat{P}_{2j}$, one establishes Majorana's result. ■

Once we are in possession of this geometric model for the pure states, we can investigate the way in which spatial rotations affect the pure-state density matrices. In so doing, we are investigating the prototype example of the action of a group [$SO(3)$, and not $SU(2)$, as we saw from Example 1] on a manifold[2] (the constellations on S^2). We shall use some of the concepts and terminology of this more general approach, as proves convenient.

The first useful concept is that of the *stability group* (also called the isotropy group or, in physics, the "little group"). The stability group, G_m, is that subgroup of G which leaves a given point m of the manifold invariant.

An *orbit* is a set of points of the manifold that are carried into each other under the group action. It can be shown that, for any two points m and m' of a given orbit, the associated stability groups are conjugates: $G_m = g^{-1} G_{m'} g$, for some element g.

The last concept to be introduced is that of a *stratum*. A stratum is defined to be a union of points whose stability groups are conjugate. A stratum is

[1] In point of fact, Majorana indicated exactly this algebraic proof in Ref. [14].

[2] The study of group actions on manifolds is a general topic of increasingly great importance in physics. For physicists a useful and clear introduction may be found in Michel [16], and references cited there.

necessarily a union of orbits; any two points on an orbit have conjugate stability groups, but points with conjugate stability groups need not lie on the same orbit.

Let us illustrate these concepts by considering as an example the pure-state density matrices for $j = 1$; the geometric model is a pair of points on the surface of the unit sphere in three-space. For a pair of points the most general configuration is that of two points separated by an arc of length Ω (with $0 \leqslant \Omega \leqslant \pi$) on the spherical surface. There are three strata:

1. For $\Omega = 0$, the two points coalesce. The stability group of this constellation is $SO(2)$, the group of rotations about an axis through the double point. The stratum consists of one orbit [isomorphic to the homogeneous space $SO(3)/SO(2)$], the surface of the sphere (since the double point may be carried to any point on the surface by some rotation). This stratum is two-dimensional.

2. For $\Omega = \pi$, the two points are diametrically opposite (antipodal). The stability group is now larger: It is $SO(2) \times Z_2$, consisting of the rotations around the axis of the two points, and an involution (rotating by π around an axis perpendicular to the diameter defined by the two points). The stratum again is a single orbit [isomorphic to $SO(3)/(SO(2) \times Z_2)$], which consists of *half* the surface of the sphere (antipodal points are identified). This stratum is also two-dimensional.

3. For $0 < \Omega < \pi$, we obtain the generic stratum. The stability group is Z_2, generated by the rotation by π, which exchanges the two points. The continuous set of orbits belonging to this stratum are labeled by Ω; each orbit is three-dimensional [$SO(3)/Z_2$]. This stratum is, accordingly, four-dimensional.

For completeness, let us note that the pure states for $j = \frac{1}{2}$ constitute a single stratum of one two-dimensional orbit [$SO(3)/SO(2)$].

It is of interest to examine the strata to which the three basis states for $j = 1$ belong. To the state $j = 1$, $m = 1$, one associates the symmetric product $|\frac{1}{2}, \frac{1}{2}\rangle \otimes |\frac{1}{2}, \frac{1}{2}\rangle$. Clearly this state belongs to stratum (1), as does $j = 1, m = -1$ on the same orbit. (These two states are often called "circularly polarized.") The state $j = 1, m = 0$ can be seen to belong to stratum (2). (This state is called "longitudinally polarized.")

Remarks. (*a*). The fact that the pure states for $m = \pm 1$ and for $m = 0$ lie in different orbits (as well as different strata) is quite significant. In particular it shows that there is, in general, *no* rotation that can transform any two pure states (each specified by a given value of m) into each other. (The contrary has been erroneously stated in some textbooks on quantum mechanics.) The *rotational inequivalence* of the set of $j = 1$ pure states is, in a sense, surprising, since the three directions $\hat{e}_1$, $\hat{e}_2$, $\hat{e}_3$ *are* equivalent, and the "z-direction" (spherical harmonic Y_{10}) does correspond to a pure state [in stratum (2)].

For the special case of $j = \frac{1}{2}$, all pure states are indeed equivalent under rotation (there is but one orbit). It is easy to see from this special case just where the confusion arose, since a pure state for $j = \frac{1}{2}$ is correlated to a direction [unit vector $\mathbf{w}$ in Eq. (7.7.13)] that corresponds to that direction along which the pure state has the m-value $\frac{1}{2}$. It is interesting to note (Ref. [12]) that there is an analog to this statement for $j = \frac{1}{2}$ that *does* generalize to all j: *For any pure state there exists a rotation such that the rotated state vector does not have a component with a specified m-value.* (For $j = \frac{1}{2}$ this statement is equivalent to the previous definition of the pure states.) The Weyl canonical labeling of state vectors in $SU(2j + 1)$ is, in fact, the systematic use of this property.

(*b*) The tensor parameters for the density matrix of a pure state having spin-j were defined in Eq. (7.7.19); these parameters all have the shift label Δ equal to zero, and it is appropriate to call them "multipole moments of the density matrix." Since there are $2j$ multipoles ($k = 1, 2, \ldots, 2j$, omitting the unit operator because of the trace condition), the number of parameters is $2j(2j + 2)$; for a pure state only $4j$ can be independent. It is a consequence of the Wigner–Eckart theorem that any measurement carried out on this system by an arbitrary (observable) tensor operator is proportional to the measurement of the corresponding multipole operator (Wigner operator). For vector operators it is not surprising that one factor of proportionality suffices (for example, the gyromagnetic ratio for the magnetic dipole operators). But it *is* surprising that for quadrupole operators, say, *one* proportionality factor suffices; classically one knows that the intrinsic (rotationally invariant) classification of quadrupole operators involves *two* parameters, and not just one. What accounts for this distinction?

The answer is that the presupposition that the state is a pure state (with sharp j-value) implies, by the uncertainty relation, that the pure states are, in effect, "cylindrically symmetric" (about the angular momentum axis, if this could be defined classically). A graphic aphorism for this is due to Weisskopf: "A football may be in a pure state of sharp j, but if so, you can't see the lacing."

(*c*). We have limited the discussion for the geometric characterization to pure states largely because accessible general results are not available. For the case $j = \frac{1}{2}$, the general density matrix is geometrically described by a single vector $\mathbf{w}$; $\|\mathbf{w}\| = 1$ implies a pure state, and $\|\mathbf{w}\| = 0$ implies an unpolarized state. The general case (spin-j) corresponds to a manifold of $[(2j + 1)^2 - 1]$ dimensions, and (as indicated in Example 2) the most general operators acting on the system are the generators of $SU(2j + 1)$.

The polarization domain of such a general structure is only recently becoming of physical interest, partly in response to the need to assess critically the validity of polarization measurements in particle physics (Doncel, *et al.* [17]).

d. The density matrix for a relativistic massive particle of spin-j. The critical distinction between the relativistic and nonrelativistic domains (Poincaré versus Galilei symmetry) is in the kinematic treatment of spin. For a particle with nonvanishing mass, the relativistic description singles out a particular frame, the rest frame, in which the description of the spin properties is precisely the same [a Hilbert space of dimension $(2j + 1)$] as the nonrelativistic description. Hence, the preceding discussion of the density matrix is valid both relativistically and nonrelativistically for a massive particle of spin-j.

It is necessary to discuss the spin density matrix, however, for a massive $(m \neq 0)$ particle viewed in an *arbitrary* frame, and not just in the rest frame alone. The reason for this is not simply the desire for generality, but a quite practical one: to be able to consider two-particle reactions. For two-particle reactions the composite system, formed in the reaction, is viewed as a "particle" that (in its rest frame defined by $\mathbf{P}_{\text{total}} = \mathbf{0}$) is characterized (Poincaré symmetry) by a mass and a spin. In order to apply angular momentum conservation to this composite system, it is necessary to be able to describe the angular momenta (hence, spin) of the incident (and also emergent) particles in a general frame (the frame for each particle that is moving with respect to the rest frame of the composite system). The kinematic aspects of the spin implied by the Poincaré symmetry are accordingly of decisive importance in discussing reactions.

Consider then a particle of mass m and spin-j. These two (Poincaré invariant) labels are the eigenvalues of the two invariant operators that we now describe. The first invariant is

$$P_\mu P^\mu = \mathbf{P} \cdot \mathbf{P} \to m^2, \tag{7.7.28}$$

where P_μ are the four-momentum operators generating the translation subgroup. The second invariant is

$$W_\mu W^\mu = \mathbf{W} \cdot \mathbf{W} \to m^2 j(j+1), \tag{7.7.29}$$

where

$$W_\alpha \equiv e_{\alpha\beta\gamma\delta} P^\beta M^{\gamma\delta}, \tag{7.7.30}$$

$$M^{\gamma\delta} \equiv \text{Lorentz subgroup generators.}$$

The four-vector operator W_μ is attributed to Pauli (see Lubanski [18]) and is often called the Pauli–Lubanski operator. A more descriptive name for W_μ would be the *polarization operator*, as will be clear shortly.

The operators P_μ and W_μ *commute*, and satisfy the operator condition that $\mathbf{P} \cdot \mathbf{W} = 0$. Both P_μ and W_μ are simultaneously observable, but the observability of W_μ (unlike that of P_μ) is restricted, since the components of W_μ do *not* commute:

$$[W_\alpha, W_\beta] = ie_{\alpha\beta\gamma\delta} P^\gamma W^\delta. \tag{7.7.31}$$

What one would like to obtain is the proper analog to the (three-vector) spin operator $\mathbf{S}$ – which exists in the rest frame, and has eigenvalue $\mathbf{S}^2 \to j(j+1)$. The way to accomplish this is to introduce – for a particle state with fixed energy-momentum $P_\mu \to p_\mu$ – the orthonormal *tetrad* $\{n_\mu^{(\alpha)}(\mathbf{p})\}$. This tetrad consists of four orthonormal four-vectors (which are functions of the numerical[1] four-vector p_μ) obeying the conditions

$$\mathbf{n}^{(\alpha)} \cdot \mathbf{n}^{(\beta)} = g^{\alpha\beta}, \tag{7.7.32}$$

$$e^{\lambda\mu\nu\rho} n_\lambda^{(\alpha)} n_\mu^{(\beta)} n_\nu^{(\gamma)} n_\rho^{(\delta)} = - e^{\alpha\beta\gamma\delta}. \tag{7.7.33}$$

The second relation expresses the condition that the tetrad is right-handed (oriented).

One now defines four-vector polarization operators with respect to the tetrad coordinates – that is,

$$W^{(\alpha)} = \mathbf{n}^{(\alpha)}(\mathbf{p}) \cdot \mathbf{W}. \tag{7.7.34}$$

These tetrad components $W^{(\alpha)}(\mathbf{p})$ satisfy the relations

$$\mathbf{W} \cdot \mathbf{W} = \sum_{\alpha\beta} W^\alpha W^\beta g_{\alpha\beta} \to m^2 j(j+1). \tag{7.7.35}$$

Using the fact that we are considering a massive particle, we may choose one tetrad vector to be a multiple of the numerical momentum four-vector – that is,

$$n_\mu^{(0)} = m^{-1} p_\mu. \tag{7.7.36}$$

It follows that $W^{(0)} = 0$. It is convenient to rename the three nonvanishing tetrad components of W_μ:

[1] It is important to realize that the tetrad consists of numerical vectors and not operators; the use of such constructs is valid only on the special subset of vectors having fixed values p_μ as eigenvalues of the four-momentum operators P_μ.

$$S^{(i)} \equiv -\mathbf{n}^{(i)} \cdot \mathbf{W}/m. \tag{7.7.37}$$

It then follows, from the commutation relations for the polarization operator **W**, that the tetrad components $S^{(i)}$ are the desired spin operators; that is,

$$[S^{(i)}, S^{(j)}] = ie_{ijk} S^{(k)}. \tag{7.7.38}$$

By means of these tetrad spin operators $\{S^{(i)}\}$, one can now obtain the set of "tetrad multipole operators," $T^k_\kappa(\mathbf{p})$, for $k = 0, 1, \ldots, 2j$. This follows, since the various products $S^{(i')} \cdots S^{(j')} \cdots S^{(i'')}$ are reducible – using the Racah coefficients – into multipole operators. Accordingly, the structure of the spin density matrix (as discussed in previous sections) can now be carried over, via the tetrad formalism, to the general case (see the Remark below).

Detailed discussion of polarization in the relativistic domain is given in the extensive work of Doncel *et al.* [17], which is too specialized to permit further discussion here.

Remark. For the Dirac equation there is a special technique available, owing to the existence of Dirac's first-order differential equation: $(\boldsymbol{\gamma} \cdot \mathbf{P} + m)\psi = 0$. One defines the (numerical) polarization four-vector S_μ (see Fradkin and Good [18a]) such that $\mathbf{S} \cdot \mathbf{p} = 0$, $\mathbf{S} \cdot \mathbf{S} = -1$. (Here $P_\mu \to p_\mu$ and $\mathbf{p} \cdot \mathbf{p} = m^2$.) Then the 4×4 density matrix (in the space of the γ-matrices) is given by the (idempotent) operator ρ:

$$\rho = (4m)^{-1}(\boldsymbol{\gamma} \cdot \mathbf{p} + m)(1 - \gamma_5 \boldsymbol{\gamma} \cdot \mathbf{S}). \tag{7.7.39}$$

(Note that the $\gamma_5 \boldsymbol{\gamma} \cdot \mathbf{S}$ commutes with $\boldsymbol{\gamma} \cdot \mathbf{p}$.)

This is an extension of a technique introduced by Casimir to facilitate calculations with the Dirac equation by using traces. It is standard now in textbooks – for example, Bjorken and Drell [19], Sakurai [20], and Merzbacher [21].

e. The special case of massless particles. For massless particles there is no rest frame, and the special tetrad used above (implying $W^{(0)} = 0$) is no longer available. The tetrad formalism still works, however. For the lightlike momentum four-vector p_μ ($\mathbf{p} \cdot \mathbf{p} = 0$, $\mathbf{p} \neq \mathbf{0}$) we take the tetrad: $\mathbf{p} \cdot \mathbf{n}^{(0)} = E = \mathbf{p} \cdot \mathbf{n}^{(3)}$. It follows from $\mathbf{P} \cdot \mathbf{W} = 0$ that $W^{(0)} + W^{(3)} = 0$, so that the tetrad component $W^{(0)}$ of the polarization vector may be eliminated.

The remaining three tetrad components $W^{(i)}$, $i = 1, 2, 3$, generate the Euclidean group $E(2)$: $W^{(1)}$ and $W^{(2)}$ generate the translations, and $W^{(3)}$ the rotations. The eigenvalue of the invariant operator $\mathbf{W} \cdot \mathbf{W}$ determines two cases: (*a*) $\mathbf{W} \cdot \mathbf{W} \to$ nonzero negative number and (*b*) $\mathbf{W} \cdot \mathbf{W} \to 0$. The first case (the "infinite" spin case) apparently does not occur physically (Wigner [22]).

In the second case, $\mathbf{W}\cdot\mathbf{W}\to 0$, one has the three invariant eigenvalue (subspace) conditions: $\mathbf{P}^2=\mathbf{P}\cdot\mathbf{W}=\mathbf{W}^2=0$, which imply that $\mathbf{W}=\lambda\mathbf{P}$. Since the polarization $\mathbf{W}$ is a pseudo-vector, and the momentum $\mathbf{P}$ is a vector, this identifies λ—the helicity—to be a pseudo-scalar. If the space reflection operator is a symmetry, then two values of λ occurs: $\pm\lambda$. Thus, for photons and gravitons, one finds ± 1 and ± 2, respectively. For neutrinos and antineutrinos, one finds $-\frac{1}{2}$ and $+\frac{1}{2}$, respectively, since parity is not a symmetry; these particles are always 100% polarized. (We have discussed the polarization properties of photons in connection with the general density matrix formulation appropriate to electromagnetic radiation in Section 6.)

There is an extensive literature on angular distributions (and polarization phenomena) in connection with weak interactions (neutrinos, etc.) that cannot be discussed here (Schopper [23]).

f. Coupling of statistical tensors. A significant application of angular momentum techniques to the density matrix, results, as discussed above, in the reduction of the density matrix by means of the characteristic operators of the angular momentum structure. This led to the concept of a statistical tensor. A second significant use of angular momentum techniques lies in the coupling of tensor operators—and hence statistical tensors—to effect the overall conservation of angular momentum of coupled systems.

To understand this coupling aspect, let us consider two distinct systems, denoted by 1 and 2, which are coupled to give a third system. Thus, one has (see Chapter 3, Section 12)

$$|(\alpha_1)j_1m_1\rangle\otimes|(\alpha_2)j_2m_2\rangle \underset{\text{Wigner coupling}}{\longrightarrow} |(\alpha_1\alpha_2j_1j_2)jm\rangle. \tag{7.7.40}$$

[Here (α_i) denotes, as usual, a set of quantum numbers that are distinct from the j_i, m_i, j, m.]

Let us suppose that we are given the two sets of statistical tensors (generalized multipole moments) that characterize systems 1 and 2. What can one conclude for the statistical tensors of the coupled system? Thus, given the coefficients, $(\alpha_i'j_i'|\rho_i|\alpha_ij_i)^{k_i}_{\kappa_i}$, in the expansion

$$\langle(\alpha_i')j_i'm_i'|\rho_i|(\alpha_i)j_im_i\rangle=\sum_{k_i\kappa_i}(\alpha_i'j_i'|\rho_i|\alpha_ij_i)^{k_i}_{\kappa_i}\,C^{j_ik_ij_i'}_{m_i\kappa_im_i'}, \tag{7.7.41}$$

we seek the coefficients in the expansion of the tensor product density matrix:

$$\langle(\alpha_1'\alpha_2'j_1'j_2')j'm'|\rho_1\otimes\rho_2|(\alpha_1\alpha_2j_1j_2)jm\rangle$$

$$=\sum_{k\kappa}(\alpha_1'\alpha_2'j_1'j_2'j'|\rho_1\otimes\rho_2|\alpha_1\alpha_2j_1j_2j)^k_\kappa\,C^{jkj'}_{m\kappa m'}. \tag{7.7.42}$$

Following the method used to derive Eq. (3.248), Chapter 3, one finds

$$(\alpha_1'\alpha_2'j_1'j_2'j'|\rho_1 \otimes \rho_2|\alpha_1\alpha_2 j_1 j_2 j)^k_\kappa$$

$$= \sum_{\substack{k_1k_2\\ \kappa_1\kappa_2}} [(2j+1)(2k+1)(2j_1'+1)(2j_2'+1)]^{\frac{1}{2}} \begin{Bmatrix} j_1 & j_2 & j \\ k_1 & k_2 & k \\ j_1' & j_2' & j' \end{Bmatrix}$$

$$\times (\alpha_1' j_1'|\rho_1|\alpha_1 j_1)^{k_1}_{\kappa_1} (\alpha_2' j_2'|\rho_2|\alpha_2 j_2)^{k_2}_{\kappa_2} C^{k_1k_2k}_{\kappa_1\kappa_2\kappa}. \tag{7.7.43}$$

The structure of the result given in Eq. (7.7.43) can be seen to be the following:

(*a*) The components $(k_1\kappa_1)$ and $(k_2\kappa_2)$ of statistical tensors for the density matrices, ρ_1 and ρ_2 of systems 1 and 2, are *Wigner-coupled* [to give the labels (k, κ), which, by index balance, are labels on the left-hand side];

(*b*) The *universal angular momentum function*, which effects the coupling of the remaining angular momentum quantum numbers, is the 9-j symbol or, what is exactly the same, the X-coefficient of Fano and Racah [24], who first emphasized this aspect of the problem.

The result expressed by Eq. (7.7.43) is the most complicated generic result needed for any application of the theory. Either Eq. (7.7.43) itself will suffice for the desired application, or, if not, a sequence of couplings [applying Eq. (7.7.43) to effect each step] will suffice.[1] From this viewpoint, the introduction of higher 3n-j coefficients is superogatory, and useful only in highly specialized circumstances.

We shall apply Eq. (7.7.43) to discuss angular correlations in Section 8.

g. Some examples illustrating the coupling formula. In order to get some physical insight into the meaning of a result as complicated as Eq. (7.7.43), let us consider here a few simplified cases.

1. To begin, let us assume that for system 1 the density matrix ρ_1 contains *minimal* information. [Note that systems 1 and 2 play fully equivalent roles in Eq. (7.7.43).] Such a state of minimal information is completely random and has for its statistical tensors the values

$$(\alpha_1' j_1'|\rho_1|\alpha_1 j_1)^{k_1}_{\kappa_1} = \delta_{k_1 0}\,\delta_{\kappa_1 0}\,\delta_{\alpha_1'\alpha_1}\,\delta_{j_1' j_1}\frac{1}{\sqrt{2j_1+1}}. \tag{7.7.44}$$

Since k_1 is restricted to be 0 by Eq. (7.7.44), the relevant 9-j symbol is reducible

[1] The coupling of one and the same system to itself leads to similar results (see discussion in Chapter 3, Section 17).

to a 6-j symbol (Racah function), using results given in Chapter 3, Eq. (3.325):

$$\begin{Bmatrix} j_1 & j_2 & j \\ 0 & k_2 & k \\ j_1 & j_2' & j' \end{Bmatrix} = (-1)^{j_1+j_2+j'+k}\,\delta_{k_2 k}\,[(2k+1)(2j_1+1)]^{-\frac{1}{2}} \begin{Bmatrix} j_1 & j_2 & j \\ k & j' & j_2' \end{Bmatrix}. \tag{7.7.45}$$

Hence, we find for the coupled statistical tensors the result

$$(\alpha_1'\,\alpha_2' j_1' j_2' j' | \rho_1 \otimes \rho_2 | \alpha_1 \alpha_2 j_1 j_2 j)^k_\kappa$$

$$= \delta_{\alpha_1'\alpha_1}\,\delta_{j_1' j_1}\,(-1)^{j_1+j_2+j'+k} \left[\frac{(2j+1)(2j_2'+1)}{(2j_1+1)}\right]^{\frac{1}{2}}$$

$$\times\,(\alpha_2' j_2' | \rho_2 | \alpha_2 j_2)^k_\kappa \begin{Bmatrix} j_1 & j_2 & j \\ k & j' & j_2' \end{Bmatrix}. \tag{7.7.46}$$

Recalling that the 6-j coefficients are real numbers with a magnitude between 0 and 1, one sees that the statistical tensors for the coupled system are (if changed) *decreased* from the corresponding values for system 2.

It is to express this interpretation that the 6-j coefficients are sometimes called *depolarization factors*. Coupling a system with angular information to a system having no information (random system) partly depolarizes the composite system. The most extensive application of such concepts is in the study of perturbed angular correlations, where the extraneous influences (for example, crystalline fields in microcrystalline systems) tend to wash out the correlation. This subject, which has extensive ramifications, is one of the most important current applications of angular momentum techniques (see Ref. [25] and the extensive literature citations).

2. As a second example illustrating the meaning of the coupling formula, let us consider the electromagnetic interaction energy of a nucleus and its surrounding cloud of electrons forming a single atomic system. Owing to the great difference in characteristic size between the nuclear and electronic systems (10^{-13} cm versus 10^{-8} cm), we may approximate the two systems as physically nonoverlapping.

To apply the coupling formula, one considers the nuclear system to be system 1 with sharp angular momentum magnitude I, and the electronic system to be system 2 with angular momentum J; the coupled system has angular momentum F, where $\mathbf{I} + \mathbf{J} = \mathbf{F}$.

Since the interaction Hamiltonian is rotationally invariant, only the statistical tensor $k = \kappa = 0$ can enter in the coupling of the two systems. By Eq. (7.7.45) (using the symmetry of the 9-j symbol), this reduces to the 6-j symbol.

Because of the simplifying physical assumption of no overlap, the two sets of statistical tensors factor into a product of coupled multipole moments.

Let us consider the interaction energy of static electric quadrupoles. The discussion of electromagnetic multipoles has been given in the preceding section; it suffices for the present (illustrative) purposes simply to take the electronic quadrupole moment (irregular at the origin) to be the reduced matrix element $e_{\mathrm{el}}\langle J\|r^{-3}P_2(\cos\theta)\|J\rangle$ and the nuclear moment (regular at the origin) to be $e_{\mathrm{nucl}}\langle I\|r^2P_2(\cos\theta)\|I\rangle$; this uses the assumption of no overlap, with the electron strictly outside the nucleus. Here $-e_{\mathrm{el}}$ is the total electronic charge and e_{nucl} is the total nuclear charge.

From the coupling formula, we obtain for the electric quadrupole interaction energy the result

$$E_{\mathrm{interaction}} = -e_{\mathrm{el}}e_{\mathrm{nucl}}\langle J\|r^{-3}P_2\|J\rangle\langle I\|r^2P_2\|I\rangle[5(2F+1)]^{\frac{3}{2}}\begin{Bmatrix} F & I & J \\ 2 & J & I\end{Bmatrix}. \tag{7.7.47}$$

The 6-j coefficient that enters here has the explicit evaluation:

$$\begin{Bmatrix} F & I & J \\ 2 & J & I\end{Bmatrix} = (-)^{I+J+F}\left[\frac{(2I-2)!(2J-2)!}{(2I+3)!(2J+3)!}\right]^{\frac{1}{2}}[6A(A+1)-8I(I+1)J(J+1)]$$

$$\text{with } A \equiv F(F+1)-I(I+1)-J(J+1). \tag{7.7.48}$$

The interest in this result lies in the fact that the term A, which enters the formula, has the significance of itself being proportional to a Racah coefficient, namely, the coefficient $\begin{Bmatrix} F & I & J \\ 1 & J & I\end{Bmatrix}$, which in the classical limit corresponds to $\hat{\mathbf{I}}\cdot\hat{\mathbf{J}}$, the cosine of the angle between $\mathbf{I}$ and $\mathbf{J}$. Correspondingly, the Racah coefficient $\begin{Bmatrix} F & I & J \\ 2 & J & I\end{Bmatrix}$ is the quantal analog to the classical Legendre coefficient $P_2(\hat{\mathbf{I}}\cdot\hat{\mathbf{J}})$ [see Eq. (5.98), Chapter 5].

The meaning of the interaction energy, as given by the coupling formula, is now clear: It is simply the quantum mechanical form of the interaction energy of two classical *cylindrically symmetric* quadrupoles, whose classical energy varies as $P_2(\cos\theta)$, θ being the angle between the two symmetry axes.

This quadrupolar interaction is significant for spectroscopy (Kopferman [26]); the energy differences—unlike the magnetic dipole hyperfine interaction—are not constant and, hence, deviate from the earlier empirical spectral rules (signaling originally the existence of nuclear quadrupolar moments).

The "quantal Legendre functions," $\begin{Bmatrix} F & I & J \\ L & J & I \end{Bmatrix}$, which enter the generic (static) multipolar interactions in spectroscopy, are tabulated in Table 7 of the Appendix of Tables.

h. The Majorana formula. The behavior of an arbitrary magnetic moment in a time-varying magnetic field was established quantum mechanically – both qualitatively and quantitatively – in a famous paper[1] by Majorana [14]. These results were of basic importance for the atomic beam experiments of I. I. Rabi, and in the recent Rabi Festschrift, Schwinger [28] discusses the Majorana formula in a retrospective paper that is most informative and engaging.[2]

The essence of the Majorana formulation, as interpreted by Schwinger, is to reduce the problem of an arbitrary spin-j to that of a superposition of $2j$ spin-$\frac{1}{2}$ systems. The motion of the superposed system of spins proves to be equivalent to that of a *single* spin-$\frac{1}{2}$ system.

To see this latter point explicitly, let us note that the Schrödinger equation for a system with magnetic moment $\mathbf{M} = g\mu\mathbf{J}$ in a magnetic field $\mathbf{B}(t)$ is

$$i\hbar\frac{\partial}{\partial t}\psi(t) = -g\mu\mathbf{J}\cdot\mathbf{B}(t)\psi(t). \tag{7.7.49}$$

The eigenvector $\psi(t)$ can be regarded as the result of a unitary operator applied to the initial eigenvector $\psi(0)$:

$$\psi(t) = U(t)\psi(0). \tag{7.7.50}$$

Since $\mathbf{J}$ can be represented as the sum of $2j$ kinematically independent spin-$\frac{1}{2}$ systems – that is,

$$\mathbf{J} = \tfrac{1}{2}\sum_{k=1}^{2j}\boldsymbol{\sigma}_k, \tag{7.7.51}$$

we find the equation of motion of the operator $U(t)$ to be

$$i\hbar\frac{\partial}{\partial t}U(t) = \left[\tfrac{1}{2}\sum_{k=1}^{2j}(-g\mu)\boldsymbol{\sigma}_k\cdot\mathbf{B}(t)\right]U(t). \tag{7.7.52}$$

The equation of motion thus factorizes, so that we may represent $U(t)$ by the product

---

[1] Güttinger [27] independently established similar results.

[2] The history of the Majorana formula is also discussed by Ramsey [29].

$$U(t) = \prod_{k=1}^{2j} U_k(t), \tag{7.7.53}$$

hence finding that each $U_k(t)$ satisfies the same equation:

$$i\hbar \frac{\partial}{\partial t} U_k(t) = -\tfrac{1}{2} g\mu\boldsymbol{\sigma}_k \cdot \mathbf{B}(t) U_k(t). \tag{7.7.54}$$

The original problem has thus been reduced to that of a system with spin-$\frac{1}{2}$ and gyromagnetic ratio g, in the time-varying magnetic field $\mathbf{B}(t)$.

One now recognizes that for this spin-$\frac{1}{2}$ problem the net effect of the magnetic field is to induce a rotation by an angle $\beta(t)$ from the initial spin direction to the final spin direction. It follows that for the desired spin-$\frac{1}{2}$ system the probability, $W(m, m', t)$, of observing the projection m' at time t for the system initially having projection m (at $t = 0$) is given by

$$W(m, m', t) = |d^j_{m'm}(\beta(t))|^2, \tag{7.7.55}$$

where $d^j_{m'm}(\beta)$ is the rotation matrix element of Chapter 3, Section 5.

Majorana gave an explicit evaluation of the rotation matrix element $d^j_{m'm}(\beta)$ appearing in Eq. (7.7.55) and related the angle $\beta(t)$ to the rotation angle of a spin-$\frac{1}{2}$ system in the time-dependent magnetic field. The "Majorana formula" is precisely Eq. (7.7.55) with Eq. (3.65) substituted for $d^j_{m'm}(\beta)$; the explicit result is unnecessary to repeat here.

Majorana's approach to this problem was based upon the remark that any state (of spin-j) can be represented by $2j$ points on the unit sphere. (This aspect has been discussed in Section c above). Schwinger recounts that it was his attempt to understand this "baffling" remark that led to his interpretation of a spin-j system as the superposition of $2j$ spin-$\frac{1}{2}$ systems (1937, unpublished) and later (in 1951) to a reinterpretation in terms of the techniques of "second quantization" (published in Biedenharn and van Dam [30]).

One can thus trace back to Majorana's formula the impetus that led to the boson calculus in angular momentum theory. The earlier, and more general, application of second quantization techniques (both fermionic and bosonic) to the general unitary group by Jordan [31] apparently played no role in this development.

In an appendix to Ref. [28], Schwinger gave another significant interpretation of Majorana's formula. To obtain Schwinger's result, we apply the Wigner product law, Eq. (3.189), Chapter 3, to the right-hand side of Eq. (7.7.55). In Schwinger's notation the result is

$$p(m, m'; \beta) = |D^j_{m'm}(\alpha\beta\gamma)|^2$$

$$= (2j+1)^{-1} \sum_{l=0}^{2j} (2l+1) P_l(j,m) P_l(\cos\beta) P_l(j,m'), \tag{7.7.56}$$

where $P_l(\cos\beta)$ is the usual Legendre function, and $P_l(j,m)$ denotes the matrix element of the unit tensor operator:

$$P_l(j,m) \equiv \langle jm| \left\langle \begin{matrix} & l & \\ 2l & & 0 \\ & l & \end{matrix} \right\rangle |jm\rangle. \tag{7.7.57}$$

(As discussed in Chapter 5, Section 8, the $P_l(j,m)$ are quantal operator analogs to the Legendre function.) Alternatively, the probability result given above can be related to the transformation properties of the Wigner operators $\left\langle \begin{matrix} & \cdot & \\ 2l & & 0 \\ & \cdot & \end{matrix} \right\rangle$.

Meckler, in discussing the Majorana formula (Meckler [32]), noted that the $P_l(j,m)$ were discrete variable Tchebichef polynomials (Erdélyi [33]). This interpretation follows from the fact that the Wigner coefficients are related to generalized hypergeometric functions, as discussed in RWA. Applications of Majorana's formula to magnetic resonance were first made by Rabi [34] (see also Abragam and Bleany [35] and Gilmore [36].

References

1. J. von Neumann, *Mathematical Foundations of Quantum Mechanics.* Princeton Univ. Press, Princeton, N. J., 1955.
2. R. P. Feynman, "Space–time approach to non-relativistic quantum mechanics," *Rev. Mod. Phys.* **20** (1948), 367–387.
3. P. Jordan and J. von Neumann, "On inner products in linear metric spaces," *Ann. of Math.* **36** (1935), 719–723.
4. J. M. Jauch, *Foundations of Quantum Mechanics.* Addison-Wesley, Reading, Mass., 1968.
5. P. Jordan, J. von Neumann, and E. Wigner, "On an algebraic generalization of the quantum mechanical formalism," *Ann. of Math.* **35** (1935), 29–64.
6. F. Gürsey, "Color quarks and octonions," in *Johns Hopkins University Workshop on Current Problems in High Energy Particle Theory* (G. Domokos and S. Kövesi-Domokos, eds.), pp. 15–42. Baltimore, Md., 1974.
7. L. P. Horwitz and L. C. Biedenharn, "Intrinsic superselection rules in algebraic Hilbert space," *Helv. Phys. Acta* **38** (1965), 385–408.
8. U. Fano, "Description of states in quantum mechanics by density matrix and operator techniques," *Rev. Mod. Phys.* **29** (1957), 74–93.
9. F. Bloch and I. I. Rabi, "Atoms in variable magnetic fields," *Rev. Mod. Phys.* **17** (1945), 237–244.
10. F. Bloch, "Nuclear induction," *Phys. Rev.* **70** (1946), 460–474.

11. I. I. Rabi, N. F. Ramsey, and J. Schwinger, "Use of rotating coordinates in magnetic resonance problems," *Rev. Mod. Phys.* **26** (1954), 167–171.
12. L. C. Biedenharn, "A note on statistical tensors in quantum mechanics," *Ann. Phys.* (*N.Y.*) **4** (1958), 104–113.
13. G. Racah, *Group Theory and Spectroscopy*, Lectures at the Institute for Advanced Study, Princeton, N. J., 1951; published in *Ergeb. Exakt. Naturw.* **37** (1965), 28–84.
14. E. Majorana, "Atomi orientati in campo magnetico variabile," *Nuovo Cimento* [8], **9** (1932), 43–50.
15. H. Bacry, "Orbits of the rotation group on spin states," *J. Math. Phys.* **15** (1974), 1686–1688.
16. L. Michel, "Nonlinear group action. Smooth action of compact Lie groups on manifolds," in *Statistical Mechanics and Field Theory* (R. N. Sen and C. Weil, eds.), pp. 133–150. Israel Universities Press, Jerusalem, Wiley-Halstead, New York, 1972.
17. M. G. Doncel, L. Michel, and P. Minnaert, "Rigorous spin tests from usual strong decays," *Nucl. Phys.* **B38** (1972), 477–528.
18. J. K. Lubanski, "Sur la theorie des particules elementaires de spin quelconque," *Physica* **9** (1942), 310–338.
18a. D. M. Fradkin and R. H. Good, Jr., "Electron polarization operators," Rev. Mod. Phys. **33** (1961), 343–352.
19. J. D. Bjorken and S. D. Drell, *Relativistic Quantum Mechanics*. McGraw-Hill, New York, 1964.
20. J. J. Sakurai, *Advanced Quantum Mechanics*. Addison-Wesley, Reading, Mass., 1967.
21. E. Merzbacher, *Quantum Mechanics*. Wiley, New York, 1961.
22. E. P. Wigner, "Relativistische Wellengleichungen," *Z. Physik* **124** (1948), 665–684.
23. H. F. Schopper, *Weak Interactions and Nuclear Beta Decay*. North-Holland, Amsterdam, 1966.
24. U. Fano and G. Racah, *Irreducible Tensorial Sets*. Academic Press, New York, 1959.
25. The very extensive literature is summarized in *The Electromagnetic Interaction in Nuclear Spectroscopy* (W. D. Hamilton, ed.). North-Holland, Amsterdam, 1975; see in particular Chapter 13, Extranuclear perturbations of angular distributions and correlations," by R. M. Steffen and K. Alder.
26. H. Kopferman, *Nuclear Moments*. Academic Press, New York, 1958.
27. P. Güttinger, "Das Verhalten von Atomen im magnetischen Drehfeld," *Z. Physik* **73** (1932), 743–765.
28. J. Schwinger, "The Majorana formula," *Trans. N. Y. Acad. Sci.* [II] **38** (1977), 170–184.
29. N. F. Ramsey, *Molecular Beams*. Oxford Univ. Press, London, 1956.
30. L. C. Biedenharn and H. van Dam (eds.), *The Quantum Theory of Angular Momentum*. Academic Press, New York, 1965.
31. P. Jordan, "Der Zusammenhang der symmetrischen und linearen Gruppen und das Mehrkörperproblem," *Z. Physik* **94** (1935), 531–535.
32. A. Meckler, "The Majorana formula," *Phys. Rev.* **111** (1958), 1447–1449.
33. A. Erdélyi, W. Magnus, F. Oberhettinger, and G. F. Tricomi, *Higher Transcendental Functions*, Vol. II, Chapter X, p. 223. McGraw-Hill, New York, 1953.
34. I. I. Rabi, "Space quantization in a gyrating magnetic field," *Phys. Rev.* **51** (1937), 652–654.
35. A. Abragam and B. Bleany, *Electron Paramagnetic Resonance of Transition Ions*. Clarendon, Oxford, 1970.
36. R. Gilmore, *Lie Groups, Lie Algebras, and Some of their Applications*. Wiley, New York, 1974.

8. Angular Correlations and Angular Distributions of Reactions

a. The nature of the angular correlation process. In order to understand intuitively the nature of the angular correlation process, it is useful to consider a schematic example to illustrate the concepts. Let us consider a nonrelativistic nuclear reaction produced by the collision of two (distinguishable) *spinless* particles that form a resonant composite state of angular momentum l decaying to a final state consisting of another (distinct) pair of spinless particles.

The reaction is described in the reference frame where the total momentum vanishes (see Section 2). As typical of quantum mechanics, the actual calculation concerns not the behavior of the particles per se but the average behavior of a statistical ensemble. In this example one prepares plane wave beams of incident particles of momentum $\mathbf{k}_i$ and then identifies by measurement the types of particles and final momentum $\mathbf{k}_f$ of the emergent beam of reaction-produced particles.

It follows at once from rotational invariance that the physical measurement (flux of emergent particles) can be a function only of the single (invariantly defined) angle: $\cos\theta = \hat{k}_i \cdot \hat{k}_f$. We may accordingly express the differential reaction probability[1] $dW/d\Omega$ as a Legendre series:

$$dW(\theta)/d\Omega = (4\pi)^{-1} \sum_{\nu} (2\nu + 1) B_\nu P_\nu(\cos\theta),$$

$$\cos\theta \equiv \hat{k}_i \cdot \hat{k}_f. \tag{7.8.1}$$

We now come to the crux of the issue: the implications of the "angular momentum information" contained in the problem. What information is there? We have two kinds of information: (1) dynamic: the information implied by the fact that the reaction proceeds through a state of definite angular momentum of magnitude l; and (2) kinematic: the angular information implied by the measurement of initial and final directions of motion, $\mathbf{k}_i$ and $\mathbf{k}_f$.

From the uncertainty relation for angular momentum,[2] we can conclude from (1) the (kinematic) implication that *the angle* χ *conjugate to* $(\mathbf{L}^2)^{\frac{1}{2}}$ *is random* (*uniformly distributed*). From the measurement made in (2), we can conclude that *the orbital angular momentum of the incident* (*or emergent*) *particles is perpendicular to the direction of motion.* (Hence, the projection of the

[1] The differential reaction probability is the flux of emergent particles within solid angle $d\Omega$ at the point (θ, ϕ) normalized to unity over 4π solid angle. In Eq. (7.8.1) this normalization implies $B_0 = 1$.

[2] This subject is developed in detail in RWA.

angular momentum, m, is zero for the z-axis taken along the direction of motion of the particle.)

This information – though seemingly rather slight, even trivial – is sufficient to yield a unique angular distribution, as we now shall show.

For simplicity, let us first take $l \gg 1$ so that we can proceed semiclassically. From the conservation of angular momentum, the angular momentum $\mathbf{L}$ of the composite (resonant) system equals the orbital angular momentum contributed by the incident beam; *thus, the direction of the angular momentum* $\mathbf{L}$ *is uniformly distributed perpendicular to the direction* $\hat{k}_i$.

The compound system, after being formed, then breaks up, emitting particles *uniformly in all directions* $\hat{k}_f$ *perpendicular to* $\mathbf{L}$. This follows from the uncertainty relation applied to (1) mentioned above.

We conclude: *The direction* $\hat{k}_f$ *is uniformly distributed over the sphere*; *that is, taking* $\hat{k}_i$ *as the axis, we find that both* θ *and* ϕ *are uniformly distributed.*[1]

The probability *in solid angle* (normalized to unity) is accordingly

$$\frac{dW(\theta)}{d\Omega} = (2\pi^2)^{-1} (\sin\theta)^{-1}. \tag{7.8.2}$$

This result (to our knowledge first obtained by Christy [1]) is of interest in its own right, for it provides a limiting form for the angular distribution of a nuclear reaction proceeding through a resonance of large angular momentum involving particles of small intrinsic spin. (Such angular distributions have been observed in heavy ion reactions.)

Note the characteristic features: There is a strong forward peaking with a width typical of large angular momentum, and there is symmetry about $\theta = 90°$, resulting from sharp parity (single orbital angular momentum involved).

Let us now reconsider this same example from a more general point of view. Any correlation where only two directions of motion are measured must be expressible as a Legendre series in the invariantly defined angle θ [see Eq. (7.8.1)].

The coefficient B_ν in Eq. (7.8.1) is to be interpreted as an average value, *the value of* $P_\nu(\hat{k}_i \cdot \hat{k}_f)$ *averaged over all possible configurations of the system*; that is,

$$B_\nu \equiv [P_\nu(\hat{k}_i \cdot \hat{k}_f)]_{\text{av}}. \tag{7.8.3}$$

If we use now the addition theorem for the Legendre polynomials [see Eq. (6.137), Chapter 6], we can formally introduce the angular momentum

---

[1] By this derivation θ ranges over 0 to 2π. The extra range guarantees that symmetry about $\theta = \pi/2$, as required by parity, will actually occur.

L as the z-axis. Recalling that the conjugate angle χ is random, we find

$$[P_\nu(\hat{k}_i \cdot \hat{k}_f)]_{\text{av}} = P_\nu(\hat{k}_i \cdot \hat{L}) P_\nu(\hat{k}_f \cdot \hat{L})$$

$$= [P_\nu(\cos \pi/2)]^2, \tag{7.8.4}$$

where in the last step we used information (2). Hence, we find

$$dW(\theta)/d\Omega = (4\pi)^{-1} \sum_\nu (2\nu + 1)[P_\nu(0)]^2 P_\nu(\cos\theta). \tag{7.8.5}$$

This result is, clearly, just the Legendre series form of Eq. (7.8.2), but the point of the derivation is that, by using the addition theorem, we are in possession of a general method.

Let us use this procedure to eliminate the requirement that $l \gg 1$. From the discussion in Chapter 5 [see Eq. (5.94)] on the classical limit of the Wigner coefficients, it will be recalled that the "proper angle functions" for the angle between a quantized (angular momentum) vector and an (unquantized) direction are just the Wigner coefficients:

$$C^{l\nu l}_{m0m} \sim P_\nu(\hat{l} \cdot \hat{k}), \tag{7.8.6}$$

where $m/l \sim \hat{l} \cdot \hat{k}$, with $\hat{l}$ a unit vector in the direction of the vector **L**. Since **L** is perpendicular to $\hat{k}$, the m-value to be used is zero: $P_\nu(\hat{k} \cdot \hat{l}) \sim C^{l\nu l}_{000}$.

The restriction of ν to even integers, which comes about in the classical case, Eq. (7.8.5), from the vanishing of $P_\nu(0)$ for odd integers, also obtains for the Wigner coefficients here, which similarly vanish for odd ν.

This limit relation suggests – and a more precise calculation confirms – that the quantum mechanical form of the angular distribution is

$$\frac{dW(\theta)}{d\Omega} = \frac{1}{4\pi} \sum_{\nu=0}^{2l} (2\nu + 1)(C^{l\nu l}_{000})^2 P_\nu(\cos\theta)$$

$$= \frac{1}{4\pi} \sum_{\nu=0}^{2l} (2l + 1)(C^{ll\nu}_{000})^2 P_\nu(\cos\theta). \tag{7.8.7}$$

The result we have obtained in Eq. (7.8.7) already displays a number of significant features that are typical of the general direction–direction angular distribution. In only one respect is this result too particular: the limitation to *spinless* incident and emergent particles.

To eliminate this restriction is, however, not quite the straightforward task as it might seem. We are, in fact, at a crucial juncture, which distinguishes the relativistic and nonrelativistic regimes. In the relativistic regime (Poincaré group), spin measurements on an elementary particle are kinematically related to the motion of the particle (Wigner [2]); this has been discussed in Section 7d. This state of affairs is in sharp contrast to the nonrelativistic situation, where the spin observables are independent of the state of motion.

Let us for simplicity particularize for the moment to the *nonrelativistic regime* (Galilei group). The spins ($\mathbf{s}_1$, $\mathbf{s}_2$) of the pair of incident particles can, with no loss of generality, each be considered in a *fixed* reference frame (identified with the center-of-mass frame). It is conventional to call the coupled spin, $\mathbf{s}_i \equiv \mathbf{s}_1 + \mathbf{s}_2$, the *channel spin* of the *i*ncident channel (a channel denotes a specific pair of particles).

The angular momentum $\mathbf{J}$ of the resonant state is accordingly the sum of the channel spin $\mathbf{s}_i$ and the orbital angular momentum $\boldsymbol{l}_i$ of the relative motion of the incident pair:

$$\mathbf{J} = \boldsymbol{l}_i + \mathbf{s}_i. \tag{7.8.8}$$

A similar relation holds for the final (exit) channel with orbital angular momentum $\boldsymbol{l}_f$ and channel spin $\mathbf{s}_f$.

The decisive point now is this: *If no observation of any spin is made, we argue that the channel spins are to be considered as randomly and uniformly oriented.* This is decisive, since from Eq. (7.8.3), the coefficient B_ν is defined to be

$$B_\nu = [P_\nu(\hat{k}_i \cdot \hat{k}_f)]_{\text{av}}, \tag{7.8.9}$$

and we can, as before, apply the addition theorem in order to average over χ (angle conjugate to $\mathbf{J}$); we thus obtain

$$B_\nu = [P_\nu(\hat{k}_i \cdot \hat{J})]_{\text{av}} \times [P_\nu(\hat{k}_f \cdot \hat{J})]_{\text{av}}. \tag{7.8.10}$$

The two required averagings in Eq. (7.8.10) can now be carried out by appealing to the random orientation of the initial and final channel spins.

Using the addition theorem on each term in Eq. (7.8.10), we find

$$B_\nu = [P_\nu(\hat{k}_i \cdot \hat{l}_i)P_\nu(\hat{l}_i \cdot \hat{J})] \times [P_\nu(\hat{k}_f \cdot \hat{l}_f)P_\nu(\hat{l}_f \cdot \hat{J})]. \tag{7.8.11}$$

The numerical coefficient B_ν is completely determined by this procedure, since Eq. (7.8.8) determines the argument $\hat{l}_i \cdot \hat{J}$ to be

$$\hat{l}_i \cdot \hat{J} = (\mathbf{J}^2 + \boldsymbol{l}_i^2 - \mathbf{s}_i^2)/(2\sqrt{\mathbf{J}^2\boldsymbol{l}_i^2}), \tag{7.8.12}$$

and (as before) $\hat{l}_i$ is perpendicular to $\hat{k}_i$. (A similar result holds for $\hat{l}_f$, $\hat{k}_f$, and $\hat{J}$.)

To obtain the quantum mechanical transcription of these results for B_ν given by Eq. (7.8.11), we recall (Chapter 5, Eq. (5.98)) that the analog to the Legendre function for the angle between two quantized angular momenta is given by the Racah coefficient:

$$P_\nu(\hat{l}_i \cdot \hat{J}) \sim (-)^\nu [(2l_i + 1)(2J + 1)]^{\frac{1}{2}} W(l_i \nu s_i J; l_i J). \qquad (7.8.13)$$

Using this result and Eq. (7.8.6) to rewrite B_ν, we now obtain from Eqs. (7.8.1) the following form for the angular distribution:

$$\frac{dW(\theta)}{d\Omega} = (-1)^{s_i - s_f}(4\pi)^{-1} \sum_\nu A_\nu(\text{initial})\, A_\nu(\text{final}) P_\nu(\cos\theta),$$

$$A_\nu(\text{initial}) \equiv (2J + 1)^{\frac{1}{2}}(2l_i + 1)\, C^{l_i l_i \nu}_{000}\, W(l_i l_i JJ; \nu s_i),$$

$$A_\nu(\text{final}) \equiv (2J + 1)^{\frac{1}{2}}(2l_f + 1)\, C^{l_f l_f \nu}_{000}\, W(l_f l_f JJ; \nu s_f). \qquad (7.8.14)$$

This result expresses the general nonrelativistic, quantum mechanical form for the angular distribution, $dW(\theta)/d\Omega$, of a reaction involving a compound state of angular momentum J with unobserved channel spins s_i and s_f, and specified orbital angular momenta l_i and l_f. (We shall shortly prove that this result is, in fact, *valid relativistically*, in the frame where the composite system has total spatial momentum zero.)

Let us discuss now the features that characterize this result:

1. The coefficients for the Legendre series *factor* into two parts, each part determined by the angular momenta associated with the measurement of a given direction of motion. This factorization property is a general feature of direction–direction correlations proceeding through an intermediate state of fixed angular momentum magnitude, even if the orbital angular momenta involved in the reaction do not have a unique value. The ultimate source of this property, as we have indicated, is the angular momentum uncertainty relation.

2. The relevant information for the correlation process comes from two facts: The orbital angular momentum $\boldsymbol{l}$ is perpendicular to the direction of motion $\mathbf{k}$, and the conservation of angular momentum fixes the "angle" [Eq. (7.8.12)] between $\boldsymbol{l}$ and $\mathbf{J}$ in the sum $\mathbf{J} = \boldsymbol{l} + \mathbf{s}$.

3. The symmetry about 90° comes from the fact that the Wigner coefficient introduced by this perpendicularity vanishes for odd ν; that is, $C^{ll\nu}_{000} = 0$ unless ν is even. This symmetry is a general feature reflecting the fact that the intermediate state has sharp parity.

4. The question as to whether the individual processes are emissions or absorptions, as well as the time sequence of the radiations, is irrelevant. This is

a general feature of the direction–direction correlation involving nuclear states of sharp angular momentum.

5. The fixing of the specific angular momenta s_i, s_f, l_i, l_f, and J is *dynamical*, and angular correlation theory thus makes no statement here, aside from conservation of angular momentum magnitude.

An important consequence of the quantization, which in fact greatly increases the utility of the correlation process as an experimental tool, is that in the Legendre series the index ν assumes only a finite number of values. There is now an upper limit on ν determined by the triangle rules in the definition of the Racah coefficients. Thus, one finds

$$\nu_{\max} \leqslant \text{minimum } (2J, 2l_i, 2l_f). \tag{7.8.15}$$

Furthermore, parity conservation, expressed by the properties of the Wigner coefficient $C^{l\nu l}_{000}$, restricts the values of ν to be even.

b. Cascades. The semiclassical construction used to derive Eq. (7.8.7), and similarly, Eq. (7.8.14), is a very direct way to construct more complicated correlations. Consider, for example, the correlation when an intermediate state of angular momentum J emits an *unobserved* radiation of fixed angular momentum l', going to a state J', that then emits the observed radiation[1] along $\hat{k}_f$. For brevity we designate this process by the quantum number sequence: $j_i(l_i)J(l')J'(l_f)j_f$. One notes now that, in addition to the randomly oriented vectors of the previous calculation, we have the additional randomly oriented vector $\boldsymbol{l}'$.

Applying the addition theorem to $[P_\nu(\hat{k}_i \cdot \hat{k}_f)]_{\text{av}}$, just as before, requires now *an extra step because of the angular momentum triangle* $(JJ'l')$. Thus, we must introduce the factor

$$[P_\nu(\hat{k}_f \cdot \hat{J})]_{\text{av}} = P_\nu(\hat{J} \cdot \hat{J}')[P_\nu(\hat{k}_f \cdot \hat{J}')]_{\text{av}}, \tag{7.8.16}$$

into Eq. (7.8.10). The additional, but unobserved, radiation introduces, therefore, another Legendre coefficient in the classical result. In the quantum analog, one thus simply inserts into Eq. (7.8.14) the extra factor:

$$[P_\nu(\hat{J} \cdot \hat{J}')]_{QM} \equiv (-)^\nu [(2J+1)(2J'+1)]^{\frac{1}{2}} W(J\nu l'J'; JJ'). \tag{7.8.17}$$

(This calculation provides a physical application of the concepts in Section 7g wherein a random subsystem partially depolarizes the composite system.)

---

[1] We now neglect recoil effects, so that the center-of-momentum frame is fixed during the complete process.

One can appreciate from this elementary derivation of the cascade correlation how direct and economical these semiclassical methods (which originated with Racah) really are! For they operate only with the *observables*, and do not introduce the cumbersome machinery of all possible observations (as the "maximal" techniques of magnetic quantum number sums necessarily do). To be sure, our derivation above was rather heuristic, but it can be made more rigorous without any essential change.

c. Stretched angular momenta. If we suppose that in a cascade an unobserved radiation of angular momentum L_1 links two states of angular momenta j_1, j_2, and that these three angular momenta are related either by $j_1 = L_1 + j_2$, or by $j_2 = L_1 + j_1$ (that is to say, the angular momentum triangle is "stretched"), then for the classical calculation we see that the extra Legendre coefficient introduced by this radiation is just unity. Thus, in this special case, the unobserved radiation has *no* influence on the correlation whatever.

The quantum result, on the other hand, does not show quite this simplicity. The reason is that the quantum equivalent to the Legendre function introduced by the unobserved radiation does not become unity (or even independent of ν) for the case of stretched angular momenta. It is clear, then, that a given unobserved radiation with stretched angular momenta does indeed generally affect the correlation in the quantum result. (The origin of this result is again to be found in the fact that the direction of an angular momentum is not an observable; the triangle formed by "stretched angular momenta" cannot have zero area.)

There is, however, one special case where even the quantum result simplifies. This is the situation where *all* the angular momenta in a given cascade are parallel, which we define as $j_n = L_n + j_{n+1}$ (the case $j_{n+1} = L_n + j_n$ is equivalent, since it corresponds to the same correlation in reverse order). For this situation, as the explicit formulas for the Racah coefficients show, the correlation in *every* case becomes the same as the double correlation for the simplest case, with the radiations having the same angular momenta as the two observed in the cascade; that is, one has the scheme $j_i(L_i)L_f(L_f)0$, with $j_i = L_i + L_f$.

d. More involved correlation processes. We consider first triple correlations. Let us take as our prototype the triple correlation of three spinless particles proceeding through sharp states of definite angular momentum—that is, the correlation process denoted by $j_i(l_0)J(l_1)J'(l_2)j_f$. The basic information for the correlation is just as for the simpler direction–direction correlation: that $\hat{l} \cdot \hat{k}$ is zero in every case, and that j_i and j_f are randomly oriented.

A classical correlation exists for this elementary case, but the necessary functions for the classical description are not very familiar. This is immediately clear when one asks, "What angle functions furnish the proper description for three vectors?" For *two* vectors, it is obvious that the Legendre functions $P_\nu(\hat{k}_1 \cdot \hat{k}_2)$ must enter, for these are the orthogonal functions of the single polar angle, defined independently of any reference system. For three vectors there are three invariantly defined angles, and the proper orthonormal angle functions hardly look classical at all. The appropriate functions have been discussed in Chapter 6, Section 17. They are the rotationally invariant functions defined by

$$P_{\nu_0\nu_1\nu_2}(\hat{k}_0\hat{k}_1\hat{k}_2) \equiv (4\pi)^{\frac{3}{2}}(2\nu_0+1)^{-\frac{1}{2}}(2\nu_1+1)^{-\frac{1}{2}}(2\nu_2+1)^{-1}(i)^{\nu_0+\nu_1-\nu_2}$$
$$\times \sum_{\alpha\beta\gamma} (-)^{\gamma} C^{\nu_0\nu_1\nu_2}_{\alpha\beta\gamma} Y_{\nu_0\alpha}(\hat{k}_0)\, Y_{\nu_1\beta}(\hat{k}_1)\, Y_{\nu_2,-\gamma}(\hat{k}_2)$$
$$= (-1)^{\nu_0-\nu_1}(i)^{\nu_0+\nu_1-\nu_2} I_{(\nu_0\nu_1\nu_2)}(\hat{k}_0,\hat{k}_1,\hat{k}_2). \tag{7.8.18}$$

[Despite the appearance of the Wigner coefficients in this formula, it is nonetheless a classical result, for the Wigner coefficients occur here in the context of a *general vector algebra*. For example, the P_{111} function is just the invariant, antisymmetric combination of three vectors; that is, $\mathbf{A}\cdot\mathbf{B}\times\mathbf{C}$ (to within an overall constant). The other functions similarly represent invariants, though of different orders, as discussed in some detail in Chapter 6, Section 17.]

The $P_{\nu_0\nu_1\nu_2}$ defined by Eq. (7.8.18) have the symmetry properties

$$P_{\nu_0\nu_1\nu_2}(\hat{k}_0\hat{k}_1\hat{k}_2) = (-1)^{\nu_0+\nu_1+\nu_2} P_{\nu_1\nu_0\nu_2}(\hat{k}_1\hat{k}_0\hat{k}_2)$$
$$= (-1)^{\nu_0+\nu_1+\nu_2} P_{\nu_0\nu_2\nu_1}(\hat{k}_0\hat{k}_2\hat{k}_1). \tag{7.8.19}$$

All other permutations may be reduced to products of these two. The antisymmetry of $\mathbf{A}\cdot\mathbf{B}\times\mathbf{C}$ thus fits the general rule (7.8.19).

The *classical* triple correlation for the reaction designated by the angular momenta $j_i(l_0)J(l_1)J'(l_2)j_f$ must be expressible by a series in terms of our orthogonal basis functions—that is,

$$W(\hat{k}_0\hat{k}_1\hat{k}_2) = (4\pi)^{-3} \sum_{\nu_0\nu_1\nu_2} (2\nu_0+1)(2\nu_1+1)(2\nu_2+1)$$
$$\times B_{\nu_0\nu_1\nu_2} P_{\nu_0\nu_1\nu_2}(\hat{k}_0\hat{k}_1\hat{k}_2). \tag{7.8.20}$$

This equation is the exact analog of Eq. (7.8.1) for the double correlation, and, in a similar fashion, the $B_{\nu_0\nu_1\nu_2}$ are to be interpreted as the average value of the $P_{\nu_0\nu_1\nu_2}$ evaluated over all possible configurations of the system consistent with our information. Thus, one has

$$B_{\nu_0\nu_1\nu_2} \equiv [P_{\nu_0\nu_1\nu_2}(\hat{k}_0\hat{k}_1\hat{k}_2)]_{\text{av}}. \tag{7.8.21}$$

To eliminate inessential detail, consider first the special case where $j_i = j_f = 0$, and the sequence of angular momenta is $0(l_0)J(l_1)J'(l_2)0$, with $J = l_0$ and $J' = l_2$.

The azimuthal angles about l_0, l_1, l_2 are each *random*. Using the form of the addition theorem given in Chapter 6, Eq. (6.23) and Eq. (7.8.18) above, it follows that

$$\begin{aligned} B_{\nu_0\nu_1\nu_2} &= [P_{\nu_0\nu_1\nu_2}(\hat{k}_0\hat{k}_1\hat{k}_2)]_{\text{av}} \\ &= P_{\nu_0}(\hat{k}_0 \cdot \hat{l}_0)[P_{\nu_0\nu_1\nu_2}(\hat{l}_0\hat{k}_1\hat{k}_2)]_{\text{av}} \\ &= P_{\nu_0}(\hat{k}_0 \cdot \hat{l}_0)P_{\nu_1}(\hat{k}_1 \cdot \hat{l}_1)[P_{\nu_0\nu_1\nu_2}(\hat{l}_0\hat{l}_1\hat{k}_2)]_{\text{av}} \\ &= P_{\nu_0}(\hat{k}_0 \cdot \hat{l}_0)P_{\nu_1}(\hat{k}_1 \cdot \hat{l}_1)P_{\nu_2}(\hat{k}_2 \cdot \hat{l}_2)P_{\nu_0\nu_1\nu_2}(\hat{l}_0\hat{l}_1\hat{l}_2). \end{aligned} \tag{7.8.22}$$

The structure of this result is apparent, for we have successively referred to $\hat{l}_0$, $\hat{l}_1$, and $\hat{l}_2$ as axes for the $P_{\nu_0\nu_1\nu_2}$ function, and averaged over the azimuth in each case. The one essentially new feature is the appearance of the $P_{\nu_0\nu_1\nu_2}(\hat{l}_0\hat{l}_1\hat{l}_2)$, which, it should be noted, is a function of three angular momentum vectors that form a definite triangle with *fixed* lengths, and *fixed* angles. The three vectors *thus lie in a plane*, and the $P_{\nu_0\nu_1\nu_2}$ function is correspondingly simpler; it is given explicitly by the relation (7.8.18) and Eq. (6.160), Chapter 6. Moreover, the symmetry of the $P_{\nu_0\nu_1\nu_2}$ requires that for $\nu_0 + \nu_1 + \nu_2 =$ odd integer the function *vanishes*.

It remains only to note now the quantum analog to the classical function $P_{\nu_0\nu_1\nu_2}$ in order to exhibit the quantum mechanical triple correlation result. This function (to within a factor) is Wigner's 9-j coefficient or, as it is also known, Fano's X-coefficient (see Chapter 3, Note 7):

$$P_{\nu_0\nu_1\nu_2}(\hat{J}\hat{l}_1\hat{J}') \sim i^{\nu_2-\nu_1-\nu_0}[(2J+1)(2J'+1)(2l_1+1)]^{\frac{1}{2}} X\begin{pmatrix} J l_1 J' \\ J l_1 J' \\ \nu_0 \nu_1 \nu_2 \end{pmatrix}. \tag{7.8.23}$$

It is interesting to note that this X-coefficient (which is special in that it has two rows the same) vanishes for $\nu_0 + \nu_1 + \nu_2$ an odd integer. This result can be interpreted as the quantum analog to the vanishing of $P_{\nu_0\nu_1\nu_2}(\hat{J}\hat{l}_1\hat{J}')$ for $\nu_0 + \nu_1 + \nu_2$ odd, since the three angular momenta must be coplanar.

It is not difficult to amend Eq. (7.8.23) to take into account initial and final states with nonzero, but unobserved, spin. In fact, the required result is readily seen to be exactly the same as was discussed for the general initial and final links in the direction–direction correlation. In both cases, it is the cylindrical averaging about $\hat{J}$ (or $\hat{J}'$) responsible for this simplification.

The result for the triple correlation of three spinless radiations with the angular momentum sequence $j_i(l_0)J(l_1)J'(l_2)j_f$ is then the following:

Classically:

$$W(\hat{k}_0\hat{k}_1\hat{k}_2) = (4\pi)^{-3} \sum_{\nu_0\nu_1\nu_2} (2\nu_0 + 1)(2\nu_1 + 1)(2\nu_2 + 1)$$

$$\times [P_{\nu_0}(\hat{k}_0 \cdot \hat{l}_0)P_{\nu_0}(\hat{l}_0 \cdot \hat{J})] \times [P_{\nu_1}(\hat{k}_1 \cdot \hat{l}_1)P_{\nu_0\nu_1\nu_2}(\hat{J}\hat{l}_1\hat{J}')]$$

$$\times [P_{\nu_2}(\hat{k}_2 \cdot \hat{l}_2)P_{\nu_2}(\hat{l}_2 \cdot \hat{J}')]P_{\nu_0\nu_1\nu_2}(\hat{k}_0\hat{k}_1\hat{k}_2). \tag{7.8.24}$$

Quantum mechanically:

$$W(\hat{k}_0\hat{k}_1\hat{k}_2) = (4\pi)^{-3}(2l_1 + 1)(-)^{j_i - j_f - J + l_1 + J'}$$

$$\times \sum_{\nu_0\nu_1\nu_2} [(2\nu_0 + 1)(2\nu_1 + 1)(2\nu_2 + 1)]^{\frac{1}{2}}\, i^{\nu_2 - \nu_1 - \nu_0}$$

$$\times \bar{Z}(l_0Jl_0J;j_i\nu_0)\bar{Z}(l_2J'l_2J';j_f\nu_2)\, C^{l_1l_1\nu_1}_{000}\, X\begin{pmatrix} Jl_1J' \\ Jl_1J' \\ \nu_0\nu_1\nu_2 \end{pmatrix}$$

$$\times P_{\nu_0\nu_1\nu_2}(\hat{k}_0\hat{k}_1\hat{k}_2). \tag{7.8.25}$$

The coefficients $\bar{Z}$ are defined in terms of universal angular momentum functions by[1]

$$\bar{Z}(lJl'J';s\nu) = [(2l + 1)(2l' + 1)(2J + 1)(2J' + 1)]^{\frac{1}{2}}\, C^{ll'\nu}_{000}\, W(lJl'J';s\nu). \tag{7.8.26}$$

Note that these coefficients are the quantal analogs to the Legendre terms occurring quite naturally in the semiclassical treatment leading to Eq. (7.8.14). [The phase factor in the summation in Eq. (7.8.25) is always ± 1, since $\nu_2 - \nu_1 - \nu_0$ is always an even integer.]

[1] Tabulations of these coefficients are cited in the Appendix of Tables.

Since our primary purpose was to illustrate the natural setting in which to comprehend the 9-j symbol intuitively, it is not useful to attempt any of the possible generalizations of this result.

We turn now to a brief discussion of parametrization techniques.

The usefulness of the angular correlation process as a technique for spectroscopic investigations is greatly enhanced by the fact that the substitution of an alternative measurement process for, say, a gamma-ray measurement in a correlation can be accomplished parametrically, by multiplying each Legendre coefficient by the corresponding *particle parameter* (Lloyd [3], Racah [4], Biedenharn and Rose [5]). In this way internal conversion angular correlations (Hager and Seltzer [6]), Coulomb excitation correlations (Alder *et al.* [7], Biedenharn and Brussaard [8]) can be encompassed by the formulation. It would take us too far afield to pursue this concept here, but it constitutes an important generalization of the correlation technique.

e. Relativistic regime. The complications introduced by relativity (Poincaré symmetry) for two-body processes are entirely restricted to spin kinematics. For two-body processes, which are defined as reactions produced by (and leading to) pairs of particles, the concept of a frame characterized by vanishing spatial momentum and sharp energy (as discussed in Section 2) can be justified fully. It follows that in this frame the composite system can be labeled by the quantum numbers E, J, and M, and the Poincaré invariant scattering matrix breaks up into well-defined submatrices S^J (de Benedetti [9], Goldberger and Watson [10]).

The kinematic effects complicating spin observations have been discussed in Sections 7d–e. Using the fact that in the chosen frame the total momentum vanishes: $\mathbf{P}_{\text{tot}} = \mathbf{P}_1 + \mathbf{P}_2 = \mathbf{0}$, we see that $\mathbf{P}_1 = -\mathbf{P}_2$, so that *the tetrads defined for each of the two particles* (in either the entrance or the exit channel) *agree in their spatial orientations.* If both particles are massive, we conclude [see Section 7d] that *there exist commuting tetrad polarization operators*, $\mathbf{S}^{(i)}(1)$, $\mathbf{S}^{(i)}(2)$ for the pair of (massive) particles, 1 and 2, so that the channel spin tetrad operator, $\mathbf{S}^{(i)} \equiv \mathbf{S}^{(i)}(1) + \mathbf{S}^{(i)}(2)$, similarly exists and is well-defined. It follows that the channel spin is a well-defined concept relativistically. Moreover, since the relative orbital angular momentum operator $\mathbf{L}$ is well-defined (for massive particles), the kinematic decomposition, using the tetrad formalism, $\mathbf{J}^{(i)} = \mathbf{L}^{(i)} + \mathbf{S}^{(i)}$, also exists.

These remarks validate for the relativistic regime the description of the scattering matrix[1], $S^{J\Pi}$, in the Wigner–Eisenbud [11] channel spin scheme: $S^J \to \langle \alpha' l' s' | S^{J\Pi} | \alpha l s \rangle$, where α denotes the pair ("alternatives") of massive

[1] We have taken parity to be valid, so that the S matrix has the labels J and Π (parity).

particles, l the orbital angular momentum, and s the channel spin. Correspondingly, we have validated the result for the angular distribution of a reaction, Eq. (7.8.14), in the relativistic regime, as stated earlier.

Using the Wigner–Eisenbud approach to two-body reactions, one can obtain a comprehensive treatment of the angular distribution of nuclear reactions (Blatt and Biedenharn [12]).

The differential cross section for the reaction in which a pair of massive particles designated by α_1 (specifically the target nucleus and incident particle) having channel spin, designated by s_1, react and form another pair of massive particles, α_2, with channel spin s_2, is given by the formula

$$d\sigma(\alpha_2 s_2;\alpha_1 s_1) = \frac{\lambda_{s_1}^2}{2s_1+1}\sum_{\nu} B_\nu(\alpha_2 s_2;\alpha_1 s_1)P_\nu(\cos\theta)\,d\Omega, \tag{7.8.27}$$

where

$$B_\nu(\alpha_2 s_2;\alpha_1 s_1) = \tfrac{1}{4}(-1)^{s_2-s_1}\sum \bar{Z}(l_{1a}J_a l_{1b}J_b;s_1\nu)\,\bar{Z}(l_{2a}J_a l_{2b}J_b;s_2\nu)$$
$$\times \operatorname{Re}\{(1-S^{J_a\Pi_a})^*_{\alpha_2 s_2 l_{2a};\alpha_1 s_1 l_{1a}} \times (1-S^{J_b\Pi_b})_{\alpha_2 s_2 l_{2b};\alpha_1 s_1 l_{1b}}\}, \tag{7.8.28}$$

in which $S^{J\Pi}$ denotes the scattering matrix, and the sum is over J_a, Π_a, J_b, Π_b, l_{1a}, l_{2a}, l_{1b}, l_{2b}. (λ_s denotes the wave number k^{-1} in the incident channel.)

The typical term of Eq. (7.8.28) is represented by the angular momentum scheme

$$s_1\begin{pmatrix} l_{1a} \\ l_{1b}\end{pmatrix}\begin{matrix} J_a \\ J_b\end{matrix}\begin{pmatrix} l_{2a} \\ l_{2b}\end{pmatrix}s_2. \tag{7.8.29}$$

For this first link in the reaction (subscript 1), we have the general coefficient $\bar{Z}(l_{1a}J_a l_{1b}J_b;s_1\nu)$; similarly, the second link introduces a general $\bar{Z}$ coefficient. These two links make a contribution to the correlation as weighted by the dynamical factor

$$\operatorname{Re}\{(1-S^{J_a\Pi_a})^*_{\alpha_2 s_2 l_{2a};\alpha_1 s_1 l_{1a}} \times (1-S^{J_b\Pi_b})_{\alpha_2 s_2 l_{2b};\alpha_1 s_1 l_{1b}}\}. \tag{7.8.30}$$

An important feature of the correlation process for pure states was the fact that the separate links *factored.* How general is this property? The angular distribution formula, Eq. (7.8.27), rather obscures the actual generality of this property, for, as written there, the links are closely tied together through the various elements of the $S^{J\Pi}$ matrix. However, the elements of the $S^{J\Pi}$ matrix are not all independent, for we have two basic restrictions: unitarity (conservation of flux), and symmetry (reciprocity), assuming time-reversal invariance. These conditions imply that the scattering matrix $S^{J\Pi}$ may be written in the form

$$S^{J\Pi} = U_{J\Pi}^{-1} \exp(2i\Delta_{J\Pi}) U_{J\Pi}, \tag{7.8.31}$$

where $U_{J\Pi}$ is real and orthogonal, and $\Delta_{J\Pi}$ is real and diagonal. The $N \times N$ matrix $U_{J\Pi}$ thus defines N real eigenvectors, U_k, where $k = 1, 2, \ldots, N$. Utilizing these eigenvectors, one may write the $S^{J\Pi}$ matrix in spectral form, and obtain the following results:

$$B_\nu(\alpha_2 s_2; \alpha_1 s_1) = (-1)^{s_2 - s_1} \sum_{J_a \Pi_a} \sum_{J_b \Pi_b} \sum_{k_a = 1}^{N} \sum_{k_b = 1}^{N} \sin(\delta_{J_a \Pi_a k_a}) \sin(\delta_{J_b \Pi_b k_b})$$
$$\times \cos(\delta_{J_a \Pi_a k_a} - \delta_{J_b \Pi_b k_b}) \times T, \tag{7.8.32}$$

where T is given by

$$T = \left[\sum_{l_{1a} l_{1b}} \bar{Z}(l_{1a} J_a l_{1b} J_b; s_1 \nu) U(J_a \Pi_a k_a)_{\alpha_1 s_1 l_{1a}} U(J_b \Pi_b k_b)_{\alpha_1 s_1 l_{1b}} \right]$$
$$\times \left[\sum_{l_{2a} l_{2b}} \bar{Z}(l_{2a} J_a l_{2b} J_b; s_2 \nu) U(J_a \Pi_a k_a)_{\alpha_2 s_2 l_{2a}} U(J_b \Pi_b k_b)_{\alpha_2 s_2 l_{2b}} \right]. \tag{7.8.33}$$

This formula, although quite complicated in appearance, is actually simpler than Eq. (7.8.28), for it has fewer subsidiary conditions on the terms that enter. (The remaining conditions are that the eigenvectors U_k be orthonormal.) We note the following features of this expression:

1. The intermediate state occurs bilinearly and each occurrence is characterized by three numbers, J, Π, and k (the latter designating the particular eigenvector of the $N \times N$ scattering matrix).

2. The two links in the correlation ("entrance" and "exit" channels in nuclear reaction terminology) enter through the T coefficient, which factors into two separate parts, and the elements of the T coefficient are all explicitly *real*. For the special case where the intermediate state is *sharp* (that is, it has a definite value of J, Π, and k), one sees that the entire coefficient B_ν factors.

There is one respect in which the formulation just described for relativistic two-body reactions is deficient. This is in the treatment of reactions where polarization measurements are to be made. It was remarked independently by several authors (Chou and Shirokov [13], Jacob and Wick [14]) that the channel spin procedure for polarization (Simon and Welton [15]) is unnecessarily complicated as compared with the more natural relativistic procedure using helicity. Such considerations have the additional advantage that massless, as well as massive, particles can be treated uniformly.

The tetrad description of spin (tetrad decomposition of the Pauli–Lubanski operator, see Sections 7d–e) is easily, and naturally, adapted to helicity. Orienting the spatial momentum along the tetrad 3-direction, one can define the helicity operator to be the tetrad operator $S^{(3)} \to \lambda$ for each of the particles (massive or massless). It follows that in the incident channel the angular momentum projection is the helicity, $\lambda_a - \lambda_b$, where a, b label the pair of oppositely moving incident particles. The composite system (angular momentum J) must have the same magnetic quantum number, $M = \lambda_a - \lambda_b$, if measured in the same frame. Similarly, the exit channel has the helicity, $\lambda_c - \lambda_d$ (c, d label the exit particles), which must agree with the composite angular momentum, *if* measured in the same frame.

Since the two frames defined by the incident and exit tetrads differ by a rotation, $R \to U$, we can conclude: *The angular distribution for the reaction $a + b \to c + d$, proceeding through the composite state J, with sharp helicities $\lambda_a \lambda_b \lambda_c \lambda_d$ is proportional to the rotation matrix element*:

$$|D^J_{\lambda_a - \lambda_b, \lambda_c - \lambda_d}(U)|^2. \tag{7.8.34}$$

This elegantly simple formulation of a relativistic angular correlation stems directly from the fundamental paper by Wigner [16].

Using the general scattering matrix formulation, this result can be given a more comprehensive form, but it is not useful to do so here.

References

1. R. F. Christy, *Proceedings of the University of Pittsburgh Conference on Nuclear Structure*, pp. 421ff., 1957, Department of Physics, University of Pittsburgh, Pittsburgh.
2. E. P. Wigner, "Relativistic invariance and quantum phenomena," *Rev. Mod. Phys.* **29** (1957), 255–268.
3. S. P. Lloyd, "The angular correlation of two successive nuclear radiations, " *Phys. Rev.* **85** (1952), 904–911; see also J. W. Gardner, "Directional correlation between successive internal-conversion electrons," *Proc. Phys. Soc.* **A62** (1949), 763–779, II. *ibid.* **A64** (1951), 238–249; "Interference in the directional correlation of conversion electrons," *ibid.* **A64** (1951), 1136–1138.
4. G. Racah, "Directional correlation of successive nuclear radiations," *Phys. Rev.* **84** (1951), 910–912.
5. L. C. Biedenharn and M. E. Rose, "Angular correlations of nuclear radiations," *Rev. Mod. Phys.* **25** (1953), 729–777.

6. R. S. Hager and E. C. Seltzer, "Internal conversion tables, Part I. *K*-, *L*-, *M*-shell conversion coefficients for $Z = 30$ to $Z = 103$," *Nuclear Data* **A4** (1968), 1–235; "Part II, Directional and polarization particle parameters for $Z = 30$ to $Z = 103$," *ibid.* **A4** (1968), 397–641; "Part III, Coefficients for analysis of penetration effects in internal conversion and E0 internal conversion," *ibid.* **A6** (1969), 1–127.
7. K. Alder, A. Bohr, T. Huus, B. Mottelson, and A. Winther, "Study of nuclear structure by electromagnetic excitation with accelerated ions," *Rev. Mod. Phys.* **28** (1956), 432–542.
8. L. C. Biedenharn and P. J. Brussaard, *Coulomb Excitation*. Clarendon, Oxford, 1965.
9. S. de Benedetti, *Nuclear Interactions*. Wiley, New York, 1964.
10. M. L. Goldberger and K. M. Watson, *Collision Theory*. Wiley, New York, 1964.
11. E. P. Wigner and L. Eisenbud, "Higher angular momenta and long range interaction in resonance reactions," *Phys. Rev.* **72** (1947), 29–41.
12. J. M. Blatt and L. C. Biedenharn, "The angular distribution of scattering and reaction cross-sections," *Rev. Mod. Phys.* **24** (1952), 258–272.
13. Chou Kuang-Chao and M. I. Shirokov, "The relativistic theory of reactions involving polarized particles," *J. Exptl. Theoret. Phys.* (*USSR*) **34** (1958), 1230–1239. (Translation in *Soviet Phys. JETP* **7** (1958), 851–857.
14. M. Jacob and G.-C. Wick, "On the general theory of collisions for particles with spin," *Ann. Phys.* (*N. Y.*) **7** (1959), 404–428.
15. A. Simon and T. A. Welton, "Polarization from isolated resonances," *Phys. Rev.* **94** (1954), 943–944; see also M. Simonius, "Determination of the scattering amplitudes from polarization measurements," *Phys. Rev. Lett.* **19** (1967), 279–281.
16. E. P. Wigner, "On unitary representations of the inhomogeneous Lorentz group," *Ann. of Math.* **40** (1939), 149–204. See also R. M. F. Houtappel, H. van Dam, and E. P. Wigner, "The conceptual basis and use of the geometric invariance principles," *Rev. Mod. Phys.* **37** (1965), 595–632, Section 5.5.

9. Some Applications to Nuclear Structure

a. Qualitative considerations. The structure of atomic nuclei was originally perceived as a many-body problem in quantum mechanics involving two types of particles, neutrons and protons (generically, nucleons), in numbers ranging from $N = 1$ (hydrogen) to $N \simeq 300$ (uranium and the man-made heavy nuclei). This difficult problem became even more intractable when it became clear that the nucleons themselves were not elementary but composite objects, belonging to large families, called hadrons, interacting by the exchange of other composite objects, the mesons. In principle, therefore, nuclear physics cannot be fully comprehended without a prior understanding of *subnuclear* physics ("particle physics"). It is clear that the problem of nuclear interactions can hardly fail to be a task of almost overwhelming complexity.

In such circumstances the only feasible procedure is phenomenological, aiming at identifying the appropriate concepts and degrees of freedom suitable to describe characteristic nuclear phenomena. Symmetry considerations, and angular momentum techniques in particular, can be expected to be—and are—of major importance in such studies.

Angular momentum techniques actually play a much more predominant role in nuclear structure than one could expect, *a priori*, as a result of the experimentally established validity of the nuclear shell model of Mayer and Jensen. The motion of the individual nucleons approximates that of independent particles moving in an average potential, and the resulting orbits are characterized primarily by angular momentum labels. It follows that nuclear structure is, in this approach, qualitatively similar to that of atomic structure; both have an *Aufbauprinzip* based largely on angular momentum. [By contrast, in subnuclear physics, angular momentum, per se, plays a less significant role (although important in labeling states), since the constituents are labeled primarily by nongeometric symmetries (isospin, strangeness, hypercharge), and for these symmetries there is no analog to the concept of orbital angular momentum.]

Contrasting to the independent particle phenomena are the phenomena involving collective degrees of freedom, based on the cooperative motions of many nucleons. Surprisingly, such phenomena can actually be viewed as compatible, in a sense, with the shell model, by ascribing the collective properties to a nonspherical distortion of the average potential. Angular momentum techniques again enter in a significant way, since the distortion is primarily quadrupolar, and can be characterized by the nuclear rotator model (rotators are discussed in Section 10).

Let us sketch the content of this section. We shall illustrate the applicability of angular momentum techniques to nuclear structure by first discussing the *shell model* and then applying this model to the properties of short-range interactions. We abstract from this discussion the concept of a *pairing interaction*, and show how this interaction is interpretable in terms of angular momentum techniques (*quasi-spin*). Let us note, for clarity, that our purpose in this section is not to review nuclear structure physics, but rather to focus attention on those special circumstances in nuclear physics where angular momentum concepts and techniques have made a significant contribution.

b. The nuclear shell model of Mayer and Jensen. It is a consequence of the basic kinematic symmetry that each of the stable atomic nuclei possesses a ground state (quantal state of minimal energy) characterized by the (exact) invariant labels of mass and spin. The approximate quantum numbers of parity and isospin (see below) complete the list of basic (model-independent) characteristics. From the coupling to the electromagnetic field, these ground states possess static electromagnetic multipole moments (charge, magnetic dipole moment, etc.) limited (see Section 6) by the magnitude of the intrinsic spin. It is clear that angular momentum theory is basic to the understanding and determination of these characteristics.

Nuclear structure theory attempts to describe nuclear systems (excited, as well as ground, states) constructively as composites made up of interacting neutrons and protons. In view of the composite nature of these constituents, this can have validity only in a limited energy domain.[1]

The nuclear shell model is, in principle, even more limited in scope, in that the primary aim is the description of the very lowest states. This model assumes that the motion of the constituent nucleons is essentially that of independent particles moving in some average potential, with the residual interparticle interactions constituting a small perturbation. It was – and is – quite astonishing that such a simple model[2] can account for the data so overwhelmingly successfully.

The wave functions for a system of independent particles is simply the (tensor) product function, $\prod_{\alpha=1}^{N} \phi_\alpha$, where $\{\phi_\alpha(\mathbf{x}_\alpha)\}$ are the wave functions of the individual particles. The Pauli exclusion principle requires that the wave function for identical particles (fermions) be antisymmetric under exchange. This requires that the physical wave function be

$$\psi(\mathbf{x}_1, \mathbf{x}_2, \ldots, \mathbf{x}_N) = \mathscr{A} \prod_{\alpha=1}^{N} \phi_\alpha(\mathbf{x}_\alpha), \tag{7.9.1}$$

where $\mathscr{A}$ is the antisymmetrizing operation. This introduces *correlations* into the "independent particle" motion, which are of essential importance physically.

To determine the wave functions of the individual particles requires specifying the average potential, which is schematized as being roughly constant over the volume of the nucleus. Two easily solvable models are (*a*) a square-well potential (with infinitely high walls at the nuclear radius for simplicity), and (*b*) an isotropic harmonic oscillator potential (in three

---

[1] A logically similar situation concerns atomic structure in which the degrees of freedom of the atomic nucleus were disregarded. The energy scale ≈ 1 eV for purely electronic structures, compared with $\approx 10^6$ eV for nuclear effects, shows the separation to be extremely favorable. For nuclei, mesonic effects involve energies ≈ 100 MeV so that the separation is less favorable.

[2] The shell model concept had been proposed much earlier by Elsasser [1] and Goldschmidt [2] to account for the earliest known data. Skepticism toward acceptance of the model was based on two characteristics of the then known "shell model systems" (the planetary system and the system of electrons in atoms), which are clearly absent in nuclear systems: First, there is the absence of a large central body, which predominantly influences the motion of the constituents. (It is, in fact, the relatively small interactions between the planets, and between the electrons in an atom, that make the motion of the constituents so nearly independent of each other.) Second, the existence of an average potential can be validated if the interaction is of long range and everywhere of the same sign, which is true of the planetary system and for electrons, but certainly not true for the nuclear forces. It was the large amount of data interpretable by the shell model that forced acceptance of the model of Mayer and Jensen [3].

dimensions). The energy levels for such potentials are labeled by the orbital angular momentum quantum number l, and are illustrated schematically in Fig. 7.4.

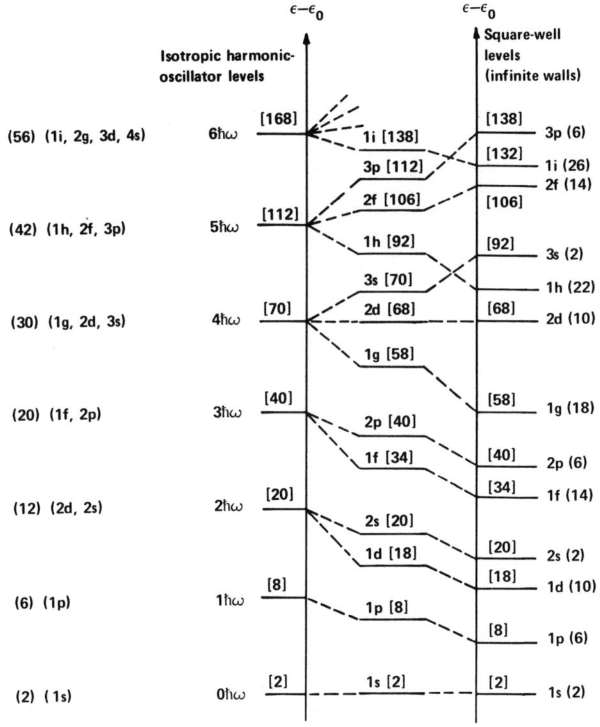

Figure 7.4. Level system of the three-dimensional isotropic harmonic oscillator (on left) contrasted with the square-well potential with infinitely high walls (on right). The degeneracy of each state is shown in parentheses, and the total number of levels up through a given state is in brackets. (From M. G. Mayer and J. H. D. Jensen, *Elementary Theory of Nuclear Shell Structure*. Wiley, New York, 1955, p. 53. Reprinted by permission.)

The data assembled in justification of the shell model showed that, experimentally, certain groups of nucleons possessed special stability [analogous to the (far more pronounced) closed shells of the inert gases in atomic structure]. These "magic numbers" are 2, 8, 20, 28, 50, 82, 126, ... From Fig. 7.4 it is clear that only the first three magic numbers are accounted for by this model. To account for the remaining "magic numbers," Mayer and Jensen (independently) suggested that the average interaction also had a spin–orbit interaction of the form[1]

$$V_{\text{spin-orbit}} = -\alpha(r)\boldsymbol{l} \cdot \mathbf{s}. \tag{7.9.2}$$

The evaluation of the resulting energy splitting involves the Racah coefficient:

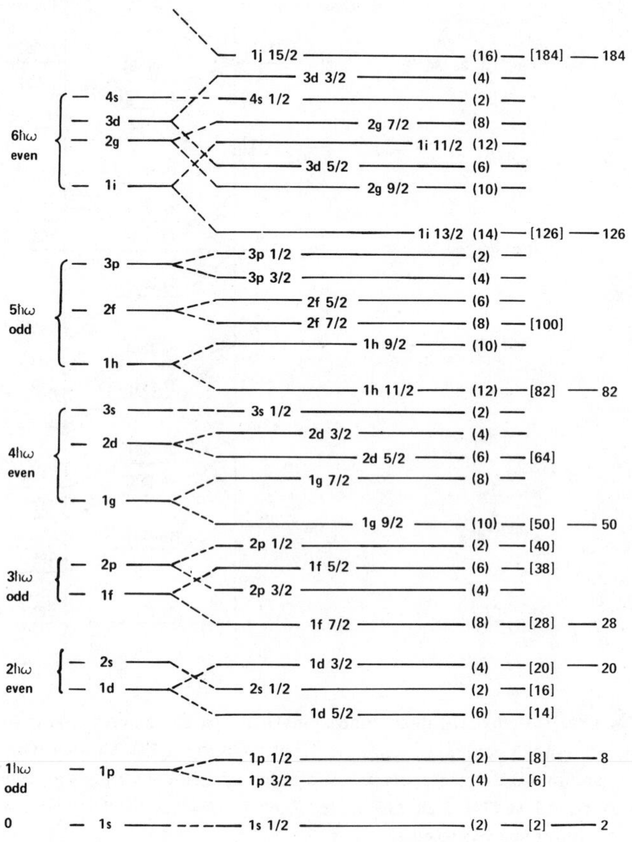

Figure 7.5. Spin-orbit coupling angular momentum eigenstates of the nuclear shell model (in middle) contrasted with the related eigenstates of the three-dimensional oscillator (on left). (From M. G. Mayer and J. H. D. Jensen, *Elementary Theory of Nuclear Shell Structure*. Wiley, New York, 1955, p. 58. Reprinted by permission.)

---

[1] In this section we use lower-case boldface letters $\boldsymbol{l}$, **s**, **j**, and **i**, to denote the orbital, spin, total ($\boldsymbol{l} + \mathbf{s}$), and isospin angular momentum, respectively, of a single nucleon. Similarly, **L**, **S**, **J**, and **I** denote the corresponding angular momenta of a physical system composed of two or more nucleons.

$$\boldsymbol{l}\cdot\mathbf{s}|nl\tfrac{1}{2};jm\rangle = -\tfrac{1}{2}[6l(l+1)(2l+1)]^{\frac{1}{2}}\, W(l1j\tfrac{1}{2};l\tfrac{1}{2})|nl\tfrac{1}{2};jm\rangle$$

$$= \tfrac{1}{2}[j(j+1) - l(l+1) - \tfrac{3}{4}]|nl\tfrac{1}{2};jm\rangle. \qquad (7.9.3)$$

Thus, the energy shifts are

$$\Delta E = -\alpha(r)\times\begin{cases} l & \text{for } j = l+\tfrac{1}{2} \\ -l-1 & \text{for } j = l-\tfrac{1}{2}. \end{cases} \qquad (7.9.4)$$

For an attractive spin–orbit coupling[1] the state $j = l + \frac{1}{2}$ lies lowest.

The resulting (schematic) single-particle energy-level system is illustrated in Fig. 7.5. The sequence of magic numbers is indicated in the figure.

Remarks. (*a*) From the point of view of the present monograph, the essential element in the Mayer–Jensen shell model is that the individual particle states are *a set of angular momentum states ordered in energy*. Thus, the problem of constructing nuclear wave functions becomes a problem in the coupling of angular momenta, and the problem of the residual interactions becomes technically a problem in tensor operator analysis. It is for this reason that textbooks on the nuclear shell model devote considerable attention to angular momentum techniques.

(*b*) In order to predict nuclear ground-state spins, for more than the nuclei in the vicinity of closed shells, it is necessary to supplement the level scheme by coupling rules (the analog to Hund's rule in atomic physics). Two empirical rules were employed:

Mayer–Jensen rule: The angular momenta of an even number of nucleons of the same type in a shell is zero; hence, for the ground state of an odd nucleus (N or Z odd), the angular momentum is that of a single nucleon.

Nordheim's rule: For odd numbers of neutrons *and* of protons, we distinguish four cases based on $j_n = l_n \pm \frac{1}{2}$ and $j_p = l_p \pm \frac{1}{2}$:

$$\begin{aligned} &+\,+ \quad \text{or} \quad -\,- \qquad J = j_n + j_p, \\ &+\,- \quad \text{or} \quad -\,+ \qquad J = |j_n - j_p|. \end{aligned} \qquad (7.9.5)$$

These rules account for essentially all the spins of the nuclear ground states, and this fact is accordingly an impressive verification of the model.

[1] There is a spin–orbit coupling (the Thomas term) in atomic physics, due to a relativistic effect. The nuclear spin–orbit term is by contrast two orders of magnitude *larger* and of the *opposite* sign, and is interpreted as a vector meson effect.

Validation of the rules themselves from properties of the nuclear interaction were developed by many methods. (We discuss below one of these validations.)

(*c*) The shell model gives a prediction for all nuclear properties dependent on single-particle matrix elements – in particular, for the electromagnetic multipole moments.

For the magnetic dipole moment, one obtains the Schmidt model. Taking the magnetic moment $\boldsymbol{\mu}$ to be linearly composed from the orbital angular moment $\boldsymbol{l}$ and the spin $\mathbf{s}$, one has

$$\boldsymbol{\mu} = -\frac{e\hbar}{2mc}(g_l \boldsymbol{l} + g_s \mathbf{s}), \tag{7.9.6}$$

where g_l and g_s are orbital and spin gyromagnetic ratios, respectively, and $-e$ and m are the charge and mass of the nucleon. Using angular momentum techniques, one obtains for the effective gyromagnetic ratio an expression involving Racah coefficients (see Section 3):

$$\boldsymbol{\mu}_j \equiv -g_j\left(\frac{e\hbar}{2mc}\right)\mathbf{j},$$

$$g_j = \frac{g_l[j(j+1) + l(l+1) - s(s+1)]}{2j(j+1)} + \frac{g_s[j(j+1) - l(l+1) + s(s+1)]}{2j(j+1)}. \tag{7.9.7}$$

According to the Schmidt model, for the magnetic moment of odd nuclei with an odd number of protons, we get the value of the magnetic moment μ_j:

$$(\text{odd } Z, \text{even } N)\colon \mu_j = \frac{e\hbar}{2m_p c}\begin{cases} 1 + \dfrac{2.29}{j}j & \text{for } \ j = l + \frac{1}{2} \\ 1 - \dfrac{2.29}{j+1}j & \text{for } \ j = l - \frac{1}{2}. \end{cases} \tag{7.9.8}$$

Correspondingly, for odd nuclei with an odd number of neutrons, we get

$$(\text{even } Z, \text{odd } N)\colon \mu_j = \frac{e\hbar}{2m_n c}\begin{cases} -\dfrac{1.91}{j}j & \text{for } \ j = l + \frac{1}{2} \\ \dfrac{1.91}{j+1}j & \text{for } \ j = l - \frac{1}{2}. \end{cases} \tag{7.9.9}$$

This simple model is surprisingly successful.[1]

The situation for the electric quadrupole moments does *not* agree with the shell model except in the vicinity of closed shells. This discrepancy was resolved by the development of the deformed nuclear model (Bohr and Mottelson [4], Rainwater [5]).

(*d*) A characteristic feature of the energy levels of the Mayer–Jensen model is the lowering of the states $j = l + \frac{1}{2}$ to complete the energy shell for $N = 50, 82,$ and 126. This has the effect of introducing states of *different* parity and *large* j into completion of a shell. A striking consequence of this effect is that excited states of large j occur relatively close above ground states having small j (and opposite parity). The de-excitations, by gamma rays, of such states accordingly have unusually long lifetimes. The occurrence of such *isomeric states* in particular regions of nucleon numbers (determined by the magic numbers – the so-called "islands of isomerism") is a striking confirmation of the shell model.

c. The isospin quantum number. The techniques of angular momentum enter into nuclear structure in a second, and quite different, way through the formal introduction of a nongeometric quantum variable called the isospin. One considers the neutron and the proton to be two instances of the same particle, the *nucleon*, defined as having isospin-$\frac{1}{2}$, with the proton having isospin projection $+\frac{1}{2}$ and the neutron having isospin projection $-\frac{1}{2}$.[2]

Linear unimodular transformations on this two-dimensional isospin basis (isospin transformations) are formally equivalent to the linear, unimodular unitary transformations on the two-dimensional spin-$\frac{1}{2}$ irrep space of angular momentum. It follows that angular momentum techniques carry over to the *SU*(2) *isospin group*.

The isospin concept was introduced by Heisenberg [6] in 1932, largely as an elegant bookkeeping device; the more fundamental status of isospin became evident in the nuclear supermultiplet model of Wigner [7] in 1937, and through the incorporation of isospin into the Gell-Mann–Ne'eman *SU*(3) model of particle physics in 1961 (Gell-Mann [8], Ne'eman [9]).

Since the two states of the nucleon are physically distinguished (by the charge of the proton and by the different masses of the neutron and proton), it is not obvious that the isospin concept has genuine content. Experimentally, the significance of the isospin concept arose from the observed equality (in the singlet spin state) of the *nn*, *np*, and *pp* nuclear interactions. [The equality of the *nn* and *pp* interactions is a less restrictive concept (charge symmetry); the

---

[1] Equations (7.9.8) and (7.9.9) use $g_l = 1$ and the empirical values of g_s for the proton and neutron.

[2] This is the standard convention at present. The original definition of isospin in nuclear physics took the neutron to have isospin projection $+\frac{1}{2}$; particle physicists perversely chose the opposite definition.

equality of all three interactions is called charge independence; see Wigner [9a] and Robson [10].] Thus, the *strong interactions* of nuclear physics are experimentally found to be invariant under isospin rotations, and accordingly they do not distinguish neutrons from protons. The electromagnetic interactions clearly do *not* respect isospin, and (presumably) this symmetry breaking accounts for the neutron–proton mass splitting.

The isospin concept allows one to express the Pauli exclusion principle in an elegant way: *Wave functions for any system of nucleons must be antisymmetric under exchange of the coordinates (spatial position, spin, and isospin) of any two nucleons.* It is essential to note that this so-called "generalized Pauli principle" is not a true generalization (involving an extra assumption) but part of a convenient formalism. The consequences of charge independence can be treated without the introduction of isospin variables or the generalized Pauli principle (see Bohr and Mottelson [4, Vol. I, p. 32]).

d. Properties of a short-range interaction. In order to understand the physical origin of the coupling rules – which implemented the shell model in predicting ground-state spins – let us consider the properties of a short-range, two-particle residual interaction in the shell model (Mayer and Jensen [3]). Since the shell model states are to be considered as correct in lowest order, we can calculate the effect of the residual interaction by first-order perturbation. This calculation, we shall find, is a nice exercise in the use of angular momentum techniques.[1]

LS-coupling. It is instructive to proceed using first *LS*-coupling, rather than *jj*-coupling, for the particles (see Section 5). Consider, then, a pair of identical nucleons in the same nuclear shell (labeled by the orbital angular momentum l). The spatial wave function for the pair of nucleons corresponding to the coupling $\mathbf{L} = \boldsymbol{l}_1 + \boldsymbol{l}_2$ of the two orbital angular momenta is[2]

$$\Phi_{(ll)LM_L}(\mathbf{x}_1, \mathbf{x}_2) = \sum_{mm'} C^{llL}_{mm'M_L} R_l(r_1) R_l(r_2) Y_{lm}(\hat{x}_1) Y_{lm'}(\hat{x}_2). \tag{7.9.10}$$

Using the symmetry properties of the Wigner coefficient (Chapter 3, Section 12), we find that under the spatial coordinate exchange, $\mathbf{x}_1 \leftrightarrow \mathbf{x}_2$, the spatial wave function has the symmetry $\Phi_{(ll)LM_L}(\mathbf{x}_2, \mathbf{x}_1) = (-1)^L \Phi_{(ll)LM_L}(\mathbf{x}_1, \mathbf{x}_2)$. For nucleons of the same type (two protons or two neutrons), the isospin has the value $I = 1$; accordingly, the antisymmetric states are given by

---

[1] The application of angular momentum techniques to validate the Nordheim rule as a property of short-range forces (with spin exchange) was first carried out by de-Shalit [11].

[2] We suppress all reference to the principal quantum number in the notation for the radial wave functions.

$$L = \text{odd: Spin function } |S = 1, M_S\rangle, M_S = 1, 0, -1;$$

$$L = \text{even: Spin function } |S = 0, M_S = 0\rangle. \tag{7.9.11}$$

Let us schematize the residual interaction by a zero-range ("delta function") interaction:

$$\begin{aligned} V_{\text{residual}}(\mathbf{x}_1, \mathbf{x}_2) &= -V_0\,\delta(\|\mathbf{x}_1 - \mathbf{x}_2\|) \\ &= -V_0 r_1^{-2}\,\delta(r_1 - r_2) \sum_{k\mu} Y_{k\mu}(\hat{x}_1) Y^*_{k\mu}(\hat{x}_2). \end{aligned} \tag{7.9.12}$$

The interaction energy, using perturbation theory, then takes the form

$$\Delta E = \langle (l^2)LM_L | V_{\text{residual}} | (l^2)LM_L \rangle = -V_0 F_0 (2l+1)^2 \begin{pmatrix} l & l & L \\ 0 & 0 & 0 \end{pmatrix}^2,$$

$$F_0 \equiv \frac{1}{4\pi} \int_0^\infty r^2\, dr (R_l(r))^4. \tag{7.9.13}$$

The 3-j coefficient $\begin{pmatrix} l & l & L \\ 0 & 0 & 0 \end{pmatrix}$ vanishes unless L is even; accordingly, the interaction energy vanishes for $S = 1$ (triplet spin states). For singlet states the energy shift ΔE can be given explicitly, using the fact that the relevant 3-j coefficient has the value [see Eqs. (3.182) and (3.194), Chapter 3]:

$$\begin{pmatrix} l & l & L \\ 0 & 0 & 0 \end{pmatrix} = \frac{(-1)^{l+\frac{1}{2}L}(l + \frac{1}{2}L)!}{(L/2)!(L/2)!(l - \frac{1}{2}L)!} \left[\frac{(2l - L)!L!L!}{(2l + L + 1)!} \right]^{\frac{1}{2}}. \tag{7.9.14}$$

(L even.)

The largest energy shift occurs for $L = 0$, for which one finds

$$(\Delta E)_{L=0} = -V_0 F_0 (2l + 1). \tag{7.9.15}$$

The energy shift decreases rapidly in magnitude as L increases. For large values of l, we have the limiting form

$$\frac{(\Delta E)_L}{(\Delta E)_0} \sim \left[\frac{(L-1)!!}{L!!} \right]^2. \tag{7.9.16}$$

This result shows that for the next energy level—that is, $L = 2$—we get only one-fourth of the $L = 0$ energy shift. For large L, the ratio $(\Delta E)_L/(\Delta E)_0$ goes to zero as $2/\pi L$.

These results appear to indicate that a zero-range interaction affects primarily the $L = 0$ state, leaving all other states only slightly perturbed. This conclusion has been stated in the literature, but it is not quite correct, since the limiting forms are valid only for $l \gg L$, and are invalid for $2l \approx L$. It is useful to pursue this result further in order to understand the correct relation.

To do so let us first obtain an asymptotic expression for the 3-j symbol that enters in the energy shift, the coefficient $\begin{pmatrix} a & b & c \\ 0 & 0 & 0 \end{pmatrix}^2$. Letting $g = (a + b + c)/2$, one has

$$\begin{pmatrix} a & b & c \\ 0 & 0 & 0 \end{pmatrix}^2 = \left[\frac{g!}{(g-a)!(g-b)!(g-c)!}\right]^2 \frac{(2g-2a)!(2g-2b)!(2g-2c)!}{(2g+1)!}. \tag{7.9.17}$$

Taking a, b, c all large compared to 1, Stirling's formula yields the asymptotic form

$$\begin{pmatrix} a & b & c \\ 0 & 0 & 0 \end{pmatrix}^2 \sim \frac{1}{2\pi[g(g-a)(g-b)(g-c)]^{\frac{1}{2}}}. \tag{7.9.18}$$

One recognizes the square root factor in this result to be just Archimedes' formula[1] for the area of a triangle having sides (a, b, c), with g being half the perimeter. Writing the area in the form Area $= (\frac{1}{2})ab \sin\theta$, where θ is the angle between a and b, we find a more suggestive form for the result:

$$\begin{pmatrix} a & b & c \\ 0 & 0 & 0 \end{pmatrix}^2 \sim (\pi ab \sin\theta)^{-1}. \tag{7.9.19}$$

Let us now apply this result to the problem at hand, the energy shift formula, Eq. (7.9.13). We note the following features:

1. The asymptotic energy shift is *symmetric about* $L = l$, where the shift reaches the minimum value $\Delta E \sim -4F_0V_0/\pi$.

2. The maximum energy shift is reached, according to the asymptotic result, for $\theta = 0$ and π ($L = 0$ and $L = 2l$). But in this region the asymptotic formula is invalid, so one should resort to the results:

[1] This formula is usually ascribed to Heron of Alexandria, but van der Waerden [12] shows this to be incorrect, and credits this formula to Archimedes.

$$(\Delta E)_{L=0} = -V_0F_0(2l+1) \simeq -2V_0F_0 l,$$

$$(\Delta E)_{L=2l} \simeq \frac{2}{\pi} V_0 F_0 l^{\frac{1}{2}}. \tag{7.9.20}$$

Hence, although the energy shift does become unlimitedly large for $l \to \infty$ for *both* $L = 0$ and $L = 2l$, the energy shift for $L = 0$ dominates.

3. There is a nice explanation for these results in physical terms. Physically, the energy shift is proportional to the overlap of the two wave functions; for large angular momenta, the wave functions become rotating "discs" perpendicular to the axis of rotation. Clearly the largest overlap occurs when the discs lie in the same plane—that is, when $\theta = 0$ or π. This is the *classical* result [asymptotic formula, Eq. (7.9.19)].

But *semiclassically* the orientation of an angular momentum vector is not precise (uncertainty principle[1]). *Only for the "trivial case,"* $L = 0$, *does the uncertainty vanish*; this is the situation that applies for $\theta = 0$, so that the overlap is in fact precise, even semiclassically. For the other maximal case, $\theta = \pi$, the uncertainty principle requires some loss of overlap, so that this energy shift consequently has a smaller magnitude than for $\theta = 0$, as shown by the discussion in (2) above. [The actual value $(\Delta E)_{L=2l} \propto l^{\frac{1}{2}}$ shows the effect to be due to fluctuations.]

4. There is a relationship between the result of this energy shift calculation and the prototype angular correlation for spinless particles (see Section 8). We shall apply this formal similarity in the *jj*-coupling calculation below, in order to derive some otherwise quite complicated results.

jj-coupling. To apply these results to the short-range interaction of two nucleons in a *jj*-coupling configuration, it is expedient to use the $(LS\text{–}jj)$ transformation [see Eq. (7.5.97)]:

$$|[(l_1\tfrac{1}{2})_{j_1}(l_2\tfrac{1}{2})_{j_2}]_{JM}\rangle = \sum_{LS} [(2S+1)(2L+1)(2j_1+1)(2j_2+1)]^{\frac{1}{2}}$$

$$\times \begin{Bmatrix} l_1 & l_2 & L \\ \frac{1}{2} & \frac{1}{2} & S \\ j_1 & j_2 & J \end{Bmatrix} |[(l_1 l_2)_L(\tfrac{1}{2}\tfrac{1}{2})_S]_{JM}\rangle. \tag{7.9.21}$$

The energy shift in the *LS*-configuration for $I = 1$ has been found above to be

[1] The uncertainty relations for angular momentum are developed in RWA.

$$(\Delta E)_{LSJ} = \begin{cases} 0 & L = \text{odd},\ S = 1 \\ -V_0F_0(2l+1)^2\begin{pmatrix} l & l & L \\ 0 & 0 & 0 \end{pmatrix}^2 & L = \text{even},\ S = 0. \end{cases} \tag{7.9.22}$$

It follows that the energy shift in the *jj*-configuration is given by

$$\Delta E(lj, lj; J) = (2j+1)^2 \sum_{\substack{S=0,1 \\ L=|J-S|,\ldots,J+S}} (2S+1)(2L+1) \times \begin{Bmatrix} l & l & L \\ \frac{1}{2} & \frac{1}{2} & S \\ j & j & J \end{Bmatrix}^2 (\Delta E)_{LSJ}. \tag{7.9.23}$$

For nucleons of the same type the isospin is $I = 1$; in this case, the only nonvanishing contributions come from $S = 0$ and $L = J$. Thus,

$$\Delta E(lj, lj; J)_{I=1} = -(2j+1)^2(2J+1)(V_0F_0)(2l+1)^2\begin{pmatrix} l & l & J \\ 0 & 0 & 0 \end{pmatrix}^2 \times \begin{Bmatrix} l & l & J \\ \frac{1}{2} & \frac{1}{2} & 0 \\ j & j & J \end{Bmatrix}^2. \tag{7.9.24}$$

This result simplifies in two ways: The 9-j symbol for one argument equal to zero [using Eq. (3.325)] is a multiple of a 6-j coefficient, here $\begin{Bmatrix} l & l & J \\ j & j & \frac{1}{2} \end{Bmatrix}$; and the product $\begin{pmatrix} l & l & J \\ 0 & 0 & 0 \end{pmatrix}\begin{Bmatrix} l & l & J \\ j & j & \frac{1}{2} \end{Bmatrix}$ is equal to $-\begin{pmatrix} j & j & J \\ \frac{1}{2} & -\frac{1}{2} & 0 \end{pmatrix}\Big/(2l+1)$ [see Eq. (A-2) in the appendix to Section 4].

Using these simplifications, one finds

$$\Delta E(lj, lj; J)_{I=1} = -\frac{(2j+1)^2}{2}(V_0F_0)\begin{pmatrix} j & j & J \\ \frac{1}{2} & -\frac{1}{2} & 0 \end{pmatrix}^2. \tag{7.9.25}$$

This result has a form quite similar to that found for the spinless case, Eq. (7.9.13), and, as we shall see, this similarity is not accidental.

The result for isospin zero could also be derived by using the same method, but this direct approach is both difficult and uninformative. Let us proceed by a different route.

Since our purpose is to illustrate angular momentum techniques, let us reconsider the *LS* calculation from the beginning. For an *LS*-coupled two-particle wave function, the angular momentum coupling coefficients provide the complete answer. For the orbital part of the wave function, one has

$$\Phi_{(l_1l_2)LM_L}(\mathbf{x}_1, \mathbf{x}_2) = \sum_{m_1m_2} C^{l_1l_2L}_{m_1m_2M_L} R_{l_1}(r_1)R_{l_2}(r_2)Y_{l_1m_1}(\hat{x}_1)Y_{l_2m_2}(\hat{x}_2). \tag{7.9.26}$$

The delta function interaction has the effect of enforcing the condition $\mathbf{x}_1 = \mathbf{x}_2$ on both the initial and the final wave functions in the matrix element, Eq. (7.9.13). Accordingly, let us investigate what effect this implies for the wave functions themselves. Applying the condition $\mathbf{x}_1 = \mathbf{x}_2$ to the orbital part of the wave function in Eq. (7.9.26), and using the product law, Eq. (3.436), for spherical harmonics of the same argument, yields

$$\Phi_{(l_1l_2)LM_L}(\mathbf{x}, \mathbf{x}) = \left\{\left[\frac{(2l_1+1)(2l_2+1)}{4\pi(2L+1)}\right]^{\frac{1}{2}} C^{l_1l_2L}_{000}\right\}$$

$$\times R_{l_1}(r)R_{l_2}(r)Y_{LM_L}(\hat{x}). \tag{7.9.27}$$

This result may be interpreted in this way: The coefficient in braces $\{\cdots\}$ is the probability amplitude that the two particles form a composite state of angular momentum (LM_L). Exactly the same probability amplitude can be seen to occur in our prototype angular correlation process for spinless particles; that is, the wave function in Eq. (7.9.27) is formally analogous (taking $\mathbf{x}_1 \to -\mathbf{x}$) to a composite state being formed by two particles of equal and opposite momenta. From this viewpoint, the zeros in the magnetic quantum number arguments of the 3-j coefficient correspond to the fact that the colliding particles are both spinless.

Let us consider now the jj-coupling wave function from this same point of view. The wave function corresponding to the coupling scheme $\mathbf{j}_i = \boldsymbol{l}_i + \mathbf{s}_i$ $(i = 1, 2)$, $\mathbf{J} = \mathbf{j}_1 + \mathbf{j}_2$, is

$$\Psi_{[(l_1\frac{1}{2})j_1(l_2\frac{1}{2})j_2]JM}(\mathbf{x}_1, \mathbf{x}_2) = \sum_{\mu_1\mu_2} C^{j_1j_2J}_{\mu_1\mu_2M}\psi_{(l_1\frac{1}{2})j_1\mu_1}(\mathbf{x}_1)\psi_{(l_2\frac{1}{2})j_2\mu_2}(\mathbf{x}_2),$$

$$\psi_{(l_i\frac{1}{2})j_i\mu_i}(\mathbf{x}_i) \equiv R_{l_i}(r_i)\sum_{m_i\nu_i} C^{l_i\frac{1}{2}j_i}_{m_i\nu_i\mu_i} Y_{l_im_i}(\hat{x}_i)|\tfrac{1}{2}, \nu_i\rangle. \tag{7.9.28}$$

If we orient the coordinate frame so that the 3-axis coincides with $\hat{x}_1$, and use

$$Y_{l_1m_1}(0, 0, 1) = \delta_{m_1,0}\left(\frac{2l_1+1}{4\pi}\right)^{\frac{1}{2}},$$

the wave function $\psi_{(l_1\frac{1}{2})j_1\mu_1}$ becomes

$$\psi_{(l_1\frac{1}{2})j_1\mu_1}(0,0,r_1) = \left(\frac{2l_1+1}{4\pi}\right)^{\frac{1}{2}} R_{l_1}(r_1) C^{l_1\frac{1}{2}j_1}_{0\mu_1\mu_1} |\tfrac{1}{2}, \mu_1\rangle. \tag{7.9.29}$$

This corresponds to the "helicity frame," where the particle has helicity $\mu_1 = \pm\frac{1}{2}$. In a general frame, the wave function undergoes a rotation so that we obtain Eq. (7.9.28) in the form:

$$\psi_{(l_1\frac{1}{2})j_1\mu_1}(\mathbf{x}_1) = \left(\frac{2l_1+1}{4\pi}\right)^{\frac{1}{2}} R_{l_1}(r_1) \sum_{\nu_1} C^{l_1\frac{1}{2}j_1}_{0\nu_1\nu_1} D^{j_1 *}_{\mu_1\nu_1}(\hat{x}_1) |\tfrac{1}{2}, \nu_1\rangle', \tag{7.9.30}$$

where $\hat{x}_1 = \mathbf{x}_1/r_1$ is related to $\hat{e}_3 = \text{col}(0,0,1)$ by the orthogonal transformation $\hat{x}_1 = R\hat{e}_3$. The special rotation matrix $D^{j_1}(\hat{x}_1)$ is then defined to be $D^{j_1}(\alpha\beta 0)$ for $\hat{x}_1 = (\cos\alpha\sin\beta, \sin\alpha\sin\beta, \cos\beta)$. The new spin function is

$$|\tfrac{1}{2}, \nu_1\rangle' \equiv \sum_{\mu_1} D^{\frac{1}{2}}_{\mu_1\nu_1}(\hat{x}_1) |\tfrac{1}{2}, \mu_1\rangle.$$

The rotation matrix occurring in Eq. (7.9.30) is not normalized with respect to the volume element $d^3\mathbf{x}$. Equation (3.138), Chapter 3, shows that we must renormalize Eq. (7.9.30) by the factor $[4\pi/(2j_1+1)]^{\frac{1}{2}}$ in order to compare with the corresponding function $\phi_{l_1\mu_1}(\mathbf{x}_1) = R_{l_1}(r_1) Y_{l_1\mu_1}(\hat{x}_1)$. Noting that the Wigner coefficients appearing in Eq. (7.9.30) have the values

$$C^{l\frac{1}{2}j}_{0\nu\nu} = \left(\frac{(2j+1)}{2(2l+1)}\right)^{\frac{1}{2}} \text{sign}(j-l-\nu), \qquad \text{sign}\, 0 \equiv +1,$$

we find that the appropriate spin-$\frac{1}{2}$ function to be compared with the spin-0 function $\phi_{l_1\mu_1}(\mathbf{x}_1)$ is

$$\phi_{(l_1\frac{1}{2})j_1\mu_1}(\mathbf{x}_1) = \frac{R_{l_1}(r_1)}{\sqrt{2}} \sum_{\nu_1} \text{sign}(j_1 - l_1 - \nu_1) D^{j_1 *}_{\mu_1\nu_1}(\hat{x}_1) |\tfrac{1}{2}, \nu_1\rangle'. \tag{7.9.31}$$

Using this result in the right-hand side of Eq. (7.9.28) and simplifying by using Eq. (3.188), we obtain the spin-$\frac{1}{2}$ analog of the spin-0 function (7.9.27):

$$\Phi_{[(l_1\frac{1}{2})j_1(l_2\frac{1}{2})j_2]JM}(\mathbf{x},\mathbf{x}) = \tfrac{1}{2} R_{l_1}(r) R_{l_2}(r)$$
$$\times \sum_{\nu_1\nu_2} (\text{phase})\, C^{j_1j_2J}_{\nu_1,\nu_2,\nu_1+\nu_2} D^{J*}_{M,\nu_1+\nu_2}(\hat{x}) |\tfrac{1}{2}, \nu_1\rangle' |\tfrac{1}{2}, \nu_2\rangle',$$

$$(\text{phase}) = \text{sign}(j_1 - l_1 - \nu_1)\,\text{sign}(j_2 - l_2 - \nu_2). \tag{7.9.32}$$

Each of the four components, $(\nu_1, \nu_2) = (\pm\frac{1}{2}, \pm\frac{1}{2})$, of this composite function is then to be considered in determining the energy shifts.

To determine how the helicities influence the components in the composite wave function, Eq. (7.9.32), note that for $\nu_1 = \nu_2 = \frac{1}{2}$ the coefficient $C^{j_1 j_2 J}_{\frac{1}{2}\frac{1}{2}1}$ vanishes unless J is odd ($j_1 = j_2$); similarly for $\nu_1 = \nu_2 = -\frac{1}{2}$. Thus, these two components correspond to isospin $I = 0$. The remaining two components $\nu_1 = -\nu_2 = \pm\frac{1}{2}$ correspond to $I = 0$ for $J =$ odd and $I = 1$ for $J =$ even.

Consider the isospin $I = 1$ case. The energy shift is immediately evaluated to be

$$\Delta E(lj, lj; J = \text{even}, I = 1) = \tfrac{1}{2}(-V_0F_0)(2j+1)^2\begin{pmatrix} j & j & J \\ \frac{1}{2} & -\frac{1}{2} & 0 \end{pmatrix}^2, \qquad (7.9.33)$$

which is precisely the result obtained earlier in Eq. (7.9.25).

To evaluate the isospin $I = 0$ case, we first determine the energy shift for each of the three components, $\nu_1 + \nu_2 = 1, 0, -1$, and sum. In this way one finds for the $I = 0$ energy shift:

$$\Delta E(lj, lj; J = \text{odd}, I = 0)$$
$$= \tfrac{1}{2}(-V_0F_0)(2j+1)^2\left[\begin{pmatrix} j & j & J \\ \frac{1}{2} & -\frac{1}{2} & 0 \end{pmatrix}^2 + \begin{pmatrix} j & j & J \\ \frac{1}{2} & \frac{1}{2} & 1 \end{pmatrix}^2\right]. \qquad (7.9.34)$$

(It is important to note in this evaluation that the $\nu_1 + \nu_2 = 0$ term contributes a factor of 2, just as for the similar case for $I = 1$.)

These results could, of course, be derived by direct computation using either Eq. (7.9.23) or the necessary angular momentum identities to simplify the matrix elements coming from the states (7.9.28). The real advantage of the procedure used above is not just simplicity, but the fact that it "explains" the physical reason for the occurrence of these particular 3-j coefficients.

The energy shift given by Eq. (7.9.33) for isospin $I = 1$ is rather different from the LS-coupled result of Eq. (7.9.13). In particular, the $I = 1$ energy shift *decreases in magnitude rapidly and monotonically*[1] *from* $J = 0$ *to* $J = 2j - 1$. By contrast, the $I = 0$ energy shift has the general shape of an inverted U (with large shifts for both $J = 1$ and $J = 2j$), and hence behaves qualitatively like the LS-coupled result discussed earlier.

It is quite remarkable that these qualitative aspects of the residual nuclear interaction, which stem from the very beginnings of the Mayer–Jensen shell

[1] There is a physical explanation for this effect in terms of a spin projection ($= \cos^2 \frac{1}{2}\theta$) multiplying the spinless (LS-coupled) result. This factor cancels the increase in the spinless result for $\theta \to \pi$.

model, have been substantially validated by the increasingly precise and much more abundant data acquired since then. A recent survey by Schiffer [12a], examining the residual nucleon–nucleon interaction for (Mayer–Jensen) shell model orbits throughout the periodic table, found that the delta function interaction was always qualitatively correct and—especially for identical particle orbits—almost quantitative.

e. The pairing interaction (*seniority*). The pairing interaction is a schematic interaction abstracted from the properties of a short-range interaction [exemplified by the delta function interaction in the singlet-even states of the previous section]. Equation (7.9.33) shows that for this case only the state $J = 0$ is substantially affected, and for $J > 0$ the effect is minor. The pairing interaction abstracts these features and is defined to affect *only* the $J = 0$ state.

Accordingly, one defines the pairing interaction operator q_{12} to be a two-particle rotationally invariant operator such that for two particles in the j^2 configuration the interaction vanishes unless the pair is coupled to $J = 0$, and has, for $J = 0$, the value $(2j + 1)$. Explicitly, one has

$$\langle jj; JM|q_{12}|jj; J'M'\rangle \equiv (2j+1)\delta_{JJ'}\,\delta_{MM'}\,\delta_{J0}. \tag{7.9.35}$$

The operator q_{12} was first introduced by Racah to define—for the f-shell ($l = 3$) in *atomic* spectroscopy—a new quantum number, "seniority" (denoted by v, from the word *vethek*, which means seniority in Hebrew). The operator q_{12} is accordingly called the seniority operator in much of the literature. Because of the close connection of this operator to short-range interactions—which are of essential importance in nuclear physics—most technical development and application of the seniority concept have been in nuclear physics, for j-shells (j half-integral).

It is important to emphasize that, although q_{12} has been abstracted from a short-range interaction, the definition shows that q_{12} is actually a *nonlocal* operator[1] and only approximates a delta function (and hence locality) in the limited function space of a single j-shell.

Two-particle configurations in a j-shell have the symmetry $(-1)^{J+1}$ under space–spin exchange (as can be seen from the symmetry of the Wigner coefficient $C^{jjJ}_{mm'M}$ under the exchange of the two j's). It follows that the seniority operator affects nucleons only in the symmetric (isospin $I = 1$) pair configuration.

The essential idea that motivates the introduction of the operator q_{12} is that pairs of particles coupled to $J = 0$ are, in a sense, "inert" elements in the many-particle wave function. In this view it is the *uncoupled* particles that are the

[1] Anderson [13] emphasized that the pairing operator violates Galilean invariance as well.

important ones. (The seniority quantum number v designates, as we shall show, the number of uncoupled particles.) This agrees qualitatively with many of the empirical results of the shell model. Clearly, adding $J = 0$ pairs to a wave function does not affect the total angular momentum; since the pairs have $I = 1$, however, the isospin can be affected. The most successful applications of the seniority concepts are, in fact, to the coupling of identical nucleons (neutrons or protons only) for which the isospin can be ignored.

In order to understand the meaning of the seniority operator, q_{12}, let us use the definition given by Eq. (7.9.35) and transform from the coupled wave function basis: $\mathbf{j}_1 + \mathbf{j}_2 = \mathbf{J}$, to the uncoupled m, m' basis corresponding to

$$|jm;jm'\rangle = \sum_J C^{jjJ}_{m,m',m+m'} |(jj)J, m + m'\rangle. \tag{7.9.36}$$

The result is

$$\langle jm';jm'''|q_{12}|jm;jm''\rangle = \delta_{m'',-m}\,\delta_{m''',-m'}\,(-1)^{m-m'}. \tag{7.9.37}$$

This result may be interpreted by saying that a pair of states $\{|jm\rangle, |jm''\rangle\}$ scatter into another pair of states $\{|jm'\rangle, |jm'''\rangle\}$ if and only if *the states in each pair are related by time reversal* (that is, $T|j,m\rangle = (-)^{j-m}|j, -m\rangle$, using results from RWA). If the pair are so related, the matrix element for the scattering has magnitude 1. If we use the time-reversal phase convention, and if we confine attention to the $(2j + 1)/2 \equiv \Omega$ independent (time-reversal related) pairs only, then the q_{12} operator may be written as a matrix of dimension Ω in the elementary form:

$$q_{12} = \begin{pmatrix} 1 & 1 & \cdots & 1 \\ 1 & 1 & \cdots & 1 \\ \vdots & \vdots & & \vdots \\ 1 & 1 & \cdots & 1 \end{pmatrix}. \tag{7.9.38}$$

This matrix is easily diagonalized, since it has the two properties: $\operatorname{tr} q_{12} = \Omega$ and $(q_{12})^2 = \Omega q_{12}$. It follows that $\Omega^{-1}q_{12}$ is a rank-one idempotent, and hence has one nonzero eigenvalue Ω, and $(\Omega - 1)$ zero eigenvalues.

The eigenvector corresponding to the eigenvalue Ω is found to be a uniformly weighted coherent combination of time-reversal related antisymmetric pairs; that is,

$$\psi^{(12)} = \sum_{m=-j}^{j} (-1)^{j-m}\,\psi^{(1)}_{jm}\,\psi^{(2)}_{j,-m}. \tag{7.9.39}$$

Two-particle states of this type are of widespread occurrence in physics – often called "coherent states," since they represent the cooperative effect of interacting particles occupying a large number of degenerate levels. [Aside from nuclear physics, the prime example is the superconducting state (described by time-reversed *momentum* pair states) (Anderson [14], Blatt [15]).]

f. Quasi-spin. Applications in nuclear physics of the pairing operator, and the seniority concept, are greatly facilitated by the fortunate circumstance that *these applications are no more and no less than the use of angular momentum techniques in yet another guise*: *quasi-spin.* (One may view this circumstance either as evidence of the pervasive importance of angular momentum, or, less grandly, as the natural human tendency to exploit familiar techniques whenever possible.)

The notion of quasi-spin is a further development in the concept that underlay the *Jordan mapping* of Chapter 5. If one has a finite space $\mathscr{H}$ – for definiteness, let us take the space $\mathscr{H}_j$ spanned by the $(2j+1)$ states $|jm\rangle$ of a j-shell – then the Jordan map allows us to associate with every operator, $\mathscr{O}$, that takes $\mathscr{H}_j$ into itself, the two operator realizations:[1]

Boson realization:

$$\mathscr{O} \to \sum_{mm'=-j}^{j} \mathscr{O}_{mm'}\, a_m \bar{a}_{m'}, \tag{7.9.40}$$

with the defining commutator relations, $[\bar{a}_m, a_{m'}] = \delta_{mm'}$, and all other commutators zero.

Fermion realization:

$$\mathscr{O} \to \sum_{mm'} \mathscr{O}_{mm'}\, b_m \bar{b}_{m'}, \tag{7.9.41}$$

with the defining *anticommutator* relations, $\{\bar{b}_m, b_{m'}\} = \delta_{mm'}$, and all other *anticommutators* zero.

From Chapter 5 we recall that the *Jordan mapping preserves all commutation relations for the family of operators* $\{\mathscr{O}\}$. Since we know the explicit tensor operators v_q^k that are a basis for the $\{\mathscr{O}\}$ – these are the unit tensor operators whose matrix elements are the Wigner coefficients $C^{j\,kj}_{mqm'}$ – we can make the Jordan map fully explicit:

---

[1] The boson and fermion operators should be denoted by a_m^α and b_m^α, where α is the particle index, designating that the boson (fermion) is associated with the space $\mathscr{H}_j$ of single-particle states that appear in position α of the tensor product space $\mathscr{H}_j \otimes \cdots \otimes \mathscr{H}_j \otimes \cdots \otimes \mathscr{H}_j$. For brevity we drop the index α but restore it when required.

$$v_q^k \to \sum_{mm'} C^{jkj}_{mqm'} \begin{cases} a_m \bar{a}_{m'} & \text{(bosons)} \\ b_m \bar{b}_{m'} & \text{(fermions).} \end{cases} \tag{7.9.42}$$

In particular, the angular momentum operator **j** is, to within an invariant factor, just v_q^1 ($q = +1, 0, -1$).

[In order to understand the physical significance of this construction, it is useful to note that the boson or fermion operators need not be considered as creating (or destroying) actual particles (as generally in field theory) but may be considered as creating–destroying *elementary excitations* of some system near its ground state (Landau [16], Migdal [17], Brown [18]). For example, one may think of (bosonic) excitations of a fluid (liquid drop nuclear model), or of (fermionic) excitations of the nuclear shell structure.[1]]

It is a natural development of these ideas to consider operators that create (or destroy) *pairs of particles*, or even more complicated clusters. This generalizes the Jordan map to the consideration of operators acting on pairs (and generally sequences) of states in tensor product spaces.[2]

Let us consider now operators that create (or destroy) a pair of particles. In order to leave the angular momentum unchanged, the pair should itself be coupled to angular momentum zero. Thus, one has (for fermions)

$$\begin{aligned} Z_+ &\equiv \tfrac{1}{2}(2j+1)^{\frac{1}{2}} \sum_m C^{jj0}_{m,-m,0}\, b_m b_{-m} \\ &= \tfrac{1}{2}\sum_m (-)^{j-m}\, b_m b_{-m}. \end{aligned} \tag{7.9.43}$$

(This assumes that $2j$ = odd integer.)

The Hermitian conjugate to Z_+ is given by

$$Z_- \equiv (Z_+)^\dagger = \tfrac{1}{2}\sum_m (-)^{j-m}\, \bar{b}_{-m} \bar{b}_m. \tag{7.9.44}$$

By construction, the operators $Z_\pm$ commute with (the Jordan map of) the angular momentum operator **j**.

Let us calculate next the commutator $[Z_+, Z_-]$. Using anticommutation rules for the fermion operators, and assuming that $2j$ = odd integer, one finds

[1] In this more general approach, the matrix elements of the operators $\{a\}$ and $\{b\}$ are called the "coefficients of fractional parentage" (see Jahn [19], Talmi [20]).

[2] This development (for physics) was initiated by Schwinger [21], who considered the (boson) pair creation–destruction operators as generating the "hyperbolic angular momentum," $SU(1,1)$ (see RWA). Further development of the idea was given by Anderson [13] (application to superconductivity) and (for seniority in nuclear physics) by Helmers [22], Kerman [23], and MacFarlane [24].

that

$$[Z_+, Z_-] = 2\sum_m (b_m\bar{b}_m - \bar{b}_m b_m)/4 \equiv 2Z_0. \tag{7.9.45}$$

To complete the commutation relations, one easily finds that

$$[Z_0, Z_\pm] = \pm Z_\pm. \tag{7.9.46}$$

Accordingly, this verifies that *the three operators* (Z_+, Z_0, Z_-) *obey the commutation rules characterizing angular momentum.* Hence the designation *quasi-spin operators* for the $\{Z\}$.

Having established that the $\{Z\}$ are formally angular momentum constructs, we can bring the full technical machinery of angular momentum to bear on applications. Before proceeding to these applications, let us first note some general features of these operators.

1. Had we used $2j$ = even integer in the definition of Z_+, the resulting operator would have vanished identically. Alternatively, one could say that the coefficients $C^{jj0}_{m,-m,0}$ define a symplectic inner product for $2j$ = odd integer.

2. Were we to use boson operators in place of the fermion operators in defining Z_+, then the situation would be reversed: The nonvanishing case requires $2j$ = even integer. Proceeding to the commutator, one finds now that

$$\begin{aligned} [Z_+, Z_-] &= -2Z_0, \\ Z_0 &\equiv \sum_m (a_m\bar{a}_m + \bar{a}_m a_m)/4 \end{aligned} \tag{7.9.47}$$

for $2j$ = even integer and boson operators.

Note the all-important sign in this commutator! It follows that *quasi-spin for bosonic operators is a formal hyperbolic angular momentum.* The hyperbolic nature of this result [which implies that faithful *unirreps* (unitary irreps) are infinite-dimensional] accords well with physical intuition, in the sense that there is no limit on the number of boson pairs that can occupy the same state, so that the action of Z_+ on the space of boson state vectors is nonterminating. Quasi-spin for *bosonic* structures has been applied in nuclear physics by Ui [25].

g. Quasi-spin wave functions (seniority label). The purpose of this section is to show how, and to what extent, the seniority quantum number v serves to label the many-particle states in a j-shell. Since the angular momentum operator $\mathbf{j}$ commutes with each quasi-spin operator Z_+, Z_0, Z_-, it follows that (besides the j, m labels) we have also the two labels furnished by Z_0 and by the invariant operator $\mathbf{Z}^2$ defined by

$$\mathbf{Z}^2 \equiv \tfrac{1}{2}(Z_-Z_+ + Z_+Z_-) + Z_0^2. \tag{7.9.48}$$

It is useful to note that the quasi-spin operator Z_0 is related to the *number operator* N (the Jordan map of the unit operator). Specifically, one has

$$Z_0 = \tfrac{1}{2}N - \tfrac{1}{4}(2j+1), \tag{7.9.49}$$

where $N \equiv \sum_m b_m\bar{b}_m$. It follows that the *vacuum* state $|0\rangle$, defined by $\bar{b}_m|0\rangle = 0$ for all m, now has the *quasi-spin property*

$$Z_0|0\rangle = -\tfrac{1}{2}\Omega|0\rangle, \qquad \Omega = (2j+1)/2, \tag{7.9.50}$$

in consequence of $N|0\rangle = 0$ (corresponding to no particles).

The basic idea behind the seniority concept is that $J = 0$ pairs in a wave function are to be "inert." This raises a question: How can one tell if a wave function is "pair-free"? Let us consider the set of two-particle ket vectors. All two-particle ket vectors are linear combinations of the basis vectors given by $\{b_mb_{m'}|0\rangle\}$. Using the vector coupling coefficients, one finds

$$|(j^2)JM\rangle = \sum_{mm'} C^{jjJ}_{mm'M} b_m b_{m'}|0\rangle, \tag{7.9.51}$$

where (from the anticommutation property of the fermion operators and the Wigner coefficient symmetry) the only nonvanishing kets have $J =$ even integer.

Clearly, the only state with a $J = 0$ pair is the state with $J = 0$; accordingly, the states $J = 2, 4, \ldots, 2j-1$ are necessarily "pair-free." Hence, if we remove a pair from these states—that is, operate with the quasi-spin operator Z_-—we must get zero:

$$Z_-|(j^2)JM\rangle = 0 \qquad \text{for} \quad J = 2, 4, \ldots, 2j-1. \tag{7.9.52}$$

(Direct evaluation verifies this result.) Since these two-particle states are pair-free, the number of uncoupled particles is 2, which defines the seniority v to be $v = 2$.

This condition that a state be pair-free also enables us to evaluate the eigenvalue of the operator $\mathbf{Z}^2$ for these same states. Thus, we find

$$\mathbf{Z}^2|(j^2)JM\rangle = z_0(z_0 - 1)|(j)^2JM\rangle, \qquad J \neq 0, \tag{7.9.53}$$

where z_0 denotes the eigenvalue of Z_0. Moreover, since N has eigenvalue v on pair-free states, we have

$$z_0 = \frac{v - \Omega}{2}. \tag{7.9.54}$$

This result shows that the quasi-spin z defined by the eigenvalue $z(z + 1)$ of $\mathbf{Z}^2$ has the value $z = (\Omega - v)/2$, so that $\mathbf{Z}^2$ has eigenvalue $\frac{1}{4}(\Omega - v)(\Omega - v + 2)$. Since $\mathbf{Z}^2$ commutes with Z_+, Z_-, and Z_0, it follows that the seniority label v is, in fact, unchanged by the number of $J = 0$ pairs. Hence, we may generate the set of states having sharp seniority v from the state having v unpaired particles:

$$|(j^n)vJM\rangle = (\text{normalization})(Z_+)^{(n-v)/2}|(j^v)vJM\rangle. \tag{7.9.55}$$

The quasi-spin eigenvalues for these states are

$$\mathbf{Z}^2 \rightarrow z(z+1) = \tfrac{1}{4}(\Omega - v)(\Omega - v + 2),$$

$$Z_0 \rightarrow z_0 = \frac{n - \Omega}{2}. \tag{7.9.55$'$}$$

In particular the ground state $|0\rangle$ is seen to be the state having $v = 0$ and quasi-spin $z = \Omega/2$ with projection $-\Omega/2$.

The calculation of the normalization factor in Eq. (7.9.55) is a straightforward exercise, equivalent to calculating the normalization of the state $(J_+)^k|j, -j\rangle$, where j is a generic angular momentum. This calculation uses only the abstract properties of angular momentum; accordingly, it is most simply accomplished by using the Jordan map of Chapter 5 for bosons. Using this map, one has

$$J_+ \rightarrow a_1\bar{a}_2, \qquad |j, -j\rangle \rightarrow \frac{(a_2)^{2j}}{\sqrt{(2j)!}}|0\rangle. \tag{7.9.56}$$

Thus, one finds

$$(J_+)^k|j, -j\rangle = \left[\frac{(2j)!}{(2j-k)!}\right]\frac{(a_1)^k(a_2)^{2j-k}}{\sqrt{(2j)!}}|0\rangle$$

$$= \left[\frac{(2j)!k!}{(2j-k)!}\right]^{\frac{1}{2}}\frac{(a_1)^k(a_2)^{2j-k}}{\sqrt{k!(2j-k)!}}|0\rangle$$

$$= \left[\frac{(2j)!k!}{(2j-k)!}\right]^{\frac{1}{2}}|j, k-j\rangle. \tag{7.9.57}$$

Transcribing now to the variables appropriate for quasi-spin—that is, $j = (\Omega - v)/2$ and $k = (n - v)/2$, we find the normalization factor to be

$$\left[\frac{(\Omega - v)!\left(\dfrac{n - v}{2}\right)!}{\left(\Omega - \dfrac{n + v}{2}\right)!}\right]^{-\frac{1}{2}}.$$

The normalized state (7.9.55) is thus given by

$$|(j^n)vJM\rangle = \left[\binom{\Omega - v}{\dfrac{n - v}{2}}\right]^{-\frac{1}{2}} \frac{(Z_+)^{(n-v)/2}|(j^v)vJM\rangle}{\left(\dfrac{n - v}{2}\right)!}. \tag{7.9.58}$$

The occurrence of the binomial factor

$$\binom{\Omega - v}{\dfrac{n - v}{2}}$$

in the normalization is interpreted physically as a "blocking effect": The *pair degeneracy* of the j-shell is given by $(2j + 1)/2 = \Omega$; the number of pairs added to the "pair-free" wave function $|(j^v)vJM\rangle$ is given by $(n - v)/2$. The number of ways of distributing $(n - v)/2$ pairs in a system having Ω possibilities is the binomial coefficient

$$\binom{\Omega}{\dfrac{n - v}{2}}$$

Comparison with the binomial coefficient obtained above shows that the "effective degeneracy" is $\Omega - v$. In other words, the v uncoupled particles have "blocked off" v possible pairs by the Pauli exclusion principle (antisymmetry).

The seniority label v is of help in labeling the states of n (identical) particles in a j-shell, but the labels n, v, J, and M are not sufficient as a general classification. For $j = \frac{1}{2}, \frac{3}{2}, \frac{5}{2}$, the seniority is superfluous; the n, J, M labels suffice. For $j = 7/2$ and $9/2$ the seniority label splits all multiplicities. For $j \geqslant 11/2$ the seniority label is helpful, but insufficient.

h. Application of quasi-spin to tensor operators. The set of all one-body tensor operators has been given above as $\{v_q^k\}$ realized by the fermionic Jordan map. To ascertain the quasi-spin tensor properties of these operators, let us first determine the quasi-spin of the fermion operators, b_m and $\bar{b}_m$, themselves. One finds

$$[Z_+, b_m] = 0, \qquad [Z_+, (-1)^{j-m}\bar{b}_{-m}] = b_m,$$

$$[Z_0, b_m] = \tfrac{1}{2} b_m, \qquad [Z_0, (-1)^{j-m}\bar{b}_{-m}] = -\tfrac{1}{2}(-)^{j-m}\bar{b}_{-m},$$

$$[Z_-, b_m] = (-)^{j-m}\bar{b}_{-m}, \qquad [Z_-, (-)^{j-m}\bar{b}_{-m}] = 0. \tag{7.9.59}$$

Thus, each pair, b_m and $(-)^{j-m}\bar{b}_{-m}$ $(m = j, j-1, \ldots, -j)$, is a quasi-spin tensor of rank $\frac{1}{2}$ with components denoted by $\frac{1}{2}$ and $-\frac{1}{2}$.

It follows from this result that the one-body operators – since they are constructed bilinearly over the basis, b_μ and $\bar{b}_{\mu'}$, and $[\frac{1}{2}] \otimes [\frac{1}{2}] = [0] \oplus [1]$ – are either quasi-spin vectors or quasi-spin scalars. Symmetry under exchange determines which is which, and one finds (as usual, from the Wigner coefficient symmetry) that the quasi-spin *scalars* are the *tensor operators with odd angular momentum*, whereas the *even* tensor operators are quasi-spin vectors.

This is a significant result. In order to appreciate its significance, let us consider the Wigner–Eckart theorem in quasi-spin space. Since the odd tensor operators are scalar in quasi-spin, the Wigner–Eckart theorem shows that these operators are diagonal in the seniority basis and *independent of the number of particles* (since the quasi-spin analog of the "magnetic quantum number" is determined by the particle number). In particular, the magnetic moment operator is a quasi-spin scalar; the implication that the magnetic moment is independent of the number of (identical) particle pairs is borne out by the shell model data. (This validates the Schmidt model calculations; see p. 498).

For the even multipole one-body tensor operators, denoted by $\mathcal{O}_{\text{even}}$, the Wigner–Eckart theorem determines that the matrix elements have the form

$$\langle(\alpha')(j^{n'})v'J'M'|\mathcal{O}_{\text{even}}|(\alpha)(j^n)vJM\rangle$$

$$= C^{\frac{\Omega-v}{2}\;\;1\;\;\frac{\Omega-v'}{2}}_{\frac{n-v}{2}\;\;z_0\;\;\frac{n'-\Omega}{2}} \langle(\alpha')(j^{v'})v'J'M'\|\mathcal{O}_{\text{even}}\|(\alpha)(j^v)vJM\rangle. \tag{7.9.60}$$

This result achieves an important goal: The explicit dependence on particle number has been determined for all one-body operators in the seniority labeling scheme (de-Shalit and Talmi [26], de-Shalit and Feshbach [26a]).

As an example of the above result, let us consider the quadrupole moment operator. As an even multipole, the operator is a quasi-spin vector. For the static quadrupole moment, the matrix elements are diagonal in both v and n.

Hence, the quadrupole moments depend on the quasi-spin analog to the Wigner coefficient $C^{j1j}_{m0m} = m/\sqrt{j(j+1)}$. One concludes from the transcription $j \rightarrow (\Omega - v)/2$ and $m \rightarrow (n - \Omega)/2$ that the quadrupole moment varies as $(n - \Omega)/2$. This implies that the quadrupole moments at the beginning of the j-shell $(n < \Omega)$ *have the opposite sign* to those near the end of the shell $(n > \Omega)$. This rule is nicely observed in the nuclear data systematics. The result we have obtained also implies that the quadrupole moment vanishes at mid-shell $(n = \Omega)$; this prediction is *not* observed in the data, and the physical explanation (deformation of the nuclear shape far from shell closings) indicates that the seniority model is limited to a few particles (or a few holes) near shell closings.

This result obtained for the quadrupole operator indicates, by example, that there is a relation between a state of n particles and a state having $2\Omega - n$ particles. This relationship—denoted "particle–hole conjugation"—is more general than the seniority labeling and is a general property of the shell model itself (Brussaard and Glaudemans [27]).

Remark. A particularly nice application of the conjugation concept is the Pandya [28] transformation, which determines from the energy spectrum of a nucleus having two particles outside closed shells the energy spectrum of the related particle–hole nucleus (assuming two-body interactions). The required transformation is the Racah transform:

$$E_J^{\text{particle-hole}} = (\text{const}) - \sum_{J'} (2J' + 1) W(j_1 j_2 j_2 j_1; JJ')\, E_{J'}^{\text{particle-particle}}, \tag{7.9.61}$$

where j_1 and j_2 denote the angular momenta of the shell model states of the two particles (with j_1 being the angular momentum of the hole state) and (constant) $= (2j_1 + 1) \sum_{J'} (2J' + 1) E_{J'}^{\text{particle-particle}}$. (This relation is derived and discussed in Ref. [27, p. 368].)

This relation is validated experimentally in the spectra of several such related pairs.

The analysis of tensor operators in terms of quasi-spin can be extended to two-body operators, these operators being quasi-spin operators with $z = 2, 1$, or 0. The essential result of this analysis is, once again, the determination of the explicit dependence of the matrix elements on particle number. Applications of such techniques to stripping and pick-up reactions (particle transfer) can be found in MacFarlane [24], Satchler [29], and Bayman [30].

i. Seniority in terms of Casimir operators. There is a very different view of the seniority operator q_{12}, which is important for the purpose of understanding angular momentum techniques and, more generally, symmetry techniques.

This view considers the operator q_{12} in terms of Casimir operators and the general two-particle exchange operator.

Let us consider a single j-shell and, for simplicity, identical nucleons filling this shell. If one assumes a definite two-particle (residual) nuclear interaction, the problem is to determine from this information the energy spectrum for the n-particle system, $n = 1, 2, \ldots, 2j + 1$.

The two-particle interaction V_{12} leads to a perturbation matrix $\langle (j^2)JM|V_{12}|(j^2)JM\rangle$, which may be parametrized in terms of the 6-j coefficients. Since the interaction V_{12} is a rotationally invariant function of the interparticle distance $r_{12} = \|\mathbf{x}_1 - \mathbf{x}_2\|$, we may introduce a Legendre polynomial expansion,

$$V_{12}(r_{12}) = \sum_{k=0}^{\infty} V_{12}^{(k)}(r_1, r_2)P_k(\hat{x}_1 \cdot \hat{x}_2), \tag{7.9.62}$$

into the perturbation matrix to obtain the form [see Eqs. (7.5.109)–(7.5.110)]

$$\langle (j^2)JM|V_{12}|(j^2)JM\rangle = \sum_k F^k(l, l) f_k[(lj)^2J]. \tag{7.9.63}$$

The parameters in this equation are as follows:

1. The Slater integrals defined by

$$F^k(l, l) = \iint r_1^2\, dr_1\, r_2^2\, dr_2\, V_{12}^{(k)}(r_1, r_2)[R_l(r_1)R_l(r_2)]^2. \tag{7.9.64}$$

2. The angular momentum factors $f_k[(lj)^2J]$ defined by

$$f_k[(lj)^2J] = (-1)^{J-1}(2j+1)^2 \begin{pmatrix} j & k & j \\ \frac{1}{2} & 0 & -\frac{1}{2} \end{pmatrix}^2 \begin{Bmatrix} j & j & J \\ j & j & k \end{Bmatrix}. \tag{7.9.65}$$

The essential point of this result is that the general residual interaction is represented in terms of a limited number of parameters (the F_k for $k = 0$ to $2j$). Since, however, the interaction determines only $(2j + 1)/2$ physical data – the energy levels of the two-particle system (which, for antisymmetry, must have $J =$ even integer) – it is clear that very different interactions may be physically equivalent (MacFarlane [24]). Optimally, the data for the two-particle system are to be parametrized in terms of $(2j + 1)/2$ basic interactions, chosen for simplicity in extending to the n-particle case. One strategy for choosing these basic interactions is to use *exchange interactions.*

In order to understand the idea of using exchange interactions, let us consider the simplest such operator, the Dirac spin exchange operator,

$$S_{12} \equiv \tfrac{1}{2}(\mathbb{1} + \boldsymbol{\sigma}_1 \cdot \boldsymbol{\sigma}_2), \tag{7.9.66}$$

where $\boldsymbol{\sigma}_i$ are the Pauli spin operators for the (kinematically independent) spins of the two particles (denoted by 1 and 2). One easily verifies that, as befits an exchange operator, $(S_{12})^2 = \mathbb{1}$, so that the eigenvalues are all ± 1. One equally easily verifies that on triplet states ($S = 1$) the eigenvalue is $+1$, and on singlet states ($S = 0$) the eigenvalue is -1.

The usefulness of the spin exchange operator is that it can be extended to n-particle systems. The appropriate definition is

$$S_{\text{exch}} \equiv \sum_{1=\alpha<\beta}^{n} S_{\alpha\beta}, \tag{7.9.67}$$

and evaluated in terms of the total spin $\mathbf{S} = \frac{1}{2}\sum_{\alpha=1}^{n} \boldsymbol{\sigma}_\alpha$:

$$\begin{aligned} S_{\text{exch}} &= \tfrac{1}{2} \sum_{\alpha<\beta} (\mathbb{1} + \boldsymbol{\sigma}_\alpha \cdot \boldsymbol{\sigma}_\beta) \\ &= \frac{n(n-1)}{4} + (\mathbf{S} \cdot \mathbf{S} - \tfrac{3}{4}n). \end{aligned} \tag{7.9.68}$$

Hence, the eigenvalue of S_{exch} is

$$\frac{n(n-4)}{4} + S(S+1). \tag{7.9.69}$$

If the total spin is a sharp quantum number (as in LS-coupling), a spin exchange interaction is a satisfactory basic interaction for parametrizing the two-body residual interactions.

These results are both simple and familiar, but, for the purpose of generalizing, it is useful to view the spin exchange in a different way. The spin-$\frac{1}{2}$ system has two states, so that the two-particle system has four states. Accordingly, the operator that exchanges the state labels (jm) of particle 1 with the corresponding state labels of particle 2 is the 4×4 operator (permutation matrix) in the tensor product space given by

$$P = \begin{pmatrix} 1 & 0 & 0 & 0 \\ 0 & 0 & 1 & 0 \\ 0 & 1 & 0 & 0 \\ 0 & 0 & 0 & 1 \end{pmatrix}. \tag{7.9.70}$$

From $P^2 = \mathbb{1}_4$, $\operatorname{tr} P = 2$, one sees that P is, in fact, equivalent to the operator S_{12}.

These observations suggest generalizing from a two-state system to a k-state system: The exchange operator $P(k)$ for two particles is then a $k^2 \times k^2$ permutation matrix. The properties $P^2(k) = \mathbb{1}_{k^2}$ and $\operatorname{tr} P(k) = k$ determine $P(k)$ to have $k(k+1)/2$ eigenvalues $+1$ and $k(k-1)/2$ eigenvalues -1. These degeneracies indicate that the exchange operator $P(k)$ assigns the eigenvalue $+1$ to the symmetric two-particle system having the Young frame $\square\square$, and the eigenvalue -1 to the antisymmetric two-particle system having the Young frame $\begin{smallmatrix}\square\\ \square\end{smallmatrix}$.

To extend the general exchange operator $P_{\alpha\beta}(k)$ to states of n particles, one evaluates the operator $\sum_{\alpha<\beta} P_{\alpha\beta}(k)$ on the set of n-particle states, each state being specified by a standard Young tableau of shape $[\lambda]$ having n boxes, restricted by having all columns of length $\leqslant k$. We state this result (without proof):

$$P([\lambda]) \overset{\text{def}}{\equiv} \langle[\lambda]|\left(\sum_{1=\alpha<\beta}^{n} P_{\alpha\beta}(k)\right)|[\lambda]\rangle = \tfrac{1}{2}\sum_{i=1}^{k} \lambda_i(\lambda_i - 2i + 1). \tag{7.9.71}$$

The previous result for the spin exchange operator is seen to be the special case $k = 2$. [This result uses $n = \lambda_1 + \lambda_2$, $S = (\lambda_1 - \lambda_2)/2$.]

Now let us show how this result is related to the Wigner operators of angular momentum theory. A basis for the set of all operators mapping the space $\mathscr{H}_j$ spanned by the basis vectors $\{|jm\rangle : m = j, j-1, \ldots, -j\}$ into itself is given by the set of multipole operators $\{v_q^k\colon k = 0, 1, \ldots, 2j;\ q = k, k-1, \ldots, -k\}$ (Wigner operators), whose matrix elements are the coefficients $\langle jm'|v_q^k|jm\rangle \equiv C_{mqm'}^{jkj}$ $(k = 0, 1, \ldots, 2j)$. Hence, the most general rotationally invariant (Hermitian) two-particle interaction (for particles in the j-shell) is given by

$$H_{12} = (2j+1)^{-1} \sum_{k=0}^{2j} A_k \mathbf{v}^k(1) \cdot \mathbf{v}^k(2),$$

$$\mathbf{v}^k(1) \cdot \mathbf{v}^k(2) \equiv \sum_q (-1)^q v_q^k(1)\, v_{-q}^k(2), \tag{7.9.72}$$

with the A_k denoting numerical (real) parameters. Taking matrix elements on two-particle states[1] yields the result

---

[1] The two-particle states $|(j^2)JM\rangle$ in this relation are for distinguishable particles so that the states exist for all $J = 0, 1, \ldots, 2j$. We require this property in order to be able to determine the coefficients A_k uniquely in terms of the interaction (see the discussion p. 518). The operator identities developed in this way [for example, Eq. (7.9.84)] are applicable (since identities) to arbitrary states (symmetric, antisymmetric, or mixed symmetry).

$$\langle (j^2)JM|H_{12}|(j^2)JM\rangle = \sum_{k=0}^{2j} (-)^{2j+J} A_k W(jjjj;kJ). \tag{7.9.73}$$

To determine the constants A_k corresponding to particularly simple interactions, let us recall the identities that exist for the Racah coefficients. The identity [Chapter 3, Eq. (3.274)]

$$\sum_f (2f+1)(-1)^{b+d-f} W(abcd;ef)W(adcb;gf) = (-1)^{e+g-a-c} W(bacd;eg) \tag{7.9.74}$$

can be specialized (take $c = b$, $d = a$, $g = 0$) to yield

$$\sum_f (2f+1)W(abba;fe) = 1, \tag{7.9.75}$$

where we have also used Eq. (3.282) and a symmetry relation in obtaining this result. Thus, choosing $A_k = (2k+1)$ in Eq. (7.9.72), we find

$$(2j+1)^{-1}\langle (j^2)JM| \sum_{k=0}^{2j} (2k+1)\mathbf{v}^k(1)\cdot\mathbf{v}^k(2)|(j^2)JM\rangle = (-1)^{J+2j}. \tag{7.9.76}$$

Since the term $(-1)^{J+2j}$ is $+1$ for symmetric two-particle states (for both j integral as well as half-integral), and -1 for antisymmetric states, it is clear that Eq. (7.9.76) *determines a Wigner operator representation for the general exchange operator*:

$$\text{Exchange operator} \equiv P_{\alpha\beta} \equiv \sum_{k=0}^{2j} \frac{(2k+1)}{(2j+1)}\mathbf{v}^k(\alpha)\cdot\mathbf{v}^k(\beta). \tag{7.9.77}$$

An equally significant interpretation of the exchange operator results from the fact that the exchange operator $P_{\alpha\beta}$ commutes with all transformations generated by the operators $\mathbf{v}^k \equiv \sum_{\alpha=1}^{n} \mathbf{v}^k(\alpha)$, $k = 0, 1, \ldots, 2j$. These operators[1] [$(2j+1)^2$ in number] generate the group $U(2j+1)$, and hence $\sum_{\alpha<\beta} P_{\alpha\beta}$ is related to the quadratic invariant of this group (known as the Casimir operator, I_2). This relation is given by

[1] In the boson realization the operators $\mathbf{v}^k$ are related to the $\mathbf{V}^k$ introduced in Section 5, Eq. (7.5.176), by $\mathbf{v}^k = [(2j+1)/(2k+1)]^{\frac{1}{2}}\,\mathbf{V}^k$.

$$\sum_{1=\alpha<\beta}^{n} P_{\alpha\beta} = \sum_{\alpha<\beta} \sum_{k=0}^{2j} \frac{(2k+1)}{(2j+1)} \mathbf{v}^k(\alpha) \cdot \mathbf{v}^k(\beta)$$

$$= \left[I_2 - \sum_{\alpha=1}^{n} I_2(\alpha) \right] \Big/ 2(2j+1), \tag{7.9.78}$$

where we have defined

$$I_2 = \sum_{k=0}^{2j} (2k+1)\mathbf{v}^k \cdot \mathbf{v}^k = \text{Casimir operator for } U(2j+1),$$

$$I_2(\alpha) = \sum_{k=0}^{2j} (2k+1)\mathbf{v}^k(\alpha) \cdot \mathbf{v}^k(\alpha). \tag{7.9.79}$$

Note that $I_2(\alpha) = I_2(1)$ for each $\alpha = 1, \ldots, n$.

It is not difficult now to obtain an expression for the seniority operator in terms of the Wigner operators. From the orthogonality relation [Chapter 3, Eq. (3.271)] for the Racah coefficients, one has the identity

$$\sum_{e} (2e+1)(2f+1) W(abcd; ef) W(abcd; eg) = \delta_{fg}. \tag{7.9.80}$$

This can be specialized (take $g = 0$) to the form

$$\sum_{k} (2k+1)(-1)^{2j+k} W(jjjj; kJ) = (2j+1)\delta_{J0}. \tag{7.9.81}$$

This result and Eq. (7.9.73) show therefore that the seniority operator, q_{12}, defined by

$$\langle (j^2)JM | q_{12} | (j^2)JM \rangle = (2j+1)\delta_{J0}, \tag{7.9.82}$$

has the Wigner operator expression

$$q_{12} = \sum_{k=0}^{2j} (-1)^k \frac{(2k+1)}{(2j+1)} \mathbf{v}^k(1) \cdot \mathbf{v}^k(2). \tag{7.9.83}$$

Using the result obtained for the exchange operator, Eq. (7.9.77), one can eliminate the even multipoles from the expression for the seniority operator

$$q_{12} = P_{12} - 2 \sum_{k\,\text{odd}} \frac{(2k+1)}{(2j+1)} \mathbf{v}^k(1) \cdot \mathbf{v}^k(2). \tag{7.9.84}$$

This expression for the seniority operator is quite significant for the group-theoretic interpretation of seniority. For both integral and half-integral j, the unitary group $U(2j+1)$ generated by the multipole operators $\{\mathbf{v}^k\}$ has a subgroup generated by the *odd* multipole operators, the group $SO(2j+1)$ for j integral, and the symplectic group $Sp(2j+1)$ for j half-integral. The expression $\sum_{k=\text{odd}}(2k+1)\mathbf{v}^k(1)\cdot\mathbf{v}^k(2)$ entering in this result shows that the seniority operator involves the quadratic invariant operator for the subgroup. Thus, we conclude that the seniority operator not only has a group-theoretic significance in the reduction $U(2j+1)\supset SO(2j+1)$ or $Sp(2j+1)$, but may also be expressed in terms of the Casimir invariant operators of the two groups.

j. Concluding remarks. The applications of angular momentum techniques in nuclear physics, which we have discussed above, are, of necessity, rather specialized and limited in scope. These examples can only hint—however inadequately—at the major (even dominant) importance of angular momentum techniques in this extensive field of research. The very existence of the nuclear shell model, classifying nuclear orbits by angular momentum quantum numbers, guarantees that extensive coupling and recoupling of individual particle angular momenta will be required in almost all calculations. The direct application of angular momentum techniques in large-scale computational attacks on nuclear structure calculations are feasible with present-day computers. The required "angular momentum technology" (based on the $3n$-j coefficients, see also Topic 12, RWA) is developed and discussed for nuclear physics in the work of Danos and Gillet [31]. The Rochester–ORNL collaboration (French *et al.* [32]) has led to the development of large-scale computer codes for nuclear structure calculations.

An alternative to large-scale numerical computation lies in truncation of the shell-model space to a manageable basis with schematic interaction Hamiltonians (Lipkin *et al.* [33], Holstein and Primakoff [34], Ginocchio [34a]). The program of solvable (schematic) interaction Hamiltonians, combined with the Landau's concept of elementary excitation modes (based, for nuclei, on angular momenta), is reviewed in Bortignon *et al.* [35]. Alternatively, truncated shell-model spaces can be viewed as carrier spaces of a larger symmetry structure, containing angular momentum symmetry as a subgroup. Such an approach is subsumed under the more general program of applying group-theoretic symmetry techniques to nuclear structure (Schiffer [12a], Hecht [36], Matsen [37], Arima and Iachello [38], Castaños [39], Parikh [40]).

We thank Prof. Hecht of the University of Michigan for a critical reading of this section.

k. Note. The concept of a Wigner operator is well-known in the physics literature (and is developed in detail in RWA); the concept of a Racah operator (discussed in Chapter 3 and in detail in RWA) is not so familiar. Equation (7.9.73) presents an excellent example of this latter concept and illustrates that: Racah operators play the same role with respect to coupled angular momentum spaces $\{|(j_1 j_2)jm\rangle$ basis$\}$ as do Wigner operators with respect to the basic angular momentum spaces $\{|jm\rangle$ basis$\}$. The Racah operators of relevance to the construction of invariant interactions acting in coupled angular momentum spaces are precisely those that effect *no shift* of the j_1, j_2 labels (see Chapter 3, Section 21):

$$\left\{\begin{matrix} & k & \\ 2k & & 0 \\ & k & \end{matrix}\right\}. \tag{7.9.85}$$

One sees from Eq. (7.9.73) that the set of operators corresponding to $k = 0, 1, \ldots, 2j$ is a basis for all invariant operator mappings of the $(2j + 1)$-dimensional space spanned by the basis $\{|(j^2)JM\rangle : J = 0, 1, \ldots, 2j\}$ into itself [see footnote, p. 520]. Thus, an alternative operator formulation of Eq. (7.9.72) is

$$H_{12} = \sum_{k=0}^{2j} a_k \left\{\begin{matrix} & k & \\ 2k & & 0 \\ & k & \end{matrix}\right\}. \tag{7.9.86}$$

Using the trace orthogonality relation [see Eq. (3.357), Chapter 3] given by

$$\mathrm{tr}\left[(2J+1)\left\{\begin{matrix} & k' & \\ 2k' & & 0 \\ & k' & \end{matrix}\right\}\left\{\begin{matrix} & k & \\ 2k & & 0 \\ & k & \end{matrix}\right\}^{\dagger}\right] = \delta_{k'k}\frac{(2j+1)^2}{(2k+1)}, \tag{7.9.87}$$

one may evaluate the expansion coefficients a_k in terms of the invariant interaction itself:

$$a_k = \frac{(2k+1)}{(2j+1)^2}\,\mathrm{tr}\left[(2J+1)H_{12}\left\{\begin{matrix} & k & \\ 2k & & 0 \\ & k & \end{matrix}\right\}^{\dagger}\right]. \tag{7.9.88}$$

Moreover, the operators (7.9.85) corresponding to $k = 0, 1, \ldots, 2j$ mutually commute, are Hermitian, and commute with $\mathbf{J}$. The set of simultaneous eigenvalues of these operators $\{(2j + 1)W(Jjjk; jj) : k = 0, 1, \ldots, 2j\}$ or, what is the same thing, the index k itself, thus provides a complete labeling of a basis of the space of invariant mappings.

The concept of a Racah operator therefore offers a significant interpretation of the expansion (7.9.73).

References

1. W. M. Elsasser, Sur le principe de Pauli dans les noyaux. II, *J. Phys. Rad.* **5** (1934), 389–397; III, *ibid.* 634–639. See also E. Feenberg, *Shell Theory of the Nucleus*. Princeton Univ. Press, Princeton, N. J., 1955, especially Chapter I.
2. V. M. Goldschmidt, "Geochemische Verteilungsgesetze der Elemente. IX. Die Mengenverhältnisse der Elemente und der Atom-Arten," *Norske Viden. Akad.* (*Oslo*) (1937), 1–148.
3. M. G. Mayer and J. H. D. Jensen, *Elementary Theory of Nuclear Shell Structure*. Wiley, New York, 1955.
4. A. Bohr and B. R. Mottelson, "Collective and individual-particle aspects of nuclear structure, *Mat. Fys. Medd. Dan. Vid. Selsk.* **27**, No. 16 (1953), 1–174. A comprehensive survey is in the monograph by A. Bohr and B. R. Mottelson *Nuclear Structure*, Vol. I. Benjamin, New York, 1969; Vol. II. 1975.
5. J. Rainwater, "Nuclear energy level argument for a spheroidal nuclear model," *Phys. Rev.* **79** (1950), 432–434.
6. W. Heisenberg, "Über den Bau der Atomkerne," *Z. Physik* **77** (1932), 1–11.
7. E. P. Wigner, "On the consequences of the symmetry of the nuclear Hamiltonian on the spectroscopy of nuclei," *Phys. Rev.* **51** (1937), 106–119.
8. M. Gell-Mann, *The Eightfold Way: A Theory of Strong Interaction Symmetry*, California Institute of Technology Report CTSL-20, 1961, unpublished; reproduced in the monograph by M. Gell-Mann and Y. Ne'eman, *The Eightfold Way*. Benjamin, New York, 1964.
9. Y. Ne'eman, "Derivation of strong interactions from a gauge invariance," *Nucl. Phys.* **26** (1961), 222–229.

9a. E. P. Wigner, "Isospin – a quantum number for nuclei," in *Proceedings of the First Robert A. Welch Foundation Conference: On the Structure of the Nucleus* (W. O. Milligan, ed.), pp. 61–81. Rice Institute, Houston, 1958.

10. D. Robson, "Theory of isobaric spin analogue resonances," *Phys. Rev.* **137** (1965), 535–546; see also A. Bohr and B. R. Mottelson, *Nuclear Structure*, Vol. I, pp. 35 ff. Benjamin, New York, 1969.
11. A. de-Shalit, "The energy levels of odd-odd nuclei," *Phys. Rev.* **91** (1953), 1479–1486.
12. B. L. van der Waerden, *Science Awakening*, see pp. 228, 277. Wiley, New York, 1963. (Translation of *Ontwakende Wetenschap*, Noordhoff).

12a. J. P. Schiffer, "The effective nucleon-nucleon interaction deduced from nuclear spectra," in *The Two-Body Force in Nuclei* (S. M. Austin and G. M. Crawley, eds.), pp. 205–227. Plenum, New York, 1972; J. P. Schiffer and W. W. True, "The effective interaction between nucleons deduced from nuclear spectra," *Rev. Mod. Phys.* **48** (1976), 191–217.

13. P. W. Anderson, "Coherent excited states in the theory of superconductivity: Gauge invariance and Meissner effect," *Phys. Rev.* **110** (1958), 827–835.
14. P. W. Anderson, "Random phase approximation in the theory of superconductivity," *Phys. Rev.* **112** (1958), 1900–1916.
15. J. M. Blatt, *Theory of Superconductivity*. Academic Press, New York, 1964.
16. L. D. Landau, "The theory of superfluidity of helium II," *J. Phys.* (*USSR*) **5** (1941), 71–90; see also *JETP* **35** (1958), 95–103. Translation in *Soviet Physics, JETP* **8** (1959), 70–74.
17. A. B. Migdal, *Nuclear Theory: The Quasi-Particle Method*. Benjamin, New York, 1968.
18. G. E. Brown, *Unified Theory of Nuclear Models*, 2nd ed. North-Holland, Amsterdam, 1967.
19. H. A. Jahn, "Theoretical studies in nuclear structure," *Proc. Roy. Soc.* **A201** (1950), 517–544.
20. I. Talmi, "Nuclear spectroscopy with harmonic oscillator wave functions," *Helv. Phys. Acta* **25** (1952), 185–234.

21. J. Schwinger, *On Angular Momentum*. Atomic Energy Commission Rept. NYO-3071, 1952, unpublished; published in L. C. Biedenharn and H. van Dam, *Quantum Theory of Angular Momentum*. Academic Press, New York, 1965.
22. K. Helmers, "Symplectic invariants and Flowers' classification of shell model states," *Nucl. Phys.* **23** (1961), 594–611.
23. A. K. Kerman, "Pairing forces and nuclear collective motion," *Ann. Phys. (N.Y.)* **12** (1961), 300–329.
24. M. H. MacFarlane, "Shell model theory of identical nucleons," in *Lectures in Theoretical Physics* (P. D. Kunz, D. A. Lind, and W. E. Brittin, eds.), Vol. VIII-C, pp. 583–677. Univ. Colorado Press, Boulder, Colo. 1966.
25. H. Ui, "*SU*(1, 1) quasi-spin formalism of the many-boson system in a spherical field," *Ann. Phys. (N.Y.)* **49** (1968), 69–92.
26. A. de-Shalit and I. Talmi, *Nuclear Shell Theory*. Academic Press, New York, 1963.
26a. A. de-Shalit and H. Feshbach, *Theoretical Nuclear Physics*. Wiley, New York, 1974.
27. P. J. Brussaard and P. W. M. Glaudemans, *Shell Model Applications in Nuclear Spectroscopy*. North-Holland, Amsterdam, 1977.
28. S. P. Pandya, "Nucleon-hole interaction in *jj* coupling," *Phys. Rev.* **103** (1956), 956–957.
29. G. R. Satchler, "Some topics in the theory of direct nuclear reactions," in *Lectures in Theoretical Physics* (P. D. Kunz, D. A. Lind, and W. E. Brittin, eds.), Vol. VIII-C, pp. 73–175. Univ. of Colorado Press, Boulder, Colo., 1966.
30. B. F. Bayman, "General properties of nuclear models," in *Proceedings of the 5th Scottish Universities Summer School: Nuclear Structure and Electromagnetic Interactions* (N. MacDonald, ed.), pp. 19–60. Oliver and Boyd, Edinburgh, 1965.
31. M. Danos and V. Gillet, *Relativistic Many-Body Bound Systems*. National Bureau of Standards Monograph 147, U. S. Government Printing Office, Washington, 1975.
32. J. B. French, E. C. Halbert, J. B. McGrory, and S. S. M. Wong, "Complex spectroscopy," in *Advances in Nuclear Physics* (M. Baranger and E. Vogt, eds.), Vol. 3, pp. 193–257. Plenum, New York, 1969.
33. H. J. Lipkin, N. Meshkov, and A. J. Glick, "Validity of many-body approximation methods for a solvable model. (I). Exact solutions and perturbation theory," *Nucl. Phys.* **62** (1965), 188–198.
34. T. Holstein and H. Primakoff, "Field dependence of the intrinsic domain magnetization of a ferromagnet," *Phys. Rev.* **58** (1940), 1098–1113.
34a. J. N. Ginocchio, "Fermion Hamiltonians with monopole and quadrupole pairing," in *Interacting Bosons in Nuclear Physics* (F. Iachello, ed.), pp. 111–119. Plenum, New York, 1978; "A schematic model for monopole and quadrupole pairing in nuclei," *Ann. Phys.* (N. Y.) **126** (1980), 234–276.
35. P. F. Bortignon, R. A. Broglia, D. R. Bès, and R. Liotta, "Nuclear field theory," *Phys. Rept.* **30C** (1977), 307–360.
36. K. T. Hecht, "Symmetries in nuclei," *Ann. Rev. Nucl. Sci.* **23** (1973), 123–161.
37. F. A. Matsen, "The unitary group and the many-body problem," in *Advances in Quantum Chemistry*, Vol. II, pp. 223–250. Academic, New York, 1978.
38. A. Arima and F. Iachello, "Interacting boson model of collective nuclear states. II. The rotational limit," *Ann. Phys. (N.Y.)* **111** (1978), 201–238.
39. O. Castaños, E. Chacón, A. Frank, and M. Moshinsky, "Group theory of the interacting boson model of the nucleus," *J. Math. Phys.* **20** (1979), 35–44.
40. J. C. Parikh, *Group Symmetries in Nuclear Structures*. Plenum, New York, 1978.

10. Body-Fixed Frames: Spectra of Spherical Top Molecules[1]

a. Introduction. The quantum theory of atomic spectra was already a sophisticated field of investigation (culminating in the publication of Condon and Shortley's *Theory of Atomic Spectra* in 1935) before the appropriate general techniques were recognized for applying the quantum theory to the interpretation of the vibration–rotation spectra of polyatomic molecules (three or more atoms).

In the late twenties and early thirties the model of a molecule as consisting of nuclei that could execute small vibrations about equilibrium positions (localized potential minima) created by the much faster motions of the electrons had been clearly recognized (Born and Oppenheimer [1]). The use of normal coordinates (Brester [2], Wigner [3]) for the description of these motions was investigated thoroughly by Wilson [4]. It was, however, only in 1934 and 1935 that Eckart [5, 6] considered methods for obtaining a general Hamiltonian that would yield an approximate separation of the over-all rotational motion of a molecule (thought of as a rigid body) and the small "internal" displacements of the nuclei away from their positions of equilibrium. (The possibility of such a separation had been suggested earlier by Casimir [7].)

The difficulty in describing the motions of the nuclei in the intuitive model sketched above was one of defining a *moving reference frame* such that the Hamiltonian, when referred to the moving frame, would fulfill Casimir's condition (small interaction between rotational and internal motions). There were subtleties in anticipating the "type" of moving reference frame that would produce the "separation of motions" that would render effective the techniques of perturbation theory so crucial to successful applications of quantum theory to many-particle systems.

In his first paper, Eckart [5] developed the classical kinetic energy expression relative to a frame defined by the *principal moments of inertia at each instant of time* (principal axes frame). The pure rotational energy term in this expression was not, however, of the classical form [see Eq. (7.10.22) below] that was to be expected if it were dominant. This same anomalous rotational energy term appears also in the Schrödinger equation for an N-particle system using Cartesian coordinates measured relative to the principal axes frame (Hirschfelder and Wigner [8]). In his second paper, Eckart [6] rejects the principal axes frame as being incompatible with a normal coordinate description of small internal motions away from fixed equilibrium positions, although Van Vleck [9] had shown how to "correct" the anomalous principal axes rotational energy to obtain the usual classical expression. (Paradoxically,

[1] The considerations of this section are confined to nonrelativistic quantum mechanics.

Eckart cites Van Vleck's paper as one of his arguments for not using the principal axes frame.[1]) Recent renewed interest in this problem (Refs. [10, 11], and references therein) may offer new insights into the use of various types of moving reference frames adapted to various physical models.

The principal axes frame and the frame introduced by Eckart are equally fundamental; each is an example of a more general kinematic concept – *the body-fixed frame*. As we shall see, this concept is intimately related to angular momentum properties.

The subject of vibration–rotation spectra of polyatomic molecules is rich in applications of angular momentum and symmetry concepts. The applications include invariance principles entering at the outset in the definition of moving frames; the development of the Hamiltonian itself; classification of interactions by their $O(3)$-tensor, G-invariant properties, where G is a subgroup of $O(3)$; determination of the level splitting associated with such interactions; the quantization of angular momentum along the various symmetry axes of a molecule; the Frobenius theory of induced representations; asymptotic behavior of Wigner coefficients; the Pauli exclusion principle; and other topics. It is the purpose of this section to develop these applications, illustrating in the last subsection their success in interpreting recent laser spectra.

This section is quite long, since we have considered in greater detail than is customary the concept of a body-fixed frame, the development and symmetries of the phenomenological molecular Hamiltonian, and the implementation of these symmetries in terms of group actions on the molecular coordinates and wave functions. We have found this detail to be necessary in order to give a consistent treatment of the structure and spectra of rigid molecules based on the techniques of standard angular momentum theory as developed in Chapter 3. The physical applications of the theory are given principally in subsections n and p–u; for the most part these subsections may be read independently of the others.

b. Definition and kinematics of a body-fixed frame. It is essential for clarity to recall the kinematical principles (discussed in Section 2) that underlie all spectroscopic considerations, in particular, as they apply to molecular spectroscopy. Based on the discussion of Section 2, we conclude that the appropriate reference frame for the quantal description of the motion of a molecule is an inertial coordinate frame for which the associated linear manifold of state vectors obeys the condition that the total linear momentum

---

[1] Eckart's intuition was, however, correct: Numerical calculations show that, in a classical model of a spherical top molecule, small displacements of the nuclei away from equilibrium positions stay small relative to Eckart's frame, but become large, even undefined, for the principal axes frame because the principal moments of inertia are nearly equal and, in fact, for some motions become equal (private communication by R. B. Walker, Los Alamos National Laboratory).

operator *vanishes* when acting on any state vector in the manifold. For such a reference frame the position of the center of mass of the molecule is (from the uncertainty principle) *uniformly probable everywhere*. Accordingly, it is meaningless, for example, to say that "the reference frame is defined such that the origin is at the center of mass of the particles."

The proper way to define the positions of the particles in this special reference frame is to introduce the *translationally invariant positions* $\vec{y}^{\alpha}$ defined by $\vec{x}^{\alpha} = \vec{y}^{\alpha} + \vec{R}$, where $\vec{R}$ denotes the center-of-mass vector defined by $\hat{R} = \sum_{\alpha} m_{\alpha}\vec{x}^{\alpha}$, in which $\vec{x}^{\alpha}$ $(\alpha = 1, 2, \ldots, N)$ denotes the position vector of particle α of mass m_{α}.

It is the translationally invariant particle positions $\{\vec{y}^{\alpha}\}$ as expressed relative to the fixed special inertial reference frame that are to be used for the description of the quantal states of a molecule. Once we recognize this conceptually important fact, the more usual language (referring incorrectly to a fixed center of mass) can be used without essential error.

Now let us turn to the concept of a body-fixed frame.

Following Eckart, we may recognize that there are many types of moving reference frames, including (1) frames whose instantaneous position depends only on some prescribed function of the time, and (2) frames whose instantaneous position and orientation depend only on the instantaneous (translationally invariant) positions of a collection of particles, etc. We single out the second class of frames and refer to them as *body-fixed frames.*[1]

As we shall prove in this section, body-fixed frames are particularly important because the *angular momentum operator that generates rotations of the frame is the total orbital angular momentum of the system of particles* (this statement will be made precise). The introduction of such a frame into the description of the motions of many-particle systems serves to define partially a *transformation from Cartesian coordinates to new coordinates* in such a way that we are assured of being able to construct states of the system having sharp total orbital angular momentum (good quantum labels LM). The difficulty with the method as a general procedure is that there exists an infinity of such body-fixed frames, and the choice of frame and of the remaining $3N - 6$ coordinates must be adapted to the given physical problem. Nonetheless, certain general properties of body-fixed frames can be developed, and we turn to this task now (we consider only Cartesian frames).

---

[1] From the discussion of Chapter 2, Section 3, it follows that *there is only one version* (the strings can be untangled after a rotation by 2π) *of a body-fixed frame associated with point particles.* This result already implies that these body-fixed frames, in contrast to those associated with solid (impenetrable) bodies, can carry only integral values of angular momenta. The mathematical verification of this result appears in the explicit results, Eqs. (7.10.10), where the angular momentum of the frame (defined such that the total linear momentum vanishes) is seen to be the *total orbital angular momentum* of the particles. Hence, the values of the total orbital angular momentum quantum number are necessarily integral.

Let the instantaneous position vectors of a collection of N point particles of masses $m_1, m_2, \ldots, m_N$ be denotes by $\vec{x}^1, \vec{x}^2, \ldots, \vec{x}^N$ relative to some fixed point O, which may be regarded as the origin of a laboratory (inertial) frame $(\hat{l}_1, \hat{l}_2, \hat{l}_3)$ (right-handed, unit triad). The definition of a body-fixed (Cartesian) reference frame is as follows:

A triad of unit vectors $(\hat{f}_1, \hat{f}_2, \hat{f}_3)$, $\hat{f}_i \cdot \hat{f}_j = \delta_{ij}$, *where each unit vector is a function,* $\hat{f}_i = \hat{f}_i(\vec{x}^1, \ldots, \vec{x}^N)$, *of the instantaneous position vectors of a collection of N particles,* $\{\vec{x}^\alpha\}$, *is said to be a body-fixed frame for the set of particles labeled by* $\{\alpha : \alpha = 1, 2, \ldots, N\}$ *if the unit vectors* $\hat{f}_i$ *satisfy the following conditions:*

(1) $\hat{f}_i(\vec{x}^1 + \vec{a}, \ldots, \vec{x}^N + \vec{a}) = \hat{f}_i(\vec{x}^1, \ldots, \vec{x}^N)$, each translation[1] $\vec{a}$;

(2) $\hat{f}_i(\mathscr{R}\vec{x}^1, \ldots, \mathscr{R}\vec{x}^N) = \mathscr{R}[\hat{f}_i(\vec{x}^1, \ldots, \vec{x}^N)]$, each rotation $\mathscr{R}$;

(3) $\left(\dfrac{\partial \hat{f}_i}{\partial x_1^\alpha}, \dfrac{\partial \hat{f}_i}{\partial x_2^\alpha}, \dfrac{\partial \hat{f}_i}{\partial x_3^\alpha}\right) \neq (0, 0, 0)$,

for at least one i and all $\alpha = 1, 2, \ldots, N$. (7.10.1)

Thus, the unit vectors are invariant under translations but rotate with the particles under a simultaneous rotation of all particles.[2] The last condition assures that the orientation of the frame depends on the instantaneous position of each particle in the collection.

This definition of body-fixed frame implies the following: *Each unit vector* $\hat{f}_k$ *constituting a body-fixed frame is a vector operator with respect to the total orbital angular momentum* $\vec{M} = \sum_\alpha \vec{x}^\alpha \times \vec{p}^\alpha$ *of the system—that is,*

$$[M_i, \hat{f}_m \cdot \hat{l}_j] = i e_{ijk} \hat{f}_m \cdot \hat{l}_k. \tag{7.10.2}$$

(The complex number $i = \sqrt{-1}$ on the right-hand side of this result is not to be confused with the indices $i, j, k, m, \ldots$, which assume the values 1, 2, 3.)

An important property of all body-fixed frames, *in contrast to arbitrary moving frames*, is that *the translationally invariant coordinates of the particles relative to a body-fixed frame are invariants under all rotation–inversions of the particles.*

---

[1] This translational invariance requirement is essential in the definition of a body-fixed frame, since all considerations can be reformulated in terms of the translationally invariant positions $\{\vec{y}^\alpha\}$.

[2] For the specific body-fixed frames discussed in this section, one finds that under inversion of all particle positions, $\vec{x}^\alpha \to -\vec{x}^\alpha$, the body-fixed frame also inverts; that is, $\hat{f}_i \to -\hat{f}_i$. It is accordingly useful to admit both left-handed and right-handed frames in the definition of a body-fixed frame. This property will be used later in discussing the invariance properties of the molecular Hamiltonian under such spatial inversions.

Proof. The coordinates y_i^α $(i = 1,2,3)$ of particle α relative to the frame $(\hat{f}_1, \hat{f}_2, \hat{f}_3)$ are the dot products $y_i^\alpha = (\vec{x}^\alpha - \vec{R}) \cdot \hat{f}_i$, where $\vec{R}$ denotes the center of mass vector, $\vec{R} \equiv \sum_\alpha m_\alpha \vec{x}^\alpha$. ■

A second important property of a body-fixed frame is that *the generator of rotations of the frame is the total orbital angular momentum of the particles*. We give the details of the proof of this significant result.

Proof. The components $M_i = \vec{M} \cdot \hat{l}_i$ of the total angular momentum operator $\vec{M}$ for a collection of N particles are defined by

$$M_i = -i \sum_\alpha \left(x_j^\alpha \frac{\partial}{\partial x_k^\alpha} - x_k^\alpha \frac{\partial}{\partial x_j^\alpha} \right), \qquad i, j, k \text{ cyclic}, \tag{7.10.3}$$

where $x_i^\alpha = \vec{x}^\alpha \cdot \hat{l}_i$ $(i = 1, 2, 3)$ denote the components of the position vector $\vec{x}^\alpha$ relative to the laboratory frame. Using $\vec{y}^\alpha$ to denote the position vector of particle α relative to the center of mass $\vec{R}$, we have

$$\vec{x}^\alpha = \vec{R} + \vec{y}^\alpha, \qquad \sum_\alpha m_\alpha \vec{y}^\alpha = \vec{0}; \tag{7.10.4}$$

that is,

$$x_i^\alpha = R_i + \sum_j C_{ij} y_j^\alpha, \tag{7.10.5}$$

where $R_i = \vec{R} \cdot \hat{l}_i$ $(i = 1,2,3)$ denote the components of $\vec{R}$ in the laboratory frame; $y_j^\alpha = \vec{y}^\alpha \cdot \hat{f}_j$ $(\alpha = 1, 2, \ldots, n;\ j = 1,2,3)$ denote the components of the particle positions relative to a body-fixed frame located at the center of mass; and

$$C_{ij} = \hat{l}_i \cdot \hat{f}_j, \qquad i, j = 1,2,3 \tag{7.10.6}$$

denote the direction cosines between the laboratory and body-fixed frames.

We next use the transformation (7.10.5) to evaluate the derivatives appearing in the definition of M_i in terms of derivatives with respect to the R_j, C_{jk}, and y_j^β:[1]

$$\frac{\partial}{\partial x_i^\alpha} = \sum_j \left(\frac{\partial R_j}{\partial x_i^\alpha} \right) \frac{\partial}{\partial R_j} + \sum_{jk} \left(\frac{\partial C_{jk}}{\partial x_i^\alpha} \right) \frac{\partial}{\partial C_{jk}} + \sum_{\beta j} \left(\frac{\partial y_j^\beta}{\partial x_i^\alpha} \right) \frac{\partial}{\partial y_j^\beta}. \tag{7.10.7}$$

Using these derivatives in Eq. (7.10.3) and the derivation property $M_i(fg) = f(M_i g) + (M_i f)g$ of the operators $\{M_i\}$, we deduce

[1] The fact that these derivatives are not independent may be accounted for explicitly, if desired, but will not alter the result, Eqs. (7.10.10).

$$M_i = \sum_j [M_i, R_j]\frac{\partial}{\partial R_j} + \sum_{jk}[M_i, C_{jk}]\frac{\partial}{\partial C_{jk}} + \sum_{\beta j}[M_i, y_j^\beta]\frac{\partial}{\partial y_j^\beta} \quad (7.10.8)$$

in which $[A, B]$ denotes the commutator $AB - BA$.

The commutators in the last result are all evaluated by noting that the operator $\vec{M}$ is the generator of the rotation $\mathscr{R}$ of the vectors $\{\vec{x}^\alpha\}$. Thus, the vectors $\{\vec{x}^\alpha\}$ as well as $\vec{R}$, $\hat{f}_1$, $\hat{f}_2$, and $\hat{f}_3$ are all *vector operators* (irreducible tensors of rank 1) with respect to $\vec{M}$, and each y_j^β is an invariant. The following list of commutators follows from this observation:

$$\begin{aligned}
[M_i, y_j^\beta] &= 0,\\
[M_i, R_j] &= ie_{ijk}R_k,\\
[M_i, \hat{f}_j] &= i(\hat{f}_j \times \hat{l}_i),\\
[M_i, C_{jm}] &= ie_{ijk}C_{km}.
\end{aligned} \quad (7.10.9)$$

Substitution of these commutators into Eq. (7.10.8) yields the result

$$\begin{aligned}
M_i &= -i\left(R_j\frac{\partial}{\partial R_k} - R_k\frac{\partial}{\partial R_j}\right) + J_i,\\
J_i &= -i\sum_l\left(C_{jl}\frac{\partial}{\partial C_{kl}} - C_{kl}\frac{\partial}{\partial C_{jl}}\right),
\end{aligned} \quad (7.10.10)$$

where i, j, k are cyclic in 1, 2, 3.[1] We now recognize that in the special inertial frame (discussed above) the action of the first term in M_i on the linear manifold of state vectors *vanishes* (since the action of the total linear momentum itself on these states was defined to give zero). Accordingly, only the second term, J_i, exists on the physical states. This justifies the statement that the total orbital angular momentum of a system of N particles is the generator of rotations of a body-fixed frame. ■

[1] If we identify $C_{ij} = \pm R_{ij} = \pm R_{ij}(\alpha\beta\gamma)$, where the R_{ij} are the elements of the Euler angle matrix (2.37), then the J_i are just the operators (3.101); that is, $J_i\Psi(C) = \mathscr{J}_i\Phi_\pm(\alpha\beta\gamma)$ for $C_{ij} = \pm R_{ij}(\alpha\beta\gamma)$, where $\Phi_\pm(\alpha\beta\gamma) = \Psi(\pm R(\alpha\beta\gamma))$. Equation (7.10.10) may be used to evaluate the operators J_i in other parametrizations, and we use the (dependent) variables C_{ij} in equations throughout this section to denote this greater generality.

Notice that Eqs. (7.10.9) and (7.10.10) are valid even for a left-handed frame $(\hat{f}_1, \hat{f}_2, \hat{f}_3)$, in which the matrix C is an improper, orthogonal matrix. Moreover, the angular momentum components (7.10.10) are invariant under the transformation $C \to -C$, this result being a consequence of the fact that angular momentum is an axial vector.

c. Form of the state vectors for isolated systems described in a body-fixed frame. In order to understand the implications of the role of body-fixed frames in the description of quantal systems, it is useful to examine the form of the state vectors, $|(\alpha)JM\rangle$, of sharp total angular momentum (J, M). Let us be quite explicit and assume the Euler angle parametrization $(\alpha\beta\gamma)$ of the matrix C [see Eq. (2.37), Chapter 2]. Then the state vectors relative to inertial frame necessarily have the form[1]

$$\langle \mathbf{y}; \alpha\beta\gamma|(\alpha)JM\rangle = \sum_K D^{J*}_{MK}(\alpha\beta\gamma)\Phi^J_{(\alpha)K}(\mathbf{y}), \tag{7.10.11}$$

where $D^J_{MK}(\alpha\beta\gamma)$ are the elements of the rotation matrix, and $\mathbf{y}$ denotes the row vector of relative coordinates:

$$\mathbf{y} = (y^1_1, y^1_2, y^1_3; \ldots ; y^N_1, y^N_2, y^N_3). \tag{7.10.12}$$

[In this case, the angular momentum operators J_i are realized by the differential operators $\mathscr{J}_i$ as given explicitly by Eqs. (3.101), Chapter 3.]

Proof. The proof of Eq. (7.10.11) is remarkably simple: It follows from the completeness of the rotation matrices, which implies that the most general solution to the diagonalization of $\mathscr{J}^2$ and $\mathscr{J}_3$ is $\sum_K a_K D^{J*}_{MK}(\alpha\beta\gamma)$, where the coefficients are *invariants* under rotations. However, the form of the transformation (7.10.5) shows that the only invariants that enter as arguments of the a_K are the $3N$ relative coordinates $\{y^\alpha_i\}$ themselves. ∎

One can also express Eq. (7.10.11) in terms of other parametrizations of the matrix C. Indeed, one need not parametrize C at all if the homogeneous polynomial forms $\mathscr{D}^J(C)$ (replace R_{ij} by C_{ij}) given by Eqs. (6.178) and (6.179), Chapter 6, are used. In this case, we write Eq. (7.10.11) as[2]

$$\langle \mathbf{y}; C|(\alpha)JM\rangle = \sum_K \mathscr{D}^{J*}_{MK}(C)\Phi^J_{(\alpha)K}(\mathbf{y}). \tag{7.10.13}$$

In this result, C may, in fact, be an orthogonal matrix; that is, it need not be proper.

The result, Eq. (7.10.11), was noted by Hirschfelder and Wigner [8] for the principal axes frame. *It is a general relation, valid for any physical system*

---

[1] We hope that the use of (α) for unspecified quantum numbers will not be confused with the Euler angle α or with the particle index α.

[2] Later [see Eq. (7.10.204)] we shall generalize the rotator wave functions in this result to include the factor $(\det C)^\sigma$, $\sigma = 0, 1$. This factor is important for considering the effect of spatial inversion.

(possessing rotational and translational symmetry) *that one chooses to describe by using a body-fixed frame.*

The significance of the result expressed by Eq. (7.10.11) is that *it solves the problem of constructing states of sharp total angular momentum for a system of N interacting particles.* One must not be misled by the apparent simplicity of this result. The complexities of the problem have clearly been placed in the determination of a body-fixed frame "suitable" for a particular physical system and in finding the "internal" wave functions $\Phi^J_{(\alpha)K}$.

Having shown the general relationship between the notion of a body-fixed frame and angular momentum theory, we now discuss, in detail, several special body-fixed frames that occur in the literature.

d. The instantaneous principal axes of inertia frame. The instantaneous moments of inertia relative to the center of mass for a system of N point particles are defined by[1]

$$Q_{ij} \equiv \sum_\alpha m_\alpha (x_i^\alpha - R_i)(x_j^\alpha - R_j). \tag{7.10.14}$$

The matrix $Q = (Q_{ij})$ is then real and symmetric (Q is often called the 3×3 mass quadrupole). The matrix Q may therefore be diagonalized by a real, proper, orthogonal matrix C:[2]

$$\tilde{C}QC = \mathrm{diag}(\lambda_1, \lambda_2, \lambda_3). \tag{7.10.15}$$

The orthogonal triad of unit vectors $(\hat{f}_1, \hat{f}_2, \hat{f}_3)$ defined by

$$\hat{f}_j = \sum_i C_{ij} \hat{l}_i \tag{7.10.16}$$

is then a body-fixed frame called the principal axes frame. (This identifies the matrix C of Eq. (7.10.15) with the matrix C of Eq. (7.10.6). Observe that the $\hat{f}_j$ depend implicitly on the x_i^α through the C_{ij}.)

The proof that the triad $(\hat{f}_1, \hat{f}_2, \hat{f}_3)$ constitutes a body-fixed frame may be given by demonstrating the defining properties (1)–(3) [Eq. (7.10.1)]. Further properties of this frame are developed in Section j below.

The principal axes frame is a well-defined concept for general motions of a collection of N particles. For physical systems where specific assumptions may be made concerning the motions of the particles, one finds in the literature

---

[1] It is the quantities $I_{ij} = \delta_{ij}(\mathrm{tr}\, Q) - Q_{ij}$ that are usually called the moments of inertia: I_{ii} is the moment of inertia about axis $\hat{l}_i$; I_{ij} $(i \neq j)$ is the moment of inertia for axes $\hat{l}_i$ and $\hat{l}_j$.

[2] The matrix C is unique only up to $\pm$ signs and permutations of its columns. Hence, *conventions* must be introduced in order that a unique C may be associated with Q. We ignore these subtleties here.

other types of body-fixed frames. Molecular physics has, for example, been developed primarily through the utilization of a body-fixed frame introduced by Eckart [6] (see also Wilson and Howard [12], Nielsen [13], Shaffer *et al.* [14], Wilson *et al.* [15]). We turn next to a description of this reference frame.

e. The Eckart molecular frame. We shall initially focus on two aspects of the description of a classical rigid molecule: (1) the description of the static model (equilibrium configuration) of a molecule; and (2) the description of the motion of the dynamic model of the molecule in space–time.

Consider first the description of the static model (the dumbbell model made up of rods and spheres) consisting of N nuclei labeled by $\alpha = 1, 2, \ldots, N$ having masses $m_1, m_2, \ldots, m_N$, where the distances between all pairs of nuclei are constant. The static model will be described in a laboratory frame L with basis vectors $(\hat{l}_1, \hat{l}_2, \hat{l}_3)$, which is a principal axes system located at the center of mass. Thus, the static model is described by a set of vectors $\mathscr{A}$,

$$\mathscr{A} = \{\vec{a}^{\alpha} : \alpha = 1, 2, \ldots, N\}, \qquad (7.10.17)$$

where $\vec{a}^{\alpha}$ is the position vector from the origin of L to the point where the atom labeled by α is located. Each vector $\vec{a}^{\alpha}$ may be expressed relative to the frame L as

$$\vec{a}^{\alpha} = a_1^{\alpha}\hat{l}_1 + a_2^{\alpha}\hat{l}_2 + a_3^{\alpha}\hat{l}_3, \qquad (7.10.18)$$

where $(a_1^{\alpha}, a_2^{\alpha}, a_3^{\alpha})$ are specified real numbers satisfying

$$\sum_{\alpha} m_{\alpha} a_i^{\alpha} = 0, \qquad (7.10.19)$$

$$\sum_{\alpha} m_{\alpha} a_i^{\alpha} a_j^{\alpha} = \delta_{ij}\lambda_i^0. \qquad (7.10.20)$$

Consider next the model for the motion of the molecule in space–time. Intuitively, we have in mind the following situation. We imagine that the rigid framework translates and rotates in space and that the nuclei execute small oscillatory motions in the neighborhood of the (moving) equilibrium points. This intuitive picture for a set of motions of N particles corresponds to our conception of the motions (based on empirical knowledge) of what is today called a "rigid" molecule. There are sufficiently many molecules of the "rigid type" to justify a careful development of such a model. (A phenomenological model of a molecule such as this one clearly ignores many aspects of a "real molecule," and one does not expect the model to have general validity – the model is designed specifically for the description of vibration–rotation motions of the nuclei, and even then for a limited energy domain.)

Even after settling on the model above, there are still many approaches that one might use to formulate a description of the motions. Let us continue the intuitive discussion. The use of a moving reference frame is suggested if one wishes to obtain a classical Hamiltonian H for the system, which, for motions in the neighborhood of the equilibrium configuration, has the approximate form

$$H \simeq T_{\text{C.M.}} + H_R + H_V, \tag{7.10.21}$$

where $T_{\text{C.M.}}$ is the kinetic energy of the center of mass, H_R is the rotational kinetic energy, and H_V is a Hamiltonian term for the kinetic and potential energies of the small motions near equilibrium. Intuition suggests the following forms for H_R and H_V:

$$H_R = \frac{\mathscr{P}_1^2}{2I_1} + \frac{\mathscr{P}_2^2}{2I_2} + \frac{\mathscr{P}_3^2}{2I_3},$$

$$H_V = \sum_{\mu=1}^{3N-6} (p_\mu^2 + \omega_\mu^2 q_\mu^2)/2. \tag{7.10.22}$$

In the definition of H_R, the symbol $\mathscr{P}_i$ $(i = 1, 2, 3)$ denotes the component of the total angular momentum $\vec{J}$ along the ith axis of the moving frame, and I_i is the principal moment of inertia of the equilibrium configuration about the ith axis of the moving frame. (We choose the moving frame to coincide with a principal axes system when the nuclei are located at their equilibrium points.) In the definition of H_V, the symbol q_μ $(\mu = 1, 2, \ldots, 3N - 6)$ denotes a normal coordinate, p_μ its conjugate linear momentum, and ω_μ the frequency of the μth normal mode of oscillation. (The normal-mode analysis of the vibrational motion may be carried out on the nonrotating molecule by several available methods.)

Let us next consider how one may formulate the approach outlined above.

The first problem to be solved is that of defining an appropriate moving reference frame. Eckart [6] solved this problem by imposing two conditions on the moving frame:

1. *Casimir's condition.* In the limit of vanishing displacements of the nuclei away from the equilibrium configuration, the term in the kinetic energy representing the Coriolis interaction between rotation and internal motions should be zero.

2. *Linearity of internal coordinates.* The internal degrees of freedom should be described by coordinates that are linear combinations (with fixed numerical coefficients) of the components of the displacements of the nuclei away from equilibrium, where the components are to be referred to the moving frame.

(The second condition is imposed to assure that the normal coordinates calculated for the nonrotating molecule can be carried over, *without change*, to the rotating, vibrating molecule.)

We now proceed directly to the definition of the Eckart frame, following the procedure of Ref. [16]. It will subsequently be shown that Casimir's condition and the linearity condition are fulfilled [see Remark (b) below]. It will be possible to verify the "zeroth-order" form (7.10.21) and (7.10.22) of the Hamiltonian only in Section m, after the complete Hamiltonian for the model of a rigid molecule has been developed.

The explicit construction of the Eckart frame may be given as follows: (1) Introduce the three vectors $\vec{F}_i$, called *Eckart vectors*, defined by

$$\vec{F}_i \equiv \sum_\alpha m_\alpha a_i^\alpha \vec{x}^\alpha = \sum_\alpha m_\alpha a_i^\alpha \vec{y}^\alpha, \tag{7.10.23}$$

where $\vec{x}^\alpha$ is the instantaneous position vector of nucleus α in the laboratory frame L, and $\vec{y}^\alpha = \vec{x}^\alpha - \vec{R}$. (2) Define a triad of unit perpendicular vectors $(\hat{f}_1, \hat{f}_2, \hat{f}_3)$ by[1]

[1] This method generalizes directly to the construction of n perpendicular vectors $x_1, x_2, \ldots, x_n$ from n given linearly independent vectors $y_1, y_2, \ldots, y_n$. It is interesting to observe that the standard Gram–Schmidt procedure is invariant under the group of triangular transformations

$$y_1 \to \alpha_{11} y_1, \qquad y_2 \to \alpha_{21} y_1 + \alpha_{22} y_2 \cdots,$$

$$y_n \to \alpha_{n1} y_1 + \alpha_{n2} y_2 + \cdots + \alpha_{nn} y_n,$$

whereas the method above is invariant under the group of permutations in the sense that any rearrangement of order of vectors in the given set $y_1, y_2, \ldots, y_n$ followed by the orthonormalization procedure produces the same rearrangement of the perpendicular vectors. The Gram–Schmidt procedure is thus invariant to the scale transformation $y_i \to \lambda_i y_i$, whereas the symmetric orthogonalization is easily shown to yield sets of orthonormal vectors depending on the λ_i. The symmetric orthogonalization thus depends on the *norms* of the vectors $\{y_i\}$ as well as on the angles between them. This latter fact was pointed out to us by B. Brandow of the Los Alamos National Laboratory. Schweinler and Wigner [17] attribute this *symmetric orthogonalization* technique to Landshoff [18]. Landshoff, however, did not give the result in its complete form: $x_i = \sum_j (G^{-\frac{1}{2}})_{ij} y_j$, $G_{ij} \equiv (y_i, y_j)$. Since Landshoff's interest was in applying perturbation theory, using overlap integrals $S_{ij} \equiv G_{ij} - \delta_{ij}$, the x_i were given only approximately in an expanded form of $(I + S)^{-\frac{1}{2}}$, which included several terms in the S_{ij}. Löwdin [19] appears to be the first to give the general construction for physical problems. Jørgensen [20] has discussed other interesting properties of this orthogonalization method. The method itself is well-known in the mathematical literature as the polar factorization of a real $n \times n$ nonsingular matrix Y into an orthogonal matrix X and a symmetric positive definite matrix $G^{\frac{1}{2}}$, that is, $Y = G^{\frac{1}{2}} X$, $G = Y\tilde{Y}$ (see Gantmacher [20a, pp. 276, 286] and the cited reference [171] published in 1929–30; C. C. MacDuffee [20b, p. 77]; and L. Autonne [20c]).

$$\hat{f}_i = \sum_j (F^{-\frac{1}{2}})_{ij} \vec{F}_j, \tag{7.10.24}$$

where F denotes the symmetric Gram matrix with elements $F_{ij} = \vec{F}_i \cdot \vec{F}_j$. (We shall assume that $\vec{F}_1$, $\vec{F}_2$, $\vec{F}_3$ are linearly independent so that F is positive definite; $F^{-\frac{1}{2}}$ is then, by definition, the positive definite matrix such that $F^{-\frac{1}{2}}F^{-\frac{1}{2}} = F^{-1}$. The positive definiteness condition is imposed so that unique roots are defined, since, in general, there are eight square roots of a 3×3 symmetric positive definite matrix.)

Observe that the Eckart vectors depend implicitly on the particle position vectors $\vec{x}^\alpha$. If we denote this result by writing $\hat{f}_i(\vec{x}^1, \ldots, \vec{x}^N)$, then one easily verifies that the defining relations, Eqs. (7.10.1), of a body-fixed frame are satisfied[1] by $(\hat{f}_1, \hat{f}_2, \hat{f}_3)$. (Note that the invariance of the constants $\{a_i^\alpha\}$ to rotations is essential to the proof that the transformation $\vec{x}^\alpha \to \mathscr{R}\vec{x}^\alpha$ implies $\vec{F}_i \to \mathscr{R}\vec{F}_i$, which, in turn, implies $\hat{f}_i \to \mathscr{R}\hat{f}_i$.)

Now that the body-fixed Eckart frame has been defined (we consider only nonplanar molecules here), we may determine the transformation from the Cartesian position vectors $\{\vec{x}^\alpha\}$ relative to the laboratory frame to "molecular coordinates." The position vector of nucleus α is given by

$$\vec{x}^\alpha = \vec{R} + \vec{u}^\alpha + \vec{\rho}^\alpha, \tag{7.10.25}$$

where

$\vec{R}$ is the instantaneous center of mass vector;

$\vec{u}^\alpha$ is the position vector of the moving equilibrium point of nucleus α relative to the center of mass and of the form

$$\vec{u}^\alpha = a_1^\alpha \hat{f}_1 + a_2^\alpha \hat{f}_2 + a_3^\alpha \hat{f}_3, \tag{7.10.26}$$

when referred to the moving Eckart frame;

$\vec{\rho}^\alpha$ is the displacement vector of nucleus α away from the equilibrium point $\vec{R} + \vec{u}^\alpha$.

There are six conditions imposed on the displacement vectors $\{\vec{\rho}^\alpha\}$:

Center of mass condition: $$\sum_\alpha m_\alpha \vec{\rho}^\alpha = \vec{0}, \tag{7.10.27}$$

Eckart conditions: $$\sum_\alpha m_\alpha \vec{u}^\alpha \times \vec{\rho}^\alpha = \vec{0}. \tag{7.10.28}$$

[1] Observe that the numerical constants $\{a_i^\alpha\}$ defining the static model enter directly into the definition of the "Eckart frame." Thus, the physical condition that there be an "average" position associated with the motion of each nuclei is an essential condition for formulating the definition of this body-fixed frame.

The second set of conditions results from an easily proved relation between the Eckart frame vectors $\hat{f}_i$ and the $\vec{F}_i$:

$$\hat{f}_1 \times \vec{F}_1 + \hat{f}_2 \times \vec{F}_2 + \hat{f}_3 \times \vec{F}_3 = \vec{0}. \tag{7.10.29}$$

We call a set of vectors $\{\vec{\rho}^\alpha : \alpha = 1, \ldots, N\}$ that satisfy Eqs. (7.10.27) and (7.10.28) *a set of displacement vectors for the frame* $(\hat{f}_1, \hat{f}_2, \hat{f}_3)$.

Remarks. (*a*) It is Eq. (7.10.25) that provides the physical (classical) model of a rigid molecule in motion: Relative to the moving Eckart frame (located at the center of mass), one "sees" nucleus α displaced away from the fixed (as viewed in the moving frame) position $\vec{u}^\alpha$ by the instantaneous displacement vector $\vec{\rho}^\alpha$. For rigid molecules the magnitude of $\vec{\rho}^\alpha$ is small, so that nucleus α remains in the neighborhood of the point defined by $\vec{u}^\alpha$. The nuclei move collectively in such a way that Eqs. (7.10.27) and (7.10.28) are satisfied. (*b*) It is conditions (7.10.28) that imply Casimir's conditions (see Ref. [21]): Differentiation of Eq. (7.10.28) with respect to time t, accounting for $d\hat{f}_i/dt = \vec{\omega} \times \hat{f}_i$ ($\vec{\omega}$ = angular velocity of the frame), yields the relation

$$\sum_\alpha m_\alpha \vec{u}^\alpha \times \vec{v}^\alpha = \vec{0}, \tag{7.10.30}$$

where $\vec{v}^\alpha$ is the velocity of nucleus α relative to the frame $(\hat{f}_1, \hat{f}_2, \hat{f}_3)$. Using now the well-known expression for the classical kinetic energy T of a collection of N particles (of total mass m) relative to a moving frame located at the center of mass,

$$\begin{aligned} 2T = m\frac{d\vec{R}}{dt} \cdot \frac{d\vec{R}}{dt} &+ \sum_\alpha m_\alpha (\vec{\omega} \times \vec{y}^\alpha) \cdot (\vec{\omega} \times \vec{y}^\alpha) \\ &+ \sum_\alpha m_\alpha \vec{v}^\alpha \cdot \vec{v}^\alpha + 2\vec{\omega} \cdot \sum_\alpha m_\alpha (\vec{y}^\alpha \times \vec{v}^\alpha), \end{aligned} \tag{7.10.31}$$

we see that the Coriolis term, $2\vec{\omega} \cdot \sum_\alpha m_\alpha (\vec{y} \times \vec{v}^\alpha)$, is zero for all $\vec{\rho}^\alpha = \vec{0}$ – that is, for $\vec{y}^\alpha = \vec{u}^\alpha$ ($\alpha = 1, 2, \ldots, N$).

The results given in this section have served to motivate the introduction of the Eckart frame using properties of a specific model of a rigid molecule. We require still further properties of this frame but shall defer these developments to subsequent sections.

f. Distinguished particle frames. In scattering problems an incoming single particle (call it particle 1) is distinguished, and one can introduce a body-fixed frame in the following manner: Let $\vec{y}^\alpha = \vec{x}^\alpha - \vec{R}$, $\alpha = 1, 2, \ldots, N$, denote the position vectors relative to the center of mass, and consider two additional particles, 2 and 3, such that the instantaneous relative position vectors, $\vec{y}^1$, $\vec{y}^2$, $\vec{y}^3$, are linearly independent (hence, $N > 3$). We may then introduce a body-fixed frame $(\hat{f}_1, \hat{f}_2, \hat{f}_3)$ by carrying out the *Gram–Schmidt orthogonalization*[1] on the vectors $\vec{y}^1$, $\vec{y}^2$, $\vec{y}^3$ ($|\ |$ denotes determinant):

$$\hat{f}_1 = \vec{y}^1 (D_1)^{-\frac{1}{2}},$$

$$\hat{f}_2 = \begin{vmatrix} \vec{y}^1 \cdot \vec{y}^1 & \vec{y}^1 \cdot \vec{y}^2 \\ \vec{y}^1 & \vec{y}^2 \end{vmatrix} (D_1 D_2)^{-\frac{1}{2}},$$

$$\hat{f}_3 = \begin{vmatrix} \vec{y}^1 \cdot \vec{y}^1 & y^1 \cdot \vec{y}^2 & \vec{y}^1 \cdot \vec{y}^3 \\ \vec{y}^2 \cdot \vec{y}^1 & \vec{y}^2 \cdot \vec{y}^2 & \vec{y}^2 \cdot \vec{y}^3 \\ \vec{y}^1 & \vec{y}^2 & \vec{y}^3 \end{vmatrix} (D_1 D_2 D_3)^{-\frac{1}{2}}, \qquad (7.10.32)$$

where D_k ($k = 1, 2, 3$) denotes the $k \times k$ Gram determinant (rotation–inversion invariants) formed from the vectors $\vec{y}^1, \ldots, \vec{y}^k$. Clearly, $(\hat{f}_1, \hat{f}_2, \hat{f}_3)$ is a body-fixed frame.[2, 3]

g. Uniform method of defining body-fixed frames. The construction (7.10.24) of the Eckart frame from the three vectors $\vec{F}_i$ ($i = 1, 2, 3$) given by Eq. (7.10.23) suggests a method of introducing body-fixed frames in a uniform manner: We replace the set of constants $\{a_i^\alpha\}$ in the definition of the $\vec{F}_i$ by a set of prescribed functions $\{u_i^\alpha = u_i^\alpha(\vec{x}^1, \ldots, \vec{x}^N)\}$ that satisfy the following conditions:

Translational invariance:

$$u_i^\alpha(\vec{x}^1 + \vec{a}, \ldots, \vec{x}^N + \vec{a}) = u_i^\alpha(\vec{x}^1, \ldots, \vec{x}^N).$$

Rotational invariance:

$$u_i^\alpha(\mathscr{R}\vec{x}^1, \ldots, \mathscr{R}\vec{x}^N) = u_i^\alpha(\vec{x}^1, \ldots, \vec{x}^N). \qquad (7.10.33)$$

Center of mass constraint:

$$\sum_\alpha m_\alpha u_i^\alpha = 0.$$

[1] One might also use the symmetric orthogonalization (see footnote, p. 537).

[2] Frames of this type were introduced into molecular scattering problems by Curtiss *et al.* [22] (see also Pack and Hirschfelder [23], Pack [24]).

[3] Curtiss, Hirschfelder, and Adler actually construct only the two vectors $\hat{f}_1$ and $\hat{f}_2$ and do not note that it is the Gram–Schmidt orthogonalization that is involved. The third vector is then *defined* by $\hat{f}_3 = \hat{f}_1 \times \hat{f}_2$.

The *generalized Eckart vectors* defined by[1,2]

$$\vec{F}_i = \sum_\alpha m_\alpha u_i^\alpha (\vec{x}^\alpha - \vec{R}) = \sum_\alpha m_\alpha u_i^\alpha \vec{y}^\alpha \tag{7.10.34}$$

are now used to define a frame $(\hat{f}_1, \hat{f}_2, \hat{f}_3)$ by the symmetric orthogonalization procedure:

$$\hat{f}_i = \sum_j (F^{-\frac{1}{2}})_{ij} \vec{F}_j. \tag{7.10.35}$$

One now easily shows that the orthonormal triad $(\hat{f}_1, \hat{f}_2, \hat{f}_3)$ is a body-fixed frame; that is, it satisfies properties (1) and (2) of Eqs. (7.10.1). (We have not attempted to formulate, in general, properties of the $\{u_i^\alpha\}$ that would imply property (3), but we shall assume that (3) is valid for any specific choice of the functions $\{u_i^\alpha\}$.)

The translationally and rotationally invariant functions $\{u_i^\alpha\}$, which lead to the body-fixed frames introduced in Sections d, e, and f, are the following:

Principal axes frame:

$$u_i^\alpha = \sum_j C_{ji}(x_j^\alpha - R_j) = \sum_j C_{ji}(\vec{y}^\alpha \cdot \hat{l}_j), \tag{7.10.36}$$

where the C_{ij} are the functions of the $\{x_i^\alpha\}$ obtained from the proper orthogonal matrix C, which diagonalizes the mass quadrupole Q [see Eq. (7.10.15)]. [This method of deriving the principal axes frame is more complicated than the direct definition, Eq. (7.10.16), but it has the advantage of unifying the derivation with other methods; furthermore, the generalized Eckart vectors (7.10.35) are themselves important, as will be shown below.]

Eckart molecular frame:

$$\{u_i^\alpha\} = \{a_i^\alpha\} = \text{equilibrium configuration parameters.} \tag{7.10.37}$$

---

[1] We assume in this general discussion that the vectors $\vec{F}_1$, $\vec{F}_2$, $\vec{F}_3$ are linearly independent for some suitable domain of definition of the vectors $\{\vec{x}^\alpha\}$.

[2] The symmetric orthogonalization map from Eckart vectors to Eckart frames is many-to-one. To show this let $(\vec{F}_1, \vec{F}_2, \vec{F}_3)$ denote a given set of Eckart vectors [specified rotational-translational invariants u_i^α in Eq. (7.10.34)] and $(\hat{f}_1, \hat{f}_2, \hat{f}_3)$ the corresponding frame. Let S denote an arbitrary 3×3 symmetric positive definite matrix whose elements are rotational invariants. Define, for each S, new Eckart vectors by $\vec{F}_i(S) \equiv \sum_\alpha m_\alpha u_i^\alpha(S) \vec{y}^\alpha$, where $u_i^\alpha(S) \equiv \sum_j (SF^{-\frac{1}{2}})_{ij} u_j^\alpha$ $(i = 1, 2, 3)$. Then, under symmetric orthogonalization, one finds $(\vec{F}_1(S), \vec{F}_2(S), \vec{F}_3(S)) \rightarrow (\hat{f}_1, \hat{f}_2, \hat{f}_3)$, each S.

Distinguished particle frame:[1]

$$u_1^\alpha = \vec{y}^\alpha \cdot \vec{y}^1 (D_1)^{-\frac{1}{2}},$$

$$u_2^\alpha = \begin{vmatrix} \vec{y}^1 \cdot \vec{y}^1 & \vec{y}^1 \cdot \vec{y}^2 \\ \vec{y}^\alpha \cdot \vec{y}^1 & \vec{y}^\alpha \cdot \vec{y}^2 \end{vmatrix} (D_1 D_2)^{-\frac{1}{2}},$$

$$u_3^\alpha = \begin{vmatrix} \vec{y}^1 \cdot \vec{y}^1 & \vec{y}^1 \cdot \vec{y}^2 & \vec{y}^1 \cdot \vec{y}^3 \\ \vec{y}^2 \cdot \vec{y}^1 & \vec{y}^2 \cdot \vec{y}^2 & \vec{y}^2 \cdot \vec{y}^3 \\ \vec{y}^\alpha \cdot \vec{y}^1 & \vec{y}^\alpha \cdot \vec{y}^2 & \vec{y}^\alpha \cdot \vec{y}^3 \end{vmatrix} (D_1 D_2 D_3)^{-\frac{1}{2}}. \tag{7.10.38}$$

The body-fixed frame $(\hat{f}_1, \hat{f}_2, \hat{f}_3)$ defined by Eq. (7.10.35) and the generalized Eckart vectors $\vec{F}_1$, $\vec{F}_2$, $\vec{F}_3$ defined by Eq. (7.10.34) always satisfy the relation

$$\hat{f}_1 \times \vec{F}_1 + \hat{f}_2 \times \vec{F}_2 + \hat{f}_3 \times \vec{F}_3 = \vec{0} \tag{7.10.39}$$

because $(F^{-\frac{1}{2}})_{ij}$ is symmetric in (i, j), whereas $\hat{F}_i \times \hat{F}_j$ is antisymmetric in (i, j). If we define $\vec{u}^\alpha$ by

$$\vec{u}^\alpha \equiv \sum_i u_i^\alpha \hat{f}_i, \tag{7.10.40}$$

then Eq. (7.10.39) may also be expressed as

$$\sum_\alpha m_\alpha \vec{u}^\alpha \times \vec{y}^\alpha = \vec{0}. \tag{7.10.41}$$

This same relation may be expressed in yet another useful way: Define the matrix $G = (G_{ij})$ by

$$G_{ij} \equiv \vec{F}_i \cdot \hat{f}_j = \text{rotational invariant}; \tag{7.10.42}$$

then Eq. (7.10.39) [or Eq. (7.10.41)] is equivalent to the relation

$$G = \tilde{G}; \tag{7.10.43}$$

that is, the matrix G is symmetric. One may also verify that

$$G^2 = F. \tag{7.10.44}$$

[1] These frames are mentioned here only for the purpose of illustration, and their properties will not be discussed further.

Thus, one also finds the following relation between the body-fixed frame vectors and the generalized Eckart vectors:

$$\hat{f}_i = \sum_j (G^{-1})_{ij} \vec{F}_j. \tag{7.10.45}$$

(This result expresses a relationship between two sets of vectors and is not to be regarded as a definition of the $\hat{f}_i$, since the G_{ij} themselves depend on the $\hat{f}_i$.)

Equations (7.10.42) and (7.10.45) show clearly the distinction between the instantaneous principal axes frame and the Eckart molecular frame.

▶ Principal axes frame:

$$G_{ij} = \delta_{ij}\lambda_i, \tag{7.10.46}$$

$$\vec{F}_i = \lambda_i \hat{f}_i. \tag{7.10.47}$$

Up to a multiplicative rotational invariant, the Eckart vectors $\vec{F}_1, \vec{F}_2, \vec{F}_3$ given by Eqs. (7.10.34) *and* (7.10.36) *define the principal axes frame itself.*

▶ Eckart molecular frame:

$$G_{ij} = G_{ji} \neq 0 \qquad \text{for } i \neq j. \tag{7.10.48}$$

In this case Eq. (7.10.41) expresses a nontrivial relationship between the vectors $\{\vec{u}^\alpha\}$ (constant components a_i^α relative to the Eckart frame) and the instantaneous position vectors $\{\vec{y}^\alpha\}$. [This was the equation that provided the physical basis for the introduction of the Eckart frame; see Remark (b), Section *e*.]

h. Internal coordinates. The key result for defining the concept of an internal coordinate is a theorem due to Weyl [25, p. 53]:

Every even rotational invariant depending on N vectors $\vec{y}^1, \vec{y}^2, \ldots, \vec{y}^N$ in three-dimensional space is expressible in terms of the N^2 scalar products $\vec{y}^\alpha \cdot \vec{y}^\beta$. Every odd invariant is a sum of terms

$$[\vec{z}^1 \cdot (\vec{z}^2 \times \vec{z}^3)] f(\vec{y}^1, \ldots, \vec{y}^N), \tag{7.10.49}$$

where $\vec{z}^1, \vec{z}^2, \vec{z}^3$ are selected from $\vec{y}^1, \ldots, \vec{y}^N$, and f is an even invariant.

In the subsequent discussion, we shall require only even invariants. Since we may write

$$\vec{y}^\alpha \cdot \vec{y}^\beta = \sum_i (\vec{y}^\alpha \cdot \hat{f}_i)(\vec{y}^\beta \cdot \hat{f}_i), \tag{7.10.50}$$

where $(\hat{f}_1, \hat{f}_2, \hat{f}_3)$ defines a body-fixed frame, Weyl's theorem implies that *every even invariant is expressible in terms of the 3N scalar products* $y_i^\alpha = \vec{y}^\alpha \cdot \hat{f}_i$, $\alpha = 1, 2, \ldots, N$; $i = 1, 2, 3$.

Let us now state the definition of internal coordinates:[1] A set of coordinates $\{q_\mu: \mu = 1, 2, \ldots, 3N - 6\}$ is said to constitute a *set of internal coordinates* for the body-fixed frame $(\hat{f}_1, \hat{f}_2, \hat{f}_3)$ if the following conditions are satisfied:

1. Each coordinate q_μ is an even invariant, and hence, by Weyl's theorem, is expressible as

$$q_\mu = h_\mu(y_1^1, y_2^1, y_3^1; \ldots; y_1^N, y_2^N, y_3^N), \tag{7.10.51}$$

where each *basic invariant* y_i^α is itself expressible in terms of the $\{q_\mu\}$ – that is, the transformation is invertible in some suitable domain:

$$y_i^\alpha = g_i^\alpha(q_1, \ldots, q_{3N-6}). \tag{7.10.52}$$

2. The center of mass conditions are satisfied:

$$\sum_\alpha m_\alpha y_i^\alpha = 0, \qquad i = 1, 2, 3. \tag{7.10.53}$$

3. The three conditions on the $\{y_i^\alpha = \vec{y}^\alpha \cdot \hat{f}_i\}$ imposed by the definition of the body-fixed frame are satisfied.

Conditions (3) become, for the two cases under discussion,

$$\text{Principal axes frame:} \qquad G_{ij} = 0, \qquad i \neq j,$$

$$\text{Eckart frame:} \qquad G \text{ symmetric.} \tag{7.10.54}$$

Internal coordinates for the Eckart frame and the principal axes frame may be introduced explicitly because of the simplicity of the conditions on the coordinates $y_i^\alpha = \vec{y}^\alpha \cdot \hat{f}_i$. Thus, one seeks $3N - 6$ internal coordinates $\{q_\mu\}$, which satisfy the defining relations, Eqs. (7.10.51)–(7.10.54), above.

i. Internal coordinates for the Eckart frame. The most significant property of the internal coordinates for the Eckart molecular frame is that they are linear combinations of the components $\{\rho_i^\alpha \equiv \vec{\rho}^\alpha \cdot \hat{f}_i\}$ of the displacement vectors with *numerical coefficients* (hence, they are linear relations in the ρ_i^α):

$$q_\mu = \sum_{\alpha i} q_{\mu i}^\alpha \rho_i^\alpha, \qquad \mu = 1, 2, \ldots, 3N - 6, \tag{7.10.55}$$

---

[1] We do not consider linear or planar molecules in this general discussion.

where

$$\rho_i^\alpha = y_i^\alpha - a_i^\alpha = (\vec{y}^\alpha - \vec{u}^\alpha)\cdot \hat{f}_i, \tag{7.10.56}$$

and the set of quantities $\{q_{\mu i}^\alpha : \alpha = 1, 2, \ldots, N;\ \mu = 1, 2, \ldots, 3N - 6;\ i = 1, 2, 3\}$ are real numbers.

Proof. In order to prove this result it is convenient to rewrite the center of mass conditions (7.10.27) and the Eckart conditions (7.10.28) in the forms (see Refs. [15] and [26])

$$\sum_\alpha \vec{s}_\lambda^{\,\alpha} \cdot \vec{\rho}^{\,\alpha} = 0, \qquad \lambda = 1, 2, \ldots, 6, \tag{7.10.57}$$

where

$$\vec{s}_i^{\,\alpha} \equiv m_\alpha \hat{f}_i / m^{\frac{1}{2}}, \qquad m = \sum_\alpha m_\alpha,$$

$$\vec{s}_{i+3}^{\,\alpha} = m_\alpha N_i^{-\frac{1}{2}} (\hat{f}_i \times \vec{u}^\alpha),$$

$$N_i = \sum_\alpha m_\alpha (\hat{f}_i \times \vec{u}^\alpha) \cdot (\hat{f}_i \times \vec{u}^\alpha). \tag{7.10.58}$$

We have normalized the vectors $\vec{s}_\lambda^{\,\alpha}$ in the sense that

$$\sum_\alpha m_\alpha^{-1} \vec{s}_\lambda^{\,\alpha} \cdot \vec{s}_\lambda^{\,\alpha} = 1. \tag{7.10.59}$$

We note also that these vectors are perpendicular; that is, they satisfy

$$\sum_\alpha m_\alpha^{-1}\, \vec{s}_\lambda^{\,\alpha} \cdot \vec{s}_{\lambda'}^{\,\alpha} = \delta_{\lambda\lambda'}. \tag{7.10.60}$$

These equations are the statement of the orthogonality of the following six row vectors in $3N$-dimensional space:

$$\vec{S}_\lambda \equiv \left(\frac{1}{\sqrt{m_1}} \vec{s}_\lambda^{\,1}, \frac{1}{\sqrt{m_2}} \vec{s}_\lambda^{\,2}, \ldots, \frac{1}{\sqrt{m_N}} \vec{s}_\lambda^{\,N} \right), \qquad \lambda = 1, 2, \ldots, 6. \tag{7.10.61}$$

If we consider that $\vec{S}_1, \ldots, \vec{S}_6$ are the first six rows of a $3N \times 3N$ orthogonal matrix, then we can find (in infinitely many ways for $N > 3$) $3N - 6$ additional rows that yield a $3N \times 3N$ orthogonal matrix. Let us denote a set of such additional rows by

$$\vec{Q}_\mu = \left(\frac{1}{\sqrt{m_1}}\vec{q}^{\,1}_\mu, \frac{1}{\sqrt{m_2}}\vec{q}^{\,2}_\mu, \ldots, \frac{1}{\sqrt{m_N}}\vec{q}^{\,N}_\mu\right), \tag{7.10.62}$$

where

$$\vec{q}^{\,\alpha}_\mu \equiv \sum_i q^\alpha_{\mu i}\hat{f}_i, \qquad \mu = 1, 2, \ldots, 3N - 6. \tag{7.10.63}$$

The conditions of orthogonality of the $\vec{Q}_\mu$ to the first six rows given explicitly by Eqs. (7.10.58) and (7.10.61) and of orthonormality among themselves become

$$\sum_\alpha \vec{q}^{\,\alpha}_\mu = \vec{0},$$

$$\sum_\alpha (\vec{u}^{\,\alpha} \times \vec{q}^{\,\alpha}_\mu) = \vec{0},$$

$$\sum_\alpha m_\alpha^{-1}(\vec{q}^{\,\alpha}_\mu \cdot \vec{q}^{\,\alpha}_\nu) = \delta_{\mu\nu} \tag{7.10.64}$$

for μ, $\nu = 1, 2, \ldots, 3N - 6$.

For definiteness, let us suppose that a specific set of vectors $\{\vec{q}^{\,\alpha}_\mu\}$ satisfying these relations (7.10.64) has been constructed.

We next introduce the $3N - 6$ rotation–inversion invariants defined by

$$q_\mu \equiv \sum_\alpha \vec{q}^{\,\alpha}_\mu \cdot \vec{\rho}^{\,\alpha}, \qquad \mu = 1, 2, \ldots, 3N - 6. \tag{7.10.65}$$

Using the orthogonality of the matrix with rows $\vec{S}_1, \ldots, \vec{S}_6, \vec{Q}_1, \ldots, \vec{Q}_{3N-6}$, this relation may be inverted to yield

$$\rho^\alpha_i = m_\alpha^{-1}\sum_\mu q^\alpha_{\mu i} q_\mu. \tag{7.10.66}$$

This equation defines a set of displacement components $\{\rho^\alpha_i\}$ in terms of $3N - 6$ rotation–inversion invariants $q_1, q_2, \ldots, q_{3N-6}$ such that the constraints (7.10.53) and (7.10.54) are satisfied. ■

A principal result obtained from the above analysis is as follows: *The rotation–inversion invariants defined by*

$$q_\mu = \sum_\alpha \vec{q}^{\,\alpha}_\mu \cdot \vec{\rho}^{\,\alpha}, \qquad \mu = 1, 2, \ldots, 3N - 6 \tag{7.10.67}$$

span the $3N - 6$ *dimensional space of the internal coordinates.* (We use the term "span" here in the sense that each internal coordinate of the form $\xi = \sum_\alpha \vec{\xi}^{\,\alpha} \cdot \vec{\rho}^{\,\alpha}$, $\vec{\xi}^{\,\alpha} = \xi^\alpha_1 \hat{f}_1 + \xi^\alpha_2 \hat{f}_2 + \xi^\alpha_3 \hat{f}_3$, has the form $\xi = \sum_\mu \xi_\mu q_\mu$. Furthermore, $\sum_\mu a_\mu q_\mu = \sum_\mu b_\mu q_\mu$ implies $a_\mu = b_\mu$.)

It is interesting to determine how the invariants [see Eq. (7.10.42)] $G_{ij} = \vec{F}_i \cdot \hat{f}_j$ fit into the scheme of internal coordinates. We may write

$$\vec{F}_i \cdot \hat{f}_j = \lambda^0_i \delta_{ij} + \sum_\alpha \vec{\xi}^{\,\alpha}_{ij} \cdot \vec{\rho}^{\,\alpha}, \tag{7.10.68}$$

where λ^0_i $(i = 1, 2, 3)$ are the principal moments of inertia for the equilibrium configuration, and

$$\vec{\xi}^{\,\alpha}_{ij} \equiv m_\alpha a^\alpha_i \hat{f}_j, \qquad i, j = 1, 2, 3. \tag{7.10.69}$$

One then finds that the three invariants

$$\sum_\alpha (\vec{\xi}^{\,\alpha}_{ij} - \vec{\xi}^{\,\alpha}_{ji}) \cdot \vec{\rho}^{\,\alpha} = 0, \qquad i < j = 1, 2, 3, \tag{7.10.70}$$

express the Eckart conditions; the three invariants

$$\xi_i = \sum_\alpha \vec{\xi}^{\,\alpha}_{ii} \cdot \vec{\rho}^{\,\alpha}, \qquad i = 1, 2, 3, \tag{7.10.71}$$

are always perpendicular internal coordinates; and the invariant $(i < j)$

$$\xi_{ij} = \sum_\alpha (\vec{\xi}^{\,\alpha}_{ij} + \vec{\xi}^{\,\alpha}_{ji}) \cdot \vec{\rho}^{\,\alpha} \tag{7.10.72}$$

is an internal coordinate if and only if $\lambda^0_i = \lambda^0_j$. (One of the invariants ξ_{12}, ξ_{13}, ξ_{23} is an internal coordinate for a symmetric top molecule; all three are internal coordinates for spherical top molecules.)

The fact that we always obtain three or more internal coordinates from the Eckart vectors themselves greatly simplifies the explicit construction of internal coordinates for specific molecules (see Ref. [16] for examples).

Let us now summarize the results giving the transformation from the $3N$ Cartesian coordinates $\{x^\alpha_i = \vec{x}^{\,\alpha} \cdot \hat{l}_i\}$ to the $3N$ coordinates consisting of (1) the three center of mass coordinates $R_i = \vec{R} \cdot \hat{l}_i$ $(i = 1, 2, 3)$; (2) the three independent parameters of the direction cosine matrix C with $C_{ij} = \hat{l}_i \cdot \hat{f}_j$; (3) the $3N - 6$ internal coordinates q_μ $(\mu = 1, 2, \ldots, 3N - 6)$:

$$x_i^\alpha = R_i + \sum_j C_{ij}(a_j^\alpha + \rho_j^\alpha) \tag{7.10.73}$$

in which

$$\rho_j^\alpha = m_\alpha^{-1} \sum_\mu q_{\mu j}^\alpha q_\mu. \tag{7.10.74}$$

This transformation is invertible for those values of the $\{x_i^\alpha\}$ for which the Eckart frame construction exists (det $F \neq 0$):

$$m\vec{R} = \sum_\alpha m_\alpha \vec{x}^\alpha,$$

$$\hat{f}_i = \sum_j (F^{-\frac{1}{2}})_{ij} \vec{F}_j, \qquad C_{ij} = \hat{l}_i \cdot \hat{f}_j,$$

$$q_\mu = \sum_\alpha \vec{q}_\mu^{\,\alpha} \cdot \vec{\rho}^{\,\alpha}. \tag{7.10.75}$$

As we have seen, there is considerable freedom (nonuniqueness) in the construction of internal coordinates $\{q_\mu : \mu = 1, 2, \ldots, 3N - 6\}$ given above. To make this more explicit, let us express the construction in matrix form.

If we introduce $\boldsymbol{\rho}$ and $\mathbf{Q}$ to denote the column vectors

$$\boldsymbol{\rho} = \mathrm{col}(\rho_1^1, \rho_2^1, \rho_3^1; \ldots; \rho_1^N, \rho_2^N, \rho_3^N),$$

$$\mathbf{Q} = \mathrm{col}(0, \ldots, 0, q_1, q_2, \ldots, q_{3N-6}), \tag{7.10.76}$$

the transformation from displacement coordinates relative to the body-fixed frame $(\hat{f}_1, \hat{f}_2, \hat{f}_3)$ to the internal coordinates q_μ is expressed by an orthogonal transformation,

$$\mathbf{Q} = S\boldsymbol{\rho}, \qquad \tilde{S}S = I_{3N} \qquad (3N \times 3N \text{ identity matrix}), \tag{7.10.77}$$

where S is, of course, not unique. Writing

$$\mathbf{q} = \mathrm{col}(q_1, q_2, \ldots, q_{3N-6}), \tag{7.10.78}$$

one sees that an arbitrary orthogonal transformation A,

$$\mathbf{q} \to A\mathbf{q} = \mathbf{q}', \tag{7.10.79}$$

effects a new transformation to internal coordinates that satisfy all the constraints. The new transformation to internal coordinates has the form

$$\mathbf{Q}' = S'\boldsymbol{\rho},$$

$$\mathbf{Q}' = \mathrm{col}(0, \ldots, 0, q'_1, q'_2, \ldots, q'_{3N-6}),$$

$$S' = \begin{pmatrix} I_6 & 0 \\ 0 & A \end{pmatrix} S, \tag{7.10.80}$$

where I_6 denotes the 6×6 unit matrix.

It is of particular importance that the new position vectors $\{\vec{x}'^{\alpha}\}$ corresponding to the new internal coordinates $\mathbf{q} \to A\mathbf{q} = \mathbf{q}'$ define the same Eckart frame $(\hat{f}_1, \hat{f}_2, \hat{f}_3)$ as do the old position vectors $\{\vec{x}^{\alpha}\}$ corresponding to the internal coordinates $\mathbf{q}$. This freedom implies that *the group $SO(3N-6)$ of real orthogonal matrices of dimension $3N-6$ is an internal transformation group that leaves the Eckart frame invariant.*[1] This results allows for the possibility of having internal transformation groups [subgroups of $SO(3N-6)$], which leave not only the frame invariant, but also the Hamiltonian.

It is an important result (as noted above) that the internal coordinates q_μ are determined in the Eckart molecular model only up to arbitrary orthogonal transformations. This additional freedom gives the model a degree of generality going beyond the specification of the parameters $\{a_i^{\alpha}\}$. *Additional physics must be put into the model before a closer determination of the matrix A in Eq.* (7.10.79) *can be given.* For example, it is often assumed that the potential energy may be approximated by a positive definite quadratic form (Wilson *et al.* [15])

$$V = \sum_{\mu\nu} u_{\mu\nu} q_\mu q_\nu, \tag{7.10.81}$$

in which the q_μ are *specified* rotation–inversion invariants of the required form

$$q_\mu = \sum_{\alpha} \vec{q}^{\,\alpha}_{\mu} \cdot \vec{\rho}^{\,\alpha}. \tag{7.10.82}$$

The matrix A is then determined (only partially if there are degenerate eigenvalues) by the requirement that $\mathbf{q}' = A\mathbf{q}$ diagonalize V.

j. Internal coordinates for the principal axes frame. The key relations for discussing internal coordinates for the principal axes frame are the center of mass relation and the Eckart vector relation, Eq. (7.10.47):

[1] Actually, this result is true for the general linear group.

$$\sum_{\alpha} m_\alpha \vec{y}^\alpha = \vec{0}, \tag{7.10.83}$$

$$\sum_{\alpha} m_\alpha y_i^\alpha \vec{y}^\alpha = \lambda_i \hat{f}_i. \tag{7.10.84}$$

Observe that the second relation yields the invariants (dot $\hat{f}_i$ into the equation)

$$\lambda_i = \sum_{\alpha} m_\alpha y_i^\alpha y_i^\alpha, \qquad i = 1, 2, 3, \tag{7.10.85}$$

as well as the constraints [dot $\hat{f}_j$ $(j \neq i)$ into the equation] on the components $\{y_i^\alpha\}$ imposed by the definition of the frame:

$$\sum_{\alpha} m_\alpha y_i^\alpha y_j^\alpha = 0, \qquad i \neq j. \tag{7.10.86}$$

The vector form of Eqs. (7.10.83) and (7.10.84) suggests that we consider other linear combinations of the position vectors $\{\vec{y}^\alpha\}$:

$$\vec{z} = \sum_{\alpha} \zeta^\alpha \vec{y}^\alpha. \tag{7.10.87}$$

With each such linear combination, we may associate a vector $\boldsymbol{\zeta}$ in N-space,

$$\boldsymbol{\zeta} = (\zeta^1, \zeta^2, \ldots, \zeta^N), \tag{7.10.88}$$

and, conversely, each such vector in N-space defines a linear combination of the position vectors.

Since the mass occurs linearly in each of Eqs. (7.10.83) and (7.10.84), it is convenient to define [as in Eq. (7.10.59)] the scalar product of two vectors $\boldsymbol{\zeta}$ and $\boldsymbol{\eta}$ in N-space by

$$\boldsymbol{\zeta} \cdot \boldsymbol{\eta} = \sum_{\alpha} m_\alpha^{-1} \zeta^\alpha \eta^\alpha. \tag{7.10.89}$$

For example, the normalized vectors in N-space associated with Eqs. (7.10.83) and (7.10.84) are given by

$$\boldsymbol{\zeta}_4 = (m_1, m_2, \ldots, m_N)/\sqrt{m}, \tag{7.10.90}$$

$$\boldsymbol{\zeta}_i = (m_1 y_i^1, m_2 y_i^2, \ldots, m_N y_i^N)/\sqrt{\lambda_i}, \qquad i = 1, 2, 3. \tag{7.10.91}$$

Observe that these vectors are already orthogonal:

$$\boldsymbol{\zeta}_\kappa \cdot \boldsymbol{\zeta}_{\kappa'} = \delta_{\kappa\kappa'}, \qquad \kappa, \kappa' = 1, 2, 3, 4. \tag{7.10.92}$$

In terms of the normalized vectors (7.10.91) and (7.10.90) the center of mass relation (7.10.83) and the Eckart vector relation (7.10.84), take the forms

$$\vec{z}_i \equiv \sum_\alpha \zeta_i^\alpha \vec{y}^\alpha = \sqrt{\lambda_i}\hat{f}_i, \qquad i = 1, 2, 3, \tag{7.10.93}$$

$$\vec{z}_4 \equiv \sum_\alpha \zeta_4^\alpha \vec{y}^\alpha = \vec{0}. \tag{7.10.94}$$

The main result for the internal coordinate problem for the principal axes frame may now be stated, using the above notations: *Let the set of vectors* $\{\zeta_\kappa : \kappa = 4, \ldots, N\}$ *together with the three vectors* ζ_i $(i = 1, 2, 3)$ *given by Eq.* (7.10.91) *be a set of orthonormal N-space vectors,* $\zeta_\kappa \cdot \zeta_{\kappa'} = \delta_{\kappa\kappa'}$ $(\kappa, \kappa' = 1, 2, \ldots, N)$. *Then*

$$\sum_\alpha \zeta_i^\alpha \vec{y}^\alpha = \sqrt{\lambda_i}\hat{f}_i, \qquad i = 1, 2, 3,$$

$$\sum_\alpha \zeta_\kappa^\alpha \vec{y}^\alpha = \vec{0}, \qquad \kappa = 4, \ldots, N. \tag{7.10.95}$$

Proof. The proof of this theorem is based on results given in Ref. [10]. It follows from the Weyl theorem (Section h) that the basic invariants $\{\vec{y}^\alpha \cdot \vec{y}^\beta\}$ are the significant invariants to consider in formulating the internal coordinate problem. This observation suggests introducing the symmetric matrix

$$S = (S^{\alpha\beta}), \tag{7.10.96}$$

where

$$S^{\alpha\beta} = \sqrt{m_\alpha m_\beta}\, \vec{y}^\alpha \cdot \vec{y}^\beta, \qquad \alpha, \beta = 1, 2, \ldots, N. \tag{7.10.97}$$

The mass factors are included in the definition of $S^{\alpha\beta}$ so that the following trace identities hold between the 3×3 mass quadrupole matrix Q [see Eq. (7.10.14)] and the $N \times N$ mass quadrupole matrix S:

$$\operatorname{tr} Q^k = \operatorname{tr} S^k, \qquad k = 1, 2, \ldots . \tag{7.10.98}$$

The trace identities (7.10.98) and the fact that S is symmetric (hence, a normal[1] matrix) imply that the Cayley–Hamilton equation for S is

[1] The properties of normal matrices used in this proof are well-known and may be found, for example, in Perlis [27].

$$S(S - \lambda_1 I_N)(S - \lambda_2 I_N)(S - \lambda_3 I_N) = 0, \tag{7.10.99}$$

where $\lambda_1, \lambda_2, \lambda_3$ are eigenvalues of S as well as of Q. This result, in turn, implies that the eigenvalues of S are λ_1, λ_2, λ_3, and 0 (repeated $N - 3$ times). Furthermore, since S is symmetric, it is diagonalizable by a real orthogonal matrix. Thus, S possesses a set of *orthonormal eigenvectors* $\{\boldsymbol{\xi}_\kappa : \kappa = 1, 2, \ldots, N\}$ such that

$$\begin{aligned} S\boldsymbol{\xi}_i &= \lambda_i \boldsymbol{\xi}_i, \qquad & i &= 1, 2, 3, \\ S\boldsymbol{\xi}_\kappa &= 0, \qquad & \kappa &= 4, \ldots, N, \end{aligned} \tag{7.10.100}$$

where the orthogonality of vectors is expressed in the usual scalar product by

$$(\boldsymbol{\xi}_\kappa, \boldsymbol{\xi}_{\kappa'}) \equiv \sum_\alpha \xi_\kappa^\alpha \xi_{\kappa'}^\alpha = \delta_{\kappa\kappa'}. \tag{7.10.101}$$

The eigenvectors $\boldsymbol{\xi}_1, \boldsymbol{\xi}_2, \boldsymbol{\xi}_3$ are already given by the Eckart vector relation, Eq. (7.10.84): We dot $\vec{y}^\beta$ into Eq. (7.10.84) to obtain the eigenvalue equation, Eq. (7.10.100), in which $\xi_i^\alpha = \zeta_i^\alpha / \sqrt{m_\alpha}$. [Also, the columns of the matrix $A \equiv \prod_{i=1}^3 (S - \lambda_i I_N)$ span the space of eigenvectors of S having eigenvalue 0, but we shall not require this result.]

Consider now the vector $\vec{z}$ defined by

$$\vec{z} = \sum_\alpha \sqrt{m_\alpha}\, \xi^\alpha \vec{y}^\alpha,$$

where $\boldsymbol{\xi} = \sum_{\kappa=4}^N a_\kappa \boldsymbol{\xi}_\kappa$ is an arbitrary (real) vector in the null space of S. Then we find

$$\vec{z} \cdot \vec{y}^\beta = 0, \qquad \beta = 1, 2, \ldots, N$$

in consequence of Eqs. (7.10.100). Since this last result must be valid for arbitrary position vectors $\{\vec{y}^\beta\}$, we must have $\vec{z} = \vec{0}$. Accounting now for the slight difference in the definitions of the scalar products (7.10.89) and (7.10.101), we obtain the proof of the stated theorem. ■

Consider now the inversion of Eqs. (7.10.95). Since $\zeta = (\zeta_\kappa^\alpha)$ is an orthogonal matrix [scalar product of rows given by Eq. (7.10.89)], it is orthogonal also in its columns:

$$\sum_\kappa \zeta_\kappa^\alpha \zeta_\kappa^\beta = m_\alpha \delta^{\alpha\beta}. \tag{7.10.102}$$

Using this result in Eqs. (7.10.95), we obtain

$$\vec{y}^{\alpha} = m_{\alpha}^{-1} \sum_{i} \mu_i \zeta_i^{\alpha} \hat{f}_i, \tag{7.10.103}$$

where

$$\mu_i \equiv \sqrt{\lambda_i}. \tag{7.10.104}$$

In this inverted result the three N-space vectors ζ_i $(i = 1, 2, 3)$ are now to be interpreted as *any general triad* $(\zeta_1, \zeta_2, \zeta_3)$ *of orthonormal vectors such that also the components of each vector satisfy* $\sum_{\alpha} \zeta_i^{\alpha} = 0$ $(i = 1, 2, 3)$ (center of mass condition). The generic triple satisfying these conditions may be parametrized explicitly in terms of $3N - 9$ angle parameters θ_{τ} $(\tau = 1, 2, \ldots, 3N - 9)$; that is,

$$\zeta_i^{\alpha} = g_i^{\alpha}(\theta_1, \ldots, \theta_{3N-9}), \tag{7.10.105}$$

but we shall not require this result in detail.

Let us now summarize the results giving the transformation from the $3N$ Cartesian coordinates $\{x_i^{\alpha} = \vec{x}^{\alpha} \cdot \hat{l}_i\}$ to the $3N$ coordinates consisting of (1) the three center of mass coordinates R_i $(i = 1, 2, 3)$; (2) the three independent parameters of the direction cosine matrix C; and (3) the $3N - 6$ internal coordinates consisting of $\mu_i = \sqrt{\lambda_i}$ $(i = 1, 2, 3)$ and the $3N - 9$ internal coordinates implicit in the triad $(\zeta_1, \zeta_2, \zeta_3)$. From $\vec{x}^{\alpha} = \vec{y}^{\alpha} + \vec{R}$ and Eq. (7.10.103), we obtain

$$x_i^{\alpha} = R_i + m_{\alpha}^{-1} \sum_{j} C_{ij} \mu_j \zeta_j^{\alpha}. \tag{7.10.106}$$

The inversion of Eqs. (7.10.106) is given by

$$\begin{aligned}
m\vec{R} &= \sum_{\alpha} m_{\alpha} \vec{x}^{\alpha}, \\
\hat{f}_i &= \sum_{j} C_{ji} \hat{l}_j, \qquad \tilde{C}QC = \operatorname{diag}(\lambda_1, \lambda_2, \lambda_3), \\
\sqrt{\lambda_i} &= \left\{ \sum_{\alpha} m_{\alpha} [(\vec{x}^{\alpha} - \vec{R}) \cdot \hat{f}_i]^2 \right\}^{\frac{1}{2}}, \\
\zeta_i^{\alpha} &= m_{\alpha} \mu_i^{-1} [(\vec{x}^{\alpha} - \vec{R}) \cdot \hat{f}_i].
\end{aligned} \tag{7.10.107}$$

Remark. Up to the freedom in the choice of parameters for describing the triad of orthonormal N-space vectors $(\zeta_1, \zeta_2, \zeta_3)$ $(\sum_\alpha \zeta_i^\alpha = 0)$, the internal coordinates for the principal axes frame are uniquely determined when one also chooses the $\mu_i = \sqrt{\lambda_i}$ $(i = 1, 2, 3)$ as three of the internal coordinates.

k. The linear momentum operators. In order to obtain the Hamiltonian for a collection of N interacting particles in which the motions are referred to a body-fixed frame, we follow Born and Oppenheimer [1] and transform directly the kinetic energy operator:

$$T = -\sum_{\alpha i} (2m_\alpha)^{-1}(\partial/\partial x_i^\alpha)^2. \tag{7.10.108}$$

For this purpose, we first transform the linear momentum $p_i^\alpha = -i(\partial/\partial x_i^\alpha)$ to the coordinates $\{R_i\}$, $\{C_{jk}\}$, $\{y_j^\alpha = \vec{y}^\alpha \cdot \hat{f}_j\}$:

$$x_i^\alpha = R_i + \sum_j C_{ij} y_j^\alpha. \tag{7.10.109}$$

Thus, we have

$$\frac{\partial}{\partial x_i^\alpha} = \sum_j \left(\frac{\partial R_j}{\partial x_i^\alpha}\right)\frac{\partial}{\partial R_j} + \sum_{jk}\left(\frac{\partial C_{jk}}{\partial x_i^\alpha}\right)\frac{\partial}{\partial C_{jk}} + \sum_{\beta j}\left(\frac{\partial y_j^\beta}{\partial x_i^\alpha}\right)\frac{\partial}{\partial y_j^\beta}. \tag{7.10.110}$$

The transformation from the $\{y_i^\alpha\}$ to internal coordinates will then be carried out subsequently.

The derivatives appearing in Eq. (7.10.110) are readily evaluated, using the relation

$$\frac{\partial \hat{f}_k}{\partial x_i^\alpha} = \hat{f}_k \times \vec{\Omega}_i^\alpha. \tag{7.10.111}$$

(The general form of this derivative follows from $\hat{f}_i \cdot \hat{f}_j = \delta_{ij}$. The explicit form is developed in Appendix A and is summarized below for the Eckart frame and the principal axes frame.) Thus, we find the following relations:

$$\frac{\partial R_j}{\partial x_i^\alpha} = \delta_{ij}(m_\alpha/m),$$

$$\frac{\partial C_{jk}}{\partial x_i^\alpha} = \hat{l}_j \cdot \frac{\partial \hat{f}_k}{\partial x_i^\alpha} = \hat{l}_j \cdot (\hat{f}_k \times \vec{\Omega}_i^\alpha) = \vec{\Omega}_i^\alpha \cdot (\hat{l}_j \times \hat{f}_k),$$

$$\frac{\partial y_j^\beta}{\partial x_i^\alpha} = \vec{\Omega}_i^\alpha \cdot (\vec{y}^\beta \times \hat{f}_j) + \delta_{\alpha\beta} C_{ij} - (m_\alpha/m) C_{ij}. \qquad (7.10.112)$$

Using these derivatives in Eq. (7.10.110) yields the following results for the various terms:

First term:

$$-i \sum_j \left(\frac{\partial R_j}{\partial x_i^\alpha}\right) \frac{\partial}{\partial R_j} = -i \frac{m_\alpha}{m} \frac{\partial}{\partial R_j}. \qquad (7.10.113)$$

Second term:

$$-i \sum_{jk} \left(\frac{\partial C_{jk}}{\partial x_i^\alpha}\right) \frac{\partial}{\partial C_{jk}} = \vec{\Omega}_i^\alpha \cdot \vec{K}, \qquad (7.10.114)$$

where $\vec{K} \equiv (\det C) \sum_i K_i \hat{f}_i$. The components (K_1, K_2, K_3) are here defined in terms of the components $\hat{f}_i \cdot \vec{J}$ of the total angular momentum $\vec{J} = \sum_i J_i \hat{l}_i$, referred to the frame $(\hat{f}_1, \hat{f}_2, \hat{f}_3)$, by[1]

$$\begin{aligned} -K_i &= (\det C)(\hat{f}_i \cdot \vec{J}) = (\det C)(\vec{J} \cdot \hat{f}_i) \\ &= (\det C) \sum_l C_{li} J_l = (\det C) \sum_l J_l C_{li} \\ &= -i \sum_l \left(C_{lj} \frac{\partial}{\partial C_{lk}} - C_{lk} \frac{\partial}{\partial C_{lj}} \right), \end{aligned} \qquad (7.10.115)$$

where i, j, k are cyclic in 1, 2, 3 (see footnote, p. 532).

Third term:

$$-i \sum_{\beta j} \left(\frac{\partial y_j^\beta}{\partial x_i^\alpha}\right) \frac{\partial}{\partial y_j^\beta} = \hat{l}_i \cdot \left\{ \vec{\pi}^\alpha - (m_\alpha/m) \sum_\beta \vec{\pi}^\beta \right\} + \Omega_i^\alpha \cdot \sum_\beta (\vec{y}^\beta \times \vec{\pi}^\beta), \qquad (7.10.116)$$

---

[1] The commutation relations for the three operators $\hat{f}_i \cdot \vec{J}$ $(i = 1, 2, 3)$ are $[(\hat{f}_i \cdot \vec{J}), (\hat{f}_j \cdot \vec{J})] = -ie_{ijk}(\det C)(\hat{f}_k \cdot \vec{J})$. The inclusion of the factor $-(\det C)$ in the definition of the K_i is thus essential in order to obtain canonical commutation relations, $[K_i, K_j] = ie_{ijk} K_k$, such that the operators K_i are invariant under inversion of all particle positions; that is, $K_i \to K_i$ under $\vec{x}^\alpha \to -\vec{x}^\alpha$. Thus, the operators (K_1, K_2, K_3) behave in all respects like the components of an angular momentum. These operators also behave correctly under time reversal; that is, $K_i \to -K_i$ under $\vec{x}^\alpha \to \vec{x}^\alpha$ and $\vec{p}^\alpha \to -\vec{p}^\alpha$. For a right-handed frame $(\hat{f}_1, \hat{f}_2, \hat{f}_3)$ the relationship of (J_1, J_2, J_3) and $(K_1, K_2, K_3) = (-\mathscr{P}_1, -\mathscr{P}_2, -\mathscr{P}_3)$ to the rotation matrices, considered as state vectors, has been discussed thoroughly in Chapter 3, Sections 7 and 8.

where $\vec{\pi}^\alpha$ is defined by

$$\vec{\pi}^\alpha = -i \sum_i \hat{f}_i \frac{\partial}{\partial y_i^\alpha}. \tag{7.10.117}$$

Using the results Eqs. (7.10.111)–(7.10.117) in Eq. (7.10.110), we obtain the following form for the linear momentum operator $-i\partial/\partial x_i^\alpha$:

$$-i\frac{\partial}{\partial x_i^\alpha} = -i\frac{m_\alpha}{m}\frac{\partial}{\partial R_i} + \hat{l}_i \cdot \left[\vec{\pi}^\alpha - (m_\alpha/m)\sum_\beta \vec{\pi}^\beta\right] + \vec{\Omega}_i^\alpha \cdot (\vec{K} + \vec{\Pi}), \tag{7.10.118}$$

where we have defined the vector $\vec{\Pi}$ by

$$\vec{\Pi} = \sum_\alpha (\vec{y}^\alpha \mathbf{x}\, \vec{\pi}^\alpha) = (\det C)\sum_i \Pi_i \hat{f}_i,$$

$$\Pi_i = -i\sum_\alpha \left(y_j^\alpha \frac{\partial}{\partial y_k} - y_k^\alpha \frac{\partial\cdot}{\partial y_j}\right), \qquad i,j,k \text{ cyclic}. \tag{7.10.119}$$

We shall see below [Eqs. (7.10.121) and (7.10.134)] that the operators Π_i depend only on the internal coordinates, and since $\mathbf{\Pi} = (\Pi_1, \Pi_2, \Pi_3)$ has the formal appearance of an angular momentum, it is often called the *internal angular momentum* of the system of particles. One must not conclude from the form (7.10.119), however, that the operators (Π_1, Π_2, Π_3) obey angular momentum commutation relations, since the coordinates $\{y_i^\alpha\}$ are not independent, and the validity or not of the commutation relations will depend on the constraints on the coordinates. (This is discussed in the Remark below.)

In the derivation of Eq. (7.10.118) no use has been made of the dependence of the $\{y_i^\alpha\}$ on the internal coordinates; this relation is therefore valid for an arbitrary body-fixed frame. In the cases of the Eckart frame and the principal axes frame, the terms in the linear momentum take on the special forms that we now summarize:

► Eckart frame:

$$\vec{\pi}^\alpha = -i\sum_\mu \vec{q}_\mu^\alpha \frac{\partial}{\partial q_\mu},$$
$$\sum_\alpha \vec{\pi}^\alpha = \vec{0}; \tag{7.10.120}$$

$$\vec{\Pi} = -i\sum_{\mu\nu} \vec{\zeta}_{\mu\nu} q_\mu \frac{\partial}{\partial q_\nu},$$

$$\vec{\zeta}_{\mu\nu} = \sum_\alpha m_\alpha^{-1}(\vec{q}_\mu^{\,\alpha} \times \vec{q}_\nu^{\,\alpha}). \tag{7.10.121}$$

The vectors $\vec{\Omega}_i^\alpha$ in Eq. (7.10.111) are of the form (see Appendix A)

$$\vec{\Omega}_i^\alpha = \sum_j \Omega_{ij}^\alpha \hat{f}_j, \tag{7.10.122}$$

where $\Omega^\alpha = (\Omega_{ij}^\alpha)$ is the 3×3 matrix given by

$$\Omega^\alpha = CA^\alpha M. \tag{7.10.123}$$

The matrices A^α and M occurring in this result have the definitions

$$A^\alpha = m_\alpha \begin{pmatrix} 0 & -a_3^\alpha & a_2^\alpha \\ a_3^\alpha & 0 & -a_1^\alpha \\ -a_2^\alpha & a_1^\alpha & 0 \end{pmatrix}, \tag{7.10.124}$$

$$M = [-G + (\operatorname{tr} G)I_3]^{-1}, \tag{7.10.125}$$

where G is the symmetric matrix (hence, M is symmetric) given by Eq. (7.10.42). In terms of internal coordinates, G and M^{-1} are given by the following expressions:

$$G = G_0 + \sum_\mu B_\mu q_\mu, \tag{7.10.126}$$

$$(G_0)_{ij} = \sum_\alpha m_\alpha a_i^\alpha a_j^\alpha, \tag{7.10.127}$$

$$(B_\mu)_{ij} = \sum_\alpha a_i^\alpha q_{\mu j}^\alpha\,; \tag{7.10.128}$$

$$M^{-1} = M_0^{-1} + \sum_\mu A_\mu q_\mu, \tag{7.10.129}$$

$$M_0^{-1} = (\operatorname{tr} G_0)I_3 - G_0 = \sum_\alpha m_\alpha^{-1}\tilde{A}^\alpha A^\alpha, \tag{7.10.130}$$

$$A_\mu = -B_\mu + (\operatorname{tr} B_\mu)I_3. \tag{7.10.131}$$

▶ Principal axes frame:

$$\pi_i^\alpha = \vec{\pi}^\alpha \cdot \hat{f}_i = -i\zeta_i^\alpha \frac{\partial}{\partial \mu_i} - i\frac{m_\alpha}{\mu_i}\frac{\partial}{\partial \zeta_i^\alpha} + i\frac{\zeta_i^\alpha}{\mu_i}\left(\sum_\beta \zeta_i^\beta \frac{\partial}{\partial \zeta_i^\beta}\right), \tag{7.10.132}$$

$$\sum_\alpha \pi_i^\alpha = -i\mu_i^{-1}\sum_\alpha m_\alpha \frac{\partial}{\partial \zeta_i^\alpha}, \tag{7.10.133}$$

$$\Pi_i = -i\left[\frac{\mu_j}{\mu_k}\left(\sum_\alpha \zeta_j^\alpha \frac{\partial}{\partial \zeta_k^\alpha}\right) - \frac{\mu_k}{\mu_j}\left(\sum_\alpha \zeta_k^\alpha \frac{\partial}{\partial \zeta_j^\alpha}\right)\right], \tag{7.10.134}$$

where i, j, k are cyclic in 1, 2, 3, and we have used the orthogonality relation, $\boldsymbol{\zeta}_i \cdot \boldsymbol{\zeta}_j = \delta_{ij}$, in obtaining this last result. The expression for $\vec{\Omega}_i^\alpha$ is still given by Eqs. (7.10.122) and (7.10.123), but now A^α and M have the definitions (see Appendix A):

$$A^\alpha = \begin{pmatrix} 0 & \mu_3\zeta_3^\alpha & \mu_2\zeta_2^\alpha \\ \mu_3\zeta_3^\alpha & 0 & \mu_1\zeta_1^\alpha \\ \mu_2\zeta_2^\alpha & \mu_1\zeta_1^\alpha & 0 \end{pmatrix}, \tag{7.10.135}$$

$$M = \begin{pmatrix} (\lambda_2-\lambda_3)^{-1} & 0 & 0 \\ 0 & (\lambda_3-\lambda_1)^{-1} & 0 \\ 0 & 0 & (\lambda_1-\lambda_2)^{-1} \end{pmatrix}. \tag{7.10.136}$$

Remark. The quantity $\vec{\Pi}$, given generally by Eq. (7.10.119) and explicitly in terms of internal coordinates by Eqs. (7.10.121) and (7.10.134) for the two cases, is often called the *internal angular momentum* associated with the internal coordinates. Since $\vec{\Pi}$ is an axial vector, its components relative to the frame $(\hat{f}_1, \hat{f}_2, \hat{f}_3)$ are defined by $\Pi_i = (\det C)\hat{f}_i \cdot \vec{\Pi}$ (under spatial inversion, $\Pi_i \to \Pi_i$). The components Π_i are not angular momentum operators, that is,

$$[\Pi_i, \Pi_j] \neq ie_{ijk}\Pi_k, \tag{7.10.137}$$

for the Eckart frame [Eq. (7.10.121)] (Watson [38]), but do satisfy angular momentum commutation relations for the principal axes frame [Eq. (7.10.134)] (Buck *et al.* [10]). Thus, although the quantity $\vec{\Pi} = \sum_\alpha (\vec{y}^\alpha \times \vec{\pi}^\alpha)$ has the formal appearance of an angular momentum, the commutation relations for the components are not valid, in general, for all body-fixed frames.

Using the transformation of the linear momentum operators, Eq. (7.10.118), one may now obtain also the transformation of the kinetic energy operator to the coordinates $\{R_i\}$, $\{C_{ij}\}$, and internal coordinates. This direct procedure requires much attention to detail, but it is, in principle, straightforward. (See

Natanson and Adamov [28], Louck [29], and Makushkin and Ulenikov [30] for the Eckart frame Hamiltonian; see Buck *et al.* [10] for the principle axes Hamiltonian.)

Introducing a potential energy function V appropriate to the physical system under investigation now yields the Hamiltonian (nonrelativistic Schrödinger equation) in terms of coordinates suitable to a body-fixed frame description.

Having indicated the generality of the body-fixed frame approach to many-particle systems, we now confine ourselves to the study of rigid molecules, using the Eckart frame.

1. The Hamiltonian for a semirigid (rigid) polyatomic molecule. The model for a semirigid molecule is one in which each mass point (nucleus) has its motion constrained to the neighborhood of an equilibrium position that is fixed relative to the equilibrium positions of the other point masses. The *numerical constants* $\{a_i^\alpha\}$ define the vector $\vec{u}^\alpha = \sum_i a_i^\alpha \hat{f}_i$, which is the *equilibrium position vector* of particle α relative to the Eckart frame.

The physical intuition for this model comes from visualizing the fast motions of the electrons as creating a potential energy function V, which has a deep minimum at each of the points $\sum_i a_i^\alpha \hat{f}_i$.

The Born–Oppenheimer [1] approximation then provides the mathematical justification for separating the motions of the electrons and the nuclei. (For a recent review of the mathematical properties of this approximation see Seiler [31]).

Since the motions are confined to the neighborhood of the equilibrium positions of the nuclei, one assumes that the potential energy may be expanded in a Taylor series about these points—that is, that one may write

$$V(Q) = \tfrac{1}{2}\sum_{\mu\nu} k_{\mu\nu} Q_\mu Q_\nu + \sum_{\mu\nu\lambda} k_{\mu\nu\lambda} Q_\mu Q_\nu Q_\lambda + \cdots \tag{7.10.138}$$

for some suitably chosen set of internal coordinates Q_μ ($\mu = 1, 2, \ldots, 3N - 6$) (see Wilson *et al.* [15]). (For equilibrium a linear term must be absent in Eq. (7.10.138), and a constant term of no physical consequence has been omitted.)

The quadratic potential energy term

$$V_0 = \tfrac{1}{2}\sum_{\mu\nu} k_{\mu\nu} Q_\mu Q_\nu \tag{7.10.139}$$

must be symmetric and positive definite. Hence, there exists an orthogonal linear transformation $\mathbf{Q} = A\mathbf{q}$, which brings the form (7.10.139) to diagonal form:

$$V_0 = \tfrac{1}{2}\sum_{\mu} k_\mu q_\mu^2, \qquad k_\mu > 0. \tag{7.10.140}$$

The internal coordinates q_μ ($\mu = 1, 2, \ldots, 3N - 6$) of the previous sections, which are appropriate to the molecular problem, are thus *normal coordinates*. (For the role of the Eckart vectors in the normal coordinate problem and other references to this problem, see the review by Louck and Galbraith [16].)

The expression for the quantum mechanical Hamiltonian operator evolved in form from the earliest days of quantum theory (Dennison [32], Casimir [7]) up to 1940, when its modern form was given by Darling and Dennison [33]. (See also Darling [34].) The earliest general derivation was given by Wilson and Howard [12] and later corrected (Wilson *et al.* [15]) to conform to the criticism of Darling and Dennison.

The steps followed in these papers in deriving the Hamiltonian may be briefly described in the following manner: (1) Develop the classical Hamiltonian for the molecular model, using the moving frame introduced by Eckart [6] (Margenau and Murphy [35]). (2) Develop the theory for transcribing a classical Hamiltonian into a quantum Hamiltonian in the case of momenta that are conjugate to coordinates (Podolsky [36], Kemble [37]). (3) Extend the theory described in (2) to the case of momenta not necessarily conjugate to coordinates.

The derivation of the quantum mechanical Hamiltonian from its classical counterpart by the procedure described above agrees with the result obtained by the direct transformation method described in the previous section.

We present the final form of the Hamiltonian in the simplified version given by Watson [38]. (The details of the derivation are too lengthy to produce here. We shall simply state the results, indicating several steps leading to the final result.)

Using the explicit transformation from the $3N$ Cartesian coordinates $\{x_i^\alpha\}$ to the $3N$ molecular coordinates $\{R_j\}$, $\{C_{ij}\}$ (containing three independent coordinates), and the normal coordinates $\{q_\mu\}$, one finds for the kinetic energy operator[1]

$$\begin{aligned} T &= -\sum_{\alpha i}(2m_\alpha)^{-1}(\partial/\partial x_i^\alpha)^2 \\ &= -(2m)^{-1}\sum_i(\partial/\partial R_i)^2 - \tfrac{1}{2}\sum_\mu(\partial/\partial q_\mu)^2 - \tfrac{1}{2}\sum_\mu \lambda_\mu(\partial/\partial q_\mu) \\ &\quad + \tfrac{1}{2}\sum_{ij}\mu_{ij}(K_i + \Pi_i)(K_j + \Pi_j) - \frac{i}{2}\sum_i \nu_i(K_i + \Pi_i). \end{aligned} \qquad (7.10.141)$$

In this result μ_{ij}, λ_μ, and ν_i have the following definitions:

---

[1] Note that H is invariant under rotation and inversion of the spatial coordinates $\vec{x}^\alpha$. This is obvious for the left-hand side of Eq. (7.10.141), and also for the right-hand side, since the internal coordinates q_μ, as well as the K_i and Π_i, are rotation–inversion invariants.

1. The matrix $\boldsymbol{\mu} = (\mu_{ij})$ is called the *effective reciprocal inertia tensor*. It is the 3×3 symmetric matrix given by

$$\boldsymbol{\mu} = \sum_{\alpha} m_{\alpha}^{-1} \tilde{\Omega}^{\alpha} \Omega^{\alpha} = M M_0^{-1} M, \tag{7.10.142}$$

where M and M_0 are defined, respectively, by Eqs. (7.10.125)–(7.10.131).

2. The real numbers λ_μ and v_j are defined by

$$\lambda_\mu \equiv \operatorname{tr}(M A_\mu),$$

$$v_j \equiv \sum_i (G_0 M H_i M)_{ij}, \tag{7.10.143}$$

where H_i is the 3×3 skew-symmetric matrix with element in row j and column k given by

$$(H_i)_{jk} = e_{ijk}. \tag{7.10.144}$$

The kinetic energy operator (7.10.141) contains linear terms in the momenta $-i(\partial/\partial q_\mu)$ and in K_i and Π_i, and these are usually removed by making a further transformation of the kinetic energy operator. This transformation is effected by the Jacobian of the transformation from the Cartesian coordinates $\{x_i^\alpha\}$ to the molecular coordinates $\{R_i\}$, $\{C_{ij}\}$, $\{q_\mu\}$:

$$x_i^\alpha = R_i + \sum_j C_{ij}\left(a_j^\alpha + m_\alpha^{-1} \sum_\mu q_\mu q_{\mu j}^\alpha\right). \tag{7.10.145}$$

The Jacobian $\mathscr{J}$ of the transformation (7.10.145) is nontrivial to calculate, but it may be shown to be

$$\mathscr{J} = \left[\left(m \Big/ \prod_\alpha m_\alpha\right)^{\frac{3}{2}}\right] (\det \Lambda)(\det M^{-1})(\det M_0)^{\frac{1}{2}}, \tag{7.10.146}$$

where $\det \Lambda$ denotes the determinant of a 3×3 matrix Λ, which we now describe: Let $(\zeta_1, \zeta_2, \zeta_3)$ denote a parametrization of the proper orthogonal matrix C. Then the derivatives $\partial C/\partial \zeta_i$ must have the form

$$\frac{\partial C}{\partial \zeta_i} = C \begin{pmatrix} 0 & \lambda_{i3} & -\lambda_{i2} \\ -\lambda_{i3} & 0 & \lambda_{i1} \\ \lambda_{i2} & -\lambda_{i1} & 0 \end{pmatrix}, \qquad i = 1, 2, 3, \tag{7.10.147}$$

for certain functions λ_{ij} of the parameters. The 3×3 *matrix* (λ_{ij}) is denoted by Λ. [For $(\zeta_1, \zeta_2, \zeta_3) = (\alpha\beta\gamma)$ = Euler angles, one has $\det \Lambda = \sin \beta$.]

The significance of the Jacobian may be understood in the following way (Kemble [37]): Let $\Psi(\mathbf{z})$ and $\Phi(\boldsymbol{\zeta}; \mathbf{q})$ denote energy eigenfunctions in terms of $3N - 3$ independent relative coordinates from the set $\{x_i^\alpha - R_i\}$ and $\{\zeta_1, \zeta_2, \zeta_3; q_1, \ldots, q_{3N-6}\}$, respectively. We impose the normalization conditions:

$$\int \Psi^*(\mathbf{z})\, \Psi(\mathbf{z}) \prod_{\alpha i} dz_i^\alpha = 1,$$

$$\int \Phi^*(\boldsymbol{\zeta}; \mathbf{q})\, \Phi(\boldsymbol{\zeta}; \mathbf{q})(\det \Lambda) \prod_i d\zeta_i \prod_\mu dq_\mu = 1. \tag{7.10.148}$$

Except for irrelevant numerical constants, this implies that

$$\Phi(\boldsymbol{\zeta}; \mathbf{q}) = \Delta^{-\frac{1}{2}}\, \Psi(\mathbf{z}), \tag{7.10.149}$$

where we have defined [see Eq. (7.10.125)]

$$\Delta = \det M. \tag{7.10.150}$$

The Hamiltonian eigenvalue equation for $H' \equiv T + V$ [see Eqs. (7.10.138) and (7.10.141)],

$$H'\Phi = E\Phi, \tag{7.10.151}$$

is thus transformed to

$$H\Phi = E\Phi, \tag{7.10.152}$$

where

$$\begin{aligned} H = \Delta^{-\frac{1}{2}} H' \Delta^{\frac{1}{2}} &= H' + \Delta^{-\frac{1}{2}} [H', \Delta^{\frac{1}{2}}] \\ &= H' + \Delta^{-\frac{1}{2}} [T, \Delta^{\frac{1}{2}}]. \end{aligned} \tag{7.10.153}$$

The additional term $\Delta^{-\frac{1}{2}}[T, \Delta^{\frac{1}{2}}]$ occurring in this last result removes the linear terms from T [Eq. (7.10.141)] containing λ_μ and v_i, yielding the molecular Hamiltonian in the following form (Watson [38]):

$$\begin{aligned} H = &-(2m)^{-1} \sum_i (\partial/\partial R_i)^2 - \tfrac{1}{2} \sum_\mu (\partial/\partial q_\mu)^2 \\ &+ \tfrac{1}{2} \sum_{ij} \mu_{ij}(K_i + \Pi_i)(K_j + \Pi_j) + U + V, \end{aligned} \tag{7.10.154}$$

where

$$U = -\sum_i \mu_{ii}/8 \qquad (7.10.155)$$

is an additional term supplied by $\Delta^{-\frac{1}{2}}[T, \Delta^{\frac{1}{2}}]$.

The theory of the vibration–rotation spectra of polyatomic molecules is to a large extent the development of techniques for determining the eigenvalues and eigenvectors of the Hamiltonian (7.10.154). Since the equation $H\Psi = E\Psi$ does not lend itself to an exact solution, much of the theory of this equation is directed toward finding effective methods for approximating solutions.

m. Approximate form of the Hamiltonian for spherical top molecules. Our primary goal here is to illustrate angular momentum techniques. The role of angular momentum theory in solving the eigenvalue problem for the molecular Hamiltonian H is nicely demonstrated for spherical top molecules, and we now particularize the Hamiltonian H to this case, referring to the literature (Nielsen [13]) for the applications to other types of molecules.

Consider the effective reciprocal inertial tensor for the case of spherical tops – that is, for the case where the principal *equilibrium* moments of inertia are all equal. (The Eckart frame coincides with the principal axes frame when all particles are at their equilibrium positions – that is, when all $q_\mu = 0$.) The reciprocal inertial tensor [see Eqs. (7.10.129)–(7.10.131), (7.10.142)] becomes

$$\boldsymbol{\mu} = I_0^{-1}\left(\mathbb{1}_3 + I_0^{-1}\sum_\mu A_\mu q_\mu\right)^{-2}, \qquad (7.10.156)$$

where the principal equilibrium moments of inertia are all equal and given by

$$I_0 = 2\sum_\alpha m_\alpha a_i^\alpha a_i^\alpha, \qquad i = 1, 2, 3. \qquad (7.10.157)$$

It is convenient at this point to restore Planck's constant $\hbar$ to the Hamiltonian (7.10.154). This is done by multiplying all terms but V by $\hbar^2$, and absorbing $\hbar$ into the definitions of K_i and Π_i [insert $\hbar$ into the right-hand sides of Eqs. (7.10.115) and (7.10.119)]. We also replace the normal coordinates q_μ [which have the dimension of $(\text{mass})^{\frac{1}{2}}(\text{length})$] by dimensionless normal coordinates q'_μ defined by

$$q'_\mu = (\omega_\mu/\hbar)^{\frac{1}{2}} q_\mu, \qquad (7.10.158)$$

where ω_μ is defined by

$$\omega_\mu \equiv (k_\mu)^{\frac{1}{2}}. \tag{7.10.159}$$

In this definition k_μ is the coefficient occurring in the quadratic form (7.10.140) for the potential energy V_0 [k_μ must have the dimension (time)$^{-2}$ in order that V_0 have the dimension (mass)(length)$^{-2}$(time)$^{-2}$ of energy]. At the same time, we introduce the standard notation

$$B \equiv \hbar/4\pi c I_0 \tag{7.10.160}$$

and define the (dimensionless) parameters ε_μ by[1]

$$\varepsilon_\mu \equiv (hcB/\hbar\omega_\mu)^{\frac{1}{2}}. \tag{7.10.161}$$

In terms of these notations, the reciprocal inertial tensor is expressed by

$$\boldsymbol{\mu} = \frac{2hcB}{\hbar^2}\left[\mathbb{1}_3 + \sum_\mu \varepsilon_\mu A'_\mu q'_\mu\right]^{-2}, \tag{7.10.162}$$

where A'_μ is the (dimensionless) matrix given by

$$A'_\mu \equiv \frac{A_\mu}{(I_0/2)^{\frac{1}{2}}}. \tag{7.10.163}$$

The parameters ε_μ^2 are essentially ratios of the $J = 1$ pure rotational energy to the ground-state vibrational energy of the ν_μ mode of motion. These ratios are typically quite small, ranging from 5/3000 for the ν_3-band of methane to 0.1/1000 for the ν_3-band of sulfur hexafluoride, yielding corresponding values of ε_μ of about 0.04 and 0.01. The elements of the matrix A'_μ are pure numbers of the order of unity. This, together with the assumption of small vibrations, justifies replacing $\boldsymbol{\mu}$ by the expansion

$$\boldsymbol{\mu} = \frac{2hcB}{\hbar^2}\sum_{n=0}^{\infty}(-1)^n(n+1)\left(\sum_\mu \varepsilon_\mu A'_\mu q'_\mu\right)^n. \tag{7.10.164}$$

Using this expansion in the Hamiltonian (7.10.154) and dropping the center of mass term and the primes on the normal coordinates $\{q'_\mu\}$, we obtain

---

[1] Planck's constant h in this equation is not canceled out in order to give ε_μ explicitly as a ratio of a rotational energy hcB to a vibrational energy $\hbar\omega_\mu$. These energy terms, hcB and $\hbar\omega_\mu$, are also kept in the succeeding equations, since it is customary to express the Hamiltonian (and the energy) in units of hc (cm^{-1}).

$$H = U + \sum_{k=0}^{\infty} H_k, \tag{7.10.165}$$

where

$$H_0 = \frac{hcB}{\hbar^2}\mathbf{J}^2 + \sum_{\mu} \frac{\hbar\omega_\mu}{2}\left(\frac{p_\mu^2}{\hbar^2} + q_\mu^2\right), \tag{7.10.166}$$

$$H_1 = \frac{2hcB}{\hbar^2}\sum_i K_i\Pi_i - \frac{2hcB}{\hbar^2}\sum_{ij} Q_{ij}K_jK_j + V_1, \tag{7.10.167}$$

$$H_{k+2} = \frac{hcB}{\hbar^2}\sum_{ij}\Big[(-1)^{k+2}(k+3)(Q^{k+2})_{ij}\,K_iK_j + (-1)^{k+1}\,2(k+2)(Q^{k+1})_{ij}\,K_i\Pi_j + (-1)^k(k+1)(Q^k)_{ij}\,\Pi_i\Pi_j\Big] + V_k,$$

$$k = 0, 1, 2, \ldots, \tag{7.10.168}$$

$$U = -\frac{hcB}{4}\sum_{k=0}^{\infty}(-1)^k(k+1)\,\mathrm{tr}\,Q^k. \tag{7.10.169}$$

We have introduced the following notations into Eqs. (7.10.165)–(7.10.169):

1. The matrix Q is the symmetric matrix (dimensionless) given by

$$Q \equiv \sum_{\mu} \varepsilon_\mu A'_\mu q_\mu, \tag{7.10.170}$$

which is linear in the normal coordinates and the parameters ε_μ.

2. The coordinate q_μ is dimensionless, and $p_\mu = -i\hbar\partial/\partial q_\mu$ is the conjugate momentum. The internal angular momentum components Π_i now take the form[1]

$$\Pi_i = \sum_{\mu<\nu} \zeta^i_{\mu\nu}\left(\sqrt{\frac{\omega_\nu}{\omega_\mu}}\,q_\mu p_\nu - \sqrt{\frac{\omega_\mu}{\omega_\nu}}\,q_\nu p_\mu\right), \tag{7.10.171}$$

$$\zeta^i_{\mu\nu} = \left[\sum_{\beta} m_\beta^{-1}\left(q^\beta_{\mu j}q^\beta_{\nu k} - q^\beta_{\mu k}q^\beta_{\nu j}\right)\right], \tag{7.10.172}$$

[1] Despite the fact that the $\zeta^i_{\mu\nu}$ appear to the components of a cross product, they are *numerical*, and thus true scalars. The form (7.10.171) of the Π_i thus shows explicitly the properties $\Pi_i \to \Pi_i$ and $\Pi_i \to -\Pi_i$, under particle inversion and time reversal, respectively.

where i, j, k are cyclic in 1, 2, 3. (Properties of the zeta coefficients have been developed in Ref. [39]).

3. The potential energy V_k denotes the term of degree $k + 2$ in the expansion of V:

$$V_0 = \tfrac{1}{2}\sum_{\mu} \hbar\omega_\mu q_\mu^2,$$

$$V_1 = hc \sum_{\mu\nu\lambda} k_{\mu\nu\lambda} q_\mu q_\nu q_\lambda, \text{ etc.,} \tag{7.10.173}$$

where the constants $k_{\mu\nu\lambda}, \ldots,$ are now expressed in units of cm^{-1}.

In defining H_k, $k = 0, 1, \ldots,$ we have collected together all terms of degree $k + 2$, where each K_i, q_μ, p_μ is assigned degree 1. We have chosen, however, to keep Watson's term U set apart. (At this point there is no implication that H_k is to be regarded as the kth order Hamiltonian in a perturbation theory.) In the expansion (7.10.165), we thus see that H is represented as a sum of polynomial forms in q_μ, p_μ, and K_i.

The eigenvalue problem for H, using the full expansion given by Eqs. (7.10.165)–(7.10.169), is still not amenable to exact treatment, and approximation techniques (perturbation theory) must be introduced. The ordering of the terms in Rayleigh–Schrödinger perturbation theory is based on (*a*) the experimental determination of the magnitude of the energy the various terms contribute to the total energy eigenvalue E; (*b*) the extent to which one can find exact solutions to the "zeroth-order" problem; and (*c*) the simplicity of the physical interpretation that one associates with the various terms.

The term we have identified as H_0 satisfies, for the most part, all the criteria above. It possesses the exact eigenfunctions[1]

$$\left(\frac{2J+1}{8\pi^2}\right)^{\frac{1}{2}} (\det C)^\sigma (-1)^{J-K} \mathscr{D}^{J*}_{M,-K}(C) \prod_{\mu} \Phi_{n_\mu}(q_\mu) \tag{7.10.174}$$

with the corresponding energy eigenvalues given by

$$E^{(0)}_{J;(n)}/hc = BJ(J+1) + \sum_{\mu} \sigma_\mu (n_\mu + \tfrac{1}{2}). \tag{7.10.175}$$

In these results Φ_{n_μ} denotes the wave function of a single harmonic oscillator in the energy eigenstate $E_{n_\mu} = hc\sigma_\mu(n_\mu + \tfrac{1}{2})$, $\sigma_\mu = \omega_\mu/2\pi c$ (units cm^{-1}), and

---

[1] The $\mathscr{D}^J_{MK}(C)$ denote the $O(3)$-representation functions defined on the elements of an orthogonal matrix $R \in O(3)$ by Eqs. (6.178) and (6.179) of Chapter 6. Not only do J_i and K_i have the standard action on these wave functions, but also the inversion operation $\mathscr{I} : \tilde{x}^\alpha \to -\tilde{x}^\alpha$ is diagonal – that is, it has eigenvalue $(-1)^{J+\sigma}$ on these states.

$n_\mu = 0, 1, \ldots$. The Hamiltonian H_0 also has a simple physical interpretation: It is the Hamiltonian for a rigid body rotating about its center of mass together with the Hamiltonian for the collective (vibrational) motions of a set of individual mass points relative to the rotating body (frame). {This complete separation of motions in zeroth-order into the two terms [cf. Eq. (7.10.22)] occurring in H_0 is one of the consequences of using the Eckart frame.} To a lesser extent, H_0 also meets criterion (*a*) above: The two energy terms (pure rotational + pure vibrational) may constitute the principal contribution to the total energy E of a molecule, but in some molecules the *Coriolis interaction term*

$$H_C \equiv \frac{2hcB}{\hbar^2} \sum_i K_i \Pi_i \tag{7.10.176}$$

may yield an energy contribution comparable to $hcBJ(J+1)$. Furthermore, the contribution from anharmonic terms in the potential energy cannot, with certainty, always be ordered by the degree of the associated term.

Despite the aforementioned difficulties, it has been the custom (Nielsen [13], Shaffer *et al.* [14], Amat *et al.* [40, and references therein]) to proceed with the formal developments using a perturbation scheme based on choosing $H_0, H_1, H_2, \ldots$ [as given by Eqs. (7.10.165)–(7.10.169)] as the zeroth-order, first-order, second-order, etc., Hamiltonians, leaving the exceptions to this ordering to be studied as special cases.

The polynomial form of each of the Hamiltonians H_k $(k = 0, 1, 2, \ldots)$ lends itself to a particularly elegant operator formulation of Rayleigh–Schrödinger perturbation theory. The idea itself originates from the earliest papers in quantum mechanics developing transformation theory (Born and Jordan [41]).

The analog in quantum mechanics to classical contact transformations is the unitary transformation

$$q'_\mu = \mathscr{U} q_\mu \mathscr{U}^{-1}, \qquad p'_\mu = \mathscr{U} p_\mu \mathscr{U}^{-1}, \tag{7.10.177}$$

which preserves the canonical commutation relations. Under this transformation, the energy eigenvalue equation $H\Psi = E\Psi$ is transformed to $H'\Psi' = E\Psi'$, where

$$H' = \mathscr{U} H \mathscr{U}^{-1}, \qquad \Psi' = \mathscr{U}\Psi. \tag{7.10.178}$$

The goal in transformation theory is to transform H to a form H', which is simpler and more manageable than H itself.

To ensure that $\mathscr{U}$ is unitary, one often takes

$$\mathscr{U} = e^{iS}, \qquad S \text{ Hermitian}. \tag{7.10.179}$$

The polynomial form of each term in the molecular Hamiltonian H suggests that one choose S to be of the form

$$S = S(q_\mu, p_\mu) = \text{polynomial in the } q_\mu \text{ and } p_\mu, \tag{7.10.180}$$

with coefficients that are themselves polynomials in the K_i. Since the transformation

$$H_k \to H'_k \equiv e^{iS} H_k e^{-iS}, \tag{7.10.181}$$

produces a new polynomial operator H'_k, the notion of simplifying H by such unitary operator transformations is feasible.

The implementation of such operator methods into perturbation theory is known as the Van Vleck [42] transformation (see Kemble [37]). Applied in the sense of Rayleigh–Schrödinger perturbation theory, where one assumes an expansion of H in terms of a small parameter λ of the form,

$$H = H_0 + \lambda H_1 + \lambda^2 H_2 + \cdots,$$

one finds, upon choosing

$$\mathscr{U} = e^{i\lambda S}, \tag{7.10.182}$$

that the transformed Hamiltonian is

$$H' = H'_0 + \lambda H'_1 + \lambda^2 H'_2 + \cdots$$

where

$$\begin{aligned} H'_0 &= H_0, \\ H'_1 &= H_1 - i[H_0, S], \\ H'_2 &= H_2 - \frac{i}{2}[H_1 + H'_1, S], \\ &\vdots \end{aligned} \tag{7.10.183}$$

The forms of the right-hand sides of Eqs. (7.10.183) suggest a method of determining S: One would like to find S such that $[H_0, S] = -iH_1$, since then $H'_1 = 0$. It is, however, in general, not possible to determine such an S, since the diagonal (in energy) matrix elements of $[H_0, S]$ are zero, whereas those of H_1 need not be zero; hence, one cannot remove the "diagonal part" of the operator H_1 by any choice of S. We shall not give any general discussion of the problem thus posed, but shall now illustrate the method for the molecular problem.

For the problem at hand, the terms of H_1 [see Eq. (7.10.167)] are of three types:

(1) $$H_1(\mu) = hch_\mu q_\mu,$$

$$h_\mu = -2B\varepsilon_\mu \sum_{ij} (A'_\mu)_{ij}(K_i K_j/\hbar^2); \tag{7.10.184}$$

(2) $$H_1(\mu\nu) = hch_{\mu\nu}\left(\sqrt{\frac{\omega_\nu}{\omega_\mu}}\, q_\mu p_\nu - \sqrt{\frac{\omega_\mu}{\omega_\nu}}\, q_\nu p_\mu\right)\Big/\hbar, \qquad \mu \neq \nu, \tag{7.10.185}$$

$$h_{\mu\nu} = 2B\left(\sum_i \zeta^i_{\mu\nu} K_i/\hbar\right); \tag{7.10.186}$$

(3) $$H_1(\mu\nu\lambda) = hck_{\mu\nu\lambda}\, q_\mu q_\nu q_\lambda. \tag{7.10.187}$$

By direct verification (Herman and Shaffer [43]), using the canonical commutation relations, $[q_\mu, p_\nu] = i\hbar\delta_{\mu\nu}$, one finds:[1]

$$[H_0, S(\mu\cdots)] = -iH_1(\mu\cdots), \tag{7.10.188}$$

where

(1) $$S(\mu) = -\frac{hch_\mu}{\hbar\omega_\mu}(p_\mu/\hbar), \tag{7.10.189}$$

(2) $$S(\mu\nu) = \frac{hch_{\mu\nu}}{\hbar(\omega_\mu\omega_\nu)^{\frac{1}{2}}(\omega_\mu^2 - \omega_\nu^2)}[(\omega_\mu^2 + \omega_\nu^2)q_\mu q_\nu + 2\omega_\mu\omega_\nu(p_\mu p_\nu/\hbar^2)], \tag{7.10.190}$$

(3) $$\begin{aligned} S(\mu\nu\lambda) = \frac{hck_{\mu\nu\lambda}}{\hbar D_{\mu\nu\lambda}}\,[&\omega_\mu(\omega_\mu^2 - \omega_\nu^2 - \omega_\lambda^2)(p_\mu q_\nu q_\lambda/\hbar) \\ &+ \omega_\nu(\omega_\nu^2 - \omega_\lambda^2 - \omega_\mu^2)(q_\mu p_\nu q_\lambda/\hbar) \\ &+ \omega_\lambda(\omega_\lambda^2 - \omega_\mu^2 - \omega_\nu^2)(q_\mu q_\nu p_\lambda/\hbar) - 2\omega_\mu\omega_\nu\omega_\lambda(p_\mu p_\nu p_\lambda/\hbar^3)], \end{aligned} \tag{7.10.191}$$

[1] Since the quantities h_μ and $h_{\mu\nu}$ commute with $\mathbf{J}^2$, one needs only calculate the commutator using the zeroth-order vibrational Hamiltonian, which itself commutes with h_μ and $h_{\mu\nu}$.

$$D_{\mu\nu\lambda} = (\omega_\mu + \omega_\nu + \omega_\lambda)(\omega_\mu + \omega_\nu - \omega_\lambda)(\omega_\mu - \omega_\nu + \omega_\lambda)(-\omega_\mu + \omega_\nu + \omega_\lambda).$$

The occurrence of the denominators $\omega_\mu^2 - \omega_\nu^2$ and $D_{\mu\nu\lambda}$ in $S(\mu\nu)$ and $S(\mu\nu\lambda)$, respectively, shows that the corresponding term $H_1(\mu\nu)$ and $H_1(\mu\nu\lambda)$ cannot be removed from H_1 by the transformation (7.10.183) if $\omega_\mu = \omega_\nu$ or $D_{\mu\nu\lambda} = 0$. The case $\omega_\mu = \omega_\nu$ $(\mu \neq \nu)$ will occur in the Coriolis interaction term (7.10.185) whenever degenerate modes of vibration (two-, three-, and higher-dimensional isotropic harmonic oscillators in the normal coordinates) are allowed by the symmetry of the molecule (see Section o). Terms of the form (7.10.185) for which $\omega_\mu = \omega_\nu$ and $\mu \neq \nu$ (degenerate oscillators) are called *Coriolis interaction terms of the first kind*, and it is impossible to remove such terms from H_1 by the transformation (7.10.183). In addition to truly degenerate vibrations, the phenomenon of *accidental degeneracy* arises in many molecules. This situation occurs for the Coriolis term $H_1(\mu\nu)$ [Eq. (7.10.185)] whenever the two frequencies ω_μ and ω_ν are approximately equal: $\omega_\mu \simeq \omega_\nu$. In this case the term introduced into H_2' by the transformation (7.10.183) may yield an energy correction to the zeroth-order levels $\hbar\omega_\mu(n_\mu + \frac{1}{2}g_\mu)$ (g_μ = degeneracy) and $\hbar\omega_\nu(n_\nu + \frac{1}{2}g_\nu)$ that is large, thus indicating a failure of the perturbation method. Terms of the form (7.10.185) for which $\omega_\mu \simeq \omega_\nu$ $(\mu \neq \nu)$ are called *Coriolis interaction terms of the second kind* and are said to produce a *Coriolis resonance*. Such cases must be given special consideration (Nielsen [13], Jahn [44]). Although degenerate modes of vibration cannot yield zero for the denominator $D_{\mu\nu\lambda}$ [Eq. (7.10.191)], one may still have accidental degeneracies. This will occur whenever two frequencies ω_μ and ω_ν are related by $\omega_\mu \simeq 2\omega_\nu$ or whenever three frequencies $\omega_\mu, \omega_\nu, \omega_\lambda$ are related by $\omega_\mu \simeq \omega_\nu + \omega_\lambda$ $(\mu \neq \nu \neq \lambda)$. In this case the S-operator (7.10.191) corresponding to the cubic anharmonic terms $hck_{\mu\nu\nu}q_\mu q_\nu^2$ or $hck_{\mu\nu\lambda}q_\mu q_\nu q_\lambda$ may introduce large terms into H_2', thus invalidating the perturbation method as in the case of a Coriolis resonance. These latter interactions originating from the cubic anharmonic terms in the potential energy are said to produce a *Fermi resonance* (Fermi [45]).

The special treatment required to deal with Coriolis and Fermi resonances in molecules is beyond the scope of the present discussion (Ref. [40]), and we henceforth ignore the possibility of such resonance phenomena. (This restriction is not as severe as might be expected, since the resulting Hamiltonian may still be applied to regions of the spectra of a given spherical top molecule where resonance effects are small.)

Under the restriction that resonance phenomena do not occur or may be ignored, we see from Eqs. (7.10.184)–(7.10.191) that all terms may be removed from H_1 by the transformation (7.10.183) except Coriolis terms of the first kind. Furthermore, symmetry considerations (see Section p) show that doubly degenerate vibrations cannot contribute a Coriolis term to the Hamiltonian of a spherical top molecule. Thus, the transformation (7.10.183) brings the

Hamiltonian H [see Eq. (7.10.165)] to the form[1]

$$H' = H'_0 + H'_1 + H'_2 + \cdots, \qquad (7.10.192)$$

where

$$H'_0 = H_0,$$

$$H'_1 = \frac{2hcB}{\hbar^2} \sum_t \zeta^t \mathbf{K} \cdot \mathbf{L}^t. \qquad (7.10.193)$$

In this result $\mathbf{L}^t$ is an internal angular momentum associated with a three-dimensional isotropic oscillator[2] (triply degenerate normal mode), and the summation is over all such degenerate normal modes ($t = 1, 2, \ldots$) of oscillation of the molecule. The term $\mathbf{K} \cdot \mathbf{L}^t$ denotes $\sum_i K_i L_i^t$, where L_i^t is expressed in terms of the normal coordinates (q_1^t, q_2^t, q_3^t) and the conjugate momenta (p_1^t, p_2^t, p_3^t) of the triply degenerate oscillator by

$$L_i^t = q_j^t p_k^t - q_k^t p_j^t, \qquad i, j, k \text{ cyclic in } 1, 2, 3. \qquad (7.10.194)$$

Finally, ζ^t denotes the ζ-coefficient for a three-dimensional oscillator, since, by symmetry, the three ζ-coefficients associated with L_1^t, L_2^t, L_3^t in Eq. (7.10.171) are equal. [We now re-index the normal coordinates q_μ ($\mu = 1, 2, \ldots, 3N - 6$) into sets describing nondegenerate vibrations, twofold degenerate vibrations, and threefold degenerate vibrations.]

The second-order Hamiltonian occurring in Eq. (7.10.192) is, in principle, calculated straightforwardly from Eqs. (7.10.183), using the S-operator [a sum of terms of types (7.10.189)–(7.10.191)] that transforms H_1 to the form H'_1. Such calculations have been carried out up to fourth-order perturbation theory by Amat and Nielsen [46], but will not be considered here. (The problem of higher-order corrections for spherical top molecules will be considered from the point of view of symmetry and tensor operators in Section p.)

Observe from Eq. (7.10.194) that each of the internal angular momenta $\mathbf{L}^t = (L_1^t, L_2^t, L_3^t)$ satisfies standard commutation relations:

$$\mathbf{L}^t \times \mathbf{L}^t = i\hbar \mathbf{L}^t. \qquad (7.10.195)$$

---

[1] The "Watson term," U [Eq. (7.10.169)], is a sum of polynomials in the normal coordinates, and we may consider it to be absorbed into the potential energy.

[2] We do not consider here the possibility of molecules possessing higher than triply degenerate normal modes (icosahedral symmetry).

Also, the internal angular momenta associated with different triply degenerate vibrations commute. Nonetheless, *the sum* $\sum_t \zeta^t \mathbf{L}^t$ *does not satisfy standard commutation relations for arbitrary parameters* ζ^t $(t = 1, 2, \ldots)$. This fact precludes any general analytic treatment of the energy correction to the zeroth-order energy [Eq. (7.10.175)] coming from H_1' (one must diagonalize a Hermitian matrix in which the unknown parameters $\zeta^1, \zeta^2, \ldots$ appear). There is, however, one specialization of the difficult general problem that can be treated fully by angular momentum techniques; this case has had considerable success in explaining a restricted region of the spectra of spherical top molecules (Hecht [47]). We discuss this in the next section.

n. First-order energy spectrum of a triply degenerate vibration in a spherical top molecule. We consider in this section the energy levels of the Hamiltonian $H_0 + H_1'$ in the case where all vibrational quantum numbers are zero except for a single vibrational quantum number n corresponding to a triply degenerate mode of vibration. In this case, except for a constant term, the energy in question is that of the Hamiltonian $H_0' \equiv H_0 + H_1'$ given by

$$H_0' = \frac{\hbar\omega}{2}\left(\frac{\mathbf{p}^2}{\hbar^2} + \mathbf{q}^2\right) + \frac{hcB}{\hbar^2}\mathbf{J}^2 + \frac{2hcB\zeta}{\hbar^2}\mathbf{K}\cdot\mathbf{L}, \qquad (7.10.196)$$

where $\mathbf{q} = (q_1, q_2, q_3)$ denotes the normal coordinates, $\mathbf{p} = (p_1, p_2, p_3)$ the conjugate linear momentum, $-\mathbf{K} = (-K_1, -K_2, -K_3)$ the components of the total angular momentum $\mathbf{J}$ referred to the Eckart frame, and $\mathbf{L} = \mathbf{q} \times \mathbf{p}$ the internal angular momentum of the triply degenerate normal mode.

The Hamiltonian (7.10.196) can be diagonalized exactly. (Thus, it is not necessary to treat the Coriolis interaction term as a perturbation.) Introducing the operator $\mathbf{R} = (R_1, R_2, R_3)$ with components defined by[1]

$$R_i = L_i + K_i = L_i \otimes \mathbb{1}(2) + \mathbb{1}(1) \otimes K_i, \qquad (7.10.197)$$

we find

$$\mathbf{K}\cdot\mathbf{L} = (\mathbf{R}^2 - \mathbf{J}^2 - \mathbf{L}^2)/2. \qquad (7.10.198)$$

Since $\mathbf{K}$ and $\mathbf{L}$ satisfy standard commutation relations, so does $\mathbf{R}$ (see Van Vleck [48]):

[1] If we define $\vec{R} \equiv \sum_i R_i \hat{f}_i$, $\vec{L} \equiv \sum_i L_i \hat{f}_i$ (right-handed frame), and recall that $-\vec{J} = \sum_i K_i \hat{f}_i$, then Eq. (7.10.197) may be written $\vec{R} = -\vec{J} + \vec{L}$, so that $-\vec{R}$ is the difference between the total angular momentum $\vec{J}$ and the internal angular momentum $\vec{L}$. Nielsen [13] calls $-\vec{R}$ the angular momentum of the rotating molecular framework. In a certain sense this is a misnomer, since we have seen that it is the total angular momentum $\mathbf{J}$ that generates rotations of the Eckart frame.

$$\mathbf{R} \times \mathbf{R} = i\hbar\mathbf{R}. \tag{7.10.199}$$

Thus, $\mathbf{R}$ is an angular momentum with eigenvalues of the two operators $\mathbf{R}^2$ and R_3 given, respectively, by $R(R+1)\hbar^2$ and $M_R\hbar$, where[1] $R = 0, 1, \ldots$ [see Eq. (7.10. 201) below] and $M_R = R, R-1, \ldots, -R$. Since the eigenvalues of $\mathbf{J}^2$ and $\mathbf{L}^2$ are, respectively, of the forms $J(J+1)\hbar^2$ and $l(l+1)\hbar^2$, we find that the eigenvalues of $2\mathbf{K}\cdot\mathbf{L}/\hbar^2$ are

$$R(R+1) - J(J+1) - l(l+1), \tag{7.10.200}$$

where, for specified $J = 0, 1, 2, \ldots$ and $l = 0, 1, 2, \ldots$, the allowed values of R are given by the addition of angular momenta rule (see Chapter 3, Section 11):

$$R = |J - l|, |J - l| + 1, \ldots, J + l. \tag{7.10.201}$$

Thus, the eigenvalues E_0' of H_0' are of the form

$$E_0'[n(lJ)R] = \hbar\omega(n + \tfrac{3}{2}) + hcBJ(J+1)$$

$$+ hcB\zeta[R(R+1) - J(J+1) - l(l+1)]. \tag{7.10.202}$$

In this energy term the ranges of the quantum numbers are the following: the total angular momentum quantum number J may assume any value $0, 1, 2, \ldots$; the vibrational quantum number n may assume any value $0, 1, 2, \ldots$; for each allowed value of n the internal angular momentum l may assume any of the values $n, n-2, \ldots, 1$ or 0; and for each allowed value of the pair (J, l) the angular momentum quantum number R may assume any of the values given by (7.10.201).

The eigenvectors of the Hamiltonian (7.10.196) may also be given exactly. We first note from Eqs. (3.123), Chapter 3, and $\mathbf{K} = -\mathscr{P}$ that $\mathbf{K}$ has the standard action (see footnote, p. 532),

$$K_\pm \Psi^{J,\sigma}_{MK} = \hbar[(J \mp K)(J \pm K + 1)]^{\frac{1}{2}} \Psi^{J,\sigma}_{MK\pm1},$$

$$K_3 \Psi^{J,\sigma}_{MK} = K\hbar\Psi^{J,\sigma}_{MK}, \tag{7.10.203}$$

on the eigenvectors $\{\Psi^{J,\sigma}_{MK}\}$ defined by

---

[1] It is customary in molecular spectroscopy to use the symbol R to denote the angular momentum quantum number corresponding to $\mathbf{R}^2$; that is, $\mathbf{R}^2 \to R(R+1)$. This usage should not be confused with the 3×3 orthogonal matrix R used later in this section.

$$\Psi^{J,\sigma}_{MK}(C) \equiv \left(\frac{2J+1}{8\pi^2}\right)^{\frac{1}{2}} (\det C)^{\sigma} (-1)^{J-K} \mathscr{D}^{J*}_{M,-K}(C). \quad (7.10.204)$$

[The components (J_1, J_2, J_3) of $\mathbf{J}$ also have the standard action on the $\{\Psi^{J,\sigma}_{MK}\}$, where M plays the role of K when $\mathbf{K}$ is replaced by $\mathbf{J}$ in Eqs. (7.10.203).] Furthermore, the three-dimensional harmonic oscillator functions $\{\Phi_{nlm}\}$ satisfy

$$\frac{1}{2}\hbar\omega\left(\frac{\mathbf{p}^2}{\hbar^2} + \mathbf{q}^2\right)\Phi_{nlm} = \hbar\omega\left(n + \frac{3}{2}\right)\Phi_{nlm},$$

$$L_{\pm}\Phi_{nlm} = \hbar[(l \mp m)(l \pm m + 1)]^{\frac{1}{2}} \Phi_{n,l,m\pm 1},$$

$$L_3\Phi_{nlm} = m\hbar\Phi_{nlm}, \quad (7.10.205)$$

where for each $n = 0, 1, 2, \ldots$ the allowed values of l are $l = n, n-2, \ldots, 1$ or 0, and for each l the allowed values of m are $m = l, l-1, \ldots, -l$. The explicit form of Φ_{nlm} is

$$\Phi_{nlm}(\mathbf{q}) = F_{nl}(q)\,\mathscr{Y}_{lm}(\mathbf{q}),$$

$$F_{nl}(q) = \sqrt{2}\left[\frac{\left(\frac{n-l}{2}\right)!}{\Gamma\left(\frac{n+l+3}{2}\right)}\right]^{\frac{1}{2}} e^{-\frac{1}{2}q^2} L^{l+\frac{1}{2}}_{(n-l)/2}(q^2), \quad (7.10.206)$$

where $q = (q_1^2 + q_2^2 + q_3^2)^{\frac{1}{2}}$, $\mathscr{Y}_{lm}(\mathbf{q})$ denotes a solid harmonic, and $L^{\alpha}_{k}(x)$ denotes an associated Laguerre polynomial [see Eq. (7.4.44)]. Since both $\mathbf{K}$ and $\mathbf{L}$ have the standard action on the eigenvectors $\{\Psi^{J,\sigma}_{MK}\}$ and $\{\Phi_{nlm}\}$, respectively, and since $\mathbf{R}$ is defined in terms of $\mathbf{K}$ and $\mathbf{L}$ in the sense of vector addition of angular momenta (see Chapter 3, Section 11), it follows that the eigenvectors of H_0' are obtained by the *standard Wigner coefficient coupling* of $\Psi^{J,\sigma}_{MK}$ and Φ_{nlm}:[1]

$$|(nM\sigma)(lJ)RM_R\rangle \equiv \sum_{Km} C^{lJR}_{mKM_R}\,\Phi_{nlm} \otimes \Psi^{J,\sigma}_{MK}. \quad (7.10.207)$$

The rotator wave functions $\{\Psi^{J,\sigma}_{MK}\}$ and the harmonic oscillator wave functions $\{\Phi_{nlm}\}$ in this coupled wave function are orthonormal functions in

[1] The subtleties of proper phase factors associated with the coupling of an angular momentum $\mathscr{P}$, satisfying anomalous commutation relations, with an ordinary angular momentum $\mathbf{L}$, and angular momentum "subtraction" in place of "addition," may be avoided completely by keeping all relations in "standard form" (see Van Vleck [48]).

the respective quantum numbers $\{\sigma JMK\}$ and $\{nlm\}$. It follows from this orthonormality that the coupled wave functions themselves are orthonormal. Explicitly, the orthonormality of the various wave functions appearing in Eq. (7.10.207) are given by the following relations:

$$\langle \sigma' J' M' K' | \sigma JMK \rangle \equiv \frac{1}{2} \int d\Omega_{C'} \, \Psi^{J',\sigma'*}_{M'K'}(C') \Psi^{J,\sigma}_{MK}(C') + \frac{1}{2} \int d\Omega_{C'} \, \Psi^{J',\sigma'*}_{M'K'}(-C') \Psi^{J,\sigma}_{MK}(-C') = \delta_{\sigma'\sigma} \delta_{J'J} \delta_{M'M} \delta_{K'K}; \tag{7.10.204'}$$

$$\langle n' l' m' | nlm \rangle \equiv \iiint_{-\infty}^{\infty} d^3\mathbf{q} \, \Phi^{*}_{n'l'm'}(\mathbf{q}) \Phi_{nlm}(\mathbf{q}) = \delta_{n'n} \delta_{l'l} \delta_{m'm}; \tag{7.10.206'}$$

$$\langle (n' M' \sigma')(l' J') R' M'_R | (nM\sigma)(lJ) RM_R \rangle = \delta_{n'n} \delta_{l'l} \delta_{\sigma'\sigma} \delta_{J'J} \delta_{M'M} \delta_{R'R} \delta_{M'_R M_R}. \tag{7.10.207'}$$

The integration in each term in Eq. (7.10.204′) is to be carried out over the set of all *proper* orthogonal matrices $\{C'\}$, where $d\Omega_{C'}$ denotes the (invariant) Haar measure for the group $SO(3)$ in some suitable parametrization;[1] for the Euler angle parametrization obtained by putting $C' = R(\alpha\beta\gamma)$ [see Eq. (2.37), Chapter 2] the measure $d\Omega_{C'}$ is given by $d\Omega_{C'} = d\alpha \, d\gamma \sin\beta \, d\beta$, and the integrals in Eq. (7.10.204′) are evaluated from the definition (7.10.204) of the rotator functions in terms of the matrix elements of the rotation matrix, $\mathscr{D}^J_{MK}(C') = (-1)^J \mathscr{D}^J_{MK}(-C') = D^J_{MK}(\alpha\beta\gamma)$ for $C' = R(\alpha\beta\gamma)$, using the orthogonality relation (3.137), Chapter 3.

The problem of finding the eigenvalues and eigenvectors of the Hamiltonian H'_0 [Eq. (7.10.196)] is thus solved completely by the results given above. The direct application of these results to particular spherical top molecules for other than small n must be approached cautiously, since one cannot always be sure that the perturbation has been treated correctly. Even when the energy

[1] More generally, the inner product, denoted (F, G), of two functions F and G, where each function is defined on the elements R_{ij} of a 3×3 orthogonal matrix R, with values denoted by $F(R)$ and $G(R)$, respectively, is defined by $(F, G) = \frac{1}{2} \int d\Omega_{R'} [F^*(R')G(R') + F^*(-R')G(-R')]$ in which the integration is carried out over all proper orthogonal matrices R'.

eigenvalues (7.10.202) are good first-order approximations, these values are not sufficiently accurate to give agreement with current high-resolution experimental spectra, and higher-order corrections must be taken into account. Electric dipole transitions from rotational sublevels of the ground vibrational state ($n = l = 0$) to rotational sublevels of the first excited vibrational state ($n = l = 1$) of triply degenerate vibrational levels[1] (fundamental band) have been studied in great detail for spherical top molecules such as methane (CH_4) and sulfur hexafluoride (SF_6), showing clearly a richness of fine structure going beyond the predictions of Eq. (7.10.202) (see Sections o–u).

The study of the fine structure of a fundamental ν-band is itself an interesting application of angular momentum techniques. It is useful here to anticipate the electric dipole selection rules for these transitions in order to exhibit the infrared absorption spectrum associated with the energy levels (7.10.202). The frequency σ of an absorption line in cm^{-1} in the ν-band is given by the Bohr frequency relation,

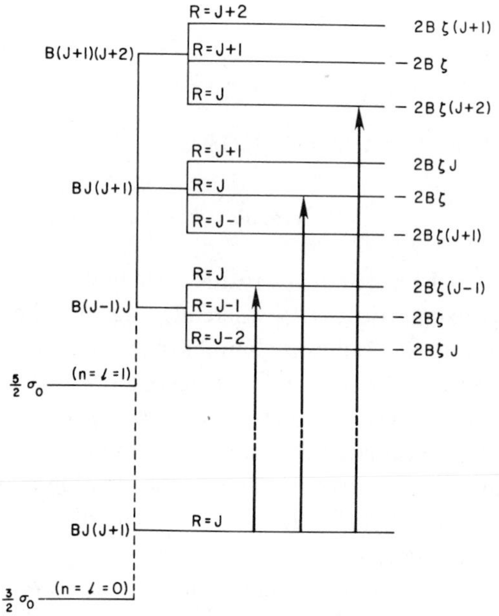

Figure 7.6. Allowed electric dipole transitions from a total angular momentum state J of the ground vibrational state ($n = l = 0$) of a triply-degenerate oscillator to the Coriolis sublevels associated with the angular momentum states $J' = J - 1$ (P-branch line), $J' = J$ (Q-branch line), $J' = J + 1$ (R-branch line) of the first excited vibrational state ($n = l = 1$).

[1] These fundamental bands are usually called ν_3-, ν_4-, etc., bands, the subscript referring to the labeling of the triply degenerate mode, and ν referring to the frequency in cm^{-1}; that is, ν denotes what we have called σ in Eq. (7.10.208).

$$\sigma = (E'_0[1;(1J')R'] - E'_0[0;(0J)R])/hc, \tag{7.10.208}$$

and the selection rules

$$\Delta J = J' - J = 0, \ \pm 1; \quad \Delta R = R' - R = 0. \tag{7.10.209}$$

[This result is refined to include parity in Eq. (7.10.346).]

These results lead to Fig. 7.6, which shows schematically the allowed transitions in a ν-band. The figure shows the ground-state energy $\frac{3}{2}\sigma_0 + BJ(J+1)$, where $\sigma_0 = \hbar\omega/hc$ denotes the frequency in cm^{-1} of the triply degenerate vibration in question, as well as the three upper-state energies $\frac{5}{2}\sigma_0 + BJ'(J'+1)$ corresponding to the allowed values $J' = J-1, J, J+1$ to which transitions can take place; each of these three upper levels ($l = 1$) is, in turn, split into three levels by the Coriolis interaction ($R = J'-1, J', J'+1$), and the Coriolis *correction* to the energy is indicated on the right-hand side of the figure. Some idea as to the scale of various portions of the figure may be obtained by noting the approximate values of the parameters in the case of the ν_3-band of SF_6 (see Section u): $\sigma_0 = 947.977$, $B = 0.0911$, $\zeta = 0.694$.

A ν-band may be divided into three distinct *branches* according to the selection rule on the total angular momentum J. By convention these three branches are called the P-branch ($\Delta J = -1$), the Q-branch ($\Delta J = 0$), and the R-branch ($\Delta J = +1$). The frequencies in cm^{-1} of the absorption lines in each of these branches is calculated from the Bohr frequency relation, Eq. (7.10.208), or may be read off Fig. 7.6 (for $J = 0$ and $J = 1$, the figure must be modified by deleting all levels corresponding to a negative value of J' or R). The results are as follows:

P-branch ($\Delta J = -1$):

$$\sigma_J = (\sigma_0 - 2B\zeta) - 2B(1-\zeta)J, \qquad J = 1, 2, 3, \ldots \tag{7.10.210}$$

Q-branch ($\Delta J = 0$):

$$\sigma_J = \sigma_0 - 2B\zeta, \qquad J = 1, 2, 3, \ldots \tag{7.10.211}$$

R-branch ($\Delta J = +1$):

$$\sigma_J = (\sigma_0 - 2B\zeta) + 2B(1-\zeta)(J+1), \qquad J = 0, 1, 2, \ldots. \tag{7.10.212}$$

(Observe in these results that J refers to the total angular momentum of the ground vibrational state.) Thus, the ν-spectrum consists, in first-order approximation, of a single line situated at $\sigma_0 - 2B\zeta$ (the Q-branch), with lines occurring to either side at equal intervals of $2B(1-\zeta)$; the P-branch lines are on the low-frequency side and are enumerated by $J = 1, 2, \ldots$; the R-branch lines are on the high-frequency side and are enumerated by $J = 0, 1, 2, \ldots$.

We have already noted that the above "ideal spectrum" of a ν-band is not accurate enough for the interpretation of a high-resolution experimental ν-band spectrum of a spherical top molecule – the problem of higher-order corrections must be addressed. Higher-order interactions may be incorporated into the theory in several ways: (1) by the direct method, which calculates H'_2 (and higher-order terms) using the transformation (7.10.183) (Shaffer *et al.* [14]); and (2) by the use of symmetry techniques that allow one to infer the "form" of all such interactions. We turn next to the development of method (2), which requires the use of the point group of a molecule.

o. The point group of a rigid molecule.[1] The set of vectors $\mathscr{A}$ given by Eq. (7.10.17), which defines the equilibrium configuration of the atoms constituting a molecule, plays the key role in the definition of the so-called point group of a rigid (semirigid) molecule. To see how this comes about, consider the partitioning of $\mathscr{A}$ corresponding to sets of identical atoms. Let $\mathscr{A}_k$ denote the subset of $\mathscr{A}$ consisting of position vectors of identical atoms of "type k." Then $\mathscr{A}$ may be written as the union of the disjoint subsets $\{\mathscr{A}_k : k \in K\}$, where K is a set indexing the distinct types of atoms:

$$\mathscr{A} = \bigcup_k \mathscr{A}_k. \tag{7.10.213}$$

The point group G of a molecule with static model given by $\mathscr{A}$ is defined to be

$$G = \{\mathscr{R} : \mathscr{A}_k \to \mathscr{A}_k, \qquad \text{each } k \in K \text{ and } \mathscr{R} \in O(3)\}, \tag{7.10.214}$$

where $O(3)$ denotes the group of rotation–inversions of the space $\mathbb{R}^3$ [Euclidean 3-space with points (x_1, x_2, x_3), which we shall describe using a laboratory frame]. Thus, G is a subgroup of the group, $O(3)$, of rotations and inversion of Euclidean 3-space, $\mathbb{R}^3$, which carries identical atoms into identical atoms in the static model.

We shall use the notation g to denote an element of the point group G.

There are two representations of the group G that play a significant role:

1. The representation of a proper rotation g as a linear transformation of the points of $\mathbb{R}^3$. In vector notation, we have from Eq. (2.1) of Chapter 2

$$g : \mathbb{R}^3 \to \mathbb{R}^3,$$

[1] A more detailed account of the results of this section may be found in Refs. [16] and [49].

$$\vec{x} \to \vec{y} = g\vec{x} = \vec{x}\cos\phi + (\hat{n}\cdot\vec{x})\hat{n}(1-\cos\phi) + (\hat{n}\times\vec{x})\sin\phi, \tag{7.10.215}$$

where g is a positive rotation (right-hand screw rule) by angle ϕ about the direction specified by the unit vector $\hat{n}$. The rotation g may also be represented by the 3×3 proper orthogonal matrix with element in row i and column j given by

$$R_{ij}(g) = \hat{l}_i \cdot g\hat{l}_j. \tag{7.10.216}$$

The inversion $\mathscr{I}$ of the space $\mathbb{R}^3$ is defined by $\mathscr{I}\vec{x} = -\vec{x}$ and is represented by $R(\mathscr{I}) \equiv -\mathbb{1}_3$, where $\mathbb{1}_3$ denotes the 3×3 unit matrix.

2. The representation of g as a linear transformation on the elements of the set $\mathscr{A}$. We may write

$$g: \mathscr{A} \to \mathscr{A}$$

$$[\vec{a}^1\vec{a}^2\cdots\vec{a}^N] \to [\vec{a}^1\vec{a}^2\cdots\vec{a}^N]\,P(g), \tag{7.10.217}$$

where we have ordered the elements of $\mathscr{A}$ and placed them in a $1\times N$ row matrix. The matrix $P(g)$ is then an $N\times N$ permutation matrix.

Observe that the group multiplication properties are satisfied:

$$g'(g\vec{x}) = (g'g)\vec{x}, \qquad \text{each } \vec{x}\in\mathbb{R}^3,$$

$$R(g')R(g) = R(g'g),$$

$$P(g')P(g) = P(g'g), \tag{7.10.218}$$

for all $g, g' \in G$. Thus, the two correspondences

$$g \to R(g) \qquad \text{and} \qquad g \to P(g), \qquad \text{each } g\in G, \tag{7.10.219}$$

are representations of the point group G.

If we denote by A the $3\times N$ matrix,

$$A = \begin{bmatrix} a_1^1 & a_1^2 & \cdots & a_1^N \\ a_2^1 & a_2^2 & \cdots & a_2^N \\ a_3^1 & a_3^2 & \cdots & a_3^N \end{bmatrix}, \tag{7.10.220}$$

then A *intertwines* the representations $\{P(g): g\in G\}$ and $\{R(g): g\in G\}$ of G; that is,

$$R(g)A = AP(g), \qquad \text{each } g\in G. \tag{7.10.221}$$

Relation (7.10.221) *is the key result obtained from the static model of a rigid molecule.*

In order to use the point group G in the description of a molecule in motion, it is necessary to *define* the action of the group G on a generic set $Y = \{\vec{y}^\alpha : \alpha = 1, 2, \ldots, N\}$ of instantaneous translationally invariant position vectors: $\vec{y}^\alpha \equiv \vec{x}^\alpha - \vec{R}$. Consider the set of linear operators

$$L(G) = \{L_g : g \in G\}, \qquad (7.10.222)$$

where $L_g : Y \to Z$ is the linear mapping of a set Y of instantaneous translationally invariant position vectors onto a second set Z of instantaneous translationally invariant position vectors given by

$$L_g : \vec{y}^\alpha \to \vec{z}^\alpha = L_g \vec{y}^\alpha \equiv \sum_\beta (g\vec{y}^\beta) P_{\alpha\beta}(g), \qquad (7.10.223)$$

in which

(1) $\{P(g) : g \in G\}$ is the $N \times N$ permutation matrix representation (7.10.217) of G;

(2) $g\vec{y}^\beta$ is the linear transformation defined for an arbitrary vector $\vec{y}$ by

$$g\vec{y} = \sum_{ij} R_{ij}(g)(\vec{y} \cdot \hat{f}_j)\hat{f}_i, \qquad (7.10.224)$$

where $\{R(g) : g \in G\}$ is the 3×3 orthogonal matrix representation (7.10.216) of G, and the Eckart frame vectors $\hat{f}_i$ ($i = 1, 2, 3$) are those corresponding to position vectors $\vec{y}^1, \ldots, \vec{y}^N$ – that is, $\hat{f}_i = \hat{f}_i(\vec{y}^1, \ldots, \vec{y}^N)$. Observe that $g\vec{y}^\alpha \cdot g\vec{y}^\beta = \vec{y}^\alpha \cdot \vec{y}^\beta$, so that g is a rotation–inversion, and that $g'(g\vec{y}) = (g'g)\vec{y}$.

An alternative expression for the transformation (7.10.223) in terms of the components $\vec{y}^\alpha \cdot \hat{f}_i$ and $(L_g\vec{y}^\alpha) \cdot \hat{f}_i$ relative to the Eckart frame vectors $\hat{f}_i = \hat{f}_i(\vec{y}^1, \ldots, \vec{y}^N)$ is

$$(L_g\vec{y}^\alpha) \cdot \hat{f}_i = \sum_{\beta j} [P(g) \otimes R(g)]_{\alpha i; \beta j} (\vec{y}^\alpha \cdot \hat{f}_j), \qquad (7.10.225)$$

where $P(g) \otimes R(g)$ is the (matrix) direct product of $P(g)$ with $R(g)$, and $[P(g) \otimes R(g)]_{\alpha i; \beta j} = P_{\alpha\beta}(g) R_{ij}(g)$ denotes the element of $P(g) \otimes R(g)$ in row αi and column βj.

The operators $\{L_g : g \in G\}$ satisfy the following relations:

(*i*) $$L_{g'}(L_g\vec{y}^\alpha) = L_{g'g}\vec{y}^\alpha \text{ for arbitrary } \vec{y}^\alpha;$$

(*ii*) $$L_g\vec{u}^\alpha = \vec{u}^\alpha, \qquad \alpha = 1, 2, \ldots, N;$$

$$(iii) \qquad L_g \hat{f}_i = \hat{f}_i, \qquad i = 1, 2, 3, \tag{7.10.226}$$

where (*iii*) denotes the property

$$(L_g \hat{f}_i)(\vec{y}^1, \ldots, \vec{y}^N) \equiv \hat{f}_i(L_g \vec{y}^1, \ldots, L_g \vec{y}^N)$$

$$= \hat{f}_i(\vec{y}^1, \ldots, \vec{y}^N).$$

The proofs of these relations are straightforward and may be found in Ref. [16].

Equation (*i*) means that the correspondence $g \to L_g$ is a *linear representation of* G; Eq. (*ii*) means that G is an *isotropy group* of the set of vectors $\{\vec{u}^\alpha : \alpha = 1, \ldots, N\}$, which define the equilibrium configuration [Eq. (7.10.221) for the static model is essential to the proof of (*ii*)]; and Eq. (*iii*) means that the Eckart frame is *invariant* under the action L_g of G (equivalently, G is an isotropy group of the Eckart frame).

Equations (7.10.226) *are the key relations for establishing the role of the point group in the dynamic molecular model* (*molecule in motion*).

A principal result is as follows: *The group of operators*

$$L(G) = \{L_g : g \in G\} \tag{7.10.227}$$

may be used to split the space of internal coordinates into subspaces that transform irreducibly under $L(G)$.

Proof. Let $\{\vec{\rho}^\alpha\}$ denote any set of displacement vectors for the frame $(\hat{f}_1, \hat{f}_2, \hat{f}_3)$. Since $L_g \vec{y}^\alpha = \vec{u}^\alpha + L_g \vec{\rho}^\alpha$, it follows that the set of vectors

$$\{L_g \vec{\rho}^\alpha : \alpha = 1, 2, \ldots, N\} \tag{7.10.228}$$

is also a set of displacement vectors for the frame $(\hat{f}_1, \hat{f}_2, \hat{f}_3)$.

Now choose any basis set $\{q_\mu : \mu = 1, \ldots, 3N - 6\}$ for the internal coordinates (see Section i). Since

$$(L_g \vec{\rho}^\alpha) \cdot \hat{f}_i = \sum_{\beta j} [P(g) \otimes R(g)]_{\alpha i; \beta j} \rho_j^\beta, \tag{7.10.229}$$

we find from Eqs. (7.10.66) and (7.10.67) that the internal coordinates $\{q'_\mu\}$ corresponding to the displacement vectors $(\vec{\rho}^\alpha)' \equiv L_g \vec{\rho}^\alpha$ $(\alpha = 1, \ldots, N)$ are

$$q'_\mu = L_g q_\mu \equiv \sum_\alpha \vec{q}_\mu^{\,\alpha} \cdot (L_g \vec{\rho}^\alpha) = \sum_\nu M_{\mu\nu}(g) q_\nu, \tag{7.10.230}$$

where

$$M_{\mu\nu}(g) = \sum_{\alpha i \beta j} m_\beta^{-1} q_{\mu i}^\alpha q_{\nu j}^\beta [P(g) \otimes R(g)]_{\alpha i;\beta j}. \tag{7.10.231}$$

The group property,[1] $L_{g'}L_g = L_{g'g}$, and the basis property, $\sum_\mu a_\mu q_\mu = \sum_\mu b_\mu q_\mu$ implies $a_\mu = b_\mu$, together imply that the set of matrices (dimension $3N - 6$)

$$\{M(g): g \in G\} \tag{7.10.232}$$

is a representation of G. The complete reduction of this representation into irreducible representations of G then defines internal coordinates of the form

$$q_\gamma^\Gamma = \sum_\mu a_{\gamma\mu}^\Gamma q_\mu, \qquad \gamma = 1, 2, \ldots, \dim \Gamma \tag{7.10.233}$$

that are transformed according to irreducible representation Γ of G under the action L_g of G. ■

We see from the above discussion that the group of operators $L(G)$ solves fully the problem of classifying the internal coordinates according to their transformation properties under the irreducible representations of the point group G.

We assume henceforth that all internal coordinates are normal coordinates that have been partitioned into sets $\{q_\gamma^\Gamma : \gamma = 1, \ldots, \dim \Gamma\}$ that transform irreducibly under the action of $L(G)$:

$$L_g : \mathbf{q}^\Gamma \to \Gamma(q)\mathbf{q}^\Gamma, \tag{7.10.234}$$

where $\mathbf{q}^\Gamma = \text{col}(q_1^\Gamma, q_2^\Gamma, \ldots)$, and $\Gamma = \{\Gamma(g): g \in G\}$ is an irreducible representation of G. The conjugate linear momenta will then also possess this transformation property:

$$L_g : \mathbf{p}^\Gamma \to \Gamma(g)\mathbf{p}^\Gamma. \tag{7.10.235}$$

The Hamiltonian H given by (7.10.165) is a function of the normal coordinates $\{q_\mu\}$ (now to be considered as re-indexed in terms of the notation q_γ^Γ), the conjugate linear momenta $\{p_\mu\}$ (similarly re-indexed), and the

---

[1] The verification of the group action $L_{g'}(L_g q_\mu) = L_{g'g} q_\mu$, where $\{q_\mu : \mu = 1, 2, \ldots, 3N - 6\}$ is an arbitrary set of internal coordinates, is given by $L_{g'}(L_g q_\mu) = \sum_\alpha \vec{q}_\mu^{\,\alpha} \cdot (L_{g'}(L_g \vec{\rho}^{\,\alpha})) = \sum_\alpha \vec{q}_\mu^{\,\alpha} \cdot (L_{g'g} \vec{\rho}^{\,\alpha}) = L_{g'g} q_\mu$. Correspondingly, the right-hand side of Eq. (7.10.230) undergoes the transformations: $q_\mu \to q'_\mu = \sum_\nu M_{\mu\nu}(g) q_\nu$ followed by $q'_\mu \to q''_\mu = \sum_{\nu'} M_{\mu\nu'}(g') q'_{\nu'} = \sum_\nu M_{\mu\nu}(g'g) q_\nu$.

components $K_i = -(\det C)\hat{f}_i \cdot \vec{J}$ of the total angular momentum $\vec{J}$. To determine the transformation properties of H under the action (7.10.223) of $L(G)$, we must still determine how $\mathbf{K}$ is transformed.

Using $\vec{J} = \sum_\alpha m_\alpha \vec{y}^\alpha \times \vec{p}^\alpha$, where $\vec{p}^\alpha$ is the linear momentum conjugate to the position vector $\vec{y}^\alpha = \vec{x}^\alpha - \vec{R}$, we find from Eqs. (7.10.223) and (7.10.224) that

$$L_g: \vec{J} \to \det R(g) \sum_{ij} (\vec{J} \cdot \hat{f}_i) R_{ij}(g) \hat{f}_i, \tag{7.10.236}$$

where we have used the fact that the Eckart frame is invariant under the action of L_g, and $R(g)$ is the matrix (7.10.216) representing $g \in G$. Thus, we find

$$L_g: K_i \to \det R(g) \sum_j R_{ij}(g) K_j. \tag{7.10.237}$$

If we define $\mathbf{K} = \mathrm{col}(K_1, K_2, K_3)$, then this result takes the following form in matrix notation:

$$L_g: \mathbf{K} \to R'(g)\mathbf{K}, \tag{7.10.238}$$

where $R'(g) = [\det R(g)]R(g)$ is the 3×3 proper orthogonal matrix representation of G obtained by multiplying the "vector" representation $R(g)$ of Eq. (7.10.216) by $\det R(g)$. [The angular momentum $\mathbf{K}$ behaves like an axial vector under the action of the group $L(G)$.]

We may now state the fundamental assumption that is made in the theory of rigid-body molecular dynamics: *The Hamiltonian H given by Eq.* (7.10.154) *is invariant under the action of the group L(G).*

This result was considered to be intuitively evident by the early spectroscopists, but its actual justification must be based on the validity of the physical approximations underlying the concept of the Eckart frame. In order to appreciate this, one must first note the clear distinction that exists between a "first principles" molecular Hamiltonian and the Eckart Hamiltonian of Eq. (7.10.154). In any Hamiltonian based on first principles, invariance under rotation-inversions is a priori valid (in the molecular physics realm of nonrelativistic quantum mechanics) as is permutational symmetry based on exchanging identical particles. Thus, for example, for the sulfur-hexaflouride molecule, SF_6, one has, a priori, the 720 element group S_6 for permuting the six (identical) flourine atoms. This is not the case for the Eckart Hamiltonian (7.10.154), which is limited to the "internal" rotation-inversion group O_h having 48 elements (octahedral group with inversions). Thus, the symmetry group of transformations $L(G)$ is distinctly smaller than the abstract (a priori) group of permutational symmetries.

The Hamiltonian for the Eckart model of a rigid molecule that is given by Eq. (7.10.154) is, accordingly, not, in general, invariant under the (a priori) permutational symmetry of identical atoms; this symmetry breaking results from the choice of a phenomenological potential energy function that keeps the (distinguished) atoms essentially at their equilibrium positions.

In the molecular model that we are describing, one assumes that the potential energy function V of a rigid molecule is the most general function of the normal coordinates[1] $\{q_\mu = 1, \ldots, 3N-6\}$ that is analytic in the neighborhood of the equilibrium configuration such that

(1) Equilibrium conditions:

$$\left.\frac{\partial V}{\partial q_\mu}\right|_{\text{equilibrium}} = 0;$$

(2) Invariance conditions:

$$V(L_g q_1, \ldots, L_g q_{3N-6}) = V(q_1, \ldots, q_{3N-6}), \qquad \text{each } g \in G. \tag{7.10.239}$$

Thus, the problem of determining which atoms are treated identically in this model is that of determining which permutations of atoms leave $V(q_1, \ldots, q_{3N-6})$ invariant. We shall see that this defines a subgroup of S_N, denoted by $P(G)$ below, that is isomorphic to $L(G)$.

In order to examine the above problem, we require a careful definition of "permutations of coordinates of identical particles" and the resulting properties of the internal coordinates under such permutations.

Let there be n_k identical atoms of type k labeled by distinct integers

$$\alpha_1(k), \alpha_2(k), \ldots, \alpha_{n_k}(k),$$

and let the position vectors of these atoms in the laboratory frame be the vectors in the set X_k given by

$$X_k = \{\vec{x}^\alpha : \alpha = \alpha_1(k), \ldots, \alpha_{n_k}(k)\}, \tag{7.10.240}$$

where $\vec{x}^\alpha$ is the position vector of the atom labeled α. *A permutation P_k of the set of position vectors of the identical atoms of type k is a mapping of X_k onto X_k.* The product of two permutations, P'_k and P_k, is then the usual composition of mappings. The set of all such mappings then forms a group isomorphic to the symmetric group S_{n_k}. Indexing the "types" of identical atoms by $k = 1, 2, \ldots, m$ $(\sum_{k=1}^{m} n_k = N)$, we see that $P_k P_{k'} = P_{k'} P_k$ $(k' \neq k)$ on the set

[1] The $\{q_\mu\}$ may, in fact, be any set of internal coordinates in the discussion that follows.

$X = \bigcup X_k$ and that each permutation $P: X \to X$ of position vectors of identical atoms of the molecule has the form

$$P = \prod_{k=1}^{m} P_k, \qquad P_k \in S_{n_k}. \tag{7.10.241}$$

Consider next the linear mapping L_g defined by Eq. (7.10.223). Then the linear mapping $P_g: Y \to Y$ defined by

$$P_g = gL_{g^{-1}}, \qquad \text{each } g \in G \tag{7.10.242}$$

is a permutation of the set of translationally invariant position vectors of identical atoms:

$$P_g: \vec{y}^\alpha \to P_g\vec{y}^\alpha = \sum_\beta \vec{y}^\beta P_{\beta\alpha}(g). \tag{7.10.243}$$

Furthermore, the set of permutations that map Y onto Y given by

$$P(G) = \{P_g: g \in G\} \tag{7.10.244}$$

forms a group under composition of mappings; hence,

$$P_{g'}(P_g\vec{y}^\alpha) = P_{g'g}\vec{y}^\alpha. \tag{7.10.245}$$

The action of the group of permutations $P(G)$ on the equilibrium position vectors $\{\vec{u}^\alpha: \alpha = 1, \ldots, N\}$ and on the set of displacement vectors $\{\vec{\rho}^\alpha: \alpha = 1, \ldots, N\}$ is given by

$$P_g\vec{u}^\alpha = g\vec{u}^\alpha = \sum_\beta \vec{u}^\beta P_{\beta\alpha}(g), \tag{7.10.246}$$

$$P_g\vec{\rho}^\alpha = gL_{g^{-1}}\vec{\rho}^\alpha = \sum_\beta \vec{\rho}^\beta P_{\beta\alpha}(g). \tag{7.10.247}$$

Observe then that Eq. (7.10.245) is also valid when $\vec{y}^\alpha$ is replaced by $\vec{u}^\alpha$ or $\vec{\rho}^\alpha$.

The following three useful properties of the permutations (7.10.244) may also be derived from the definition (7.10.242) or (7.10.243):

(1) $$\hat{f}_i(P_g\vec{y}^1, \ldots, P_g\vec{y}^N) = g[\hat{f}_i(\vec{y}^1, \ldots, \vec{y}^N)]. \tag{7.10.248}$$

This result states that the Eckart frame corresponding to the permuted position vectors $P_g\vec{y}^1, \ldots, P_g\vec{y}^N$ is the rotation-inversion g of the frame F corresponding to the position vectors $\vec{y}^1, \ldots, \vec{y}^N$.

(2)
$$P_g q_\mu = L_{g^{-1}} q_\mu. \tag{7.10.249}$$

In this relation, $P_g q_\mu$ is defined to be the result obtained from q_μ by applying the permutation (7.10.243) to the displacement vectors $\{\vec{\rho}^\alpha\}$. Thus,

$$P_g q_\mu \equiv \sum_{\alpha i} q^\alpha_{\mu i}(g\hat{f}_i) \cdot (P_g\vec{\rho}^\alpha)$$

$$= \sum_{\alpha i} q^\alpha_{\mu i}\hat{f}_i \cdot (g^{-1}P_g\vec{\rho}^\alpha) = \sum_\alpha \vec{q}^\alpha_\mu \cdot L_{g^{-1}}\vec{\rho}^\alpha = L_{g^{-1}} q_\mu.$$

(3)
$$P_{g'}(P_g q_\mu) = P_{g'g} q_\mu = (g^{-1}L_{g'^{-1}}g)(L_{g^{-1}}q_\mu). \tag{7.10.250}$$

This last relation is proved as follows:

$$P_{g'}(P_g q_\mu) = \sum_{\alpha i} q^\alpha_{\mu i}\,[g'(g\hat{f}_i)] \cdot [P_{g'}(P_g\vec{\rho}^\alpha)] = P_{g'g} q_\mu$$

$$= \sum_\alpha \vec{q}^\alpha_\mu \cdot g^{-1}g'^{-1}P_{g'}(P_g\vec{\rho}^\alpha) = \sum_\alpha \vec{q}^\alpha_\mu \cdot (g^{-1}L_{g'^{-1}}g)(L_{g^{-1}}\vec{\rho}^\alpha)$$

$$= (g^{-1}L_{g'^{-1}}g)(L_{g^{-1}}q_\mu).$$

We can now state the important result on the permutation symmetry of the potential energy function $V(q_1, \ldots, q_{3N-6})$ of a rigid molecule: *The potential energy function $V(q_1, \ldots, q_{3N-6})$ is invariant under the group $P(G)$*:

$$V(P_g q_1, \ldots, P_g q_{3N-6}) = V(q_1, \ldots, q_{3N-6}), \qquad \text{each } g \in G. \tag{7.10.251}$$

Proof. See Eqs. (7.10.249) and (7.10.239). ■

It follows from Eq. (7.10.248) and the invariance of H to rotations and inversion of $\mathbb{R}^3$, and from Eq. (7.10.251), that *the Eckart model Hamiltonian* (7.10.154) *for a rigid molecule is invariant under the action of the group $P(G)$*. It is this result that allows us to conclude: *Atoms that are treated as being identical in this model are precisely those that are transformed into one another by the action of the point group G on the static model.* [This is an important result for calculating the so-called statistical weight of a level (see Section t).]

It is clear from the above results that one could have developed the symmetry properties of the rigid-molecule Hamiltonian by starting with an appropriate group of permutations of identical particles,[1] ignoring the point group concept. A principal problem is that of deciding which atoms are to be treated as identical in the model (the problem of selecting a potential energy function that predicts with suitable accuracy the phenomenon in question). An advantage of using $P(G)$ is that it accords more closely with our notions of the basic symmetries of nature; an advantage of using $L(G)$ is that it is a true invariance group of the internal motions [$L(G)$ leaves the Eckart frame invariant, whereas the isomorphic group $P(G)$ does not [see Eq. (7.10.248)].

With this brief description of the invariance properties of the Hamiltonian of a rigid molecule, we now return to the problem of second-order corrections to the Hamiltonian (7.10.196).

p. Higher-order corrections: phenomenological Hamiltonian. Let us first observe that the Hamiltonian (7.10.196) is invariant under the transformations[2]

$$\mathbf{K} \to (\det R)R\mathbf{K}, \qquad \mathbf{q} \to R\mathbf{q}, \qquad \text{each } R \in O(3), \qquad (7.10.252)$$

since the second transformation implies also

$$\mathbf{L} \to (\det R)R\mathbf{L}. \qquad (7.10.253)$$

These results are to be compared with the transformations $L(G)$ induced by the point group G:

$$\mathbf{K} \to [\det R(g)]R(g)\mathbf{K},$$

$$\mathbf{q} \to A(g)R(g)\mathbf{q}, \qquad \text{each } g \in G, \qquad (7.10.254)$$

where $\{A(g): g \in G\}$ is a one-dimensional irrep of G, with $A(g) = \pm 1$. In obtaining the second result we have used the fact that the vector representation $\{R(g): g \in G)\}$ is irreducible for each point group G of a spherical top molecule ($G = O_h, O, T_d, T_h, T$; see footnote, p. 589), and, furthermore, under the action of $L(G)$ the normal coordinates of each triply degenerate vibration transform according to an irreducible representation $\{\Gamma(g) = A(g)R(g)\}$ of G [see Ref. [12], Appendix A).

---

[1] This viewpoint has been developed and extended to nonrigid molecules, principally by Longuet-Higgens [50], Hougen [51, 52], and Bunker [53, 54]. A somewhat more general viewpoint has also been developed by Harter *et al.* [55], Harter [55a], and Bauder *et al.* [55b].

[2] Observe that these "internal" rotation–inversions are independent of rotation–inversions of the molecule relative to a laboratory frame (the two types of rotations commute). This property is discussed in greater detail in Ref. [16].

The group–subgroup structure observed above,

$$O(3) \supset G, \tag{7.10.255}$$

is significant, for it allows us to characterize definitively the perturbation problem for the Hamiltonian (7.10.196) in terms of the general concepts of level splitting associated with group–subgroup restriction (see Refs. [56–58]). *All higher-order interaction terms to the Hamiltonian* H'_0 *are polynomial operators in* **K**, **q**, *and* **p** *that may be classified as irreducible tensor operators under the action* (7.10.252) *of* $O(3)$ *and invariants under the action* (7.10.254) *of* G.

Let us now tabulate interactions that can contribute to the first-order vibration–rotation energy spectrum of a triply degenerate vibration as described by the Hamiltonian H'_0 [Eq. (7.10.196)].

We shall make the simplifying assumption that different vibrational levels (different n) are far enough apart so that only those matrix elements of the perturbation that are diagonal in n will be important. In this case one has first the general result[1] for harmonic oscillator matrix elements:

$$\langle nl'm'|q_i^s|nlm\rangle = \langle nl'm'|(p_i/\hbar)^s|nlm\rangle \tag{7.10.256}$$

for $s = 1, 2, \ldots$, where, in fact, the matrix elements for odd s are zero. The second general result, pertaining again to harmonic oscillator matrix elements, is as follows: If $P(\mathbf{q}, \mathbf{p})$ is an $O(3)$-invariant *polynomial*, then the matrix element $\langle nl'm'|P(\mathbf{q}, \mathbf{p})|nlm\rangle$ has the form $\delta_{l'l}\delta_{m'm}\langle nl\|P'(h_{\text{osc}}, \mathbf{L}^2)\|nl\rangle$, where P' is a polynomial in the harmonic oscillator Hamiltonian,

$$h_{\text{osc}} \equiv \tfrac{1}{2}(\mathbf{p}\cdot\mathbf{p}/\hbar^2 + \mathbf{q}\cdot\mathbf{q}), \tag{7.10.257}$$

and in the angular momentum, $\mathbf{L}^2$.

Using these two results, let us next list all $O(3)$-invariant operators, up to degree 5, that have independent matrix elements diagonal in n. In the degree classification, one counts each q_i, p_i, or K_i to be of degree 1, and L_i to be of degree 2. Furthermore, it is necessary to keep all invariants of lower degree, since terms like $q'^2 h_{\text{osc}}$, where q' is a normal coordinate of another mode of vibration, contribute a correction, $(\#)h_{\text{osc}}$, to the Hamiltonian (7.10.196). In order to indicate this latter property, a symbolic λ^r $(r = 0, 1, \ldots)$ multiplies a given invariant in the list below, so that r is to be added to the degree of the invariant in **q** and **K** in obtaining the total degree in all normal coordinates and momenta from which the invariant originated.

---

[1] Because of this relationship between matrix elements, it is not necessary below to consider tensor operators in **p** alone.

$O(3)$-invariant interactions up to degree 5:

Degree 4: $\lambda^2 h_{\text{osc}}, \lambda^2 \mathbf{J}^2, h_{\text{osc}}^2, \mathbf{J}^4, h_{\text{osc}}\mathbf{J}^2, \mathbf{L}^2,$

$$T_0^{(22)0} \equiv \frac{2}{15}\sqrt{4\pi}\sum_{\mu}(-1)^{\mu}\mathscr{Y}_{2,\mu}(\mathbf{q})\mathscr{T}_{2,\mu}(\mathbf{K}); \tag{7.10.258}$$

Degree 5: $\lambda^2 \mathbf{K}\cdot\mathbf{L}, (\mathbf{K}\cdot\mathbf{L})\mathbf{J}^2, (\mathbf{K}\cdot\mathbf{L})h_{\text{osc}}.$

In the definition of $T_0^{(22)0}$, the solid harmonics in $\mathbf{q}$ are denoted by $\mathscr{Y}_{k\mu}$, and the tensor harmonics in $\mathbf{K}$ by $\mathscr{T}_{k\mu}$ (see Tables 4 and 5 in the Appendix of Tables). (One replaces J_i^{λ} by $\hbar^{k-\lambda}K_i^{\lambda}$ to restore $\hbar$ to the tensor harmonics listed in Table 5.)

It is not obvious that the list (7.10.258) is complete, and one must show, for example, that the $O(3)$-invariant $(\mathbf{q}\cdot\mathbf{K})^2$ is a linear combination of $\mathbf{q}^2\mathbf{J}^2$ and $T_0^{(22)0}$, noting that $\mathbf{q}^2\mathbf{J}^2$ is accounted for by $h_{\text{osc}}\mathbf{J}^2$ in the list (7.10.258).

It is also necessary to consider $O(3)$-irreducible tensor operators, which are invariant under the action of the point group G (G-invariants). Up to degree 5 there are only two generic G-invariants to consider for the cubic groups[1] $G = T, T_h, T_d, O,$ or O_h. These are of the form[2] (Jahn [44])

$$G^3 = T_2^3 - T_{-2}^3, \tag{7.10.259}$$

$$G^4 = \frac{5}{4}\left(\sqrt{\frac{14}{5}}\,T_0^4 + T_4^4 + T_{-4}^4\right), \tag{7.10.260}$$

where T_{μ}^k is an $O(3)$-irreducible tensor operator.

For the case at hand, the tensor operators T_{μ}^k are all of the coupled form [see Eq. (3.233) of Chapter 3] given by

---

[1] The cubic groups O_h, O, T_d, T_h, and T may be described succinctly in terms of groups of transformations of an arbitrary vector $\mathbf{x} = (x_1, x_2, x_3) \in \mathbb{R}^3$ that correspond to rotation-inversions of the regular cube into itself. The group O_h is isomorphic to the group of forty-eight transformations $(x_1, x_2, x_3) \to (\sigma_1 x_{i_1}, \sigma_2 x_{i_2}, \sigma_3 x_{i_3})$, where (i_1, i_2, i_3) is any permutation of $(1, 2, 3)$ and each σ_i may be $+1$ or -1. The groups O, T_d, T_h, and T are isomorphic to subgroups of this O_h-transformation group obtained by imposing additional conditions: O contains those elements having (i_1, i_2, i_3) even and $\sigma_1\sigma_2\sigma_3 = +1$ or (i_1, i_2, i_3) odd and $\sigma_1\sigma_2\sigma_3 = -1$ (twenty-four elements); T_d contains those elements having $\sigma_1\sigma_2\sigma_3 = +1$ (twenty-four elements); T_h contains those elements having (i_1, i_2, i_3) even (twenty-four elements); and T contains those elements having (i_1, i_2, i_3) even and $\sigma_1\sigma_2\sigma_3 = +1$ (twelve elements). These faithful irreducible representations of the cubic groups by 3×3 orthogonal matrices are called vector representations.

[2] The factor 5/4 in G^4 has been chosen to obtain the agreement with Hecht [47] given by Eqs. (7.10.268) below, using the definitions of tensor operators given by Eqs. (7.10.265).

$$T_\mu^k = T_\mu^{(k_1k_2)k} = \sum_{\mu_1\mu_2} C^{k_1k_2k}_{\mu_1\mu_2\mu}\, T_{k_1\mu_1}(\text{vib})\, T_{k_2\mu_2}(\text{rot}). \quad (7.10.261)$$

In this expression, $T_{k\mu}(\text{rot})$ denotes the $O(3)$-tensor operator,

$$T_{k\mu}(\text{rot}) \equiv [k!k!/(2k)!]^{\frac{1}{2}}\, \mathscr{T}_{k\mu}(\mathbf{K}). \quad (7.10.262)$$

For this part of the discussion, $T_{k\mu}(\text{vib})$ denotes the $O(3)$-tensor operator,

$$T_{k\mu}(\text{vib}) \equiv (-2)^k \sqrt{\frac{4\pi}{2k+1}} \left[\frac{k!k!}{(2k)!}\right]^{\frac{1}{2}} \mathscr{Y}_{k\mu}(\mathbf{q}), \quad (7.10.263)$$

but will later be generalized to include more general irreducible tensor operators constructed from vibrational quantities. [The normalization in the definitions (7.10.262) and (7.10.263) is such that $T_{kk}(\text{rot}) = (-K_+)^k$ and $T_{kk}(\text{vib}) = q_+^k$. This normalization then gives the agreement with the tensor operators introduced by Hecht [47], as shown by Eqs. (7.10.268) below.]

The G-invariant linear combination (7.10.260) of these $O(3)$-tensor operators is now denoted by

$$G^{(k_1k_2)4} \equiv \frac{5}{4}\left(\sqrt{\frac{14}{5}}\, T_0^{(k_1k_2)4} + T_4^{(k_1k_2)4} + T_{-4}^{(k_1k_2)4}\right). \quad (7.10.264)$$

The cubic interaction term, $T_2^3 - T_{-2}^3$, constructed from the tensor operators $T_\mu^{(21)3}$ defined by Eqs. (7.10.261)–(7.10.263) (these are the only cubic terms that are quadratic in $\mathbf{q}$), cannot occur in the Hamiltonian (7.10.192). This result may be seen from the transformation (7.10.254): In order to construct such a G-invariant, we must construct three quantities, $\Lambda_k = \sum_{ij} a_{ij}^{(k)} q_i q_j$ $(k = 1, 2, 3)$, quadratic in the components of $\mathbf{q}$, such that $\mathbf{\Lambda}$ transforms as a vector under the action of G (transforms in the same way as $\mathbf{K}$). Then, and only then, can we construct a G-invariant from $\mathbf{\Lambda}$ and $\mathbf{K}$. But the unique quantity $\mathbf{\Lambda}$ with the required transformation property is $\mathbf{\Lambda} = \mathbf{q} \times \mathbf{q} = \mathbf{0}$.

The fourth-degree tensor operators (7.10.261) that are of even degree in $\mathbf{q}$ and that define G-invariant interactions (7.10.264) are

$$T_\mu^{(40)4} = \frac{16}{3}\sqrt{\frac{4\pi}{70}}\, \mathscr{Y}_{4,\mu}(\mathbf{q}),$$

$$T_\mu^{(04)4} = \frac{1}{\sqrt{70}}\, \mathscr{T}_{4,\mu}(\mathbf{K}),$$

$$T_{\mu}^{(22)4} = \frac{2}{3}\sqrt{\frac{4\pi}{5}} \sum_{\mu_1\mu_2} C^{224}_{\mu_1\mu_2\mu} \mathscr{Y}_{2,\mu_1}(\mathbf{q})\mathscr{T}_{2,\mu_2}(\mathbf{K}). \qquad (7.10.265)$$

The fifth-degree tensor operators of even degree in **q** and **p** that define G-invariant interactions are again obtained from Eq. (7.10.261). There are two such operators, denoted by $T_{\mu}^{(13)4}$ and $T_{\mu}^{(31)4}$, and defined in terms of Eq. (7.10.261) by choosing

$$T_{1,\mu_1}(\text{vib}) \equiv -\frac{1}{\sqrt{2}}\mathscr{T}_{1,\mu_1}(\mathbf{L}),$$

$$T_{3,\mu_1}(\text{vib}) \equiv -2\sqrt{\frac{4\pi}{15}} \sum_{\nu_1\nu_2} C^{123}_{\nu_1\nu_2\mu_1} \mathscr{T}_{1,\nu_1}(\mathbf{L})\mathscr{Y}_{2,\nu_2}(\mathbf{q}). \qquad (7.10.266)$$

The G-invariant operators $G^{(13)4}$ and $G^{(31)4}$ are then defined by Eq. (7.10.264) in terms of $T_{\mu}^{(13)4}$ and $T_{\mu}^{(31)4}$.

Again, one must prove that the five G-invariant interactions

$$\{G^{(k_1k_2)4}: (k_1k_2) = (40), (04), (22), (13), (31)\} \qquad (7.10.267)$$

include all $O(3)$-tensor interactions, up to degree 5,[1] that have independent matrix elements diagonal in n. (We omit this proof, which may be given by exhausting all cases.)

The G-invariant interactions (7.10.267) have been chosen such that the following relations to Hecht's [47] definitions are valid:

$$O_{33}(\text{tensor}) = G^{(40)4},$$

$$O_{PPPP}(\text{tensor}) = G^{(04)4},$$

$$O_{PPP3}(\text{tensor}) = G^{(13)4},$$

$$O_{PP33}(\text{tensor}) = G^{(22)4},$$

$$O_{P333}(\text{tensor}) = G^{(31)4}. \qquad (7.10.268)$$

(Hecht's tensor operators are written out in terms of the components q_i, L_i, and $K_i = -\mathscr{P}_i$. This form is very useful for the explicit verification of their invariance under G.)

---

[1] The Hamiltonian (7.10.154) is invariant under the time-reversal transformation $\hat{f}_i \to \hat{f}_i$, $\mathbf{q} \to \mathbf{q}$, $\mathbf{p} \to -\mathbf{p}$, $\mathbf{K} \to -\mathbf{K}$, and this invariance precludes interaction terms that contain tensors of the type $\mathbf{T}^{(23)4}$ that are of odd degree in **K**.

The effective rotational parameters of a vibration–rotation band[1] are determined by the scalar perturbation. For the Hamiltonian H'_0 [Eq. (7.10.196)], the scalar perturbation (to the approximation being considered) is a linear combination of the $O(3)$-invariant interactions (7.10.258), using real numerical coefficients. The effect of this scalar perturbation is to produce a slight shift in the first-order levels given by Eq. (7.10.202) – that is, $E'_0[n(lJ)R]$ is replaced by[2]

$$\frac{E_s[n(lJ)R]}{hc} = \frac{E_{nl}(\text{vib})}{hc} + \left[B' + e\left(n + \frac{3}{2}\right)\right]J(J+1)$$

$$+ \left[(B\zeta)' + aJ(J+1) + b\left(n + \frac{3}{2}\right)\right]$$

$$\times\ [R(R+1) - J(J+1) - l(l+1)]$$

$$-\ d[J(J+1)]^2 + f\langle n(lJ)R\|T_0^{(22)0}\|n(lJ)R\rangle,$$

$$\frac{E_{nl}(\text{vib})}{hc} = \sigma'\left(n + \frac{3}{2}\right) + g\left(n + \frac{3}{2}\right)^2 + ul(l+1). \tag{7.10.269}$$

In these two equations σ', B', and $(B\zeta)'$ denote shifted values (by small amounts) of the parameters $\sigma = \omega/2\pi c$, B, and $B\zeta$ occurring in the first-order energy expression (7.10.202), these shifts being due to the $O(3)$-invariant interactions $\lambda^2 h_{\text{osc}}$, $\lambda^2\mathbf{J}^2$, and $\lambda^2(\mathbf{K}\cdot\mathbf{L})$ arising from ground-state matrix elements of other modes of vibration. The parameters a, b, d, e, f, g, and u are the coefficients of the remaining $O(3)$-invariant interactions (7.10.258) and are presumably small compared to the appropiate σ', B', or $(B\zeta)'$.

---

[1] For our purpose a vibration absorption *band* may be defined to be the set of all observable frequencies of radiation that a molecule absorbs in making transitions from any rotational substate of a specified pure vibrational energy state E_{nl}(vib) to any rotational substate of a second specified pure vibrational energy state $E_{n'l'}$(vib) $(n' > n)$.

[2] These energy levels are independent of the quantum numbers M and σ in consequence of the invariance of the general Hamiltonian to rotation-inversions of Euclidean three-space; they are independent of the quantum number M_R in consequence of the invariance of the *approximate Hamiltonian*, H'_0 [Eq. (7.10.196)] plus linear combinations of the $O(3)$-invariants (7.10.258), to the *internal* $O(3)$ transformations (7.10.252). It is this latter internal symmetry (corresponding to the degeneracy of the internal quantum number M_R) that is broken to the level of the point group G by the $O(3)$-tensor, G-invariant interactions (see Section q). We emphasize again that "internal symmetries" are consequences of the model and are sometimes called approximate symmetries – they are usually inferred from empirical properties of specific physical systems. (We often suppress the labels σ, M, and M_R in the notation for matrix elements.)

The scalar energy term becomes explicit with the evaluation of the diagonal matrix elements of the $O(3)$-invariant $T_0^{(22)0}$ in the coupled representation (7.10.207). This is a standard application of Eq. (3.260) of Chapter 3. It is, however, more instructive to observe that the diagonal matrix elements[1] of $T_0^{(22)0}$ are the same as those of the $O(3)$-invariant defined by

$$T_0^{(22)0}(\text{eff}) \equiv -\frac{8}{15} h_{\text{osc}} \frac{[3\mathbf{K}\cdot\mathbf{L}(2\mathbf{K}\cdot\mathbf{L}+1) - 2\mathbf{J}^2\mathbf{L}^2]}{4\mathbf{L}^2 - 3}. \quad (7.10.270)$$

We thus find yet another type of invariant operator making its appearance: The numerator factor in Eq. (7.10.270) is the "polynomial part" of the Racah invariant operator

$$\begin{Bmatrix} & 2 & \\ 4 & & 0 \\ & 2 & \end{Bmatrix}$$

(see Section 21 of Chapter 3 and the operator P_0^2 in Table 7 in the Appendix of Tables). [The result, Eq. (7.10.270), may be verified by direct computation, using Eq. (3.260) from Chapter 3, the reduced matrix elements given in Table 8 in the Appendix of Tables, and the appropriate 6-j coefficient from Table 2.]

Remark. The general form of the scalar part of the Hamiltonian may be given. As we have discussed at the end of Section 9 (see also Chapter 3, Section 21), the set of commuting Hermitian Racah operators,

$$\left\{\begin{Bmatrix} & k & \\ 2k & & 0 \\ & k & \end{Bmatrix} : k = 0, 1, \ldots, 2R\right\},$$

is a basis of all invariant operator mappings (operators commuting with $\mathbf{R}$) of the space spanned by the orthonormal basis $\{|(nM\sigma)(lJ)RM_R\rangle : R = |l - J|, \ldots, l + J\}$ into itself. This result implies that the most general form of the scalar Hamiltonian is

$$H_{\text{scalar}} = \sum_{k=0}^{2R} A_k(h_{\text{osc}}, \mathbf{L}^2, \mathbf{J}^2) \begin{Bmatrix} & k & \\ 2k & & 0 \\ & k & \end{Bmatrix},$$

[1] The invariant $T_0^{(22)0}$ also has off-diagonal matrix elements in l (the final state may have $l' = l + 2, l, l - 2$), but the possibility of mixing of states of the same n and different l will not be considered here.

where A_k is, in general, a rational function of the operators h_{osc}, $\mathbf{L}^2$, and $\mathbf{J}^2$. (Observe that, since the Racah operators above effect no shift in the labels n, l, and J, they commute with h_{osc}, $\mathbf{L}^2$, and $\mathbf{J}^2$.) To our knowledge this approach to molecular spectroscopy, using Racah operators, is not in the literature.

We shall not consider further the details of the distortions of the relatively simple first-order spectrum (7.10.202) that are introduced by the $O(3)$-invariant interactions above, although this is an important problem for sorting out the values of molecular parameters from experimental spectra and has been worked out in detail (Hecht [47], Fox [59], Moret-Bailly [60]).

We turn next to a discussion of the effect on a fixed level $E_s[n(lJ)R]$ of the $O(3)$-tensor, G-invariant interactions $G^{(k_1k_2)4}$.

q. Splitting patterns. In general, an $O(3)$-tensor, G-invariant interaction $[O(3) \supset G]$ will split an eigenvalue (level) of a Hamiltonian having $O(3)$ invariance alone into a number of distinct eigenvalues (sublevels). It is the purpose of this section to discuss the manner in which the $O(3)$-tensor, G-invariant interaction

$$V \equiv \sum_{k_1k_2} t_{k_1k_24}\, G^{(k_1k_2)4} \tag{7.10.271}$$

splits a specified scalar energy level, $E_s[n(lJ)R]$, into sublevels having G symmetry. [The summation in Eq. (7.10.271) is to be taken over the pairs (40), (04), (22), (13), (31), and the $t_{k_1k_24}$ are numerical coefficients.]

In our initial discussion below, it is useful to employ the bra–ket notation of Chapter 3, since the results to be presented are quite general.

The general features of such splitting patterns are well-known and may be formulated in terms of the reduction of an irreducible representation (irrep) $\mathscr{D}^{j,\sigma}$ ($j = 0, 1, \ldots;\ \sigma = 0, 1$),

$$\mathscr{D}^{j,\sigma} = \{(\det R)^{\sigma}\mathscr{D}^{j}(R) \colon R \in O(3)\},$$

of the group $O(3)$ into irreps Γ^{α} ($\alpha = 1, 2, \ldots, k$),

$$\Gamma^{\alpha} = \{\Gamma^{\alpha}(g) \colon g \in G\}, \tag{7.10.272}$$

of the point group G. This reduction is described abstractly by the group–subgroup reduction series (direct sum):

$$\mathscr{D}^{j,\sigma} = \sum_{\alpha} \oplus\, m_j(\Gamma^{\alpha})\Gamma^{\alpha}, \tag{7.10.273}$$

where $m_j(\Gamma^\alpha)$ denotes the multiplicity (number of occurrences) of Γ^α in $\mathscr{D}^{j,\sigma}$. Corresponding to this reduction, one has also the dimension formula

$$2j+1=\sum_\alpha m_j(\Gamma^\alpha)\dim\Gamma^\alpha. \qquad (7.10.274)$$

In many physical applications, a more detailed description of the group–subgroup reduction process is required. Thus, let $\{|jm\rangle: m=j, j-1,\ldots,-j\}$ denote a set of orthonormal basis vectors of a Hilbert space $\mathscr{H}_{j,\sigma}$, which transforms irreducibly under a specified action of $O(3)$ (for simplicity we suppress σ in the notation for basis vectors):

$$R:|jm\rangle\to\sum_{m'}\mathscr{D}^{j,\sigma}_{m'm}(R)|jm'\rangle, \qquad \text{each } R\in O(3). \qquad (7.10.275)$$

Then the group–subgroup problem for $O(3)\supset G$ is one of determining all linear combinations of the form

$$|\kappa;\Gamma^\alpha\gamma\rangle=\sum_m\langle jm|\Gamma^\alpha\gamma\rangle_\kappa|jm\rangle \qquad (7.10.276)$$

such that the functions and coefficients in this relation have the following properties:

1. The indices γ and κ range over the values

$$\gamma=1,2,\ldots,\dim\Gamma^\alpha; \qquad \kappa=1,2,\ldots,m_j(\Gamma^\alpha) \qquad (7.10.277)$$

and enumerate linearly independent functions for each irrep Γ^α ($\alpha=1,2,\ldots,k$). These functions are usually taken to be orthonormal:

$$\langle\kappa';\Gamma^{\alpha'}\gamma'|\kappa;\Gamma^\alpha\gamma\rangle=\delta_{\alpha'\alpha}\,\delta_{\gamma'\gamma}\,\delta_{\kappa'\kappa}. \qquad (7.10.278)$$

2. The set of functions

$$\{|\kappa;\Gamma^\alpha\gamma\rangle:\gamma=1,2,\ldots,\dim\Gamma^\alpha\} \qquad (7.10.279)$$

transforms irreducibly under the action of G:

$$g:|\kappa;\Gamma^\alpha\gamma\rangle\to\sum_{\gamma'}\Gamma^\alpha_{\gamma'\gamma}(g)|\kappa;\Gamma^\alpha\gamma'\rangle, \qquad \text{each } g\in G. \qquad (7.10.280)$$

3. The coefficients $\{\langle jm|\Gamma^\alpha\gamma\rangle_\kappa\}$ are called subduction coefficients for the reduction of $O(3)$ to G and are the elements of a unitary matrix of dimension

$2j+1$. Thus, Eq. (7.10.276) is a unitary transformation of the space $\mathscr{H}_{j,\sigma}$ such that the new basis $\{|\kappa;\Gamma^\alpha\gamma\rangle\}$ carries a unitary irrep of $O(3)$, which is equivalent to $\mathscr{D}^{j,\sigma}$, and also such that the space $\mathscr{H}_{j,\sigma}$ has been split into subspaces that carry the irreps of G.

The occurrence of the multiplicity index κ in the linear combinations (7.10.276) means that there are $m_j(\Gamma^\alpha)$ perpendicular subspaces of $\mathscr{H}_{j,\sigma}$, each of which carries the irrep Γ^α of G. One cannot determine explicitly the reduction of irrep $\mathscr{D}^{j,\sigma}$ of $O(3)$ into irreps Γ^α of G without at the same time specifying how the multiplicity enumerated by κ is to be broken, since arbitrary unitary transformations of the form $U=(u_{\kappa'\kappa})$,

$$U\colon|\kappa;\Gamma^\alpha\gamma\rangle\to\sum_{\kappa'}u_{\kappa'\kappa}|\kappa';\Gamma^\alpha\gamma\rangle, \tag{7.10.281}$$

will yield a new basis of the space $\mathscr{H}_{j,\sigma}$ with the same transformation properties under g as the old basis. Methods for dealing with this general "multiplicity problem" have been discussed in the literature (see Fox and Ozier [61], Moret-Bailly *et al.* [62], and references therein).

In physical problems the "degeneracy" in κ described above is often broken by introducing an operator A with the following properties:

(1) $$A\colon\mathscr{H}_{j,\sigma}\to\mathscr{H}_{j,\sigma},$$

(2) $$A \text{ is Hermitian on } \mathscr{H}_{j,\sigma},$$

(3) $$A \text{ is a } G\text{-invariant.} \tag{7.10.282}$$

It follows that A may be diagonalized on the space $\mathscr{H}_{j,\sigma}$ and that each distinct eigenvalue defines a subspace of $\mathscr{H}_{j,\sigma}$ that is invariant under the action of G. Under favorable circumstances, these invariant subspaces are also irreducible. In such cases, A serves the same role in the group–subgroup reduction from $O(3)$ to G that J_3 serves in the standard $O(3)\supset O(2)$ reduction. (The classic example of this structure in molecular physics is the diagonalization of the asymmetric rotator Hamiltonian operator $A=aJ_1^2+bJ_2^2+cJ_3^2$ on the space $\mathscr{H}_{j,\sigma}$; in this case G is isomorphic to Klein's *Vierergruppe* and has the invariant action $J_i\to\lambda_iJ_i$, $\lambda_i=\pm1$, $\lambda_1\lambda_2\lambda_3=1$ on A. The asymmetric rotator problem has been discussed from many points of view (see Nielsen [13], Wilson *et al.* [15], Amat *et al.* [40], and references therein). A recent discussion emphasizing the viewpoint above is given by Patera and Winternitz [63].)

The molecular problem with the perturbation (7.10.271) is of the type described above. To show this, let G^4 be defined by Eq. (7.10.260), where, for the present discussion, T_μ^4 may denote any irreducible tensor operator acting on the Hilbert space of a physical system that has been split by the total angular momentum $\mathbf{J}$ and parity σ into a direct sum of spaces $\{\mathscr{H}_{j,\sigma}\}$. We require only that G^4 be a Hermitian operator with nonvanishing matrix elements diagonal in j: $\langle j, m + \mu | G^4 | jm \rangle \not\equiv 0$. The operator A defined by

$$A|jm\rangle = \sum_{m'} \langle jm'|G^4|jm\rangle |jm'\rangle \tag{7.10.283}$$

satisfies the conditions (7.10.282) for the cubic groups $G = O_h, O, T_d$. Note, then, that up to an inessential multiplicative reduced matrix element the operator A is

$$A = \sqrt{\frac{14}{5}}\mathscr{T}_{40}(\mathbf{J}) + \mathscr{T}_{4,4}(\mathbf{J}) + \mathscr{T}_{4,-4}(\mathbf{J}). \tag{7.10.284}$$

The crucial result required now has been proved by Kramer and Moshinsky [64]: *The eigenvalues of the operator A are distinct, and each eigenvalue defines a subspace of $\mathscr{H}_{j,\sigma}$ that is irreducible under the action of G* (G-irreducible) *for $G = O_h$, O, or T_d.*[1]

Let us next relate the molecular problem with the perturbation (7.10.271) to the above results. The calculation of the matrix elements of the perturbation T [Eq. (7.10.271)] in the coupled basis (7.10.207) is a standard application of the Wigner–Eckart theorem (Section 15, Chapter 3) using the reduced matrix elements given by Eq. (3.248). The results are (suppressing quantum numbers M and σ):

$$\langle n(l'J')R'M'_R|V|n(lJ)RM_R\rangle = \delta_{J'J}\langle n(l'J)R'\|V\|n(lJ)R\rangle \times \left(\sqrt{\frac{14}{5}}C^{R4R'}_{M_R0M'_R} + C^{R4R'}_{M_R4M'_R} + C^{R4R'}_{M_R,-4,M'_R}\right), \tag{7.10.285}$$

[1] The interaction A is invariant under the action of O_h – this interaction breaks the $O(3)$ symmetry only to the level of the group O_h. However, it splits the spaces $\mathscr{H}_{j,0}$ and $\mathscr{H}_{j,1}$ in exactly the same manner (we assume that σ is always a sharp quantum number for the system in question). Thus, for the study of the splitting patterns themselves, we may assume without loss of generality that $j + \sigma$ is even – that is, that we are dealing with $SO(3) \supset O$. Since the irreps of O_h are irreducible when restricted to the subgroup O (or T_d), the irreps of O_h contained in an irrep $\mathscr{D}^{j,\sigma}$ of $O(3)$ are obtained from the irreps of O contained in an irrep $\mathscr{D}^j$ of $SO(3)$ by assigning the extra g (gerade) or u (ungerade) label according to whether $j + \sigma$ is even or odd. Moreover, since only the 2×2 E-type irreps of O_h are reducible when O_h is restricted to either of the subgroups T or T_h, and then the reduction into two one-dimensional irreps is unique, we see also that the reduction of an $O(3)$ irrep into irreps of T and T_h is also essentially solved by the diagonalization of A.

where the reduced matrix element is given by

$$\langle n(l'J)R' \| V \| n(lJ)R\rangle = \frac{5}{4}\sum_{k_1k_2} t_{k_1k_24}\langle n(l'J)R' \| \mathbf{T}^{(k_1k_2)4} \| n(lJ)R\rangle, \quad (7.10.286)$$

in which

$$\langle n(l'J)R' \| \mathbf{T}^{(k_1k_2)4} \| n(lJ)R\rangle$$
$$= [9(2l'+1)(2J+1)(2R+1)]^{\frac{1}{2}} \begin{Bmatrix} l & J & R \\ k_1 & k_2 & 4 \\ l' & J & R' \end{Bmatrix}$$
$$\times \langle nl' \| \mathbf{T}_{k_1}(\text{vib}) \| nl\rangle\langle J \| \mathbf{T}_{k_2}(\text{rot}) \| J\rangle. \quad (7.10.287)$$

Again in the spirit of first-order perturbation theory [the Hamiltonian (7.10.196) and the $O(3)$-invariant interactions (7.10.258), $T_0^{(22)0}$ excepted, are diagonal in the coupled basis (7.10.207)], one considers, initially, only those matrix elements (7.10.285) that are diagonal in l and R ($l' = l$, $R' = R$). (This approximation is called the dominant approximation.) Thus, in the dominant approximation, which includes all $O(3)$-tensor, G-invariant interactions up to the fifth degree, we obtain the following energy for a triply degenerate oscillation in interaction with the rotational motion in a spherical top molecule:

$$E[n(lJ)R; \Gamma^\alpha\kappa] = E_s[n(lJ)R] + t[n(lJ)R]E(R\Gamma^\alpha\kappa), \quad (7.10.288)$$

where

$$t[n(lJ)R] \equiv \left[\frac{24}{5}(2R+1)\right]^{\frac{1}{2}} \langle n(lJ)R \| V \| n(lJ)R\rangle, \quad (7.10.289)$$

and $E(R\Gamma^\alpha\kappa)$ is an eigenvalue of a $(2R+1) \times (2R+1)$ real symmetric matrix A with elements given by

$$A_{M_R'M_R} = \left[\frac{5}{24(2R+1)}\right]^{\frac{1}{2}} \left(\sqrt{\frac{14}{5}}\, C^{R4R}_{M_R0M_R'} + C^{R4R}_{M_R4M_R'} + C^{R4R}_{M_R,-4,M_R'}\right). \quad (7.10.290)$$

Now, putting for brevity

$$|RM_R\rangle = |(nM\sigma)(lJ)RM_R\rangle, \qquad M_R = R, \ldots, -R, \quad (7.10.291)$$

we see that the matrix elements of A are just

$$A_{M'_R M_R} = \langle RM'_R | T | RM_R \rangle, \qquad (7.10.292)$$

where[1,2]

$$T \equiv \left[\frac{5}{24(2R-3)_9}\right]^{\frac{1}{2}} \left(\sqrt{\frac{14}{5}}\,\mathscr{T}_{40}(\mathbf{R}) + \mathscr{T}_{44}(\mathbf{R}) + \mathscr{T}_{4,-4}(\mathbf{R})\right) \qquad (7.10.293)$$

in which $\mathscr{T}_{k\mu}(\mathbf{R})$ is a tensor harmonic (Table 5 in the Appendix of Tables, with $\mathbf{J} = \mathbf{R}$). Thus, the problem of diagonalizing A is the same as that of diagonalizing the Hermitian operator T on the $(2R+1)$-dimensional space spanned by the basis vectors (7.10.291). (The operator T has been defined so that it coincides exactly with the operator T in Ref. [58] upon putting $\mathbf{R} = \mathbf{J}$; the eigenvalues of T have been tabulated by Krohn [65] for all values $R = 1, 2, \ldots, 100$; they coincide up to a phase $(-1)^R$ with the $F^{(4)}$-coefficients introduced by Moret-Bailly [60].)

Using now the Kramer–Moshinsky result above, we conclude: *The $O(3)$-tensor, G-invariant interaction V given by Eq.* (7.10.271) *splits each level $E_s[n(lJ)R]$ into a unique set of sublevels, and each sublevel is described by linear combinations of the basis vectors $\{|RM_R\rangle : M_R = R, \ldots, -R\}$ that are G-irreducible for $G = O_h$, O, or T_d.*

This result then implies that the eigenvalues of T may be uniquely denoted by the notation $E(R\Gamma^\alpha\kappa)$, where κ enumerates the multiplicity of irrep Γ^α of G in irrep $\mathscr{D}^{R,\sigma}$ of $O(3)$, each such irrep now being distinguished by the eigenvalue $E(R\Gamma^\alpha\kappa)$ itself.

Notice next from Eq. (7.10.288) that each scalar level $E_s[n(lJ)R]$ is split into sublevels in exactly the same way by the set of eigenvalues

$$\{E(R\Gamma^\alpha\kappa) : \alpha = 1, 2, \ldots, k; \kappa = 1, 2, \ldots, m_R(\Gamma^\alpha)\}, \qquad (7.10.294)$$

the splitting for different J and l being a scaling of this same basic eigenvalue set[3].

[1] The notation $(x)_a$ denotes the rising factorial $(x)_a = x(x+1)\cdots(x+a-1)$.

[2] Since $\mathbf{R} = \mathbf{L} + \mathbf{K}$, it follows from Eqs. (7.10.253) and (7.10.254) that the action of the point group G on $\mathbf{R}$ is $L_g : \mathbf{R} \to [\det R(g)]R(g)\mathbf{R}$, each $g \in G$. In particular, the vector representation of O_h (see footnote p. 589) yields the explicit transformations of $\mathbf{R}$ of the form $(\sigma_1 R_{i_1}, \sigma_2 R_{i_2}, \sigma_3 R_{i_3})$ with (i_1, i_2, i_3) even or odd and $\sigma_1\sigma_2\sigma_3 = +1$ or -1, respectively. Since the operator T may be written in Cartesian form as $R_1^4 + R_2^4 + R_3^4$, up to a multiplying constant and a sum of $O(3)$ invariants, one easily verifies explicitly the O_h invariance of T. Note also that the transformation $\mathbf{R} \to -\mathbf{R}$ is a consequence of time reversal and not of an internal coordinate transformation.

[3] The interpretation of laser-induced absorption vibration–rotation bands requires detailed knowledge of the scaling factor $t[n(lJ)R]$, which is seen from Eqs. (7.10.289) and (7.10.286) to depend on known reduced matrix elements (see Table 8 in the Appendix of Tables). It is the values of these factors, $t[n(lJ)R]$, that are to be inferred from experiment (Hecht [47]) (see Section u). It is beyond the scope of the present discussion to give these details.

Table 7.2. Eigenvalues (multiplied by 100) of the operator T for $j = 100$. Symmetry type is denoted by the letter to the left of each eigenvalue.

A_1	5.125 690 090 113 115 680 256 257 41
F_1	5.125 690 090 113 115 680 256 246 55
E	5.125 690 090 113 115 680 256 241 11
F_2	4.613 143 588 753 669 695 986 086 41
F_1	4.613 143 588 753 669 695 984 526 12
E	4.123 813 565 974 013 576 249 958 69
F_2	4.123 813 565 974 013 576 195 545 78
A_2	4.123 813 565 974 013 576 086 719 95
F_2	3.657 180 034 191 149 605 188 983 55
F_1	3.657 180 034 191 149 600 281 883 37
A_1	3.212 732 649 388 184 644 879 787 44
F_1	3.212 732 649 388 184 484 121 005 93
E	3.212 732 649 388 184 403 741 615 18
F_2	2.789 971 626 892 359 674 156 371 85
F_1	2.789 971 626 892 355 597 481 874 43
E	2.388 408 909 335 097 267 599 682 75
F_2	2.388 408 909 335 055 641 500 819 30
A_2	2.388 408 909 334 972 389 303 092 32
F_2	2.007 569 673 405 330 918 805 171 95
F_1	2.007 569 673 403 924 645 927 347 27
A_1	1.646 994 299 230 227 427 363 868 59
F_1	1.646 994 299 210 200 870 576 139 99
E	1.646 994 299 200 187 592 180 041 52
F_2	1.306 240 983 446 697 745 268 577 70
F_1	1.306 240 983 202 868 013 774 524 95
E	0.984 889 268 710 685 073 571 903 90
F_2	0.984 889 267 428 308 849 092 921 45
A_2	0.984 889 264 863 556 326 927 654 06
F_2	0.682 544 915 774 415 105 086 555 01
F_1	0.682 544 892 285 740 302 042 659 88
A_1	0.398 846 926 458 295 844 008 468 22
F_1	0.398 846 738 104 829 771 571 132 97
E	0.398 846 643 927 897 998 386 189 49
F_2	0.133 477 415 121 653 930 832 118 01
F_1	0.133 476 087 600 358 713 277 347 41
E	− 0.113 820 550 714 971 537 012 721 75
F_2	− 0.113 824 670 725 801 047 764 678 17
A_2	− 0.113 832 911 516 736 218 647 037 97
F_2	− 0.343 211 700 124 708 016 123 815 95
F_1	− 0.343 256 749 021 544 314 879 348 59
A_1	− 0.554 586 765 960 791 560 014 224 37
F_1	− 0.554 802 947 346 126 586 805 606 85
E	− 0.554 911 306 144 353 572 766 778 27
F_2	− 0.747 916 422 095 131 911 371 734 66
F_1	− 0.748 829 443 283 038 391 070 882 10
E	− 0.921 740 940 781 814 669 346 563 49
F_2	− 0.923 367 440 228 536 798 770 416 95
A_2	− 0.926 750 544 469 145 034 187 247 51
F_2	− 1.073 780 721 465 328 536 413 833 16
F_1	− 1.083 217 941 412 453 593 557 705 35
A_1	− 1.190 130 139 104 612 221 122 307 78
F_1	− 1.211 465 541 115 730 083 575 430 55
E	− 1.223 807 792 495 174 406 634 504 30
F_2	− 1.298 669 531 856 416 780 417 344 07
F_1	− 1.357 463 645 331 767 352 225 532 58
E	− 1.398 953 665 318 508 955 615 312 30
F_2	− 1.413 942 186 561 829 616 697 199 63
A_2	− 1.530 407 059 651 272 556 104 560 10
F_2	− 1.542 737 626 348 819 454 833 532 19
F_1	− 1.552 860 244 399 062 344 812 493 06
A_1	− 1.561 835 047 007 528 733 480 626 01
F_1	− 1.735 872 930 017 428 548 567 057 65
E	− 1.738 329 462 409 720 394 247 929 18
F_2	− 1.740 792 637 826 024 220 948 537 19
F_1	− 1.958 800 917 925 736 438 246 769 67
E	− 1.959 211 917 221 652 421 431 964 77
F_2	− 1.959 617 474 398 551 574 555 539 92
A_2	− 2.205 690 537 106 250 549 362 472 70
F_2	− 2.205 740 107 216 926 575 826 673 52
F_1	− 2.205 789 633 426 881 222 180 206 50
A_1	− 2.205 839 115 846 819 559 087 037 52
F_1	− 2.475 887 153 282 221 645 230 334 37
E	− 2.475 891 508 594 447 061 256 675 41
F_2	− 2.475 895 863 794 787 859 379 167 96
F_1	− 2.768 492 514 561 974 034 171 497 12
E	− 2.768 492 779 390 955 664 450 115 70
F_2	− 2.768 493 044 217 690 926 871 150 73
A_2	− 3.082 960 347 082 066 430 850 404 83
F_2	− 3.082 960 357 090 862 681 561 542 41
F^1	− 3.082 960 367 099 656 707 919 444 58
A_1	− 3.082 960 377 108 448 509 925 335 12
F_1	− 3.418 969 545 015 500 357 195 826 09
E	− 3.418 969 545 193 238 111 742 763 24
F_2	− 3.418 969 545 370 975 865 769 605 86

Attention is thus shifted to the determination of the eigenvalues of the matrix A, and this problem itself [see Eq. (7.10.290)] is a purely algebraic one based on the properties of Wigner coefficients and the group–subgroup reduction $O(3) \supset G$. (See Appendix B for further discussion of this latter group-theoretic problem.)

Since one might not, at first, expect much regularity in the spectrum (7.10.294), it came somewhat as a surprise to find that numerical calculation of these levels showed a great deal of regularity, not only in the spacing of levels (clusters of nearly degenerate levels) but also in the ordering. Table 7.2 shows the spectrum of the operator T (eigenvalues of the matrix A) for $R(=j) = 100$, where A_1, A_2, E, F_1, and F_2 denote[1] the irrep labels $\{\Gamma^{\alpha}: \alpha = 1, \ldots, 5\}$ of the octahedral group O (see footnote p. 597). Reading downward from the top of Table 7.2, one sees that the levels are combined into sixfold, nearly degenerate levels of symmetry types (direct sum)

$$A_1 + F_1 + E, \qquad F_2 + F_1, \qquad E + F_2 + A_2, \qquad F_2 + F_1, \tag{7.10.295}$$

this grouping of levels by common approximate eigenvalue becoming rather bad after about five repetitions of the four levels (7.10.295) of O. On the other hand, reading upward from the bottom of Table 7.2, one sees that the levels are combined into eightfold, nearly degenerate levels of symmetry types

$$F_2 + E + F_1, \qquad A_1 + F_1 + F_2 + A_2, \qquad F_2 + E + F_1, \tag{7.10.296}$$

this grouping of levels by common approximate eigenvalue becoming rather bad after two full repetitions of the three levels (7.10.296) of O, and partial repetition consisting in the first two (approximate) levels in (7.10.296). Notice also that the sum of the irreps occurring in either (7.10.295) or (7.10.296) is just the regular representation of O.

A study of this cluster phenomenon for different values of $j(R)$ shows that the "near-degeneracy" in a given cluster improves with increasing j, but is already evident for $j = 10$ (Ref. [58]). Cluster degeneracies of the type illustrated in Table 7.2 were first noticed in computer calculations for $j = 2, \frac{5}{2}, \ldots, 8$ by Lea *et al.* [66], and again by Dorney and Watson [67] ($j = 1, \ldots, 20$) [clusters may also be inferred from the table ($j = 1, \ldots, 21$) given by Moret-Bailly *et al.* [62], but they were not noted explicitly.] The recent availability of high-resolution spectra of spherical top molecules (Aldridge *et al.* [68]) and the interpretation of such spectra (McDowell *et al.* [69, 70]) required for the first time an understanding of the spectrum of T for

[1] The relation of these irrep labels of O to those for the isomorphic symmetric group S_4 is given in the footnote, p. 611.

large angular momentum j. Interest was thus renewed in understanding in greater detail the cluster phenomenon. (The most extensive computer calculations have been carried out by Krohn [65]).

As was emphasized by Harter and Patterson [71], the physical explanation of clustering is due to *the localization of the values of the state vectors of a cluster around a classically stable axis of symmetry of a molecule*. The suggestion that molecules in states of high angular momentum would tend to rotate around a stable symmetry axis of the molecule was made by Dorney and Watson [67] (the classical result is due to Klein and Sommerfeld [72]; a stable axis of a perturbed rotator is a direction from the origin to a point that minimizes or maximizes the perturbation). This idea has been developed into a valuable calculational tool by Harter and Patterson [73, 74] and by Galbraith *et al.* [75]. The general applicability of this concept to other areas of physical phenomena (crystal field theory, etc.) has also been emphasized by Harter and Patterson [73] and by Judd [73a].

r. Symmetry axes and induced representations. The mathematical description of the physical intuition described above requires, then, that we consider the angular momentum state vectors of a physical system corresponding to different choices of *quantization axes* (different z-axes). If we let $\{|jm\rangle\}$ denote the standard angular momentum state vectors of a physical system referred to an arbitrary Cartesian frame $(\hat{e}_1, \hat{e}_2, \hat{e}_3)$ $(J_3 \to m)$, then the angular momentum state vectors $\{|jm;\hat{a}\rangle : m = j, \ldots, -j\}$ of this same physical system referred to the quantization axis $\hat{a}$ are given by

$$|jm;\hat{a}\rangle = \sum_{m'} \mathscr{D}^j_{m'm}(R^{\hat{a}})|jm'\rangle, \tag{7.10.297}$$

where $R^{\hat{a}} = (R^{\hat{a}}_{ij})$ is an orthogonal matrix that describes the orientation of axis $\hat{a}$ relative to the frame $(\hat{e}_1, \hat{e}_2, \hat{e}_3)$:

$$\hat{a} = R^{\hat{a}}_{13}\hat{e}_1 + R^{\hat{a}}_{23}\hat{e}_2 + R^{\hat{a}}_{33}\hat{e}_3. \tag{7.10.298}$$

For example, in the molecular problem $|jm\rangle$ denotes the state vector $|RM_R\rangle$ [Eq. (7.10.291)], the frame $(\hat{e}_1, \hat{e}_2, \hat{e}_3)$ is the moving Eckart frame $(\hat{f}_1, \hat{f}_2, \hat{f}_3)$ – it is the internal quantization axes that are of interest –, and $\hat{a}$ is a symmetry axis of the molecule having a fixed (numerical) orientation relative to the Eckart frame.

We return now to the generic angular momentum notation in order to emphasize the generality of results. As we shall show below, the explanation of the regularities in the spectrum of T combines angular momentum techniques (transformation properties, asymptotic properties, etc.), the induced representation theory of Frobenius, and physical intuition. (Our treatment is based on that on Harter and Patterson [73, 74].)

In order to establish the notation, it is convenient to recall some elementary properties of symmetry axes of a point group G (identity element e). [The group G may be any subgroup of $O(3)$ in the discussion that follows, where the elements of $O(3)$ have an action defined on a Euclidean 3-space].

Symmetry axes of G. An axis $\hat{a}$ is called a symmetry axis of a point group G if there exists at least one $g \in G$, $g \neq e$, such that $g\hat{a} = \hat{a}$. Thus, the set A of symmetry axes of G is

$$A = \{\hat{a}\colon g\hat{a} = \hat{a} \text{ for at least one } g \in G \text{ and } g \neq e\}. \qquad (7.10.299)$$

The group G is then a transformation group on A; that is, $g\colon A \to A$, each $g \in G$ (the set A is a G-space).

Equivalence of symmetry axes. Two symmetry axes $\hat{a}$ and $\hat{b}$ are equivalent, $\hat{a} \sim \hat{b}$, if there exists a $g \in G$ such that $g\hat{a} = \hat{b}$.

G-orbits of A. The G-orbit containing $\hat{a}$ is the subset of A defined by

$$G_{\text{orb}}(\hat{a}) = \{\text{all distinct axes } g\hat{a}\colon g \in G\}. \qquad (7.10.300)$$

Thus, $G_{\text{orb}}(\hat{a})$ contains all symmetry axes of G that are equivalent to axis $\hat{a}$.

Isotropy group $G^{\hat{a}}$. The isotropy group $G^{\hat{a}}$ of axis $\hat{a}$ is the subgroup of G defined by

$$G^{\hat{a}} = \{g \in G\colon g\hat{a} = \hat{a}\}. \qquad (7.10.301)$$

Coset of a subgroup H. Let H be a subgroup of G. A left coset of H in G is the set of elements of G given by

$$gH = \{gh\colon h \in H\}.$$

Two cosets are either identical or disjoint, and accordingly G may be partitioned (uniquely) into disjoint left cosets $S_1, S_2, \ldots, S_\sigma$, where $S_i = g_i H$, $g_i \in G$, $S_i \neq S_j$, and $\sigma = |G|/|H|$, in which $|S|$ denotes the number of elements in a set S. The notation G/H denotes the *set* whose elements are the left cosets of H in G:

$$G/H = \{S_1, S_2, \ldots, S_\sigma\}. \qquad (7.10.302)$$

(Similar results hold for right cosets.)

Three principal results, relating the sets introduced above, are important for the discussion of quantization axes:

1. If g transforms symmetry axis $\hat{a}$ to symmetry axis $\hat{b}$ – that is, $\hat{b} = g\hat{a}$ – then (see Chapter 2, Section 2)

$$gG^{\hat{a}}g^{-1} = G^{\hat{b}} = G^{g\hat{a}}. \tag{7.10.303}$$

Thus, axes in the same G-orbit have conjugate, hence isomorphic, isotropy groups.

2. The number of left cosets in the set $G/G^{\hat{a}}$ is equal to the number of (equivalent) axes in the G-orbit, $G_{\text{orb}}(\hat{a})$, containing $\hat{a}$. It follows from this result that the cosets $S_1^{\hat{a}}, S_2^{\hat{a}}, \ldots$ in the set $G/G^{\hat{a}}$ may be enumerated by the notation

$$G/G^{\hat{a}} = \{S_{\hat{b}}^{\hat{a}} : \hat{b} \in G_{\text{orb}}(\hat{a})\}. \tag{7.10.304}$$

Thus, the number of elements in $G_{\text{orb}}(\hat{a})$ is $|G|/|G^{\hat{a}}|$.

3. Let $s_{\hat{b}}$ be a fixed representative of the coset $S_{\hat{b}}^{\hat{a}}$ of $G^{\hat{a}}$. Then

$$\phi_{\hat{b}} : g \to (s_{g\hat{b}})^{-1} g s_{\hat{b}} \equiv \phi_{\hat{b}}(g) \tag{7.10.305}$$

is a mapping of G into the subgroup $G^{\hat{a}}$ for each $\hat{b} \in G_{\text{orb}}(\hat{a})$. Furthermore, this mapping satisfies

$$\phi_{g\hat{b}}(g')\phi_{\hat{b}}(g) = \phi_{\hat{b}}(g'g). \tag{7.10.306}$$

Let us now consider the "choice of quantization axis problem" for the case when the quantization axes are taken to be the elements of the set A of symmetry axes of a point group G. Using the one-to-one correspondence between the elements of $G_{\text{orb}}(\hat{a})$ and $G/G^{\hat{a}}$ given by (2) above, it is now convenient to denote the set of angular momentum state vectors of the type (7.10.297), corresponding to equivalent axes, by the notation:

$$|jm; S_{\hat{b}}^{\hat{a}}\rangle \equiv |jm; \hat{b}\rangle, \tag{7.10.307}$$

each $\hat{b} \in G_{\text{orb}}(\hat{a})$, and hence each $S_{\hat{b}}^{\hat{a}} \in G/G^{\hat{a}}$. The scalar product of two basis vectors (7.10.307) is obtained from the definition (7.10.297) and the property $R^{g\hat{a}} = R^{\hat{a}}\tilde{R}(g)$:

$$\langle jm'; S_{g'\hat{a}}^{\hat{a}} | jm; S_{g\hat{a}}^{\hat{a}}\rangle = \mathscr{D}^j_{m'm}(\tilde{R}(g')R(g)). \tag{7.10.308}$$

Thus, the set of basis vectors

$$\{|jm; S_{\hat{b}}^{\hat{a}}\rangle : \hat{b} \in G_{\text{orb}}(\hat{a})\}, \qquad j, m, \text{ and } \hat{a} \text{ specified} \tag{7.10.309}$$

is not orthonormal. We shall, however, assume that the vectors in this set are linearly independent.

We next construct a representation of G using the angular momentum state vectors (7.10.307) and a representation of the isotropy subgroup $G^{\hat{a}}$. We shall, however, assume, for simplicity, that $G = G_+$ is a group of pure rotations. In this case the isotropy group of an axis $\hat{a}$ is always one of the cyclic abelian groups denoted by C_n with elements generated by the rotation $\mathscr{R}_n \equiv \mathscr{R}(2\pi/n, \hat{a})$ about $\hat{a}$ by angle $2\pi/n$.

Specifically, the isotropy group $G^{\hat{a}}$ is the group $C_n(\hat{a})$ of rotations about direction $\hat{a}$ by angles $2\pi s/n$ $(s = 0, 1, \ldots, n-1)$:

$$G^{\hat{a}} = C_n(\hat{a}) = \left\{\mathscr{R}_n^s = \left[\mathscr{R}\left(\frac{2\pi}{n}, \hat{a}\right)\right]^s : s = 0, 1, \ldots, n-1\right\}. \tag{7.10.310}$$

Under the action of $\mathscr{R}_n^s$, the angular momentum state vector undergoes the unitary transformation:

$$|jm; \hat{a}\rangle \to \exp\left[\frac{-2\pi i s}{n}(\hat{a} \cdot \mathbf{J})\right]|jm; \hat{a}\rangle$$

$$= (e^{-2\pi i m/n})^s |jm; \hat{a}\rangle. \tag{7.10.311}$$

Thus, the state vector $|jm; \hat{a}\rangle$ will transform according to irrep k_n of $C_n(\hat{a})$ given by

$$k_n \equiv \{(e^{-2\pi i k/n})^s : s = 0, 1, \ldots, n-1\}, \qquad k = 0, 1, \ldots, n-1, \tag{7.10.312}$$

if and only if

$$m \equiv k (\text{mod}\, n), \qquad m \geqslant 0. \tag{7.10.313}$$

We may now obtain a representation of G_+ using the basis of (7.10.309) by the following well-known procedure. We define the linear operator $L(g)$, each $g \in G_+$, by giving its action on each basis vector in the set (7.10.309):

$$L(g)|jm; S_{\hat{b}}^{\hat{a}}\rangle = \chi_{k_n}(\phi_{\hat{b}}(g))|jm; gS_{\hat{b}}^{\hat{a}}\rangle, \tag{7.10.314}$$

for each $\hat{b} \in G_{\text{orb}}(\hat{a})$. In this definition, m may be any integer $m \equiv k (\text{mod}\, n)$, where k is a specified integer from the set $\{0, 1, \ldots, n-1\}$, and $\chi_{k_n}(\mathscr{R}_n^s)$ denotes the character of the rotation $\mathscr{R}_n^s$ in the irrep k_n. Thus,

$$\chi_{k_n}(\mathscr{R}_n^s) = (e^{-2\pi ik/n})^s \in k_n. \tag{7.10.315}$$

Notice that, since $\phi_{\hat{b}}(g) \in C_n(\hat{a})$, the character appearing in Eq. (7.10.314) belongs to irrep k_n of $C_n(\hat{a})$, so that $\chi_{k_n}(\phi_{\hat{b}}(g))$ is defined.

The group multiplication for representations,

$$L(g')L(g) = L(g'g), \qquad \text{all } g', g \in G_+, \tag{7.10.316}$$

may now be verified from Eq. (7.10.314) by using the following properties of cosets and characters:

$$gS_{\hat{b}}^{\hat{a}} = S_{g\hat{b}}^{\hat{a}}, \tag{7.10.317}$$

$$\begin{aligned} \chi_{k_n}(\phi_{g\hat{b}}(g'))\chi_{k_n}(\phi_{\hat{b}}(g)) &= \chi_{k_n}(\phi_{g\hat{b}}(g')\,\phi_b(g)) \\ &= \chi_{k_n}(\phi_{\hat{b}}(g'g)). \end{aligned} \tag{7.10.318}$$

Thus,

$$L(G_+) = \{L(g): g \in G_+\} \tag{7.10.319}$$

is a representation of G_+. The dimension of $L(G_+)$ is $|G_+|/|G_+^{\hat{a}}|$, since only the basis vectors in the set (7.10.309), with $m \equiv k (\text{mod}\, n)$ specified, enter into the transformation $L(g)$.

The representation of G_+ obtained above by using an irrep k_n of C_n is often denoted by

$$L(G_+) = k_n \uparrow G_+. \tag{7.10.320}$$

Observe that $L(G_+)$ does not depend on which of the isomorphic isotropy groups $G_+^{\hat{b}} \cong G_+^{\hat{a}}$, $\hat{b} \in G_{\text{orb}}(\hat{a})$, is used in the construction – the same basis vectors $\{|jm; \hat{b}\rangle : b \in G_{\text{orb}}(\hat{a})\}$ enter in the transformation $L(g)$ with the same coefficients. The general method of constructing a representation of a group G from a representation $\Delta(H)$ of a subgroup H was discovered by Frobenius. The representation $\Delta(H) \uparrow G$ is called the representation of G induced from the representation $\Delta(H)$ of H.

One of the most important results for representations of a group obtained by inducing from a subgroup is the *Frobenius reciprocity theorem*, which states: *The multiplicity of irrep Γ^α of G in the representation $\Delta(H) \uparrow G$ of G induced from* *an irrep $\Delta(H)$ of a subgroup H equals the multiplicity of irrep $\Delta(H)$ in the reduction of Γ^α into irreps of H.*

Since the basis vectors (7.10.309) are formed from linear combinations of the original angular momentum states $\{|jm\rangle : m = j, \ldots, -j\}$ [see Eq. (7.10.297)], the preceding construction, using induced representation theory, allows one to construct numerous subspaces of $\mathscr{H}_{j,\sigma}$ [see Eq. (7.10.275)] that are invariant under G_+. We can construct such a subspace not only for each irrep of any one of the isotropy groups C_n, but also for each value of $m \equiv k(\bmod\, n)$. Clearly, there are too many such spaces – they cannot be independent. The sorting out of which of these subspaces are important in a physical problem is provided by physical intuition and will here be justified by using asymptotic properties of the Wigner coefficients.

This physical intuition was discussed earlier, but it may now be stated more precisely: (1) Any physical G_+-invariant interaction V will determine various classically stable axes of rotation; (2) the angular momentum state vectors $|jm; \hat{a}_0\rangle$ for each such stable axis $\hat{a}_0$ will, for large j, and for m such that $\cos\theta_m = m/\sqrt{j(j+1)}$ is near unity, have values that are localized in the vicinity of the axis $\hat{a}_0$; and (3) each state vector in the set

$$\{|jm; \hat{a}\rangle : \hat{a} \in G_{\text{orb}}(\hat{a}_0)\}, \qquad j, m \text{ specified; } \cos\theta_m \simeq 1, \quad (7.10.321)$$

will be an approximate eigenstate of V with eigenvalue E that is independent of $\hat{a} \in G_{\text{orb}}(\hat{a}_0)$:

$$V|jm; \hat{a}\rangle \simeq E|jm; \hat{a}\rangle. \qquad (7.10.322)$$

(The eigenvalue E may, of course, depend on j, m as well as on the choice of stable axis $\hat{a}_0$.)

It is subspaces of the type (7.10.321), associated with classically stable rotation axes and values of m near j, that one expects to be useful in the construction of G_+-invariant subspaces for a physical problem.

In order to implement the method outlined above, it is necessary to have the form $V(\hat{a})$ of the interaction V referred to an arbitrary symmetry axis. Letting $\mathscr{U}(\hat{a})$ denote the unitary transformation (7.10.297) – that is, $\mathscr{U}(\hat{a})|jm\rangle = |jm; \hat{a}\rangle$ – one has

$$V(\hat{a}) = \mathscr{U}(\hat{a}) V \mathscr{U}^{-1}(\hat{a}). \qquad (7.10.323)$$

In particular, since V has the general form

$$V = \sum_{k\mu} \alpha_{k\mu} T^k_\mu, \qquad (7.10.324)$$

$V(\hat{a})$ has the form

$$V(\hat{a}) = \sum_{k\mu} \alpha_{k\mu} T^k_\mu(\hat{a}), \tag{7.10.325}$$

where the matrix elements in the original basis $\{|jm\rangle\}$ of the tensor operator $T^k_\mu(\hat{a})$ referred to the axis $\hat{a}$ are

$$\langle j'm'|T^k_\mu(\hat{a})|jm\rangle = \mathscr{D}^k_{m'-m,\mu}(R^{\hat{a}})\langle j'm'|T^k_{m'-m}|jm\rangle. \tag{7.10.326}$$

The importance of the form (7.10.323) for the present problem is that, since V and $V(\hat{a})$ are unitarily equivalent, they have the same spectrum. Hence, any of the forms $V(\hat{a})$, $\hat{a} \in A$, may be used in determining the eigenvalues and eigenvectors of V. We shall see below that, to obtain approximate eigenvalues and eigenvectors of V for large j, one needs to utilize all the forms $V(\hat{a})$, where $\hat{a}$ is a stable axis of rotation. As an example, we have the following results for the interaction T [Eq. (7.10.293), $\mathbf{R} = \mathbf{J}$]. The stable axes for this interaction are identified by using a mathematically equivalent formulation of the perturbation problem, which replaces the tensor harmonics $\mathscr{T}_{k\mu}(\mathbf{J})$ by the solid harmonics $\mathscr{Y}_{k\mu}(\mathbf{x})$. Up to an $O(3)$-invariant term, the interaction T is thus replaced by $f(\mathbf{x}) = x_1^4 + x_2^4 + x_3^4$. The stable axes of rotation of the perturbation T are the axes corresponding to the extremal points of $f(\mathbf{x})$ subject to $x_1^2 + x_2^2 + x_3^2 = 1$: Thus, the stable symmetry axes of the octahedrally (octahedral group O) invariant interaction T are determined to be

$$O_{\text{orb}} \equiv \{\hat{e}_1, \hat{e}_2, \hat{e}_3, -\hat{e}_1, -\hat{e}_2, -\hat{e}_3\}, \tag{7.10.327}$$

corresponding to points that maximize $f(\mathbf{x})$; and

$$O'_{\text{orb}} \equiv \left\{\frac{1}{\sqrt{3}}(\alpha\hat{e}_1 + \beta\hat{e}_2 + \gamma\hat{e}_3): \alpha, \beta, \gamma = \pm 1\right\}, \tag{7.10.328}$$

corresponding to points that minimize $f(\mathbf{x})$.

The interaction T, as given by Eq. (7.10.293) (replace $\mathbf{R}$ by $\mathbf{J}$), is already referred to any of the O-equivalent axes belonging to the maximizing orbit O_{orb}; its form referred to any of the equivalent axes belonging to the minimizing orbit O'_{orb} is

$$T' = -\frac{2}{3}\left[\frac{5}{24(2j-3)_9}\right]^{\frac{1}{2}}\left(\sqrt{\frac{14}{5}}\mathscr{T}_{40}(\mathbf{J}) - 2\mathscr{T}_{4,3}(\mathbf{J}) + 2\mathscr{T}_{4,-3}(\mathbf{J})\right) \tag{7.10.329}$$

with matrix elements

$$\langle jm'|T'|jm\rangle = \frac{-2}{3}\left[\frac{5}{24(2j+1)}\right]^{\frac{1}{2}}\left(\sqrt{\frac{14}{5}}C^{j4j}_{m0m'} - 2C^{j4j}_{m3m'} + 2C^{j4j}_{m,-3,m'}\right). \tag{7.10.330}$$

s. High angular momentum effects.[1] An important result for the study of physical phenomena for large values of the angular momentum j is the asymptotic relationship (see Chapter 5, Section 8):

$$C^{jkj+\Delta}_{m,\mu,m+\mu} \sim d^k_{\mu\Delta}(\theta), \tag{7.10.331}$$

where $\cos\theta = m/\sqrt{j(j+1)}$. If we put $m = j - \alpha$ and take the limit for infinite j, we obtain [see Eq. (3.305), Chapter 3]

$$\lim_{j\to\infty} C^{jkj+\Delta}_{j-\alpha,\mu,j-\alpha+\mu} = \delta_{\mu\Delta}, \tag{7.10.332}$$

where we have used $\cos\theta \to 1$, $d^k_{\mu\Delta}(0) = \delta_{\mu\Delta}$.

Applying these results to the Wigner coefficients appearing in the interaction T [see Eqs. (7.10.290) and (7.10.293)] and those appearing in T' [see Eqs. (7.10.329) and (7.10.330)], we find

$$C^{j4j}_{m0m} \sim d^4_{00}(\theta) = P_4(\cos\theta_m) \simeq 1.$$

$$C^{j4j}_{m,\mu,m+\mu} \sim d^4_{0\mu}(\theta_m) \simeq 0 \qquad \text{for } \cos\theta_m \simeq 1, \mu \neq 0. \tag{7.10.333}$$

Thus, we may expect that for large j approximate eigenvalues of the interaction T are given by the diagonal matrix elements of T itself and its unitary equivalent form T':

$$A_{mm} = \sqrt{\frac{7}{24(2j+1)}}\,C^{j4j}_{m0m},$$

$$A'_{mm} = -\tfrac{2}{3}A_{mm}, \tag{7.10.334}$$

where $m = j, j-1, j-2, \ldots, j_0$ for some undetermined value $m = j_0$ such that $j_0/\sqrt{j(j+1)} \simeq 1$. (The exact eigenvalues of T have been denoted earlier in the discussion on p. 599 by $E(R\Gamma^\alpha\kappa)$.]

[1] Because of the general validity of the results in this subsection, we formulate the results in terms of generic angular momentum quantum numbers j $(= R)$ and m $(= M_R)$.

For states of rotation about the "hard" axes belonging to the maximizing orbit O_{orb} [see Eq. (7.10.327)], one expects the corresponding energies to be largest, whereas for the "soft" axes belonging to the minimizing orbit O'_{orb} [see Eq. (7.10.328)] the corresponding energies should be smallest. Furthermore, since $P_4(\cos\theta_m)$ is monotone decreasing for $m = j, j-1, \ldots$, in the neighborhood of $\cos\theta_m \simeq 1$, the values A_{mm}, $m = j, j-1, j-2, \ldots$, should approximate the eigenvalues of T at the high end of the spectrum (top of Table 7.2), and the values A'_{mm}, $m = j, j-1, j-2, \ldots$, should approximate those at the low end of the spectrum (bottom of Table 7.2). These qualitative features of the spectrum of T have been verified by explicit calculations (Ref. [74]), the values of A_{mm} and A'_{mm} for $m = j, j-1, \ldots$, enumerating, in order, the cluster energies at the high and low ends, respectively (e.g., 5.125, 4.613, 4.123, ... at the top, and -3.418, -3.082, -2.768, ... at the bottom of Table 7.2).

One can also estimate the point in the spectrum of T where the "crossover" between the values given by A_{mm} (hard axes) and those given by A'_{mm} (soft axes) will occur (Refs. [67, 74]). This crossing occurs near a value of the diagonal matrix elements of the interaction T referred to one of the twofold symmetry axes (saddle point of $x_1^4 + x_2^4 + x_3^4$):

$$O''_{\text{orb}} = \{(\alpha\hat{e}_i + \beta\hat{e}_j)/\sqrt{2} : \alpha, \beta = \pm 1; (i, j) = (1,2), (1,3), (2,3)\}. \quad (7.10.335)$$

The form of T referred to an axis belonging to the transition orbit O''_{orb} is

$$T'' = -\frac{1}{4}\left[\frac{5}{24(2j+1)}\right]^{\frac{1}{2}}\left(\sqrt{\frac{14}{5}}\,\mathscr{T}_{40}(\mathbf{J}) - 2\sqrt{7}\,[\mathscr{T}_{4,2}(\mathbf{J}) + \mathscr{T}_{4,-2}(\mathbf{J})] - 3[\mathscr{T}_{4,4}(\mathbf{J}) + \mathscr{T}_{4,-4}(\mathbf{J})]\right) \quad (7.10.336)$$

with diagonal matrix element

$$A''_{mm} = -\tfrac{1}{4}A_{mm}. \quad (7.10.337)$$

For large j the crossing occurs at about $A''_{jj} = -\frac{1}{4}A_{jj}$.

The description of the approximate state vectors having the eigenvalues A_{mm} and A'_{mm} may also be given, using the construction of the last section. We list below the basis states that are the bases of carrier spaces [see the construction, Eq. (7.10.314)] of the representations of O denoted by $k_4 \uparrow O$ (high end of spectrum) and $k_3 \uparrow O$ (low end of spectrum):

▶ States for levels at the high end of the spectrum of T:

$$k_4 \uparrow O\ [j \equiv k(\mathrm{mod}\,4)]: \{|jj;\hat{a}\rangle : \hat{a} \in O_{\mathrm{orb}}\}$$

$$k'_4 \uparrow O\ [j-1 \equiv k'(\mathrm{mod}\,4)]: \{|jj-1;\hat{a}\rangle : \hat{a} \in O_{\mathrm{orb}}\}$$

$$k''_4 \uparrow O\ [j-2 \equiv k''(\mathrm{mod}\,4)]: \{|jj-2;\hat{a}\rangle : \hat{a} \in O_{\mathrm{orb}}\}$$

$$k'''_4 \uparrow O\ [j-3 \equiv k'''(\mathrm{mod}\,4)]: \{|jj-3;\hat{a}\rangle : \hat{a} \in O_{\mathrm{orb}}\}$$

$$\vdots$$

▶ States for levels at the low end of the spectrum of T:

$$k_3 \uparrow O\ [j \equiv k(\mathrm{mod}\,3)]: \{|jj\rangle;\hat{a}\rangle : \hat{a} \in O'_{\mathrm{orb}}\}$$

$$k'_3 \uparrow O\ [j-1 \equiv k'(\mathrm{mod}\,3)]: \{|jj-1;\hat{a}\rangle : \hat{a} \in O'_{\mathrm{orb}}\}$$

$$k''_3 \uparrow O\ [j-2 \equiv k''(\mathrm{mod}\,3)]: \{|jj-2;\hat{a}\rangle : \hat{a} \in O'_{\mathrm{orb}}\}$$

$$\vdots$$

For example, for $j = 100$, the representations of O associated with the clusters at the high end of the spectrum are, in order, $0_4 \uparrow O$, $3_4 \uparrow O$, $2_4 \uparrow O$, $1_4 \uparrow O; 0_4 \uparrow O, 3_4 \uparrow O, 2_4 \uparrow O, 1_4 \uparrow O; \ldots$. Similarly, those for the low end are, in order, $1_3 \uparrow O$, $0_3 \uparrow O$, $2_3 \uparrow O$; $1_3 \uparrow O$, $0_3 \uparrow O$, $2_3 \uparrow O; \ldots$. The Frobenius reciprocity theorem quickly yields the results:[1]

$$\begin{array}{ll}
0_4 \uparrow O = A_1 + E + F_1 & \qquad \vdots \\
1_4 \uparrow O = F_1 + F_2 & \qquad 0_3 \uparrow O = A_1 + A_2 + F_1 + F_2 \\
2_4 \uparrow O = A_2 + E + F_2 & \qquad 1_3 \uparrow O = E + F_1 + F_2 \\
3_4 \uparrow O = F_1 + F_2 & \qquad 2_3 \uparrow O = E + F_1 + F_2. \\
\qquad \vdots &
\end{array} \tag{7.10.338}$$

Note the complete agreement with the results given in Table 7.2.

[1] The symbols A, E, F for irreps are standard for molecular spectroscopy; in terms of the standard tableau notation for irreps of the symmetric group, the correspondence between irreps of the octahedral group O (or the tetrahedral group T_d) and the symmetric group S_4 is

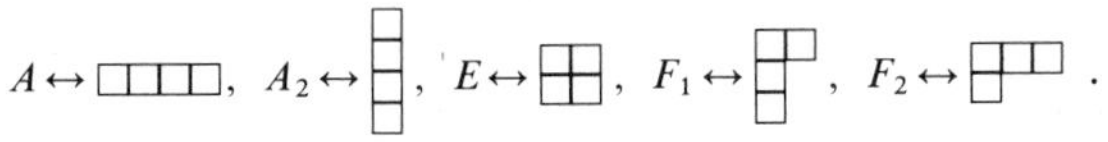

The qualitative explanation of level splitting based on induced representations presented in this and the preceding section does not predict the order and relative splittings *within* a given cluster. Furthermore, the approximate eigenvalues and eigenvectors described here need to be improved to achieve the high resolution accuracy required for the interpretation of spectra (such a spectrum is discussed in Section u). A model based on tunneling between the local axes angular momentum states $\{|jm;a\rangle : \hat{a} \in G_{\text{orb}}(\hat{a})\}$ has been proposed in Ref. [73]. This model accounts qualitatively for the observed splitting ratios within a cluster (Ref. [58]). The physical picture is that the molecule spins on one internal axis, "tumbles" over to another axis, spins for a while, tumbles again, spins, tumbles, etc. (Ref. [71]).

The eigenvalue expressions (7.10.334) and corresponding eigenvectors may be improved further by treating the off-diagonal terms in the interactions (7.10.293) and (7.10.329) as a perturbation (Refs. [74, 75]). Calculations of this sort yield expressions for the transition frequencies in a triply degenerate fundamental ν-band of a "heavy" spherical top molecule that are accurate to current spectroscopic resolution. This result obviates the need in the case of high angular momentum for the calculation of cubic vector coupling coefficients [Eq. (7.10.276)] and the diagonalization of large matrices (Ref. [75]).

We conclude that *angular momentum concepts lead to an understanding of certain aspects of precision laser spectroscopy in an unsuspected and intuitively satisfying manner.*

t. Selection rules and statistical weights. The basic expression for the intensity of an absorption line for a transition from an energy state E to an energy state E' ($E' > E$) when a gas sample of molecules is irradiated by light depends on several parameters: the frequency of the incident radiation, the Bohr frequency formula, a Boltzmann temperature factor, the density of the gas, a transition amplitude factor, and a statistical weight (degeneracy of a state) factor (see Dennison [32], Tolman [76]). Of all these factors, only the last two are of interest here: the transition amplitude and the statistical weight.

The transition probability, $B(E \to E')$, for electric dipole transitions is given by

$$B(E \to E') \equiv \sum_{i=1}^{3} \sum_{\gamma'\gamma} |\langle E';\gamma'|D_i|E;\gamma\rangle|^2. \tag{7.10.339}$$

In this expression $|E;\gamma\rangle$ denotes an eigenvector of the energy state E in question, in which γ denotes a set of quantum numbers that enumerate the degeneracy of level E, and D_i is the component of the electric dipole moment $\vec{D}$ of the molecule along the laboratory axis $\hat{l}_i$, that is, $D_i = \hat{l}_i \cdot \vec{D}$. The nonvanishing of the transition amplitude, $\langle E';\gamma'|D_i|E;\gamma\rangle$, determines the *selection rules* for electric dipole transitions from an energy state E to an energy state E'.

For a molecule, where the states are described relative to an internal frame (Eckart frame in our development), the components of the electric dipole moment referred to the moving frame, $d_i = \hat{f}_i \cdot \vec{D}$, are functions of the internal coordinates, which we take to be normal coordinates $\mathbf{q}$. The two sets of components are related by

$$D_i(\mathbf{q}) = \sum_j C_{ij} d_j(\mathbf{q}), \tag{7.10.340}$$

where C is the direction cosine matrix specifying the orientation of the Eckart frame relative to the laboratory frame. Equivalently, relation (7.10.340) is expressed in terms of the spherical components of a vector (see Appendix D, Chapter 3) by

$$D_\mu(\mathbf{q}) = \sum_\nu \mathscr{D}^{1*}_{\mu\nu}(C)\, d_\nu(\mathbf{q}). \tag{7.10.341}$$

A first step in the evaluation of the electric dipole moment transition amplitude is the calculation of the matrix elements of $\mathscr{D}^{1*}_{\mu\nu}$ with respect to the rotator states

$$\langle C|\sigma JMK\rangle \equiv \sqrt{\frac{2J+1}{8\pi^2}}\,(\det C)^\sigma(-1)^{J-K}\mathscr{D}^{J*}_{M,-K}(C). \tag{7.10.342}$$

These matrix elements are obtained from Eq. (3.189), Chapter 3, using the normalization given by Eq. (3.137) and symmetry relation (3.331):[1]

$$\langle\sigma'J'M'K'|\mathscr{D}^{1*}_{\mu\nu}|\sigma JMK\rangle$$
$$= (-1)^{\nu-1}\tfrac{1}{2}[1-(-1)^{J'+\sigma'+J+\sigma}]\sqrt{\frac{2J+1}{2J'+1}}\,C^{J1J'}_{M\mu M'}\,C^{J1J'}_{K,-\nu,K'} \tag{7.10.343}$$

For general vibration–rotation states of the form

$$|(\text{vib})\sigma JM\rangle \equiv \sum_K A_K|(\text{vib})K\rangle \otimes |\sigma JMK\rangle, \tag{7.10.344}$$

one thus finds

---

[1] These matrix elements are evaluated using the inner product defined in the footnote, p. 575.

$$\langle(\text{vib})'\sigma' J' M'|D_\mu|(\text{vib})\sigma JM\rangle = \tfrac{1}{2}[1-(-1)^{J'+\sigma'+J+\sigma}]$$
$$\times \sqrt{\frac{2J+1}{2J'+1}}\, C^{J1J'}_{M\mu M'} \sum_{K\nu}(-1)^{\nu-1} A_K A^*_{K+\nu}$$
$$\times\, C^{J1J'}_{K,\nu,K+\nu} \langle(\text{vib})'K+\nu|d_{-\nu}|(\text{vib})K\rangle. \tag{7.10.345}$$

This relation shows that the selection rules on the total angular momentum and parity are

$$\Delta J = 0, \qquad \Delta\sigma = \pm 1; \qquad \Delta J = \pm 1, \qquad \Delta\sigma = 0, \tag{7.10.346}$$

where $\Delta J = J' - J$ and $\Delta\sigma = \sigma' - \sigma$. *These selection rules are independent of the internal states.* They could have been obtained more directly from the fact that $\vec{D}$ is a vector operator with respect to rotations of physical space generated by the total angular momentum $\vec{J}$ and changes sign under spatial inversion.

Without further assumptions about the internal states $|(\text{vib})K\rangle$, little more can be said about the transition probability (7.10.339). For the case of a triply degenerate mode of oscillation [wave functions given by Eq. (7.10.207)], the explicit evaluation of the transition amplitude may be given. Although this calculation is a special case of Eq. (7.10.345), we repeat it in order to illustrate directly the angular momentum concepts involved.

Let us rewrite Eq. (7.10.341) in the form

$$D_\mu(\text{vib}, \text{rot}) = \sum_\nu (-1)^{\nu-1} T^1_\nu(\text{vib}) T^1_{\mu;-\nu}(\text{rot}). \tag{7.10.347}$$

In this result, we have defined

$$T^1_\nu(\text{vib}) = d_\nu(\mathbf{q}),$$
$$T^1_{\mu;\nu}(\text{rot}) = (-1)^{1-\nu} \mathscr{D}^{1*}_{\mu,-\nu}(C), \tag{7.10.348}$$

in which $d_\nu(\mathbf{q})$ is now a function of the normal coordinates $\mathbf{q} = (q_1, q_2, q_3)$ of the triply degenerate oscillation. As the notation indicates, $T^1_\nu(\text{vib})$ is a tensor operator of rank 1 with respect to the internal angular momentum $\mathbf{L}$, and $T^1_{\mu;\nu}(\text{rot})$ is a tensor operator of rank 1 in the index ν with respect to the angular momentum $\mathbf{K} = (K_1, K_2, K_3)$. Thus, $D_\mu(\text{vib, rot})$ *is an invariant with respect to* $\mathbf{R} = \mathbf{L} + \mathbf{K}$. [This observation already foretells the selection rule $\Delta R = 0$ obtained below (Galbraith [77]).]

The evaluation of the matrix elements of $D_\mu(\text{vib, rot})$ is now just a standard application of the Wigner–Eckart theorem [Eq. (3.260) of Chapter 3]:

$$\langle (n'M'\sigma')(l'J')R'M'_R|D_\mu|(nM\sigma)(lJ)RM_R\rangle$$

$$= \delta_{R'R}\,\delta_{M'_RM_R}(-1)^{J'+l+R}\tfrac{1}{2}[(-1)^{J'+\sigma'+J+\sigma}-1]$$

$$\times\,[(2l'+1)(2J+1)]^{\frac{1}{2}}\langle n'l'\|\mathbf{T}^1(\text{vib})\|nl\rangle\, C^{J1J'}_{M\mu M'}\begin{Bmatrix} l & J & R\\ J' & l' & 1\end{Bmatrix}. \quad (7.10.349)$$

In obtaining this result, we have used the fact [see Eq. (7.10.343)] that the reduced matrix element of $T^1_{\mu;\nu}$, which is a tensor operator of rank 1 (in the index ν) with respect to $\mathbf{K}$, is given by $\frac{1}{2}[1-(-1)^{J'+\sigma'+J+\sigma}]\times[(2J+1)/(2J'+1)]^{\frac{1}{2}}C^{J1J'}_{M\mu M'}$.

Consider next the evaluation of the reduced vibrational matrix elements in Eq. (7.10.349). The components $d_\nu(\mathbf{q})$ of the electric dipole moment must transform as a vector with respect to the internal angular momentum $\mathbf{L}$ (Wilson [15, p. 157]). This implies two results: (1) A triply degenerate normal mode $\mathbf{q}$ will have an associated electric dipole moment only if $\mathbf{q}$ transforms according to the vector representation of the point group G (see footnote, p. 589; these are the irreps F_{1u} of O_h, F_2 of T_d, F_1 of O, F of T, F_u of T_h); and (2) the general form of the expansion of $d_\nu(\mathbf{q})$ into the normal coordinates (q_1, q_2, q_3) of a triply degenerate (vector) mode has the form $d_\nu(\mathbf{q}) = I(q^2)q_\nu$, where (q_{+1}, q_0, q_{-1}) are the spherical components of $\mathbf{q}$, and $I(q^2)$ is an analytic function (polynomial, if terminated) in the invariant $q^2 = \mathbf{q}\cdot\mathbf{q}$ (invariant with respect to $\mathbf{L}$). Using $\mathscr{Y}_{1\nu}(\mathbf{q}) = (3/4\pi)^{\frac{1}{2}}q_\nu$ and Eq. (7.10.206) for the harmonic oscillator wave functions, and Eqs. (3.226) and (3.437), we obtain the following expression for the reduced vibrational matrix elements in Eq. (7.10.349):

$$\langle n'l'\|\mathbf{T}^1(\text{vib})\|nl\rangle = \langle n'l'\|\mathbf{d}\|nl\rangle$$

$$= [(2l+1)/(2l'+1)]^{\frac{1}{2}}\, C^{l1l'}_{000}\int_0^\infty q^2\,dq\, R_{n'l'}(q)qI(q^2)R_{nl}(q), \quad (7.10.349')$$

in which $R_{nl}(q) = q^lF_{nl}(q)$ (and similarly for $n'l'$).

If we approximate $I(q^2)$ by a polynomial of degree k in q^2, then the radial integral[1] in the reduced matrix element (7.10.349′) vanishes unless $\Delta n = n'-n = \pm 1, \pm 3, \ldots, \pm(2k+1)$.

[1] It is interesting to note that the evaluation of these radial integrals can be given by angular momentum techniques. This is developed in RWA.

We now obtain from Eqs. (3.10.349) and (3.10.349′) the selection rules for electric dipole transitions from the Coriolis energy state $E_0[n(lJ)R]$ to the Coriolis energy state $E_0[n'(l'J')R'] > E_0[n(lJ)R]$ for a triply degenerate oscillation (vector type):

$$\Delta n = 1, 3, \ldots; \quad \Delta l = \pm 1; \quad \Delta J = 0, \qquad \Delta\sigma = \pm 1; \quad \Delta R = 0$$

$$\Delta J = \pm 1, \quad \Delta\sigma = 0; \tag{7.10.350}$$

Since the leading (constant) term in the expansion of $I(q^2)$ is usually much larger than the remaining terms, it is the $\Delta n = 1$ transitions that are the most intense (and, hence, observed). We discuss in Section u below a fundamental band of the molecule SF_6 (sulfur hexafloride) in which the selection rules (7.10.350) are experimentally obeyed. (General $\Delta n = 1$ transitions have been considered by Galbraith [77]; transitions from the ground vibrational states ($n = 0$) to the $n' = 3$ vibrational states ($\Delta n = 3$) in SF_6 – the so-called $3\nu_3$ absorption spectrum – have also recently been analyzed by Ackerhalt *et al.* [77a] and by Patterson *et al.* [77b].)

We emphasize again that for electric dipole transitions the selection rules (7.10.346) for the angular momentum quantum number J and the (parity) quantum number σ are *exact* – the general molecular Hamiltonian H [Eq. (7.10.154)] is invariant under rotation-inversions of physical space, and each eigenvalue of H is correspondingly $2(2J + 1)$-fold degenerate; the additional selection rules in Eq. (7.10.350), although exact for electric dipole transitions between Coriolis levels (7.10.202), are, in fact, applicable only if these levels are a reasonably accurate description of actual energy states of the molecule. Even if these levels are valid (a result we shall assume), a high resolution spectra may show finer detail corresponding to electric dipole transitions between the (superfine) levels given by Eq. (7.10.288), or between cluster levels.

To verify the validity (or not) of the model Hamiltonian (7.10.196), as corrected by the higher-order interactions given in Section p, it thus becomes necessary to refine the selection rules (7.10.350) by including the "quantum numbers" Γ^{α} and κ that appear in the energy expression $E[n(lJ)R; \Gamma^{\alpha}\kappa]$ given by Eq. (7.10.288). These selection rules are determined from the point group symmetry of the levels as follows: electric dipole transitions between two levels $E[n'(l'J')R'; \Gamma^{\alpha'}\kappa'] > E[n(lJ)R; \Gamma^{\alpha}\kappa]$ occur only if the quantum numbers in these levels obey the selection rules (7.10.350) and, in addition, the direct product representation of the point group G given by $\Gamma^{\alpha'} \otimes \Gamma_{\text{vector}} \otimes \Gamma^{\alpha}$ contains the identity irrep [here $\Gamma_{\text{vector}}(g) = R(g)$ = vector representation of G]. Using this result, one can, by superposition, also determine the selection rules for transitions between cluster levels.

It is a sizeable task to carry out the detailed construction of wave functions for the levels $E[n(lJ)R;\Gamma^{\alpha}\kappa]$ — or even of the wave functions for the cluster levels — and to determine their symmetry. We refer to the recent literature (as already cited) for these details.

We turn now to the second factor in the transition probability that is of interest for illustrating angular momentum techniques — the statistical weight.

The statistical weight factor entering into the intensity formula for absorption is the degeneracy of the lower molecular energy level taking into account the nuclear spin and the Pauli exclusion principle (Wilson [78]). In applying the Pauli principle, it is important to recognize that the Hamiltonian we have developed for describing a rigid molecule treats equivalently only those nuclei that are permuted under the action of the point group G, as discussed earlier on p. 586. This group has been denoted by $P(G)$ in Section o, and it is an invariance group of the molecular Hamiltonian. *Precisely these permutations of nuclei are to be considered in applying the Pauli principle*. If the nuclei are labeled initially by $1, 2, \ldots, N$ and the permutation $1 \to 1'$, $2 \to 2', \ldots, N \to N'$ is not an element of $P(G)$, then one obtains a new Hamiltonian and a new permutation group $P'(G)$ that leads to an equivalent description of the molecule and to the concept of framework (Wilson [78]).

For the purpose of this discussion, we shall assume that the lower state is the ground electronic and ground vibrational state, and, furthermore, that these states are invariant under all permutations $P_g \in P(G)$. Our problem is then simplified to a determination of the transformation properties, under permutations P_g, of the product states:

$$|\sigma JMK\rangle \otimes |SK_S\rangle, \tag{7.10.351}$$

in which $|\sigma JMK\rangle$ denotes the rotator state vector (7.10.204),

$$\langle C|\sigma JMK\rangle \equiv \Psi^{J,\sigma}_{MK}(C), \tag{7.10.352}$$

and $|SK_S\rangle$ denotes a total spin state, which we now describe[1] (see footnote, p. 621).

Let $\mathbf{S} = (S_1, S_2, S_3)$ denote the total spin with components S_i ($i = 1, 2, 3$) referred to the laboratory (inertial) frame $(\hat{l}_1, \hat{l}_2, \hat{l}_3)$. These operators then

[1] We choose the Eckart frame axis $\hat{f}_3$ as quantization axis for the spin states, which are denoted by $|SK_S\rangle$. This is convenient in the construction of rotator-spin states that obey the Pauli principle, since the permutation group $P(G)$ induces transformations [see Eq. (7.10.360)] on the internal quantum number K, which also refers to quantization axis $\hat{f}_3$. The spin states, $|SM_S\rangle_{\text{lab}}$, which have the laboratory axis $\hat{l}_3$ as quantization axis, are related to the internal spin states by $|SM_S\rangle_{\text{lab}} = \sum_{K_S} D^S_{K_S M_S}(U)|SK_S\rangle$, where $\pm U \to (\det C)C$ in the homomorphism of $SU(2)$ onto $SO(3)$ [see Eq. (7.10.297)].

satisfy standard commutation relations

$$[S_i, S_j] = ie_{ijk}S_k. \tag{7.10.353}$$

The definition (see footnote, p. 555) of the spin operators referred to the body-fixed frame $(\hat{f}_1, \hat{f}_2, \hat{f}_3)$ is

$$T_i = +(\det C)\hat{f}_i \cdot \mathbf{S}, \qquad i = 1, 2, 3. \tag{7.10.354}$$

The operators $\{T_i\}$ then satisfy the standard commutation relations

$$[T_i, T_j] = ie_{ijk}T_k, \tag{7.10.355}$$

since each of the spin operators S_i commutes with each of the frame vectors $\hat{f}_i$. Moreover, each T_i is a rotation–inversion invariant (since $\mathbf{S} \to \mathbf{S}$ under inversion) and changes sign under time reversal (since $\mathbf{S} \to -\mathbf{S}$ under time reversal). (Notice that the operators $\{S_i\}$ *do not commute* with the operators $\{T_i\}$ – the total spin functions do not have the structure of rotator functions, see footnote, p. 617.) The basis vectors $\{|SK_S\rangle : K_S = S, S-1, \ldots, -S\}$ are the simultaneous eigenvectors of $\mathbf{S}^2 = \mathbf{T}^2 = T_1^2 + T_2^2 + T_3^2$ and T_3, and the operators $\{T_i\}$ have the standard action on these basis vectors:

$$T_\pm |SK_S\rangle = [(S \mp K_S)(S \pm K_S + 1)]^{\frac{1}{2}} |S, K_S \pm 1\rangle,$$

$$T_3 |SK_S\rangle = K_S |SK_S\rangle. \tag{7.10.356}$$

Thus, S refers to the total spin and K_S to the projection quantum number associated with the operator T_3, which refers to the Eckart frame axis $\hat{f}_3$ as quantization axis.

Consider next the effect of the group of permutations $P(G)$ on the product state vectors (7.10.351). We have from Eqs. (7.10.248) and (7.10.243) that the Eckart frame undergoes the transformation

$$P_g : \hat{f}_i \to \sum_j R_{ji}(g)\hat{f}_j, \qquad \text{each } P_g \in P(G), \tag{7.10.357}$$

where $R(g)$ is the orthogonal matrix given by Eq. (7.10.216) and determined by the static molecular model. Hence, one has

$$P_g : C \to CR(g). \tag{7.10.358}$$

Thus, under the action of each $P_g \in P(G)$, we have the following transformations of angular momentum and spin components:

$$P_g\colon \mathbf{K} \to [\det R(g)]\tilde{R}(g)\mathbf{K},$$

$$\mathbf{T} \to [\det R(g)]\tilde{R}(g)\mathbf{T}, \qquad (7.10.359)$$

where $\mathbf{K} = \text{col}(K_1, K_2, K_3)$ and $\mathbf{T} = \text{col}(T_1, T_2, T_3)$ in these matrix transformations.

We thus find that under the action of a permutation $P_g \in P(G)$ the rotator functions undergo the transformation

$$P_g\colon |\sigma JMK\rangle \to [\det R(g)]^\sigma \sum_{K'} \mathscr{D}^J_{K'K}(R(g))|\sigma JMK'\rangle. \qquad (7.10.360)$$

The rotator basis thus carries the reducible representation,

$$P_g \to [\det R(g)]^\sigma \mathscr{D}^J(R(g)), \qquad (7.10.361)$$

of the permutation group $P(G)$. Let us denote this representation of $P(G) \cong G$ by $\Gamma^{J,\sigma}(G)$.

Recall now that the permutation group $P(G)$ and the (isomorphic) transformation group $L(G)$ are both invariance groups of the molecular Hamiltonian (and, in particular, of the Hamiltonian (7.10.196), including all higher-order interactions). It follows from this fact that the subspace of state vectors corresponding to a given energy level $E[n(lJ)R; \Gamma^\alpha\kappa]$ [see Eq. (7.10.288)] is irreducible under the action of the permutation group $P(G)$ as well as that of the transformation group $P(G)$. In particular, the orthogonal matrix that diagonalizes the interaction matrix A [see Eq. (7.10.290)] for $R = J$ must completely reduce the representation $\Gamma^{J,\sigma}(G)$. [Indeed, the representation of $P(G)$ and $L(G)$ carried by the rotator functions (7.10.352) are identical—to obtain this result set $l = 0$, $R = J$, and $A(g) = 1$ in Eq. (B-5) of Appendix B and compare the result with Eq. (7.10.361).]

Let us assume that the explicit reduction of the representation $\Gamma^{J,\sigma}(G)$ of $P(G)$ has been carried out. For use below [see Eqs. (7.10.371)–(7.10.373)] it is convenient to write the corresponding abstract reduction series as

$$\Gamma^{J,\sigma}(G) = A_\sigma \otimes \left(\sum_\alpha \oplus k_\alpha \Gamma^\alpha\right), \qquad (7.10.361')$$

where A_σ denotes the one-dimensional irrep of G defined by $A_\sigma(g) = [\det R(g)]^\sigma$, $\sigma = 0, 1$. Accordingly, the term $\sum_\alpha \oplus k_\alpha \Gamma^\alpha$ is the reduction series for the representation $\Gamma^{J,0}(G) = \{\mathscr{D}^J(R(g))\colon g \in G\}$ in which $\sigma = 0$.

Consider next the transformation properties of the spin basis

$$\{|SK_S\rangle\colon K_S = S, S-1, \ldots, -S\} \qquad (7.10.362)$$

under permutations. In general, these functions will be linear combinations of basis vectors in the tensor product space:

$$\{|s\mu_1\rangle \otimes |s\mu_2\rangle \otimes \cdots \otimes |s\mu_n\rangle : \text{each } \mu_k = s, s-1, \ldots, -s\}, \tag{7.10.363}$$

where s is the spin of a nucleus in a set of n identical nuclei and μ_i is the projection quantum number referred to the Eckart frame axis $\hat{f}_3$. Thus, the explicit construction of the spin states in Eq. (7.10.351) would require finding those linear combinations of the basis vectors (7.10.363) that transform irreducibly under the action of the total spin operators (T_1, T_2, T_3) and irreducibly under the action of permutations P_g of the spin states.

In general, the effect of a permutation $P \in S_n$ on a state vector in the set (7.10.363) is to permute the single-nucleus spin states to the various positions in the tensor product. In the present case, however, a permutation P_g not only has the action of moving these single-nucleus spin states into various positions in the tensor product, it also transforms the individual spin components according to Eq. (7.10.359). Thus, the action of a permutation

$$P_g = \begin{pmatrix} 1 & 2 & \cdots & n \\ i_1(g) & i_2(g) & \cdots & i_n(g) \end{pmatrix} \tag{7.10.364}$$

on a basis vector in the set (7.10.363) is

$$\begin{aligned} P_g\colon & |s\mu_1\rangle \otimes |s\mu_2\rangle \otimes \cdots \otimes |s\mu_n\rangle \\ & \to \mathscr{U}_g|s\mu_{i_1(g)}\rangle \otimes \mathscr{U}_g|s\mu_{i_2(g)}\rangle \otimes \cdots \otimes \mathscr{U}_g|s\mu_{i_n(g)}\rangle, \end{aligned} \tag{7.10.365}$$

where $\mathscr{U}_g$ is the unitary transformation of single-nucleus spin states given by

$$\mathscr{U}_g|s\mu\rangle = \sum_{\mu'} D^s_{\mu'\mu}(U(g))|s\mu'\rangle. \tag{7.10.366}$$

In this result $\pm U(g)$ is the unitary matrix corresponding to the *proper orthogonal* transformation [det $R(g)$] $R(g)$ occurring in Eq. (7.10.359) [see Eq. (2.40), Chapter 2].

In general, the first problem in the explicit construction of the nuclear spin functions is to determine those linear combinations of the basis vectors (7.10.363) that transform irreducibly under the action of P_g. Let us illustrate this construction in the case of $s = \frac{1}{2}$.

In the case of spin-$\frac{1}{2}$, that is, $s = \frac{1}{2}$, we gave in the Appendix to Section 5 the explicit construction of the spin states that transform irreducibly under the action of a permutation $P \in S_n$. These are the double standard tableau basis vectors corresponding to the Young frame of shape

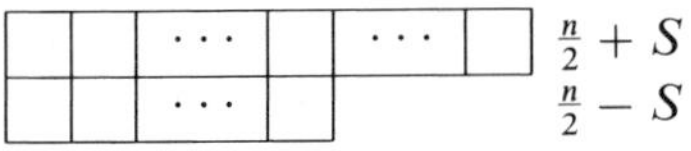

These basis vectors are denoted by

$$|(T_{U_2}|T_{S_n})\rangle = |(y_1 y_2 \cdots y_n)SK_S\rangle, \tag{7.10.367}$$

where (S, K_S) are the spin quantum numbers of the standard T_{U_2} tableau, and $(y_1 y_2 \cdots y_n)$ is the Yamanouchi symbol of the standard T_{S_n} tableau.[1]

Under the action of P_g the basis vectors (7.10.367) are transformed according to

$$P_g\colon |(y)SK_S\rangle \to \sum_{(y')K'_S} D^S_{K'_S K_S}(U(g))\,\Gamma^{[(n/2)+S,\,(n/2)-S]}_{(y')(y)}(P_g)|(y')SK'_S\rangle. \tag{7.10.368}$$

Thus, the vector space spanned by the double standard tableau basis for fixed S (fixed shape of the Young frame) carries the (reducible) representation

$$P_g \to D^S(U(g)) \otimes \Gamma^{[(n/2)+S,\,(n/2)-S]}(P_g), \qquad \text{each } g \in G \tag{7.10.369}$$

of the permutation group $P(G) \cong G$. Thus, the problem of constructing explicit spin states ($s = \frac{1}{2}$) that transform irreducibly under the permutation group $P(G) \cong G$ is that of reducing the direct product representation (7.10.369) of $P(G)$ into irreps of the point group G. One must, of course, carry this procedure out for each spin value $S = \frac{n}{2}, \frac{n}{2} - 1, \ldots, 1$ or 0.

[For cases of nuclei having spin other than $\frac{1}{2}$, the procedure is analogous to that described above, but more difficult to describe in the same detail, and will not be considered here.]

Let $\Gamma^{n,S}(G)$ denote the representation (7.10.369) of the point group G that is carried by the spin functions (7.10.367) under the action of the permutation group $P(G)$ [the dimension of $\Gamma^{n,S}(G)$ is $(2S+1) \times \dim[\frac{n}{2} + S\ \frac{n}{2} - S]$— see Eq. (A-18), Appendix A, Section 5]. We denote the reduction series for the representation $\Gamma^{n,S}(G)$ of G by

$$\Gamma^{n,S}(G) = \sum_{\alpha} \oplus\, d_\alpha \Gamma^\alpha. \tag{7.10.370}$$

[1] Observe then that for n spin-$\frac{1}{2}$ nuclei each spin vector in Eq. (7.10.351) will also be labeled by a Yamanouchi symbol.

It is difficult, for general n and S, to give the construction of explicit spin states that transform according to each of the irreps Γ^α of G occurring in the reduction series (7.10.370), although it can, in principle, be carried out. We assume, for this discussion, that this reduction has been effected.

Combining the splitting of the rotator state vector space [Eq. (7.10.361′)] into subspaces that carry irrep $A_\sigma \otimes \Gamma^\alpha$ of $P(G)$ with the splitting of the spin state vector space [Eq. (7.10.370)] into subspaces that carry irrep $\Gamma^{\alpha'}$ of $P(G)$ will lead to sets of direct product vectors of the form

$$|(\sigma JM)\kappa; \Gamma^\alpha \gamma\rangle \otimes |(nS)\lambda; \Gamma^{\alpha'}\gamma'\rangle, \tag{7.10.371}$$

where the vectors enumerated by $\gamma = 1, \ldots, \dim \Gamma^\alpha$ and $\gamma' = 1, \ldots, \dim \Gamma^{\alpha'}$ are a basis of the space that carries irrep

$$A_\sigma \otimes \Gamma^\alpha \otimes \Gamma^{\alpha'} \tag{7.10.372}$$

of the permutation group $P(G) \cong G$. [The index κ has the significance of enumerating the eigenvalues $E(J\Gamma^\alpha\kappa)$ in Eq. (7.10.294), and λ is a "multiplicity" index enumerating the number of occurrences, $d_{\alpha'}$, of the irrep $\Gamma^{\alpha'}$ in the reduction series (7.10.370).] One now completes the construction of rotator-spin states transforming irreducibly under the action of $P(G)$ by reducing the representation (7.10.372) by forming the appropriate linear combinations (over the indices γ and γ') of the basis vectors (7.10.371). [As discussed below, the only states of physical interest in this reduction are the "Pauli states," which transform as the one-dimensional "antisymmetric" irrep of G.]

We have given considerable detail above in the discussion of the transformation properties of rotator and spin basis vectors under the action of the permutation group $P(G)$—and the reduction of the corresponding matrix representations—, since there is considerable confusion on this subject in the literature.

Let us consider now the implementation of the Pauli principle (for spin-$\frac{1}{2}$ nuclei). Let $\Gamma_{\text{Pauli}}(G)$ denote the one-dimensional representation of the point group G defined by $\Gamma_{\text{Pauli}}(g) = +1$ or -1 according to whether or not the permutation $P_g \leftrightarrow g$ is even or odd. We define a Pauli state of sharp angular momentum (J, M), sharp parity $J + \sigma$, and sharp spin S to be a linear combination of the direct product states $|\sigma JKM\rangle \otimes |(y)SK_S\rangle$ that transforms like the irrep $\Gamma_{\text{Pauli}}(G)$ under the action of the permutation group $P(G)$. (The general linear combination is taken over all $2J + 1$ values of K, over all $2S + 1$ values of K_S, and over all Yamanouchi symbols corresponding to the standard Young tableaux of shape $[\frac{n}{2} + S\ \frac{n}{2} - S]$.)

The statistical weight of the ground vibrational (superfine) level $E[0(0J)0; \Gamma^\alpha\kappa]$ [see Eq. (7.10.288)] is the number of orthogonal Pauli states that possess this energy eigenvalue, accounting for all spin values $S = \frac{n}{2}, \frac{n}{2} - 1, \ldots, \frac{1}{2}$ or 0, and for the two parity values $\sigma = 0, 1$ (since the energy eigenvalue is independent of both these quantum numbers). The statistical weight of the level $E[0(0J)0; \Gamma^{\alpha}\kappa]$ is therefore the total number of states that will be found to belong to irrep $\Gamma_{\text{Pauli}}(G)$ at the end of the reduction procedure described on pp. 619–622. It is not necessary to carry out the reduction procedure explicitly to count the number of (orthogonal) Pauli states: If we define

$$\Gamma_{\text{spin}} = \sum_{S} \Gamma^{n,S}(G), \tag{7.10.373}$$

then the statistical weight of the level $E[0(0J)0; \Gamma^{\alpha}\kappa]$ is clearly the number of occurrences of $\Gamma_{\text{Pauli}}(G)$ in the *two* direct products

$$A_{\sigma} \otimes \Gamma^{\alpha} \otimes \Gamma_{\text{spin}}, \qquad \sigma = 0, 1, \tag{7.10.374}$$

where we recall that A_{σ} denotes the one-dimensional irrep of G given by $A_{\sigma}(g) = [\det R(g)]^{\sigma}$, and Γ^{α} is an irrep of G contained in $\Gamma^{J,0}$.

Let us now apply the above results to the determination of the statistical weights of an octahedrally (O_h) symmetric molecule XY_6 in which the Y-nuclei have spin-$\frac{1}{2}$ and $\Gamma_{\text{Pauli}}(G) = A_{2u}$. The statistical weight of each O_h-irreducible energy level $E(\Gamma^{\alpha})$, where Γ^{α} denotes an irrep of O_h, is obtained from the rule (Cantrell and Galbraith [79]): *The statistical weight of $E(\Gamma^{\alpha})$ equals the number of irreps A_{2u} that appear in the reduction of*

$$A_{1g} \otimes \Gamma^{\alpha} \otimes \Gamma_{\text{spin}} \qquad \text{and} \qquad A_{1u} \otimes \Gamma^{\alpha} \otimes \Gamma_{\text{spin}}.$$

$$[\Gamma^{\alpha} = A_{1\tau}, A_{2\tau}, E_{\tau}, F_{1\tau}, F_{2\tau}; \tau = g(\text{gerade}), \text{ and } \tau = u(\text{ungerade}).] \tag{7.10.375}$$

The reduction of the 2^6 spin wave functions into irreps of O_h is given by (see Galbraith [80] for the generalization of this result to arbitrary spin and Weber [80a] for detailed tabulations of statistical weights):

$$\Gamma_{\text{spin}} = 10A_{1g} + A_{2g} + A_{2u} + 8E_g + 6F_{1u} + 3F_{2g} + 3F_{2u}.$$

The number of A_{2u} appearing in the direct products (7.10.375) is easily determined by using the multiplication of irreps (algebra of representations) given by

	A_1	A_2	E	F_1	F_2
A_1	A_1	A_2	E	F_1	F_2
A_2	A_2	A_1	E	F_2	F_1
E	E	E	$A_1 + A_2 + E$	$F_1 + F_2$	$F_1 + F_2$
F_1	F_1	F_2	$F_1 + F_2$	$A_1 + E + F_1 + F_2$	$A_2 + E + F_1 + F_2$
F_2	F_2	F_1	$F_1 + F_2$	$A_2 + E + F_1 + F_2$	$A_1 + E + F_1 + F_2$

$$g \times g = g, \qquad g \times u = u \times g = u, \qquad u \times u = g.$$

The results of this calculation are (Ref. [79])

Level $E(\Gamma^\alpha)$:	A_1	A_2	E	F_1	F_2
Statistical weight:	2	10	8	6	6

(The statistical weight is independent of the labels g and u.)

Since the sublevels in a cluster are not resolved, it is also useful to note the statistical weights for the cluster levels:

Cluster symmetry	Statistical weight
$A_1 + E + F_1$	16
$F_1 + F_2$	12
$A_2 + E + F_2$	24
$A_1 + A_2 + F_1 + F_2$	24
$E + F_1 + F_2$	20

u. Spectra of fundamental transitions of SF_6. We conclude by illustrating briefly the applicability of the conceptual developments of this section to real spectra, noting, in particular, the accuracy between theoretical predictions and experimental observations.

In the dominant approximation to the interaction between a triply degenerate oscillation and the rotation of a spherical top molecule, the transition frequencies in a fundamental band are obtained from Eq. (7.10.288). For the P-, Q-, and R-branches, the standard forms (Moret-Bailly [60]) for expressing these energy terms (in cm^{-1}) are

$$P(J) = m - nJ + pJ^2 - qJ^3 + t_{JJ} F^{(4JJ)}_{A_1pp},$$

$$Q(J) = m + vJ(J+1) + t_{JJ} F^{(4JJ)}_{A_1pp},$$

$$R(J) = m + n(J+1) + p(J+1)^2 + q(J+1)^3 + t_{JJ} F^{(4JJ)}_{A_1pp}, \qquad (7.10.376)$$

where $(-1)^J F^{(4JJ)}_{A_1pp}$ enumerates, for various values of a discrete index p, the eigenvalues of the operator T [Eq. (7.10.293)]. The value of the total angular momentum of the ground state is J. The tensor factor t_{JJ} is different for each branch. The three parameters t_{JJ} are related to two numerical parameters, denoted by g and h, and to J-dependent factors as follows:

$$(-1)^J t_{JJ} = \begin{cases} (g + hJ)[(2J-3)_9]^{\frac{1}{2}}/2J(2J-1), & (P) \\ -2g[(2J-3)_9]^{\frac{1}{2}}/2J(2J+2), & (Q) \\ [g - h(J+1)][(2J-3)_9]^{\frac{1}{2}}/(2J+2)(2J+3) & (R). \end{cases} \quad (7.10.377)$$

For the ν_3 fundamental band of SF_6, the seven parameters appearing in these expressions are known (Galbraith *et al.* [75], Loëte *et al.* [81]). Expressed in cm^{-1}, they are

$$\begin{aligned} m &= 947.976\,575\,9 \pm 0.000\,004\,3 & v &= (-6.998\,70 \pm 0.000\,18)10^{-5} \\ n &= (5.581\,760 \pm 0.000\,014)10^{-2} & g &= (-2.458\,283 \pm 0.000\,082)10^{-5} \\ p &= (-1.618\,642 \pm 0.000\,022)10^{-4} & h &= (-5.63 \pm 0.12)10^{-10} \\ q &= (1.038\,9 \pm 0.003\,8)10^{-8} \end{aligned}$$

These data for the ν_3-band of SF_6 were obtained in saturated absorption, using CO_2 lasers. Such techniques are subdoppler and are limited only by the homogeneous line width.

For the ν_4 fundamental band of SF_6, these same seven parameters have also been determined (Kim *et al.* [82]). They are

$$\begin{aligned} m &= 615.024\,7 \pm 0.000\,4 & v &= (-1.91 \pm 0.02)\,10^{-5} \\ n &= 0.221\,514 \pm 0.000\,013 & g &= (2.79 \pm 0.02)\,10^{-6} \\ p &= (-2.414 \pm 0.014)\,10^{-5} & h &= (6.3 \pm 0.3)\,10^{-9} \\ q &= (-4.4 \pm 0.3)\,10^{-8} \end{aligned}$$

These data may be used to obtain the approximate values of B and ζ-factors given by (Kim *et al.* [82])

$$B = (0.09111 \pm 0.00005),$$

$$\zeta_3 = 0.6937 \pm 0.0002,$$

$$\zeta_4 = -0.2156 \pm 0.0007.$$

The ν_4-band data were obtained by direct absorption using laser diodes and kilometer path spectrometers. The method of determining the constants consists in fitting (by a least-squares method) the expressions (7.10.376) to a large number of measured line frequencies having unambiguous sharp peaks.

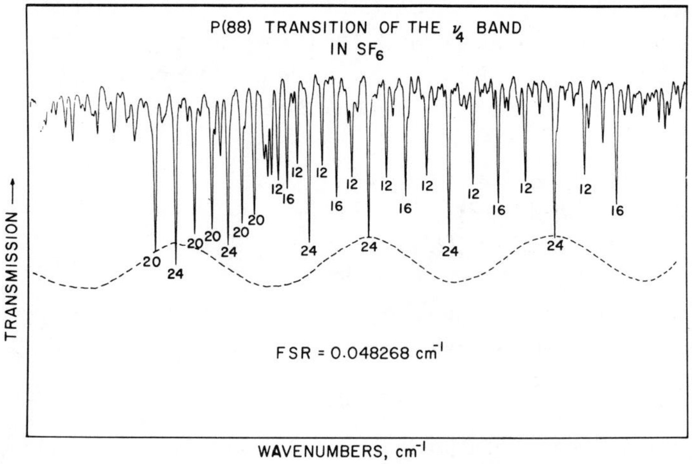

Figure 7.7. Diode spectrum of the $P(88)$ transition in the ν_4-band of SF_6 at 595.4 cm^{-1}. Numbers at resonances are spin statistical weights for the cluster levels [see Eq. (3.74)]. The dashed curve represents an interference pattern taken from a one-inch germanium etalon, used for wave number calibration – the periodicity is the free spectral range (FSR). (From K. C. Kim, W. B. Person, D. Seitz, and B. J. Krohn, "Analysis of the ν_4 (615 cm^{-1}) region of the Fourier transform and diode laser spectra of SF_6," *J. Mol. Spectrosc.* **76** (1979), 322–340. – Reprinted by permission.)

To illustrate this latter data for the fundamental ν_4-band of SF_6, we refer to Fig. 7.7, which shows a high-resolution spectrum. The lines identified by the integers 12, 16, 20, and 24 are the absorption lines for transitions from the ground vibrational state with total angular momentum $J = 88$ to the cluster sublevels into which the upper ($n = 1$) Coriolis level (see Fig. 7.6) having $R = 88$ and total angular momentum $J' = 87$ (P-branch, $\Delta J = -1$) is split by the octahedrally invariant interaction T. The integers 12, 16, 20, and 24 are themselves the statistical weights of the clusters (see Section t). Since $88 \equiv 1$ (mod 3) and $88 \equiv 0$ (mod 4), the clusters are predicted to have the relative intensities occurring in the sequence 20, 24, 20, (repeat), ... reading from left to right at the left end (low energy) of the spectrum (Fig. 7.7) and in the sequence 16, 12, 24, 12, (repeat), ... reading from right to left at the right end (high energy) of the spectrum. The "crossover" region should be located at a distance from the left end of the spectrum that is about one-quarter the length of the spectrum. Note the remarkable agreement.

v. Appendices.

Appendix A. Calculation of $\vec{\Omega}_i^\alpha$. The matrix expression, Eq. (7.10.123), for Ω^α is derived in this Appendix.

The solution to Eq. (7.10.111), which defines $\vec{\Omega}_i^\alpha$, is given in component form by

$$\Omega_{ij}^{\alpha} \equiv \vec{\Omega}_i^{\alpha} \cdot \hat{f}_j = \hat{f}_k \cdot \frac{\partial \hat{f}_l}{\partial x_i^{\alpha}} = -\hat{f}_l \cdot \frac{\partial \hat{f}_k}{\partial x_i^{\alpha}}, \tag{A-1}$$

where j, k, l are cyclic in 1, 2, 3.

To evaluate the right-hand side of Eq. (A-1), we use Eq. (7.10.45) for the Eckart frame case and Eq. (7.10.15) for the principal axes case.

▶ *Eckart frame.* We rewrite Eq. (7.10.45) in the form

$$\sum_k \hat{f}_k G_{jk} = \vec{F}_j = \sum_\beta m_\beta a_j^\beta \vec{x}^\beta, \tag{A-2}$$

and differentiate to obtain first

$$\sum_k \frac{\partial \hat{f}_k}{\partial x_i^\alpha} G_{jk} + \sum_k \hat{f}_k \frac{\partial G_{jk}}{\partial x_i^\alpha} = m_\alpha a_j^\alpha \hat{l}_i, \tag{A-3}$$

and then

$$\sum_k \hat{f}_l \cdot \frac{\partial \hat{f}_k}{\partial x_i^\alpha} G_{jk} + \frac{\partial G_{jl}}{\partial x_i^\alpha} = m_\alpha a_j^\alpha C_{il}. \tag{A-4}$$

We obtain a second equation from Eq. (A-4) by interchanging j and l, noting that $G_{jl} = G_{lj}$. Subtracting this new equation from Eq. (A-4) eliminates the derivative term in G_{jl}. Thus,

$$\sum_k \hat{f}_l \cdot \frac{\partial \hat{f}_k}{\partial x_i^\alpha} G_{jk} - \sum_k \hat{f}_j \cdot \frac{\partial \hat{f}_k}{\partial x_i^\alpha} G_{lk} = m_\alpha (a_j^\alpha C_{il} - a_l^\alpha C_{ij}). \tag{A-5}$$

Using Eq. (A-1), this last result takes the matrix form:

$$\begin{pmatrix} -G_{11} + \operatorname{tr} G & -G_{12} & -G_{13} \\ -G_{21} & -G_{22} + \operatorname{tr} G & -G_{23} \\ -G_{31} & -G_{32} & -G_{33} + \operatorname{tr} G \end{pmatrix} \begin{pmatrix} \Omega_{i1}^\alpha \\ \Omega_{i2}^\alpha \\ \Omega_{i3}^\alpha \end{pmatrix} = -(A^\alpha \tilde{C})_i, \tag{A-6}$$

where the subscript i on the right designates the ith column of the matrix product $A^\alpha \tilde{C}$.

Letting $i = 1, 2, 3$ in Eq. (A-6), we obtain

$$[-G + (\operatorname{tr} G)\mathbb{1}_3]\tilde{\Omega}^\alpha = -A^\alpha \tilde{C}, \tag{A-7}$$

where Ω^α is the 3×3 matrix having Ω_{ij}^α in row i and column j. We thus find

$$\Omega^\alpha = C A^\alpha M, \tag{A-8}$$

where M is the symmetric matrix

$$M = [-G + (\operatorname{tr} G)\mathbb{1}_3]^{-1}. \tag{A-9}$$

▶ *Principal axes.* Using $\hat{f}_l = \sum_m C_{ml}\hat{l}_m$ and the definition (A-1), we obtain

$$\Omega^\alpha_{ij} = \left(\tilde{C}\frac{\partial C}{\partial x^\alpha_i}\right)_{kl}, \tag{A-10}$$

where j, k, l are cyclic in 1, 2, 3. Differentiation of the defining relation, $QC = C\Lambda$, $\Lambda = \text{diag}(\lambda_1, \lambda_2, \lambda_3)$ and multiplication of the resulting expression from the left by $\tilde{C}$ yields

$$\tilde{C}\frac{\partial Q}{\partial x^\alpha_i}C - \frac{\partial \Lambda}{\partial x^\alpha_i} = \tilde{C}\frac{\partial C}{\partial x^\alpha_i}\Lambda - \Lambda\tilde{C}\frac{\partial C}{\partial x^\alpha_i}. \tag{A-11}$$

Combining this result with Eq. (A-10), we obtain

$$\Omega^\alpha_{ij} = (\lambda_k - \lambda_l)^{-1}\left(\tilde{C}\frac{\partial Q}{\partial x^\alpha_i}C\right)_{kl}, \tag{A-12}$$

where j, k, l, are cyclic in 1, 2, 3. Using $Q_{jk} = \sum_\alpha m_\alpha(x^\alpha_j - R_j)(x^\alpha_k - R_k)$ and the transformation to internal coordinates, Eq. (7.10.106), we find

$$\frac{\partial Q_{jk}}{\partial x^\alpha_i} = \delta_{ij}\sum_l C_{kl}\mu_l\zeta^\alpha_l + \delta_{ik}\sum_l C_{jl}\mu_l\zeta^\alpha_l. \tag{A-13}$$

Substituting this result into Eq. (A-12) now leads to

$$\Omega^\alpha = CA^\alpha M, \tag{A-14}$$

where A^α and M are defined, respectively, by Eqs. (7.10.135) and (7.10.136).

Appendix B. Global formulation of the $O(3) \supset G$ reduction problem for spherical top molecules. It is useful to give the explicit irrep of the internal $O(3)$ transformation group that is carried by the basis $|RM_R\rangle = |(nM\sigma)(lJ)RM_R\rangle$, $M_R = R, \ldots, -R$ [see Eq. (7.10.207)]. We find:[1]

$$\langle S^{-1}\mathbf{q}; CS|RM_R\rangle = (\det S)^{\sigma+l+J+R}\sum_{M'_R}\mathcal{D}^R_{M'_RM_R}(S)|RM'_R\rangle, \tag{B-1}$$

each $S \in O(3)$. For $S \in SO(3)$ this result follows from the fact that $\mathbf{L}$, $\mathbf{K}$, and $\mathbf{R} = \mathbf{L} + \mathbf{K}$ have the standard action on the uncoupled and coupled bases, $\{\Phi_{nlm} \otimes \Psi^{J,\sigma}_{MK}\}$ and $\{|RM_R\rangle\}$; for $S \in O(3)$ the result follows from $-S \in SO(3)$ and the homogeneous polynomial properties of the solid harmonics and of the $\Psi^{J,\sigma}_{MK}$. Alternatively, we may establish the above result by carrying out explicitly the transformations $S^{-1}\mathbf{q}$ and CS in

$$\sum_{mK} C^{lJR}_{mKM_R}\Phi_{nlm}(S^{-1}\mathbf{q})\Psi^{J,\sigma}_{MK}(CS) \tag{B-2}$$

[1] In this appendix the symbol S denotes a 3×3 orthogonal matrix.

by using the transformation properties of the solid harmonics and $O(3)$ representation functions as given by Eqs. (6.23), (7.10.204), (6.182), (6.32), and (3.188). Thus, we find the desired transformation of the basis $\{|RM_R\rangle\}$ under the action of each $S \in O(3)$:

$$S: |RM_R\rangle \to \sum_{M'_R} \mathscr{D}^{R,\tau}_{M'_R M_R}(S)|RM'_R\rangle, \tag{B-3}$$

where $\tau \equiv (\sigma + l + J + R)(\text{mod } 2)$ and

$$\mathscr{D}^{R,\tau}_{M'_R M_R}(S) = (\det S)^\tau \mathscr{D}^{R}_{M'_R M_R}(S). \tag{B-4}$$

It follows from Eqs. (7.10.254) that the space with basis $\{|RM_R\rangle: M_R = R, \ldots, -R\}$ is the carrier space of the reducible representation

$$\{\mathscr{D}^{R,\tau}(A(g)R(g)): g \in G\} \tag{B-5}$$

of the cubic group $G = O_h, O, T_d, T_h$, or T. The reduction of this representation of G is effected fully by the orthogonal matrix that diagonalizes the matrix A given by Eq. (7.10.290) for $G = O_h, O$, or T_d, and almost fully for $G = T_h$ or T (see footnote p. 597).

The representation of G given by Eq. (B-5) may be simplified further by using the property $\mathscr{D}^R(\lambda S) = \lambda^R \mathscr{D}^R(S)$ of the $O(3)$ representation functions. Using $R(g) = [\det R(g)]R'(g)$, where $R'(g) = [\det R(g)]R(g) \in SO(3)$, and the fact that $A(g)$ is a one-dimensional irrep of G, we find:

$$\mathscr{D}^{R,\tau}(A(g)R(g)) = A'(g)\mathscr{D}^R(R'(g)), \tag{B-6}$$

where $A'(g) = [A(g) \det R(g)]^{R+\tau} = \pm A(g) \det R(g)$ is again a one-dimensional irrep of G [each $A'(g)$ is $+1$ or -1]. Using the reduction of $\mathscr{D}^R(R'(g))$ into irreps of G as determined by the operator A, we may use Eq. (B-6) and the multiplication of irreps of G to determine the corresponding reduction of the irreps $\mathscr{D}^{R,\tau}(S)$ of $O(3)$ into irreps of G.

References

1. M. Born and J. Oppenheimer, "Zur Quantentheorie der Molekeln," *Ann. Physik* **84** (1927), 457–484.
2. J. C. Brester, *Kristallsymmetrie und Reststrahlen*. Dissertation, Utrecht, 1923; *Z. Physik* **24** (1924), 324–344.
3. E. P. Wigner, "Über die elastischen Eigenschwingungen symmetrischer Systeme," *Göttinger Nachrichten* (1930), 133–146. A translation and discussion of this paper among others appears in the monograph by A. P. Cracknell, *Applied Group Theory*. Pergamon, New York, 1968.
4. E. B. Wilson, "The degeneracy, selection rules, and other properties of the normal vibrations of certain polyatomic molecules," *J. Chem. Phys.* **2** (1934), 432–439.
5. C. Eckart, "The kinetic energy of polyatomic molecules," *Phys. Rev.* **46** (1934), 383–387.
6. C. Eckart, "Some studies concerning rotating axes and polyatomic molecules," *Phys. Rev.* **47** (1935), 552–558.
7. H. B. G. Casimir, *Rotation of a Rigid Body in Quantum Mechanics*. Thesis, University of Leyden, Wolters, Groningen, 1931. [*Koninkl. Ned. Akad. Wetenschap. Proc.* **34** (1931), 844].

8. J. O. Hirschfelder and E. Wigner, "Separation of rotational coordinates from the Schrödinger equation for n particles," *Proc. Amer. Acad. Sci.* **21** (1935), 113–119.
9. J. H. Van Vleck, "The rotational energy of polyatomic molecules," *Phys. Rev.* **47** (1935), 487–493.
10. B. Buck, L. C. Biedenharn, and R. Y. Cusson, "Collective coordinates for the description of rotational motion in many particle systems," *Nucl. Phys.* **A317** (1979), 205–241.
11. F. Jørgensen, *On the Molecular Vibration–Rotation Problem.* Thesis, University of Copenhagen, 1978.
12. E. B. Wilson and J. B. Howard, "The vibration–rotation energy levels of polyatomic molecules," *J. Chem. Phys.* **4** (1936), 260–268.
13. H. H. Nielsen, "The vibration–rotation energies of molecules," *Rev. Mod. Phys.* **23** (1951), 90–136.
14. W. H. Shaffer, H. H. Nielsen, and L. H. Thomas, "The rotation–vibration energies of tetrahedrally symmetric pentatomic molecules. I," *Phys. Rev.* **56** (1939), 895–907.
15. E. B. Wilson, J. C. Decius, and P. C. Cross, *Molecular Vibrations. The Theory of Infrared and Raman Vibrational Spectra.* McGraw-Hill, New York, 1955.
16. J. D. Louck and H. W. Galbraith, "Eckart vectors, Eckart frames, and polyatomic molecules," *Rev. Mod. Phys.* **48** (1976), 69–106.
17. H. C. Schweinler and E. P. Wigner, "Orthogonalization methods," *J. Math. Phys.* **11** (1970), 1693–1694.
18. R. Landshoff, "Quantenmechanische Berechnung des Verlaufes der Gitterenergie des Na-Cl-Gitters in Abhängigkeit vom Gitterabstand," *Z. Physik* **102** (1936), 201–228.
19. P.-O. Löwdin, "On the non-orthogonality problem connected with the use of atomic wave functions in the theory of molecules and crystals," *J. Chem. Phys.* **18** (1950), 365–375; "Correlation problem in many electron quantum mechanics, I. Review of different approaches and discussion of some current ideas," in *Advances in Chemical Physics* (I. Prigogine, ed.), pp. 207–322. Interscience New York, 1970.
20. F. Jørgensen, "Orientation of the Eckart frame in a polyatomic molecule by symmetric orthonormalization," *Int. J. Quant. Chem.* **14** (1978), 55–63.
20a. F. R. Gantmacher, *The Theory of Matrices*, Vol. 1, Chelsea, New York, 1959.
20b. C. C. MacDuffee, *The Theory of Matrices*, Chelsea, New York, 1946.
20c. L. Autonne, "Sur les groupes linéaires, réels et orthogonaux," *Societe Math. de France, Bulletin* **30** (1902), 121–134.
21. S. M. Ferigle and A. Weber, "The Eckart conditions for a polyatomic molecule," *Amer. J. Phys.* **21** (1953), 102–107.
22. C. F. Curtiss, J. O. Hirschfelder, and F. T. Adler, "The separation of the rotational coordinates from the n-particle Schrödinger equation," *J. Chem. Phys.* **18** (1950), 1638–1642.
23. R. T. Pack and J. O. Hirschfelder, "Separation of rotational coordinates form the n-electron diatomic Schrödinger equation," *J. Chem. Phys.* **49** (1968), 4009–4020.
24. R. T. Pack, "Space-fixed vs body-fixed axes in atom-diatomic molecule scattering. Sudden approximation," *J. Chem. Phys.* **60** (1974), 633–639.
25. H. Weyl, *The Classical Groups.* Princeton Univ. Press, Princeton, N. J., 1946.
26. R. J. Malhiot and S. M. Ferigle, "Eckart conditions in Wilson's treatment of molecular vibrations," *J. Chem. Phys.* **22** (1954), 717–719.
27. S. Perlis, *Theory of Matrices.* Addison-Wesley, Cambridge, Mass., 1952.
28. G. A. Natanson and M. N. Adamov, "Hamiltonian of a polyatomic molecule. I. General representation of Hamiltonian in terms of impulse and momentum operators," *Vestn. Leningr. Univ. Ser. Fiz. i Khim.* **4** (1973), 28–35; "II. The application of the Eckart–Sayvetz conditions," *ibid.* **2** (1974), 24–32 (in Russian).

29. J. D. Louck, "Derivation of the molecular vibration–rotation Hamiltonian from the Schrödinger equation for the molecular model," *J. Mol. Spectrosc.* **61** (1976), 107–137.
30. Yu. S. Makushkin and O. N. Ulenikov, "On the transformation of the complete electron-nuclear Hamiltonian of a polyatomic molecule to the intramolecular coordinates," *J. Mol. Spectrosc.* **68** (1977), 1–20.
31. R. Seiler, "The Born-Oppenheimer approximation," presented at Conference on Mathematical Properties of Wave Functions, Bielefeld, Germany, 1978.
32. D. M. Dennison, "The infrared spectra of polyatomic molecules, I.", *Rev. Mod. Phys.* **3** (1931), 280–345; II., *ibid.* **12** (1940), 175–214.
33. B. T. Darling and D. M. Dennison, "The water vapor molecule," *Phys. Rev.* **57** (1940), 128–139.
34. B. T. Darling, "Quantum mechanical Hamiltonian in group operator form and the molecular Hamiltonian," *J. Mol. Spectrosc.* **11** (1963), 67–78.
35. H. Margenau and G. M. Murphy, *The Mathematics of Physics and Chemistry*, Chapter 9. Van Nostrand, New York, 1943.
36. B. Podolsky, "Quantum-mechanically correct form of Hamiltonian function for conservative systems," *Phys. Rev.* **32** (1928), 812–816.
37. E. C. Kemble, *The Fundamental Principles of Quantum Mechanics*. McGraw-Hill, New York, 1937; reissued by Dover, New York, 1958.
38. J. K. G. Watson, "Simplification of the molecular vibration–rotation Hamiltonian," *Mol. Phys.* **15** (1968), 479–490.
39. J. H. Meal and S. R. Polo, "Vibration–rotation in polyatomic molecules. I. The zeta matrices," *J. Chem. Phys.* **24** (1956), 119–1125; II. The determination of Coriolis coupling coefficients," *ibid.*, 1126–1138.
40. G. Amat, H. H. Nielsen, and G. Tarrago, *Rotation–Vibration Spectra of Polyatomic Molecules*, Dekker. New York, 1971.
41. M. Born and P. Jordan, "Zur Quantenmechanik," *Z. Physik* **34** (1925), 858–888.
42. J. H. Van Vleck, "On σ-type doubling and electron spin in the spectra of diatomic molecules," *Phys. Rev.* **35** (1929), 467–506.
43. R. C. Herman and W. H. Shaffer, "The calculation of perturbation energies in vibrating rotating polyatomic molecules," *J. Chem. Phys.* **16** (1948), 453–465.
44. H. A. Jahn, "A new Coriolis perturbation in the methane spectra. I. Vibrational–rotational Hamiltonian and wave functions," *Proc. Roy. Soc.* **A128** (1938), 469–495; "II. Energy levels," *ibid.*, 495–518.
45. E. Fermi, "Über den Ramaneffekt des Kohlendioxyds," *Z. Physik* **71** (1931), 250–259.
46. G. Amat and H. H. Nielsen, "Higher order rotation–vibration energies of molecules, IV," *J. Chem. Phys.* **29** (1958), 665–672.
47. K. T. Hecht, "The vibration–rotation energies of tetrahedral XY_4 molecules. Part I. Theory of spherical top molecules," *J. Mol. Spectrosc.* **5** (1960), 355–389; "Part II. The fundamental ν_3 of CH_4," *ibid.*, 390–404.
48. J. H. Van Vleck, "The coupling of angular momentum in molecules," *Rev. Mod. Phys.* **23** (1951), 213–227.
49. J. D. Louck, "Relationship between the feasible group and the point group of a rigid molecule," *Lecture Notes in Chemistry* (J. Hinze, ed.), Vol. V, pp. 57–76. Springer, Berlin, 1979.
50. H. C. Longuet-Higgens, "The symmetry groups of non-rigid molecules," *Mol. Phys.* **6** (1963), 445–460.
51. J. T. Hougen, "Calculation of rotational energy levels for symmetric-top molecules," *J. Chem. Phys.* **37** (1962), 1433–1441.
52. J. T. Hougen, "Classification of rotational levels. II," *J. Chem. Phys.* **39** (1963), 358–365.

53. P. R. Bunker, "Practically everything you ought to know about the molecular symmetry group," in *Vibrational Spectra and Structure* (J. R. Durig, ed.), Vol. III, Chapter 1. Dekker, New York, 1975.
54. P. R. Bunker, *Molecular Symmetry and Spectroscopy*. Academic Press, New York, 1979.
55. W. G. Harter, C. W. Patterson, and F. J. da Paixano, "Frame transformation relations and multipole transitions in symmetric polyatomic molecules," *Rev. Mod. Phys.* **50** (1978), 37–83.
55a. W. G. Harter, "Theory of hyperfine and superfine levels in symmetric polyatomic molecules. Trigonal and tetrahedral molecules: Elementary spin-$\frac{1}{2}$ cases in vibronic ground states," *Phys. Rev.* **A19** (1979), 2277–2303.
55b. A. Bauder, R. Meyer, and Hs. H. Günthard, "The isometric group of non-rigid molecules. A new approach to symmetry groups of non-rigid molecules," *Mol. Phys.* **28** (1974), 1305–1343.
56. L. C. Biedenharn and A. Gamba, "The splitting of a degenerate level under the action of a symmetry breaking Hamiltonian," *Rev. Bras. Fis.* **2** (1972), 319–333.
56a. E. De Vries and A. J. van Zanten, "The splitting of a degenerate level under the action of a symmetry-breaking Hamiltonian," *J. Phys.* **A7** (1974), 807–817.
57. J. D. Louck, "Algebra of level splitting," *Phys. Rev.* **A9** (1974), 2273–2281.
58. K. Fox, H. W. Galbraith, B. J. Krohn, and J. D. Louck, "Theory of level splitting: Spectrum of the octahedrally invariant fourth-rank tensor operator," *Phys. Rev.* **A15** (1977), 1363–1381.
59. K. Fox, "Vibration–rotation interactions in infrared active overtone levels of spherical top molecules; $2\nu_3$ and $2\nu_4$ of CH_4, $2\nu_3$ of CD_4," *J. Mol. Spectrosc.* **9** (1962), 381–420.
60. J. Moret-Bailly, "Sur l'interprétation des spectres de vibration–rotation des molécules à symétrie tétraédrique or octahédrique," *Cah. Phys.* **15** (1961), 237–314; " Calculations of the frequencies of the lines in a threefold degenerate fundamental band of a spherical top molecule," *J. Mol. Spectrosc.* **15** (1965), 344–354.
61. K. Fox and I. Ozier, "Construction of tetrahedral harmonics," *J. Chem. Phys.* **52** (1970), 5044–5056.
62. J. Moret-Bailly, L. Gautier, and J. Montagutelli, "Clebsch–Gordan coefficients adapted to cubic symmetry," *J. Mol. Spectrosc.* **15** (1965), 355–377.
63. J. Patera and P. Winternitz, "A new basis for the representations of the rotation group. Lamé and Heun polynomials," *J. Math. Phys.* **14** (1973), 1130–1139.
64. P. Kramer and M. Moshinsky, "Group theory of harmonic oscillators. III. States with permutational symmetry," *Nucl. Phys.* **82** (1966), 241–273.
65. B. J. Krohn, *Diagonal $F^{(4)}$ and $F^{(6)}$ Coefficients for Spherical-Top Molecules in Angular-Momentum States up to $J = 100$*. Los Alamos Scientific Laboratory Report LA-6554-MS, October, 1976.
66. K. R. Lea, M. J. M. Leask, and W. P. Wolf, "The raising of angular momentum degeneracy of *f*-electron terms by cubic crystal fields," *J. Phys. Chem. Solids* **23** (1962), 1381–1405.
67. A. J. Dorney and J. K. G. Watson, "Forbidden rotational spectra of polyatomic molecules," *J. Mol. Spectrosc.* **42** (1972), 135–148.
68. J. P. Aldridge, H. Filip, H. Flicker, R. F. Holland, R. S. McDowell, and N. G. Nereson, "Octahedral fine-structure splittings in ν_3 of SF_6," *J. Mol. Spectrosc.* **58** (1975), 165–168.
69. R. S. McDowell, H. W. Galbraith, B. J. Krohn, C. D. Cantrell, and E. D. Hinkley, "Identification of the SF_6 transitions pumped by a CO_2 laser," *Optics Commun.* **17** (1976), 178–182.
70. R. S. McDowell, H. W. Galbraith, C. D. Cantrell, N. G. Nereson, and E. D. Hinkley, "The ν_3 Q branch of SF_6 at high resolution: Assignment of the levels pumped by P(16) of the CO_2 laser," *J. Mol. Spectrosc.* **68** (1977), 288–298.
71. W. G. Harter and C. W. Patterson, "Advances in understanding of laser spectra of symmetric molecules," *Dimension* **62,** No. 6 (1978), 23–26.

72. F. Klein and A. Sommerfeld, *Über die Theorie des Kreisels.* Teubner, Stuttgart, 1965, Heft I, Kapitel II, § 8.
73. W. G. Harter and C. W. Patterson, "Orbital level splitting in octahedral symmetry and SF_6 rotational spectra. I. Qualitative features of high J levels," *J. Chem. Phys.* **66** (1977), 4872–4885.
73a. B. R. Judd, "Group theory in atomic and molecular physics," in *Symmetries in Science* (B. Gruber and R. S. Millman, eds.), pp. 151–160. Plenum, New York, 1980.
74. C. W. Patterson and W. G. Harter, "Orbital level splitting in octahedral symmetry and SF_6 rotational spectra. II. Quantitative treatment of high J levels," *J. Chem. Phys.* **66** (1977), 4886–4892.
75. H. W. Galbraith, C. W. Patterson, B. J. Krohn, and W. G. Harter, "Line frequency expressions for triply degenerate fundamentals of spherical top molecules appropriate for large angular momentum," *J. Mol. Spectrosc.* **73** (1978), 475–493.
76. R. C. Tolman, *The Principles of Statistical Mechanics.* Oxford Univ. Press, London, 1938.
77. H. W. Galbraith, "Single photon transition moments in excited states of spherical top molecules," *Optics Lett.* **3** (1978), 154–155.
77a. J. R. Ackerhalt, H. Flicker, H. W. Galbraith, J. King, and W. B. Person, "Analysis of $3\nu_3$ in SF_6," *J. Chem. Phys.* **69** (1978), 1461–1464.
77b. C. W. Patterson, B. J. Krohn, and A. S. Pine, "Interacting band analysis of the high-resolution spectrum of the $3\nu_3$ manifold of SF_6," *J. Mol. Spectrosc.* **88** (1981), 133–166.
78. E. B. Wilson, Jr., "The statistical weights of the rotational levels of polyatomic molecules, including methane, ammonia, benzene, cyclopropane, and ethylene," *J. Chem. Phys.* **3** (1935), 276–285.
79. C. D. Cantrell and H. W. Galbraith, "Statistical weights for octahedral spherical-top molecules," *J. Mol. Spectrosc.* **58** (1975), 158–164.
80. H. W. Galbraith, "$SU(2s+1) \times \zeta_r$ and spin statistical weights for tetrahedral XY_4, trigonal bipyramidal XY_5, and octahedral XY_6, *J. Chem. Phys.* **68** (1978), 1677–1682.
80a. A. Weber, "Ro-vibronic species, overall allowed species, and nuclear spin statistical weights for symmetric top molecules belonging to the D_{nd} and D_{nh} ($n \leqslant 6$) point-groups," *J. Chem. Phys.* **73** (1980), 3952–3972.
81. M. Löete, A. Clairon, A. Frichet, R. S. McDowell, H. W. Galbraith, J. C. Hilico, J. Moret-Bailly, and L. Henry, "Constantes spectrales de la band ν_3 de la molécule $^{32}SF_6$ calculées à partir du spectre d'absorption saturée," *C. R. Acad. Sci. Paris* **B285** (1977), 175–178.
82. K. C. Kim, W. B. Person, D. Seitz, and B. J. Krohn, "Analysis of the ν_4 ($615\ cm^{-1}$) region of the Fourier transform and diode laser spectra of SF_6," *J. Mol. Spectrosc.* **76** (1979), 322–340.

Appendix of Tables

Tables of algebraic coefficients and other quantities needed frequently in applications of angular momentum theory to physical problems are given in this appendix. The tables included are:

Table 1. Algebraic tables of $j_2 = \frac{1}{2}, 1, \frac{3}{2}, 2, \frac{5}{2}$, and 3 Wigner coefficients.[1]
Table 2. Algebraic tables of $l = \frac{1}{2}, 1, \frac{3}{2}$, and 2 Racah coefficients.[2]
Table 3. Algebraic tables of $j_2 = \frac{1}{2}$ and 1 Racah coefficients in the pattern calculus form.
Table 4. The solid and spherical harmonics for $l = 0, 1, 2, 3$, and 4.
Table 5. The tensor harmonics.
Table 6. The $\Delta = [00]$ Wigner operators.
Table 7. The Racah operators.
Table 8. Reduced matrix elements.

The use of the tables is self-explanatory.

Tables 1, 2, 4, 5, and 8 consist of standard material from the general literature (see footnotes 1 and 2, and the Bibliography of Tables below). The most extensive tabulations of algebraic formulas for the Wigner coefficients and Racah coefficients may be found in Varshalovich *et al.* and Ishidzu *et al.*, respectively.

Tables 3, 6, and 7 consist of formulas which illustrate the algebraic structure underlying angular momentum theory as developed in the present volume. Table 7, in particular, summarizes properties of a class of Racah operators which had previously been discussed only in the special case $\mu = 0$, and then not within the present algebraic formulation (see Refs. [37] and [55] in Chapter 3).

[1] The tables for $j_2 = \frac{1}{2}, 1, \frac{3}{2}, 2$ are from Condon and Shortley; for $j_2 = \frac{5}{2}$ from Melvin and Swamy; for $j_2 = 3$ from Falkoff *et al.* Reprinted by permission (see Bibliography of Tables).

[2] These tables are from Biedenharn *et al.* Reprinted by permission (see Bibliography of Tables).

Table 1. Algebraic tables of $j_2 = \frac{1}{2}, 1, \frac{3}{2}, 2, \frac{5}{2}$, and 3 Wigner coefficients

$$C^{j_1\ \frac{1}{2}\ j}_{m-m_2\ m_2\ m} = (-1)^{j_1-\frac{1}{2}+m}(2j+1)^{1/2}\begin{pmatrix} j_1 & \frac{1}{2} & j \\ m-m_2 & m_2 & -m \end{pmatrix}$$

$j=$	$m_2=\frac{1}{2}$	$m_2=-\frac{1}{2}$
$j_1+\frac{1}{2}$	$\sqrt{\frac{j_1+m+\frac{1}{2}}{2j_1+1}}$	$\sqrt{\frac{j_1-m+\frac{1}{2}}{2j_1+1}}$
$j_1-\frac{1}{2}$	$-\sqrt{\frac{j_1-m+\frac{1}{2}}{2j_1+1}}$	$\sqrt{\frac{j_1+m+\frac{1}{2}}{2j_1+1}}$

$$C^{j_1\ 1\ j}_{m-m_2\ m_2\ m} = (-1)^{j_1-1+m}(2j+1)^{1/2}\begin{pmatrix} j_1 & 1 & j \\ m-m_2 & m_2 & -m \end{pmatrix}$$

$j=$	$m_2=1$	$m_2=0$	$m_2=-1$
j_1+1	$\sqrt{\frac{(j_1+m)(j_1+m+1)}{(2j_1+1)(2j_1+2)}}$	$\sqrt{\frac{(j_1-m+1)(j_1+m+1)}{(2j_1+1)(j_1+1)}}$	$\sqrt{\frac{(j_1-m)(j_1-m+1)}{(2j_1+1)(2j_1+2)}}$
j_1	$-\sqrt{\frac{(j_1+m)(j_1-m+1)}{2j_1(j_1+1)}}$	$\frac{m}{\sqrt{j_1(j_1+1)}}$	$\sqrt{\frac{(j_1-m)(j_1+m+1)}{2j_1(j_1+1)}}$
j_1-1	$\sqrt{\frac{(j_1-m)(j_1-m+1)}{2j_1(2j_1+1)}}$	$-\sqrt{\frac{(j_1-m)(j_1+m)}{j_1(2j_1+1)}}$	$\sqrt{\frac{(j_1+m+1)(j_1+m)}{2j_1(2j_1+1)}}$

Table 1. Algebraic tables of $j_2 = \frac{1}{2}, 1, \frac{3}{2}, 2, \frac{5}{2}$, and 3 Wigner coefficients (continued)

$$C^{j_1\ \frac{3}{2}\ j}_{m-m_2\ m_2\ m} = (-1)^{j_1-\frac{3}{2}+m}(2j+1)^{1/2}\begin{pmatrix} j_1 & \frac{3}{2} & j \\ m-m_2 & m_2 & -m \end{pmatrix}$$

$j=$	$m_2 = \frac{3}{2}$	$m_2 = \frac{1}{2}$
$j_1+\frac{3}{2}$	$\sqrt{\frac{(j_1+m-\frac{1}{2})(j_1+m+\frac{1}{2})(j_1+m+\frac{3}{2})}{(2j_1+1)(2j_1+2)(2j_1+3)}}$	$\sqrt{\frac{3(j_1+m+\frac{1}{2})(j_1+m+\frac{3}{2})(j_1-m+\frac{3}{2})}{(2j_1+1)(2j_1+2)(2j_1+3)}}$
$j_1+\frac{1}{2}$	$-\sqrt{\frac{3(j_1+m-\frac{1}{2})(j_1+m+\frac{1}{2})(j_1-m+\frac{3}{2})}{2j_1(2j_1+1)(2j_1+3)}}$	$-(j_1-3m+\frac{3}{2})\sqrt{\frac{j_1+m+\frac{1}{2}}{2j_1(2j_1+1)(2j_1+3)}}$
$j_1-\frac{1}{2}$	$\sqrt{\frac{3(j_1+m-\frac{1}{2})(j_1-m+\frac{1}{2})(j_1-m+\frac{3}{2})}{(2j_1-1)(2j_1+1)(2j_1+2)}}$	$-(j_1+3m-\frac{1}{2})\sqrt{\frac{j_1-m+\frac{1}{2}}{(2j_1-1)(2j_1+1)(2j_1+2)}}$
$j_1-\frac{3}{2}$	$-\sqrt{\frac{(j_1-m-\frac{1}{2})(j_1-m+\frac{1}{2})(j_1-m+\frac{3}{2})}{2j_1(2j_1-1)(2j_1+1)}}$	$\sqrt{\frac{3(j_1+m-\frac{1}{2})(j_1-m-\frac{1}{2})(j_1-m+\frac{1}{2})}{2j_1(2j_1-1)(2j_1+1)}}$

$j=$	$m_2 = -\frac{1}{2}$	$m_2 = -\frac{3}{2}$
$j_1+\frac{3}{2}$	$\sqrt{\frac{3(j_1+m+\frac{3}{2})(j_1-m+\frac{1}{2})(j_1-m+\frac{3}{2})}{(2j_1+1)(2j_1+2)(2j_1+3)}}$	$\sqrt{\frac{(j_1-m-\frac{1}{2})(j_1-m+\frac{1}{2})(j_1-m+\frac{3}{2})}{(2j_1+1)(2j_1+2)(2j_1+3)}}$
$j_1+\frac{1}{2}$	$(j_1+3m+\frac{3}{2})\sqrt{\frac{j_1-m+\frac{1}{2}}{2j_1(2j_1+1)(2j_1+3)}}$	$\sqrt{\frac{3(j_1+m+\frac{3}{2})(j_1-m-\frac{1}{2})(j_1-m+\frac{1}{2})}{2j_1(2j_1+1)(2j_1+3)}}$
$j_1-\frac{1}{2}$	$-(j_1-3m-\frac{1}{2})\sqrt{\frac{j_1+m+\frac{1}{2}}{(2j_1-1)(2j_1+1)(2j_1+2)}}$	$\sqrt{\frac{3(j_1+m+\frac{1}{2})(j_1+m+\frac{3}{2})(j_1-m-\frac{1}{2})}{(2j_1-1)(2j_1+1)(2j_1+2)}}$
$j_1-\frac{3}{2}$	$-\sqrt{\frac{3(j_1+m-\frac{1}{2})(j_1+m+\frac{1}{2})(j_1-m-\frac{1}{2})}{2j_1(2j_1-1)(2j_1+1)}}$	$\sqrt{\frac{(j_1+m-\frac{1}{2})(j_1+m+\frac{1}{2})(j_1+m+\frac{3}{2})}{2j_1(2j_1-1)(2j_1+1)}}$

Table 1. Algebraic tables of $j_2 = \frac{1}{2}$, 1, $\frac{3}{2}$, 2, $\frac{5}{2}$, and 3 Wigner coefficients (continued)

$$C^{j_1\ 2\ j}_{m-m_2\ m_2\ m} = (-1)^{j_1-2+m}(2j+1)^{1/2}\begin{pmatrix} j_1 & 2 & j \\ m-m_2 & m_2 & -m \end{pmatrix}$$

$j =$	$m_2 = 2$	$m_2 = 1$
j_1+2	$\sqrt{\frac{(j_1+m-1)(j_1+m)(j_1+m+1)(j_1+m+2)}{(2j_1+1)(2j_1+2)(2j_1+3)(2j_1+4)}}$	$\sqrt{\frac{(j_1-m+2)(j_1+m+2)(j_1+m+1)(j_1+m)}{(2j_1+1)(j_1+1)(2j_1+3)(j_1+2)}}$
j_1+1	$-\sqrt{\frac{(j_1+m-1)(j_1+m)(j_1+m+1)(j_1-m+2)}{2j_1(j_1+1)(j_1+2)(2j_1+1)}}$	$-(j_1-2m+2)\sqrt{\frac{(j_1+m+1)(j_1+m)}{2j_1(2j_1+1)(j_1+1)(j_1+2)}}$
j_1	$\sqrt{\frac{3(j_1+m-1)(j_1+m)(j_1-m+1)(j_1-m+2)}{(2j_1-1)\,2j_1(j_1+1)(2j_1+3)}}$	$(1-2m)\sqrt{\frac{3(j_1-m+1)(j_1+m)}{(2j_1-1)\,j_1(2j_1+2)(2j_1+3)}}$
j_1-1	$-\sqrt{\frac{(j_1+m-1)(j_1-m)(j_1-m+1)(j_1-m+2)}{2(j_1-1)\,j(j_1+1)(2j_1+1)}}$	$(j_1+2m-1)\sqrt{\frac{(j_1-m+1)(j_1-m)}{(j_1-1)\,j_1(2j_1+1)(2j_1+2)}}$
j_1-2	$\sqrt{\frac{(j_1-m-1)(j_1-m)(j_1-m+1)(j_1-m+2)}{(2j_1-2)(2j_1-1)\,2j_1(2j_1+1)}}$	$-\sqrt{\frac{(j_1-m+1)(j_1-m)(j_1-m-1)(j_1+m-1)}{(j_1-1)(2j_1-1)\,j_1(2j_1+1)}}$

$j =$	$m_2 = 0$	$m_2 = -1$
j_1+2	$\sqrt{\frac{3(j_1-m+2)(j_1-m+1)(j_1+m+2)(j_1+m+1)}{(2j_1+1)(2j_1+2)(2j_1+3)(j_1+2)}}$	$\sqrt{\frac{(j_1-m+2)(j_1-m+1)(j_1-m)(j_1+m+2)}{(2j_1+1)(j_1+1)(2j_1+3)(j_1+2)}}$
j_1+1	$m\sqrt{\frac{3(j_1-m+1)(j_1+m+1)}{j_1(2j_1+1)(j_1+1)(j_1+2)}}$	$(j_1+2m+2)\sqrt{\frac{(j_1-m+1)(j_1-m)}{j_1(2j_1+1)(2j_1+2)(j_1+2)}}$
j_1	$\frac{3m^2-j_1(j_1+1)}{\sqrt{(2j_1-1)\,j_1(j_1+1)(2j_1+3)}}$	$(2m+1)\sqrt{\frac{3(j_1-m)(j_1+m+1)}{(2j_1-1)\,j_1(2j_1+2)(2j_1+3)}}$
j_1-1	$-m\sqrt{\frac{3(j_1-m)(j_1+m)}{(j_1-1)\,j_1(2j_1+1)(j_1+1)}}$	$-(j_1-2m-1)\sqrt{\frac{(j_1+m+1)(j_1+m)}{(j_1-1)\,j_1(2j_1+1)(2j_1+2)}}$
j_1-2	$\sqrt{\frac{3(j_1-m)(j_1-m-1)(j_1+m)(j_1+m-1)}{(2j_1-2)(2j_1-1)\,j_1(2j_1+1)}}$	$-\sqrt{\frac{(j_1-m-1)(j_1+m+1)(j_1+m)(j_1+m-1)}{(j_1-1)(2j_1-1)\,j_1(2j_1+1)}}$

$j =$	$m_2 = -2$
j_1+2	$\sqrt{\frac{(j_1-m-1)(j_1-m)(j_1-m+1)(j_1-m+2)}{(2j_1+1)(2j_1+2)(2j_1+3)(2j_1+4)}}$
j_1+1	$\sqrt{\frac{(j_1-m-1)(j_1-m)(j_1-m+1)(j_1+m+2)}{j_1(2j_1+1)(j_1+1)(2j_1+4)}}$
j_1	$\sqrt{\frac{3(j_1-m-1)(j_1-m)(j_1+m+1)(j_1+m+2)}{(2j_1-1)\,j_1(2j_1+2)(2j_1+3)}}$
j_1-1	$\sqrt{\frac{(j_1-m-1)(j_1+m)(j_1+m+1)(j_1+m+2)}{(j_1-1)\,j_1(2j_1+1)(2j_1+2)}}$
j_1-2	$\sqrt{\frac{(j_1+m-1)(j_1+m)(j_1+m+1)(j_1+m+2)}{(2j_1-2)(2j_1-1)\,2j_1(2j_1+1)}}$

Table 1. Algebraic tables of $j_2 = \frac{1}{2}, 1, \frac{3}{2}, 2, \frac{5}{2}$, and 3 Wigner coefficients (continued)

$$C^{j_1\ \frac{5}{2}\ j}_{m-m_2\ m_2\ m} = (-1)^{j_1-\frac{5}{2}+m}(2j+1)^{1/2}\begin{pmatrix} j_1 & \frac{5}{2} & j \\ m-m_2 & m_2 & -m \end{pmatrix}$$

$j \diagdown m_2$	$\frac{5}{2}$
$j_1+\frac{5}{2}$	$\left[\dfrac{(j_1+m+\frac{5}{2})(j_1+m+\frac{3}{2})(j_1+m+\frac{1}{2})(j_1+m-\frac{1}{2})(j_1+m-\frac{3}{2})}{(2j_1+5)(2j_1+4)(2j_1+3)(2j_1+2)(2j_1+1)}\right]^{\frac{1}{2}}$
$j_1+\frac{3}{2}$	$-\left[\dfrac{5(j_1-m+\frac{5}{2})(j_1+m+\frac{3}{2})(j_1+m+\frac{1}{2})(j_1+m-\frac{1}{2})(j_1+m-\frac{3}{2})}{(2j_1+5)(2j_1+3)(2j_1+2)(2j_1+1)(2j_1)}\right]^{\frac{1}{2}}$
$j_1+\frac{1}{2}$	$\left[\dfrac{10(j_1+m+\frac{1}{2})(j_1+m-\frac{1}{2})(j_1+m-\frac{3}{2})(j_1-m+\frac{5}{2})(j_1-m+\frac{3}{2})}{(2j_1+4)(2j_1+3)(2j_1+1)(2j_1)(2j_1-1)}\right]^{\frac{1}{2}}$
$j_1-\frac{1}{2}$	$-\left[\dfrac{10(j_1+m-\frac{1}{2})(j_1+m-\frac{3}{2})(j_1-m+\frac{5}{2})(j_1-m+\frac{3}{2})(j_1-m+\frac{1}{2})}{(2j_1+3)(2j_1+2)(2j_1+1)(2j_1-1)(2j_1-2)}\right]^{\frac{1}{2}}$
$j_1-\frac{3}{2}$	$\left[\dfrac{5(j_1-m+\frac{5}{2})(j_1-m+\frac{3}{2})(j_1-m+\frac{1}{2})(j_1-m-\frac{1}{2})(j_1+m-\frac{3}{2})}{(2j_1+2)(2j_1+1)(2j_1)(2j_1-1)(2j_1-3)}\right]^{\frac{1}{2}}$
$j_1-\frac{5}{2}$	$-\left[\dfrac{(j_1-m+\frac{5}{2})(j_1-m+\frac{3}{2})(j_1-m+\frac{1}{2})(j_1-m-\frac{1}{2})(j_1-m-\frac{3}{2})}{(2j_1+1)(2j_1)(2j_1-1)(2j_1-2)(2j_1-3)}\right]^{\frac{1}{2}}$

$j \diagdown m_2$	$\frac{3}{2}$
$j_1+\frac{5}{2}$	$\left[\dfrac{5(j_1+m+\frac{5}{2})(j_1+m+\frac{3}{2})(j_1+m+\frac{1}{2})(j_1+m-\frac{1}{2})(j_1-m+\frac{5}{2})}{(2j_1+5)(2j_1+4)(2j_1+3)(2j_1+2)(2j_1+1)}\right]^{\frac{1}{2}}$
$j_1+\frac{3}{2}$	$[-3j_1+5m-(15/2)]\left[\dfrac{(j_1+m+\frac{3}{2})(j_1+m+\frac{1}{2})(j_1+m-\frac{1}{2})}{(2j_1+5)(2j_1+3)(2j_1+2)(2j_1+1)(2j_1)}\right]^{\frac{1}{2}}$
$j_1+\frac{1}{2}$	$(2j_1-10m+9)\left[\dfrac{(j_1+m+\frac{1}{2})(j_1+m-\frac{1}{2})(j_1-m+\frac{3}{2})}{2(2j_1+4)(2j_1+3)(2j_1+1)(2j_1)(2j_1-1)}\right]^{\frac{1}{2}}$
$j_1-\frac{1}{2}$	$(2j_1+10m-7)\left[\dfrac{(j_1+m-\frac{1}{2})(j_1-m+\frac{3}{2})(j_1-m+\frac{1}{2})}{2(2j_1+3)(2j_1+2)(2j_1+1)(2j_1-1)(2j_1-2)}\right]^{\frac{1}{2}}$
$j_1-\frac{3}{2}$	$-[3j_1+5m-(9/2)]\left[\dfrac{(j_1-m+\frac{3}{2})(j_1-m+\frac{1}{2})(j_1-m-\frac{1}{2})}{(2j_1+2)(2j_1+1)(2j_1)(2j_1-1)(2j_1-3)}\right]^{\frac{1}{2}}$
$j_1-\frac{5}{2}$	$\left[\dfrac{5(j_1+m-\frac{3}{2})(j_1-m+\frac{3}{2})(j_1-m+\frac{1}{2})(j_1-m-\frac{1}{2})(j_1-m-\frac{3}{2})}{(2j_1+1)(2j_1)(2j_1-1)(2j_1-2)(2j_1-3)}\right]^{\frac{1}{2}}$

Table 1. Algebraic tables of $j_2 = \frac{1}{2}$, 1, $\frac{3}{2}$, 2, $\frac{5}{2}$, and 3 Wigner coefficients (continued)

$$C^{j_1\ \frac{5}{2}\ j}_{m-m_2\ m_2\ m} = (-1)^{j_1-\frac{5}{2}+m}(2j+1)^{1/2}\begin{pmatrix} j_1 & \frac{5}{2} & j \\ m-m_2 & m_2 & -m \end{pmatrix}$$

$j \diagdown m_2$	$\frac{1}{2}$
$j_1+\frac{5}{2}$	$\left[\dfrac{10(j_1+m+\frac{5}{2})(j_1+m+\frac{3}{2})(j_1+m+\frac{1}{2})(j_1-m+\frac{5}{2})(j_1-m+\frac{3}{2})}{(2j_1+5)(2j_1+4)(2j_1+3)(2j_1+2)(2j_1+1)}\right]^{\frac{1}{2}}$
$j_1+\frac{3}{2}$	$(-j_1+5m-\frac{5}{2})\left[\dfrac{2(j_1+m+\frac{3}{2})(j_1+m+\frac{1}{2})(j_1-m+\frac{3}{2})}{(2j_1)(2j_1+5)(2j_1+3)(2j_1+2)(2j_1+1)}\right]^{\frac{1}{2}}$
$j_1+\frac{1}{2}$	$(-2j_1^2-4j_1m+10m^2+\frac{3}{2}-8m-2j_1)\left[\dfrac{(j_1+m+\frac{1}{2})}{(2j_1+4)(2j_1+3)(2j_1+1)(2j_1)(2j_1-1)}\right]^{\frac{1}{2}}$
$j_1-\frac{1}{2}$	$(2j_1^2-10m^2-4j_1m+2j_1+4m-\frac{3}{2})\left[\dfrac{(j_1-m+\frac{1}{2})}{(2j_1+3)(2j_1+2)(2j_1+1)(2j_1-1)(2j_1-2)}\right]^{\frac{1}{2}}$
$j_1-\frac{3}{2}$	$(j_1+5m-\frac{3}{2})\left[\dfrac{2(j_1+m-\frac{1}{2})(j_1-m+\frac{1}{2})(j_1-m-\frac{1}{2})}{(2j_1+2)(2j_1+1)(2j_1)(2j_1-1)(2j_1-3)}\right]^{\frac{1}{2}}$
$j_1-\frac{5}{2}$	$-\left[\dfrac{10(j_1+m-\frac{1}{2})(j_1-m+\frac{1}{2})(j_1+m-\frac{3}{2})(j_1-m-\frac{1}{2})(j_1-m-\frac{3}{2})}{(2j_1+1)(2j_1)(2j_1-1)(2j_1-2)(2j_1-3)}\right]^{\frac{1}{2}}$

$j \diagdown m_2$	$-\frac{1}{2}$
$j_1+\frac{5}{2}$	$\left[\dfrac{10(j_1+m+\frac{5}{2})(j_1+m+\frac{3}{2})(j_1-m+\frac{5}{2})(j_1-m+\frac{3}{2})(j_1-m+\frac{1}{2})}{(2j_1+5)(2j_1+4)(2j_1+3)(2j_1+2)(2j_1+1)}\right]^{\frac{1}{2}}$
$j_1+\frac{3}{2}$	$(j_1+5m+\frac{5}{2})\left[\dfrac{2(j_1+m+\frac{3}{2})(j_1-m+\frac{3}{2})(j_1-m+\frac{1}{2})}{(2j_1+5)(2j_1+3)(2j_1+2)(2j_1+1)(2j_1)}\right]^{\frac{1}{2}}$
$j_1+\frac{1}{2}$	$(-2j_1^2+10m^2+4j_1m-2j_1+8m+\frac{3}{2})\left[\dfrac{(j_1-m+\frac{1}{2})}{(2j_1+4)(2j_1+3)(2j_1+1)2j_1(2j_1-1)}\right]^{\frac{1}{2}}$
$j_1-\frac{1}{2}$	$(-2j_1^2+10m^2-4j_1m-2j_1+4m+\frac{3}{2})\left[\dfrac{(j_1+m+\frac{1}{2})}{(2j_1+3)(2j_1+2)(2j_1+1)(2j_1-1)(2j_1-2)}\right]^{\frac{1}{2}}$
$j_1-\frac{3}{2}$	$(j_1-5m-\frac{3}{2})\left[\dfrac{2(j_1+m+\frac{1}{2})(j_1+m-\frac{1}{2})(j_1-m-\frac{1}{2})}{(2j_1+2)(2j_1+1)(2j_1)(2j_1-1)(2j_1-3)}\right]^{\frac{1}{2}}$
$j_1-\frac{5}{2}$	$\left[\dfrac{10(j_1+m+\frac{1}{2})(j_1+m-\frac{1}{2})(j_1+m-\frac{3}{2})(j_1-m-\frac{1}{2})(j_1-m-\frac{3}{2})}{(2j_1+1)(2j_1)(2j_1-1)(2j_1-2)(2j_1-3)}\right]^{\frac{1}{2}}$

Table 1. Algebraic tables of $j_2 = \frac{1}{2}, 1, \frac{3}{2}, 2, \frac{5}{2}$, and 3 Wigner coefficients (continued)

$$C^{j_1 \ \frac{5}{2} \ j}_{m-m_2 \ m_2 \ m} = (-1)^{j_1 - \frac{5}{2} + m} (2j+1)^{1/2} \begin{pmatrix} j_1 & \frac{5}{2} & j \\ m-m_2 & m_2 & -m \end{pmatrix}$$

$j \diagdown m_2$	$-\frac{3}{2}$
$j_1+\frac{5}{2}$	$\left[\frac{5(j_1+m+\frac{5}{2})(j_1-m+\frac{5}{2})(j_1-m+\frac{3}{2})(j_1-m+\frac{1}{2})(j_1-m-\frac{1}{2})}{(2j_1+5)(2j_1+4)(2j_1+3)(2j_1+2)(2j_1+1)}\right]^{\frac{1}{2}}$
$j_1+\frac{3}{2}$	$[3j_1+5m+(15/2)]\left[\frac{(j_1-m+\frac{3}{2})(j_1-m+\frac{1}{2})(j_1-m-\frac{1}{2})}{(2j_1+5)(2j_1+3)(2j_1+2)(2j_1+1)(2j_1)}\right]^{\frac{1}{2}}$
$j_1+\frac{1}{2}$	$(2j_1+10m+9)\left[\frac{(j_1+m+\frac{3}{2})(j_1-m+\frac{1}{2})(j_1-m-\frac{1}{2})}{2(2j_1+4)(2j_1+3)(2j_1+1)(2j_1)(2j_1-1)}\right]^{\frac{1}{2}}$
$j_1-\frac{1}{2}$	$(-2j_1+10m+7)\left[\frac{(j_1+m+\frac{3}{2})(j_1+m+\frac{1}{2})(j_1-m-\frac{1}{2})}{2(2j_1+3)(2j_1+2)(2j_1+1)(2j_1-1)(2j_1-2)}\right]^{\frac{1}{2}}$
$j_1-\frac{3}{2}$	$[-3j_1+5m+(9/2)]\left[\frac{(j_1+m+\frac{3}{2})(j_1+m+\frac{1}{2})(j_1+m-\frac{1}{2})}{(2j_1+2)(2j_1+1)(2j_1)(2j_1-1)(2j_1-3)}\right]^{\frac{1}{2}}$
$j_1-\frac{5}{2}$	$-\left[\frac{5(j_1+m+\frac{3}{2})(j_1+m+\frac{1}{2})(j_1+m-\frac{1}{2})(j_1+m-\frac{3}{2})(j_1-m-\frac{3}{2})}{(2j_1+1)(2j_1)(2j_1-1)(2j_1-2)(2j_1-3)}\right]^{\frac{1}{2}}$

$j \diagdown m_2$	$-\frac{5}{2}$
$j_1+\frac{5}{2}$	$\left[\frac{(j_1-m+\frac{5}{2})(j_1-m+\frac{3}{2})(j_1-m+\frac{1}{2})(j_1-m-\frac{1}{2})(j_1-m-\frac{3}{2})}{(2j_1+5)(2j_1+4)(2j_1+3)(2j_1+2)(2j_1+1)}\right]^{\frac{1}{2}}$
$j_1+\frac{3}{2}$	$\left[\frac{5(j_1+m+\frac{5}{2})(j_1-m+\frac{3}{2})(j_1-m+\frac{1}{2})(j_1-m-\frac{1}{2})(j_1-m-\frac{3}{2})}{(2j_1+5)(2j_1+3)(2j_1+2)(2j_1+1)(2j_1)}\right]^{\frac{1}{2}}$
$j_1+\frac{1}{2}$	$\left[\frac{10(j_1+m+\frac{5}{2})(j_1+m+\frac{3}{2})(j_1-m+\frac{1}{2})(j_1-m-\frac{1}{2})(j_1-m-\frac{3}{2})}{(2j_1+4)(2j_1+3)(2j_1+1)(2j_1)(2j_1-1)}\right]^{\frac{1}{2}}$
$j_1-\frac{1}{2}$	$\left[\frac{10(j_1+m+\frac{5}{2})(j_1+m+\frac{3}{2})(j_1+m+\frac{1}{2})(j_1-m-\frac{1}{2})(j_1-m-\frac{3}{2})}{(2j_1+3)(2j_1+2)(2j_1+1)(2j_1-1)(2j_1-2)}\right]^{\frac{1}{2}}$
$j_1-\frac{3}{2}$	$\left[\frac{5(j_1+m+\frac{5}{2})(j_1+m+\frac{3}{2})(j_1+m+\frac{1}{2})(j_1+m-\frac{1}{2})(j_1-m-\frac{3}{2})}{(2j_1+2)(2j_1+1)2j_1(2j_1-1)(2j_1-3)}\right]^{\frac{1}{2}}$
$j_1-\frac{5}{2}$	$\left[\frac{(j_1+m+\frac{5}{2})(j_1+m+\frac{3}{2})(j_1+m+\frac{1}{2})(j_1+m-\frac{1}{2})(j_1+m-\frac{3}{2})}{(2j_1+1)2j_1(2j_1-1)(2j_1-2)(2j_1-3)}\right]^{\frac{1}{2}}$

Table 1. Algebraic tables of $j_2 = \frac{1}{2}, 1, \frac{3}{2}, 2, \frac{5}{2}$, and 3 Wigner coefficients (continued)

$$C^{j\ \ 3\ \ J}_{M-m\ \ m\ \ M} = (-1)^{j-3-M}(2J+1)^{1/2}\begin{pmatrix} j & 3 & J \\ M-m & m & -M \end{pmatrix}$$

Column 1 (m = 3)

$J = j+3$	$\sqrt{\dfrac{(j+M-2)(j+M-1)(j+M)(j+M+1)(j+M+2)(j+M+3)}{8(j+1)(j+2)(j+3)(2j+1)(2j+3)(2j+5)}}$
$j+2$	$-\sqrt{\dfrac{3(j+M-2)(j+M-1)(j+M)(j+M+1)(j+M+2)(j-M+3)}{8j(j+1)(j+2)(j+3)(2j+1)(2j+3)}}$
$j+1$	$\sqrt{\dfrac{15(j+M-2)(j+M-1)(j+M)(j+M+1)(j-M+2)(j-M+3)}{8j(j+1)(j+2)(2j+1)(2j+5)(2j-1)}}$
j	$-\sqrt{\dfrac{5(j+M-2)(j+M-1)(j+M)(j-M+1)(j-M+2)(j-M+3)}{4j(j+1)(j+2)(j-1)(2j+3)(2j-1)}}$
$j-1$	$\sqrt{\dfrac{15(j-M+3)(j-M+2)(j-M+1)(j-M)(j+M-1)(j+M-2)}{8(j+1)j(j-1)(2j+1)(2j-3)(2j+3)}}$
$j-2$	$-\sqrt{\dfrac{3(j-M+3)(j-M+2)(j-M+1)(j-M)(j-M-1)(j+M-2)}{8(j+1)j(j-1)(j-2)(2j+1)(2j-1)}}$
$j-3$	$\sqrt{\dfrac{(j-M+3)(j-M+2)(j-M+1)(j-M)(j-M-1)(j-M-2)}{8j(j-1)(j-2)(2j+1)(2j-1)(2j-3)}}$

Column 2 (m = 2)

$J = j+3$	$\sqrt{\dfrac{3(j-M+3)(j+M-1)(j+M)(j+M+1)(j+M+2)(j+M+3)}{4(j+1)(j+2)(j+3)(2j+1)(2j+3)(2j+5)}}$
$j+2$	$-\left(j-\frac{3}{2}M+3\right)\sqrt{\dfrac{(j+M-1)(j+M)(j+M+1)(j+M+2)}{j(j+1)(j+2)(j+3)(2j+1)(2j+3)}}$
$j+1$	$(j-3M+4)\sqrt{\dfrac{5(j-M+2)(j+M-1)(j+M)(j+M+1)}{4j(j+1)(j+2)(2j+1)(2j+5)(2j-1)}}$
j	$(M-1)\sqrt{\dfrac{15(j+M)(j-M+1)(j-M+2)(j+M-1)}{2j(j+1)(j+2)(j-1)(2j+3)(2j-1)}}$
$j-1$	$-(j+3M-3)\sqrt{\dfrac{5(j+M-1)(j-M+2)(j-M+1)(j-M)}{4(j+1)j(j-1)(2j+1)(2j-3)(2j+3)}}$
$j-2$	$\left(j+\frac{3}{2}M-2\right)\sqrt{\dfrac{(j-M+2)(j-M+1)(j-M)(j-M-1)}{(j+1)j(j-1)(j-2)(2j+1)(2j-1)}}$
$j-3$	$-\sqrt{\dfrac{3(j+M-2)(j-M+2)(j-M+1)(j-M)(j-M-1)(j-M-2)}{4j(j-1)(j-2)(2j+1)(2j-1)(2j-3)}}$

Column 3 (m = 1)

$J = j+3$	$\sqrt{\dfrac{15(j-M+2)(j-M+3)(j+M)(j+M+1)(j+M+2)(j+M+3)}{8(j+1)(j+2)(j+3)(2j+1)(2j+3)(2j+5)}}$
$j+2$	$-(j-3M+3)\sqrt{\dfrac{5(j-M+2)(j+M)(j+M+1)(j+M+2)}{8j(j+1)(j+2)(j+3)(2j+1)(2j+3)}}$
$j+1$	$-\left\{j(j+7)+10(M-1)\left(j-\frac{3}{2}M+1\right)\right\}\sqrt{\dfrac{(j+M)(j+M+1)}{8j(j+1)(j+2)(2j+1)(2j+5)(2j-1)}}$
j	$\left\{j(j+1)-5M(M-1)-2\right\}\sqrt{\dfrac{3(j-M+1)(j+M)}{4j(j+1)(j+2)(j-1)(2j+3)(2j-1)}}$
$j-1$	$-\left\{(j+1)(j-6)-10(M-1)\left(j+\frac{3}{2}M\right)\right\}\sqrt{\dfrac{(j-M+1)(j-M)}{8(j+1)j(j-1)(2j+1)(2j-3)(2j+3)}}$
$j-2$	$-(j+3M-2)\sqrt{\dfrac{5(j+M-1)(j-M+1)(j-M)(j-M-1)}{8(j+1)j(j-1)(j-2)(2j+1)(2j-1)}}$
$j-3$	$\sqrt{\dfrac{15(j+M-1)(j+M-2)(j-M+1)(j-M)(j-M-1)(j-M-2)}{8j(j-1)(j-2)(2j+1)(2j-1)(2j-3)}}$

Table 1. Algebraic tables of $j_2 = \frac{1}{2}, 1, \frac{3}{2}, 2, \frac{5}{2}$, and 3 Wigner coefficients (continued)

$$C^{j\ \ 3\ \ J}_{M-m\ \ m\ \ M} = (-1)^{j-3-M}(2J+1)^{1/2}\begin{pmatrix} j & 3 & J \\ M-m & m & -M \end{pmatrix}$$

Column 4 (m = 0)

$J = j+3$	$\sqrt{\frac{5(j-M+1)(j-M+2)(j-M+3)(j+M+1)(j+M+2)(j+M+3)}{2(j+1)(j+2)(j+3)(2j+1)(2j+3)(2j+5)}}$
$j+2$	$M\sqrt{\frac{15(j-M+1)(j-M+2)(j+M+1)(j+M+2)}{2j(j+1)(j+2)(j+3)(2j+1)(2j+3)}}$
$j+1$	$-\left\{j(j+2)-5M^2\right\}\sqrt{\frac{3(j+M+1)(j-M+1)}{2j(j+1)(j+2)(2j+1)(2j+5)(2j-1)}}$
j	$-M\left\{3j(j+1)-5M^2-1\right\}\sqrt{\frac{1}{j(j+1)(j+2)(j-1)(2j+3)(2j-1)}}$
$j-1$	$(j^2-5M^2-1)\sqrt{\frac{3(j-M)(j+M)}{2(j+1)j(j-1)(2j+1)(2j-3)(2j+3)}}$
$j-2$	$M\sqrt{\frac{15(j+M)(j+M-1)(j-M)(j-M-1)}{2(j+1)j(j-1)(j-2)(2j+1)(2j-1)}}$
$j-3$	$-\sqrt{\frac{5(j+M)(j+M-1)(j+M-2)(j-M)(j-M-1)(j-M-2)}{2j(j-1)(j-2)(2j+1)(2j-1)(2j-3)}}$

Column 5 (m = − 1)

$J = j+3$	$\sqrt{\frac{15(j+M+2)(j+M+3)(j-M)(j-M+1)(j-M+2)(j-M+3)}{8(j+1)(j+2)(j+3)(2j+1)(2j+3)(2j+5)}}$
$j+2$	$(j+3M+3)\sqrt{\frac{5(j+M+2)(j-M)(j-M+1)(j-M+2)}{8j(j+1)(j+2)(j+3)(2j+1)(2j+3)}}$
$j+1$	$-\left\{j(j+7)-10(M+1)\left(j+\frac{3}{2}M+1\right)\right\}\sqrt{\frac{(j-M)(j-M+1)}{8j(j+1)(j+2)(2j+1)(2j+5)(2j-1)}}$
j	$-\left\{j(j+1)-5M(M+1)-2\right\}\sqrt{\frac{3(j+M+1)(j-M)}{4j(j+1)(j+2)(j-1)(2j+3)(2j-1)}}$
$j-1$	$-\left\{(j+1)(j-6)+10(M+1)\left(j-\frac{3}{2}M\right)\right\}\sqrt{\frac{(j+M+1)(j+M)}{8(j+1)j(j-1)(2j+1)(2j-3)(2j+3)}}$
$j-2$	$(j-3M-2)\sqrt{\frac{5(j-M-1)(j+M+1)(j+M)(j+M-1)}{8(j+1)j(j-1)(j-2)(2j+1)(2j-1)}}$
$j-3$	$\sqrt{\frac{15(j-M-1)(j-M-2)(j+M+1)(j+M)(j+M-1)(j+M-2)}{8j(j-1)(j-2)(2j+1)(2j-1)(2j-3)}}$

Column 6 (m = − 2)

$J = j+3$	$\sqrt{\frac{3(j+M+3)(j-M-1)(j-M)(j-M+1)(j-M+2)(j-M+3)}{4(j+1)(j+2)(j+3)(2j+1)(2j+3)(2j+5)}}$
$j+2$	$\left(j+\frac{3}{2}M+3\right)\sqrt{\frac{(j-M-1)(j-M)(j-M+1)(j-M+2)}{j(j+1)(j+2)(j+3)(2j+1)(2j+3)}}$
$j+1$	$(j+3M+4)\sqrt{\frac{5(j+M+2)(j-M-1)(j-M)(j-M+1)}{4j(j+1)(j+2)(2j+1)(2j+5)(2j-1)}}$
j	$(M+1)\sqrt{\frac{15(j-M)(j+M+1)(j+M+2)(j-M-1)}{2j(j+1)(j+2)(j-1)(2j+3)(2j-1)}}$
$j-1$	$-(j-3M-3)\sqrt{\frac{5(j-M-1)(j+M+2)(j+M+1)(j+M)}{4(j+1)j(j-1)(2j+1)(2j-3)(2j+3)}}$
$j-2$	$-\left(j-\frac{3}{2}M-2\right)\sqrt{\frac{(j+M+2)(j+M+1)(j+M)(j+M-1)}{(j+1)j(j-1)(j-2)(2j+1)(2j-1)}}$
$j-3$	$-\sqrt{\frac{3(j-M-2)(j+M+2)(j+M+1)(j+M)(j+M-1)(j+M-2)}{4j(j-1)(j-2)(2j+1)(2j-1)(2j-3)}}$

Table 1. Algebraic tables of $j_2 = \frac{1}{2}, 1, \frac{3}{2}, 2, \frac{5}{2},$ and 3 Wigner coefficients (continued)

$$C^{j\ \ 3\ \ J}_{M-m\ \ m\ \ M} = (-1)^{j-3-M}(2J+1)^{1/2}\begin{pmatrix} j & 3 & J \\ M-m & m & -M \end{pmatrix}$$

Column 7 ($m = -3$)

$J = j + 3$	$\sqrt{\dfrac{(j-M-2)(j-M-1)(j-M)(j-M+1)(j-M+2)(j-M+3)}{8(j+1)(j+2)(j+3)(2j+1)(2j+3)(2j+5)}}$
$j + 2$	$\sqrt{\dfrac{3(j-M-2)(j-M-1)(j-M)(j-M+1)(j-M+2)(j+M+3)}{8j(j+1)(j+2)(j+3)(2j+1)(2j+3)}}$
$j + 1$	$\sqrt{\dfrac{15(j-M-2)(j-M-1)(j-M)(j-M+1)(j+M+2)(j+M+3)}{8j(j+1)(j+2)(2j+1)(2j+5)(2j-1)}}$
j	$\sqrt{\dfrac{5(j-M-2)(j-M-1)(j-M)(j+M+1)(j+M+2)(j+M+3)}{4j(j+1)(j+2)(j-1)(2j+3)(2j-1)}}$
$j - 1$	$\sqrt{\dfrac{15(j+M+3)(j+M+2)(j+M+1)(j+M)(j-M-1)(j-M-2)}{8(j+1)j(j-1)(2j+1)(2j-3)(2j+3)}}$
$j - 2$	$\sqrt{\dfrac{3(j+M+3)(j+M+2)(j+M+1)(j+M)(j+M-1)(j-M-2)}{8(j+1)j(j-1)(j-2)(2j+1)(2j-1)}}$
$j - 3$	$\sqrt{\dfrac{(j+M+3)(j+M+2)(j+M+1)(j+M)(j+M-1)(j+M-2)}{8j(j-1)(j-2)(2j+1)(2j-1)(2j-3)}}$

Table 2. Algebraic tables of $l = \frac{1}{2}$, 1, $\frac{3}{2}$, and 2 Racah coefficients

$$W(l_1 J_1 l_2 J_2 ; \tfrac{1}{2}, L) = (-1)^{l_1+J_1+l_2+J_2} \begin{Bmatrix} l_1 & J_1 & \frac{1}{2} \\ J_2 & l_2 & L \end{Bmatrix}$$

	$l_1 = J_1 + \frac{1}{2}$	$l_1 = J_1 - \frac{1}{2}$
$l_2 = J_2 + \frac{1}{2}$	$(-1)^{J_1+J_2-L}\left[\frac{(J_1+J_2+L+2)(J_1+J_2-L+1)}{(2J_1+1)(2J_1+2)(2J_2+1)(2J_2+2)}\right]^{\frac{1}{2}}$	$(-1)^{J_1+J_2-L}\left[\frac{(L-J_1+J_2+1)(L+J_1-J_2)}{(2J_1)(2J_1+1)(2J_2+1)(2J_2+2)}\right]^{\frac{1}{2}}$
$l_2 = J_2 - \frac{1}{2}$	$(-1)^{J_1+J_2-L}\left[\frac{(L+J_1-J_2+1)(L-J_1+J_2)}{(2J_1+1)(2J_1+2)(2J_2)(2J_2+1)}\right]^{\frac{1}{2}}$	$(-1)^{J_1+J_2-L-1}\left[\frac{(J_1+J_2+L+1)(J_1+J_2-L)}{2J_1(2J_1+1)(2J_2)(2J_2+1)}\right]^{\frac{1}{2}}$

Table 2. Algebraic tables of $l = \frac{1}{2}, 1, \frac{3}{2}$, and 2 Racah coefficients (continued)

$$W(l_1 J_1 l_2 J_2\ ;\ 1\ ,\ L) = (-1)^{l_1+J_1+l_2+J_2} \begin{Bmatrix} l_1 & J_1 & 1 \\ J_2 & l_2 & L \end{Bmatrix}$$

$l_2 = J_2 + 1$	
$l_1 = J_1 + 1$	$(-1)^{J_1+J_2-L}\left[\dfrac{(L+J_1+J_2+3)(L+J_1+J_2+2)(-L+J_1+J_2+2)(-L+J_1+J_2+1)}{4(2J_1+3)(J_1+1)(2J_1+1)(2J_2+3)(J_2+1)(2J_2+1)}\right]^{1/2}$
$l_1 = J_1$	$(-1)^{J_1+J_2-L}\left[\dfrac{(L+J_1+J_2+2)(-L+J_1+J_2+1)(L-J_1+J_2+1)(L+J_1-J_2)}{4J_1(2J_1+1)(J_1+1)(2J_2+1)(J_2+1)(2J_2+3)}\right]^{1/2}$
$l_1 = J_1 - 1$	$(-1)^{J_1+J_2-L}\left[\dfrac{(L+J_1-J_2)(L+J_1-J_2-1)(L-J_1+J_2+2)(L-J_1+J_2+1)}{4(2J_1+1)(2J_1-1)(J_1)(J_2+1)(2J_2+1)(2J_2+3)}\right]^{1/2}$
$l_2 = J_2$	
$l_1 = J_1 + 1$	$(-1)^{J_1+J_2-L}\left[\dfrac{(L+J_1+J_2+2)(L+J_1-J_2+1)(J_1+J_2-L+1)(L-J_1+J_2)}{4(2J_1+1)(J_1+1)(2J_1+3)(J_2)(J_2+1)(2J_2+1)}\right]^{1/2}$
$l_1 = J_1$	$(-1)^{J_1+J_2-L-1}\left[\dfrac{J_1(J_1+1)+J_2(J_2+1)-L(L+1)}{[4J_1(J_1+1)(2J_1+1)(J_2)(J_2+1)(2J_2+1)]^{1/2}}\right]$
$l_1 = J_1 - 1$	$(-1)^{J_1+J_2-L-1}\left[\dfrac{(L+J_1+J_2+1)(-L+J_1+J_2)(L+J_1-J_2)(L-J_1+J_2+1)}{4(2J_1+1)(J_1)(2J_1-1)(J_2)(2J_2+1)(J_2+1)}\right]^{1/2}$

Table 2. Algebraic tables of $l = \frac{1}{2}$, 1, $\frac{3}{2}$, and 2 Racah coefficients (continued)

$$W(l_1 J_1 l_2 J_2 \; ; \; 1 \, , \, L) = (-1)^{l_1+J_1+l_2+J_2} \begin{Bmatrix} l_1 & J_1 & 1 \\ J_2 & l_2 & L \end{Bmatrix}$$

	$l_2 = J_2 - 1$
$l_1 = J_1 + 1$	$(-1)^{J_1+J_2-L}\left[\dfrac{(L-J_1+J_2)(L-J_1+J_2-1)(L+J_1-J_2+2)(L+J_1-J_2+1)}{4(2J_1+1)(J_1+1)(2J_1+3)(2J_2-1)(J_2)(2J_2+1)}\right]^{1/2}$
$l_1 = J_1$	$(-1)^{J_1+J_2-L-1}\left[\dfrac{(L+J_1+J_2+1)(L+J_1-J_2+1)(L+J_2-J_1)(J_1+J_2-L)}{4J_1(2J_1+1)(J_1+1)(J_2)(2J_2+1)(2J_2-1)}\right]^{1/2}$
$l_1 = J_1 - 1$	$(-1)^{J_1+J_2-L}\left[\dfrac{(L+J_1+J_2+1)(L+J_1+J_2)(-L+J_1+J_2)(-L+J_1+J_2-1)}{4(2J_1+1)(J_1)(2J_1-1)(2J_2+1)(J_2)(2J_2-1)}\right]^{1/2}$

Table 2. Algebraic tables of $l = \frac{1}{2}, 1, \frac{3}{2}$, and 2 Racah coefficients (continued)

$$W(l_1 J_1 l_2 J_2 ; \tfrac{3}{2}, L) = (-1)^{l_1+J_1+l_2+J_2} \begin{Bmatrix} l_1 & J_1 & \frac{3}{2} \\ J_2 & l_2 & L \end{Bmatrix}$$

$l_1 = J_1 + 3/2$	
$l_2 = J_2 + \frac{3}{2}$	$(-1)^{J_1+J_2-L}\left[\dfrac{(L+J_1+J_2+4)(L+J_1+J_2+3)(L+J_1+J_2+2)(-L+J_1+J_2+3)(-L+J_1+J_2+2)(-L+J_1+J_2+1)}{(2J_1+4)(2J_1+3)(2J_1+2)(2J_1+1)(2J_2+4)(2J_2+3)(2J_2+2)(2J_2+1)}\right]^{\frac{1}{2}}$
$l_2 = J_2 + \frac{1}{2}$	$(-1)^{J_1+J_2-L}\left[\dfrac{3(L+J_1+J_2+3)(L+J_1+J_2+2)(L+J_1-J_2+1)(L-J_1+J_2)(-L+J_1+J_2+2)(-L+J_1+J_2+1)}{(2J_1+4)(2J_1+3)(2J_1+2)(2J_1+1)(2J_2+3)(2J_2+2)(2J_2+1)(2J_2)}\right]^{\frac{1}{2}}$
$l_2 = J_2 - \frac{1}{2}$	$(-1)^{J_1+J_2-L}\left[\dfrac{3(L+J_1+J_2+2)(L+J_1-J_2+2)(L+J_1-J_2+1)(L-J_1+J_2)(L-J_1+J_2-1)(J_1+J_2-L+1)}{(2J_1+4)(2J_1+3)(2J_1+2)(2J_1+1)(2J_2+2)(2J_2+1)(2J_2)(2J_2-1)}\right]^{\frac{1}{2}}$
$l_2 = J_2 - \frac{3}{2}$	$(-1)^{J_1+J_2-L}\left[\dfrac{(L-J_1+J_2)(L-J_1+J_2-1)(L-J_1+J_2-2)(L+J_1-J_2+3)(L+J_1-J_2+2)(L+J_1-J_2+1)}{(2J_1+4)(2J_1+3)(2J_1+2)(2J_1+1)(2J_2+1)(2J_2)(2J_2-1)(2J_2-2)}\right]^{\frac{1}{2}}$
$l_1 = J_1 + \frac{1}{2}$	
$l_2 = J_2 + \frac{3}{2}$	$(-1)^{J_1+J_2-L}\left[\dfrac{3(L+J_1+J_2+3)(L+J_1+J_2+2)(-L+J_1+J_2+2)(-L+J_1+J_2+1)(L-J_1+J_2+1)(L+J_1-J_2)}{(2J_1+3)(2J_1+2)(2J_1+1)(2J_1)(2J_2+4)(2J_2+3)(2J_2+2)(2J_2+1)}\right]^{\frac{1}{2}}$
$l_2 = J_2 + \frac{1}{2}$	$(-1)^{J_1+J_2-L-1}\dfrac{[(L+J_1+J_2+2)(-L+J_1+J_2+1)]^{\frac{1}{2}}[(L+J_1+J_2+3)(-L+J_1+J_2)-2(L-J_1+J_2)(L+J_1-J_2)]}{[(2J_1+3)(2J_1+2)(2J_1+1)(2J_1)(2J_2+3)(2J_2+2)(2J_2+1)(2J_2)]^{\frac{1}{2}}}$
$l_2 = J_2 - \frac{1}{2}$	$(-1)^{J_1+J_2-L-1}\dfrac{[(L+J_1-J_2+1)(L-J_1+J_2)]^{\frac{1}{2}}[2(L+J_1+J_2+2)(-L+J_1+J_2)-(L-J_1+J_2-1)(L+J_1-J_2)]}{[(2J_1+3)(2J_1+2)(2J_1+1)(2J_1)(2J_2+2)(2J_2+1)(2J_2)(2J_2-1)]^{\frac{1}{2}}}$
$l_2 = J_2 - \frac{3}{2}$	$(-1)^{J_1+J_2-L-1}\left[\dfrac{3(L+J_1+J_2+1)(L-J_1+J_2)(L-J_1+J_2-1)(L+J_1-J_2+2)(L+J_1-J_2+1)(-L+J_1+J_2)}{(2J_1+3)(2J_1+2)(2J_1+1)(2J_1)(2J_2+1)(2J_2)(2J_2-1)(2J_2-2)}\right]^{\frac{1}{2}}$

Table 2. Algebraic tables of $l = \frac{1}{2}$, 1, $\frac{3}{2}$, and 2 Racah coefficients (continued)

$$W(l_1 J_1 l_2 J_2 \,;\, \tfrac{3}{2}, L) = (-1)^{l_1+J_1+l_2+J_2} \begin{Bmatrix} l_1 & J_1 & \frac{3}{2} \\ J_2 & l_2 & L \end{Bmatrix}$$

$l_1 = J_1 - \frac{1}{2}$	
$l_2 = J_2 + \frac{3}{2}$	$(-1)^{J_1+J_2-L}\left[\dfrac{3(L+J_1+J_2+2)(-L+J_1+J_2+1)(L-J_1+J_2+2)(L-J_1+J_2+1)(L+J_1-J_2)(L+J_1-J_2-1)}{(2J_1+2)(2J_1+1)(2J_1)(2J_1-1)(2J_2+4)(2J_2+3)(2J_2+2)(2J_2+1)}\right]^{\frac{1}{2}}$
$l_2 = J_2 + \frac{1}{2}$	$(-1)^{J_1+J_2-L}\dfrac{[(L-J_1+J_2+1)(L+J_1-J_2)]^{\frac{1}{2}}[(L-J_1+J_2)(L+J_1-J_2-1)-2(L+J_1+J_2+2)(-L+J_1+J_2)]}{[(2J_1+2)(2J_1+1)(2J_1)(2J_1-1)(2J_2+3)(2J_2+2)(2J_2+1)(2J_2)]^{\frac{1}{2}}}$
$l_2 = J_2 - \frac{1}{2}$	$(-1)^{J_1+J_2-L}\dfrac{[(L+J_1+J_2+1)(-L+J_1+J_2)]^{\frac{1}{2}}[(L+J_1+J_2+2)(-L+J_1+J_2-1)-2(L+J_1-J_2)(L-J_1+J_2)]}{[(2J_1+2)(2J_1+1)(2J_1)(2J_1-1)(2J_2+2)(2J_2+1)(2J_2)(2J_2-1)]^{\frac{1}{2}}}$
$l_2 = J_2 - \frac{3}{2}$	$(-1)^{J_1+J_2-L}\left[\dfrac{3(L+J_1+J_2+1)(L+J_1+J_2)(L-J_1+J_2)(L+J_1-J_2+1)(-L+J_1+J_2)(-L+J_1+J_2-1)}{(2J_1+2)(2J_1+1)(2J_1)(2J_1-1)(2J_2+1)(2J_2)(2J_2-1)(2J_2-2)}\right]^{\frac{1}{2}}$
$l_1 = J_1 - 3/2$	
$l_2 = J_2 + \frac{3}{2}$	$(-1)^{J_1+J_2-L}\left[\dfrac{(L+J_1-J_2)(L+J_1-J_2-1)(L+J_1-J_2-2)(L-J_1+J_2+3)(L-J_1+J_2+2)(L-J_1+J_2+1)}{(2J_1+1)(2J_1)(2J_1-1)(2J_1-2)(2J_2+4)(2J_2+3)(2J_2+2)(2J_2+1)}\right]^{\frac{1}{2}}$
$l_2 = J_2 + \frac{1}{2}$	$(-1)^{J_1+J_2-L-1}\left[\dfrac{3(L+J_1+J_2+1)(-L+J_1+J_2)(L+J_1-J_2)(L+J_1-J_2-1)(L-J_1+J_2+2)(L-J_1+J_2+1)}{(2J_1+1)(2J_1)(2J_1-1)(2J_1-2)(2J_2+3)(2J_2+2)(2J_2+1)(2J_2)}\right]^{\frac{1}{2}}$
$l_2 = J_2 - \frac{1}{2}$	$(-1)^{J_1+J_2-L}\left[\dfrac{3(L+J_1+J_2+1)(L+J_1+J_2)(-L+J_1+J_2)(-L+J_1+J_2-1)(L+J_1-J_2)(L-J_1+J_2+1)}{(2J_1+1)(2J_1)(2J_1-1)(2J_1-2)(2J_2+2)(2J_2+1)(2J_2)(2J_2-1)}\right]^{\frac{1}{2}}$
$l_2 = J_2 - \frac{3}{2}$	$(-1)^{J_1+J_2-L-1}\left[\dfrac{(L+J_1+J_2+1)(L+J_1+J_2)(L+J_1+J_2-1)(-L+J_1+J_2)(-L+J_1+J_2-1)(-L+J_1+J_2-2)}{(2J_1+1)(2J_1)(2J_1-1)(2J_1-2)(2J_2+1)(2J_2)(2J_2-1)(2J_2-2)}\right]^{\frac{1}{2}}$

$$W(l_1 J_1 l_2 J_2\,;\,2\,,\,L) = (-1)^{l_1+J_1+l_2+J_2}\begin{Bmatrix} l_1 & J_1 & 2 \\ J_2 & l_2 & L \end{Bmatrix}$$

$l_1 = J_1 + 2$

$l_2 = J_2+2$	$(-1)^{L-J_1-J_2}\left[\dfrac{(L+J_1+J_2+5)(L+J_1+J_2+4)(L+J_1+J_2+3)(L+J_1+J_2+2)(-L+J_1+J_2+4)(-L+J_1+J_2+3)(-L+J_1+J_2+2)(-L+J_1+J_2+1)}{(2J_1+5)(2J_1+4)(2J_1+3)(2J_1+2)(2J_1+1)\cdot(2J_2+5)(2J_2+4)(2J_2+3)(2J_2+2)(2J_2+1)}\right]^{1/2}$
$l_2 = J_2+1$	$(-1)^{L-J_1-J_2}\left[\dfrac{4(L+J_1+J_2+4)(L+J_1+J_2+3)(L+J_1+J_2+2)(L+J_1-J_2+1)(-L+J_1+J_2+3)(-L+J_1+J_2+2)(-L+J_1+J_2+1)(L-J_1+J_2)}{(2J_1+5)(2J_1+4)(2J_1+3)(2J_1+2)(2J_1+1)\cdot(2J_2+4)(2J_2+3)(2J_2+2)(2J_2+1)(2J_2)}\right]^{1/2}$
$l_2 = J_2$	$(-1)^{L-J_1-J_2}\left[\dfrac{6(L+J_1+J_2+3)(L+J_1+J_2+2)(L+J_1-J_2+2)(L+J_1-J_2+1)(-L+J_1+J_2+2)(-L+J_1+J_2+1)(L-J_1+J_2)(L-J_1+J_2-1)}{(2J_1+5)(2J_1+4)(2J_1+3)(2J_1+2)(2J_1+1)\cdot(2J_2+3)(2J_2+2)(2J_2+1)(2J_2)(2J_2-1)}\right]^{1/2}$
$l_2 = J_2-1$	$(-1)^{L-J_1-J_2}\left[\dfrac{4(L+J_1+J_2+2)(L-J_1+J_2)(L-J_1+J_2-1)(L-J_1+J_2-2)(L+J_1-J_2+3)(L+J_1-J_2+2)(L+J_1-J_2+1)(-L+J_1+J_2+1)}{(2J_1+5)(2J_1+4)(2J_1+3)(2J_1+2)(2J_1+1)\cdot(2J_2+2)(2J_2+1)(2J_2)(2J_2-1)(2J_2-2)}\right]^{1/2}$
$l_2 = J_2-2$	$(-1)^{L-J_1-J_2}\left[\dfrac{(L-J_1+J_2)(L-J_1+J_2-1)(L-J_1+J_2-2)(L-J_1+J_2-3)(L+J_1-J_2+4)(L+J_1-J_2+3)(L+J_1-J_2+2)(L+J_1-J_2+1)}{(2J_1+5)(2J_1+4)(2J_1+3)(2J_1+2)(2J_1+1)\cdot(2J_2+1)(2J_2)(2J_2-1)(2J_2-2)(2J_2-3)}\right]^{1/2}$

$l_1 = J_1 + 1$

$l_2 = J_2+2$	$(-1)^{L-J_1-J_2}\left[\dfrac{4(L+J_1+J_2+4)(L+J_1+J_2+3)(L+J_1+J_2+2)(L-J_1+J_2+1)(L+J_1-J_2)(-L+J_1+J_2+3)(-L+J_1+J_2+2)(-L+J_1+J_2+1)}{(2J_1+4)(2J_1+3)(2J_1+2)(2J_1+1)(2J_1)\cdot(2J_2+5)(2J_2+4)(2J_2+3)(2J_2+2)(2J_2+1)}\right]^{1/2}$
$l_2 = J_2+1$	$(-1)^{L-J_1-J_2-1}\left[\dfrac{(L+J_1+J_2+3)(L+J_1+J_2+2)(-L+J_1+J_2+2)(-L+J_1+J_2+1)}{(2J_1+4)(2J_1+3)(2J_1+2)(2J_1+1)(2J_1)\cdot(2J_2+4)(2J_2+3)(2J_2+2)(2J_2+1)(2J_2)}\right]^{1/2}\cdot 4\cdot[(J_1+1)(J_1-J_2)-L(L+1)+J_2(J_2+2)]$
$l_2 = J_2$	$(-1)^{L-J_1-J_2-1}\left[\dfrac{6(L+J_1+J_2+2)(L-J_1+J_2)(L+J_1-J_2+1)(-L+J_1+J_2+1)}{(2J_1+4)(2J_1+3)(2J_1+2)(2J_1+1)(2J_1)\cdot(2J_2+3)(2J_2+2)(2J_2+1)(2J_2)(2J_2-1)}\right]^{1/2}\cdot 2\cdot[J_1(J_1+2)+J_2(J_2+1)-L(L+1)]$
$l_2 = J_2-1$	$(-1)^{L-J_1-J_2-1}\left[\dfrac{(L-J_1+J_2)(L-J_1+J_2-1)(L+J_1-J_2+2)(L+J_1-J_2+1)}{(2J_1+4)(2J_1+3)(2J_1+2)(2J_1+1)(2J_1)\cdot(2J_2+2)(2J_2+1)(2J_2)(2J_2-1)(2J_2-2)}\right]^{1/2}\cdot 4\cdot[J_1(J_1+2)+J_2(J_2+1)+J_1J_2-L(L+1)]$
$l_2 = J_2-2$	$(-1)^{L-J_1-J_2-1}\left[\dfrac{4(L+J_1+J_2+1)(-L+J_1+J_2)(L-J_1+J_2)(L-J_1+J_2-1)(L-J_1+J_2-2)(L+J_1-J_2+3)(L+J_1-J_2+2)(L+J_1-J_2+1)}{(2J_1+4)(2J_1+3)(2J_1+2)(2J_1+1)(2J_1)\cdot(2J_2+1)(2J_2)(2J_2-1)(2J_2-2)(2J_2-3)}\right]^{1/2}$

Table 2. Algebraic tables of $l = \frac{1}{2}$, 1, $\frac{3}{2}$, and 2 Racah coefficients (continued)

$$W(l_1 J_1 l_2 J_2 \; ; \; 2, L) = (-1)^{l_1+J_1+l_2+J_2} \begin{Bmatrix} l_1 & J_1 & 2 \\ J_2 & l_2 & L \end{Bmatrix}$$

$l_1 = J_1$

$l_2 = J_2+2$	$(-1)^{L-J_1-J_2}\left[\dfrac{6(L+J_1+J_2+3)(L+J_1+J_2+2)(L+2-J_1+J_2)(L+1-J_1+J_2)(L+J_1-J_2)(L+J_1-J_2-1)(-L+J_1+J_2+2)(-L+J_1+J_2+1)}{(2J_1+3)(2J_1+2)(2J_1+1)(2J_1)(2J_1-1)\cdot(2J_2+5)(2J_2+4)(2J_2+3)(2J_2+2)(2J_2+1)}\right]^{1/2}$
$l_2 = J_2+1$	$(-1)^{L-J_1-J_2-1}\left[\dfrac{6(L+J_1+J_2+2)(L+J_1-J_2)(L-J_1+J_2+1)(-L+J_1+J_2+1)}{(2J_1+3)(2J_1+2)(2J_1+1)(2J_1)(2J_1-1)\cdot(2J_2+4)(2J_2+3)(2J_2+2)(2J_2+1)(2J_2)}\right]^{1/2}\cdot 2\cdot[J_1(J_1+1)+J_2(J_2+2)-L(L+1)]$
$l_2 = J_2$	$(-1)^{L-J_1-J_2}\left[\dfrac{1}{(2J_1+3)(2J_1+2)(2J_1+1)(2J_1)(2J_1-1)\cdot(2J_2+3)(2J_2+2)(2J_2+1)(2J_2)(2J_2-1)}\right]^{1/2}\cdot 6\cdot[A(A+1)-\frac{4}{3}J_1(J_1+1)J_2(J_2+1)]$
	where $A = L(L+1)-J_1(J_1+1)-J_2(J_2+1)$
$l_2 = J_2-1$	$(-1)^{L-J_1-J_2}\left[\dfrac{6(L+J_1+J_2+1)(L-J_1+J_2)(L+J_1-J_2+1)(-L+J_1+J_2)}{(2J_1+3)(2J_1+2)(2J_1+1)(2J_1)(2J_1-1)\cdot(2J_2+2)(2J_2+1)(2J_2)(2J_2-1)(2J_2-2)}\right]^{1/2}\cdot 2\cdot[J_1(J_1+1)-L(L+1)+J_2^2-1]$
$l_2 = J_2-2$	$(-1)^{L-J_1-J_2}\left[\dfrac{6(L+J_1+J_2+1)(L+J_1+J_2)(L-J_1+J_2)(L-J_1+J_2-1)(-L+J_1+J_2)(-L+J_1+J_2-1)(L+J_1-J_2+2)(L+J_1-J_2+1)}{(2J_1+3)(2J_1+2)(2J_1+1)(2J_1)(2J_1-1)\cdot(2J_2+1)(2J_2)(2J_2-1)(2J_2-2)(2J_2-3)}\right]^{1/2}$

$l_1 = J_1 - 1$

$l_2 = J_2+2$	$(-1)^{L-J_1-J_2}\left[\dfrac{4(L+J_1+J_2+2)(-L+J_1+J_2+1)(L-J_1+J_2+3)(L-J_1+J_2+2)(L-J_1+J_2+1)(L+J_1-J_2)(L+J_1-J_2-1)(L+J_1-J_2-2)}{(2J_1-2)(2J_1-1)(2J_1)(2J_1+1)(2J_1+2)\cdot(2J_2+1)(2J_2+2)(2J_2+3)(2J_2+4)(2J_2+5)}\right]^{1/2}$
$l_2 = J_2+1$	$(-1)^{L-J_1-J_2-1}\left[\dfrac{(L-J_1+J_2+2)(L-J_1+J_2+1)(L+J_1-J_2)(L+J_1-J_2-1)}{(2J_1-2)(2J_1-1)(2J_1)(2J_1+1)(2J_1+2)\cdot(2J_2)(2J_2+1)(2J_2+2)(2J_2+3)(2J_2+4)}\right]^{1/2}\cdot 4\cdot[(J_1-1)(J_1+J_2+2)-(L+J_2+2)(L-J_2-1)]$
$l_2 = J_2$	$(-1)^{L-J_1-J_2}\left[\dfrac{6(L+J_1+J_2+1)(L+J_1-J_2)(-L+J_1+J_2)(L-J_1+J_2+1)}{(2J_1-2)(2J_1-1)(2J_1)(2J_1+1)(2J_1+2)\cdot(2J_2-1)(2J_2)(2J_2+1)(2J_2+2)(2J_2+3)}\right]^{1/2}\cdot 2\cdot[J_1^2-1-(L+J_2+1)(L-J_2)]$
$l_2 = J_2-1$	$(-1)^{L-J_1-J_2-1}\left[\dfrac{(L+J_1+J_2+1)(L+J_1+J_2)(-L+J_1+J_2)(-L+J_1+J_2-1)}{(2J_1-2)(2J_1-1)(2J_1)(2J_1+1)(2J_1+2)\cdot(2J_2-2)(2J_2-1)(2J_2)(2J_2+1)(2J_2+2)}\right]^{1/2}\cdot 4\cdot[(J_1-1)(J_1-J_2+1)-(L+J_2)(L-J_2+1)]$
$l_2 = J_2-2$	$(-1)^{L-J_1-J_2-1}\left[4(L+J_1+J_2+1)(L+J_1+J_2)(L+J_1+J_2-1)(L+J_1-J_2+1)(-L+J_1+J_2)(-L+J_1+J_2-1)(-L+J_1+J_2-2)(L-J_1+J_2)\right]^{1/2}$

Table 2. Algebraic tables of $l = \frac{1}{2}, 1, \frac{3}{2}$, and 2 Racah coefficients (continued)

$$W(l_1 J_1 l_2 J_2 ; 2, L) = (-1)^{l_1+J_1+l_2+J_2} \begin{Bmatrix} l_1 & J_1 & 2 \\ J_2 & l_2 & L \end{Bmatrix}$$

	$l_1 = J_1 - 2$
$l_2 = J_2 + 2$	$(-1)^{L-J_1-J_2}\left[\dfrac{(L+4-J_1+J_2)(L+3-J_1+J_2)(L+2-J_1+J_2)(L+1-J_1+J_2)(L+J_1-J_2)(L+J_1-J_2-1)(L+J_1-J_2-2)(L+J_1-J_2-3)}{(2J_1-3)(2J_1-2)(2J_1-1)(2J_1)(2J_1+1)\cdot(2J_2+5)(2J_2+4)(2J_2+3)(2J_2+2)(2J_2+1)}\right]^{\frac{1}{2}}$
$l_2 = J_2 + 1$	$(-1)^{L-J_1-J_2-1}\left[\dfrac{4(L+1+J_1+J_2)(L+3-J_1+J_2)(L+2-J_1+J_2)(L+1-J_1+J_2)(L+J_1-J_2)(L+J_1-J_2-1)(L+J_1-J_2-2)(-L+J_1+J_2)}{(2J_1-3)(2J_1-2)(2J_1-1)(2J_1)(2J_1+1)\cdot(2J_2+4)(2J_2+3)(2J_2+2)(2J_2+1)(2J_2)}\right]^{\frac{1}{2}}$
$l_2 = J_2$	$(-1)^{L-J_1-J_2}\left[\dfrac{6(L+1+J_1+J_2)(L+J_1+J_2)(L+2-J_1+J_2)(L+1-J_1+J_2)(L+J_1-J_2)(L+J_1-J_2-1)(-L+J_1+J_2)(-L+J_1+J_2-1)}{(2J_1-3)(2J_1-2)(2J_1-1)(2J_1)(2J_1+1)\cdot(2J_2+3)(2J_2+2)(2J_2+1)(2J_2)(2J_2-1)}\right]^{\frac{1}{2}}$
$l_2 = J_2 - 1$	$(-1)^{L-J_1-J_2-1}\left[\dfrac{4(L+1+J_1+J_2)(L+J_1+J_2)(L+J_1+J_2-1)(L+1-J_1+J_2)(L+J_1-J_2)(-L+J_1+J_2)(-L+J_1+J_2-1)(-L+J_1+J_2-2)}{(2J_1-3)(2J_1-2)(2J_1-1)(2J_1)(2J_1+1)\cdot(2J_2+2)(2J_2+1)(2J_2)(2J_2-1)(2J_2-2)}\right]^{\frac{1}{2}}$
$l_2 = J_2 - 2$	$(-1)^{L-J_1-J_2}\left[\dfrac{(L+1+J_1+J_2)(L+J_1+J_2)(L+J_1+J_2-1)(L+J_1+J_2-2)(-L+J_1+J_2)(-L+J_1+J_2-1)(-L+J_1+J_2-2)(-L+J_1+J_2-3)}{(2J_1-3)(2J_1-2)(2J_1-1)(2J_1)(2J_1+1)\cdot(2J_2+1)(2J_2)(2J_2-1)(2J_2-2)(2J_2-3)}\right]^{\frac{1}{2}}$

Table 3. Algebraic tables of $j_2 = \frac{1}{2}$ and 1 Racah coefficients in the pattern calculus form

$$W^{j_1 \ \frac{1}{2} \ j}_{m-m_2 \ m_2 \ m}(J)$$

$j =$	$m_2 = \frac{1}{2}$	$m_2 = -\frac{1}{2}$
$j_2+\frac{1}{2}$	$\left[\frac{(j_1+m+\frac{1}{2})(2J+j_1-m+\frac{3}{2})}{(2j_1+1)(2J+1)}\right]^{\frac{1}{2}}$	$\left[\frac{(j_1-m+\frac{1}{2})(2J-j_1-m+\frac{1}{2})}{(2j_1+1)(2J+1)}\right]^{\frac{1}{2}}$
$j_2-\frac{1}{2}$	$-\left[\frac{(j_1-m+\frac{1}{2})(2J-j_1-m+\frac{1}{2})}{(2j_1+1)(2J+1)}\right]^{\frac{1}{2}}$	$\left[\frac{(j_1+m+\frac{1}{2})(2J+j_1-m+\frac{3}{2})}{(2j_1+1)(2J+1)}\right]^{\frac{1}{2}}$

Table 3. Algebraic tables of $j_2 = \frac{1}{2}$ and 1 Racah coefficients in the pattern calculus form (continued)

$$W^{j_1\ \ 1\ \ j}_{m-m_2\ m_2\ m}(J)$$

$j =$	$m_2 = 1$	$m_2 = 0$
j_1+1	$\left[\frac{(j_1+m)(j_1+m+1)(2J+j_1-m+1)(2J+j_1-m+2)}{(2j_1+1)(2j_1+2)(2J)(2J+1)}\right]^{\frac{1}{2}}$	$\left[\frac{(j_1-m+1)(j_1+m+1)(2J-j_1-m)(2J+j_1-m+2)}{(2j_1+1)(j_1+1)(2J)(2J+1)}\right]^{\frac{1}{2}}$
j_1	$-\left[\frac{(j_1+m)(j_1-m+1)(2J+j_1-m+1)(2J-j_1-m)}{2j_1(j_1+1)(2J)(2J+1)}\right]^{\frac{1}{2}}$	$\frac{m(2J+1)-m^2}{[4j_1(j_1+1)J(J+1)]^{1/2}}$
j_1-1	$\left[\frac{(j_1-m)(j_1-m+1)(2J-j_1-m)(2J-j_1-m+1)}{2j_1(2j_1+1)(2J)(2J+1)}\right]^{\frac{1}{2}}$	$-\left[\frac{(j_1-m)(j_1+m)(2J+j_1-m+1)(2J-j_1-m+1)}{j_1(2j_1+1)(2J)(2J+2)}\right]^{\frac{1}{2}}$

Table 3. Algebraic tables of $j_2 = \frac{1}{2}$ and 1 Racah coefficients in the pattern calculus form (continued)

$j =$	$m_2 = 1$
j_1+1	$\left[\frac{(j_1-m)(j_1-m+1)(2J-j_1-m)(2J-j_1-m+1)}{(2j_1+1)(2j_2+2)(2J+1)(2J+2)}\right]^{\frac{1}{2}}$
j_1	$\left[\frac{(j_1-m)(j_1+m+1)(2J+j_1-m+2)(2J-j_1-m+1)}{2j_1(j_1+1)(2J+1)(2J+2)}\right]^{\frac{1}{2}}$
j_1-1	$\left[\frac{(j_1+m)(j_1+m+1)(2J+j_1-m+1)(2J+j_1-m+2)}{2j_1(2j_1+1)(2J+1)(2J+3)}\right]^{\frac{1}{2}}$

Table 4. The solid and spherical harmonics for $l = 0, 1, 2, 3$, and 4

ℓ	m	$\sqrt{4\pi}\, y_{\ell m}(\vec{r})$	$\sqrt{4\pi}\, Y_{\ell m}(\theta,\varphi)$
0	0	1	1
1	±1	$\mp\sqrt{\frac{3}{2}}\,(x \pm iy)$	$\mp\sqrt{\frac{3}{2}}\, e^{\pm i\varphi} \sin\theta$
	0	$\sqrt{3}\, z$	$\sqrt{3}\cos\theta$
2	±2	$\frac{1}{2}\sqrt{\frac{15}{2}}\,(x \pm iy)^2$	$\frac{1}{2}\sqrt{\frac{15}{2}}\, e^{\pm 2i\varphi} \sin^2\theta$
	±1	$\mp\sqrt{\frac{15}{2}}\,(x \pm iy)z$	$\mp\sqrt{\frac{15}{2}}\, e^{\pm i\varphi} \sin\theta\cos\theta$
	0	$\frac{1}{2}\sqrt{5}\,(3z^2 - r^2)$	$\frac{1}{2}\sqrt{5}\,(3\cos^2\theta - 1)$
3	±3	$\mp\frac{1}{4}\sqrt{35}\,(x \pm iy)^3$	$\mp\frac{1}{4}\sqrt{35}\, e^{\pm 3i\varphi} \sin^3\theta$
	±2	$\frac{1}{2}\sqrt{\frac{105}{2}}\,(x \pm iy)^2 z$	$\frac{1}{2}\sqrt{\frac{105}{2}}\, e^{\pm 2i\varphi} \sin^2\theta\cos\theta$
	±1	$\mp\frac{1}{4}\; 21\,(x \pm iy)(5z^2 - r^2)$	$\mp\frac{1}{4}\sqrt{21}\, e^{\pm i\varphi} \sin\theta\,(5\cos^2\theta - 1)$
	0	$\frac{1}{2}\sqrt{7}\,(5z^2 - 3r^2)z$	$\frac{1}{2}\sqrt{7}\,(5\cos^2\theta - 3)\cos\theta$
4	±4	$\frac{3}{16}\sqrt{70}\,(x \pm iy)^4$	$\frac{3}{16}\sqrt{70}\, e^{\pm 4i\varphi} \sin^4\theta$
	±3	$\mp\frac{3}{4}\sqrt{35}\,(x \pm iy)^3 z$	$\mp\frac{3}{4}\sqrt{35}\, e^{\pm 3i\varphi} \sin^3\theta\cos\theta$
	±2	$\frac{3}{8}\sqrt{10}\,(x \pm iy)^2(7z^2 - r^2)$	$\frac{3}{8}\sqrt{10}\, e^{\pm 2i\varphi} \sin^2\theta\,(7\cos^2\theta - 1)$
	±1	$\mp\frac{3}{4}\sqrt{5}\,(x \pm iy)(7z^2 - 3r^2)z$	$\mp\frac{3}{4}\sqrt{5}\, e^{\pm i\varphi} \sin\theta\,(7\cos^2\theta - 3)\cos\theta$
	0	$\frac{15}{8}\,(7z^4 - 6z^2r^2 + \frac{3}{5}r^4)$	$\frac{15}{8}\,(7\cos^4\theta - 6\cos^2\theta + \frac{3}{5})$

Table 5. The tensor harmonics

$$T^k_\mu \equiv \left\langle k \begin{matrix} 0 \\ \mu \end{matrix} -k \right\rangle \left[\prod_{s=1}^{k} (4J^2 + 1 - s^2) \right]^{1/2}$$

$$T^k_k = (-1)^k [(2k)!/k!k!]^{1/2} J^k_+$$

$$T^{k\dagger}_\mu = (-1)^\mu T^k_{-\mu}$$

$$\sum_\mu (-1)^\mu T^k_\mu T^k_{-\mu} = \prod_{s=1}^{k} (4J^2 + 1 - s^2)$$

k	μ	T^k_μ
0	0	1
1	1	$-\sqrt{2}\,J_+$
	0	$2J_3$
	-1	$\sqrt{2}\,J_-$
2	2	$\sqrt{6}\,J_+^2$
	1	$-\sqrt{6}\,J_+(2J_3 + 1)$
	0	$2(3J_3^2 - J^2)$
	-1	$\sqrt{6}\,J_-(2J_3 - 1)$
	-2	$\sqrt{6}\,J_-^2$

Table 5. The tensor harmonics (continued)

k	μ	T^k_μ
3	3	$-2\sqrt{5}\,J_+^3$
	2	$2\sqrt{30}\,J_+^2\,(J_3+1)$
	1	$-2\sqrt{3}\,J_+(5J_3^2-J^2+5J_3+2)$
	0	$4(5J_3^2-3J^2+1)J_3$
	-1	$2\sqrt{3}\,J_-\,(5J_3^2-J^2-5J_3+2)$
	-2	$2\sqrt{30}\,J_-^2\,(J_3-1)$
	-3	$2\sqrt{5}\,J_-^3$
4	4	$\sqrt{70}\,J_+^4$
	3	$-2\sqrt{35}\,J_+^3\,(2J_3+3)$
	2	$2\sqrt{10}\,J_+^2\,(7J_3^2-J^2+14J_3+9)$
	1	$-\sqrt{5}\,J_+(28J_3^3-12J^2J_3+42J_3^2-6J^2+38J_3+12)$
	0	$70J_3^4-60J^2J_3^2+6(J^2)^2+50J_3^2-12J^2$
	-1	$\sqrt{5}\,J_-(28J_3^3-12J^2J_3-42J_3^2+6J^2+38J_3-12)$
	-2	$2\sqrt{10}\,J_-^2(7J_3^2-J^2-14J_3+9)$
	-3	$2\sqrt{35}\,J_-^3(2J_3-3)$
	-4	$\sqrt{70}\,J_-^4$

Table 6. The $\Delta = [00]$ Wigner operators

$$\left\langle \begin{matrix} 2j & & 0 \\ & j+m+\mu & \end{matrix} \right| \left\langle \begin{matrix} & 0 & \\ k & & -k \\ & \mu & \end{matrix} \right\rangle \left| \begin{matrix} 2j & & 0 \\ & j+m & \end{matrix} \right\rangle = C^{j\ k\ \ j}_{m\ \mu\ m+\mu} = (-1)^k\, C^{\ j\ \ \ k\ \ \ \ j}_{-m,-\mu,-m-\mu}$$

$$= (-1)^k \left[\frac{(k+\mu)!}{k!}\frac{(k-\mu)!}{k!}\right]^{1/2} \left[\frac{[j-m]_\mu}{[2j]_k}\frac{(j+m+1)_\mu}{(2j+2)_k}\right]^{1/2}$$

$$\times \sum_{s=0}^{k-\mu} (-1)^s \binom{k}{s}\binom{k}{\mu+s} [j+m]_s [j-m-\mu]_{k-\mu-s}$$

$$\left\langle \begin{matrix} & 0 & \\ k & & -k \\ & \mu & \end{matrix} \right\rangle = (-1)^k \left[\frac{(k+\mu)!}{k!}\frac{(k-\mu)!}{k!}\right]^{1/2} J_+^\mu\, P^k_\mu(J^2, J_3) \prod_{s=1}^{k} \left[4J^2 + 1 - s^2\right]^{-1/2}$$

where $\mu = 0, 1, \ldots k$; $k = 0, 1, \ldots, 2j$

$$\langle jm | P^k_\mu(J^2, J_3) | jm \rangle \equiv \sum_{s=0}^{k-\mu} (-1)^s \binom{k}{s}\binom{k}{\mu+s} [j+m]_s [j-m-\mu]_{k-\mu-s}$$

$$P^k_k(J^2, J_3) = 1$$

$$P^k_{k-1}(J^2, J_3) = k(1 - k - 2J_3), \quad k \geqslant 1$$

$$P^k_{k-2}(J^2, J_3) = 3(k-1)\binom{k}{3} - kJ^2 + \binom{2k}{2}\left[(k-2)J_3 + J_3^2\right], \quad k \geqslant 2$$

$$P^k_{k-3}(J^2, J_3) = -8\binom{k-1}{2}\binom{k}{4} + 2\binom{k}{2}(k-3)J^2$$

$$+ \binom{k}{2}\left[-\frac{2}{3}(3k^3 - 18k^2 + 32k - 13) + 4J^2\right]J_3$$

$$- \binom{2k}{3}\left[\frac{3}{2}(k-3)J_3^2 + J_3^3\right], \quad k \geqslant 3$$

$$P^k_{k-4}(J^2, J_3) = 30\binom{k-1}{3}\binom{k}{5} - \binom{k}{2}(k^3 - 9k^2 + 26k - 22)J^2 + \binom{k}{2}(J^2)^2$$

$$+ \frac{1}{3}\binom{k}{2}(k-4)[2k^4 - 17k^3 + 51k^2 - 59k + 21) - 6(2k-3)J^2]J_3$$

$$+ \frac{1}{3}\binom{k}{2}[(6k^4 - 57k^3 + 182k^2 - 211k + 69) - 6(2k-3)J^2]J_3^2$$

$$+ \binom{2k}{4}[2(k-4)J_3^3 + J_3^4], \quad k \geqslant 4$$

Table 7. The Racah operators

DEFINITION AND GENERAL PROPERTIES

Matrix element form:

$$\langle j_1+\Delta, j_2+\Delta', j| \begin{Bmatrix} & \Delta & \\ k & & -k \\ & \Delta' & \end{Bmatrix} |j_1 j_2 j\rangle$$

$$= \langle j, m, \mu+\Delta-\Delta', \kappa+\Delta-\Delta' | \begin{Bmatrix} & \Delta & \\ k & & -k \\ & \Delta' & \end{Bmatrix} |jm\rangle$$

$$= [(2j_1+2\Delta+1)(2j_2+1)]^{1/2} W(j, j_1, j_2+\Delta', k; j_2, j_1+\Delta)$$

$$= (-1)^{j+j_1+j_2+\Delta'+k} [(2j_1+2\Delta+1)(2j_2+1]^{1/2} \begin{Bmatrix} j & j_1 & j_2 \\ j_2+\Delta' & k & j_1+\Delta \end{Bmatrix} ,$$

where $\mu=j_1-j_2$ and $\kappa=j_1+j_2+1$ with ranges $\mu=j,j-1,\ldots,-j;\kappa=j+1,j+2,\ldots$.

Operator form:

$$\begin{Bmatrix} & \Delta' & \\ k & & -k \\ & \Delta' & \end{Bmatrix} = N^k_{\Delta'\Delta}(K,H) P^k_{\Delta'\Delta}(\vec{J}^2, K_3, H_3) \; [\bar{D}^k_\Delta(H_3+K_3) \underline{D}^k_{-\Delta'}(H_3-K_3)]^{-1} ,$$

where

$$N^k_{\Delta'\Delta}(K,H) = (K_\delta)^{|\Delta-\Delta'|}(H_{\delta'})^{|\Delta+\Delta'|}$$

$$\delta = \mathrm{sign}(\Delta-\Delta'), \delta' = \mathrm{sign}(\Delta+\Delta')$$

$$\bar{D}^k_\Delta(H_3+K_3) = [(H_3+K_3-k+\Delta)_{k+\Delta}(H_3+K_3+2\Delta+1)_{k-\Delta}]^{1/2}$$

$$\underline{D}^k_{-\Delta'}(H_3-K_3) = [(H_3-K_3+1)_{k+\Delta'}(H_3-K_3-k+\Delta')_{k-\Delta'}]^{1/2}$$

Special cases of the polynomials $P^k_{\Delta'\Delta}$ are given below.

Operator actions:

$$J_+|jm\mu\kappa\rangle=[(j-m)(j+m+1)]^{1/2}|j,m+1,\mu,\kappa\rangle$$

$$J_-|jm\mu\kappa\rangle=[(j+m)(j-m+1)]^{1/2}|j,m-1,\mu,\kappa\rangle$$

$$J_3|jm\mu\kappa\rangle=m|jm\mu\kappa\rangle$$

$$K_+|jm\mu\kappa\rangle=[(j-\mu)(j+\mu+1)]^{1/2}|j,m,\mu+1,\kappa\rangle$$

$$K_-|jm\mu\kappa\rangle=[(j+\mu)(j-\mu+1)]^{1/2}|j,m,\mu-1,\kappa\rangle$$

$$K_3|jm\mu\kappa\rangle=\mu|jm\mu\kappa\rangle$$

$$H_+|jm\mu\kappa\rangle=[(\kappa-j)(\kappa+j+1)]^{1/2}|j,m,\mu,\kappa+1\rangle$$

$$H_-|jm\mu\kappa\rangle=[(\kappa+j)(\kappa-j-1)]^{1/2}|j,m,\mu,\kappa-1\rangle$$

$$H_3|jm\mu\kappa\rangle=\kappa|jm\mu\kappa\rangle$$

The operators in the sets $\{J_+,J_3,J_-\}$ and $\{K_+,K_3,K_-\}$ satisfy standar angular momentum commutation relations; those in the set $\{H_+,H_3,H_-\}$ satisfy hyperbolic commutation rules. Operators from different set are mutually commuting.

Relations between operators:

$$\vec{J}^2=J_3(J_3+1)+J_-J_+=K_3(K_3+1)+K_-K_+$$

$$= H_3(H_3-1)-H_+H_-$$

$$H_3+K_3=(4\vec{J}_1^2+1)^{1/2}\rightarrow 2j_1+1$$

$$H_3-K_3=(4\vec{J}_2^2+1)^{1/2}\rightarrow 2j_2+1$$

$$2\vec{J}_1\cdot\vec{J}_2=\vec{J}^2-(H_3^2+K_3^2-1)/2$$

Table 7. The Racah operators (continued)

SPECIAL CASES OF THE POLYNOMIALS ($\Delta_1=k+\Delta$, $\Delta_2=k-\Delta$)

$$P^k_{k\Delta}=1,\quad \Delta=k,k-1,\ldots,-k$$

$$P^k_{k-1,\Delta}=-2k\vec{J}^2+\Delta_1K_3(K_3-\Delta_2+1)+\Delta_2(H_3-1)(H_3+\Delta_1)$$

$$-(k-1)\leqslant\Delta\leqslant k-1$$

$$\begin{aligned}P^k_{k-2,\Delta}&=k(2k-1)\vec{J}^4-(2k-1)\vec{J}^2[\Delta_1K_3(K_3-\Delta_2+2)\\&+\Delta_2H_3(H_3+\Delta_1-2)-2\Delta_1\Delta_2+2k]+\tfrac{1}{2}\Delta_1(\Delta_1-1)K_3^4\\&-\Delta_1(\Delta_1-1)(\Delta_2-1)K_3^3+\tfrac{1}{2}\Delta_1(\Delta_1-1)(\Delta_2^2-7\Delta_2+5)K_3^2\\&+\tfrac{1}{2}\Delta_1(\Delta_1-1)(3\Delta_2^2-7\Delta_2+2)K_3+\tfrac{1}{2}\Delta_2(\Delta_2-1)H_3^4\\&+(\Delta_1-2)\Delta_2(\Delta_2-1)H_3^3+\tfrac{1}{2}\Delta_2(\Delta_2-1)(\Delta_1^2-7\Delta_1+5)H_3^2\\&-\tfrac{1}{2}\Delta_2(\Delta_2-1)(3\Delta_1^2-7\Delta_1+2)H_3+\Delta_1\Delta_2K_3^2H_3^2\\&-\Delta_1\Delta_2(\Delta_2-2)K_3H_3^2+\Delta_1\Delta_2(\Delta_1-2)K_3^2H_3\\&-\Delta_1(\Delta_1-2)\Delta_2(\Delta_2-2)K_3H_3+\Delta_1(\Delta_1-1)\Delta_2(\Delta_2-1)\end{aligned}$$

$$-(k-2)\leqslant\Delta\leqslant k-2$$

Table 7. The Racah operators (continued)

The polynomials $P_0^k \equiv P_{0,0}^k$:

Invariant construction:

$$P_0^k = \sum_\mu (-1)^\mu T_\mu^k(\vec{J}_1) T_{-\mu}^k(\vec{J}_2)$$

Recursion relation:

$$P_0^{k+1} = \frac{2k+1}{k+1}[4\vec{J}_1\cdot\vec{J}_2 + k(k+1)]P_0^k$$

$$-\frac{k}{k+1}[4\vec{J}_1^2 - (k^2-1)][4\vec{J}_2^2 - (k^2-1)]P_0^{k-1}, \quad P_0^0 = 1$$

$$P_0^1 = 4\vec{J}_1\cdot\vec{J}_2$$

$$P_0^2 = 4[6(\vec{J}_1\cdot\vec{J}_2)^2 + 3\vec{J}_1\cdot\vec{J}_2 - 2\vec{J}_1^2\vec{J}_2^2]$$

$$P_0^3 = 4^2[10(\vec{J}_1\cdot\vec{J}_2)^3 + 20(\vec{J}_1\cdot\vec{J}_2)^2$$

$$-2(\vec{J}_1\cdot\vec{J}_2)(3\vec{J}_1^2\vec{J}_2^2 - \vec{J}_1^2 - \vec{J}_2^2 - 3) - 5\vec{J}_1^2\vec{J}_2^2]$$

$$P_0^4 = 4^2[70(\vec{J}_1\cdot\vec{J}_2)^4 + 350(\vec{J}_1\cdot\vec{J}_2)^3 + 2(\vec{J}_1\cdot\vec{J}_2)^2(-30\vec{J}_1^2\vec{J}_2^2 + 25\vec{J}_1^2$$

$$+25\vec{J}_2^2 + 195) + 2(\vec{J}_1\cdot\vec{J}_2)(-85\vec{J}_1^2\vec{J}_2^2 + 30\vec{J}_1^2 + 30\vec{J}_2^2 + 45)$$

$$+3\vec{J}_1^2\vec{J}_2^2(2\vec{J}_1^2\vec{J}_2^2 - 4\vec{J}_1^2 - 4\vec{J}_2^2 - 27)$$

Table 8. Reduced matrix elements

GENERAL PROPERTIES

Generic matrix elements of a tensor operator T^k with respect to angular momentum $\vec{J}$ with basis states $\{|\alpha jm\rangle\}$:

$$\langle \alpha' j'm'|T^k_\mu|\alpha jm\rangle = \langle \alpha' j'\|T^k\|\alpha j\rangle\, C^{j\;k\;j'}_{m\;\mu\;m'}$$

$$= (-1)^{j'-m'}\left(\alpha' j'\|T^k\|\alpha j\right)\begin{pmatrix} j' & k & j \\ -m' & \mu & m\end{pmatrix} ,$$

where

$$\left(\alpha' j'\|T^k\|\alpha j\right) \equiv (-1)^{2k}(2j'+1)^{1/2}\langle \alpha' j'\|T^k\|\alpha j\rangle$$

SPECIAL CASES

Spherical harmonics Y_k [tensor operator with respect to $\vec{L} = \vec{r}\times\vec{p}$ with basis states $\langle r\theta\varphi|\alpha\ell m\rangle = R_{\alpha\ell}(r)Y_{\ell m}(\theta\varphi)$]:

$$\langle \alpha'\ell'\|F(r)Y_k\|\alpha\ell\rangle = \left[\frac{(2\ell+1)(2k+1)}{4\pi(2\ell'+1)}\right]^{1/2} C^{\ell\;k\;\ell'}_{0\;0\;0}\int_0^\infty r^2dr\, R^*_{\alpha'\ell'}(r)\,F(r)\,R_{\alpha\ell}(r)$$

Position vector $\vec{r}$:

$$\langle \alpha'\ell'\|r\|\alpha\ell\rangle = \int_0^\infty r^2dr\, R^*_{\alpha'\ell'}(r)\,r\,R_{\alpha\ell}(r)\times\begin{cases}\left[\dfrac{\ell+1}{2\ell+3}\right]^{1/2}, & \ell'=\ell+1\\ 0, & \ell'=\ell\\ -\left[\dfrac{\ell}{2\ell-1}\right]^{1/2}, & \ell'=\ell-1\end{cases}$$

Linear momentum $\vec{p}$:

$$\langle \alpha'\ell'\|p\|\alpha\ell\rangle = -i\hbar\langle \alpha'\ell'\|r\|\alpha\ell\rangle\int_0^\infty r^2dr\, R^*_{\alpha'\ell'}\left[\frac{dR_{\alpha\ell}}{dr}+\frac{\ell(\ell+1)-\ell'(\ell'+1)+2}{r}R_{\alpha\ell}\right]$$

Runge-Lenz-Pauli vector $\vec{A}$:

$$\langle n'\ell'\|A\|n\ell\rangle = \delta_{n'n}\times\begin{cases}\left[\dfrac{(\ell+1)(n+\ell+1)(n-\ell-1)}{(2\ell+3)}\right]^{1/2}, & \ell'=\ell+1\\ 0, & \ell'=\ell\\ -\left[\dfrac{\ell(n-\ell)(n+\ell)}{2\ell-1}\right]^{1/2}, & \ell'=\ell-1\end{cases}$$

Table 8. Reduced matrix elements (continued)

Vector operators $\vec{J}_1$ and $\vec{J}_2$ ($\vec{J} = \vec{J}_1 + \vec{J}_2$):

$$\langle j_1 j_2 j + 1 \| J_1 \| j_1 j_2 j \rangle = -\langle j_1 j_2 j + 1 \| J_2 \| j_1 j_2 j \rangle$$

$$= \left[\frac{[(j_1 + j_2 + 1)^2 - (j + 1)^2][(j + 1)^2 - (j_1 - j_2)^2]}{4(j + 1)(2j + 3)}\right]^{1/2}$$

$$\langle j_1 j_2 j \| J_1 \| j_1 j_2 j \rangle = \langle j_2 j_1 j \| J_2 \| j_2 j_1 j \rangle$$

$$= \frac{j(j + 1) + j_1(j_1 + 1) - j_2(j_2 + 1)}{[j(j + 1)]^{1/2}}$$

$$\langle j_1 j_2 j - 1 \| J_1 \| j_1 j_2 j \rangle = -\langle j_1 j_2 j - 1 \| J_2 \| j_1 j_2 j \rangle$$

$$= -\left[\frac{[(j_1 + j_2 + 1)^2 - j^2][j^2 - (j_1 - j_2)^2]}{4j(2j - 1)}\right]^{1/2}$$

Tensor operators T^k (Table 6):

$$\langle j \| T^k \| j \rangle = \left[[2j]_k (2j + 2)_k\right]^{1/2}$$

Tensor product of tensor operators (S^{k_1} and T^{k_2} are tensor operators with respect to $\vec{J}$):

$$\left[S^{k_1} \times T^{k_2}\right]^{k}_{\mu} \equiv \sum_{\mu_1 \mu_2} C^{k_1 k_2 k}_{\mu_1 \mu_2 \mu} \, S^{k_1}_{\mu_1} T^{k_2}_{\mu_2}$$

Table 8. Reduced matrix elements (continued)

$$\left\langle \alpha' \; j' \| [S^{k_1} \times T^{k_2}]^{k} \| \; \alpha \; j \right\rangle$$

$$= (-1)^{j'+k+j} \sum_{\alpha'' j''} [(2j''+1)(2k+1)]^{1/2} \begin{Bmatrix} j' & k_1 & j'' \\ k_2 & j & k \end{Bmatrix}$$

$$\cdot \left\langle \alpha' \; j' \| S^{k_1} \| \; \alpha'' \; j'' \right\rangle \left\langle \alpha'' \; j'' \| T^{k_2} \| \; \alpha \; j \right\rangle$$

Tensor product of tensor operators $T^{k_1}(1)$ and $T^{k_2}(2)$ acting in different spaces:

$$[T^{k_1}(1) \times T^{k_2}(2)]^{k}_{\mu} \equiv \sum_{\mu_1 \mu_2} C^{k_1 k_2 \; k}_{\mu_1 \mu_2 \; \mu} \; T^{k_1}_{\mu_1}(1) \otimes T^{k_2}_{\mu_2}(2)$$

$$\left\langle \alpha_1' \alpha_2' j_1' j_2' \; j' \| [T^{k_1}(1) \times T^{k_2}(2)]^{k} \| \; \alpha_1 \alpha_2 j_1 j_2 \; j \right\rangle$$

$$= \begin{bmatrix} j_1 & j_2 & j \\ k_1 & k_2 & k \\ j_1' & j_2' & j' \end{bmatrix} \left\langle \alpha_1' \; j_1' \| T^{k_1}(1) \| \; \alpha_1 \; j_1 \right\rangle \left\langle \alpha_2' \; j_2' \| T^{k_2}(2) \| \; \alpha_2 \; j_2 \right\rangle$$

$$\begin{bmatrix} j_1 & j_2 & j \\ k_1 & k_2 & k \\ j_1' & j_2' & j \end{bmatrix} \equiv [(2j_1'+1)(2j_2'+1)(2j+1)(2k+1)]^{1/2} \begin{Bmatrix} j_1 & j_2 & j \\ k_1 & k_2 & k \\ j_1' & j_2' & j \end{Bmatrix}$$

Special cases:

$$\left\langle \alpha_1' \alpha_2' j_1' j_2' \; j' \| T^{k_1}(1) \otimes I^{(j_2)} \| \; \alpha_1 \alpha_2 j_1 j_2 \; j \right\rangle$$

$$= (-1)^{j_1'+k_1+j_2+j} [(2j_1'+1)(2j+1)]^{1/2} \begin{Bmatrix} j' & j_2 & j_1' \\ j_1 & k_1 & j \end{Bmatrix} \delta_{\alpha_2' \alpha_2} \; \delta_{j_2' j_2} \left\langle \alpha_1' \; j_1' \| T^{k_1}(1) \| \; \alpha_1 \; j_1 \right\rangle$$

Table 8. Reduced matrix elements (continued)

$$\left\langle \alpha_1'\alpha_2'j_1'j_2'\, j'\| \; I^{(j_1)} \otimes T^{k_2}(2)\| \; \alpha_1\alpha_2 j_1 j_2\, j \right\rangle$$

$$= (-1)^{j'+k_2+j_1+j_2} [(2j_2'+1)(2j+1)]^{1/2} \begin{Bmatrix} j' & j_1 & j_2' \\ j_2 & k_2 & j \end{Bmatrix} \delta_{\alpha_1'\alpha_1} \; \delta_{j_1'j_1} \; \left\langle \alpha_2' \; j_2' \| T^{k_2}(2)\| \; \alpha_2 \; j_2 \right\rangle$$

$$\left\langle \alpha' \; j \,\|\, [S^k \times T^k]^0 \,\|\, \alpha \; j \right\rangle$$

$$= \sum_{\alpha'' \; j''} (-1)^{k+j''-j} \left[\frac{(2j''+1)}{(2k+1)(2j+1} \right]^{1/2}$$

$$\times \; \varepsilon_{kjj''} \; \left\langle \alpha' \; j\| S^k \| \; \alpha'' \; j'' \right\rangle \; \left\langle \alpha'' \; j'' \| T^k \| \; \alpha \; j \right\rangle$$

$$[S^k \times T^k]_0^0 = \sum_\mu \frac{(-1)^{k-\mu}}{(2k+1)^{1/2}} \; S_\mu^k \; T_{-\mu}^k$$

$$\left\langle \alpha_1'\alpha_2'j_1'j_2' \; j\| [T^k(1) \times T^k(2)]^0 \| \; \alpha_1\alpha_2 j_1 j_2 \; j \right\rangle = (-1)^{j_2'+k+j_1+j} \left[\frac{(2j_1'+1)(2j_2'+1)}{2k+1} \right]^{1/2}$$

$$\times \begin{Bmatrix} j_1 & j_2 & j \\ j_2' & j_1' & k \end{Bmatrix} \left\langle \alpha_1' \; j_1'\| T^k(1)\| \; \alpha_1 \; j_1 \right\rangle \; \left\langle (\alpha_2') j_2' \| T^k(2) \| \; \alpha_2 \; j_2 \right\rangle$$

$$[T^k(1) \times T^k(2)]_0^0 = \sum_\mu \frac{(-1)^{k-\mu}}{(2k+1)^{1/2}} \; T_\mu^k(1) \otimes T_{-\mu}^k(2)$$

Bibliography of Tables

The abbreviation $x = a(b)c$ means that x runs from a to c, inclusive, in increments of b.

Clebsch–Gordan (Wigner) Coefficients

The standard form referred to is $(j_1 j_2 j_3 m_3 | j_1 m_1 j_2 m_2) = C^{j_1 j_2 j_3}_{m_1 m_2 m_3}$.

K. Alder, *Helv. Phys. Acta*, **25**, 235 (1952). Table of Clebsch–Gordan coefficients in rational fractions. $j_1 = j_2 = 1(1)4$; $j_3 = 0(2)6$; $m_1 = -m_2 = 0(1)2$.

E. U. Condon and G. H. Shortley, *The Theory of Atomic Spectra*, Cambridge University Press (1935). Tables of formulas of the Clebsch–Gordan coefficients. $j_2 = \frac{1}{2}(\frac{1}{2})2$.

A. R. Edmonds, *Angular Momentum in Quantum Mechanics*, Princeton University Press, Princeton, N. J. (1957). Algebraic Table of the 3-j Symbol. $j_3 = 0(\frac{1}{2})2$.

D. L. Falkoff, G. S. Holladay, and R. E. Sells, *Can. J. Phys.*, **30**, 253 (1952). Tables of formulas of the Clebsch–Gordan coefficients. $j_2 = 3$.

A. P. Jucys, I. B. Levinson, and V. V. Vanagas, *Mathematical Apparatus of The Theory of Angular Momentum*, Russian edition, 1960; translation by A. and A. R. Sen, Jerusalem (1962). Algebraic Formulas for the Clebsch–Gordan coefficients. $j_2 = \frac{1}{2}(\frac{1}{2})4$.

M. A. Melvin and N. V. V. J. Swamy, *Phys. Rev.*, **107**, 186 (1957). Tables of formulas of the Clebsch–Gordan coefficients. $j_2 = \frac{5}{2}$.

M. E. Rose, *Elementary Theory of Angular Momentum*, John Wiley & Sons, Inc., New York (1957). Table of formulas of Clebsch–Gordan coefficients. $j_2 = \frac{1}{2}$, 1.

M. Rotenberg, R. Bivins, N. Metropolis, and J. K. Wooten, Jr., *The 3-j and 6-j Symbols*. M.I.T. Press, Cambridge, Massachusetts (1959). Tables of Selected 3-j Symbols. $m_1 = m_2 = 0$; $j_1, j_2, j_3 = 0(1)16$. Tables of the 3-j Symbol. $j_1, j_2, j_3 = 0(\frac{1}{2})8$.

R. Saito and M. Morita, *Progr. Theoret. Phys.* (*Japan*), **13**, 540 (1955). Table of formulas of Clebsch–Gordan coefficients. $j_2 = \frac{5}{2}$.

B. J. Sears and M. G. Radtke, *Algebraic Tables of Clebsch–Gordan Coefficients*, TPI-75, Atomic Energy of Canada Limited, Chalk River, Ontario (1954). $j_2 = \frac{1}{2}(\frac{1}{2})3$.

W. T. Sharp, J. M. Kennedy, B. J. Sears, and M. G. Hoyle, *Tables of Coefficients for Angular Distribution Analysis*, CRT-556, AECL No. 97, Atomic Energy of Canada Limited, Chalk River, Ontario (1954). Tables of the Clebsch–Gordan coefficients. Powers-of-primes notation. j_1, $j_2 = 0(1)6$; $j_3 = 0(1)12$; $m_1 = m_2 = m_3 = 0$. Also j_1, $j_2 = 1(1)3$; $j_3 = 0(1)6$; $m_1 = -1$; $m_2 = 1$; $m_3 = 0$.

A. Simon, *Numerical Table of the Clebsch–Gordan Coefficient*. ORNL-1718, Oak Ridge National Laboratory (1954). Ten-place decimal fractions. Any j-value $0(\frac{1}{2})\frac{9}{2}$.

D. A. Varshalovich, A. N. Moskalev, and V. K. Khersonskii, *Quantum Theory of Angular Momentum*. Nauka, Leningrad (1975) (in Russian). Algebraic Tables of Clebsch–Gordan Coefficients: $j_2 = \frac{1}{2}(\frac{1}{2})5$. Numerical Tables of Clebsch–Gordan Coefficients: $j_1, j_2, j_3 = 0(\frac{1}{2})3$.

M. Yamada and L. Morita, *Prog. Theor. Phys.* **8**, 431 (1952). On the β-ray angular correlation. Algebraic Table of Transformation Coefficients. $j_2 = 3$.

[1] Adapted from M. Rotenberg, R. Bivins, N. Metropolis, and J. K. Wooten, Jr. *The 3-j and 6-j Symbols*. M.I.T. Press, Cambridge, Massachusetts (1959).

Racah Coefficients

The standard form referred to is $W(a\,b\,c\,d; e\,f)$.

K. Alder, *Helv. Phys. Acta*, **25**, 235 (1952). Table of formulas of Racah coefficients. $a = f$; $c = 2(2)6$; $d = e = 1(1)4$; b ranges over permitted values. Also formulas for $e = d + 1 = 1(1)4$.

L. C. Biedenharn, *Tables of the Racah Coefficients*. ORNL-1098, Oak Ridge National Laboratory (1952). Tables of formulas of Racah coefficients for $e = \frac{1}{2}(\frac{1}{2})2$. Table of W^2 in rational fractions. $e = \frac{1}{2}(\frac{1}{2})3$; $f = 0(1)8$; $a, c = 0(1)4$; $b, d = 0(\frac{1}{2})4$.

L. C. Biedenharn, J. M. Blatt, and M. E. Rose, *Revs. Mod. Phys.*, **24**, 249 (1952). Tables of formulas of Racah coefficients. $e = \frac{1}{2}(\frac{1}{2})2$.

S. F. Boys and R. C. Sahni, *Phil. Trans. Roy. Soc.* (*London*), *A*, **246**, 463 (1954). Nine-place decimal fractions $a, \ldots, d = \frac{1}{2}(\frac{1}{2})\frac{3}{2}$; $e, f \leqslant 5$.

A. R. Edmonds, *op. cit.* Algebraic table of the 6-*j* Symbol. $a = 0(\frac{1}{2})2$.

M. Goldstein and C. Kazek, *Table of the Racah and Z-Coefficients*, LAMS-1739, Los Alamos Scientific Laboratory (1954). Eight-place decimal fractions. $b = d$; $e = 0(\frac{1}{2})\frac{13}{2}$; $a = c = 0(\frac{1}{2})\frac{17}{2}$; $f = 0(2)16$.

K. M. Howell, *Tables of the Wigner 6-j Symbols*, Research Report US 58-1, University of Southampton, Southampton, England (1958). Squares of the 6-*j* symbols in prime factors, with sign. Listed in order of largest triangular sum Δ. $\Delta = 0(1)17$, with no parameters beyond $\frac{17}{2}$. Also *Revised Tables of the 6-j Symbols*, Research Report US 59-1, University of Southampton, Southampton, England (1959). More extensive than the previous *Tables* in that it takes advantage of the Regge symmetries.

T. Ishidzu, H. Horie, S. Obi, M. Sato, Y. Tanabe, and S. Yanagawa, *Tables of the Racah coefficients*, Pan-Pacific Press, Tokyo (1960). Algebraic Formulas for the Racah coefficients $W(abcd;ef)$ for $e = \frac{1}{2}(\frac{1}{2})7$. Numerical tables of Racah coefficients in prime factors: Tables 1–29, all integral a and b consistent with $e, f = 0(1)7$, $e \geqslant f$; $c \leqslant 7$, $d \leqslant 7$, $c + d + f \leqslant 15$; Tables 30–79, all half-integral a and integral b consistent with $e = \frac{1}{2}(1)\frac{13}{2}$, $f = 0(1)7$, $c \leqslant \frac{13}{2}$, (c half-integral), $d \leqslant 7$, $c + d + f \leqslant \frac{29}{2}$; Tables 80–108, all half-integral a, b, c, d consistent with $e, f = 0(1)7$, $c, d \leqslant \frac{13}{2}$, $c + d + f \leqslant 14$; Numerical tables of some Wigner coefficients: $m_1 = m_2 = 0$, $j_2 = 0(1)6$, $j_3 = 0(1)12$, all possible $j_1 \geqslant j_2$; $m_1 = -m_2 = \frac{1}{2}$, all j_1 such that $j_2 = \frac{1}{2}(1)\frac{13}{2}$, $j_3 = 0(1)(13)$, all possible $j_1 \geqslant j_2$. [Published in *Ann. Tokyo Astro. Obs.*, Second Series, Vol. III, No. 3 (1953); Vol. IV, No. 1 (1954); Vol. IV, No. 2 (1955); Vol. V, (1958).]

H. A. Jahn, *Proc. Roy. Soc.* (*London*), *A*, **205**, 192 (1951). Tables of formulas of the Racah coefficients. $e = \frac{1}{2}(\frac{1}{2})2$.

M. E. Rose, *op. cit.* Table of formulas of the Racah coefficients. $e = \frac{1}{2}, 1$.

M. Rotenberg, R. Bivins, N. Metropolis, and J. K. Wooten, Jr., *op. cit.* Tables of the 6-*j* Symbols. Powers-of-primes notation. $a, b, \ldots, f = 0(\frac{1}{2})8$.

M. Sato, *Progr. Theoret. Phys.* (*Tokyo*), **13**, 405 (1955). Tables of formulas of the Racah coefficients. $e = 3(\frac{1}{2})\frac{9}{2}$.

W. T. Sharp, J. M. Kennedy, B. J. Sears, and M. G. Hoyle, *Tables of Coefficients for angular Distribution Analysis*, DRT-556, AECL-97, Atomic Energy Commission of Canada Limited, Chalk River, Ontario (1953). Tables of Racah coefficients. $a, c = 0(1)4$; $b = d = 0(\frac{1}{2})5$; $e = 0(\frac{1}{2})4$.

A. Simon, J. H. Van der Sluis, and L. C. Biedenharn, *Tables of the Racah Coefficients*, ORNL-1679, Oak Ridge National Laboratory (1952). Ten-place decimal fractions of $W(abcd;ef)$. $a, c = 0(\frac{1}{2})\frac{15}{2}$; $b, d = 0(\frac{1}{2})\frac{9}{2}$; $e = 0(\frac{1}{2})3$; $f = 0(1)8$.

D. A. Varshalovich, A. N. Moskalev, and V. K. Khersonskii, *op. cit.* Algebraic Tables of the 6-*j* Symbols: $d = \frac{1}{2}(\frac{1}{2})4$. Numerical Tables of the 6-*j* Symbols: all a, b, c, d consistent with $e = 1(1)(3)$, $f = 0(\frac{1}{2})3$.

9-j Symbols

The standard form referred to is $\begin{Bmatrix} a & b & c \\ a' & b' & c' \\ d & e & f \end{Bmatrix}$

J. M. Kennedy, B. J. Sears, and W. T. Sharp, *Tables of X Coefficients*, CRT-569, AECL-106, Atomic Energy of Canada Limited, Chalk River, Ontario (1954). Five tables of the X-coefficient in powers-of-primes notation.

Table I: $a = a'$; $b = b'$; $c = c'$; $a = 1, 2$; $b, c = 1(1)5$; $d, e, f = 0(2)8$.

Table II: $a = 1$; $a' = 2$; $b = b'$; $c = c'$; $d, e, f = 0(2)8$.

Table III: $a = a'$; $b = b'$; $c = c'$; $a = 1, 2$; $b, c = \frac{1}{2}(1)\frac{9}{2}$; $d, e, f = 0(2)8$.
$a = 3, 4$; $b = \frac{1}{2}(1)\frac{5}{2}$; $c = \frac{3}{2}(1)\frac{9}{2}$; $d, e, f = 0(2)8$.

Table IV: $a = 1$; $a' = 2$; $b = b'$; $c = c'$; $b, c = \frac{1}{2}(\frac{1}{2})\frac{9}{2}$; $d, e, f = 0(2)8$.

Table V: $a = a'$; $b = \frac{1}{2}$; $c = c'$; $2 = 1(1)4$; $b = \frac{3}{2}, \frac{5}{2}$; $c = \frac{1}{2}(\frac{1}{2})\frac{9}{2}$; $d, e, f = 0(2)8$.

H. Matsunobu and H. Takebe, *Progr. Theoret. Phys.* (*Japan*), **14**, 589 (1955). Tables of the 9-j symbol (both formulas and numbers) in prime factors. $a', b' = \frac{1}{2}(1)\frac{7}{2}$; $c, c' = 0(1)5$; $f = 0, 1$. a and b are limited to integers that range over the admissible values.

K. Smith and J. W. Stephenson, *A Table of the Wigner 9-j Coefficients for Integral and Half-Integral Values of the Parameters*, ANL-5776, Argonne National Laboratory, Lemont, Ill. (1957). Ten-place decimal fractions. $a = \frac{1}{2}, \frac{3}{2}$; $b, c = \frac{1}{2}, \frac{3}{2}, \frac{5}{2}$; $a' = \frac{1}{2}(\frac{1}{2})\frac{7}{2}$; $b', c', d, e, f = 0(1)7$. Also a *Supplement* to these tables:

K. Smith, *Supplement to a Table of Wigner 9-j Coefficients for Integral and Half-Integral Values of the Parameters*, Parts 1 and 2, ANL-5860, Argonne National Laboratory, Lemont, Ill. (1958). Ten-place decimal fractions. $a, b, c, a' = \frac{1}{2}(1)\frac{7}{2}$; $b', c', d, e, f = 0(1)7$.

D. A. Varshalovich, A. N. Moskalev, and V. K. Khersonskii, *op. cit.* Algebraic Tables of the 9-j Symbols: all a, b, c, a', d', c', and d consistent with $f = 0, 1$ and $e = \frac{1}{2}(\frac{1}{2})2$. Numerical Tables of the 9-j Symbols: all $a, b, a', b', c' \leqslant 4$ consistent with $d = e = \frac{1}{2}$; $f = 0, 1$; $c = 0(\frac{1}{2})4$.

Z-Coefficients

L. C. Biedenharn, *Revised Z Tables of the Racah Coefficients*, ORNL-1501, Oak Ridge National Laboratory (1953). Rational fractions for $Z^2(l_1 j_1 l_2 j_2; sL)$; $s = \frac{1}{2}(\frac{1}{2})3$; $L = 0(1)8$; $l_1, l_2 = 0(1)4$; $j_1, j_2 = 0(\frac{1}{2})4$.

M. Goldstein and C. Kazek, *Table of the Racah and Z Coefficients*, LAMS-1739, Los Alamos Scientific Laboratory (1954). Eight-place decimal fractions of $Z(ljlj; sL)$, $j, s = 0(\frac{1}{2})\frac{13}{2}$; $l = 0(\frac{1}{2})\frac{17}{2}$; $L = 0(2)16$.

W. T. Sharp, J. M. Kennedy, B. J. Sears, and M. G. Hoyle, *Tables of Coefficients for Angular Distribution Analysis*, *op. cit.* Powers-of-primes notation. $s = 0(\frac{1}{2})4$; $l_1, l_2 = 0(1)4$; $j_1 = j_2 = 0(\frac{1}{2})5$. Both Z- and Z_1-coefficients.

List of Symbols

Many of the symbols used in this volume are given below together with their meanings. The list has been compiled mostly from Chapters 1–6. These "standard" notations have been used in the applications given in Chapter 7 whenever they are consistent with those found in that specialized area of the physics literature.

Signs of Relation

$\equiv$	definition; identically equal to; congruent (in modulo relations)
$\simeq$	approximately equal to
$\approx$	order of magnitude
$\rightarrow$	tends to; approaches; yields; is replaced by
$\sim$	asymptotically equal to, approximately equal to for a large value of a variable; equivalent to (in equivalence relations)
$\cong$	isomorphic to
$\propto$	proportional to
$<$	less than
$>$	greater than
$\leqslant$	less than or equal to
$\geqslant$	greater than or equal to
$\subset$	a subset of, contained in
$\supset$	contains as a subset
$\in$	an element of
$\leftrightarrow$	exchange; one-to-one correspondence
$\Rightarrow$	implies
$\Leftrightarrow$	if and only if
$\perp$	perpendicular

Operations

$\times$	times; direct product of groups
$\mathbf{x}$	cross (vector) product of vectors in $\mathbb{R}^3$; coupling of irreducible tensor operators
$\cdot$	dot or scalar product of vectors
$\pm$	plus or minus
$\mp$	minus or plus
$\lvert z \rvert$	absolute value
$\oplus$	direct sum of vector spaces

$\otimes$	tensor product of vector spaces or operators; direct or Kronecker product of matrices
$\circ$	composition of functions
$\cup$	union of sets
$\cap$	intersection of sets
$\blacksquare$	conclusion of proof
A^*	complex conjugation of a matrix A
$\tilde{A}$	transposition of a matrix A
$A^\dagger$	Hermitian conjugation of a matrix A
A^{-1}	inverse of a matrix A
$\operatorname{tr} A$	trace of a matrix A
$\det A$	determinant of a matrix A
∂	partial derivative
$\boldsymbol{\nabla}$	gradient operator in $\mathbb{R}^3$
$\boldsymbol{\nabla}^2$	Laplacian in $\mathbb{R}^3$
$\boldsymbol{\nabla}_4$	gradient operator in $\mathbb{R}^4$
$\boldsymbol{\nabla}_4^2$	Laplacian in $\mathbb{R}^4$

Sets

$\mathbb{Z}$, $\mathbb{Z}^+$	set of all integers, set of nonnegative integers
$\mathbb{Q}$	set of rational numbers
$\mathbb{R}$	set of real numbers
$\mathbb{C}$	set of complex numbers
$\mathbb{R}^n$	set of real n-tuples; n-dimensional real vector space
$\mathbb{C}^n$	set of complex n-tuples; n-dimensional complex vector space
$\hat{\mathbb{C}}^{2j+1}$	projective space corresponding to $\mathbb{C}^{2j+1}$
$E(3)$	Euclidean 3-space
$\emptyset$	empty set
S^2	unit sphere in $\mathbb{R}^3$
S^3	unit sphere in $\mathbb{R}^4$
$SO(n)$	group of $n \times n$ real, orthogonal matrices of determinant $+1$, $n = 2, 3, \ldots$
$O(n)$	group of $n \times n$ real, orthogonal matrices, $n = 2, 3, \ldots$
$SU(n)$	group of $n \times n$ unitary matrices of determinant $+1$, $n = 2, 3, \ldots$
$U(n)$	group of $n \times n$ unitary matrices, $n = 1, 2, \ldots$

S_n	group of permutations on n objects, symmetric group
G	generic group
G_+	finite subgroup of pure rotations
G/K	right cosets of G with respect to a subgroup K
Z_2	two element group containing the inversion and the identity
T	group of rotations mapping a regular tetrahedron into itself
T_h	direct product group $T \times Z_2$
T_d	group of rotation – inversions mapping a regular tetrahedron into itself
O	group of rotations mapping a regular octahedron into itself
O_h	group of rotation – inversions mapping a regular octahedron into itself
C_n	cyclic group of rotations about an n-fold symmetry axis
$\mathbf{T}^J$	irreducible tensor operator of rank J with elements indexed by $M = J, \ldots, -J$
$\{x \in X: P(x)\}$	set of x in X with property $P(x)$
$f: X \to Y$	function (or mapping) f with domain X and range in Y
$f: x \to y = f(x)$	function f mapping $x \in X$ to $y \in Y$

Vectors and Operations in Euclidean Three-Space

$\vec{a}, \vec{b}, \ldots$	generic vectors with common origin in Euclidean 3-space, $E(3)$
$\vec{r}, \vec{s}, \vec{x}, \vec{y}, \ldots$	position vectors in $E(3)$ ($\vec{r}$ is used synonymously with $\vec{x}$, z-axis with x_3-axis)
$\vec{p}$	linear momentum vector in $E(3)$
$\hat{n}, \hat{e}_i, \ldots$	unit vectors in $E(3)$
$(\hat{e}_1, \hat{e}_2, \hat{e}_3)$	right-handed triad of orthogonal unit vectors in $E(3)$, inertial frame
$\vec{x} = \sum_i x_i \hat{e}_i$	representation of a vector $\vec{x}$ in terms of the components (x_1, x_2, x_3) and the frame $(\hat{e}_1, \hat{e}_2, \hat{e}_3)$
$\mathbf{x} = (x_1, x_2, x_3)$	representation of a vector $\vec{x}$ as a 3-tuple in $\mathbb{R}^3$ ($\mathbf{x}$ is also called a vector)
$\vec{a} \cdot \vec{b} = \mathbf{a} \cdot \mathbf{b} = a_1 b_1 + a_2 b_2 + a_3 b_3$	dot or scalar product of two vectors

$\vec{a} \times \vec{b} = (a_2b_3 - a_3b_2)\hat{e}_1 + (a_3b_1 - a_1b_3)\hat{e}_2 + (a_1b_2 - a_2b_1)\hat{e}_1$	cross or vector product of vector $\vec{a}$ with vector $\vec{b}$
$\mathbf{a} \times \mathbf{b} = (a_2b_3 - a_3b_2, a_3b_1 - a_1b_3, a_1b_2 - a_2b_1)$	cross or vector product of vector $\mathbf{a}$ with vector $\mathbf{b}$
$a = \|\mathbf{a}\| = (\mathbf{a} \cdot \mathbf{a})^{\frac{1}{2}}$	length of vector $\mathbf{a}$
$\vec{L} = \vec{x} \times \vec{p} = \vec{r} \times \vec{p}, \mathbf{L} = \mathbf{x} \times \mathbf{p} = \mathbf{r} \times \mathbf{p}$	angular momentum about point O of a (point) particle located at position $\vec{x} = \vec{r}$ and having linear momentum $\vec{p}$ (equivalently at position $\mathbf{x}$ with linear momentum $\mathbf{p}$)
$\mathscr{R}(\phi, \hat{n}): \vec{r} \to \vec{r}' = \mathscr{R}(\phi, \hat{n})\vec{r}$	action of a rotation by angle ϕ about direction $\hat{n}$ on a vector $\vec{r}$
$R: \mathbf{x} \to \mathbf{x}' = R\mathbf{x}$	action of a rotation $R \in SO(3)$ on a vector $\mathbf{x} \in \mathbb{R}^3$
$\mathscr{I}: \mathbf{x} \to \mathbf{x}' = -\mathbf{x}$	action of spatial inversion $\mathscr{I}$ on a vector $\mathbf{x} \in \mathbb{R}^3$
$\boldsymbol{\xi} = \begin{pmatrix} \xi_0 \\ \xi_1 \end{pmatrix}$	Cartan spinor associated with an isotropic vector $\mathbf{x} \in \mathbb{C}^3$, $\mathbf{x} \cdot \mathbf{x} = 0$
$R: \boldsymbol{\xi} \to \boldsymbol{\xi}' = \pm U\xi$	rotation of a Cartan spinor corresponding to the rotation $\mathbf{x}' = R\mathbf{x}$ of $\mathbb{R}^3$
$(\hat{\xi}_{+1}, \hat{\xi}_0, \hat{\xi}_{-1}), (\hat{\eta}_{+1}, \hat{\eta}_0, \hat{\eta}_{-1})$	spherical basis vectors in $\mathbb{R}^3$
$\Gamma_{\vec{A}}\mathbf{x}$	operation of forming the vector product of a given vector $\vec{A}$ with an arbitrary vector
$(\hat{f}_1, \hat{f}_2, \hat{f}_3)$	triad of unit vectors in $E(3)$ constituting a body-fixed reference frame; vector operators with respect to the total orbital angular momentum operator

Rotations and Rotation Operators

$\mathscr{R}$	generic rotation about a common origin in $E(3)$
$\mathscr{R}^{-1}$	inverse rotation to $\mathscr{R}$
$\mathscr{R}(\phi, \hat{n})$	rotation by angle ϕ about direction $\hat{n}$
E	identity rotation
$\mathscr{R}_{2\pi}, \mathscr{R}_{4\pi}$	rotations by 2π and 4π about any axis
$R = (R_{ij})$	generic 3×3 real, proper orthogonal matrix corresponding to rotation $\mathscr{R}$
$C = (C_{ij}) = (\hat{l}_i \cdot \hat{f}_j)$	direction cosine matrix between laboratory fame $(\hat{l}_1, \hat{l}_2, \hat{l}_3)$ and body-fixed frame $(\hat{f}_1, \hat{f}_2, \hat{f}_3)$
$R(\phi, \hat{n})$	parametrization of R in terms of angle of rotation ϕ about direction $\hat{n}$

$R(\alpha\beta\gamma)$	parametrization of R in terms of Euler angles
$R(\alpha_0, \boldsymbol{\alpha})$	parametrization of R in terms of Euler-Rodrigues parameters $(\alpha_0, \boldsymbol{\alpha}) \in S^3$
$U = (u_{ij})$	generic 2×2 unitary or unitary unimodular matrix; element of quantal rotation group $SU(2)$ corresponding to rotation $R \in SO(3)$
$U(\psi, \hat{n})$	parametrization of U (unimodular) in terms of angle of rotation ψ about direction $\hat{n}$
$U(\alpha\beta\gamma)$	parametrization of U (unimodular) in terms of Euler angles
$U(\alpha_0, \boldsymbol{\alpha})$	parametrization of U (unimodular) in terms of Euler-Rodrigues parameters $(\alpha_0, \boldsymbol{\alpha}) \in S^3$
$\mathscr{U}$	rotation operator; unitary operator realization in Hilbert space of rotation $U \in SU(2)$
$\mathscr{U}^{-1}$	inverse rotation operator
$\mathscr{U}(\psi, \hat{n}) = e^{-i\psi\hat{n}\cdot\mathbf{J}}$	parametrization of $\mathscr{U}$ in terms of angle of rotation ψ about direction $\hat{n}$
$\mathscr{U}(\alpha\beta\gamma)$	parametrization of $\mathscr{U}$ in terms of Euler angles
$\mathscr{U}(\alpha_0, \boldsymbol{\alpha})$	parametrization of $\mathscr{U}$ in terms of Euler-Rodrigues parameters $(\alpha_0, \boldsymbol{\alpha}) \in S^3$

Domains of Parameters of Rotations

$\{(\phi, \hat{n}) : 0 \leqslant \phi \leqslant \pi,\ \hat{n}\cdot\hat{n} = 1,\ (\pi, \hat{n}) \equiv (\pi, -\hat{n})\}$ parameters of a real, proper orthogonal rotation of vectors in $\mathbb{R}^3$ in terms of the angle of rotation ϕ about direction $\hat{n}$

$\{(\psi, \hat{n}) : 0 \leqslant \psi \leqslant 2\pi, \hat{n}\cdot\hat{n} = 1\}$ parameters of a unitary unimodular rotation of spinors in $\mathbb{C}^2$ in terms of the angle of rotation ψ about direction $\hat{n} \in \mathbb{R}^3$

$\{(\alpha\beta\gamma) : 0 \leqslant \alpha < 2\pi,\ 0 \leqslant \beta \leqslant \pi,\ 0 \leqslant \alpha < 2\pi\}$ Euler angle parameters of a rotation of vectors in $\mathbb{R}^3$

$\{(\alpha\beta\gamma) : 0 \leqslant \alpha < 2\pi,\ 0 \leqslant \beta \leqslant \pi$ or $2\pi \leqslant \beta \leqslant 3\pi,\ 0 \leqslant \alpha < 2\pi\}$ Euler angle parameters of a unitary unimodular rotation of spinors in $\mathbb{C}^2$

$\{(\alpha_0, \boldsymbol{\alpha}) : \alpha_0^2 + \alpha_1^2 + \alpha_2^2 + \alpha_3^2 = 1, (-\alpha_0, -\boldsymbol{\alpha}) \equiv (\alpha_0, \boldsymbol{\alpha})\}$ Euler-Rodrigues parameters of a real, proper orthogonal rotation of vectors in $\mathbb{R}^3$

$\{(\alpha_0, \boldsymbol{\alpha}) : \alpha_0^2 + \alpha_1^2 + \alpha_2^2 + \alpha_3^2 = 1\}$ Euler-Rodrigues parameters of a unitary unimodular rotation of spinors in $\mathbb{C}^2$

Hilbert Space

$\mathscr{H}$	generic Hilbert space
$\mathscr{H}_j$	$(2j + 1)$-dimensional Hilbert space, which is a carrier space for an irreducible representation of $SU(2)$
$\mathscr{H}^{(2j)}$	Hilbert space of homogeneous polynomials of degree $2j$ in two bosons a_1 and a_2
$\|(\alpha)jm\rangle, \|jm\rangle$	orthonormal basis of $\mathscr{H}_j$ on which the angular momentum $\mathbf{J}$ has the standard action
$\|\psi\rangle, \|\Psi\rangle, \|\phi\rangle, \|\Phi\rangle, \ldots$	ket vector notation for vectors in Hilbert space
$\langle\psi\|, \langle\Psi\|, \langle\phi\|, \langle\Phi\|, \ldots$	bra vector notation for vectors in the dual space
$\psi, \Psi, \phi, \Phi, \ldots$	function notation for vectors in Hilbert space
$(\psi, \phi), \langle\psi, \phi\rangle, \langle\psi\|\phi\rangle, \ldots$	inner or scalar product of vectors in Hilbert space
$\\|\psi\\| = (\psi, \psi)^{\frac{1}{2}}$	norm of a vector (function) ψ
$\langle\mathbf{x}\|\psi\rangle = \psi(\mathbf{x})$	bra-ket vector notation for the value of a ket vector $\|\psi\rangle$ with domain $\mathbb{R}^3$
$\langle(\alpha')j'm'\|(\alpha)jm\rangle$	bra-ket notation for inner product of angular momentum basis vectors
$\mathscr{O}$	generic operator in Hilbert space
$A^{\text{op}}, A = \mathbf{r}, \mathbf{x}, \mathbf{p}, \mathbf{L}, \ldots$	preliminary notation for operator in Hilbert space corresponding to classical observable A
$\mathbb{1}, \mathbf{1}, 1$	unit operator
$\langle(\alpha')j'm'\|\mathscr{O}\|(\alpha)jm\rangle$	matrix elements of an operator, $\mathscr{O}$, in the angular momentum basis
$\sum_j \oplus \mathscr{H}_j$	direct sum of Hilbert spaces $\mathscr{H}_{\frac{1}{2}}, \mathscr{H}_1, \ldots$
$(j, m), (J, M), (l, m),$ $(L, M), (S, M_S), (k, \mu), \ldots$	angular momentum labels of vectors or operators in Hilbert space
$\mathscr{H} \otimes \mathscr{K}$	tensor product of generic Hilbert spaces
$\phi \otimes \psi$	tensor product of vectors $\phi \in \mathscr{H}$ and $\psi \in \mathscr{K}$
$S \otimes T$	tensor product of operator S acting in $\mathscr{H}$ with operator T acting in $\mathscr{K}$
$\mathscr{H}_{j_1}(1) \otimes \mathscr{H}_{j_2}(2)$	tensor product of angular momentum carrier spaces $\mathscr{H}_{j_1}(1)$ and $\mathscr{H}_{j_2}(2)$
$\mathbb{1}(i)$	unit operator in $\mathscr{H}_{j_i}(i)$

$|j_1m_1;j_2m_2\rangle = |j_1m_1\rangle|j_2m_2\rangle = |j_1m_1\rangle \otimes |j_2m_2\rangle$ uncoupled angular momentum basis of $\mathscr{H}_{j_1}(1) \otimes \mathscr{H}_{j_2}(2)$

$|(j_1j_2)jm\rangle$ coupled angular momentum basis of $\mathscr{H}_{j_1}(1) \otimes \mathscr{H}_{j_2}(2)$

$\langle \Psi|\mathcal{O}|\Phi\rangle$ generic matrix element of an operator $\mathcal{O}$ between states $|\Psi\rangle$ and $|\Phi\rangle$

$e^{-i\phi\hat{n}\cdot\mathbf{L}}\Psi(\mathbf{x}) = \Psi(R^{-1}(\phi, \hat{n})\mathbf{x})$ action of orbital rotation operator on analytic functions defined on $\mathbb{R}^3$

Angular Momentum Operators

$\mathbf{L}^{\text{op}} = (L_1^{\text{op}}, L_2^{\text{op}}, L_3^{\text{op}})$ preliminary notation for orbital angular momentum operators

$\mathbf{L} = (L_1, L_2, L_3)$ orbital angular momentum operator with action defined on vectors in Hilbert space; generators of the group $SO(3)$; basis of the Lie algebra of $SO(3)$; total orbital angular momentum operator

$\mathbf{L}^2 = L_1^2 + L_2^2 + L_3^2 = \mathbf{L}\cdot\mathbf{L}$ square of the orbital angular momentum; Casimir operator for the group $SO(3)$

$\mathbf{J} = (J_1, J_2, J_3)$ generic angular momentum operator with action defined on vectors in Hilbert space; generators of the group $SU(2)$; basis of the Lie algebra of $SU(2)$; total angular momentum

$\mathbf{J}^2 = J_1^2 + J_2^2 + J_3^2 = \mathbf{J}\cdot\mathbf{J}$ square of the angular momentum $\mathbf{J}$; Casimir operator for the group $SU(2)$

$\mathbf{L}\times\mathbf{L} = i\hbar\mathbf{L}$ vector form of the commutation relations for orbital angular momentum $\mathbf{L}$

$\mathbf{J}\times\mathbf{J} = i\mathbf{J}$ vector form of the commutation relations for the dimensionless generators of the quantum rotation group, $SU(2)$

$\mathscr{L} = (\mathscr{L}_1, \mathscr{L}_2, \mathscr{L}_3), \mathscr{L}^2$ differential operator realizations of orbital angular momentum

$\mathscr{J} = (\mathscr{J}_1, \mathscr{J}_2, \mathscr{J}_3)$
$\mathscr{K} = (\mathscr{K}_1, \mathscr{K}_2, \mathscr{K}_3)$
$\mathscr{P} = (\mathscr{P}_1, \mathscr{P}_2, \mathscr{P}_3)$ differential operator realization of angular momentum

$J_\pm = J_1 \pm iJ_2, L_\pm = L_1 \pm iL_2,$ etc. raising and lowering operators; complex extension of the Lie algebra of $SO(3)$ or $SU(2)$

$J_{+1} \equiv -\frac{1}{\sqrt{2}}J_+, J_0 \equiv J_3, J_{-1} \equiv \frac{1}{\sqrt{2}}J_-$ spherical components of angular momentum (see Appendix D, Chapter 3)

$\mathbb{1}^{(j)}$ — $(2j+1) \times (2j+1)$ unit matrix

$\sigma_0 = \mathbb{1}^{(\frac{1}{2})}$ — 2×2 unit matrix

$\boldsymbol{\sigma} = (\sigma_1, \sigma_2, \sigma_3)$ — Pauli matrices

$\mathbf{x} \cdot \boldsymbol{\sigma} = x_1\sigma_1 + x_2\sigma_2 + x_3\sigma_3$, — Cartan map $\mathbf{x} \to \mathbf{x} \cdot \boldsymbol{\sigma}$ of points $\mathbf{x} \in \mathbb{R}^3$ onto 2×2 traceless Hermitian matrices

$\hat{n} \cdot \boldsymbol{\sigma}/2$ — component of spin $\boldsymbol{\sigma}/2$ in direction $\hat{n}$

$\mathbf{a} \cdot \mathbf{J} = a_1J_1 + a_2J_2 + a_3J_3$ — general element in the Lie algebra of $SU(2)$; dot product of a vector $\mathbf{a} \in \mathbb{R}^3$ with the angular momentum operator $\mathbf{J}$

$\mathbf{a} \times \mathbf{J} = (a_2J_3 - a_3J_2, a_3J_1 - a_1J_3, a_1J_2 - a_2J_1)$ vector product of a vector $\mathbf{a} \in \mathbb{R}^3$ and the angular momentum operator $\mathbf{J}$

$\mathbf{J}^2|jm\rangle = j(j+1)|jm\rangle$
$J_3|jm\rangle = m|jm\rangle$
$J_\pm|jm\rangle = [(j \mp m)(j \pm m + 1)]^{\frac{1}{2}}|jm \pm 1\rangle$ — standard action of angular momentum operators in the space $\mathscr{H}_j$

$\mathbf{J}(i)$ — angular momentum of "part" i of a composite system, $i = 1, 2, \ldots$

$T^{k_1}_{\mu_1}(1) \otimes \mathbb{1}(2), \mathbb{1}(1) \otimes T^{k_2}_{\mu_2}(2)$ — tensor product of operators with action in the tensor product space $\mathscr{H}_{j_1}(1) \otimes \mathscr{H}_{j_2}(2)$

$E^{(j)}_m$ — principal idempotent matrix corresponding to the angular momentum matrix $J^{(j)}_2$

$\mathscr{T}^k_\mu$ — tensor harmonic; μth component of an irreducible tensor operator of rank k belonging to the enveloping algebra of the Lie algebra of $SU(2)$

C_i, Ω — operations of forming the commutator of J_i, respectively, $\mathbf{J}^2$ with a tensor operator

$J^{(j_1)}_i \otimes \mathbb{1}^{(j_2)}, \mathbb{1}^{(j_1)} \otimes J^{(j_2)}_i$ — matrix representations of the operators $J_i \otimes \mathbb{1}(2)$ and $\mathbb{1}(1) \otimes J_i$ in the space $\mathscr{H}_{j_1}(1) \otimes \mathscr{H}_{j_2}(2)$

(V_{+1}, V_0, V_{-1}) — spherical components of a vector operator $\mathbf{V}$

$\mathbf{T}^J, \mathbf{S}^J$ — generic symbol for irreducible tensor operator of rank J

$T^J_M: M = J, J-1, \ldots, -J$ — components of $\mathbf{T}^J$ (also called irreducible tensor operators)

$\bar{\mathbf{T}}^J, \bar{T}^J_M$ — generic symbol for a conjugate irreducible tensor operator of rank J

$[\mathbf{S}^{k_1} \mathbf{x}\, \mathbf{T}^{k_2}]^k$	irreducible tensor operator of rank k obtained by Wigner coupling of two irreducible tensor operators having a common Hilbert space as domain
$[\mathbf{S}^{k_1} \mathbf{x}\, \mathbf{T}^{k_2}]^k_\mu$	μth component of $[\mathbf{S}^{k_1} \mathbf{x}\, \mathbf{T}^{k_2}]^k$
$\mathbf{T}^{k_1}(1) \otimes \mathbf{T}^{k_2}(2)$	tensor product of two irreducible tensor operator components with action in $\mathscr{H}_{j_1}(1) \otimes \mathscr{H}_{j_2}(2)$
$\langle(\alpha')j'm'\lvert T^J_M \rvert(\alpha)jm\rangle$	matrix element in an angular momentum basis of an irreducible tensor operator
$\langle(\alpha')j'\lVert\mathbf{T}^J\rVert(\alpha)j\rangle$	Condon-Shortley notation for a reduced matrix element; matrix element of the $SU(2)$ invariant operator corresponding to $\mathbf{T}^J$
$\left\langle \begin{matrix} & J+\Delta & \\ 2J & & 0 \\ & J+M & \end{matrix} \right\rangle$	$SU(2)$ unit tensor operator, Wigner operator; action defined in angular momentum representation space $\mathscr{H} = \mathscr{H}_0 \oplus \mathscr{H}_{\frac{1}{2}} \oplus \mathscr{H}_1 \oplus \cdots$
$\mathbf{W}^{abc}_{\rho\sigma\tau}$	Racah invariant operator in $\mathscr{H}$ corresponding to coupling of unit tensor operators
$\left\{ \begin{matrix} & a+\rho & \\ 2a & & 0 \\ & a+\sigma & \end{matrix} \right\}$	Racah operator with action defined in the tensor product space $\mathscr{H} \otimes \mathscr{H}$
$\begin{bmatrix} abc \\ \rho\sigma\tau \end{bmatrix}$	9-j invariant operator with action defined in tensor product space $\mathscr{H} \otimes \mathscr{H}$

Functions

$D^j_{m'm}(\psi, \hat{n})$	matrix elements of rotation matrix D^j of the quantal rotation group $SU(2)$ expressed in terms of the rotation angle ψ and the direction $\hat{n}$
$D^j_{m'm}(U)$	functions defined by irreps ("irrep functions") of $SU(2)$ realized as homogeneous polynomials in the elements u_{ij} of $U \in SU(2)$
$D^j_{m'm}(\alpha\beta\gamma)$	irrep functions of $SU(2)$ expressed in terms of Euler angles
$D^j_{m'm}(\alpha_0, \boldsymbol{\alpha})$	irrep functions of $SU(2)$ expressed in terms of Euler-Rodrigues parameters
$D^j_{m'm}(x_0, \mathbf{x})$	extension of irrep functions of $SU(2)$ to arbitrary points $(x_0, \mathbf{x}) \in \mathbb{R}^4$ (quaternionic variables)

$\mathscr{D}^{l}_{m'm}(R)$	irrep functions of $O(3)$ expressed as homogeneous polynomials in the elements R_{ij} of $R \in O(3)$
$d^{j}_{m'm}(\beta)$	irrep functions of $SU(2)$ depending only on the Euler angle parameter β
$P_n^{(\alpha,\beta)}(x)$	Jacobi polynomials in x
$Y_{lm}(\beta\alpha)$	spherical harmonics on S^2
$\mathscr{Y}_{lm}(x)$	solid harmonics in $\mathbf{x} \in \mathbb{R}^3$
$P_l^m(\cos\beta)$	associated Legendre polynomials in $\cos\beta$
$P_l(\cos\beta)$	Legendre polynomials in $\cos\beta$
$D^{j*}_{m'm}(U), D^{j*}_{m'm}(\alpha\beta\gamma), D^{j*}_{m'm}(x_0, \mathbf{x})$	complex conjugated set of values of irrep functions of $SU(2)$ (see Section 6, Chapter 3)
$D^j(U), D^j(\psi, \hat{n}), D^j(\alpha\beta\gamma)$	$(2j+1) \times (2j+1)$ unitary matrix irrep of $SU(2)$
$D^{j_1}(U) \otimes D^{j_2}(U)$ $D^{j_1}(\psi, \hat{n}) \otimes D^{j_2}(\psi, \hat{n})$	matrix direct product of irreps of $SU(2)$
$\boldsymbol{\mathscr{Y}}^{(ls)jm}(\mathbf{x})$	tensor solid harmonics in $\mathbf{x} \in \mathbb{R}^3$
$\chi^j(U) = \operatorname{tr} D^j(U)$	trace (character) of irrep j of $SU(2)$
$L_k^{\alpha}(\xi)$	associated Laguerre polynomials in ξ
$C_n^{(\alpha)}(x)$	Gegenbauer polynomials in x
$I_l(\mathbf{x}, \mathbf{y}), I_{(l_1 l_2 l_3)}(\mathbf{x}^1, \mathbf{x}^2, \mathbf{x}^3)$	rotationally invariant polynomials in two and three vectors, respectively (see Section 17, Chapter 6)

Angular Momentum Coefficients

ε_{abc}	characteristic functions defined by the Clebsch-Gordan series [see Eq. (3.272), Chapter 3]
$C^{j_1 j_2 j}_{m_1 m_2 m}$	Wigner coefficient; Clebsch-Gordan coefficient; vector addition coefficient [see Section 12, Chapter 3]

$$\begin{pmatrix} j_1 & j_2 & j \\ m_1 & m_2 & -m \end{pmatrix} \equiv (-1)^{j_1 - j_2 + m}(2j+1)^{-\frac{1}{2}} C^{j_1 j_2 j}_{m_1 m_2 m}$$

3-j coefficient [see Eq. (3.182), Chapter 3]

$$\begin{bmatrix} j_1 + m_1 & j_2 + m_2 & j - m \\ j_1 - m_1 & j_2 - m_2 & j + m \\ j - j_1 + j_2 & j + j_1 - j_2 & -j + j_1 + j_2 \end{bmatrix}$$

Regge array for a Wigner coefficient [see Eq. (3.181), Chapter 3]

$\Delta(abc)$	triangle coefficient [see Eq. (3.276), Chapter 3]

$W(abcd;ef)$ — Racah coefficient [see Eq. (3.290), Chapter 3]

$\begin{Bmatrix} a & b & e \\ d & c & f \end{Bmatrix} = (-1)^{a+b+c+d} W(abcd;ef)$ — 6-j coefficient

$W^{abc}_{\rho\sigma\tau}(j) \equiv [(2c+1)(2j-2\sigma+1)]^{\frac{1}{2}} W(j-\tau, a, j, b; j-\sigma, c)$ — Racah coefficient with dimension factors

$\begin{Bmatrix} j_1 & j_2 & j \\ k_1 & k_2 & k \\ j_1' & j_2' & j' \end{Bmatrix}$ — 9-j coefficient [see Eq. (3.251), Chapter 3]

$\begin{bmatrix} j_1 & j_2 & j \\ k_1 & k_2 & k \\ j_1' & j_2' & j' \end{bmatrix}$ — 9-j coefficient with dimension factors [see Eq. (3.250), Chapter 3]

Boson Calculus

H	harmonic oscillator Hamiltonian; generic Hamiltonian
p, q	conjugate linear momentum and position operators for harmonic oscillator satisfying $[p,q] = -i\hbar$
a, $\bar{a}$	creation and destruction operators, respectively, for harmonic oscillator satisfying $[\bar{a},a] = 1$; boson and conjugate boson operators
b, $\bar{b}$	fermion creation and destruction operators, respectively, satisfying $\bar{b}b + b\bar{b} = 1$ (see Section 9, Chapter 7)
m, ω	mass, angular frequency parameters of a harmonic oscillator
$N = a\bar{a}$	number operator
Δp, Δq	dispersion in momentum, position
$\mathscr{U}(\xi,\eta)$, $\mathscr{U}(\zeta)$	unitary displacement operator of bosons
$\mathscr{H} = \mathscr{H}_1$	Hilbert space of states of the harmonic oscillator
$\{\|\hat{e}_\zeta\rangle : \zeta \in \mathbb{C}\}$	family of unit vectors in $\mathscr{H}$ on which the action of $\bar{a}$ is multiplication by ζ
$\langle \hat{e}_\alpha \| \hat{e}_\beta \rangle$	scalar or inner product in $\mathscr{H}$
$\{\|\mathbf{e}_\zeta\rangle = e^{\frac{1}{2}\|\zeta\|^2}\|\hat{e}_\zeta\rangle : \zeta \in \mathbb{C}\}$	family of principal vectors in $\mathscr{H}$
$\mathscr{F}$	Hilbert space of entire analytic functions in a complex variable ζ
(f,g)	inner product in $\mathscr{F}$
$\mathscr{K}(\omega,\zeta)$	reproducing kernel in $\mathscr{F}$

$\mathscr{F}_n = \mathscr{F} \otimes \cdots \otimes \mathscr{F}$	Hilbert space of entire analytic functions in n complex variables $(\zeta_1, \ldots, \zeta_n)$
$\mathbf{a} = (a_1, a_2, \ldots, a_n)$	n-component boson operator $\mathbf{a}$
$\mathscr{H}_n = \mathscr{H}_1 \otimes \cdots \otimes \mathscr{H}_1$	Hilbert space of state vectors in n bosons $(a_1, \ldots, a_n)$ which is isomorphic to $\mathscr{F}_n$
$X \to \mathscr{L}_X = \sum_{i,j=1}^n X_{ij} a_i \bar{a}_j$	Jordan-Schwinger map of an $n \times n$ matrix X into a boson operator $\mathscr{L}_X$ in $\mathscr{H}_n$
$\mathscr{E} = \mathscr{L}_1 = \sum_{i=1}^n a_i \bar{a}_i$	Euler operator in $\mathscr{H}_n$; number operator
$\lvert 0\rangle$	vector (vacuum ket) in $\mathscr{H}$ such that $\bar{a}_i \lvert 0\rangle = 0$, $i = 1, 2, \ldots, n$
$\langle 0 \lvert F^*(\bar{a}) G(a) \rvert 0\rangle = \langle F \vert G\rangle = [F^*(\partial/\partial\zeta) G(\zeta)]_{\zeta=0}$	inner product in $\mathscr{H}$ of state vectors $\lvert G\rangle = G(a)\lvert 0\rangle$ and $\lvert F\rangle = F(a)\lvert 0\rangle$
$J_i = \mathscr{L}_{\frac{1}{2}\sigma_i} = \frac{1}{2}(\tilde{a}\sigma_i \bar{a})$	Jordan-Schwinger boson operator realization of generator $\mathbf{J} = (J_1, J_2, J_3)$ of $SU(2)$
$\mathscr{H}^{(2j)}$	subspace of vectors in $\mathscr{H}_2$ of the form $P(a_1, a_2)\lvert 0\rangle$, where $P(a_1, a_2)$ is a homogeneous polynomial of degree $2j$
$\lvert jm\rangle = P_{jm}(a_1, a_2)\lvert 0\rangle$ $(m = j, \ldots, -j)$	standard orthonormal angular momentum ket vectors spanning $\mathscr{H}^{(2j)}$
$\mathscr{T}_U$	unitary operator in $\mathscr{H}_2$ giving the standard unitary irrep $\mathscr{T}_U \to D^j(U)$ of $SU(2)$ when acting in the invariant subspace $\mathscr{H}^{(2j)}$
$D^j(Z)$	extension of irreps of $SU(2)$ to irreps of $GL(2, C)$
$A = (a_i^j)$ $(i, j = 1, \ldots, n)$	$n \times n$ matrix with n^2 bosons as elements; matrix boson
$X \to \mathscr{L}_X = \text{trace}(\tilde{A} X \bar{A})$	generalized Jordan map of an $n \times n$ matrix X into boson operator acting in $\mathscr{H}_{n^2}$; generator of left translations
E_{ij} $(i, j = 1, \ldots, n)$	basis of the set of boson operator maps $\{\mathscr{L}_X : X = (x_{ij}), x_{ij} \in \mathbb{C}\}$; left boson operator realization of the Weyl basis of the Lie algebra of $U(n)$
$Y \to \mathscr{R}_Y = \text{trace}(\tilde{Y} \tilde{A} \bar{A})$	generalized Jordan map of an $n \times n$ matrix Y into boson operator acting in $\mathscr{H}_{n^2}$; generator of right translations
$E^{\alpha\beta}$ $(\alpha, \beta = 1, 2, \ldots, n)$	basis of the set of boson operator maps $\{\mathscr{R}_Y : Y = (y_{\alpha\beta}), y_{\alpha\beta} \in \mathbb{C}\}$; right boson operator realization of the Weyl basis of the Lie algebra of $U(n)$
$\det A = a_{12\cdots n}^{12\cdots n}$	determinant of the $n \times n$ matrix boson A

$L_U: A \to UA = L_U(A)$, $R_V: A \to A\tilde{V} = R_V(A)$	left and right translations of the $n \times n$ matrix boson A by unitary matrices U and V
$\mathscr{S}_U, \mathscr{T}_V$	unitary operator realizations in $\mathscr{H}_{n^2}$ of left and right translations L_U and R_V
$J_+ = E_{12}, J_- = E_{21}$, $J_3 = \frac{1}{2}(E_{11} - E_{22})$	generators of unimodular left translations $\mathscr{S}_U$ in $\mathscr{H}_4$, $U \in SU(2)$
$K_+ = E^{12}, K_- = E^{21}$, $K_3 = \frac{1}{2}(E^{11} - E^{22})$	generators of unimodular right translations $\mathscr{T}_V$ in $\mathscr{H}_4$, $V \in SU(2)$
$\mathscr{E} = \sum_k E^{kk}$	generator of phase transformations in $\mathscr{H}_4$; number operator, Euler operator
$D^j_{m'm}(A)$	boson polynomial obtained from the representation functions $D^j_{m'm}(U)$ by the replacement $u_{ij} \to a^j_i$
$\|k, j; m, m'\rangle$	boson basis vector in $\mathscr{H}_4$ on which the generators $\mathbf{J}$ and $\mathbf{K}$ have the standard action
$\mathbf{a}^i = (a^i_1, a^i_2, \ldots, a^i_n)$ $(i = 1, 2, \ldots, n)$	n-component boson operators
$\mathbf{J}(i)$, $i = 1, 2$	generators of $SU(2)$ corresponding to the 2-component boson $\mathbf{a}^i$
$\mathscr{H}^{(j_1, j_2)}$	vector space of state vectors of the form $P(\mathbf{a}^1, \mathbf{a}^2)\|0\rangle$, where $P(\mathbf{a}^1, \mathbf{a}^2)$ is polynomial homogeneous of degree $2j_1$ in $\mathbf{a}^1$ and $2j_2$ in $\mathbf{a}^2$
$\|j_1 m_1; j_2 m_2\rangle = P_{j_1 m_1}(\mathbf{a}^1) P_{j_2 m_2}(\mathbf{a}^2)\|0\rangle$, $m_1 = j_1, j_1 - 1, \ldots, -j_1$, $m_2 = j_2, j_2 - 1, \ldots, -j_2$	orthonormal basis of the space $\mathscr{H}^{(j_1, j_2)}$ on which the generators $\mathbf{J}(i)$ have the standard action
$\mathbf{J} = \mathbf{J}(1) + \mathbf{J}(2)$	generator of the diagonal subgroup of $SU(2) \times SU(2)$
$\mathbf{K} = \mathbf{J}(1) - \mathbf{J}(2)$	vector operator with respect to $\mathbf{J}$
$\|j_1 j_2; jm\rangle, j = j_1 + j_2, j_1 + j_2 - 1, \ldots, \|j_1 - j_2\|, m = j, j-1, \ldots, -j$	basis of the space $\mathscr{H}^{(j_1, j_2)}$ on which the $\mathbf{J}$ and $\mathbf{K}$ have the standard action
$(\pm\sqrt{k})^s \to (\pm 1)^s [k!/(k-s)!]^{\frac{1}{2}} = (\pm 1)^s([k]_s)^{\frac{1}{2}}$	map from ordinary powers to falling factorials that carries irrep functions into Wigner coefficients in $SU(2)$
$[\lambda] = [\lambda_1 \lambda_2 \cdots \lambda_n]$, $[m] = [m_{1n} m_{2n} \cdots m_{mn}]$	partitions of an integer $N \geqslant 0$ into n parts, including 0
$Y_{[\lambda]}, Y_{[m]}$	Young frames of shape $[\lambda]$, $[m]$
$\begin{pmatrix} [m] \\ (m) \end{pmatrix}$	Gel'fand pattern for $U(n)$

$(W) = (w_1, w_2, \ldots, w_n)$	weight of a Gel'fand pattern; content of a standard Weyl tableau
$(y) = (y_1, y_2, \ldots, y_n)$	Yamanouchi symbol of a standard Young tableau
$\begin{pmatrix} 2j & & 0 \\ & j+m & \end{pmatrix}$	Gel'fand pattern for $SU(2)$
$\begin{pmatrix} m_{12} & & m_{22} \\ & m_{11} & \end{pmatrix}$	Gel'fand pattern for $U(2)$
$\begin{pmatrix} [m] \\ (m) \end{pmatrix}$	boson basis vector in a carrier space of irrep $[m]$ of $U(n)$
$H^{[\lambda]}$	hook product of a standard tableau of shape $[\lambda]$
$\dim[\lambda] = n!/H^{[\lambda]}$	dimension of irrep $[\lambda]$ of S_n
$\mathrm{Dim}[\lambda] = \prod_{ij}(n+j-i)/H^{[\lambda]}$	dimension of irrep $[\lambda]$ of $U(n)$
$P\begin{pmatrix} & m'_{11} & \\ m_{12} & & m_{22} \\ & m_{11} & \end{pmatrix}(A)$	boson polynomial corresponding to a double standard tableau for $U(2) * U(2)$
$B\begin{pmatrix} & m'_{11} & \\ m_{12} & & m_{22} \\ & m_{11} & \end{pmatrix}(A)$	boson polynomial corresponding to a double Gel'fand pattern for $U(2) * U(2)$
$\left\| \begin{pmatrix} & \mu_{11} & \\ m_{12} & & m_{22} \\ & m_{11} & \end{pmatrix} \right\rangle$	orthonormal boson basis vector in the carrier space of irrep $[m_{12}m_{22}]$ of $U(2) * U(2)$
$\begin{pmatrix} (m') \\ [m] \\ (m) \end{pmatrix}$	double Gel'fand pattern for $U(n) * U(n)$
$P\begin{pmatrix} (m') \\ [m] \\ (m) \end{pmatrix}(A)$	boson polynomial corresponding to a double standard Weyl tableau; nonunitary irrep function under the replacement $A \to U$
$B\begin{pmatrix} (m') \\ [m] \\ (m) \end{pmatrix}(A)$	boson polynomial corresponding to a double Gel'fand pattern; unitary irrep function of $U(n)$ under the replacement $A \to U$
$\left\| \begin{pmatrix} (m') \\ [m] \\ (m) \end{pmatrix} \right\rangle = [\mathscr{M}([m])]^{-\frac{1}{2}} B\begin{pmatrix} (m') \\ [m] \\ (m) \end{pmatrix}(A)\|0\rangle$	orthonormal boson basis vector in the carrier space of irrep $[m]$ of $U(n) * U(n)$
$\mathscr{M}([m]) = \prod_{i=1}^{n} p_{in}! \Big/ \prod_{i<j}(p_{in} - p_{jn})$	normalization factor for boson polynomials; measure of a highest-weight tableau

$\mathscr{L}_k^{(\lambda)}, \mathscr{U}_k^{(\lambda)}, \lambda = 1, \ldots, n; k = 1, \ldots, \lambda$	complete set of mutually commuting Hermitian operators in the space $\mathscr{H}_{n^2}$
$\mathscr{R}, \mathscr{C}, \mathscr{T}, \mathscr{K}$	generators of a finite group of symmetries of the $SU(2)$ representation functions
$D\begin{pmatrix} & \mu_{11} & \\ m_{12} & & m_{22} \\ & m_{11} & \end{pmatrix}(U)$	unitary irrep function of $U(2)$

Miscellaneous Mathematical and Physical Symbols

$\hbar$	Dirac's notation for Planck's constant h divided by 2π; basic unit of angular momentum
$e = -1.60 \times 10^{-19}$ coulomb	charge of the electron in rationalized MKS units
$\delta_{ij}, \delta_{i'j}, \delta_{m'm}$	Kronecker delta
e_{ijk}	$+1$ (-1) for ijk an even (odd) permutation of 123; otherwise 0
$e_{\alpha\beta\gamma\delta}, e^{\alpha\beta\gamma\delta}$	$+1$ (-1) for $\alpha\beta\gamma\delta$ an even (odd) permutation of 0123; otherwise zero
$(a_1, a_2, \ldots, a_n)$	ordered n-tuple
$\mathbb{1}$, $\mathbf{1}$, 1	unit operators
$\mathbb{1}_n$	$n \times n$ unit matrix
$\mathbb{1}^{(j)}$	$(2j+1) \times (2j+1)$ unit matrix
$[A, B] = AB - BA$	commutator of operators or matrices A and B
$[A, B]_{(k)} \equiv [A, [A, B]_{(k-1)}]$ $[A, B]_{(0)} \equiv B$	multiple commutator of operator A with operator B
$[A, B]_{PB} = \sum_i \left(\frac{\partial A}{\partial q_i} \frac{\partial B}{\partial p_i} - \frac{\partial A}{\partial p_i} \frac{\partial B}{\partial p_i} \right)$	Poisson bracket of observables A and B
$\binom{z}{a} = z(z-1) \cdots (z-a+1)/a!$	binomial function
$(z)_a = z(z+1) \cdots (z+a-1)$	Pochhammer's symbol for a rising factorial
$[z]_a = z(z-1) \cdots (z-a+1)$	falling factorial
$1, i, j, k$	quaternionic basis elements
$q = q_0 1 + q_1 i + q_2 j + q_3 k$	general quaternion
$\bar{q} = q_0 1 - q_1 i - q_2 j - q_3 k$	conjugate quaternion to q
$N(q) = \bar{q}q = q_0^2 + q_1^2 + q_2^2 + q_3^2$	norm of a quaternion q
$N(q) = 1$	unimodular quaternion
$dS(\hat{n})$	differential surface area of S^2 at the point $\hat{n} \in S^2$

$d\omega = d\phi \sin\theta \, d\theta$	differential surface area of S^2 at the point $\hat{n} = (\sin\theta \sin\phi, \sin\theta \sin\phi, \cos\theta)$
$d\Omega, d\Omega_U$	differential surface area of S^3 at the point $(x_0, \mathbf{x}) \leftrightarrow U(x_0, \mathbf{x}) = U \in SU(2)$
$A^\dagger$, A^*, $\bar{A}$	conjugate of an operator A
$z^*, \bar{z}$	complex conjugate of $z \in \mathbb{C}$
$\mathbf{T}_0, -\mathbf{T}, \mathbf{T}^c, \mathbf{T}_\pi, \mathbf{E}$	special turns: identity, inverse, conjugate, length π, length $\pi/2$, respectively
$\mathbf{T}_1 + \mathbf{T}_2$	sum of two turns (see Figure 4.4, Chapter 4)
$\lVert\mathbf{T}\rVert$	length of a turn
$T(s, \mathbf{v})$	turn parametrized in terms of quaternionic coordinates $(s, \mathbf{v})$
$L_g(\tau) = g + \tau$	left translation of a turn τ by a turn g
$R_g(\tau) = \tau + g^{-1} = \tau + (-g)$	right translation of a turn τ by a turn g
$(\mathscr{L}_g f)(\tau) = f(L_{g^{-1}}(\tau))$ $(\mathscr{R}_g f)(\tau) = f(R_{g^{-1}}(\tau))$	unitary operator realizations in Hilbert space of left and right translations of turns
$\delta(\mathbf{x} - \mathbf{x}')$	Dirac delta function
$U(\mathbf{a})f(\mathbf{x}) = f(\mathbf{x} - \mathbf{a})$	action of the unitary displacement operator $U(\mathbf{a}) = e^{-i\mathbf{a}\cdot\mathbf{p}/\hbar}$ on a function $f: \mathbf{x} \to f(\mathbf{x})$
$\Delta L_i, \langle L_i \rangle$	dispersion in the angular momentum operator L_i; expectation value of L_i

Author Index

Subject Index

About the authors

[illegible]

[illegible]

About the book

[illegible]

Addison-Wesley Publishing Company
Advanced Book Program
Reading, Massachusetts